RESTORATION AND MANAGEMENT OF LAKES AND RESERVOIRS

THIRD EDITION

RESTORATION AND MANAGEMENT OF LAKES AND RESERVOIRS

THIRD EDITION

G. DENNIS COOKE
EUGENE B. WELCH
SPENCER A. PETERSON
STANLEY A. NICHOLS

Taylor & Francis
Taylor & Francis Group

Boca Raton   London   New York   Singapore

A CRC title, part of the Taylor & Francis imprint, a member of the
Taylor & Francis Group, the academic division of T&F Informa plc.

Published in 2005 by
CRC Press
Taylor & Francis Group
6000 Broken Sound Parkway NW, Suite 300
Boca Raton, FL 33487-2742

International Standard Book Number-10: 1-56670-625-4 (Hardcover)
International Standard Book Number-13: 978-1-5667-0625-4 (Hardcover)
Library of Congress Card Number 2004062816

Library of Congress Cataloging-in-Publication Data

Restoration and management of lakes and reservoirs / edited by G. Dennis Cooke
  … [et al.].—3rd ed.
    p. cm.
  Includes bibliographical references and index.
  ISBN 1-56670-625-4 (alk. paper)
  1. Restoration ecoology. 2. Water quality management. I. Cooke, G. Dennis
  (George Dennis), 1937- II. Title.

QH541.15.R45R49 2005
628.1'68—dc22   2004062816

Taylor & Francis Group
is the Academic Division of T&F Informa plc.

Visit the Taylor & Francis Web site at
http://www.taylorandfrancis.com

and the CRC Press Web site at
http://www.crcpress.com

Preface

Environmental problems usually develop from the interactions of people, consumption, and resources. Increasing population, increasing consumption and limited resources exacerbate these problems. One concern that heads the list of critical problems is the availability of clean, fresh, surface water. It is the basis of the existence of human societies and economies. Fresh water is essential for many forms of life, is required by humans for drinking, agriculture, and most industrial processes, and plays a prominent role in our recreational activities.

Since we completed the second edition of this book in 1992 (Cooke et al., 1993), hundreds of millions of people have been added to the human population, each of them exerting demands and impacts on a finite supply of fresh water. As noted in our Introduction, the overall quality of lakes and reservoirs in many areas of the United States, southern Canada, and Europe continues to deteriorate. In some areas, fresh water resources are so polluted that economic systems and human health are impaired. Although there are several urgent global environmental problems, scientists, environmentalists, and policy makers must focus much more attention on the human population explosion and its well-known relationship to fresh water pollution. Certainly all nations should be taking significant steps to reduce the likelihood of global climate change and to limit additional water pollution and aquatic habitat destruction.

Lake and reservoir management and restoration methods are new technologies that have developed over the last 35 years, and are ones that promise to be of great significance in protecting and improving fresh water systems. We hope that our book will be a useful addition. Every lake or reservoir utilized by humans requires management. This may involve only monitoring to assure that it is not degraded, or it may require regular efforts to maintain it, perhaps with equipment or techniques that have been adopted to enhance or protect the system. Restoration of impaired lakes and reservoirs, in the strict sense, is not possible, but the term is applied to procedures to return the system to some approximation of an earlier, less disturbed condition. We are just beginning to learn the art and science of management and restoration.

Applied limnology developed as an extension of basic sciences. There is a great need to understand fresh water systems if we are to provide for their competent protection, management, and restoration for current and future generations. Long-term funding to support basic and applied limnology must be greatly expanded, and this must be recognized by politicians, administrators, and others who support science through policies and appropriations. We strongly endorse the work of the North American Lake Management Society (NALMS), and other professional and environmental organizations, which together have been so consistent in delivering this message to scientists, appropriate legislators, and citizens.

Our goals in this book are to describe the eutrophication process, outline methods for developing a pre-management and restoration diagnosis-feasibility study, and to provide detailed descriptions of scientifically sound management and restoration methods. Each chapter includes an introduction to the scientific basis of the problem, a description of the method's procedures, and presents some case histories. Potential negative impacts and costs, where known, also are noted. The chapters are updated and extensively referenced, and three new chapters have been added to this edition. Our book will be useful as a classroom text, as a reference manual, and as a general guide for interested lake users.

This book is certainly not the last word on the topic. It is our sincere hope that it will stimulate new and improved perspectives and ideas in lake and reservoir management and restoration. The content of this book is a product of the study, input, and concurrence of all of the authors, as well as a product of our combined years of field and laboratory research in limnology.

Where appropriate and possible, we report costs in 2002 U.S. dollars by correcting for inflation. This was done by using year-to-year increases in the Consumer Price Index (CPI) to correct costs reported for earlier years to their present values. We thank Dr. Thomas S. Lough (Sonoma State University, Rohnert Park, California) for the use of his CPI scale to correct for inflation.

The contributions to this book by Spencer A. Peterson, an employee of the U.S. Environmental Protection Agency (EPA), were made on his own time, with Agency permission. However, the research and writing were independent of USEPA employment and have not been subjected to the Agency's peer and administrative review. Therefore, the conclusions and opinions stated are solely those of the author and should not be construed to reflect the views of the USEPA.

Specific chapter authorship is: G. Dennis Cooke (Chapters 5, 9, 10, 13, 15, and 17), Eugene B. Welch (Chapters 3, 4, 6, 7, 18, and 19), Spencer A. Peterson (Chapter 20), Stanley A. Nichols (Chapters 11, 12, 14, and 16), G. Dennis Cooke and Spencer A. Peterson (Chapters 1 and 2), and Eugene B. Welch and G. Dennis Cooke (Chapter 8).

Acknowledgments

Numerous and often unnamed, our colleagues and students have provided a rich array of ideas, articles, books, theses, and reports from which to draw materials to write the book. Many have spent years in the field and in the laboratory, collecting data and studying lakes. The many stimulating discussions with them have been invaluable as well. We dedicate this book to them.

We thank Dr. Brent Bruot, Chair, Department of Biological Sciences, Kent State University, for invaluable facilities and support during the book's preparation, and Dr. Gertrud Cronberg and the late Dr. Gunnar Andersson for permission to use unpublished figures and photographs, respectively, in Chapter 20. We also thank Chris Lind and the General Chemical Corporation for permission to use a figure in Chapter 8, and Drs. Richard Lathrop, William Walker and Jacob Kann for permission to use unpublished figures in Chapter 3. We thank Tetra Tech, Inc. (Seattle, Washington) for its general office and computer assistance to Eugene Welch. We thank the U.S. Environmental Protection Agency for authorizing Spencer Peterson to write this book on his own time, but also to have occasional use of his computing and graphic arts contractor (Computer Sciences Corporation), especially Suzanne M. Pierson, for drafting some new figures for this third edition.

We gratefully acknowledge the technical assistance of the Wisconsin Geological and Natural History Survey Staff, especially Susan Hunt and Mindy James, for graphic, editorial, and computer support for chapters prepared by Stanley Nichols.

CRC Press has been a joy to work with. We are especially grateful to Patricia Roberson for her exceptional assistance during the book's preparation, Jill Jurgensen and Sylvia Wood for their able editorial work, and our editor, Matt Lamoreaux, for his continuous support. Suzanne Pierson and Spencer Peterson assisted CRC's Shayna Murry to design, compose and select colors for the book cover.

G. Dennis Cooke
Eugene B. Welch
Spencer A. Peterson
Stanley A. Nichols
January 2005

BOOK COVER PHOTO CREDITS

Front Cover

Top: A partitioned pond phosphorus inactivation experiment at Cline's Pond, Oregon (top half of left pond untreated; bottom half of left pond treated with zirconium tetrachloride; right pond is untreated reference pond). Courtesy of Spencer Peterson (1974).

Bottom: Whole lake phosphorus inactivation at Dollar Lake, Ohio (left, small round lake treated with alum in 1974), West Twin Lake, Ohio (right, round lake treated with alum in 1975) and reference lake (center, irregularly shaped East Twin Lake). Courtesy of Dennis Cooke (1976).

Background cover photo is an enlargement of the Twin Lakes photo by Dennis Cooke.

Back Cover

Left to right, row 1:
1. Shoreline of West Twin Lake, Ohio. Courtesy of Dennis Cooke (1976).
2. IR photo of Lilly Lake, Wisconsin prior to dredging. Courtesy of Spencer Peterson (1977).

3. Sewer pipe installation around Liberty Lake, Washington prior to phosphorus inactivation with alum. Courtesy of Spencer Peterson (1977).
4. Dredge pipeline in Lake Trummen, Sweden. Gunnar Andersson (1969), University of Lund, Lund, Sweden. With permission.
5. Milman Mudcat dredge on Lake Jarnsjon, Sweden. Ellicott, Division of Baltimore Dredges LLC (1993), Baltimore, MD. With permission.

Left to right, row 2:
1. Mudcat dredge in Mexico (nd). Ellicott, Division of Baltimore Dredges LLC, Baltimore, Maryland. With permission.
2. Aerator installation in Lake Stevens, Washington (nd). Courtesy of Harry Gibbons, Tetra Tech, Inc., Seattle, Washington.
3. Grass carp or white amur (*Ctenopharyngodon idella* Val.) (1987). Courtesy of Dennis Cooke.
4. Aquatic plant harvester on Lake Sallie, Minnesota. Courtesy of Spencer Peterson (1969).
5. Aquatic plant harvester. Courtesy of Dennis Cooke (1980).

Left to right, bottom:
1. Alum application barge on Green Lake, Washington. Courtesy of Eugene Welch (nd).
2. Alum application at Medical Lake, Washington. Courtesy of Spencer Peterson (1977).
3. Waldo Lake, Oregon. Courtesy of Spencer Peterson (1982).

Bottom of back cover, bar graph figure:
Biomass before dredging and over a more than 30-year history following dredging in Lake Trummen, Sweden. Gertrud Cronberg, University of Lund, Lund, Sweden. With permission.

Authors

G. Dennis Cooke is Emeritus Professor of Biological Sciences and Member of the Water Resources Research Institute at Kent State University, Kent, Ohio. He was a founding member and the first President of the North American Lake Management Society and also served two terms as a board member. He is also a founding member of the Ohio Lake Management Society and served as its president and as a board member. Dr. Cooke is the author of several books, including *Reservoir Management for Water Quality and THM Precursor Control*, and many articles and reports on limnology and lake and reservoir management.

Eugene B. Welch is Emeritus Professor of Civil and Environmental Engineering at the University of Washington, Seattle, and is a consultant with Tetra Tech, Inc., in Seattle. He is Past President of the North American Lake Management Society (1992–93 term), was a founding member of the Society, and served on its first Board of Directors. Dr. Welch is author of two other books, including *Pollutant Effects in Fresh Water: Applied Limnology*, and many reports and articles on applied limnology and lake and reservoir management.

Spencer A. Peterson is a Senior Research Ecologist with the USEPA's Environmental Monitoring and Assessment Program at the National Health and Ecological Effects Research Laboratory, Western Ecology Division, Corvallis, Oregon, and affiliate Professor of Civil and Environmental Engineering, University of Washington, Seattle. Dr. Peterson is a founding member of the North American Lake Management Society and the author of many articles on lake management, contaminated sediments, and non-point source and hazardous waste assessment.

Stanley A. Nichols is Emeritus Professor of Environmental Sciences at the University of Wisconsin-Extension in Madison. For most of his career he worked at the Environmental Resources Center and the Wisconsin Geological and Natural History Survey. His initial efforts in lake restoration and management began more than 30 years ago as a member of the Inland Lake Renewal and Demonstration Project in Wisconsin and the Lake Wingra International Biological Program team. He has published widely in the areas of aquatic plant ecology and management, lake protection, exotic species control, habitat restoration, and lake sampling. He is a past member of the North American Lake Management Society and the Aquatic Plant Management Society. Presently, he consults and writes on aquatic plants, lake management, and habitat restoration issues.

Contents

SECTION I   *Overview*

Chapter 1   Introduction..3

1.1   The Hydrologic Cycle and the Quantity of Fresh Water 3
1.2   Status of Fresh Water in the United States 7
1.3   Sources of Lake and Reservoir Problems 11
1.4   Restoration and Management of Lakes and Reservoirs...................................13
1.5   History of Lake Restoration and Management 15
References 17

Chapter 2   Basic Limnology 23

2.1   Introduction 23
2.2   Lakes and Reservoirs 23
2.3   Basic Limnology 26
         2.3.1   Physical–Chemical Limnology...26
2.4   Biological Limnology 28
2.5   Limiting Factors 30
2.6   The Eutrophication Process 31
2.7   Characteristics of Shallow and Deep Lakes..33
2.8   Ecoregions and Attainable Lake Conditions 34
2.9   Summary 41
References 41

Chapter 3   Lake and Reservoir Diagnosis and Evaluation...47

3.1   Introduction 47
3.2   Diagnosis/Feasibility Studies...47
         3.2.1   Watershed...47
         3.2.2   In-Lake...53
         3.2.3   Data Evaluation 57
                      3.2.3.1   Example 1 70
                      3.2.3.2   Example 2 71
3.3   Selection of Lake Restoration Alternatives 73
         3.3.1   Algal Problems 73
                      3.3.1.1   Nutrient Diversion/Advanced Waste Treatment 73
                      3.3.3.2   P Inactivation 73
                      3.3.3.3   Dilution/Flushing 74
                      3.3.3.4   Lake Protection From Urban Runoff 74
                      3.3.3.5   Hypolimnetic Withdrawal 74
                      3.3.3.6   Artificial Circulation 74
                      3.3.3.7   Food-Web Manipulations...74
                      3.3.3.8   Copper Sulfate Treatment 74

      3.3.4   Macrophyte Problems 74
              3.3.4.1   Harvesting 75
              3.3.4.2   Biological Controls 75
              3.3.4.3   Lake-Level Drawdown 75
              3.3.4.4   Sediment Covers 75
              3.3.4.5   Sediment Removal 75
              3.3.4.6   Hypolimnetic Aeration 75
3.5   Guidelines for Choosing Lake Restoration Alternatives 76
3.6   The Lake Improvement Restoration Plan 78
References 80

SECTION II   Algal Biomass Control Techniques Directed toward Control of Plankton Algae

Chapter 4   Lake and Reservoir Response to Diversion and Advanced Wastewater
                 Treatment 89
4.1   General 89
4.2   Techniques for Reducing External Nutrient Loads 90
4.3   Recovery of World Lakes 91
4.4   Lake Washington, Washington 95
4.5   Lake Sammamish, Washington 98
4.6   Lake Norrviken, Sweden 100
4.7   Shagawa Lake, Minnesota 101
4.8   Madison Lakes, Wisconsin 103
4.8   Lake Zürich, Switzerland 104
4.9   Lake Søbygaard, Denmark 105
4.10   Costs 106
4.11   In-Lake Treatment Following Diversion 107
4.12   Summary 108
References 109

Chapter 5   Lake and Reservoir Protection From Non-Point Pollution 113
5.1   Introduction 113
5.2   In-Stream Phosphorus Removal 114
5.3   non-point Nutrient Source Controls: Introduction 116
5.4   non-point Source Controls: Manure Management 119
5.5   non-point Nutrient Source Controls: Ponds and Wetlands 122
      5.5.1   Introduction 122
      5.5.2   Dry And Wet Extended Detention (ED) Ponds 122
      5.5.3   Constructed Wetlands 124
5.6   Constructed Wetlands: Case Histories 127
5.7   Pre-Dams 130
5.8   Riparian Zone Rehabilitation: Introduction 131
5.9   Riparian Zone Rehabilitaton Methods 132
5.10   Reservoir Shoreline Rehabilitation 135
5.11   Lakeshore Rehabilitation 137

5.12   Summary 140
References 140

Chapter 6   Dilution and Flushing...149

6.1   Introduction 149
6.2   Theory and Predictions 150
6.3   Case Studies 151
       6.3.1   Moses Lake 152
       6.3.2   Green Lake 158
       6.3.3   Lake Veluwe..160
6.4   Summary: Effects, Applications, and Precautions 161
References 162

Chapter 7   Hypolimnetic Withdrawal...165

7.1   Introduction 165
7.2   Test Cases..167
       7.2.1   General Trends..167
       7.2.1   Specific Cases...170
                7.2.1.1   Mauen See...170
                7.2.1.2   Austrian Lakes 170
                7.2.1.3   U.S. Lakes...171
                7.2.1.4   Canada...172
7.3   Costs..173
7.4   Adverse Effects 173
7.5   Summary 173
References 174

Chapter 8   Phosphorus Inactivation and Sediment Oxidation 177

8.1   Introduction 177
8.2   Chemical Background..178
       8.2.1   Aluminum 178
       8.2.2   Iron and Calcium 180
8.3   Dose Determination and Application Techniques 182
       8.3.1   Aluminum 182
       8.3.2   Iron and Calcium 191
       8.3.3   Application Techniques for Alum 191
8.4   Effectiveness and Longevity of P Inactivation 195
       8.4.1   Introduction...195
       8.4.2   Stratified Lake Cases 195
                8.4.2.1   Mirror and Shadow Lakes, Wisconsin (WI) 200
                8.4.2.2   West Twin Lake (WTL), Ohio....................................201
                8.4.2.3   Kezar Lake, New Hampshire.....................................203
                8.4.2.4   Lake Morey, Vermont 204
       8.4.3   Shallow, Unstratified Lake Cases..206
                8.4.3.1   Long Lake, Kitsap County, Washington 208
                8.4.3.2   Campbell and Erie Lakes, Washington 209
                8.4.3.3   Green Lake, Washington..210

8.4.4   Reservoirs 211
8.4.5   Ponds..211
8.4.6   Iron Applications 212
8.4.7   Calcium Applications to Hardwater Lakes 213
8.5   Problems that Limit Effectiveness of P Inactivation...215
8.6   Negative Aspects 216
8.7   Costs 224
8.8   Sediment Oxidation..224
8.8.1   Equipment and Application Rates...225
8.8.2   Lake Response 225
8.8.3   Costs...228
8.8.4   Prospectus 228
References 230

Chapter 9   Biomanipulation 239

9.1   Introduction 239
9.2   Trophic Cascade..239
9.3   Basic Trophic Cascade Research..243
9.4   Biomanipulation 244
9.5   Shallow Lakes 245
9.6   Biomanipulation: Shallow Lakes 247
9.6.1   Cockshoot Broad (UK)..247
9.6.2   Lake Zwemlust (and Other Dutch Lakes) 248
9.6.3   Lake Vaeng (and other Danish Lakes)..250
9.6.4   Lake Christina, Minnesota 250
9.7   Biomanipulation: Deep Lakes 252
9.7.1   Lake Mendota, Wisconsin 252
9.7.2   Bautzen Reservoir And Grafenheim Experimental Lakes (Germany).............253
9.8   Costs 255
9.9   Summary and Conclusions 255
References 256

Chapter 10   Copper Sulfate 263

10.1   Introduction 263
10.2   Principle of Copper Sulfate Applications...263
10.3   Application Guidelines 265
10.4   Effectiveness of Copper Sulfate 266
10.5   Negative Effects of Copper Sulfate 267
10.6   Costs of Copper Sulfate 269
References 270

SECTION III   Macrophyte Biomass Control

Chapter 11   Macrophyte Ecology and Lake Management..275

11.1   Introduction 275
11.2   Planning and Monitoring for Aquatic Plant Management 275
11.2.1   Case Study: White River Lake Aquatic Plant Management Plan 276

11.3	Species and Life-Form Considerations		280
11.4	Aquatic Plant Growth and Productivity		281
	11.4.1	Light	281
	11.4.2	Nutrients	282
	11.4.3	Dissolved Inorganic Carbon (DIC), pH, and Oxygen (O_2)	283
	11.4.4	Substrate	284
	11.4.5	Temperature	284
11.5	Plant Distribution within Lakes		285
11.6	Resource Allocation and Phenology		285
11.7	Reproduction and Survival Strategies		286
11.8	Relationships with Other Organisms		287
11.9	The Effects of Macrophytes on Their Environment		289
References			291

Chapter 12 Plant Community Restoration 295

12.1	Introduction		295
12.2	The "Do Nothing" Approach		296
	12.2.1	Case history: Lake Wingra, "Doing Nothing"	297
12.3	The Habitat Alteration Approach		298
	12.3.1	Case History: No-Motor, Slow-No-Wake Regulations	299
		12.3.1.1 Long and Big Green Lakes: Heavily Used Recreational Lakes in Southeastern Wisconsin	299
		12.3.1.2 Active Habitat Manipulation: Engineering and Biomanipulation Case Studies	300
12.4	Aquascaping		307
12.5	The Founder Colony: A Reasonable Restoration Approach		315
	12.5.1	Case Studies	316
		12.5.1.1 Founder Colonies in North Lake, Lake Lewisville, and Lake Conroe, Texas and Guntersville Reservoir, Alabama	316
		12.5.1.2 Cootes Paradise Marsh: Volunteers in Action	317
		12.5.1.3 Rice Lake at Milltown, Wisconsin: Lessons Learned	317
12.6	Concluding Thoughts		321
References			321

Chapter 13 Water Level Drawdown 325

13.1	Introduction		325
13.2	Methods		325
13.3	Positive and Negative Factors of Water Level Drawdown		330
13.4	Case Studies		332
	13.4.1	Tennessee Valley Authority (TVA) Reservoirs	332
	13.4.2	Louisiana Reservoirs	333
	13.4.3	Florida	333
	13.4.4	Wisconsin	334
	13.4.5	Connecticut	337
	13.4.6	Oregon	337
13.5	Fish Management with Water Level Drawdown		337
13.6	Case Histories		337
13.7	Summary		339
References			339

Chapter 14 Preventive, Manual, and Mechanical Methods..343

14.1   Introduction 343
14.2   Preventive Approaches...343
        14.2.1   The Probabilities of Invasion 344
        14.2.2   Education, Enforcement, and Monitoring as Preventive Approaches 346
        14.2.3   Barriers and Sanitation 346
14.3   Manual Methods and Soft Technologies 348
14.4   Mechanical Methods 349
        14.4.1   The Materials Handling Problem 349
        14.4.2   Machinery and Equipment 350
        14.4.3   Cutting 352
                 14.4.3.1   Case Study: Water chestnut (*Trapa natans*) Management in
                             New York, Maryland, and Vermont..352
                 14.4.3.2   Case Study: Pre-Emptive Cutting to Manage Curly-Leaf
                             Pondweed (*Potamogeton crispus*) in Minnesota 353
                 14.4.3.3   Case Study: Deep Cutting, Fish Lake, Wisconsin 353
                 14.4.3.4   Case Study: Cutting the Emergents, Cattails (*Typha* spp.) and
                             Reeds (*Phragmites* spp.)...355
        14.4.4   Harvesting 355
                 14.4.4.1   Efficacy, Regrowth, and Change in Community Structure..............355
                 14.4.4.2   The Nutrient Removal Question..358
                 14.4.4.3   Environmental Effects...362
                 14.4.4.4   Operational Challenges 365
        14.4.5   Shredding and Crushing 365
        14.4.6   Diver-Operated Suction Dredges..366
        14.4.7   Hydraulic Washing 367
        14.4.8   Weed Rollers: Automated, Untended Aquatic Plant Control Devices...............367
        14.4.9   Mechanical Derooting 368
        14.4.10 Costs and Productivity...369
14.5   Concluding Remarks..374
References 375

Chapter 15 Sediment Covers and Surface Shading for Macrophyte Control 381

15.1   Introduction 381
15.2   Comparison of Synthetic Sediment Covers...381
        15.2.1   Polyethylene..381
        15.2.2   Polypropylene 382
        15.2.3   Aquascreen..383
        15.2.4   Burlap..383
15.3   Application Procedures for Sediment Covers 384
15.4   Shading of Macrophytes with Surface Covers..384
References 385

Chapter 16 Chemical Controls 387

16.1   Introduction 387
16.2   Effective Concentration — Dose, Time Considerations, Active Ingredients,
        Site-Specific Factors, and Herbicide Formulation 387
16.3   Types of Chemicals..388
        16.3.1   Contact vs. Systemic 389

16.3.2 Broad-spectrum vs. Selective Herbicides..390
16.3.3 Persistent vs. Non-Persistent 390
16.3.4 Tank Mixes 390
16.3.5 Plant Growth Regulators (PGRs) 390
16.3.6 Adjuvants 391
16.4   Increasing Herbicide Selectivity 391
16.5   Environmental Impacts, Safety and Health Considerations..393
16.5.1 Herbicide Fate in the Environment 393
16.5.2 Toxic Effects 394
16.5.2.1 Direct Effects 395
16.5.2.2 Indirect Impacts 397
16.5.2.3 What Should a Lake Manager or Concerned Citizen Do?..............399
16.6   Ways of Minimizing Environmental Risks 399
16.7   Case Studies 401
16.7.1 Plant Management with Fluridone in the Northern United States...................401
16.7.1.1 Minnesota Experiences 401
16.7.1.2 Wisconsin Experiences — Potters and Random Lakes 404
16.7.1.3 Michigan Experiences...406
16.7.1.4 Vermont Experiences — Lake Hortonia and Burr Pond 409
16.7.1.5 Increasers and Decreasers 412
16.7.2 2,4-D in Cayuga Lake, New York and Loon Lake, Washington State.............412
16.7.2.1 Cayuga Lake 412
16.7.2.2 Loon Lake 414
16.7.3 Triclopyr in Pend Oreille River, Washington State and Lake Minnetonka,
Minnesota..415
16.7.3.1 Pend Oreille River 415
16.7.3.2 Lake Minnetonka 417
16.8   Costs 418
16.9   Concluding Remarks 418
References 420

Chapter 17 Phytophagous Insects, Fish, and Other Biological Controls....................................425
17.1   Introduction 425
17.2   Hydrilla (*Hydrilla verticillata*) 426
17.3   Water Hyacinth (*Eichhornia crassipes*)..427
17.4   Alligatorweed (*Alternanthera philoxeroides*)...429
17.5   Eurasian Watermilfoil (*Myriophyllum spicatum*) 430
17.6   Grass Carp..433
17.6.1 History and Restrictions 433
17.6.2 Biology of Grass Carp..434
17.6.3 Reproduction of Grass Carp...435
17.6.4 Stocking Rates 438
17.6.5 Case Histories 440
17.6.5.1 Deer Point Lake, Florida...440
17.6.5.2 Lake Conway, Florida...442
17.6.5.3 Lake Conroe, Texas 442
17.6.5.4 Smaller Lakes and Ponds 443
17.6.6 Water Quality Changes...443
17.7   Other Phytophagous Fish...445
17.8   Developing Areas of Macrophyte and Algae Management 446

17.8.1 Fungal Pathogens..446
17.8.2 Water hyacinth...446
17.8.3 Hydrilla..447
17.8.4 Eurasian Watermilfoil..447
17.8.5 Allelopathic Substances..447
17.8.6 Plant Growth Regulators 448
17.8.7 Barley Straw 448
17.8.8 Reducing Algae Growth with Bacteria 448
17.8.9 Viruses for Blue-Green Algae Management 449
References 449

SECTION IV   Multiple Benefit Treatments

Chapter 18 Hypolimnetic Aeration and Oxygenation...459

18.1 Introduction 459
18.2 Description and Operation of Units 459
18.3 Unit Sizing 464
18.4 Beneficial Effects and Limitations 465
18.5 Undesirable Effects 470
18.6 Costs 470
18.7 Summary 470
References 471

Chapter 19 Artificial Circulation...475

19.1 Introduction 475
19.2 Devices and Air Quantities 475
19.3 Theoretical Effects of Circulation 483
    19.3.1 Dissolved Oxygen (DO)...483
    19.3.2 Nutrients..483
    19.3.3 Physical Control of Phytoplankton Biomass 484
    19.3.4 Effects on Phytoplankton Composition...488
19.4 Effects of Circulation on Trophic Indicators..490
19.5 Undesirable Effects 494
19.6 Costs...494
19.7 Summary and Recommendations 496
References 496

Chapter 20 Sediment Removal..503

20.1 Introduction 503
20.2 Objectives of Sediment Removal 503
    20.2.1 Deepening 503
    20.2.2 Nutrient Control..503
    20.2.3 Toxic Substances Removal...504
    20.2.4 Rooted Macrophyte Control 504
20.3 Environmental Concerns 505
    20.3.1 In-Lake Concerns..505
    20.3.2 Disposal Area Concerns 506

20.4   Sediment Removal Depth 507
20.5   Sediment Removal Techniques...508
    20.5.1   Mechanical Dredges 509
    20.5.2   Hydraulic Dredges..509
    20.5.3   Special-Purpose Dredges..513
    20.5.4   Pneumatic Dredges 514
20.6   Suitable Lake Conditions...514
20.7   Dredge Selection and Disposal Area Design 516
    20.7.1   Dredge Selection...517
        20.7.1.1   Plan to Optimize the Available Disposal Area..................517
        20.7.1.2   Analyze the Production Capacity of Available Dredging
                  Equipment 518
        20.7.1.3   Compute Dredging Days Required to Complete the Job 520
        20.7.1.4   Determine the Required Head Discharge Characteristics of the
                  Main Pump When Pumping Material with the Specific Gravity
                  of Lake Sediment (Approximately 1.20).........................520
        20.7.1.5   Determine Minimum Head Conditions When Pumping to the
                  Nearest Disposal Area 528
        20.7.1.6   Analyze Booster Pump Requirements for Pumping to Distances
                  Beyond the Capacity of the Main Pump...........................529
    20.7.2   Disposal Area Design 535
        20.7.2.1   Flocculent Settling Procedure..535
        20.7.2.2   Zone/Compression Settling Test Procedure 536
        20.7.2.3   Design Procedures 537
20.8   Case Studies 543
    20.8.1   Lake Trummen, Sweden...544
    20.8.2   Lilly Lake, Wisconsin..548
        20.8.2.1   Initial Diagnosis and Results..548
        20.8.2.2   Long-Term Effects 550
        20.8.2.3   Other WDNR Dredging Experiences 553
    20.8.3   Lake Springfield, Illinois 554
        20.8.3.1   Sediment Removal Guidelines..555
        20.8.3.2   Sediment Removal Techniques and Disposal Site Selection...........556
        20.8.3.3   Permits...557
        20.8.3.4   Disposal Site 558
        20.8.3.5   Sediment Removal 558
    20.8.4   Lake Järnsjön, Sweden 559
20.9   Costs...562
20.10 Summary 567
References 567

Index 575

Section I

Overview

1 Introduction

"The frog does not drink up the pond in which it lives."

Sandra Postel (1995)

This brief sentence, an Inca proverb according to Dr. Postel, aptly describes the predicament facing humanity and the rest of Earth's biota. Everyone is aware that humans and other terrestrial animals, as well as plants and a huge diversity of aquatic species, are completely dependent on adequate and sustainable fresh water supplies. Yet many humans behave as if the amount of clean fresh water is infinite and the lives and activities of aquatic species are insignificant.

This introductory chapter illustrates our dependence on fresh water and the condition or quality of these waters, and argues that protection, restoration, and management of them is an increasingly vital activity. The goal of this chapter is to provide every reader with a sense of urgency, and with an understanding of the history, significance, and need for studying restoration ecology and biology of fresh water habitats.

1.1 THE HYDROLOGIC CYCLE AND THE QUANTITY OF FRESH WATER

The amount of fresh water on the Earth is finite. Unlike fossil fuels, the other backbone of modern human economies, it has no substitute. It is essential for plant and animal metabolism, habitat for many species, and it is the fluid of the Earth's circulatory system. Water evaporating from land and water surfaces returns to the Earth as precipitation. It replenishes aquifers, flows across the land filling lakes, ponds, wetlands, and streams, and finally discharges to the oceans, bringing nutrients and organic matter that subsidize marine food chains. All life, and all human economies and cultures, are dependent on this hydrologic cycle.

Most fresh water is in ice caps, and 99% of liquid fresh water is in underground aquifers (Table 1.1). About 75% of groundwater has a residence time much greater than 100 years, and therefore is not considered renewable (Jackson et al., 2001). Although the amount of water in streams and lakes is small, it is renewed rapidly. Therefore, these habitats are the primary sustainable supplies of fresh water for most regions. Their protection, rational use, and restoration where needed, should be paramount in the water policies of every state and nation.

How much of this finite resource is available for current and future supplies to aquatic habitats and to human economies? Table 1.2 is a balance sheet of global fresh water runoff, including renewable groundwater, and a list of global water uses. The approximate total annual runoff is 40,700 km^3. When remote flows and uncaptured floodwaters are subtracted, the remainder (accessible runoff) is 12,500 km^3/year or 31% of total runoff. The estimated annual human use of accessible fresh water is 6,780 km^3/year (54%). Of this, 4,430 km^3/year is withdrawn (2880 km^3 by agriculture). About 65% of agricultural withdrawal is consumed via evapotranspiration (Postel et al., 1996; Postel, 2000). Less than half of accessible runoff remains for future human use and for support of aquatic ecosystems. More than 70% of accessible runoff may be appropriated by

TABLE 1.1
Water in the Biosphere

	Volume (thousands of km³)	Percentage of total	Renewal time
Oceans	1,370,000	97.61	3,100 years
Polar ice, glaciers	29,000	2.08	16,000 years
Groundwater (actively exchanged)	4,067	0.295	300 years
Fresh Water lakes	126	0.009	1–100 years
Saline lakes	104	0.008	10–1000 years
Soils and subsoil moisture	67	0.005	280 days
Rivers	1.2	0.00009	12–20 days
Atmospheric water vapor	14	0.0009	9 days

Source: Wetzel, R.G. 2001. *Limnology. Lake and River Ecosystems,* 3rd Edition. With permission.

TABLE 1.2
Global Runoff, Withdrawals, and Human Appropriations of Fresh Water Supply

Parameter	Fresh Water (km³/yr)
Total global runoff	40,700
Remote flow, total	7,800
Amazon basin	5,400
Zaire–Congo basin	660
Remote northern rivers	1,740
Uncaptured floodwater	20,400
Accessible runoff	12,500
Total human appropriation	6,780
Global water withdrawals, total	4,430
Agriculture	2,880
Industry	975
Municipalities	300
Reservoir losses	275
Instream uses	2,350

Note: Remote flow refers to river runoff that is geographically inaccessible for human use, estimated to include 95% of runoff in the Amazon basin, 95% of remote northern North American and Eurasian river flows, and 50% of the Zaire–Congo basin runoff. Runoff estimates also include renewable groundwater. An estimated 18% (2285 km³/year) of accessible runoff is consumed, compared with an estimated appropriation (including withdrawals and instream uses) of 6780 km³/year (54%). Water that is withdrawn but not consumed is not always returned to the same river or lake from which it was taken, nor does it always provide the same natural ecosystem functions.

Source: Jackson, R.B. et al., 2001. *Ecol. Appl.* 11: 1027–1045. With permission. Data shown are from Postel, S.L. et al., 1996. *Science* 271: 785–788.

humans by 2025 (Postel et al., 1996). It will be difficult to meet this water demand without great reductions in pollution and a shift in attitudes towards sustainable water use.

This balance sheet (Table 1.2) is deceptive because it provides the appearance that fresh water is abundant. But, precipitation is not evenly spread over the Earth's surface. Some regions are rich in fresh water (e.g., Canada) and others face chronic drought (e.g., U.S. Southwest, North Africa). A more revealing statistic of water scarcity is the per capita supply of a nation or region. This is the water supply for all activities, including food production, industry, waste disposal, and habitat for the rest of Earth's biota. "Water stressed" and "water scarce" countries and regions are those with less than 1,700 m^3 per person per year, and less than 1,000 m^3 per person per year, respectively (Postel, 1996). Many nations and regions are below these thresholds, and others soon will be.

Several interacting factors assure falling per capita water supplies, with impacts to aquatic ecosystems and human affairs extending far into the future. These factors, discussed in the following paragraphs, provide a persuasive rationale for fresh water protection, judicious use, and restoration.

The global human population has an annual net increase of more than 70 million, or a projected net increase of about 1.5 billion by 2025 (MacDonald and Nierenberg, 2003). Globally, human population growth is very rapid in some water-stressed nations (e.g., Egypt's population will double in less than 25 years) (Postel, 1992). Also, Egypt is an example of a nation dependent upon water originating outside its borders, forcing it to respond to any decision to reduce that supply. The most rapid human population growth in the United States may continue to be in states with rapidly declining water supplies (e.g., Florida, California, Arizona). Per capita supply must fall most rapidly where supplies are lowest and population growth highest.

Pollution and aquatic habitat destruction, directly linked to human economies and population growth, are increasing and "consume" water, thereby reducing per capita supply. Wetzel (2001) calls the combined impact of population and technology "demotechnic growth."

Climatologists have warned that significant human-induced climate change is occurring. Mean global temperature is increasing, perhaps as much as 1.5–5.8°C in this century, with earlier snowmelt and runoff that lead to altered flow regimes (e.g., winter and spring floods, summer droughts), and with projected changes in biota (including possibly severe species extinctions), abrupt climate shifts, falling lake levels, and more runoff and eutrophication from intense storms (Houghton et al., 2001; Jackson et al., 2001; Stefan et al., 2001; Poff et al., 2002; Parmesan and Yohe, 2003; Thomas et al., 2004). Changing climate is likely to have major impacts on aquatic systems and may contribute greatly to falling per capita supply.

Agriculture uses 65% of all water removed from lakes, reservoirs, and rivers, and most of this is then lost to evapo-transpiration (Postel, 1996). Animal agriculture requires huge amounts of water to produce feed grain (about 1000 metric tons per ton of grain). Unsustainable over-pumping of non-renewable groundwater for irrigation of grain land is common. Declining levels of the Ogallala Aquifer (Western High Plains), the source of water irrigating 20% of U.S. irrigated land, is one example. In California, groundwater over-pumping exceeds recharge by 1.6 billion m^3/year, mostly in the Central Valley that grows half of the U.S. fruits and vegetables (Postel, 1999). New sources of fresh water to meet the growing U.S. food demands are not evident. Worldwide, water demand for food production continues to escalate. In many places, water shortages must be met by importing grain, placing even greater pressure on surface and groundwater resources of the grain-producing regions.

The impact of cities on water supplies is increasing. By 2025, 61% of the global population will live in cities, requiring water previously used by agriculture (Postel, 2000). Figure 1.1 illustrates the diverse demands on Lake Biwa and the Yodo River, the water supplies for metropolitan Otsu, Kyoto, and Osaka, Japan. About 56% is used for power generation and is thus not consumed. Of the 44% that is consumed, agriculture uses about 35%. The fastest growing consumptive use, tap water, uses about 43%, and will produce a shortage of water for agriculture that must be made up by importing water in the form of grain and other food commodities. Thus, some other region or nation subsidizes this population with its own water.

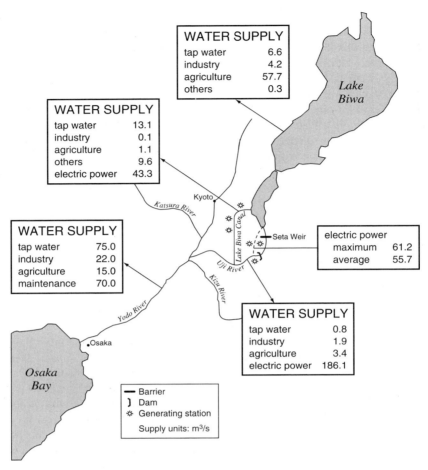

FIGURE 1.1 Water uses in the Lake Biwa-Yodo River basin. Numbers in boxes represent the relative water uses (m³/s) in various reaches of the Yodo River, Japan. Water uses changed dramatically from 1972–1992. Agricultural use rose 42%. (Redrawn from Ohkubo, 2000. With permission.)

These and other factors contribute to growing fresh water limitations. Conflicts between nations, states, regions, and cities are certain to intensify. For example, peace in the Middle East will not occur unless there are agreements among all parties about water in this water-scarce area of rapid population growth and political conflict (Hillel, 1994). In the U.S. there are many disputes over water, including current and future attempts to divert water from the Great Lakes to water-poor states of the West and Southwest (Beeton, 2002). The conflict at Klamath Lake (Oregon) between irrigators and the Klamath Indian Tribe's fish production is an example of a "water war," pitting economic and political interests against environmental and cultural needs (Service, 2003).

The needs of fresh water species for clean water and undisturbed habitats, and our reliance on processes of aquatic ecosystems for sustainable human economies, are often forgotten in our human-centric culture. Valuations of these services are difficult, but exceed ten billion dollars annually in the U.S. (e.g., Wilson and Carpenter, 1999). Despite this value, the species extinction rate in fresh water ecosystems is higher than in terrestrial systems (Postel, 2000), suggesting that additional ecological deficits may be developing, leading to unpredicted changes in these ecosystems and their "services" to humans.

Restoration of fresh water systems is an essential way of adding to a sustainable, high-quality water supply and to the beginnings of stabilizing or increasing the per capita supply (e.g., Cairns et al., 1992; Baron et al., 2002). Restoration and protection of fresh water systems is often directed toward impaired recreational sites. This is an important need and one of the primary topics in this book. We focus on reservoirs and natural lakes, and on streams to the extent that they transport particulate and dissolved organic and inorganic materials to these water bodies. However, lakes and reservoirs have values well beyond their recreational attributes. They are primary sources of raw potable water and irrigation water worldwide, they are habitat for thousands of species, and they contribute to ecosystem sustainability in many ways, including water and nutrient retention and storage. Because 75% of groundwater is not renewable (Jackson et al., 2001), the significance of surface waters, and the need to protect and restore them, will increase as impacts of social, economic, and political forces on fresh waters intensifies and per capita supply dwindles. Thus, human economic and personal security may become more dependent on our ability to restore impaired fresh water habitats. The "politics of scarcity" (Postel, 1996) will become increasingly important, and water wars seem inevitable. *The Last Oasis. Facing Water Security* (Postel, 1992) is "must reading" for every limnologist for its assessment of threats to fresh water, its forecast of future human demands, and its suggestions for remedies.

In this introductory chapter, we examine the condition of U.S. aquatic systems, with emphasis on lakes and reservoirs, and then focus on the characteristics of the restoration process and upon the history of lake restoration. This leads, in subsequent chapters, to discussions of the principles of limnology as they apply to restoration, to problem diagnosis and selection of restoration methods, and to detailed descriptions of methods to protect and restore lakes and reservoirs that have been impacted by eutrophication and exotic plants.

1.2 STATUS OF FRESH WATER IN THE UNITED STATES

The continued and growing need for sustainable supplies of high-quality fresh water is apparent. The U.S. Clean Water Act (Section 305 b) requires that states assess their water on a biennial basis to determine how well water quality standards are being met. This report to Congress and the nation is an analysis of our current water quality, and a reflection of the values we place on this resource. Local, state, and federal officials are the public's stewards of environmental quality and of the enactment and enforcement of laws to protect it. As citizens who use fresh water and as stewards who protect and regulate it, how well are we doing to assure its high quality?

The condition of fresh waters in the U.S., and elsewhere, is often described in reports listing lake, wetland, and stream conditions separately. This approach provides focus, but aquatic habitats are linked to each other, and to the land and air as well. Thus, degraded stream quality produces degraded lake or reservoir quality. Wetland destruction or alteration means reduced water storage or increased sediment and nutrient transport. Air pollution leads to lake acidification, and to their contamination with pollutants such as mercury. These and many other links between systems, often on very large scales, mean that restoration or protection of one unit, such as an individual lake, will most likely fail unless restoration and protection are linked to contiguous aquatic communities and their terrestrial environs.

Streams and groundwater are the primary sources of water to most lakes and reservoirs. The 1998 305(b) report (USEPA, 2000) for streams indicated that all 50 states and nine American Indian tribes had assessed 23% (1,355,463 km) of their total streams and rivers, a 21% increase over the 1996 report. Of these assessed streams, 35% (468,642 km) were impaired, primarily by siltation, pathogens, and nutrients. Other substances that would "impair" water quality, such as pesticides and PCBs, were not included in this analysis. Further, this 35% figure cannot be reliably extrapolated to all of the nation's streams and rivers because only a few states used a statistical design for sampling.

Worldwide, streams and rivers are the planet's most polluted ecosystems (Malmqvist and Rundle, 2002). The deposition of mercury in streams and other aquatic systems, followed by biomagnification, is one of the most serious and developing concerns everywhere (e.g., Peterson et al., 2002), and is a main reason for the growing list of fish consumption advisories issued by U.S. state and federal agencies (see http://map1.epa.gov/html/federaladv.html). Presently, the United States Environmental Protection Agency (USEPA) advises pregnant or nursing women to limit consumption of fresh water fish to one meal per week. The principal sources of mercury are industries, mining, trash incineration, and coal-fired power plants.

Despite the importance of flowing waters to our economic needs, their condition appears to be deteriorating. One measure of this is the number of imperiled lotic species. The annual extinction rates of North American fresh water fauna is about five times higher than terrestrial rates and 1000 times higher than historical rates. Extinction rates average about 4% per decade, approaching those of tropical rainforests. Projected percent species losses per decade are: fishes 2.4%, crayfish 3.9%, mussels 6.4%, and amphibians 3.0% (Ricciardi and Rasmussen, 1999). These rates signal deteriorating U.S. stream conditions.

Groundwater assessments were provided by 31 states and ten tribes for 146 aquifers or hydrogeologic settings in the 1998 305(b) report (USEPA, 2000), and overall groundwater quality was described as "good." These data are not representative and may be skewed towards a conclusion that groundwater quality is better than it actually is. Elevated levels of contaminants have been regularly reported, and water quantity is a concern because most aquifers contain non-renewable or "fossil" water.

Enormous wetland destruction continues worldwide. Most of it involves conversion to agricultural uses (e.g., Andreas and Knoop, 1992). About half of the original wetlands in the 48 contiguous states are gone, with a net loss of about 40,000 ha annually (Dahl, 1990; Dahl et al., 1991). This rate will increase following the 2001 U.S. Supreme Court ruling (SWANNC vs. U.S. Army Corps of Engineers) that held that non-navigable, isolated, intrastate waters cannot be protected by the Clean Water Act based on the use of these waters by migratory birds (the Migratory Bird Rule; Nadeau and Leibowitz, 2003). This decision greatly reduces the area of wetlands subject to federal regulation.

Only 11 states and tribes reported wetland data for the 1998 305(b) report (USEPA, 2000), and about 4% of the nation's wetlands were assessed (73% of that assessment was in North Carolina). Sedimentation, followed by draining, was the most widespread cause of wetland loss and pollution.

Wetland losses directly affect lake and reservoir quality because wetlands retain runoff of nutrients and particulate material. Wetland losses directly influence water quantity because they are significant storage sites and sources of groundwater recharge. For example, flooding in the Missouri and Mississippi River watersheds in 1993 produced billions of dollars in damages and many human injuries and deaths. Flooding would have been minimal if just half of the extirpated wetlands in these drainage basins had been left intact (Hey and Phillipi, 1995). A large amount of fresh water is lost annually to the oceans as floodwater.

The 48 contiguous states have about 100,000 lakes of 40 ha or more, and a total of 1.6×10^6 ha of lakes, ponds, and reservoirs. Forty-two states, Puerto Rico, the District of Columbia, and two tribes assessed 42% of their lake areas for the 1998 305(b) report (USEPA, 2000), a significant increase from the 1989 report when only 33 states and one territory reported. Hawaii, Idaho, Minnesota, New Jersey, Ohio, Pennsylvania, Wyoming, and Washington, did not submit lake data for the 1998 305(b) report to Congress.

The condition of U.S. lakes and reservoirs from the 1998 305(b) report is summarized in Table 1.3. The data supplied by states are not consistent from year to year for sample site selection, variables selected for analysis, analysis procedures, and frequency of sampling. They are not reliably comparable across states, or from year to year. Because these data were developed from non-statistically selected sampling sites, they pertain only to the lakes from which the samples were drawn, and no inference regarding the quality of the entire population of lakes in a state or region can be made. The survey serves primarily as a point-in-time snapshot of water quality.

Only 27% of the total lake area reported by states was actually assessed for its ability to support aquatic life. Not all states reported data in all condition classes. Based on our calculations from

TABLE 1.3
Lake Condition Relative to the Support of Habitat Suitable for Protection and Propagation of Desirable Fish, Shellfish, and Other Aquatic Organisms in Various States and one Territory

State	Total Reported (ha)[a]	Amount Assessed (ha)	Fully Support (%)[b]	Threatened (%)	Partially Support (%)	Not Support (%)	Not Attainable (%)
Alabama	198,494	187,422	67	15	17	2	0
Alaska	5,174,980	1,909	0	—[c]	100	—	—
Arizona	142,692	31,203	18	48	32	1	—
Arkansas	233,827	161,988	100	—	0	0	—
California	760,569	310,672	25	8	48	19	—
Colorado	66,382	24,144	88	—	11	1	—
Connecticut	26,294	10,970	88	10	1	0	0
Delaware	1,195	1,195	70	—	16	14	—
Dist. Columb.	96	96	57	0	0	43	0
Florida	843,848	259,992	46	7	35	12	—
Georgia	172,152	161,596	73	—	25	2	—
Hawaii	877	—	—	—	—	—	—
Idaho	283,290	—	—	—	—	—	—
Illinois	125,189	76,125	42	10	46	3	—
Indiana	57,871	5,445	50	50	0	0	0
Iowa	65,304	16,897	32	32	35	0	—
Kansas	73,387	73,387	0	51	47	2	0
Kentucky	92,427	88,014	74	24	2	<1	—
Louisiana	436,279	15,180	8	2	68	23	—
Maine	399.955	399,955	74	16	10	0	—
Maryland	31,552	8,502	37	—	63	0	—
Massachusetts	61,179	11,737	6	2	88	1	2
Michigan	360,002	3,299	—	—	—	100	—
Minnesota	1,331,503	—	—	—	—	—	—
Mississippi	202,254	11,108	66	32	2	0	0
Missouri	118,254	118,254	99	—	<1	1	—
Montana	341,189	322,622	14	—	86	1	—
Nebraska	113,316	49,262	68	13	19	<1	—
Nevada	215,818	85,932	74	—	8	18	—
N. Hampshire	68,8 02	65,344	97	—	2	1	0
N. Jersey	9,712	—	—	—	—	—	—
N. Mexico	403,674	50,517	11	—	89	<1	0
N. York	320,029	320,029	94	1	4	1	—
N. Carolina	125,957	125,957	68	30	2	<1	—
N. Dakota	267,141	259,247	24	72	4	0	—
Ohio	76,270	—	—	—	—	—	—
Oklahoma	421,650	245,275	21	35	38	5	—
Oregon	250,482	53,541	<1	35	0	65	—
Pennsylvania	65,336	—	—	—	—	—	—
Puerto Rico[d]	4,901	4,901	18	0	0	82	0
Rhode Island	8,620	6,436	43	43	11	3	—
S. Carolina	148,353	85,578	92	—	2	5	—
S. Dakota	303,525	53,484	16	—	26	58	0
Tennessee	217,725	217,752	90	—	3	7	—
Texas	1,240,648	533,685	89	0	7	4	0

TABLE 1.3 (Continued)
Lake Condition Relative to the Support of Habitat Suitable for Protection and Propagation of Desirable Fish, Shellfish, and Other Aquatic Organisms in Various States and one Territory

State	Total Reported (ha)[a]	Amount Assessed (ha)	Fully Support (%)[b]	Threatened (%)	Partially Support (%)	Not Support (%)	Not Attainable (%)
Utah	194,918	186,389	65	0	34	1	0
Vermont	92,641	6,614	23	35	24	18	—
Virginia	60,697	56,689	94	6	0	0	—
Washington	100,882	—	—	—	—	—	—
W. Virginia	9,054	8,710	11	21	60	8	0
Wisconsin	397,478	26,107	37	3	55	6	—
Wyoming	131,546	—	—	—	—	—	—
Totals	16,850,216	4,382,797[c]					

[a] Hectares × 2.47 = acres.

[b] Percentage fully supported, threatened, etc., is the percentage of the amount assessed, not the total reported. Percentages might not equal 100% due to rounding and/or a state not reporting in all categories.

[c] State did not report these data.

[d] U.S. Territory.

[e] Note that only about 29.5% of the total reported lake area has been assessed for this category by states who assessed lake conditions.

Source: USEPA, 2000. *National Water Quality Inventory. 1998 Report to Congress.* USEPA 841-R-00-001.

Table 1.3, about 53% of the assessed lake area fully supports aquatic life criteria (42 states reporting), 21% was threatened (29 states reporting), 12% partially supported (42 states reporting), and 12% did not support aquatic life criteria (43 states reporting).

The 305(b) report also described the ability of assessed lakes and reservoirs to support fish for consumption, as well as their condition for swimming, boating, drinking water supply, and agricultural uses. (Designated Uses; Table 1.4). These data are valuable, but in some cases are nonquantitative and often are based on opinion and perception.

Surface drinking water supplies in some regions are threatened by eutrophication, and by toxic materials that include pesticide residues from agriculture, and mercury. Waters from eutrophic reservoirs may have poor taste, odor, and color, and some have high concentrations of naturally occurring organic molecules that might form carcinogenic and mutagenic trihalomethanes and other by-products of raw drinking water disinfection with chlorine (Palmstrom et al., 1988; Cooke and Carlson, 1989; Cooke and Kennedy, 2001). Also, human gastrointestinal disorders have been associated with water consumption from reservoirs with cyanobacteria blooms (Kotak et al., 2000; Carmichael et al., 2001).

Two thirds of the U.S. population obtains drinking water from surface water sources, and of the 600 largest public utilities (serving more than 50,000 customers each), 68% obtain raw drinking water from lakes and reservoirs (Cooke and Carlson, 1989). These facts, combined with evidence of widespread surface water deterioration, suggest that there could be problems on the horizon with regard to drinking water quality and human health.

The status of lakes and reservoirs other than those in the U.S. is less well known. Canada has the largest lake area of any country, and an inventory, much less an assessment of their trophic status is not feasible at this time. Most are oligotrophic, though some lakes in southern Canadian provinces are impaired from domestic and agricultural runoff. Eutrophication is widespread in

TABLE 1.4
Proportion of Lake, Reservoir, and Pond Areas Assessed by States for Each Category of Designated Use

Designated Use	Assessed Area (ha)[a]	Fully Support (%)	Threatened (%)	Partially Support (%)	Not Support (%)	Not Attainable (%)
Aquatic life support	4,955,662	58	13	23	6	<1
Fish consumption	3,172,195	54	5	35	6	<1
Swimming	5,833,293	69	11	15	5	<1
Boating	2,963,548	78	8	10	4	<1
Drinking water	3,406,880	82	4	9	5	0
Agricultural use	1,904,246	89	4	3	4	0

[a] Hectares × 2.47 = acres.

Source: USEPA, 2000. *National Water Quality Inventory. 1998 Report to Congress.* USEPA 841-R-00-001.

Europe, but reports such as the U.S. 305(b) report are not available from all nations. Accounts of extensive soil erosion and massive siltation of reservoirs throughout the world, coupled with the absence of wastewater treatment in many regions, indicate that eutrophication is a worldwide problem, especially in developing nations (Bronmark and Hansson, 2002). Rapid in-filling of major reservoirs in developing nations is particularly troubling in view of their need for irrigation water, potable supplies, and flood control.

These reports about the quality of fresh waters are not encouraging. A significant fraction of our fresh water systems have not been assessed, and of those that have been, a disturbing percentage are impaired. Some states in the U.S. have not reported at all, and others make the unrealistic statement, perhaps based on selective sampling, that 99–100% of their waters fully support aquatic life conditions (Table 1.3). The answer to the question that began this section, "if fresh water is so valuable, what is its quality?" might be answered as "not as good as expected or needed." In view of the projected increases in demand for clean fresh water, restoration and protection of lakes and reservoirs will become increasingly important.

1.3 SOURCES OF LAKE AND RESERVOIR PROBLEMS

The preceding paragraphs indicate that many streams, lakes, and reservoirs in the United States, Europe, and elsewhere have serious water quality problems. The causes and correctives of some of these problems are the subject of this book. The next sections of this chapter are an introduction to the sources of these problems and to topics we address in subsequent chapters (see Figures 1.2 and 1.3).

Control of point source nutrients and toxic materials, such as wastewater or industrial discharges, was the primary focus of efforts to protect and improve streams and lakes in the 1970s and 1980s. Laws were enacted and enforced, including bans on phosphorus (P) in detergents, leading to a significant decline in point source loading to aquatic habitats in the U.S. Now the chief contaminant and nutrient sources to streams and lakes are "non-point" sources (NPS) such as agricultural runoff, erosion from urban or deforested areas, surface mining, or atmospheric depositions (Table 1.4).

According to the 305(b) report (USEPA, 2000), 45% of assessed lakes and reservoirs were impaired by nutrients, primarily from agriculture (Table 1.4). Agriculture has become dependent on fertilizers, manure, and pesticides to meet growing and changing food demands, but it has not

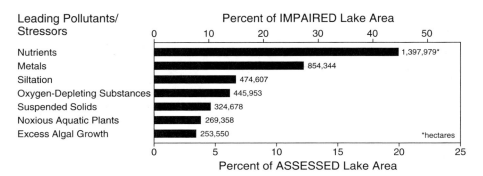

FIGURE 1.2 Leading lake pollutant/stressor types. (Redrawn from USEPA, 2000. *National Water Quality Inventory. 1998 Report to Congress.*)

made a sustained and effective effort to prevent runoff. About 11 million metric tons of nitrogen (N) and about 2 million metric tons of P are applied to the land annually as fertilizer. Another 6.4 million metric tons of N and 2 million metric tons of P are applied as manure (U.S. Geological Survey, 1999). A large fraction of these applications runs off to streams. In addition, poultry, beef, and pork production consume huge amounts of water in the form of grain production, and are increasingly centered in streamside feedlots that may be major non-point nutrient sources.

Managing non-point sources, and restoring lakes and reservoirs exposed to them, are among the major topics of this book and represent a difficult environmental problem to solve. One consequence of external nutrient loading to lakes is that nutrients settle to the lake bottom, often as detritus, and are recycled and sustain impaired lake conditions. Treatment of this internal non-point source is often as significant as the treatment of runoff, and many lakes require attention to both sources. Control of non-point "internal loading" is a major feature of this text.

A problem at least equal to excessive nutrient loading is the establishment of exotic species. (NALMS, 2002; Mack et al., 2000). One example is the common carp (*Cyprinus carpio*), an enormously destructive fish introduced to the United States in the late 1800s as a food source, and distributed by the millions from rail cars to lakes and streams along the rights-of-way (Bright, 1998). The Great Lakes provide other examples. At least 141 exotic species are established in them, and the invasion rate is one new species annually (Mills et al., 1998; cited in Bright, 1998). Perhaps the most widely known are the zebra (*Dreissena polymorpha*) and quagga mussels (*D. bugensis*). Zebra mussels spread from the Great Lakes into the Ohio, Mississippi, Illinois, Tennessee, and Hudson River drainages, with great economic and ecological impacts. Zebra mussels have the potential to bring about the extinction of nearly half of the 300 species of fresh water mussels in North America (Abramovitz, 1996), may alter fish production (Baldwin et al. 2002), increase biofouling of potable and cooling water intakes, and stimulate production of cyanobacteria (Garton, 2002). The speed of successful exotic species invasions has increased, partly from the "stock and see" (Bain, 1993) attitude of agencies, agriculturalists, and horticulturalists, exemplified by the introductions of common carp and white amur (*Ctenopharyngodon idella*), and the propagation of purple loosestrife (*Lythrum salicaria*) by the horticulture industry.

Throughout this text, methods to manage and control exotic organisms in lakes, primarily plants such as Eurasian watermilfoil (*Myriophyllum spicatum*), are discussed. Eradication of exotics, in most cases, is impossible, and our increasingly "borderless world" (Bright, 1998) promises that more and more of them will be successful and may dwarf our concerns about excessive nutrients (Bronmark and Hansson, 2002). How many cases like Lake Victoria (Uganda, Nigeria, Tanzania) will there be, where more than 200 native fish species were extirpated by the introduction of the Nile perch? The "stock and see" attitude may be over, but human activities continue to encourage the spread of exotic species to lakes (Hall and Mills, 2000).

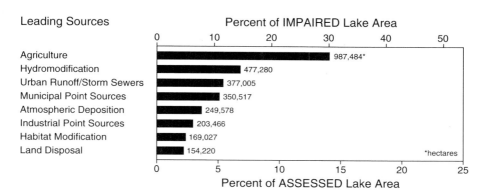

FIGURE 1.3 Leading lake pollutant sources. (Redrawn from USEPA, 2000. *National Water Quality Inventory. 1998 Report to Congress.*)

Lakes and reservoirs, as well as other aquatic habitats, are essential resources at growing risk from non-point pollution and exotic species invasions, as well as from factors like excessive water withdrawals, climate change, and human population increases. All of these forces may reduce per capita water availability in some U.S. and world regions to levels that threaten human well-being, human economies, and biotic integrity. These facts are well known to many citizens, policy makers and politicians, and industrial/agricultural water users. Some of these large-scale problems of water quality and quantity, especially in food- and water-rich nations, seem to be willful and created without concern for the future.

The U.S. National Research Council (Cairns et al., 1992) identified fresh water as a crucial, irreplaceable resource that is threatened everywhere and in decline in many areas. They concluded (p. 37): "To withstand the possible compound stresses from increasing population, and increased demands for aquatic ecosystem services, prudence requires that the nation adopt a national aquatic ecosystem restoration agenda." Aquatic restoration is the subject of this book. While we will often emphasize restoration of recreational lakes, we focus consistently on abatement of the causes of lake impairment that are common to all lakes and reservoirs.

Pollution and destruction of fresh water habitats is a global problem and serious ecological deficits have occurred. We need to learn as much as possible about these ecosystems, including studies of their fundamental properties and behaviors. In many areas of the Earth, even the most rudimentary research has yet to be performed. This basic science approach provides the understanding necessary to develop protection, management, and restoration plans. Thus the training of new limnologists, water scientists, engineers, including those interested in the science of fresh water management and restoration, is a critical task.

1.4 RESTORATION AND MANAGEMENT OF LAKES AND RESERVOIRS

The most obvious, persistent, and pervasive global water quality problem at this time is eutrophication. Lakes and reservoirs have deteriorated through excessive loading of plant nutrients, organic matter, and silt, that cause increased primary producer biomass, reduced water clarity, good growing conditions for nuisance species, and usually decreased lake or reservoir volumes. Eutrophic water bodies lose much of their beauty, their attractiveness for recreation, and their usefulness and safety as industrial and domestic water supplies. Their degraded condition reflects, in large part, the effects of point and especially non-point loading. The symptoms of eutrophication, including impaired potable supplies, dissolved oxygen depletions, and fish kills, can bring about economic losses in the forms of decreased property values, high cost treatments of raw drinking water, illness, depressed recreation industries, need to build new reservoirs, and the costs for management and restoration (Pretty et al., 2003).

The first and most obvious step towards improving lake or reservoir water quality is to limit, divert, or treat excessive external loading. Unlike streams, which respond quickly to pollution abatement unless toxic materials or nutrients contaminate their sediments, lakes and reservoirs trap and recycle materials. Elimination or curtailment of external loading, an expensive and necessary step, may be insufficient to produce immediate or long-lasting improvements due to the presence of extensive shallow areas and to internal recycling of toxics or nutrients. These lake characteristics allow continued production of algae and macrophytes. Several authors (e.g., Rich and Wetzel, 1978; Carpenter, 1981; Moss et al., 1996; Sondergaard et al., 2001) identified feedback loops in eutrophic systems that maintain the eutrophic state after loading has been curtailed, including macrophyte growth–death–decay cycles, nutrient release from sediments, and bioturbation. A second step, a technique to manipulate or alter one or more internal chemical, biological or physical processes or conditions, may be needed to restore the water body.

In the strict sense, *restoration* means returning something to its original form. With the possible exception of dredging, which removes nutrients, excessive vegetation, and sediment, hopefully toward a former state, other lake treatments are not true restoration activities. Restoration (*sensu strictu*) is defined (Webster, 1972) as "the bringing back to a former or normal condition, as by repairing, rebuilding, and/or altering." Cairns et al. (1992) defined restoration as (p.18) "the return of an ecosystem to a close approximation of its condition prior to disturbance." They recognized that actual restoration is not possible. The pre-disturbance species list and the chemical, biological, and physical interactions for a lake and its watershed are not known for any lake.

A more accurate term for what we, as restoration ecologists, are attempting, is *rehabilitation* (Cooke, 1999). Rehabilitation means re-establishment of important missing or altered processes, habitats, concentrations, and species. It is a well-understood idea in the medical community where treatment of the severely ill or injured can only return patients to a condition allowing the most important and vital functions to take place. There is no attempt to make the patient exactly as before the injury or illness. This concept emphasizes return of degraded systems to attainable approximations of pre-disturbance conditions, and the establishment of protections against future disturbances. Although the term "restoration" is commonly used, and we use it in this book, what is referred to in most cases is "rehabilitation."

There are some procedures that are not considered to be "restoration/rehabilitation" activities, because problem causes are not addressed and protections are not implemented. They include harvesting, herbicide/algicide application, and stocking of plant-eating exotic fish. However, there are circumstances where these procedures are appropriate.

All lake protection and restoration activities involve *management*. There is essential day-to-day management such as maintenance of an air compressor. Nature is not static, and human actions and behaviors are very dynamic, meaning that effectiveness must be monitored and changes made when necessary. A sedimentation basin, or fish removal project, as examples, require long-term management that might include project re-evaluation and additional efforts.

Reservoirs are different from lakes in many ways but they can be rehabilitated as well. Their drainage basins are usually much larger than lakes, and many are located in watersheds with extensive agricultural activities. Reservoirs therefore typically lose volume to siltation. Sediment removal, coupled with land management and the construction of devices to trap silt, are examples of reservoir restoration and protection. Practically, however, many reservoir rehabilitation projects, like some lake treatments, primarily involve the treatment of symptoms because the impacts of land activities are usually very large.

We have not included a discussion of every lake problem. Excluded are topics such as institutional arrangements, the formation and operation of a lake association, sports fishery management, and other related problems that are described in publications such as Moore and Thornton (1988) or Olem and Flock (1990).

The focus of this book is on the management and restoration of eutrophic lakes and reservoirs. We examine the most common in-lake techniques, and the procedures to reduce loading, with regard to their scientific basis, application methods, effectiveness, feasibility, drawbacks, and where known, costs. We also identify further research and development.

1.5 HISTORY OF LAKE RESTORATION AND MANAGEMENT

This book would be incomplete without providing some understanding of the short history of lake restoration. Hasler (1947) was among the first to recognize that restoration of eutrophic lakes would be difficult. He stated: "The problem is especially serious because there is no way at present for reversing the process of eutrophy." Twenty years later, most of the techniques described in this book had been suggested, some had been tried, and many were discussed in symposia. By the late 1960s, publication of several of these national (American Association for the Advancement of Science, 1970) and international (U.S. National Academy of Sciences, 1969) proceedings and literature reviews (e.g., Hasler, 1969) began to inform a wider audience concerning lake eutrophication and lake restoration.

In the 1970s, stimulated and supported in part by the USEPA Clean Lakes Program (CLP) (Section 314 of the Clean Water Act), research into and application of these techniques intensified. Spencer Peterson of USEPA's Office of Research and Development led the lake restoration evaluation portion of the CLP. Conferences were held, initiated at the University of Wisconsin by Lowell Klessig, Jim Peterson, Doug Knauer, George Gibson, and many others. The United States joined with other nations of the Organization for Economic Cooperation and Development (OECD) to study the problem (USEPA, 1977). Numerous reviews and conference proceedings were published, among them Peterson et al. (1974), Dunst et al. (1974), USEPA (1979; 1980a, b; 1983), Ryding (1981), Golterman and de Oude (1991), Lee and Jones (1991), Welch (1992), Moss (1996), Smith et al. (1999), and Gulati and van Donk (2002).

In 1980, as a product of some of these conferences and the intense concern about eutrophication and its remedies, a new professional society, the North American Lake Management Society (NALMS, 4513 Vernon Blvd., Suite 100, Madison, WI 53705-5443, U.S.) was founded at the Portland, Maine International Symposium on Inland Waters and Lake Restoration. This was a jointly sponsored USEPA and OECD meeting. A history of the movement to organize this new society, with notes from past presidents to celebrate its 20th anniversary, was published by NALMS (2000a, b). NALMS has, for over 20 years, continued to act as an information conduit among lake scientists, the public, consultants, and the U.S. Congress on matters concerning eutrophication and lake management. NALMS was instrumental in urging Congress to fund Section 314 (CLP) of the Clean Water Act for several years. Its dedicated members and leadership have been very effective in developing and providing information needed by technical and political/regulatory communities regarding the plight of aquatic systems and their protection and restoration. No other organization has filled this role.

Despite termination of federal funding for the CLP, money for lake projects is still available through Section 319 (non-point Source Program). USEPA guidance to states concerning NPS funds encourages their use for lake and reservoir restoration and protection. There is also encouragement to request Clean Water Act State Revolving Funds for lake restoration and protection, including drinking water supply reservoirs. In addition, USEPA encourages use of Safe Drinking Water Act funds for priority lake management projects. Funds through the NPS program are available for all phases of a lake restoration project, including diagnosis, implementation, and post-treatment monitoring. There are qualifications associated with NPS program funding, including the stipulation that a lake(s) must be included in the state NPS management program and only up to 20% of state NPS allocations can be used to update state NPS assessment programs, including Phase I diagnostic-feasibility studies and state-wide lake water-quality assessments. Despite these limitations, potentially there is much more money available under the current USEPA guidelines than was available in CLP appropriations (D. Weitman, Director of USEPA NPS Program, personal communication).

Since 1988, $900 million has been directed to NPS projects. Using the 20% guideline, up to $180 million could have been applied to lake projects.

In 1985, NALMS began publication of a new professional periodical, *Lake and Reservoir Management*, in 1986 the first edition of this book (Cooke et al., 1986) was published, and in 1988 a book written specifically for interested citizens and lake managers by NALMS members, *The Lake and Reservoir Restoration Guidance Manual*, updated in 1990, was published by the USEPA CLP (Moore and Thornton, 1988; Olem and Flock, 1990). A third edition was published in 2001 (NALMS and Terrene Institute, 2001).

There was rapid development and testing of lake and reservoir management technologies in Europe, Canada, and the United States in the last two decades. The role of biology in lake management was intensely explored, following the suggestion of Shapiro et al. (1975) that fish activities could be important determinants of algal biomass. This led to new hypotheses and discoveries about lake productivity controls (e.g., Carpenter and Kitchell, 1993; Benndorf et al., 2002). The use of phytophagous insects (e.g., Van Driesche et al., 2002) for macrophyte management became widespread in southern waters, and grass carp were introduced throughout the continent (Leslie et al., 1987). A major advance was the development of software and an applications manual, based upon a large database from reservoirs, to guide managers in determining reservoir conditions and their responses to changed external or internal P loading (Walker, 1987). The small lake or pond owner was not neglected. Small scale, low cost, and effective procedures were described in two useful books (McComas 1993, 2003), written specifically for consultants and small lake owners.

Nowhere has the development of lake management and restoration technology been more rapid than in Western Europe, particularly Sweden, Denmark, Germany, U.K., and The Netherlands. European limnologists (e.g., Moss et al., 1996; Hosper, 1997; Jeppesen et al., 1999; Nienhuis and Gulati, 2002) were among the leaders in developing an understanding of shallow lake ecology and management.

The U.S. Army Corps of Engineers, concerned about water quality in the hundreds of reservoirs it manages, produced a volume on eutrophication and reservoir management (Cooke and Kennedy, 1989), and reports on sampling, data management, and modeling (e.g., Walker, 1987; Kennedy, 1999a, b, 2001).

Eutrophication poses special problems for potable water supplies, particularly the production of disinfection by-products and the appearance of taste and odor compounds and toxins produced by algae, fungi, and bacteria. The American Water Works Association Research Foundation, NALMS, and European investigators, among others, have been interested in this problem (e.g., Cooke and Carlson, 1989; Heinzman and Sarfert, 1995; Cooke and Kennedy, 2001), and with watershed management to protect water supply lakes and reservoirs (Robbins et al., 1991; Pannetter, 1991; Smith et al., 2002).

The history of lake restoration and protection has not been without controversy. Based on limited evidence it was assumed that P, and to a lesser extent N, were the essential nutrients that limited algal biomass in fresh waters. Therefore, lake and reservoir management was directed towards reducing the concentration of these elements by limiting loading from wastewater and by bans on P-containing detergents. Phosphorus limitation of algal biomass was challenged in the mid-1970s by proponents of the hypothesis that carbon was the limiting nutrient (the so-called limiting nutrient controversy). It is now clear that P is most often the limiting nutrient, on a long-term basis (Schindler, 1974; Guildford and Hecky, 2000).

These findings about P limitation of algal biomass led many limnologists and lake managers to base their approach to the eutrophication problem on controlling P concentration. While this emphasis led directly to significant lake and reservoir improvements, it often restricted the view of eutrophication, and its control, to simply that of external nutrient loading and algal biomass. Most lakes and many reservoirs are small and/or shallow, with extensive wetland and littoral zones, macrophyte development, and high ratios of bottom sediment to lake volume (Wetzel, 1990). Lakes and reservoirs, and their associated wetland and littoral zones, contain interacting food webs and dynamic stores of nutrients in their sediments that directly affect lake condition. Lakes are not simply reaction vessels containing water, nutrients, and algae. Recognition of the significance of sediments and highly productive shallow areas, and the roles of biological interactions and feedback processes, is now standard in lake restoration. We emphasize these views throughout, and discourage the erroneous belief that eutrophication is simply the result of excessive external nutrient loading and algal responses. These views were strongly influenced by Robert Wetzel's limnology text (Wetzel, 2001).

Our book is divided into four units: principles of limnology and problem diagnosis/evaluation, methods of algal biomass control, methods of macrophyte biomass control, and multiple benefit treatments. It is our contention that we must learn as much as we can about fresh water ecosystems, emphasizing ecosystem-level questions. We agree with Moss (1999) that reductionistic research is unlikely to provide many useful advances in understanding the functioning of large systems such as lakes, or how to manage and rehabilitate them. While our emphasis is on applied limnology, which too often leads to "quick-fix" approaches, we wish also to emphasize that these methods and procedures, present and future, must have a foundation in the basic science of limnology.

Aquatic macrophyte biology is poorly understood by many applied limnologists, in part because many modern limnology texts and courses essentially ignore macrophytes. Therefore, we greatly expanded the sections on plant ecology and restoration of plant communities.

We did not include herbicide use in the last edition (Cooke et al., 1993) due to possible harsh secondary effects of some herbicides, though we did acknowledge that herbicide treatment might be the only reasonable approach in some cases (e.g., southern waters where exotics like water hyacinth have become established). Also in 1993, relatively comprehensive coverage of herbicide applications was available (e.g., Ross and Lembi, 1985; Westerdahl and Getsinger, 1988). Herbicides in popular use now are less toxic to non-target organisms, decompose fairly quickly, apply more readily, and some are specific to selected plant species thereby reducing the potential for complete plant eradication and adverse secondary effects. We give much greater coverage to herbicide U.S.ge in this edition.

There are other important topics on lake and reservoir management and restoration, such as problems of toxic metals and organics, but we have chosen not to address them, except for their removal via dredging (Chapter 20), because they have been covered extensively elsewhere or are beyond the scope of this book.

It is our sincere desire that we have retained the good, discarded the bad, and added enough of the new, that this book is even more helpful than previous editions. It represents nearly 175 years of our combined knowledge and experience, but it can only begin to meet the needs in this field. We hope it will stimulate additional research and applications that improve our understanding of lakes and of what does and does not work, and at the same time will serve as a useful guide to limnologists, students, consultants, engineers, lake managers, and others who are trying to solve lake and reservoir problems.

REFERENCES

Abramovitz, J.N. 1996. Imperiled Waters, Impoverished Future: The Decline of Fresh Water Ecosystems. WorldWatch Paper 128. WorldWatch Institute. Washington, DC.

American Association for the Advancement of Science. 1970. *Lake Restoration.* Proceedings of a Symposium. Washington, DC (tapes 67–70. Sessions 1 and 2, tapes 2-3730–2-3722).

Andreas, B.K. and J.D. Knoop. 1992. 100 years of changes in Ohio peatlands. *Ohio J. Sci.* 92: 130–138.

Bain, M.B. 1993. Assessing impacts of introduced species: Grass carp in large systems. *Environ. Manage.* 17: 211–224.

Baldwin, B.S., M.S. Mayer, J. Dayton, N. Pau, J. Mendilla, M. Sullivan, A. Moore, A. Ma and E.L. Mills. 2002. Comparative growth and feeding in zebra and quagga mussels (*Dreissena polymorpha* and *Dreissena bugensis*): Implications for North American lakes. *Can. J. Fish. Aquatic Sci.* 59: 680–694.

Baron, J.S., N.L. Poff, P.L. Angermeier, C.N. Dahm, P.H. Glieck, N.G. Hairston, Jr., R.B. Jackson, C.A. Johnston, B.D. Richter and A.D. Steinman. 2002. Meeting ecological and societal needs for fresh water. *Ecol. Appl.* 12: 1247–1260.

Beeton, A.M. 2002. Large fresh water lakes: Present state, trends, and future. *Environ. Conserv.* 29: 21–38.

Benndorf, J., W. Boing, J. Koop and I. Neubauer. 2002. Top-down control of phytoplankton: The role of time scale, lake depth and trophic state. *Freshwater Biol.* 47: 2282–2295.

Bright, C. 1998. *Life Out of Bounds. Bioinvasion in a Borderless World.* W.W. Norton & Co., New York, NY.

Bronmark, C. and L.-A. Hansson. 2002. Environmental issues in lakes and ponds: Current state and perspectives. *Environ. Conserv.* 29: 290–306.

Cairns, J., et al. (National Research Council). 1992. *Restoration of Aquatic Ecosystems. Science, Technology, and Public Policy.* National Academy Press, Washington, DC.

Carmichael, W.W., S.M.F.O. Azevedo, J.S. An, R.J.R. Molica, E.M. Jochimsen, S. Lau, K.L. Rinehart, G.R. Shaw and G.K. Eaglesham. 2001. Human fatalities from Cyanobacteria: Chemical and biological evidence for cyanotoxins. *Environ. Health Perspect.* 109: 663–668.

Carpenter, S.R. 1981. Submersed vegetation: An internal factor in lake ecosystem succession. *Am. Nat.* 118: 372–383.

Carpenter, S.R. and J.F. Kitchell. 1993. *The Trophic Cascade in Lakes.* Cambridge University Press. Cambridge, U.K.

Cooke, G.D. 1999. Ecosystem rehabilitation. *Lake and Reservoir Manage.* 15: 1–4.

Cooke, G.D. and R.E. Carlson. 1989. *Reservoir Management for Water Quality and THM Precursor Control.* American Water Works Association Research Foundation. Denver, CO.

Cooke, G.D. and R.H. Kennedy. 1989. Water Quality Management for Reservoirs and Tailwaters. Report I. In-Reservoir Water Quality Management Techniques. Tech. Rept. E-89-1. U.S. Army Corps Engineers. Vicksburg, MS.

Cooke, G.D. and R.H. Kennedy. 2001. Managing drinking water supplies. *Lake and Reservoir Manage.* 17: 157–174.

Cooke, G.D., E.B. Welch, S.A. Peterson and P.R. Newroth. 1986. *Lake and Reservoir Restoration.* Butterworth, Stoneham, MA.

Cooke, G.D., E.B. Welch, S.A. Peterson and P.R. Newroth. 1993. *Restoration and Management of Lakes and Reservoirs, 2nd Edition.* Lewis Publishers and CRC Press. Boca Raton, FL.

Dahl, T.E. 1990. *Wetland Losses in the United States, 1970s to 1980s.* U.S. Department of the Interior. Fish and Wildlife Service, Washington, DC.

Dahl, T.E., C.E. Johnson and W.E. Frayer. 1991. *Status and Trends of Wetlands in the Conterminous United States, mid-1970s to mid-1980s.* U.S. Department of the Interior. Fish and Wildlife Service, Washington, DC.

Dunst, R.D., S.M. Born, P.D. Uttomark, S.A. Smith, S.A. Nichols, J.O. Peterson, D.R. Knauer, S.L. Serns, D.R. Winter and T.L. Wirth. 1974. Survey of Lake Rehabilitation Techniques and Experiences. Tech. Bull. 75. Wisconsin Dep. Nat. Res., Madison, WI.

Garton, D.W. 2002. Ecological consequences of zebra mussels in North American lakes. North American Lake Management Society. *LakeLine* 22(1): 48–51.

Golterman, H.L. and N.T. deOude. 1991. Eutrophication of lakes, rivers, and coastal seas. In: O. Hutzinger (Ed.). *The Handbook of Environmental Chemistry.* Vol. 5, Part O. Springer-Verlag, Berlin. pp. 79–124.

Guildford, S.J. and R.E. Hecky. 2000. Total nitrogen, total phosphorus, and nutrient limitation in lakes and oceans: Is there a common relationship? *Limnol. Oceanogr.* 45: 1213–1223.

Gulati, R.D. and E. van Donk. 2002. Lakes in The Netherlands, their origin, eutrophication and restoration: State-of-the-art review. In: P.H. Nienhuis and R.D. Gulati (Eds.), *Ecological Restoration of Aquatic and Semi-Aquatic Ecosystems in the Netherlands (NW Europe).* Kluwer Academic Publishers, Boston, MA. Reprinted from *Hydrobiologia*, Volume 478, 2002. pp. 73–106.

Hall, S.R. and E.L. Mills. 2000. Exotic species in large lakes of the world. *Aquatic Ecosystem Health Manage.* 3: 105–135.

Hasler, A.D. 1947. Eutrophication of lakes by domestic drainage. *Ecology* 28: 383–395.

Hasler, A.D. 1969. Cultural eutrophication is reversible. *BioScience* 19: 425–431.

Heinzman, B. and F. Sarfert. 1995. An integrated water management concept to ensure a safe drinking water supply and high drinking water quality on an ecologically sound basis. *Water Sci. Technol.* 31: 281–291.

Hey, D.L. and N.S. Phillipi. 1995. Flood reduction through wetland restoration: The Upper Mississippi River Basin as a case history. *Restor. Ecol.* 3: 4–17.

Hillel, D. 1994. *Rivers of Eden. The Struggle for Water and the Quest for Peace in the Middle East.* Oxford University Press, New York, NY.

Hosper, H. 1997. *Clearing Lakes. An Ecosystem Approach to the Restoration and Management of Shallow Lakes in The Netherlands.* RIZA, Lelystad, The Netherlands.

Houghton, J.T., Y. Ding, D.J. Griggs, M. Noguer, P.J. van der Linden and V. Xiaosu. 2001. Intergovernmental Panel on Climate Change: Working Group 1. Cambridge University Press. Cambridge, U.K.

Jackson, R.B., S.R. Carpenter, C.N. Dahm, D.M. McKnight, R.J. Naiman, S.L. Postel and S.W. Running. 2001. Water in a changing world. *Ecol. Appl.* 11: 1027–1045.

Jeppesen, E., J.P. Jensen, M. Sondergaard and T. Lauridsen. 1999. Trophic dynamics in turbid and clearwater lakes with special emphasis on the role of zooplankton for water clarity. *Hydrobiologia* 408/409: 217–231.

Kennedy, R.H. 2001. Considerations for establishing nutrient criteria for reservoirs. *Lake and Reservoir Manage.* 17: 175–187.

Kennedy, R.H. 1999a. Basin-wide considerations for water quality management: Importance of phosphorus retention by reservoirs. *Int. Rev. ges. Hydrobiol.* 84: 557–566.

Kennedy, R.H. 1999b. Reservoir design and operation: Limnological implications and management opportunities. In: J.G. Tundisi and M. Straskraba (Eds.), *Theoretical Reservoir Ecology and its Applications.* International Institute of Ecology, Brazilian Academy of Sciences and Backhuys Publishers. pp. 1–28.

Kotak, B.G., A.K. Lam, E.E. Prepas and S.E. Hurley. 2000. Role of chemical and physical variables in regulating microcystin-LR concentration in phytoplankton of eutrophic lakes. *Can. J. Fish. Aquatic Sci.* 57: 1584–1593.

Lee, G.F. and R.A. Jones. 1991. Effects of eutrophication on fisheries. *Rev. Aquatic Sci.* 5: 287–305.

Leslie, A.J., Jr., J.M. VanDyke, R.S. Hestand, III and B.Z. Thompson. 1987. Management of aquatic plants in multi-use lakes with grass carp (*Ctenopharyngodon idella*). *Lake and Reservoir Manage.* 3: 266–276.

MacDonald, M. and D. Nierenberg. 2003. Linking population, women, and biodiversity. In: L. Starke (Ed.), *State of the World 2003.* W.W. Norton & Co., New York, NY. Chapter 3.

Mack, R.N., D. Simberloff, W.M. Lonsdale, H. Evans, M.Clout and F.A. Bazzaz. 2000. Biotic invasions: Causes, epidemiology, global consequences, and control. *Ecol. Appl.* 10: 689–710.

Malmqvist, B. and S. Rundle. 2002. Threats to the running water ecosystems of the world. *Environ. Conserv.* 29: 134–153.

McComas, S. 1993. *Lake Smarts. The First Lake Maintenance Handbook.* Terrene Institute, Washington, DC and U.S. Environmental Protection Agency, Office of Water, Washington, DC.

McComas, S. 2003. *Lake and Pond Management Guidebook.* Lewis Publishers and CRC Press, Boca Raton, FL.

Mills, E.L., S.R. Hall and N.K. Pauliukonis. 1998. Exotic species in the Laurentian Great Lakes. *Great Lakes Res. Rev.* February 1998.

Moore, L. and K. Thornton (Eds.). 1988. *Lake and Reservoir Restoration Guidance Manual.* USEPA 440/5-88-002.

Moss, B. 1996. A land awash with nutrients — the problem of eutrophication. *Chem. Ind.* 3: 407–411.

Moss, B. 1999. Ecological challenges for lake management. *Hydrobiologia* 395/396: 3–11.

Moss, B., J. Madgwick and G. Phillips. 1996. *A Guide to the Restoration of Nutrient-Enriched Shallow Lakes.* Broads Authority, Norfolk, U.K.

Nadeau, T.-L. and S.G. Leibowitz. 2003. Isolated wetlands: An introduction to the special issue. *Wetlands* 23: 471–474.

Nienhuis, P.H. and R.D. Gulati. 2002. *Ecological Restoration of Aquatic and Semi-aquatic Ecosystems in The Netherlands.* Kluwer Academic, Dordrecht, The Netherlands and Norwell, MA,

North American Lake Management Society. 2000a. History. *LakeLine* 20(2): 17–18.

North American Lake Management Society. 2000b. Past presidents of NALMS. *LakeLine* 20(2): 20–35.

North American Lake Management Society and Terrene Institute. 2001. *Managing Lakes and Reservoirs.* NALMS. Madison, WI.

North American Lake Management Society. 2002. Exotic species. *LakeLine* 22(1): 17–57.

Olem, H. and G. Flock (Eds.). 1990. *Lake and Reservoir Restoration Guidance Manual, 2nd Edition*. USEPA 440/4-90-006.

Ohkubo, T. 2000. Lake Biwa. In: M. Okada and S. Peterson (Eds.), *Water Pollution Control Policy and Management: The Japanese Experience*. Gyosei Publishers, Tokyo, Japan. Chapter 14. pp. 188–217.

Palmstrom, N.S., R.E. Carlson and G.D. Cooke. 1988. Potential links between eutrophication and the formation of carcinogens in drinking water. *Lake and Reservoir Manage.* 4(2): 1–15.

Pannetter, E. 1991. Watershed protection and compliance with the Safe Drinking Water Act amendments. *Lake and Reservoir Manage.* 7: 120–123.

Parmesan, C. and G. Yohe. 2003. A globally coherent fingerprint of climate change. Impacts across natural systems. *Nature* 421: 37–42.

Peterson, J.O., S.M. Born and R.D. Dunst. 1974. Lake rehabilitation techniques and experiences. *Water Resourc. Bull.* 10: 1228–1245.

Peterson, S.A., A.T. Herlihy, R.M. Hughes, K.L. Motter and J.M. Robbins. 2002. Level and extent of mercury contamination in Oregon, lotic fish. *Environ. Toxicol. Chem.* 21: 2157–2164.

Poff, N.L., M.M. Brinson and J.W. Day. 2002. *Aquatic Ecosystems and Global Climate Change*. Pew Center on Global Climate Change. Arlington, VA.

Postel, S. 1992. *Last Oasis: Facing Water Scarcity*. W.W. Norton & Co. New York, NY.

Postel, S. 1995. Where have all the rivers gone? *WorldWatch* 8: 9–19.

Postel, S. 1996. Dividing the Waters: Food Security, Ecosystem Health and the New Politics of Scarcity. WorldWatch Paper No. 132. WorldWatch Institute, Washington, DC.

Postel, S. 1999. When the world's wells run dry. *WorldWatch* Sept/Oct: 30–38.

Postel, S. 2000. Entering an era of water scarcity: The challenges ahead. *Ecol. Appl.* 10: 941–948.

Postel, S.L., G.C. Dailey and P.R. Ehrlich. 1996. Human appropriation of renewable fresh water. *Science* 271: 785–788.

Pretty, J.N., C.F. Mason, D.B. Nedwell, R.E. Hine, S. Leaf and R.Dils. 2003. Environmental costs of eutrophication in England and Wales. *Environ. Sci. Technol.* 37: 201–208.

Rich, P.H. and R.G. Wetzel. 1978. Detritus in the lake ecosystem. *Am. Nat.* 112: 57–71.

Ricciardi, A. and J.B. Rasmussen. 1999. Extinction rates of North American fresh water fauna. *Conserv. Biol.* 13: 1220–1222.

Robbins, R.W., D.M. Glicker and B.M. Niss. 1991. *Effective Watershed Management*. American Water Works Association Research Foundation, Denver, CO.

Ross, M. and C.A. Lembi. 1985. *Applied Weed Science*. Burgess Publishers. Minneapolis, MN.

Ryding, S.O. 1981. Reversibility of man-induced eutrophication – experiences of a lake recovery study in Sweden. *Int. Rev. ges. Hydrobiol.* 66: 449–503.

Schindler, D.W. 1974. Eutrophication and recovery in experimental lakes: Implications for lake management. *Science* 184: 897–899.

Service, R.F. 2003. "Combat biology" on the Klamath. *Science* 300: 36–39.

Shapiro, J., V. LaMarra and M. Lynch. 1975. Biomanipulation: An ecosystem approach to lake restoration. In: P.L. Brezonik and J.L. Fox (Eds.), *Proceedings of a Symposium on Water Quality Management and Biological Control*. University of Florida, Gainesville, FL. pp. 85–96.

Smith, V.H., G.D. Tilman and J.C. Nekola. 1999. Eutrophication: Impacts of excess nutrient inputs on fresh water, marine, and terrestrial ecosystems. *Environ. Pollut.* 100: 179–196.

Smith, V.H., J. Sieber-Denlinger, F. deNoyelles, Jr., S. Campbell, S. Pan, S.J. Randtke, G.T. Blain and V.A. Strasser. 2002. Managing taste and odor in a eutrophic drinking water reservoir. *Lake and Reservoir Manage.* 18: 319–323.

Sondergaard, M., J.P. Jensen and E. Jeppesen. 2001. Retention and internal loading of phosphorus in shallow, eutrophic lakes. *Sci. World* 1: 427–442.

Stefan, H.G., X. Fang and J.G. Eaton. 2001. Simulated fish habitat changes in North American lakes in response to projected climate warming. *Trans. Am. Fish. Soc.* 130: 459 477.

Thomas, C.D., A. Cameron, R.E. Green, M. Bakkenes, L.J. Beaumont, Y.C. Collingham, et al., 2004. Extinction risk from climate change. *Nature* 427: 145–148.

U.S. Environmental Protection Agency. 1977. *North American Project — A Study of U.S. Water Bodies*. USEPA 600/3-77-086.

U.S. Environmental Protection Agency. 1979. *Lake Restoration*. USEPA 440/5-79-001.

U.S. Environmental Protection Agency. 1980a. *Restoration of Lakes and Inland Waters*. USEPA 440/5-81-010.

U.S. Environmental Protection Agency. 1980b. *Clean Lakes Program Guidance Manual.* USEPA 440/5-81-003.

U.S. Environmental Protection Agency. 1983. *Lake Restoration, Protection and Management.* USEPA 440.5-83-001.

U.S. Environmental Protection Agency. 2000. National Water Quality Inventory. 1998 Report to Congress. USEPA 841-R-00-001.

U.S. Geological Survey. 1999. *The Quality of Our Nation's Waters: Nutrients and Pesticides.* U.S. Department of Interior. USGS Circular 1225.

U.S. National Academy of Sciences. 1969. *Eutrophication: Causes, Consequences, Correctives.* Proceedings of a Symposium. National Academy of Sciences. Washington, DC.

Van Driesche, R., S. Lyon, B. Blossey, M. Hoddle and R. Reardon (technical coordinators). 2002. *Biological Control of Invasive Plants in the Eastern United States.* U.S. Department of Agriculture. Forest Service. FHTET-2002-04. Morgantown, WV.

Walker, W.E. Jr. 1987. Empirical Methods for Predicting Eutrophication in Impoundments. Report 4. Phase III. Applications Manual. Tech. Rept. E-81-9. U.S. Army Corps Engineers, Vicksburg, MS.

Webster. 1972. *New World Dictionary.* World Publishing Co., New York, NY.

Welch, E.B. 1992. *Ecological Effects of Wastewater. Applied Limnology and Pollutant Effects.* 2nd edition. Chapman and Hall, New York, NY.

Westerdahl, H.E. and K.D. Getsinger (Eds.). 1988. Aquatic Plant Identification and Herbicide Use Guide. Volume II. Aquatic Plants and Susceptibility to Herbicides. Tech. Rept. A-88-9. U.S. Army Corps Engineers, Vicksburg, MS.

Wetzel, R.G. 1990. Land–water interfaces: Metabolic and limnological regulators. *Verh. Int. Verein. Limnol.* 24: 6–24.

Wetzel, R.G. 2001. *Limnology. Lake and River Ecosystems.* 3rd edition. Academic Press, New York, NY.

Wilson, M.A. and S.R. Carpenter. 1999. Economic valuation of freshwater ecosystems services in the United States. *Ecol. Appl.* 9: 772–783.

2 Basic Limnology

2.1 INTRODUCTION

Lake managers, students, consultants, and others interested in lake and reservoir restoration should have a thorough understanding of limnology. The next two chapters outline some basic principles of limnology that are significant to restoration and management decisions. A brief comparison of lakes and reservoirs is presented in this chapter, along with a description of regional lake conditions and the forces, both external and internal, that promote lake and reservoir problems. Procedures to obtain the data necessary to diagnose lake condition, select a restoration alternative, and prepare a project report are described in the next chapter.

Readers familiar with the fundamentals of limnology could go directly to sections on restoration methods. While these next two chapters cannot substitute for the in-depth understanding of limnology required to make competent and effective decisions, they do provide a review or guide to some basic principles. The reader is referred to Hutchinson (1957, 1967, 1975), Cole (1994), Horne and Goldman (1994), Lampert and Sommer (1997) and Scheffer (1998) for thorough discussions of limnology. Welch and Jacoby (2004), Wetzel (2001), and Kalff (2002) are especially useful for their holistic viewpoints, and for their coverage of macrophyte biology and stream and reservoir ecology.

2.2 LAKES AND RESERVOIRS

The physics, chemistry, and biology of dimictic (deeper lakes that thermally stratify in summer and winter) natural lakes have dominated limnological literature and the training of many limnologists. This bias reflects the fact that there are many of these lakes in North America and Europe. It is also a result of the emergence of "limnology schools" located primarily in North American and European areas dominated by deep lakes. But, shallow lakes are far more common than deep lakes (Wetzel, 1992), and limnology programs emphasizing them are now emerging, particularly in Europe.

Reservoirs are as important as natural lakes for recreation, but have additional values for flood control, hydropower generation, and water supply. While both lakes and reservoirs are subject to silt, organic, and nutrient loadings, reservoirs are more likely to have water quality problems due to their usually large watersheds and their morphometric configurations. Reservoirs are a vital part of the economy of many nations. The U.S. Army Corps of Engineers (USCOE) manages approximately 783 reservoirs with a combined surface area of 27,000 km^2 (Kennedy and Gaugush, 1988). Despite their abundance and importance, most limnology texts only mention them, or incorrectly imply that they are functionally equivalent to natural lakes and that no distinction is necessary.

While natural lakes and reservoirs have biotic and abiotic processes in common, they have important differences. Both have similar habitats (pelagic, benthic, profundal, and littoral zones), organisms, and processes, but it is their differences, summarized by Thornton et al. (1980), Walker (1981), Kennedy et al. (1982, 1985), Søballe and Kimmel (1987), Thornton et al. (1990), and Kennedy (1999, 2001) (Table 2.1), that also must be understood to successfully manage them. These fundamental reports are important supplements to most texts in limnology. A brief comparison of lakes and reservoirs is presented here.

TABLE 2.1
Comparison of Geometric Means (Probability That Means for Each Comparison Are < 0.0001) of Selected Variables of Natural Lakes and Army Corps of Engineers Reservoirs

Variable	Natural Lakes ($N = 309$)	Reservoirs ($N = 107$)
Drainage area (km²)	222.00	3228.00
Surface area (km²)	5.60	34.50
Maximum depth (m)	10.70	19.80
Mean depth (m)	4.50	6.90
Hydraulic residence time (yr)	0.74	0.37
Areal water load (m/yr)	6.50	19.00
Drainage/surface area	33.00	93.00
P loading (gm/m²/yr)	0.87	1.70
N loading (gm/m²/yr)	18.00	28.00

Source: Modified from Thornton, K.W. et al., 1980. *Symposium on Surface Water Impoundments.* Proceedings Am. Soc. Civil Eng. pp. 654–661. With permission.

Reservoirs differ from lakes in their geologic history and setting, basin morphology, and hydrologic factors (Kennedy et al., 1985; Kennedy, 2001). When natural lakes and USCOE reservoirs are compared, it is apparent that reservoirs are located primarily where flooding may occur or where water shortages require water storage. Reservoirs thus dominate the middle latitudes of the U.S. (Walker, 1981). Reservoirs are also used for hydropower generation. Very small reservoirs for recreation and farming operations are found at all latitudes.

Lakes of North America are also located in distinct regions. They are: (1) the continental glacial lakes in the mesic northeast, Canada and upper midwest, (2) the mostly alpine glacial lakes in Alaska and the mountainous west, (3) the coastal plain and karst (solution) lakes of the southeast, especially Florida, and (4) scattered small regions of playas, potholes and sandhill lakes in arid and semi arid areas (J.M. Omernik, USEPA, personal communication). Further discussions of these lake distributions are found in Hutchinson (1957) and Frey (1966).

Latitudinal differences in climate and geology have a major influence on the quality and rates of materials loaded to lakes and reservoirs, and on their degree of thermal stratification and mixing. The average reservoir watershed area is nearly an order of magnitude greater than the average lake's watershed, a factor accounting for the much higher average areal water (and contaminant) loadings to reservoirs (Table 2.1). Some lakes also have large watersheds and thus, like reservoirs, have high water loads. Reservoirs can become distinctly "lake-like" during summer low flow periods. Therefore, it should be noted that the values in Table 2.1 are averages, and that the ranges of lake and reservoir characteristics overlap.

Natural lakes are more likely to be located centrally in a fairly symmetrical drainage area, whereas reservoirs are elongated and dendritic, and usually at the downstream boundary of the watershed. The deep zone of a reservoir is normally at the dam; in lakes there may be several "deep holes."

Average nutrient and sediment loads are much higher for reservoirs and this material may have undergone a far longer period of in-stream processing than material loaded to natural lakes. Water often enters lakes via smaller streams that are likely to traverse wetland or littoral areas, whereas reservoirs may have characteristics of a river for long distances into the reservoir.

While natural lake outflows are at the surface, or occasionally through the ground, reservoirs usually have multiple depth, constructed outlets, leading to in-reservoir mixing processes and to discharge of water that might be anoxic, enriched with soluble nutrients, or high in hydrogen sulfide, methane, and reduced metals. Lake levels vary with precipitation, evaporation, and surface outflows, but it is uncommon for the amplitude to be large or to change quickly. An exception is the wind-induced displacement of some of the water mass, creating a to and fro "sloshing" of water in the lake basin, sometimes with amplitudes of one meter or more (surface and internal seiches). A well-known example is the occasional seiche in Lake Erie (U.S.-Canada). Reservoirs, however, can have rapid and significant changes in levels due to management decisions and these changes in level may eliminate or greatly reduce the littoral community of rooted aquatic plants.

Unlike lakes, reservoirs are operated to store and release water, and these operations profoundly influence their limnological characteristics (Kennedy, 2001; Cooke and Kennedy, 2001). For example, when deep waters are released, heat is stored. When surface waters are released heat is dissipated. These actions greatly alter thermal structure, including depth of the metalimnion (layer of water with a sharp thermal gradient) and retention or loss of materials.

Lakes and reservoirs represent a continuum of ecological conditions (Canfield and Bachmann, 1981). Kimmel and Groeger (1984) and Søballe and Kimmel (1987) viewed this continuum as one ordered by water residence time (volume divided by outflow rate), and indicated that reservoirs and natural lakes with similar residence times have similar ecological attributes. In rapidly flushed systems, for example, algal abundance is less likely to depend on nutrient concentrations than on flushing rate (Chapter 6). Therefore, despite features that might separate lakes and reservoirs as classes of aquatic habitats, convergence can occur when water residence times are similar.

Geographic location of a reservoir determines the quantity and timing of inflow. For example, inflows in some areas of California are in spring to mid-summer, whereas peak inflows in the Pacific Northwest and southeastern U.S. are in winter to early spring. Similarities and differences between lakes and reservoirs based on water residence times are thus modified by location (Kennedy, 1999).

Figure 2.1 illustrates the expected gradient in reservoir characteristics of a main-stem reservoir (dam on the stream) from the river entrance to the dam. Unlike many natural lakes where water enters from several smaller tributaries draining comparatively small sub-watersheds, reservoirs have a distinct riverine zone dominated by flow and mixing, followed by a transition zone where inflow velocity slows, rapid sedimentation begins, and water clarity increases. When inflowing river water is colder than surface water of the reservoir, a "plunge point" is found where the colder, heavier water loses velocity and descends to a depth equal to its density, creating a distinct inter- or underflow (Figure 2.2). Unlike lakes, where it is often assumed that nutrient loads are completely mixed with lake waters, loading to a reservoir might not mix with upper waters at all, but instead might be carried through the reservoir via an inter- or underflow, greatly altering standard loading model assumptions (Chapter 3) (Kimmel and Groeger, 1984; Gaugush, 1986; Walker, 1987). The lacustrine zone near the dam is the most lake-like, with thermal stratification and a higher probability that algae growth is nutrient limited. Some natural lakes in narrow valleys, with large inflow rivers and low water residence times, have many reservoir characteristics. This gradient of conditions along the length of a reservoir means that reservoir characterization requires multiple sampling stations. The same is true for large natural lakes, and lakes with distinct pelagic and littoral zones.

Reservoir basin design also influences hydrodynamic features. For example, a tributary reservoir and a main stem reservoir receiving identical water loads and having identical basin volumes are likely to have different responses. Main stem reservoirs have low capacity to store excess volume and thus have water residence times that fluctuate with water loading events, whereas tributary reservoirs have much higher storage capacity and are used for flood control. In these reservoirs, the hypolimnion may be large whereas the main stem reservoir may be longer and shallower and greatly influenced by interactions between sediments and overlying water (Kennedy, 1999).

Reservoirs are important sources of fresh water for potable, irrigation, and industrial purposes. Their protection and management requires that more traditional views of sampling, correlations between loading and responses of biota, and choice of restoration techniques be modified to take these, and the basic differences between natural lakes and reservoirs, into account.

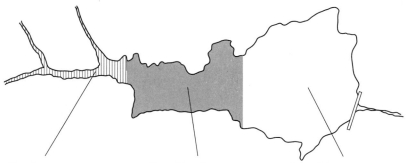

Riverine zone
- Narrow basin
- High flow
- High susp. solids, low light
- High nutrients, advective supply
- Light limited photosynthesis
- Algal cell loss by sedimentation
- Organic matter supply allochthonous
- More "eutrophic"

Transitional zone
- Broader, deeper basin
- Reduced flow
- Lower susp. solids, more light
- Advective nutrient supply reduced
- High photosynthesis
- Algal cell loss by sedimentation, grazing
- Intermediate

Lacustrine zone
- Broad, deep, lake-like
- Little flow
- Clearer
- Internal nutrient recycling, low nutrients
- Nutrient limited Photosynthesis
- Algal cell loss by grazing
- Organic matter supply autochthonous
- More "oligotrophic"

FIGURE 2.1 Longitudinal zonation in environmental factors that control primary productivity, phytoplankton biomass, and trophic state within reservoir basins. Changes in shading indicate decline in turbidity. (From Kimmel, B.C. and A.W. Groeger, 1984. *Lake and Reservoir Management.* USEPA 440/5-84-001. pp. 277–281.

2.3　BASIC LIMNOLOGY

2.3.1　PHYSICAL–CHEMICAL LIMNOLOGY

Some lakes and reservoirs stratify thermally during summer months into an upper warm, well-mixed zone termed the epilimnion. Below this is a zone of rapidly decreasing temperature with depth, the metalimnion, followed by a deep, colder, often dark bottom layer, the hypolimnion. This phenomenon, brought about by wind mixing, solar input, and by large differences in water density between cold and warm waters, is a primary determinant of summer physical, chemical, and biological interactions. During ice cover lake water temperature inversely stratifies, with colder water at the surface. This happens because water's maximum density is at 4°C, and water colder than this temperature, including ice, is lighter and floats above this slightly warmer layer. Lakes with two mixing periods (spring and fall) and two stratified periods (summer, winter) are dimictic and are typical of deep lakes and reservoirs of north temperate latitudes (Wetzel, 2001). Details of mechanisms leading to this and other types of thermal stratification are found in all basic limnology texts. Figure 2.3 illustrates the characteristics of the three thermal layers in a dimictic lake, or in the lacustrine zone (Figure 2.1) of a reservoir during summer months. The figure also illustrates typical summer temperature and dissolved oxygen (DO) profiles with depth in a stratified eutrophic lake or reservoir.

Polymictic lakes are more common than dimictic lakes. Because polymictic lakes are shallow, they may mix continuously, or stratify briefly (hours, days) in calm, hot weather, followed by renewed complete mixing. Polymictic lakes are found at all latitudes.

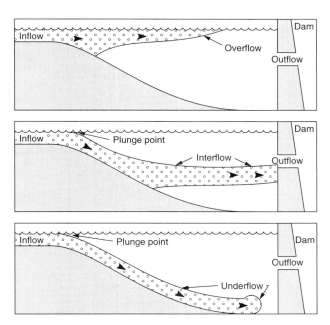

FIGURE 2.2 Density flows in reservoirs. The upper panel illustrates an "overflow" of warm incoming waters, the middle panel illustrates an "interflow," and the bottom panel shows an "underflow." (From Moore, L. and K. Thornton (Eds.). 1988. *Lake and Reservoir Restoration Guidance Manual.* USEPA 440/5-88-002.

There can be significant modifications of temperature regimes, particularly in reservoirs. River inflows to reservoirs may have very different temperatures than reservoir waters, producing under- inter-, or overflows of incoming water (Figure 2.2). The upper reaches of the reservoir, like the wave washed littoral zone of a lake, may exhibit little thermal stratification except during hot, calm, low-flow periods. In the transition zone (Figure 2.1), where mixing and sedimentation processes are dominant, the volume of a reservoir's hypolimnion may be small. Only in the deep lacustrine zone is the temperature stratification similar to natural lakes, though the hypolimnion is likely to be less stable due to underflows and withdrawals of deep water at the dam.

The shape of a reservoir or lake's basin affects its productivity, kinds of organisms, water chemistry, and the choices available to manage and restore it. Most natural lakes are small in area and shallow (mean depth $\leq$ 3 m). Rooted plants, and algae associated with leaf and sediment surfaces, can have very high primary productivity, biomass, and areal distribution, unless the lake is turbid from silt loading, wind mixing, or algal blooms that cause rooted plants to be light-limited. Also, as noted in later sections and other chapters, the large area of shallow, warm sediments and the small hypolimnetic volumes associated with (mainly) polymictic shallow lakes and reservoirs provide ideal circumstances for processes that allow sediment nutrient release (actually recycling or "internal loading") and transport to the water column. This can greatly stimulate algal productivity. Internal loading processes may be biological (e.g., microbial activities and temporary anoxia, and sediment disturbance by methane release or by burrowing animals), chemical (e.g., high pH from photosynthesis), and physical (e.g., turbulence from the wind) in nature (Chapter 3). Because of these processes, lake productivity is often negatively correlated with mean depth (Wetzel, 2001) and with the ratio of mean to maximum depth (Carpenter, 1983). Therefore, many shallow water bodies will have more algae or rooted plants than the less common steep-sided, deep lakes and reservoirs.

High macrophyte growth is to be expected in shallow lakes. A hypsograph (a representation describing the relationship between lake area and depth) is useful in explaining this. Figure 2.4 compares the area–depth relationship for two hypothetical lakes with different areas of shallow

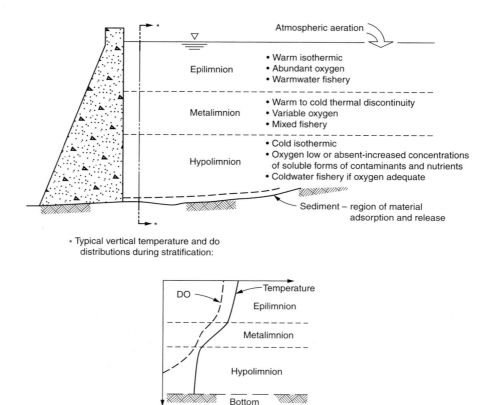

FIGURE 2.3 Cross section of a thermally stratified reservoir indicating location and characteristics of the epilimnion, metalimnion, and hypolimnion, and typical summer temperature-dissolved oxygen distributions in the lacustrine zone of a eutrophic reservoir. (From Gunnison, D. and J.M. Brannon. 1981. Characterization of Anaerobic Chemical Processes in Reservoirs: Problem Description and Conceptual Model Formulation. Tech. Rept. E-81-6. U.S. Army Corps Engineers, Vicksburg, MS.

water. Both lakes could have nuisance algal blooms if nutrient concentrations were high. Only the shallow one has the potential to have a large area with rooted plants because of the extensive shallow, well-lighted sediment area. Physical factors, particularly waves, transparency, and the slope of the littoral zone (amount of stable sediment area exposed to light) are among the determinants of maximum macrophyte biomass and maximum depth of plant colonization (Canfield et al., 1985; Duarte and Kalff, 1986, 1988) (Chapter 11). The development of a hypsograph is an important first step in lake and reservoir problem diagnosis.

2.4   BIOLOGICAL LIMNOLOGY

Lakes and reservoirs have three distinct and interacting biotic communities (Figure 2.5): (1) the wetland-littoral zone, and its sediments, (2) the open water pelagic zone, and (3) the benthic or deep water (profundal) zone and sediments. Problems or characteristics appearing in one zone (e.g., deep water oxygen depletions, littoral zone aquatic plants, pelagic zone algal blooms) directly or indirectly affect other zones, meaning that successful lake restoration requires a holistic view of lake and watershed processes. For example, nutrients causing algal blooms may come from lake sediments and decomposition of littoral plants, as well as from external loading. All sources might require attention to solve the problem.

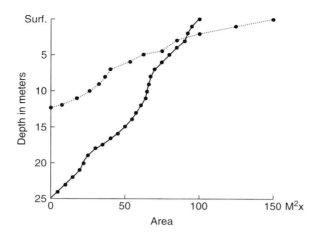

FIGURE 2.4 Depth-area hypsographs. Solid line illustrates the less common deep lake with a small littoral zone; dotted line illustrates the more common shallow lake with extensive littoral area and volume. (From Cooke, G.D., E.B. Welch, S.A. Petersen, and P.R. Newroth. 1993. *Restoration and Management of Lakes and Reservoirs*, 2nd Edition. Lewis Publishers and CRC Press, Boca Raton, FL.

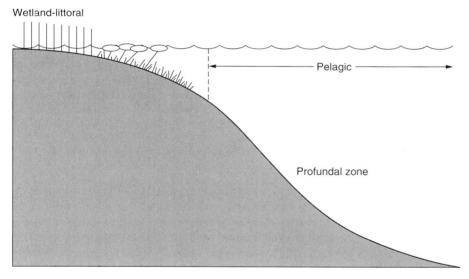

FIGURE 2.5 Biotic communities in lakes and reservoirs.(From Cooke, G.D., E.B. Welch, S.A. Petersen, and P.R. Newroth. 1993. *Restoration and Management of Lakes and Reservoirs*, 2nd Edition. Lewis Publishers and CRC Press, Boca Raton, FL.)

Rooted emergent, floating, and submersed vascular plants, collectively called macrophytes, and their attached flora and fauna, dominate the wetland-littoral. These plants are distinctly different from the microscopic, floating (planktonic) cells, colonies, and filaments of algae, often seen as surface "scums" in some eutrophic systems. Macrophytes are usually vascular plants and are found in shallow water. They may have large masses of filamentous (string or hair-like) algae attached to them as thick mats. Shallow, lighted sediments often have a highly productive epilithic, epipelic, and epiphytic flora (algae growing on surfaces of rocks, sediments and vascular plants). Macrophyte biology is described in Chapter 11.

The littoral zone often has high species diversity, and is commonly the site where fish reproduction and development occurs. It is also an important waterfowl habitat. Littoral zone plant biomass replaces itself two or more times per summer in productive lakes, leading to inputs of non-living dissolved and particulate organic matter, termed "detritus," to the water column and sediments. Detritus, whether from watershed drainage or from in-lake productivity, is a stable energy and nutrient source to the lake's food webs, especially to microbial flora and plankton (Wetzel, 1992, 1995). Many lakes, especially those surrounded by dense forest, are actually heterotrophic (photosynthetic rate is less than total respiration rate), and depend upon organic carbon from terrestrial sources to subsidize their food webs (Cole, 1999). Therefore lakes are strongly linked to the land, not only through nutrient and silt loading, but through detritus imports.

Macrophytes, in addition to being a significant energy source and habitat, stabilize littoral zone sediments from the impacts of wind and boat-generated waves, thus reducing internal P loading and sediment resuspension (Bachmann et al., 2000; Anthony and Downing, 2003; Horppila and Nurminen, 2003).

Macro- and microplankton, and the fish and invertebrates grazing on them, dominate the pelagic zone. The plankton includes algae that produce unsightly "blooms" and low water clarity, and bacteria, fungi, Protozoa, and filter-feeding crustaceans like *Bosmina* and *Daphnia*. The pelagic community obtains energy from sunlight and from detritus transported to it from stream inflows and the littoral zone. The plankton of most enriched lakes and reservoirs is dominated by one or a few species of highly adapted algae and bacteria, particularly nuisance blue-green algae (cyanobacteria). *Bosmina*, *Daphnia*, and other planktonic microcrustacea are significant grazers of detritus, bacteria, and some algae species, though their abundance may be regulated by complex interactions with predators such as fish and insects (Chapter 9).

The profundal benthic community receives nutrients and energy from organic matter loaded to or produced in the lake or reservoir and deposited on the sediments. Inorganic forms of nutrients may be added to the sediments in the form of precipitates. This pelagic-benthic coupling is a fundamental feature of lakes (Vadeboncoeur et al., 2002). In productive lakes and reservoirs, large areas of the sediment community in deep water are continuously anoxic during thermal stratification due to intense microbial respiration that is stimulated by deposits of detritus. Anoxic conditions provide conditions favoring high rates of nutrient release to the water column (Figure 2.3).

2.5 LIMITING FACTORS

Nuisance densities of algae or macrophytes, and associated water quality problems, are conditions managed by manipulating or altering their biomass or by manipulating one or more of the factors controlling their abundance. Macrophyte density, while in part related to sediment type and composition, and to nutrient factors, is often determined by light availability (Duarte and Kalff, 1986; Canfield et al., 1985; Barko et al., 1986; Smith and Barko, 1990). Long-term control of algal biomass requires significant water column nutrient reduction. Phosphorus (P) is most frequently targeted because it is usually the nutrient in shortest supply relative to demands by algae (the limiting nutrient). Phosphorus does not have a gaseous phase so the atmosphere is not a significant source, unlike nitrogen or carbon. Lake P concentration, therefore, can be lowered significantly by reducing loading from land and in-lake sources.

A significant reduction in external nutrient loading is an essential, but not necessarily sufficient, step toward reducing lake P concentrations. Internal loading from aerobic and anaerobic sediments, groundwater seepage, decomposing macrophytes, sediment resuspension, and organism activities might add more nutrients to the lake than external loading during some times of the year.

The shape of a lake's basin (Figure 2.4) has an important bearing on the amount of internal loading. Most of the variance in algal productivity among some Ontario lakes was explained by the ratio of sediment area in contact with the epilimnion to epilimnion volume. Steep-sided, deep lakes have a low ratio, producing less influence on overlying water (Fee, 1979). Epilimnetic sediments are warm, leading to increased microbial decomposition rates and to nutrient release (Jensen and Andersen, 1992). Extensive littoral areas, typical of shallower lakes, may have distinct day-night cycles of high and low DO concentrations that stimulate nighttime P releases, especially under dense macrophyte beds (Frodge et al., 1991). Wind mixing and convective currents may scour sediments or entrain nutrient-rich littoral or bottom waters of shallow lakes, especially those with low macrophyte density, thus transporting nutrients to the pelagic zone.

The hypolimnion may or may not be a P source to the epilimnion. When thermal stratification occurs, hypolimnetic waters are isolated from the atmosphere and are usually too deep to permit sufficient light penetration for photosynthetic oxygen generation. Respiration in deep waters leads to DO depletion or elimination, to reducing conditions, and to the associated release of P from sediment iron complexes. High sulfate concentrations may lead to ferrous sulfide (FeS) production under reducing conditions, and loss of Fe control of sediment P (Caraco et al., 1989; Golterman, 1995; Gächter and Müller, 2003). In stratified lakes with low resistance to mixing (large surface area relative to depth), summer winds either briefly destratify the lake (polymixis), or force vertical entrainment of P-rich hypolimnetic water to the epilimnion. In either case, surface water P concentration increases, stimulating an algal bloom. For example, Stauffer and Lee (1973) calculated that all of the summer algal blooms in Lake Mendota, Wisconsin could be accounted for by transport of P from the metalimnion to the epilimnion.

This internal P source to the epilimnion may not be significant in lakes that are deep relative to area of lake surface exposed to wind mixing. This type of lake offers greater resistance to the force of summer wind (Osgood, 1988). The best predictor of vertical P transport to the epilimnion appears to be the vertical gradient of P concentration, not lake morphometry (Mataraza and Cooke, 1997). These ideas are explored in Chapters 3 and 4 with respect to model predictions, and in Chapter 8 where sediment treatment with P inactivating chemicals is discussed.

Macroscopic animals play major roles in nutrient releases from lake sediments. Common carp digestive activities release P at rates similar to external loading (La Marra, 1975). Bioturbation (sediment disturbance) by fish and insects and high rates of sloughing of vascular plant tissues are also nutrient sources to the epilimnion. Reviews of internal recycling include Carlton and Wetzel (1988), Marsden (1989), Welch and Cooke (1995), Pettersson (1998), and Søndergaard et al. (2001). These characteristics of littoral and pelagic zones mean that expensive nutrient diversion projects may not meet expectations for reduced algal biomass until internal nutrient sources are addressed (Chapters 4 and 8).

Other factors affecting algal biomass include flushing rate, light availability, pH, and zooplankton grazing. These factors can be manipulated as part of a management plan, though significant reduction of external and internal nutrient loading remains the central part of plans for long term improvement of excessive algae problems.

2.6 THE EUTROPHICATION PROCESS

A eutrophic lake or reservoir is rich in nutrients and organic materials, and those enriched by human activities are said to be culturally eutrophic. We have expanded the definition of the eutrophication process to include the loading of silt and organic matter, as well as nutrients. Thus, we define the eutrophication process as the loading of inorganic and organic dissolved and particulate matter to lakes and reservoirs at rates sufficient to increase the potential for high biological production, decrease basin volume, and deplete DO. This concept of eutrophication is more complete because it includes all materials that produce the eutrophic condition. The eutrophication process and associated major in-lake interactions are summarized in Figure 2.6.

Traditionally, eutrophication referred only to nutrient loading, its eventual high concentrations in the water column, and the high productivity and biomass of algae that could occur. Organic matter loading may lead to sediment enrichment and loss of volume. Organic matter, whether added to the water column from external or internal sources, also leads to increased nutrient availability via direct mineralization, or through release from sediments when respiration is stimulated by this organic matter and DO is depleted. Net internal P loading appears to increase exponentially with

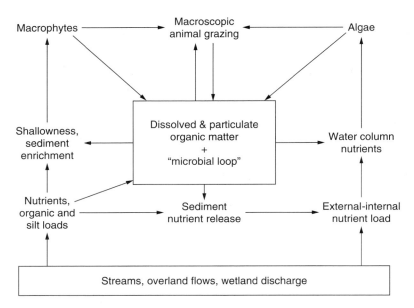

FIGURE 2.6 Loadings and primary interactions in lakes and reservoirs. (From Cooke, G.D., E.B. Welch, S.A. Petersen, and P.R. Newroth. 1993. *Restoration and Management of Lakes and Reservoirs*, 2nd Edition. Lewis Publishers and CRC Press, Boca Raton, FL.)

increasing dissolved organic carbon content of the lake (Ryding, 1985). Allochthonous organic matter contains molecules producing changes in algal and microbial metabolism independently of effects of added nutrients (e.g., Franko and Wetzel, 1981). Finally, organic matter added to a lake or reservoir contains energy that is incorporated, in both dissolved and particulate forms, into plant and animal biomass, leading directly to increased living biomass (the microbial loop). Dissolved and particulate organic matter entering the lake or reservoir from streams, wetlands, and from macrophytes, is of great significance to lake metabolism. These ideas are developed in Wetzel (1995, 2001) and Cole (1999).

Silt may be rich in organic matter and in nutrients sorbed to surfaces of particulate matter. These may become available to algae or macrophytes immediately or at some later time. Silt loading also contributes directly to volume loss and to an increase in shallow sediment area. Whether volume loss is produced by silt deposition or by the build-up of refractory organic matter from terrestrial and aquatic sources, the development of shallow areas fosters further spread of macrophytes and their attendant epiphytic algae. Ultimately these plants promote further losses of DO and release of organic molecules and nutrients as they decay (Carpenter, 1980, 1981, 1983) (Figure 2.6).

Thus, silt and organic loadings have effects on lakes that are additional to their nutrient content, and cannot be excluded when defining the eutrophication process. This view is not meant to downplay or negate the fundamental importance of high nutrient loading in stimulating lake productivity. Instead, following Odum's (1971) holistic view, it is meant as a more complete description of the process.

Excessive nutrient loading creates potential for eutrophic conditions but does not guarantee increased productivity. Figure 2.6 does not account for the "oligotrophication" effects of high rates of lake flushing and dilution, the effects of organisms in stimulating nutrient release from sediments, or the effects of grazing (or lack of grazing) on algae biomass.

Lakes and reservoirs that are naturally eutrophic, or have become so, have characteristics separating them from less enriched and oligotrophic ("poorly nourished") water bodies. Eutrophic lakes have algal "blooms," often of monospecific blue-green (cyanobacteria) populations. Some also have macrophytes, though exotic macrophyte infestations are not a symptom of the eutrophic condition because large populations can develop in oligotrophic waters. Eutrophic lakes and reservoirs also have colored water (green/brown), and low or zero DO levels in the deepest areas (Figure 2.3). Warmwater fish production is likely to be high (Jones and Hoyer, 1982). Fish can be limited by low DO and high pH (Welch and Jacoby, 2004), and lakes may be dominated by less desirable fish species or stunted fish populations.

An oligotrophic lake or reservoir is low in nutrients and productivity because organic matter and nutrient loadings are low or large basin water volumes and short water residence times dilute or pass material through the lake. In addition, high water hardness may foster co-precipitation of calcium carbonate (e.g., marl lakes) and essential nutrients, rendering them unavailable to algae. Oligotrophic lakes are often deep and steep-sided, with nutrient-poor sediments, few macrophytes, usually no nuisance cyanobacteria, and large amounts of DO in deep water. Water clarity is high, as is phytoplankton diversity, but total algal biomass is low.

Low biological productivity is not always perceived as a benefit when, for example, a sports fishery is desired. Some lakes, Lake Mead, Nevada, for example (Axler et al., 1988), have been fertilized in an attempt to develop more fish biomass. The words "eutrophic" and "oligotrophic" therefore do not represent "bad" and "good," but are only descriptive of the state or condition of a lake or reservoir. Perceived quality is a judgment based upon needs and expectations.

2.7 CHARACTERISTICS OF SHALLOW AND DEEP LAKES

Shallow lakes and reservoirs (< 3 m mean depth) are more common than larger, deeper ones, and many are eutrophic or heavily impacted by siltation and high turbidity. Their problems, and solutions to those problems, are reflected in their characteristics. Most lake and reservoir restoration techniques and paradigms were developed from research and testing on less common deep lakes and may not be entirely suitable for shallow lakes. Throughout this text, we attempt to emphasize applicability of methods to both classes of lakes. Table 2.2 is a comparison of the characteristics of deep and shallow lakes, primarily based on European research (e.g., Moss et al., 1996; Jeppesen, 1998; Scheffer, 1998; Havens et al., 1999; Cooke et al., 2001; NALMS, 2003).

Shallow lakes are less sensitive to significant reductions in external nutrient loading because benthic-pelagic interactions tend to maintain high nutrient levels. Nutrients released from bottom sediments of shallow lakes affect the entire water column, in contrast to stratified, deep lakes. In shallow lakes, nutrient release may be very high from bioturbation, wind disturbance, the effects of gas bubbles, high pH from intense photosynthesis, and from DO deficits at the sediment-water interface. Diversion of external nutrient loading, while necessary, may not be sufficient to rehabilitate a shallow lake and a sediment treatment may be necessary.

Shallow lakes are more likely to exist in one of two alternative and often stable states (Chapter 9). The algae-dominated turbid state is almost a certainty at high nutrient concentrations, whereas the clear water state, possibly with macrophytes across the well-lighted sediments, will occur at low concentrations. Between these extremes, either the clear or turbid water state can exist, largely based on biotic interactions. Lakes with dense populations of planktivorous and benthivorous fish (e.g., grass carp, common carp, shad), and lakes with large populations of herbivorous birds, are likely to have few phytoplankton grazers (large-bodied zooplankton), high internal P loading, turbid water, and little chance of extensive establishment of native submersed plants. Canopy-forming plants such as Eurasian watermilfoil (*Myriophyllum spicatum*) may be successful in these lakes. In contrast, shallow lakes with dominance by piscivorous fish and birds (e.g., largemouth bass, northern pike, Great Blue Heron) may have abundant algae grazers, stable sediments, clear water, and populations of submersed plants, even at nutrient concentrations identical to the algae-dominated lake (Moss et al., 1996). In most cases, a shallow lake will have either a community of macrophytes or turbid water with phytoplankton. A shallow lake that is free of both aquatic plants

TABLE 2.2
Characteristics of Shallow and Deep Lakes

Characteristic	Shallow	Deep
1. Likely size of drainage area to lake area	Large	Smaller
2. Responsiveness to diversion of external loading	Less	More
3. Polymictic	Often	Rarely
4. Benthic–pelagic coupling	High	Low
5. Internal loading impact on photic zone	High	Lower
6. Impact of benthivorous fish on nutrients/turbidity	High	Lower
7. Fish biomass per unit volume	Higher	Lower
8. Fish predation on zooplankton	Higher	Lower
9. Nutrient control of algal biomass	Lower	Higher
10. Responsiveness to strong biomanipulation	More	Lesser
11. Chance of turbid state with plant removal	Higher	Lower
12. Probability of fish winterkill	Higher	Lower
13. % Area/volume available for rooted plants	High	Low
14. Impact of birds/snails on lake metabolism	Higher	Lower
15. Chance of macrophyte-free clear water	Low	Higher

Source: Modified from Cooke, G.D. et al. 2001. *LakeLine (NALMS)* 21: 42–46. With permission.

and algae is uncommon, and it is unrealistic to expect such a lake to occur without a large investment of money and energy.

Shallow lakes are more susceptible than deep lakes to strong biomanipulations, such as fish addition or removal, leading to a switch in stable states. Adding grass carp at densities sufficient to eliminate macrophytes, for example, is almost certain to switch a clear lake to a turbid, algae-dominated one. A fish winterkill may create conditions leading to clear water.

Lake management requires consideration of the differences between deep and shallow lakes. The consequences of using a particular technique in one lake type may be different from using it in another type. For example, an alum application to a deep lake may have little effect on epilimnetic P concentrations if there is no substantial vertical P transport or a steep P gradient. The impact of an alum treatment on a shallow lake is likely to be dramatic.

2.8 ECOREGIONS AND ATTAINABLE LAKE CONDITIONS

A lake's geographic location has an important bearing on its attainable condition or trophic state, a concept valuable to lake managers and lake association members. It places realistic boundaries or expectations on achievable lake conditions, and provides limits on the types and amounts of treatment or management that reasonably might be imposed to achieve the desired lake quality.

There was little formal recognition of regional water quality limits until publication (with map supplement) of "Ecoregions of the Conterminous United States" (Omernik, 1987), and subsequent articles (Rohm et al., 1995; Omernik, 1995; Omernik and Bailey, 1997; Griffith et al., 1997a; Griffith et al., 1999; Bryce et al., 1999; Omernik et al., 2000; Rohm et al., 2002). The original 76 ecoregions were delineated from integrative factors such as land use, and were "based on hypotheses that ecosystems and their components display regional patterns that are reflected in spatially variable combinations of causal factors, including climate, mineral availability (soils and geology), vegetation, and physiography" (Omernik, 1987; Figure 2.7).

The purpose for describing ecoregions was to assist resource managers in understanding aquatic ecosystem regional patterns of nutrient concentrations, biotic assemblages, and lake trophic state,

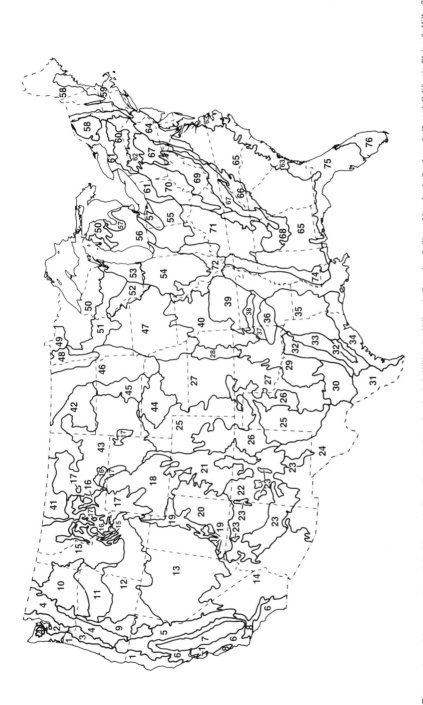

FIGURE 2.7 Ecoregions of the conterminous United States. 1, Coast Range; 2, Puget Lowland; 3, Willamette Valley; 4, Cascades; 5, Sierra Nevada; 6, Southern & Central California Plains & Hills; 7, Central California Valley; 3, Southern California Mountains; 9, Eastern Cascades Slopes & Foothills; 10, Columbia Basin; 11, Blue Mountains; 12, Snake River Basin/High Desert; 13, Northern Basin & Range; 14, Southern Basin & Range; 15, Northern Rockies; 16, Montana Valley & Foothill Prairies; 17, Middle Rockies; 18, Wyoming Basin; 19, Wasatch and Uinta Mountains; 20, Colorado Plateaus; 21, Southern Rockies; 22, Arizona/New Mexico Plateau; 23, Arizona/New Mexico Mountains; 24, Southern Deserts; 25, Western High Plains; 26, Southwestern Tablelands; 27, Central Great Plains; 28, Flint Hills; 29, Central Oklahoma/Texas Plains; 30, Central Texas Plateau; 31, Southern Texas Plains; 32, Texas Blackland Prairies; 33, East Central Texas Plains; 34, Western Gulf Coastal Plain; 35, South Central Plains; 36, Ouachita Mountains; 37, Arkansas Valley; 38, Boston Mountains; 39, Ozark Highlands; 40, Central Irregular Plains; 41, Northern Montana Glaciated Plains; 42, Northwestern Glaciated Plains; 43, Northwestern Great Plains; 44, Nebraska Sand Hills; 45, Northeastern Great Plains; 46, Northern Glaciated Plains; 47, Western Corn Belt Plains; 48, Red River Valley; 49, Northern Minnesota Wetlands; 50, Northern Lakes and Forests; 51, North Central Hardwood Forests; 52, Driftless Area; 53, Southeastern Wisconsin Till Plains; 54, Central Corn Belt Plains; 55, Eastern Corn Belt Plains; 56, Southern Michigan/Northern Indiana Till Plains; 57, Huron/Erie Lake Plain; 58, Northeastern Highlands; 59, Northeastern Coastal Zone; 60, Northern Appalachian Plateau and Uplands; 62, Erie/Ontario Lake Plain; 62, North Central Appalachians; 63, Middle Atlantic Coastal Plain; 64, Northern Piedmont; 65, Southeastern Plains; 66, Blue Ridge Mountains; 67, Central Appalachian Ridges and Valleys; 68, Southwestern Appalachians; 69, Central Appalachians; 70, Western Allegheny Plateau; 71, Interior Plateau; 72, Interior River Lowland; 73, Mississippi Alluvial Plain; 74, Mississippi Valley Loess Plains; 75, Southern Coastal Plain; 76, Southern Florida Coastal Plain. (From Omernik, J.M., 1987. *Ann. Assoc. Am. Geogr.* 77: 118–125. With permission.)

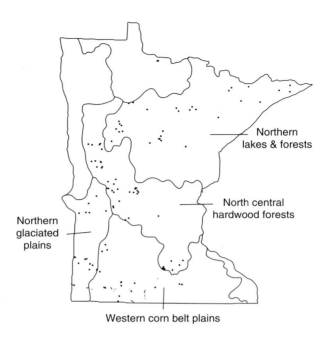

Northern
lakes & forests

North central
hardwood forests

Northern
glaciated
plains

Western corn belt plains

FIGURE 2.8 Minnesota's lake ecoregions and spatial distribution of representative lakes. These lakes comprise the "ecoregion data base." (From Wilson, C.B. and W.W. Walker, Jr. 1989. *Lake and Reservoir Manage.* 5(2): 11–22. With permission.)

thus allowing informed management decisions and reasonable expectations about attainable conditions. For example, the concentrations of silt, organic matter, and nutrients in streams within an ecoregion, although variable, are likely to be more similar to each other than to concentrations of these constituents in streams of adjacent ecoregions with different soils, vegetation types, and runoff potentials. It follows that the condition of lakes within an ecoregion that has nutrient-rich, erodible soils and reduced vegetation cover is likely to be different from lakes in a nearby ecoregion with sandy soils, flat relief, and dense tree cover. These expectations have been verified for streams of Arkansas, Kansas, Minnesota, Ohio, and Oregon and for lakes of Michigan, Minnesota, Ohio, and Wisconsin (Hawkes et al., 1986; Hughes and Larsen, 1988; Larsen et al., 1988; Omernik et al., 1988; Wilson and Walker, 1989; Fulmer and Cooke, 1990).

Minnesota uses the ecoregion concept to manage lakes. While there are seven ecoregions that extend into Minnesota (Figure 2.8), 98% of the state's 12,000 lakes over 10 ha are found in only four of them. Table 2.3 lists characteristics of the lakes and their streams and watersheds for the four ecoregions (Heiskary et al., 1987; Heiskary and Wilson, 1989; Wilson and Walker, 1989). There are substantial differences between the North Central Hardwood Forest (NCHF) and Northern Lakes and Forests (NLF) ecoregions versus the Western Corn Belt Plains (WCBP) and Northern Glaciated Plains (NGP) ecoregions. Lake users in the NCHF and NLF ecoregions should expect their lakes to be clear and essentially free of algal blooms. This is the attainable lake condition for these ecoregions, and management of problem lakes is directed toward achieving that realistic lake condition. Lakes a few hundred kilometers to the south, in the WCBP ecoregion, usually have macrophytes, algal blooms, low transparency, and anoxic hypolimnia to some degree. Lake management in this ecoregion cannot produce lakes with mean depths, chlorophyll levels, or water transparencies like those in the NCHF or NLF without extraordinary expenditures. Therefore lakes of a particular ecoregion should be managed for reasonable attainability and realistic expectations relative to overall regional or ecoregional lake quality.

The concept of regional lake quality can be both misleading and helpful in management decisions. For example, Shagawa Lake, Minnesota is located in the NLF ecoregion where mean

TABLE 2.3
Summary of Land Use and Water Quality Data for Four Ecoregions in Minnesota

Variable	Units	NCHF	NLF	NGP	WCBP
Number of lakes	36	30	8	11	
Land uses					
Cultivated	%	34.8	1.8	73.0	60.6
Pasture	%	18.0	3.9	9.2	5.9
Urban	%	0.7	0.0	2.0	1.5
Residential	%	6.4	4.8	0.4	9.9
Forested	%	16.4	66.2	0.0	7.0
Marsh	%	2.5	2.1	0.6	1.2
Water	%	20.9	20.9	14.4	13.6
Watershed area	ha	4670	2140	2464	756
Lake area	ha	364	318	218	107
Mean depth	m	6.6	6.3	1.6	2.5
Total phosphorus (P)	mg/L	33	21	156	98
Chlorophyll A	mg/L	14	6	61	67
Secchi disc	m	2.5	3.5	0.6	0.9
Total P load	kg/yr	1004	305	1943	590
Inflow P	mg/L	183	58	5666	564
Areal P load	kg/km^2/yr	276	96	891	551
Outflow	km^3/yr	6.2	5.3	0.9	1.0
Water residence time	yr	9.3	5.0	36.2	4.8
Stream total P	mg/L	148	52	1500	570

Note: Ecoregions: NCHF, Northern Central Hardwood Forests; NLF, Northern Lakes and Forests; NGP, Northern Glaciated Plains; WCBT, Western Corn Belt Plains. Data are listed as averages.

Source: From Wilson, C.B. and W.W. Walker, Jr. 1989. *Lake and Reservoir Manage.* 5(2): 11–22. With permission.

summer lake total P (TP) concentration is about 20–25 µg P/L (mesotrophic). However, summer TP in Shagawa Lake averaged between 50 and 60 µg P/L (eutrophic), an anomaly for the region, due to loading from wastewater treatment at Ely, Minnesota (Peterson, et al. 1995). Advanced wastewater treatment was expected to reduce effluent concentration to less than 20 µg P/L and the lake was expected to return to a normal TP level for the region. However, what was unknown when the treatment plant upgrade was proposed was that sediment nutrient recycling would maintain high summer P concentrations in the lake. Although annual TP levels decreased (Chapter 4), internal loading kept summer TP concentration high and algae bloomed. Lake managers were misled by regional lake conditions, thinking that a major reduction in inflow nutrients would return the lake to a mesotrophic or oligotrophic status. The importance of internal loading in maintaining high summer TP in Shagawa Lake was demonstrated with a dynamic TP model (Larsen et al., 1979).

Another example is that of the Fairmont Lakes, Minnesota in the WCBP ecoregion. The TP concentration in the lake's hypereutrophic surface waters ranged from 30 to 150 µg P/L (Stefan and Hanson, 1981) that differed little from the WCBP ecoregion mean of about 130 µg P/L. This regional average lake trophic state information was not generally available or well understood by limnologists when lake management began, and lake managers spent nearly 60 years of well-intended but futile efforts to change the Fairmont Lakes to a low algal biomass lake through chemical treatments (Chapter 10) and dredging. It is now clear that the quality of the Fairmont Lakes is

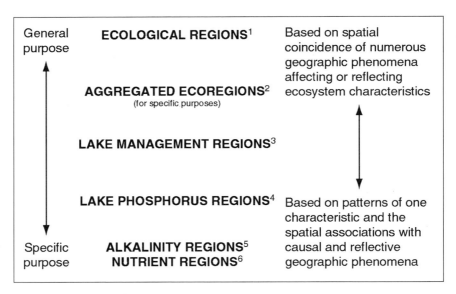

General   **ECOLOGICAL REGIONS**[1]   Based on spatial
purpose   coincidence of numerous
                                               geographic phenomena
                                               affecting or reflecting
              AGGREGATED ECOREGIONS[2]   ecosystem characteristics
                  (for specific purposes)

              LAKE MANAGEMENT REGIONS[3]

              LAKE PHOSPHORUS REGIONS[4]   Based on patterns of one
                                                characteristic and the
                                                spatial associations with
Specific   **ALKALINITY REGIONS**[5]   causal and reflective
purpose   **NUTRIENT REGIONS**[6]   geographic phenomena

FIGURE 2.9 Continuum of regional frameworks for lake assessment and management. [1]Omernik, 1987, 1995; [2]USEPA, 1998; [3]Griffith et al, 1997b; [4]Omernik et al., 1998, Rohm et al., 1995; [5]Omernik and Powers, 1983, Omernick and Griffith, 1986; [6]Omernik, 1977. (From Griffith, G.E. et al. 1999. *J. Soil Water Conserv.* 54: 666–677. With permission.)

"normal" for this region. These lakes will always resemble others of the region unless there are very expensive in-lake manipulations and major changes in land uses.

Ecoregions were not established to regionalize a specific characteristic, but were meant as "spatial tools" to work with the quality of the "aggregate of environmental resources" of a region (Omernik and Bailey, 1997, p. 939; Figure 2.9). Misunderstanding of this concept can lead to misinterpretations of regional data by resource managers. For example, Secchi disk transparency data collected by volunteers as part of the Great American Dip-In (Carlson et al., 1977) may not be suitable for extrapolation to lakes in the region because the data might not be representative of all lakes in the region. Peterson (1997) cautioned potential Dip-In data users about possible biases resulting from the data provided by volunteers who sampled lakes they liked, disliked, lived on, or were willing to sample. For example, median Secchi disk transparency in northeastern U.S. lakes, based on data from Dip-In volunteers, was reported as 4.2 m (Lee et al., 1997). However, a random sampling of lakes in the northeastern U.S. demonstrated that median transparency was only 2.4 m and that lakes sampled by the volunteers were nearly nine times larger than the general population of lakes in this region (Figures 2.10 and 2.11; Peterson et al., 1999). This demonstrates how extrapolations from non-representative lake data can be misleading. In another example, Dip-In transparency data for the upper Midwestern U.S. region, which were not obtained from a statistics-based sampling program, were extrapolated as continuous three dimensional plots for the region in an attempt to indicate overall regional lake quality (Lee et al., 1997). Unless attainable resource conditions for an ecoregion are based on representative data, resource conditions can be misrepresented and misunderstood, and may lead to erroneous conclusions and perhaps to erroneous management decisions (Peterson, 1997; Peterson et al., 1999; Omernik and Bailey, 1997).

Some states rank lakes and reservoirs on the basis of their current trophic state (Chapter 3), and on public use and other similar considerations, in order to establish funding priorities for restoration. Typically the most eutrophic lakes are given top priority. The differences in stream quality among ecoregions within a state suggests another approach that ranks lakes according to their potential for improvement, thereby directing limited public funds to the best candidates for restoration. Fulmer and Cooke (1990) examined this idea for 19 Ohio reservoirs in four adjacent

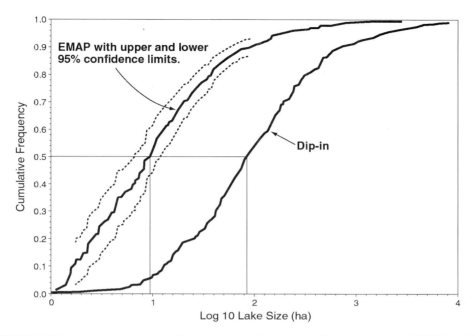

FIGURE 2.10 Lake size cumulative distribution function (cumulative frequency) for the EMAP and Dip-In data sets across the northeastern United States (vertical lines show median lake sizes for each data set). (From Peterson, S.A. et al., 1999. *Environ. Sci. Technol.* 33: 1559–1565. With permission.)

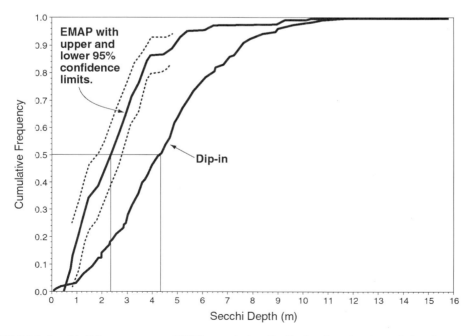

FIGURE 2.11 Secchi disk transparency (SDT) cumulative distribution function (cumulative frequency) for EMAP and Dip-In data sets across the northeastern United States. (vertical lines show median SDT's for each data set). (From Peterson, S.A. et al., 1999. *Environ. Sci. Technol.* 33: 1559–1565. With permission.)

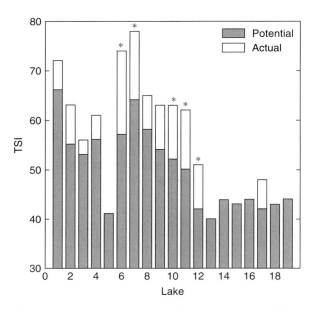

FIGURE 2.12 Potential phosphorus reduction determined by comparing the 1989 actual lake trophic state to the estimated attainable trophic state. Lakes with the greatest potential for improvement are 6, 7, 10, 11, and 12. (From Fulmer, D.G. and G.D. Cooke. 1990. *Lake and Reservoir Manage.* 6: 197–206. With permission).

ecoregions. Stream TP concentrations were selected that represented the 25th percentile concentration (the mean concentration in the lowest one-quarter of all concentrations) in the least impacted streams in the Ohio ecoregion in which each reservoir is found (from Larsen et al., 1988). These concentrations, along with hydrologic and morphometric data, were used with the Canfield and Bachmann (1981) loading model (Chapter 3). The model predicted the attainable steady state P concentration in the lacustrine deep-water zone of each reservoir. The 25th percentile concentration was not arbitrarily chosen. It was considered to be a concentration that can be reached in impacted streams through technologically feasible remediations in the watershed, including advanced waste treatment, detention of feedlot runoff, and improved agricultural practices.

Fulmer and Cooke (1990) compared predicted reservoir TP concentrations to measured concentrations. Five reservoirs were identified that had higher actual lacustrine zone TP concentrations than expected or predicted for that ecoregion. These five were not the most eutrophic of the 19 reservoirs, but were those deviating the most from expected conditions and were, therefore, those that should exhibit the greatest improvement if stream TP loading is reduced to the 25th percentile P concentration of the least impacted streams in the ecoregions (Figure 2.12). Final selections of lakes for management or restoration also involve considerations such as intended uses, proximity to users, number of other lakes in the area, and an in-depth analysis of nutrient dynamics in relation to lake trophic state.

The use of ecoregions for lake management continues to be evaluated. A special purpose map of summer total TP in lakes was developed for Minnesota, Wisconsin and Michigan (Omernik et al., 1988, 1991). The Wisconsin Department of Natural Resources currently is developing an approach for setting lake quality standards based in part on these lake TP maps. Although ecoregions are helpful in establishing regional expectations with regard to lake quality, and in evaluating relationships between lake quality and morphometric and landscape characteristics, lake TP maps explain more clearly some of the regional patterns. Work on TP maps has been completed for the northeastern U.S. (Rohm et al., 1995), Florida (Griffith et al., 1997a), and the entire U.S. (Rohm et al., 2002).

The use of ecoregions to assist with lake restoration and management decisions differs from approaches that might, for example, advocate national lake quality standards, or the use of a median TP concentration as a statewide, or even a nationwide, standard. Statewide or national lake standards are likely to fail to protect lakes in ecoregions where stream quality should be very high and could encourage large expenditures in cases where improved lake and stream quality would be difficult. The use of an ecoregional assessment approach brings a sense of regional ecological reality to lake management and restoration goal setting. The USEPA's Science Advisory Board stated (USEPA, 1991): "the ecoregion approach is a defensible classification technique for large areas (covering one or more states) that is superior to the classification methods that are currently used by most environmental managers."

2.9 SUMMARY

The purpose of this chapter is to familiarize the reader with some of the basic concepts of limnology that are important to lake and reservoir management. Reservoirs and lakes have fundamental differences, but the characteristics of many of their most important features converge when water residence times are similar. The most important determinants of lake and reservoir productivity are the amount of shallow water that can support wetland/littoral plants and the concentration of a limiting nutrient in the water column. Lakes and reservoirs of some ecoregions are expected to be clear, deep, and comparatively unproductive, and deviations from this should respond to management. In other areas, moderate or even high aquatic productivity is to be expected, and far more costly or complicated management methods will be needed to achieve any improvements in condition beyond this level.

REFERENCES

Anthony, J.L. and J.A. Downey. 2003. Physical impacts of wind and boat traffic on Clear Lake, Iowa. *Lake and Reservoir Manage.* 19: 1–14.

Axler, R., L. Paulson, P. Vaux, P. Sollberger and D.H. Baepler. 1988. Fish Aid — the Lake Mead fertilization project. *Lake and Reservoir Manage.* 4(2): 125–135.

Bachmann, R.W., M.V. Hoyer and D.E. Canfield, Jr. 2000. The potential for wave disturbance in shallow Florida lakes. *Lake and Reservoir Manage.* 16: 281-291.

Barko, J.W., M.S. Adams and N.L. Clesceri. 1986. Environmental factors and their consideration in the management of submersed aquatic vegetation: A review. *J. Aquatic Plant Manage.* 24: 1–10.

Bryce, S.A., J.M. Omernik and D.P. Larsen. 1999. Ecoregions: A geographic framework to guide risk characterization and ecosystem management. *Environ. Practice* 1: 141-155.

Canfield, D.E., Jr. and R.W. Bachmann. 1981. Prediction of total phosphorus concentrations, chlorophyll a, and Secchi depths in natural and artificial lakes. *Can. J. Fish. Aquatic Sci.* 38: 414–423.

Canfield, D.E. Jr., K.A. Langeland, S.B. Linda and W.T. Haller. 1985. Relations between water transparency and maximum depth of macrophyte colonization in lakes. *J. Aquatic Plant Manage.* 23: 25–28.

Caraco, N.F., J.J. Cole and G.E. Likens. 1989. Evidence for sulphate-controlled phosphorus release from sediments of aquatic systems. *Nature* 341: 316–318.

Carlson, R.E., J. Lee and D. Waller. 1997. The 1995 and 1996 Great American Secchi Dip-In: A report to the volunteers. *Lakeline* 17(2): 32a–32d.

Carlton, R.G. and R.G. Wetzel. 1988. Phosphorus flux from lake sediments: effect of epiphytic algal oxygen production. *Limnol. Oceanogr.* 33: 562–570.

Carpenter, S.R. 1980. Enrichment of Lake Wingra, Wisconsin, by submersed macrophyte decay. *Ecology* 61: 1145–1155.

Carpenter, S.R. 1981. Submersed vegetation: an internal factor in lake ecosystem succession. *Am. Nat.* 118: 372–383.

Carpenter, S.R. 1983. Lake geometry: implications for production and sediment accretion rates. *J. Theor. Biol.* 105: 273–286.

Cole, G.A. 1994. *Textbook of Limnology.* 4th edition Waveland Press, Prospect Heights, IL.

Cole, J.J. 1999. Aquatic microbiology for ecosystem scientists: New and recycled paradigms in ecological microbiology. *Ecosystems* 2: 215–225.

Cooke, G.D. and R.H. Kennedy. 2001. Managing drinking water supplies. *Lake and Reservoir Manage.* 17: 157–174.

Cooke, G.D., P. Lombardo and C. Brant. 2001. Shallow and deep lakes: Determining successful management options. *LakeLine (NALMS)* 21: 42–46.

Cooke, G.D., E.B. Welch, S.A. Petersen, and P.R. Newroth. 1993. *Restoration and Management of Lakes and Reservoirs,* 2nd Edition. Lewis Publishers and CRC Press, Boca Raton, FL.

Duarte, C.M. and J. Kalff. 1986. Littoral slope as a predictor of the maximum biomass of submerged macrophyte communities. *Limnol. Oceanogr.* 31: 1072–1080.

Duarte, C.M. and J. Kalff. 1988. Influence of lake morphometry on the response of submerged macrophytes to sediment fertilization. *Can. J. Fish. Aquatic Sci.* 45: 216–221.

Fee, E.J. 1979. A relation between lake morphometry and primary productivity and its use in interpreting whole-lake eutrophication experiments. *Limnol. Oceangr.* 24: 401–416.

Franko, D.A. and R.G. Wetzel. 1981. Synthesis and release of cyclic adenosine $3':5'$-monophosphate by aquatic macrophytes. *Physiol. Plant.* 52: 33–36.

Frey, D.G. Ed. 1966. *Limnology in North America.* University of Wisconsin Press, Madison.

Frodge, J.D., G.I. Thomas and G.B. Pauley. 1991. Sediment phosphorus loading beneath dense canopies of aquatic macrophytes. *Lake and Reservoir Manage.* 7: 61–71.

Fulmer, D.G. and G.D. Cooke. 1990. Evaluating the restoration potential of 19 Ohio reservoirs. *Lake and Reservoir Manage.* 6: 197–206.

Gächter, R. and B. Müller. 2003. Why the phosphorus retention of lakes does not necessarily depend on the oxygen supply to their sediment surface. *Limnol. Oceanogr.* 48: 929–933.

Gaugush, R.F. 1986. Statistical methods for reservoir water quality investigations. Instr. Rept. E-86-2. U.S. Army Corps Engineers, Vicksburg, MS.

Golterman, H.L. 1995. The role of the iron-hydroxide-phosphate-sulfide system in the phosphorus exchange between sediments and the overlying water. *Hydrobiologia* 297: 43–54.

Griffith, G.E., J.M. Omernik and A.J. Kinney. 1997. Interpreting patterns of lake alkalinity in the upper Midwest region. *Lake and Reservoir Manage.* 10(6): 329–336.

Griffith, G.E., D. E. Canfield Jr., C.A. Horsburgh, and J.M. Omernik. 1997b. Lake regions of Florida. USEPA R97/127.

Griffith, G.E., J.M. Omernik and A.J. Woods. 1999. Ecoregions, watersheds, basins, and HUC's: How state and federal agencies frame water quality. *J. Soil Water Conserv.* 54: 666–677.

Gunnison, D. and J.M. Brannon. 1981. Characterization of Anaerobic Chemical Processes in Reservoirs: Problem Description and Conceptual Model Formulation. Tech. Rept. E-81-6. U.S. Army Corps Engineers, Vicksburg, MS.

Havens, K.E., H.J. Carrick, E.F. Lowe and M.F. Coveney. 1999. Contrasting relationships between nutrients, chlorophyll a and secchi transparency in two shallow subtropical lakes: Lakes Okeechobee and Apopka (Florida). *Lake and Reservoir Manage.* 15: 298–309.

Hawkes, C.L., D.L. Miller and W.G. Layther. 1986. Fish ecoregions of Kansas: stream fish assemblage patterns and associated environmental correlates. *Environ. Biol. Fish.* 17: 267–279.

Heiskary, S.A. and C.B. Wilson. 1989. The regional nature of water quality across Minnesota: An analysis for improving resource management. *J. Minn. Acad. Sci.* 55(1): 71–77.

Heiskary, S.A., C.B. Wilson and D.P. Larsen. 1987. Analysis of regional pattern in lake water quality: Using ecoregions for lake management in Minnesota. *Lake and Reservoir Manage.* 3: 337–344.

Horppila, J. and L. Nurminen. 2003. Effects of submerged macrophytes on sediment resuspension and internal phosphorus loading in Lake Hiidenvesi (southern Finland). *Water Res.* 37: 4468–4474.

Horne, A.J. and C.R. Goldman. 1994. *Limnology.* 2nd edition. McGraw-Hill, New York, NY.

Hughes, R.M. and D.P. Larsen. 1988. Ecoregions: an approach to surface water protection. J. *Water Pollut. Contr. Fed.* 60: 486–493.

Hutchinson, G.E. 1957. *A Treatise on Limnology. Volume I. Geography, Physics, and Chemistry.* John Wiley & Sons, New York.

Hutchinson, G.E. 1967. *A Treatise on Limnology. Volume II. Introduction to Lake Biology and the Limno-plankton.* John Wiley & Sons, New York.

Hutchinson, G.E., 1975. *A Treatise on Limnology. Volume III. Limnological Botany.* John Wiley & Sons, New York.

Jensen, H.S. and F.O. Andersen. 1992. Importance of temperature, nitrate, and pH for phosphate release from aerobic sediments of four shallow, eutrophic lakes. *Limnol. Oceanogr.* 37: 577–589.

Jeppesen, E. 1998. *The Ecology of Shallow Lakes. Trophic Interactions in the Pelagial.* National Environmental Research Institute, Silkeborg, Denmark.

Jones, J.R. and M.V. Hoyer. 1982. Sportfish harvest predicted by summer chlorophyll a concentration in Midwestern lakes and reservoirs. *Trans. Am. Fish. Soc.* 111: 176–179.

Kalff, J. 2002. *Limnology. Inland Water Ecosystems.* Prentice-Hall, Upper Saddle, NJ.

Kennedy, R.H. 1999. Reservoir design and operation: Limnological implications and management In: J.G. Tundisi and M. Straskraba (Eds.), *Theoretical Reservoir Ecology and Its Applications.* International Institute of Ecology, Brazilian Academy of Sciences. Backhuys Publishers, Leiden, The Netherlands. pp. 1–28.

Kennedy, R.H., 2001. Considerations for establishing nutrient criteria for reservoirs. *Lake and Reservoir Manage.* 17: 175–187.

Kennedy, R.H. and R.F. Gaugush, 1988. Assessment of water quality in Corps of Engineers reservoirs. *Lake and Reservoir Manage.* 4(2): 253–260.

Kennedy, R.H., K.W. Thornton and R.C. Gunkey. 1982. The establishment of water quality gradients in reservoirs. *Can. Water Res.* J. 7: 71–87.

Kennedy, R.H., K.W. Thornton and D.E. Ford. 1985. Characterization of the reservoir ecosystem. In: D. Gunnison. (Ed.), *Microbial Processes in Reservoirs.* Junk Publishers, The Hague, Netherlands. pp. 27–38.

Kimmel, B.C. and A.W. Groeger. 1984. Factors controlling primary production in lakes and reservoirs: a perspective. In: *Lake and Reservoir Management.* USEPA 440/5-84-001. pp. 277–281.

LaMarra, V.J., Jr. 1975. Digestive activities of carp as a major contributor to the nutrient loading of lakes. *Verh. Int. Verein. Limnol.* 19: 2461–2468.

Lampert, W. and U. Sommer. 1997. *Limnoecology. The Ecology of Lakes and Streams.* Oxford University Press, New York, NY.

Larsen, D.P., D.R. Dudley and R.M. Hughes. 1988. A regional approach for assessing attainable surface water quality: an Ohio case study. *J. Soil Water Conserv.* 43: 171–176.

Larsen, D.P., J. VanSickle, K.W. Malueg and P.D. Smith. 1979. The effect of wastewater phosphorus removal on Shagawa Lake, Minnesota: Phosphorus supplies, lake phosphorus, and chlorophyll-a. *Water Res.* 13:1259–1272.

Lee, J., M.R. Binkley and R.E. Carlson. 1997. The Great American Secchi Dip-In: GIS contributes to national snapshot of lake-water quality. *GIS World* August: 42–44.

Marsden, M.W. 1989. Lake restoration by reducing external phosphorus loading: the influence of sediment phosphorus release. *Freshwater Biol.* 21: 139–162.

Mataraza, L.K. and G.D. Cooke. 1997. A test of a morphometric index to predict vertical phosphorus transport in lakes. *Lake and Reservoir Manage.* 13: 328–337.

Moore, L. and K. Thornton, Eds. 1988. *Lake and Reservoir Restoration Guidance Manual.* USEPA 440/5-88-002.

Moss, B., J. Madgwick and G. Phillips. 1996. *A Guide to the Restoration of Nutrient-Enriched Shallow Lakes.* Broads Authority. Norwich, Norfolk, UK.

North American Lake Management Society (NALMS). 2003 Shallow Lakes. *LakeLine* 23(1).

Odum. E.P. 1971. *Fundamentals of Ecology.* 3rd Edition. W.B. Saunders, Philadelphia.

Omernik, J.M. 1977. Nonpoint source stream nutrient level relationships: A nationwide study. USEPA 600/3-77/105.

Omernik, J.M. 1987. Ecoregions of the conterminous United States. *Ann. Assoc. Am. Geogr.* 77: 118–125.

Omernik, J.M. 1995. Ecoregions: A framework for managing ecosystems. *George Wright Forum* 12: 35–51.

Omernik, J.M. and R.G. Bailey. 1997. Distinguishing between watersheds and ecoregions. J. *Am. Water Resour. Assoc.* 33: 935–949.

Omernik, J.M. and G.E. Griffith. 1986. Total alkalinity of surface waters: A map of the Upper Midwest region of the United States. *Environ Manage.* 10: 829–839.

Omernik, J.M. and C.F. Powers. 1983. total alkilinity of surface waters: A national map. *Ann. Assoc. Amer. Geogr.* 73: 133–136.

Omernik, J.M., D.P. Larsen, C.M. Rohm and S.E. Clarke. 1988. Summer total phosphorus in lakes: a map of Minnesota, Wisconsin, and Michigan. *Environ. Manage.* 12: 815–825.

Omernik, J.M., C.M. Rohm, R.N. Lillie and N. Mesner. 1991. Usefulness of natural regions for lake management: analysis of variation among lakes in northwestern Wisconsin. *Environ. Manage.* 15: 281–293.

Omernik, J.M.S.S. Chapman, R.A. Lillie and R.T. Dumke. 2000. Ecoregions of Wisconsin. *Trans. Wisc. Acad. Sci.* 88: 77–103.

Osgood, R.A. 1988. Lake mixes and internal phosphorus dynamics. *Arch. Hydrobiol.* 113: 629–638.

Peterson, S.A. 1997. Liked volunteerism coverage, not cover; cautions against Dip-In generalization. *Lakeline* 17(3): 4–5.

Peterson, S.A., R.M. Hughes, D.P. Larsen, S.G. Paulsen and J.M. Omernik. 1995. Regional lake quality patterns: Their relationship to lake conservation and management decisions. *Lakes Reservoirs: Res. Manage.* 1:163–167.

Peterson, S.A., D.P. Larsen, S.G. Paulsen and N.S. Urquhart. 1998. Regional lake trophic patterns in the northeastern United States: Three approaches. *Environ. Manage.* 22(5): 789–801.

Peterson, S.A., N.S. Urquhart and E.B. Welch. 1999. Sample representativeness: A must for reliable regional lake condition estimates. *Environ. Sci. Technol.* 33: 1559–1565.

Pettersson, K. 1998. Mechanisms for internal loading of phosphorus in lakes. *Hydrobiologia* 373/374: 21–25.

Rohm, C.M., J.M. Omernik and C.W. Kiilsgaard. 1995. Regional patterns of total phosphorus in lakes of the northeastern United States. *Lake and Reservoir Manage.* 11(1): 1–14 (color map).

Rohm, C.M., J.M. Omernik, A.J. Woods and J.L. Stoddard. 2002. Regional characteristics of nutrient concentrations in streams and their application to nutrient criteria development. *J. Am. Water Resour. Assoc.* 38(1): 1–27.

Ryding, S.O. 1985. Chemical and microbiological processes as regulators of the exchange of substances between sediments and water in shallow eutrophic lakes. *Int. Rev. ges. Hydrobiol.* 70: 657–702.

Scheffer, M. 1998. *Ecology of Shallow Lakes.* Kluwer Academic Publishers, Norwell, MA.

Smith, C.S. and J.W. Barko. 1990. Ecology of Eurasian watermilfoil. *J. Aquat. Plant Manage.* 28: 55–63.

Søballe, D.M. and B.C. Kimmel. 1987. A large-scale comparison of factors influencing phytoplankton abundance in rivers, lakes, and impoundments. *Ecology* 68: 1943–1954.

Søndergaard, M., J.P. Jensen and E. Jeppesen. 2001. Retention and internal loading of phosphorus in shallow, eutrophic lakes. *Sci. World* 1: 427–442.

Stauffer, R.E. and G.F. Lee. 1973. The role of thermocline migration in regulating algal blooms. In: E.J. Middlebrooks, D.H. Falkenborg, and T.E. Maloney (Eds.), *Modeling the Eutrophication Process*, Utah State University, Water Resources Center, Logan, UT. pp. 73–82.

Stefan, H.G. and M.J. Hanson. 1981. Phosphorus recycling in five shallow lakes. *J. Environ. Eng. Div. ASCE*, 107(EE4): 713–730.

Thornton, K.W., R.H. Kennedy, J.H. Carroll, W.W. Walker, R.C. Gunkey and S. Ashby. 1980. Reservoir sedimentation and water quality — an heuristic model. In: *Symposium on Surface Water Impoundments*. Proceedings Am. Soc. Civil Eng. pp. 654–661.

Thornton, K.W., B.L. Kimmel and F.E. Payne (Eds.). 1990. *Reservoir Limnology: Ecological Perspectives.* John Wiley & Sons, New York, pp. ix and 246.

U.S. Environmental Protection Agency (USEPA) Science Advisory Board. 1991. Evaluation of the ecoregion concept. Report of the Ecoregions Subcommittee of the Ecological Processes and Effects Committee. USEPA-SAB-EPEC-91-003. Washington, DC.

U.S. Environmental Protection Agency (USEPA). 1998. National strategy for the development of regional nutrient criteria. USEPA 822/R98–002,

Vadeboncoeur, Y., M.J. Vander Zanden and D.M. Lodge. 2002. Putting the lake back together: Reintegrating benthic pathways into lake food web models. *BioScience* 52: 44–54.

Walker, W.W., Jr. 1981. Empirical Methods for Predicting Eutrophication in Impoundments. Report I. Phase II: Date Base Development. Tech. Rept. E-81-9. U.S. Army Corps Engineers, Vicksburg, MS.

Walker, W.W., Jr. 1987. Empirical Methods for Predicting Eutrophication in Impoundments. Report 4. Phase III: Applications Manual. Tech. Rept. E-81-9. U.S. Army Corps Engineers, Vicksburg, MS.

Welch, E.B. and J.M. Jacoby. 2004. *Pollutant Effects in Freshwater: Applied Limnology.* 3rd Edition. Spon Press, New York.

Welch, E.B. and G.D. Cooke. 1995. Internal phosphorus loading in shallow lakes: Importance and control. *Lake and Reservoir Manage.* 11: 273-281.

Wetzel, R.G. 1992. Gradient-dominated ecosystems: sources and regulatory functions of dissolved organic matter in fresh water ecosystems. *Hydrobiologia* 229: 181–198.

Wetzel, R.G. 1995. Death, detritus, and energy flow in aquatic ecosystems. *Freshwater Biol.* 33: 83–89.

Wetzel, R.G. 2001. *Limnology. Lake and River Ecosystems.* 3rd edition. Academic Press, New York.

Wilson, C.B. and W.W. Walker, Jr. 1989. Development of lake assessment methods based upon the aquatic ecoregion concept. *Lake and Reservoir Manage.* 5(2): 11–22.

3 Lake and Reservoir Diagnosis and Evaluation

3.1 INTRODUCTION

The success of efforts to restore and/or improve the quality of lakes and reservoirs depends on the thoroughness of the diagnosis and evaluation prior to initiating restoration measures. Thorough diagnosis with appropriate predictive methods allows realistic expectations. This chapter describes the following: (1) the constituents and variables that should be determined in the watershed and in the lake and its sediment; (2) the sample number needed and their frequency; (3) ways to express the data collected; (4) the levels of constituents that indicate trophic state; and (5) how to determine the limiting nutrient. Also, it covers aspects of phosphorus modeling, how to predict the response to treatment and how to choose a treatment(s) based on predicted response, past success, and cost.

There have been many mistakes made in the name of lake restoration and management. Techniques that are the correct choice in some situations have been used in the wrong circumstances, sometimes for political reasons, but sometimes because the diagnosis and evaluation were inadequate (Peterson et al., 1995). Techniques, such as external controls on nutrient input and in-lake controls, such as drawdown to control macrophytes, were implemented without the benefit of a complete prerestoration diagnosis/evaluation. Improvement in water quality or an acceptable control of macrophytes did not occur because certain factors/conditions were not considered fully, such as: (1) the relative unimportance of external nutrient sources, compared to internal sources, (2) the uncertainty of drawdown as a macrophyte control under the particular climatic conditions (e.g., Long Lake, Washington, Chapter 13), or (3) the "natural" condition of other lakes in the region, i.e., unreasonable expectations (Peterson et al., 1999). In other instances, in-lake nutrient control measures were initiated where the major inputs were external and similarly, improvements in water quality did not result (e.g., Riplox in Long Lake, Minnesota, Chapter 8).

Lake and reservoir restoration has progressed markedly in its relatively short history, but a proven "track record" for some techniques is lacking. Thus, there is still uncertainty in estimating cost effectiveness of some techniques. For that reason, a thorough prerestoration diagnosis/evaluation is an absolute requirement, not only for the increased assurance of success, but also to contribute new knowledge that benefits future projects.

3.2 DIAGNOSIS/FEASIBILITY STUDIES

3.2.1 WATERSHED

Lake and reservoir quality, or trophic state, is a direct result of their location within the landscape and nutrients and sediment that enter them from their watersheds. Thus, a thorough understanding of the watershed's characteristics (soils, slope, vegetation, tributaries, wetlands, unique non-point nutrient sources, etc.) is necessary to explain the condition of the lake/reservoir. Where the lake fits within the population of lakes in the region is also important (Peterson et al., 1999; Heiskary and Wilson, 1989; Chapter 2). For many areas, some of these characteristics can be determined using geographical information systems (GIS).

Initially, detailed maps must be obtained. Tributaries and wells for surface and groundwater (GW) nutrient content and flow determinations must be located. These are usually indicated on U.S. Geological Survey quadrangle maps. These maps also have contour lines so watershed boundaries for the main basin, as well as sub-basins, can be drawn. While these maps are usually complete, they probably do not include stormwater pipes if the lake is in a developing urban area. Hydrologic changes may have occurred since the map was drawn, so ground reconnaissance is absolutely necessary. For example, 45 inflow sources were identified for 2000 ha Lake Sammamish in 1971, and most were stormwater pipes not on the quadrangle map. From that information, 13 minor tributaries were selected, along with the major inflow that contributed 70% of the water, to construct water and nutrient budgets (Moon, 1973; Welch et al., 1980). Location and sampling of inputs becomes an increasing problem as lake size increases.

Watershed area, lake area and lake volume are often known, but if not, must be determined from maps. Sub-watershed (sub-basins) delineation may be important if development varies from one part of the watershed to another. Nutrient yield coefficients (mg/m2 per yr) vary with the density of development, and therefore, are of value in developing control strategies. Sub-basins can be further subdivided into land use types, such as forest, agricultural and urban (commercial and single family) for purposes of proportioning sub-basin nutrient loading to land use.

Lake depth contours are necessary to calculate lake volume and for locating water/sediment sampling sites. If existing contour maps are old, new soundings may be necessary, especially for reservoirs with large inflows from erosive watersheds. Soundings should be made with electronic methods to improve accuracy if soft (high water content) sediments are present. Depth–area (or depth–volume) hypsographic curves should be constructed to illustrate the lake's morphometry (Figure 2.4).

Construction of an accurate water budget is the first step in diagnosing a lake's problem(s), because the substances that determine quality, or trophic state, originally are transported by water from the watershed. Major tributaries can be selected from a reconnaissance survey of water discharges. Continuous gauge recording is recommended to determine flow in major tributaries, because high flows are the most important segment of the water budget and large volume influxes are accompanied by high substance concentrations, especially in urban areas. From subsequent continuous records of flow in the major tributaries and the outflow(s), an annual water budget is constructed so that measured/estimated inflows equal outflows with correction for lake storage. The water budget formulation is:

$$SF_i + GW + DP + WW = SF_o + EVP + EXF + WS \pm \Delta STOR \qquad (3.1)$$

SF_i is stream flow in and out, GW is groundwater in (includes deep and subsurface seepage), DP is direct precipitation on the lake surface, WW is wastewater, if any, EVP is evaporation, EXF is exfiltration, WS is removal for water supply, if any, and $\Delta STOR$ is change in lake volume. There may be other sources/losses than those designated above. Winter (1981) has described the methods, uncertainties, and problems in estimating a lake's water budget. A brief description of procedures to determine the values for Equation 3.1 follows.

Stream flow (SF) is estimated by taking velocity measurements over a known cross section of stream. SF, or discharge, is:

$$SF \ (m^3/s) = velocity \ (m/s) \times cross\text{-sectional area} \ (m^2) \qquad (3.2)$$

A staff gauge may be installed and calibrated over the full range of measured discharge rates, so that observations of water level are used to estimate discharge from a regression equation. Discrete observations are inadequate if discharge is so variable that high rates are missed if observations are made weekly, twice monthly, etc. The greatest accuracy in annual stream flow estimates is by

TABLE 3.1
Comparison of Hydraulic Input as Calculated by Five Commonly Used Methods (Seven Streams on Harp Lake, Ontario, January–December 1977)

Data	Stream Discharge Calculation Method	Mean Absolute % Error	Range in % Error
Discharge calculated from contin. stage records	Integration of continuous discharge vs. time plot	0	0
	Integration of discrete discharge vs. time plot	12	−19 to + 35
Discharge measured at discrete time intervals	Three-point running mean of discrete discharge	35	−15 to + 130
No measured discharge	Long-term unit runoff (Pentland, 1968)	18	−2 to + 68
	Precipitation-evapotranspiration (Morton, 1976)	36	+12 to + 91

Source: From Scheider W.A. et al., 1979. *Lake Restoration*. USEPA 440/5-79-001. p. 77.

automatic continuous discharge with a stage-height recorder. Estimates of SF_i from discrete discharge measurements and calculated values from runoff maps and precipitation-evaporation records had errors ranging from 12% to 36% compared with those from continuous gauge-height records (Scheider et al., 1979; Table 3.1).

If the project cannot afford continuous gauge-height recording, an alternative, capable of intermediate accuracy, is as follows. SF_i is separated into base flow and storm flow, with the former being estimated from discrete observations and the latter from continuous (manual) observations during several storm events during the year. Discharge during other storm events is estimated by a relationship with precipitation, which is not always satisfactory due to varying antecedent dry periods, or with a continuous flow record from a nearby stream (e.g., one equipped with a USGS station). Runoff can also be estimated using contour maps developed with existing runoff data for broad regions (Rochelle et al., 1989).

Outlet SF_o is typically less complicated than inflows, because there is usually one outlet stream and the lake dampens flow variation. In reservoirs, overflow from a uniform spillway may simplify measurement procedures. For many reservoirs, records of continuous outflows are available.

Precipitation directly on the lake surface (DP) is determined with a collector installed preferably at the lake and on the water rather than the shore. A constantly open collector is recommended so that dry fall, as well as precipitation, is obtained. Events should be collected separately, as with stormwater, due to the variability from one event to another. Several collectors may be needed at a large lake or reservoir. The relative importance of precipitation in the total budget increases as the ratio of total watershed area to lake area decreases. For example, for Ontario lakes, precipitation amounted to only 3% of the total phosphorus (TP) load for a watershed to lake area ratio of 100:1, 9% for a ratio of 30:1, and 23% for a ratio of 10:1 (Rigler, 1974).

Wastewater (WW) contributions are determined in the same way as SF, but are usually more constant so discrete observations may be adequate. Those data are usually collected as part of plant operations. Urban stormwater (and agricultural) runoff may contain suspended solids and nutrient concentrations nearly as high as wastewater. In some instances, estimations from paved areas based on precipitation may be adequate (Arnell, 1982; Brater and Sherrill, 1975).

Groundwater may be an important component and comprise 50% or more of the total influx. Some lakes receive very little GW. However, this cannot be assumed. GW is by far the most difficult influx to estimate (Winter, 1978, 1980, 1981). The most common, but usually least adequate method to estimate GW is to treat it as the residual term in Equation 3.2. The accuracy of this approach depends on the accuracy of all the other terms in the equation. La Baugh and Winter (1984) found that the residual term was of the same magnitude as the measurement errors of the other terms in the water budget for a Colorado reservoir.

A direct method for groundwater estimation is to calculate it in a flow net using the following equation:

$$Q = KIA \tag{3.3}$$

Q is groundwater discharge, K is hydraulic conductivity, I is hydraulic gradient, and A is cross-sectional area through which flow occurs. This procedure requires establishing nests of piezometers to determine the hydraulic gradient of the water table (and substance concentration), measuring hydraulic conductance through pump tests, and establishing hydrogeologic boundaries for flow.

Another direct method is the use of seepage meters (Lee, 1977; Lee and Hynes, 1978; Barwell and Lee, 1981). These are constructed of plastic barrel halves, inverted over the lake bottom so that GW flows into an attached collecting bag, the contents of which represent the total net flow per unit barrel area over the collection time. An adequate sampling design is necessary with this method, because they measure flow at a discrete site and flow can vary greatly among sites. Also, the need for SCUBA gear to sample the barrels limits their use to ice-free periods in northern latitudes. Although they have proven to be a convenient and useful tool for detecting the direction and quantity of GW flow, they are not as reliable in determining nutrient transport via GW. Enclosure of the surficial sediments within the meter promotes anaerobic conditions. Hence, determination of nutrient content in that water can lead to substantial overestimates in transport rates (Belanger and Mikutel, 1985). To characterize the GW quality entering a lake, Mitchell et al. (1989) have demonstrated the usefulness of a modified hydraulic potentiomanometer to sample interstitial pore water in the littoral. Also, to obtain accurate estimates of water input, the seepage meter bags should be partially pre-filled to prevent an anomaly of an excessive initial influx (Shaw and Prepas, 1989).

Evaporation (EVP) is a water-loss term estimated by several methods, all with potentially significant errors. EVP pan is the most common method, but no standard pan technique exists, and there are problems in extrapolation from the pan to the lake. Pan EVP rates are often obtained from the nearest National Weather Service station and multiplied by 0.7 to estimate lake EVP, based on a class A pan. However, this coefficient is based on annual averages and will be incorrectly applied if used for monthly values (Siegel and Winter, 1980).

Finally, the lake level, or storage (volume) term, is determined from a gauge-height recorder or discrete observations of a staff gauge. Records of level are often available for reservoirs. Errors in lake level measurement are largely attributable to lake area and volume estimates, and to seiches in large lakes and reservoirs. Exfiltration (EXF) is very difficult to determine and is usually assumed to be nil. Some indication of EXF may be obtained by observing changes in storage during periods of low GW influx.

The nutrient budget is constructed by multiplying each term (except EVP) in the water budget by a representative concentration. While concentrations tend to be less variable than flow, frequent observations are nonetheless desirable. A suggested minimum frequency is twice monthly. Scheider et al. (1979) used discrete observations of TP concentration and continuous SF as the absolute estimate in comparing eight methods of computing TP loading (Table 3.2). Estimates of inputs from urban (and rural agricultural) stormwater runoff, where TP concentration is normally high at the beginning of a storm event, and declines as the storm continues, may require far more frequent observations of concentration during storms or, preferably, the use of flow-activated automatic sampling.

Concentrations in GW, DP, and WW are less variable and usually need not be observed so frequently. Direct precipitation can often represent a substantial fraction and affect the in-lake N:P ratio, especially for oligotrophic lakes (Jassby et al., 1994).

TABLE 3.2
Comparison of Phosphorus Input Calculation by Nine Commonly Used Methods (Seven Streams on Harp Lake, Ontario, January–December 1977)

Data	Phosphorus Input Calculation Method	Mean Absolute % Error	Range in % Error
Discharge calculated from continuous stage records; [P] measured at discrete time intervals	1. Product of integrated discharge vs. time plot and [P] at midpoint of time interval	0	0
	2. Product of integrated discharge vs. time plot and mean of [P] at end point of time interval	3	–4 to + 5
	3. Product of integrated discharge vs. time plot and mean of [P] at midpoint of time intervals	11	–19 to + 11
	4. Product of integrated discharge vs. time plot and [P] at endpoints of time interval	14	–25 to + 16
	5. Product of discharge as calculated by three-point running mean and [P] at midpoint of time interval	30	–19 to + 92
Discharge and [P] measured at discrete time intervals	6. Integration of the plot of the product of discharge and [P] vs. time	10	–19 to + 8
	7. Three-point running mean of product of discharge and [P]	27	–14 to + 57
	8. Product of total monthly discharge (Pentland, 1968) and [P]	49	–4 to + 85
No measured discharge and [P] measured monthly	9. Product of total monthly discharge (precipitation-evapotranspiration) and [P]	71	–19 to + 111

Source: From Scheider, W.A. et al. 1979. *Lake Restoration.* USEPA 440/5-79-001. p. 77.

A minimum of bi-monthly computations of the TP budget is recommended in order to determine the among- and within-seasonal variation in sources and sinks. The mass balance, in units of kilograms per whole lake or milligrams per square meter of lake area, is as follows:

$$\Delta TP_l = TP_{in} - TP_{out} - TP_{sed} \tag{3.4}$$

where TP_l is whole-lake content, TP_{in} is all external inputs. TP_{out} is the output and TP_{sed} is sedimentation in the lake. Internal loading of P from anoxic (or oxic) sediment release or decomposition of macrophytes can be estimated by solving for TP_{sed} in Equation 3.4:

$$TP_{sed} = TP_{in} - TP_{out} - \Delta TP_l \tag{3.5}$$

where a negative TP_{sed} indicates that TP_{out} and/or ΔTP_l exceeds the external input of TP_{in} and, thus, there is net internal loading. That is, the gross rate of sediment release exceeds the gross rate of sedimentation. The gross rate of sediment release may be estimated by independent measurements in cores in the laboratory or by estimation of the gross sedimentation rate by means of sediment traps in the lake (if not too shallow). The gross release rate may be estimated by calibration of a mass balance model as will be described later. If TP_{sed} is positive, gross sedimentation exceeds gross release, which is the case on a long-term basis in all lakes. However, during short-term periods of anoxia, high temperature, or wind action, or for several years following reduction of external

TABLE 3.3
Watershed TP Yield Coefficients

Land Use	Yield Coefficient (mg/m² per yr)
Forest	2–45
Precipitation	15–60
Agriculture	10–300
Urban	50–500
Septic-tank drain fields	0.3–1.8 kg/cap per yr

Source: From Reckhow, K.H. and S.C. Chapra. 1983. *Engineering Approaches for Lake Management: Vol. I. Data Analysis and Empirical Modeling.* Butterworths, Boston, MA. With permission.

inputs, net internal loading can be highly significant. Estimation of net internal loading on an annual basis will underestimate its importance, because algal problems occur in summer when internal loading may be the largest P source (Welch and Jacoby, 2001). Restoration attempts by controlling external inputs have often been unsuccessful, or unexpected, because internal sources were either underestimated or not estimated at all.

Sedimentation rates from traps agreed with TP retention on an annual basis in Eau Galle Reservoir, Wisconsin, but exceeded retention during summer indicating additional internal P sources (James and Barko, 1997). Trap data were helpful in estimating a settling rate for a TP model for Lake Sammamish, Washington (Perkins et al., 1997).

External nutrient loading may also be estimated indirectly using published yield (or export) coefficients, preferably calibrated to local conditions. The procedure was originally developed to estimate the capacity of a lake to accommodate development of summer homes around its shore (Dillon and Rigler, 1975). The approach allows a consultant or lake manager to estimate the current mean lake TP concentration and compare it to a predicted post-development concentration of TP, transparency, and algal biomass. Lake TP concentration is obtained by summing the yields from the land-use areas (urban, agricultural and forest), including that from precipitation and from cultural sources, such as septic tank drain fields. Water flow is estimated from runoff maps and lake volume and area from topographic maps or direct measurement.

The potential for large errors with this approach is great. A procedure for estimating uncertainty for each separate estimate of TP yield, as well as providing improved yield coefficients, was described by Reckhow and Simpson (1980). Also, a method of error analysis appropriate when prediction of a new steady state TP concentration is desired for a change in land use was developed (Reckhow, 1983). Existing lake quality data are used, eliminating the need to project all land-use impacts. Suggested ranges in TP yield coefficients are shown in Table 3.3.

Rast and Lee (1978) also developed TP yield coefficients for three land-use types (wetlands were assumed to have no net yield) plus precipitation, based on data from 473 sub-drainage areas in the eastern U.S. (USEPA, 1974) and data from Uttormark et al. (1974) and Sonzogni and Lee (1974). These coefficients are single values and fall toward the lower end of the ranges shown in Table 3.3 (Table 3.4), which may be reasonable since data of this type tend to be log normally distributed. Rast and Lee (1978) considered that the coefficients in Table 3.4 would approximate the true load from a watershed by ± 100%. There was good agreement between the loading computed from their export coefficients and the loading rate empirically determined for 38 U.S. water bodies.

Estimated N and P export coefficients exist for Wisconsin lakes (Clesceri et al., 1986; Omernik, 1977), Lake Mendota, Wisconsin (Soranno et al., 1996); Lake Okeechobee, Florida (Fluck et al., 1992) and for Canadian Shield lakes (Nürnberg and LaZerte, 2004). The latter were used in a modeling approach that predicted the effect of development on internal as well as external TP loading.

TABLE 3.4
Watershed TP Yield Coefficients

Land Use	Yield Coefficient (mg/m² per yr)
Forest	10
Precipitation	20
Agriculture/rural	50
Urban	100
Dry fall	80

Source: From Rast, W. and G.F. Lee. 1978. *Summary analysis of the North American (U.S. Portion) OECD eutrophication project: nutrient loading-lake response relationships and trophic state.* USEPA 600/3-78-008.

Yield coefficients can provide a reasonable estimate of TP (and N; Rast and Lee, 1978) loading to a lake, and at relatively low cost. However, the degree of uncertainty should be computed, and field verification would reduce that uncertainty. To use this indirect method of loading estimation to predict effects of increased development, an annual water budget must be available, and one preferably determined directly. However, the only estimate possible using coefficients is for an annual loading, which is not as useful for estimating internal loading as a seasonal budget analysis.

Yield coefficients may have their greatest value in estimating lake quality changes from planned development near water bodies with complete water and nutrient budgets that were determined directly. Although direct measurement of sub-basin loading is most reliable, it gives no information on the distribution of that loading among land-use types. Thus, by using the ratios among yields in Tables 3.3 or 3.4, together with information on the areas devoted to the respective land uses in each sub-basin, the known load can be partitioned among land uses. In that way, the effect of future changes in land use can be more reliably determined for a particular lake (Shuster et al., 1986). Yield coefficients were calibrated to local conditions to develop estimates of loading for a set of Massachusetts lakes (Matson and Isaac, 1999). A significant forecasting problem using yield coefficients is the uncertainty due to changing SF_i. Because future loading is estimated from calibrated yield coefficients, they would not include the effect of changing SF_i. When estimated loads are superimposed on a range of SF_i possibilities, lower inflow TP concentrations result from high flow and higher concentrations, the opposite of that expected in urbanizing watersheds. Normally, increased runoff in urbanized watersheds produces higher TP concentrations. Therefore, some adjustment is necessary.

3.2.2 IN-LAKE

The data needs for a lake or reservoir are more varied than those from the watershed (nutrients, solids and water flow). In-lake data are used to describe a lake's trophic state (quality), help understand why that trophic state exists (Peterson et al., 1995, 1999), and provide clues as to its restoration potential. The data needed include physical, chemical, and biological variables.

Temperature profiles determine the extent of thermal (density) stratification and mixing, which are important to understanding the distribution of chemical/biological characteristics. Temperature should be determined at 1 m intervals with depth, at a minimum (Figure 3.1). Usually, one profile at the deepest point is adequate if the water body is relatively small, but more sites may be necessary if the water body is large and there are multiple basins or embayments, such as in reservoirs, where wind and flushing can produce differing effects on water column stability. Wind speed and direction may be useful for explaining the seasonal (and diurnal) variability in chemical/biological characteristics. Seasonal changes in water column stability are especially important in shallow polymictic

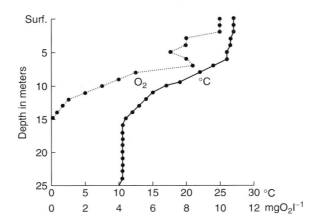

FIGURE 3.1 Distribution of temperature (solid line) and dissolved oxygen (dotted line) during summer thermal stratification of a eutrophic lake. (From Cooke, G.D., E.B. Welch, S.A. Petersen, and P.R. Newroth. 1993. *Restoration and Management of Lakes and Reservoirs*, 2nd Edition. Lewis Publishers and CRC Press, Boca Raton, FL.)

lakes (Jones and Welch, 1990). Temperature (density) profiles help determine if density interflows are important and several profiles distributed longitudinally along the reservoir may be necessary for that purpose. Inflows to reservoirs often dive to some intermediate depth, due to density differences, and that may result in incoming nutrients being unavailable to phytoplankton in the photic zone. Some more complicated hydrodynamic modeling approach, other than a completely mixed assumption, may be needed.

Water transparency, determined with a Secchi disc, is one of the most reliable, frequently used, and meaningful indicators of lake quality. The depth of transparency is the path length in the Beer's law equation through which light is scattered and absorbed as a function of particle concentration in the water. As the concentration increases, transparency depth decreases exponentially. However, transparency is usually related to particle concentration, whether those particles are algae or other suspended solids. The measurement is easy and is used by lakeshore residents to monitor lake quality. There may be more horizontal variability in transparency than with temperature, especially if buoyant blue-green algae are abundant in the lake and are distributed unevenly by the wind. Measurements at more than one site, even in small lakes, are recommended. Plot transparency against time for each sampling site.

Suspended solids (TSS) determined by gravimetric analysis may be useful, especially in highly-flushed reservoirs in watersheds subject to erosion. Turbidity, determined by light scattering (nephelometry), is an indirect measure of suspended solids and may be useful information. If there is a sizable influx of solids to the lake/reservoir, a horizontal gradient in concentration can be expected as water velocity decreases upon entry to the water body and deposition occurs. These variables are not as useful to indicate trophic state as is transparency.

The chemical variables that should be determined are nutrients (TP and total nitrogen [TN] and the soluble fractions NO_3, NH_4 and SRP), pH, dissolved oxygen (DO), total dissolved solids (specific conductance) and ANC (acid neutralizing capacity or alkalinity). Biochemical oxygen demand (BOD) may be useful when assessing DO demands and sources. Nutrients, pH, and dissolved solids should be determined at several depths at the deep-water site, at least three depths in the epilimnion and three in the hypolimnion. Fewer sampling depths are needed when the water column is completely mixed. Surface samples may be sufficient in shallow lakes (Brown et al., 1999). The purpose here is to insure that respective water layers are adequately represented for computing whole-lake mean concentrations. To check for variation in horizontal distribution, integrated (tube) samples could be collected at other sites. Again, if the lake/reservoir has multiple basins/embayments, additional sampling sites may be necessary. Whole-lake mean concentrations (sum of the products of depth-interval volumes and concentrations) or epilimnetic water column means are useful for assessing long-term change and the nutrient budget and models. Profile plots of TP, SRP, DO, and temperature for several dates in the summer may also be instructive to illustrate the effects of stratification and DO depletion on sediment P release. Volume weighted hypolimnetic TP plotted against time can be used to calculate a release rate from sediments.

DO and temperature should be determined at 1 m intervals, sampling as close to the bottom as possible to detect DO depletion at the sediment/water interface, especially in shallow, unstratified lakes. DO sensors are easy to use and can be located at discrete depths, as opposed to 0.5 m sampling. DO should be determined by the standard wet chemical method (APHA, 2003) at a minimum of 10% of the depths sampled, including depths with DO < 1 mg/L, to verify the probe-determined values. Unreliable values from depth in the water column may occur with sensors that operated satisfactorily in the laboratory. All sensors, except microelectrode sensors, are unreliable for DO < 1 mg/L, or for steep gradients, such as the sediment-water interface or at metabolic boundaries (Wetzel and Likens, 1991). The vertical temperature-DO data should be plotted on a depth-time graph, with isopleths of values represented rather than a separate graph for each sampling date, to illustrate periods of stratification and DO loss from the hypolimnion and/or supersaturation in the lighted zone.

A twice-monthly sampling frequency during May through September and monthly for the remainder of the year is recommended for temperate waters. Monthly during summer may miss algal blooms completely and result in underestimated means for trophic state indices. Twice-monthly sampling is also recommended for nutrient budgets. ANC and BOD need not be sampled as frequently or at as many sites. ANC does not change appreciably, but is used to calculate CO_2, which changes with pH in response to diurnal cycles of photosynthesis/respiration, and alum dose (Chapter 8). DO is usually correlated with pH and inversely with CO_2. These variables influence nutrient cycling and blue-green algal buoyancy (see Chapter 19), which can affect trophic state. Except in highly enriched lakes, BOD is usually not significant, and oxygen deficit rate (AHOD, Chapter 18) determinations from hypolimnetic DO data are more relevant.

Sediment cores from the deepest site are useful to determine the chronology of cultural eutrophication, the character of P (fractions), its release rate in and from the sediments and alum dose (Chapter 8). Vertical changes in the concentration of stable or radioactive lead are used to date depths in the core, providing inferences about the history of P and organic loading. Figure 3.2 is an example showing the increase in stable lead at about 20 cm (circa 1930; the start of leaded gasoline use) and decrease again around 1972 (started unleaded gasoline use). In this case, two sedimentation rates could be determined. Anomalies, such as the value at 15 cm, often occur. That value could not be explained and was ignored in estimating sedimentation rates. Chronology may not always be clear.

The question is often asked, "Is lake quality being restored to an earlier state or has quality always been poor and is simply being improved?" Historical chronology from core data can answer that question with evidence on sedimentation rate, productivity, nutrient loading, and plankton species composition over time. Some of the specific indicators are algal pigments, chiromomid midge head capsules and P-fraction content (Wetzel, 1983; Welch, 1989). Total chlorophyll, myxoxanthophyll (cyanobacteria) and diatom-inferred TP and chl a, showed the chronology of eutrophication of Lake Haines, Florida, with dating by lead-210 (Whitmore and Riedinger-Whitmore, 2004). Pollen analysis is also useful for establishing historical markers, although it does not indicate lake trophic state.

Cores can be incubated under conditions of constant temperature and oxic or anoxic conditions, in order to measure P release rates. These may be comparable to those occurring in the lake. Cores can also be sectioned and P fractions determined, such as loosely bound P, iron-bound P, aluminum-bound P, and organic P, which may give insight into the process of P cycling from sediments and prospects for restoration (Boström et al., 1982; Psenner et al., 1988). Sediment release rates deter-

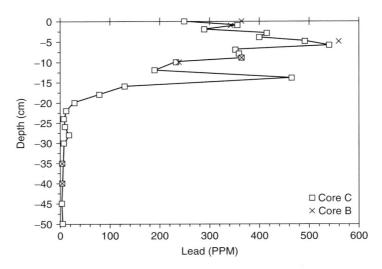

FIGURE 3.2 Content of stable lead in two cores from the deep station (15.5 m) in Silver Lake, Washington. (From Cooke, G.D., E.B. Welch, S.A. Petersen, and P.R. Newroth. 1993. *Restoration and Management of Lakes and Reservoirs*, 2nd Edition. Lewis Publishers and CRC Press, Boca Raton, FL.)

mined in the laboratory can be used, in conjunction with observed rates of hypolimnetic P increase, to characterize internal loading for constructing P budgets or calibrate mass balance models.

The usual biological variables are phytoplankton, zooplankton, macrophytes, if present, and benthic invertebrates and fish in certain circumstances. Water samples for phytoplankton analysis should be collected from two to three depths in the epilimnion and preserved with Lugol's solution. Samples from the metalimnion and even hypolimnion may show separate populations from those in the epilimnion and that possibility should be examined. Phytoplankton can be simply counted or their taxa biovolumes determined. Taxonomic separation can be by species or genera, with the latter being adequate for separation of biovolumes into diatoms, greens, and blue-greens and/or determining diversity.

Chl *a* is a conventional method to estimate phytoplankton biomass and is used more often than biovolume to indicate trophic state. It is a reliable indicator despite its dependence (per unit cell) on nutrient status, light, and species composition. Cell chl content can vary by a factor of two or more with the above variables. Again, some sampling time and site combination of data plotted against time is an appropriate display. An illustration of when, where, and how much blue-green algae is often useful.

Zooplankton can be sampled from discrete depths by filtering water bottle (e.g., Van Dorn type) collections through appropriate size nets, by vertical net hauls through all or part (closing net) of the water column, or by horizontal tows at particular depth intervals with a Clarke–Bumpus sampler. The Schindler–Patalas trap technique is also useful. Taxonomic separations can be crude (cladocerans, copepods, etc.) or by species or genera, although at least genera is desirable. A useful separation for display may be the abundance (No./m^3) of large daphnids, which are the important grazers, vs. the smaller forms.

Macrophyte distribution can be determined by several methods ranging from satellite imagery to depth-interval, stratified, random design sampling for biomass (g dry weight/m^2). The latter is most desirable to determine whole-lake and species-specific biomass, but is also most expensive and time consuming. Plants for areal dry weight can be conveniently collected by SCUBA using a device to delimit a unit area. Sample size can be determined from known measures of plant-species variability within each depth interval. Samples can also be collected using SCUBA, with sites spaced randomly along shore-to-depth transects or by less quantitative means along such transects. One sample collection per year may be all that is necessary to characterize the macrophyte crop. The annual mean biomass in each plant zone (emergent, floating-leaved and submersed) can be predicted from a measure of the maximum biomass in each zone, determined once per year by one of the above sample collection techniques (Canfield et al., 1990). A map showing abundance in relation to lake depth, as well as depth of visibility, is a useful method for illustration. Floristic quality of macrophyte communities was related to ecoregional and lake-type differences (Nichols, 1999). Satellite imagery may be more cost effective for monitoring long-term trends, but is generally inadequate for assessing specific biomass levels that can be used in nutrient budget computations.

This discussion of sampling, analytical techniques, and data display is rather superficial and the reader is referred to Wetzel and Likens (1991), Standard Methods (APHA, 2003), Golterman (1969), Edmondson and Winberg (1971) and Vollenweider (1969a).

3.2.3 Data Evaluation

Lake assessment for management usually requires a model that adequately predicts P in the lake/reservoir in question. Mass balance models for P are based on the kinetics of continuously stirred tank reactors (CSTR), which are commonly used in chemical engineering (Reckhow and Chapra, 1983). By continuously mixing the volume in such a reactor, holding that volume constant, and maintaining the water inflow rate equal to the water outflow rate, the following mass balance equation applies with units of mass/time:

$$\mathrm{d}CV/\mathrm{d}t = C_i Q - CQ + KCV \tag{3.6}$$

where C is the concentration of a substance in the reactor and C_i is concentration in the inflow, Q is flow rate, V is reactor volume, and K is the reaction rate coefficient. If K is assumed to represent a first order depletion reaction (rate of decrease dependent on concentration) and both sides are divided by V, so that $Q/V = \rho$ (the flushing rate in $1/t$), the equation becomes

$$\frac{\mathrm{d}C}{\mathrm{d}t} = \rho C_i - \rho C - KC \tag{3.7}$$

At steady state, the equation becomes

$$C = \frac{C_i}{1 + K/\rho} = \frac{\rho C_i}{\rho + K} \tag{3.8}$$

That is essentially the same as the TP mass balance proposed by Vollenweider (1969b) for lakes:

$$\frac{\mathrm{d}\,\mathrm{TP}}{\mathrm{d}t} = \frac{L}{z} - \rho\,\mathrm{TP} - \sigma\,\mathrm{TP} \tag{3.9}$$

where L is TP areal loading in mg/m² per yr ($L/\overline{z} = \rho C_i$ in 3.7, 3.8), z is mean depth (m), ρ is flushing rate (1/yr), and σ is the sedimentation rate coefficient (1/yr). The steady state equation is

$$\mathrm{TP} = \frac{L}{\overline{z}(\rho + \sigma)} \tag{3.10}$$

which is equivalent to Equation 3.8 because $L/\overline{z} = \rho\,\mathrm{TP}_i$.

According to Equation 3.9, each new concentration of TP entering the lake is immediately mixed throughout the lake producing a new concentration after a fraction leaves through the outlet and a fraction sediments to the lake bottom, both of which are a function of the new, slightly changed concentration. According to Equation 3.10, over the long term, the lake will equilibrate to the given loading. If the loading is changed then some time interval will be required for equilibration to the new loading. Assuming a first order rate reaction, the time interval to 50% (100/50) and 90% (100/10) of equilibrium will be, respectively:

$$t_{50} = \frac{\ln 2}{\rho + \sigma}$$

$$t_{90} = \frac{\ln 10}{\rho + \sigma}$$

(3.11)

The principal limitation with these models is determining the sedimentation rate coefficient. All other variables can be determined directly. Thus, for a lake with known loading, mean annual TP concentration, and flushing rate, σ could be estimated from Equation 3.10 according to:

$$\sigma = \frac{L}{\text{TP}\,\bar{z}} - \rho$$

(3.12)

However, to develop a model to describe a large number of lakes, it is useful to have some general way to estimate sedimentation. One approach is to use a unitless retention coefficient RTP (Vollenweider and Dillon, 1974; Dillon and Rigler, 1974a), which can be derived from Equation 3.10 by multiplying the numerator and denominator by ρ (Ahlgren et al., 1988):

$$\text{TP} = \frac{L}{\rho\bar{z}} * \frac{\rho}{\rho + \sigma}$$

(3.13)

where $L/\rho\bar{z}$ is the inflow concentration and $\rho/(\rho + \sigma)$ is a dimensionless reduction term equal to $1 - R_{TP}$, the retention coefficient for TP. Thus:

$$R_{TP} = 1 - \frac{\rho}{\rho + \sigma} = \frac{\sigma}{\sigma + \rho}$$

(3.14)

There is still the difficulty with estimating σ, but Vollenweider (1976) found that σ could be approximated by $10/\bar{z}$, where 10 has the dimensions of m/yr and is considered to be an apparent settling velocity for TP. If the numerator and denominator in Equation 3.13 are multiplied by $\bar{z}$ and substituting $10/\bar{z}$ for σ, it becomes:

$$R_{TP} = \frac{10}{\rho\bar{z} + 10}$$

(3.15)

The surface hydraulic loading, $\rho\bar{z}$ in m/yr, is designated as q_s in many formulations, and the retention coefficient is described as

$$R_{TP} = \frac{v}{q_s + v} \qquad (3.16)$$

where v is the settling velocity. Several estimates of v exist in the literature, e.g., 16 m/yr from Chapra (1975) and see Nürnberg (1984) for others.

R_{TP} can also be determined directly for an individual lake according to

$$R_{TP} = 1 - \frac{TP}{TP_i} \qquad (3.17)$$

where TP_i is inflow concentration and TP is the lake concentration if assumed to equal the outflow concentration. From Equations 3.13 and 3.14 it is clear that (see Vollenweider and Dillon, 1974):

$$TP = \frac{L(1 - R_{TP})}{\bar{z}\rho} \qquad (3.18)$$

R_{TP} has been related to hydraulic variables in several empirical formulations, one of which is $1/(1 + \rho^{0.5})$ (Larsen and Mercier, 1976; Vollenweider, 1976). With this and other such relationships (Equation 3.16), R_{TP} decreases as flushing rate increases. A R_{TP}–flushing rate relation may be relatively constant with loading change (Edmondson and Lehman, 1981), or vary with loading (Kennedy, 1999). There are several forms of the steady state Equation 3.10 that are based on this dependence of retained TP on flushing rate. Using $TP_i = L/\bar{z}\rho$ for simplicity, three such equations, in sequence, are

$$TP = TP_i(1 - R_{TP}) = \frac{TP_i}{1 + \frac{1}{\rho^{0.5}}} = \frac{L}{\bar{z}(\rho + \rho^{0.5})} \qquad (3.19)$$

The negative relation between flushing rate and R_{TP} is logical. That is, as flushing rate increases there is less time for TP to settle, so R_{TP} decreases accordingly. Seemingly in contrast, the sedimentation rate coefficient is positively related to the flushing rate ($\sigma = \rho0.5$). However, to calculate actual sedimentation, which is flux rate to the sediment, R_{TP} must be multiplied by L, while σ must be multiplied by lake TP. Therefore, it is readily apparent that if L is held constant, increasing the flushing rate will give increasingly smaller TP_i. As a result, σ must increase in order that the flux rate to the sediments does not decrease too rapidly. Ahlgren et al. (1988) modified the relationship found by Canfield and Bachmann (1981) that shows such a relationship between σ and both flushing rate and TP_i:

$$\sigma = 0.129 \, (TP_i \, \rho)^{0.549} \qquad (3.20)$$

The steady state mass balance model illustrated by Equation 3.18 has been verified for a large population of lakes (Chapra and Reckhow, 1979). This suggests that the general form of the sedimentation term is reasonable, although the error for predicting the TP content in any given lake may be quite large (about ± 50 μg/L).

If internal loading is important, as may be the case in either oxic or anoxic lakes, then the model may need to be modified to account for the two sources. Nürnberg (1984) formulated the following model to account for internal load (L_{int}):

$$TP = TP_i(1 - R_{pred}) + \frac{L_{int}}{\bar{z}\rho} \qquad (3.21)$$

where R_{pred} in 54 oxic lakes was best represented by:

$$R_{pred} = \frac{15}{18 + \bar{z}\rho} \qquad (3.22)$$

Internal loading can also be added to Equations 3.18 and 3.19. However, there was no attempt to separately treat oxic and anoxic lakes in the development of those models.

Solving Equation 3.21 for L_{int}, using observed TP, allows calibration of Nürnberg's model for a particular stratified, anoxic lake. L_{int} can then be compared with other estimates of internal loading for the lake/reservoir in question, such as sediment P release rates determined from cores incubated in the laboratory or by the observed rate of increase in hypolimnetic P concentrations. These two methods of estimating internal loading in anoxic lakes have shown rather good agreement (Nürnberg, 1987). Sediment release rate in anoxic cores also has been directly related to iron-bound P (BD-P) in sediment (Nürnberg, 1988). Lake-wide internal loading can be estimated as the product of anoxic release rate and anoxic factor (Nürnberg and LaZerte, 2004). Such good agreement among these different estimates of internal loading for a particular lake indicates that the model is verified for that lake. If the agreement is poor, then an error in the estimate for sedimentation may exist and a different modeling approach must be taken. Agreement may be poor if the lake is not in equilibrium with its external loading.

Even if verification of a particular steady state model is satisfactory, problems are encountered using the steady state version. First, an appropriate time interval (most often annual), when the lake mean TP represents a steady state, is often difficult to determine, especially if flushing rate is much greater than 1/yr. Second, internal loading usually occurs during the summer and may contribute proportionately more to growing season TP and biomass than external loading, especially if the lake is unstratified and external loading occurs primarily during the non productive period (e.g., winter in the Pacific Northwest). These problems may be averted by calibrating and verifying a transient version of Equation 3.9 including L_{int}:

$$\frac{d\,TP}{d\,t} = \frac{L_{ext}}{\bar{z}} - \rho\,TP - \sigma\,TP + \frac{L_{int}}{\bar{z}} \qquad (3.23)$$

Because sedimentation is a function of TP concentration resulting from both L_{ext} and L_{int} at each time step in Equation 3.23, L_{int} is a gross rate. In this case, the numerator in Equation 3.10 would be $L_{ext} + L_{int}$.

The transient version usually requires no more data, because as recommended above, TP loading and lake concentration data are collected twice monthly as a minimum. With the steady state approach, the data are usually reduced to annual means (or some interval consistent with ρ), whereas TP is computed for each time interval with the transient version. Weekly time steps are preferred in the modeling process to obtain a more realistically smooth curve even if less frequent data were available. The model can be calibrated by determining the sedimentation rate coefficient (σ) that gives the best fit between predicted and observed TP for the oxic period. Larsen et al. (1979) used a constant σ among years in Shagawa Lake, Minnesota with good success. However, the model could be verified year to year in Lake Sammamish, Washington only if σ were allowed to vary as a function of flushing rate, i.e., $\sigma = \rho^x$, where $0 < x < 1$ (Shuster et al., 1986; Welch et al., 1986). That is analogous to Equation 3.19 where $x = 0.5$. A formulation such as $\sigma = y\rho^x$, where $y < 1$, may

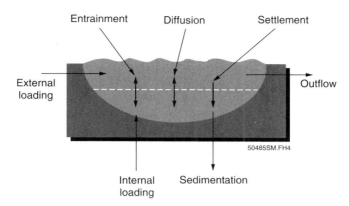

FIGURE 3.3 TP fluxes in a stratified lake. (From Perkins, W.W. 1995. *Lake Sammamish Phosphorus Model.* King County Surface Water Manage., Seattle, WA.)

be necessary if sedimentation rates are low, because as *x* approaches zero in the previous formulation, the sedimentation rate remains around 1.0 regardless of the flushing rate.

There still may be a problem with using the transient model for stratified lakes even if it can be verified for whole-lake TP. From predicted TP, chl *a* and transparency are usually predicted as biological and physical factors defining trophic state and lake quality, and are a function of TP in the productive zone (i.e., epilimnion) and not of whole-lake TP. Usually, epilimnetic TP declines during the stratified period while hypolimnetic TP increases. Thus, either the epilimnion and hypolimnion must be modeled separately with diffusion between the two strata included to account for exchange of TP, or mean epilimnetic TP must be estimated from a relationship between that and whole lake TP. The latter may be satisfactory, because relationships among chl *a*, TP, and transparency are usually based on summer means, which are in turn most often used for management purposes (Shuster et al., 1986).

The use of a two-layer mass balance TP model for stratified lakes is routine. The earlier TP modeling work for Lake Sammamish described above was considered inadequate to separate the effects of urban runoff from internal loading. The model of Auer et al. (1997) was developed for Lake Onondaga and later applied to Lake Sammamish (Perkins et al., 1997). While internal loading from anoxic sediments represented a substantial fraction of the annual and, especially, summer total loading, availability of hypolimnetic P via entrainment and diffusion to the epilimnion for algae production was much less important than external loading. A two-layer model is based on representing the transfers shown in Figure 3.3. A quantitative estimate of the magnitude of internal loading availability has become very important in judging the probable cost-effectiveness of in-lake treatment techniques.

There are qualitative procedures to indicate the importance of internally-loaded P availability in stratified lakes. The Osgood Index of mixing (OI = mean depth/$\sqrt{km^2}$; Osgood, 1988) is a measure of the lake volume in relation to wind fetch. As the ratio decreases, the chance for mixing hypolimnetic with epilimnetic water increases. Based on data from 96 lakes in central Minnesota, those with an OI < 6–7 had summer surface water TP that exceeded the concentration predicted from external loading. All of these lakes were continuously mixed, polymictic, or weakly stratified dimictic lakes. Dimictic lakes with OI values > 8 were strongly stratified with summer surface water TP concentrations that conformed to values predicted from external loading.

This index works in some stratified lakes, but not others. Where wind mixing is effective low OIs are consistent with significant transport of hypolimnetic P to surface water. Shagawa Lake is a case in point. The eastern basin (OI = 3.6) is smaller and more wind-sheltered and was shown to have less vertical transport than the west basin (OI = 2.3 – see Chapter 4; Larsen et al., 1981; Stauffer and Lee, 1973; Stauffer and Armstrong, 1984). Also, no transport was consistent with a

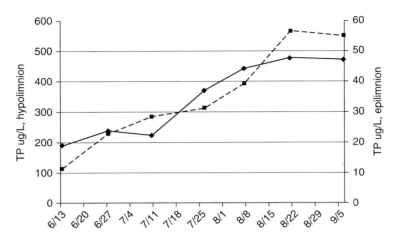

FIGURE 3.4 Epilimnetic and hypolimnetic TP concentration in McDonald Lake, Washington.

high OI (36.7) in Third Sister Lake, Michigan (Lehman and Naumoski, 1986). But in others, the OI is unreliable. Where wind mixing is less important and diffusion dominates due to a large TP concentration gradient between hypolimnion and epilimnion, transport of P may be significant in spite of a high OI (26; Dollar Lake, Mataraza and Cooke, 1998). This is also shown for Lake McDonald, a similarly small (7.2 ha), relatively deep (7 m mean depth) lake with an OI of 26, the same as Dollar Lake (2 ha, 3.9 m mean depth). TP at Z_{max} reached about 800 µg/L and over 1000 µg/L during the stratified period in the hypolimnia of McDonald and Dollar, respectively. Surface TP (0–2 m) increased during the summer in proportion to the increase in hypolimnetic TP (mean of depths 9 and 13 m) in spite of continued water column thermal stability (Figure 3.4). Surface chl a also increased from about 6 µg/L in mid June to 32 µg/L in mid August while TP increased from 12 µg/L to 56 µg/L. Nürnberg (1985) calculated a transport to the epilimnion via eddy diffusion in Lake Magog (OI = 4.4) equaling 30% of gross internal loading to the hypolimnion and cited three other examples ranging from 50–100%. In contrast, surface TP in Lakes Sammamish (OI = 3.9) and Onondaga (OI = 3.15) remained rather constant during summer, until fall turnover approached, despite increasing hypolimnetic TP. These data can be used to indicate the availability of internal loading and its effect on lake trophic state. Given that hypolimnetic P can be effectively transported to the epilimnion either by wind mixing in lakes with low OIs or diffusion across large concentration gradients, internal loading is likely to affect trophic state in most stratified lakes. This is demonstrated with alum-treated lakes in Chapter 8.

Incorporating internal loading into a two-layer TP model is usually straightforward because sediment release is typically rather constant during the anoxic period. That is, the increase in hypolimnetic TP is usually linear with time. Sediment cores incubated under anoxic conditions have rates comparable to those derived from hypolimnetic TP–time plots (Nürnberg, 1987). However, Penn et al. (2000) observed seasonal variation in core release in Lake Onondaga. While the P release rate in stratified lakes may not always be dependent solely on iron redox reactions (Gächter and Meyer, 1993; Gächter and Müller, 2003; Golterman, 2001; Søndergaard et al., 2002), the pattern of release is usually consistent from year-to-year and can be reasonably simulated for a given lake. The iron cycle usually controls sediment P release in stratified anoxic lakes, as indicated by a strong correlation between sediment P release in anoxic cores and the Fe-P (as BD-P, indicating the extraction reagent) fraction in sediment (Nürnberg, 1988). Release rates determined by the increase in hypolimnetic TP have varied some from year-to-year in Lake Sammamish (Figure 3.5), although the area of anoxia (< 1 mg/L DO) remained relatively constant. Nevertheless, post-diversion rates were similar for most years allowing the use of an average value for long-term modeling (Perkins

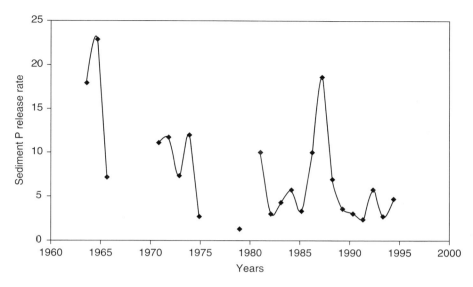

FIGURE 3.5 Sediment P release rate (mg/m² per day) in Lake Sammamish, Washington. (Data from Perkins, W.W. 1995. *Lake Sammamish Phosphorus Model.* King County Surface Water Manage., Seattle, WA.)

et al., 1997). Mechanisms become important, however, when determining the effectiveness of in-lake controls, especially hypolimnetic aeration (Chapter 18).

Simulating internal loading in shallow polymictic lakes is more difficult than in stratified lakes because several mechanisms may operate simultaneously and the pattern of sediment P release may not be similar among years. Moreover, macrophyte senescence and/or anoxic conditions under macrophyte beds may provide an additional source to the sediment-water exchange processes (Frodge et al., 1990; Stephen et al. 1997). Macrophytes may also decrease resuspension and thus internal loading (Welch et al., 1994; Christiansen et al., 1997). However, there is not the issue of P availability to algae in shallow lakes, because P entering the water column from the sediment is readily available in the lighted zone. Internal loading in a shallow lake can occur through any or all of the following processes (Boström et al., 1982; Welch and Cooke, 1995; Søndergaard et al., 1999):

- Photosynthetically caused high pH dissolving Al- and Fe-bound P
- Wind-induced entrainment of soluble P released from anoxic sediment during calm, temporarily stratified conditions
- Temperature-driven mineralization of organic P by microbial metabolism
- Soluble P release from bacterial cells or via metabolism of organic P excreted from algal cells in sediment
- Soluble P desorption from wind-caused resuspended particles via high particle-water concentration gradient enhanced by high pH
- Macrophyte senescence and bioturbation (e.g., benthic fish activity)

Several of these processes may occur simultaneously and their within- and between-year variations can be great. Much of that variation is due to changing wind speed and its effect on water-column stability and sediment resuspension. For example, net P internal loading varied from year-to-year by ± 100% and was strongly related to RTRM (relative thermal resistance to mixing) over a 12-year period in largely polymictic Moses Lake, Washington (Jones and Welch, 1990). Wind-caused mixing was a good predictor of resuspension and TP in several, large shallow lakes (Søndergaard, 1988; Kristensen et al., 1992; Koncsos and Somlyody, 1994). High pH can enhance desorption of P from resuspended particles (Lijklema, 1980; Koski-Vähälä and Hartikainen, 2001;

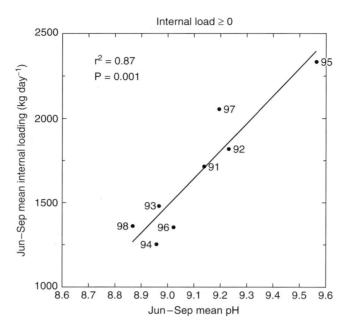

FIGURE 3.6 Net internal P loading versus pH in Upper Klamath Lake, Oregon. (From J. Kann, Aquatic Ecosystem Sci., Ashland, OR 97520, personal communication.)

Duras and Hejzlar, 2001; Van Hullebusch et al., 2003). Photosynthetically caused high pH was apparently the dominating factor resulting in high internal loading in large (270 km²), shallow (2 m mean depth) Upper Klamath Lake, Oregon (Figure 3.6). TP was not related to calculated particle resuspension, possibly due to the dependence of pH on algal biomass (Welch et al., 2004; Kann and Smith, 1999). Because of the year-to-year variability in timing and magnitude of internal loading, a time-dependent constant internal loading rate, such as used for Lake Sammamish and other stratified lakes, could not be used in a non-steady state mass balance TP model for shallow Upper Klamath Lake.

There are several approaches for dealing with the uncertainty in TP predictions for individual lakes. For example, in using Equation 3.23 to predict future TP concentrations resulting from increased development in the Lake Sammamish watershed, uncertainty was included by choosing a range in land use yield coefficients and the 5 and 95% flow probabilities for the principal inflow stream (Shuster et al., 1986). TP sedimentation was a function of ρ and increased/decreased flow resulted in, respectively, dilution/concentration of the estimated TP loading. By this procedure, the prediction of 31 µg/L TP with future development had a ± 10% error due to land use yield and a ± 20% error due to flow. Most of the year-to-year variation in loading was due to surface inflow.

Another approach is to use first order error analysis to calculate uncertainty in loading and TP predictions based on low, high, and most likely loading estimates from yield coefficients (Reckhow and Chapra, 1983). For a model of the type of Equation 3.22, Reckhow and Chapra (1983) determined an error of ± 30%, which is added to the loading uncertainty. By summing those uncertainties, confidence intervals for a single model estimate for TP can be calculated. To evaluate small changes in TP, predicted from relatively small changes in loading, uncertainty can be applied to the TP concentration change, rather than the before and after concentration as noted earlier.

These mass balance models described above do not predict the long-term response of lake TP to input reduction (Chapter 4). If the lake has not yet reached equilibrium to the new reduced loading, they may under-predict lake TP (Havens and James, 1997). Long-term response can be predicted by including a mass balance on sediment P (Chapra and Canale, 1991; Pollman, personal communication; Walker, personal communication). However, such predictions have not yet been verified.

Criteria exist to describe the quality and trophic state of a lake. They include the concentration and loading rate of nutrients, which are the cause, as well as physical and biological indices, which are the effect, as noted above. Numerical criteria allow precise definition of a lake's quality or classification. Criteria are used to accurately chart the course of a lake as it becomes more or less eutrophic and to judge if the lake is suitable or unsuitable for recreational or water supply use.

The literature is replete with indices to classify trophic state and lake quality. Porcella et al. (1980) listed 30 different sources for trophic state criteria and there are still others. Also, goals for lake quality may be in conflict. The aesthetically pleasing, clear, blue water of ultra-oligotrophic lakes is usually associated with low fish production (but not necessarily small size). Compromises may be needed between lake quality that is more favorable to fish production (meso, meso-eutrophic, or even eutrophic) and that preferred for swimming, boating, and aesthetics. However, for cold-water fish species in lakes with epilimnetic temperature that exceeds preferred levels, there may be little difference between trophic state criteria appropriate for fisheries and recreational use.

The most commonly determined biological variable to define trophic state and lake quality is chl a, and several empirical relationships between chl a and TP exist (see Ahlgren et al., 1988; Downing and McCauley, 1992; Jones et al., 1998; Seip et al., 2000). Probably the two most often used are by Dillon and Rigler (1974b) and Jones and Bachmann (1976), which, respectively, are

$$\log \text{chl } a = 1.449 \log \text{TP} - 1.136 \tag{3.24}$$

$$\log \text{chl } a = 1.46 \log \text{TP} - 1.09 \tag{3.25}$$

The Dillon and Rigler data set contained TP values from spring turnover and mean summer chl a while the Jones and Bachmann set was composed of summer means for both variables. The equations agree rather closely, despite of the difference in data averaging times. Ahlgren et al. (1988) compared seven different TP–chl a relationships, which yield a wide range in predictions. Some of the variability in prediction is due to the variation in cellular chl a (0.5–2% of dry weight), due to such factors as light and nutrition, but also some of the measured TP may not be in cells. This is an explanation why the ratio of chl a to TP and hence, the slope of the regression line, can be expected to vary between 1.0 and 0.5. Some relationships had slopes below 0.5, presumably because measured non-cellular P was high. Zooplankton grazing also reduces the chl a:TP ratio if large-bodied *Daphnia* are abundant, and thus improves transparency relative to TP (e.g., Lake Washington, Chapter 4) (Lathrop et al., 1999).

Because most TP–chl a relationships using large data sets are usually log-log, the accuracy of prediction for a single lake is not great. With Equation 3.24, for example, a chl a concentration of 5.6 µg/L (10 µg/L TP) has a prediction error of ± 60–170% and 30–40% for 95% and 50% confidence. respectively. The high correlation coefficients between TP and chl a tend to mask the accuracy problem, which may be due to lake-to-lake and seasonal variations in cellular chl a, zooplankton grazing (Chapter 9), and other limiting factors such as light and nitrogen (Ahlgren et al., 1988; Jones et al., 2003). Developing a relationship for the individual lake of interest that provides much greater accuracy of prediction is recommended where data are sufficient (Smith and Shapiro, 1981). However, data may be insufficient for a reliable relationship so a published relationship that provides the best agreement with the individual lake data would be preferable.

Summer means for chl a and TP are most often used to define lake trophic state, so sampling intensively throughout the non-growing season to only determine trophic state is unjustified. Although TP may be higher when inflows are greater during the winter and spring, the summer mean represents the residual after sedimentation and, therefore, should be most closely related to P in algal biomass.

The TP–chl a relationship was used by Carlson (1977) to develop a numerical trophic state index (TSI). This is probably the most commonly used index, which includes three variables: TP,

TABLE 3.5
Completed Trophic State Index (TSI) and Its Associated
Parameters

TSI	Secchi Disc (m)	Surface Phosphorus (mg/m³)	Surface Chlorophyll (mg/m³)
0	64	0.75	0.04
10	32	1.5	0.12
20	16	3	0.12
30	8	6	0.94
40	4	12	2.6
50	2	24	7.3
60	1	48	20
70	0.5	96	56
80	0.25	192	154
90	0.12	384	427
100	0.062	768	1183

Source: From Carlson, R.E. 1977. *Limnol. Oceanogr.* 22: 361–368. With permission.

chl *a* and Secchi transparency. Carlson's TSI and Porcella's (1980) LEI (Lake Evaluation Index) (Porcella et al., 1980) reduce lake trophic state to one or more numbers, in an attempt to remove the subjectivity inherent in the terms oligotrophic, mesotrophic, and eutrophic. Instead, they emphasize the degree of eutrophication within each classification. To classify a lake as eutrophic encompasses a wide range of lake conditions and just how eutrophic is not specified by the term itself, although use of those terms in communications about lake quality is still necessary.

Carlson's TSIs (and LEIs) represent absolute values for chl *a*, TP, and transparency (SD) applicable to any lake (with minimal nonalgal turbidity), in contrast to indices for which values are relative and confined to uses with a particular data set. Transformations of the data to $\log_2$ were used to interrelate these three indices within a scale of 0 to 100, so that a doubling in TP is related to a reduction by half in SD. Representative values for TP, chl *a*, and SD, calculated from the following equations for TSI are shown in Table 3.5.

$$TSI = 10(6 - \log_2 SD) \tag{3.26}$$

$$= 10(6 - \log_2 7.7/\text{chl } a^{0.68}) \tag{3.27}$$

$$= 10(6 - \log_2 48/TP) \tag{3.28}$$

If annual mean values are used for TP in Equation 3.28, then 64.9 is used as the numerator instead of 48. Note that the greatest change in SD occurs below a chl *a* concentration of about 30 μg/L. Above 30 μg/L, there is relatively little change in transparency with increasing chl *a*. Thus, more TP must be removed from a highly eutrophic lake to see benefits in transparency than in a moderately eutrophic or mesotrophic lake. For example, the range between 40 and 50 is most often associated with mesotrophy. Between 40 and 50, TP concentration doubles and SD halves (4 m at TSI of 40 and 2 m at TSI of 50), which is a change that would be obvious to lake users through changes such as blue-green algal blooms and oxygen deficits. On the other hand, if a management strategy proposed for a P-limited lake with a TSI of 70 will only cut the concentration in half, then the lake users may not notice the small (0.5 m) improvement in transparency (Table 3.5).

The Carlson index has been misused, particularly in lakes with high nonalgal turbidity or with extensive macrophyte populations. It makes no sense to locate a sampling boat over the only macrophyte-free patch of water and measure trophic state based on water column values for TP, chl a, and SD. The lake could be classified as oligotrophic from these measurements, while anyone viewing the lake would consider it highly eutrophic and unusable due to the extensive macrophyte cover (Bachmann et al., 2001). Another problem often occurs in reservoirs where transparency is determined primarily by nonalgal turbidity or color (Lind, 1986). In this case, the effect of nonalgal turbidity can be determined by comparing the calculated TSIs for each of the three variables (Havens, 2000).

Insight into nutrient limitation is also possible using the TSI. If TSI values for TP, chl a and SD are nearly identical, this is evidence that algal biomass is P-limited and that chl a is the primary determinant of transparency. But suppose the chl TSI is much smaller (i.e., oligotrophic) than the TP TSI. That suggests algal biomass limitation by other factors, such as zooplankton grazing or N limitation.

Canfield et al. (1983) proposed an index for classification of lakes largely covered by macrophytes. The total biomass of submersed macrophytes is determined and its P content, as determined by tissue analysis, is then multiplied by the total biomass estimate of each species. The sum for all species gives an amount of P associated with macrophytes. Then the P content of the water (whole-lake mean) is added to that of macrophytes to give a total, whole-lake mean, which is then used in the Carlson index. Canfield et al. found that macrophytes had little effect on trophic state when they were less than 25% of the total whole-lake TP, and when mean macrophyte biomass is less than 1 g dry wt/m^2.

The LEI includes SD, TP, TN, chl a, DO, and macrophytes (Porcella et al., 1980). Water-column Carlson TSIs are essentially the same as LEIs if P is limiting, but the advantage of the LEI is if N is limiting, macrophytes are abundant, and/or if DO is important, whether stratified or unstratified (see net DO below). Walker (1980, 1984) noted that some lakes and many reservoirs may deviate in several ways from Carlson's equations, perhaps due to N limitation or nonalgal turbidity. Walker (1984) developed a 2-dimensional classification system, which appears to be preferable to the Carlson index for reservoirs. The consultant/manager must choose the appropriate index for the lake/reservoir in question.

Expressing lake trophic state on a probability basis may be more realistic (OECD, 1982; Chapra and Reckhow, 1979). This approach recognizes that a high degree of uncertainty exists in trophic state criteria. From the OECD model, for example, an annual mean TP of 40 μg/L has a 38% chance of representing eutrophy, a 56% chance of mesotrophy, and a 6% chance for oligotrophy. With this model, the generally accepted TP threshold for eutrophy of 25 μg/L represents a lake with a high probability of being mesotrophic, but has a low and equal chance of being either oligotrophic or eutrophic. A meso-eutrophic threshold by the OECD model would be a lake with equal chance of being either eutrophic or mesotrophic, i.e., a TP concentration of almost 50 μg/L. Although overlap and uncertainty in trophic state are realities, a threshold value of 50 μg/L represents a condition that is far too degraded from the standpoint of recreational and water supply use to be interpreted as mesotrophy. This represents more than a doubling in chl a from a generally accepted eutrophic threshold (Porcella et al., 1980; Nürnberg, 1996). Carlson suggested a TSI of 40–50 for mesotrophy; 50 μg/L TP is a TSI of 60. Rast and Holland (1988) apparently recognized that problem, recommending 35 μg/L as a meso-eutrophic threshold while advocating the OECD model. Nevertheless, a meso-eutrophic threshold of 25 μg/L is the most frequently used criterion (Nünberg, 1996).

TN has been used infrequently as a trophic state indicator. Except for unique cases (e.g., Lake Tahoe, Goldman, 1981), the use of TN as an indicator would normally be pertinent only in highly eutrophic lakes where N availability could be expected to control productivity. Smith (1982) has presented a TP–TN–chl a predictive equation:

TABLE 3.6
Trophic State Boundary Values

Trophic State Indicator	o-m	m-e	e-h
TP (µg/L)	10	25	100
Chl a (µg/L)	3.5	9	25
Secchi (m)	4	2	1
AHOD (mg/m² per day)	250	400	550
AF (days)	20	40	60
Net DO (mg/L)	4.5	5.0	
Min. DO (mg/L)	7.2	6.2	
TN (µg/L)	350.0	650	1200

Note: o-m, oligotrophic–mesotrophic; m-e, mesotrophic–eutrophic; e-h, eutrophic–hypereutrophic. TP, TN, Chl a and Secchi transparency are summer means. AF = anoxic factor.

Source: After Nünberg, G.K. 1996. *Lake and Reservoir Manage.* 12: 432–447. With permission.

$$\log \text{chl } a = 0.6531 \log \text{TP} + 0.548 \log \text{TN} - 1.517 \qquad (3.29)$$

Equation 3.29 may be more useful in highly eutrophic systems than a TP-chl a relationship alone. For example, it predicted a chl a concentration of 21 ± 9 µg/L in Moses Lake, Washington, while Equation 3.25, based only on TP, predicted 50 ± 23 µg/L. The observed chl a value in that N-limited lake was 23 ± 11 µg/L. Once mean TP concentration decreased below about 50 µg/L, Equation 3.25 was a good predictor of mean chl a (Welch et al., 1989; Chapter 6). To take into account the effect of different TN to TP ratios, Prairie et al. (1989) developed separate chl a–TP relationships over a range in TN to TP ratios from 5 to 60. However, Prairie et al. (1995) suggested that variation was probably due to TP and not an N fertilizing effect. Moreover, a long-term data set from 184 Missouri reservoirs show that N had a minor effect on TP–chl relations with an average summer chl:TP ratio of 0.33, agreeing with other global TP–chl relations, so long as spring nonalgal turbidity events are avoided. (Jones and Knowlton, in press). While N limitation may cause Equation 3.25 to over predict chl a in some cases, reduction of P is still the most appropriate approach to controlling eutrophication. Several relationships that show chl a is dependent on TP at concentrations up to 200 µg/L support that contention (Seip, 1994; Scheffer, 1998; Welch and Jacoby, 2004).

Indices of DO have included; (1) areal hypolimnetic oxygen deficit rate (AHOD) in mg/m² per day, (2) net DO, (3) minimum DO, and (4) anoxic factor (AF). Neither minimum DO nor net DO has been correlated with TP loading or concentration, but AHOD was related to TP retention (Cornett and Rigler, 1979) and oxygen deficit rate (ODR) to TP loading (Welch and Perkins, 1979). AHOD is the DO index most often used for trophic state (Nürnberg, 1996; Table 3.6), and its significance to fish was reviewed by Welch (1989) and Welch and Jacoby (2004).

AHOD is usually calculated as the slope of the linear plot of mean hypolimnetic DO against time, multiplied by the hypolimnetic mean depth. DO sensors are not recommended, or at least should be verified by adequate wet chemical method values, for AHOD calculation, due to the index's sensitivity to low DO concentrations. In some highly enriched lakes, DO may disappear too rapidly to give an accurate estimate of AHOD, even if lake sampling is twice per month. Thus, twice-weekly sampling in late spring and early summer may be necessary. Calculated AHODs can vary depending on the time interval chosen, which should be held constant from year-to-year and

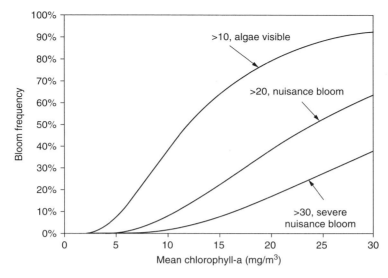

FIGURE 3.7 Frequency of algal blooms greater than 10, 20 and 30 µg/L chl *a* related to summer mean chl *a*; calibrated to Corps of Engineers Reservoirs. (From Walker, W.W., Jr. 1985. Empirical methods of predicting eutrophication in impoundments. Applications Manual. EWQOS Program, U.S. Army Corps Eng., Vicksburg, MS; and personal communication.)

encompass the whole stratified period or until DO at the bottom reaches 1 mg/L. Although time interval and hypolimnetic depth were held constant, AHODs in Lakes Sammamish and Washington varied year to year over 20–30 years (King County, 2002; personal communication).

Net DO may answer the dilemma of a suitable DO index for unstratified lakes, because AHOD (Chapter 18) is appropriate for stratified lakes only. Porcella et al. (1980) developed net DO, for stratified and unstratified lakes alike, with values ranging from 0 to 10. Net DO is defined as the absolute difference from an equilibrium condition (saturation) and is calculated by summing those differences (equilibrium DO – measured DO) over intervals of depth, thus incorporating the increasing tendency of supersaturation as well as deficiency in response to eutrophication.

Anoxic factor (AF) is equal to $(\Sigma t_i \bullet a_i)/A_o$, where t is days of detectable anoxic conditions, a is the sediment area and A_o is surface area, both in m^2 (Nürnberg, 1995a,b). AF is a measure of the lake bottom area covered by ≤ 1 mg/L DO and is more useful than AHOD in determining the extent of conditions suitable for P internal loading and bottom area (habitat) inaccessible to bottom-feeding fish. Year-to-year variation in AF was much less than AHOD in Lake Sammamish, Washington (Perkins, 1995).

Boundary or threshold values for trophic states can be useful in communicating lake quality conditions and as general management goals. Nürnberg (1996) reviewed these values and developed, in most cases, new ones (except for minimum and net DO) based on regression equations that related one variable to another (Table 3.6). Boundary values for SD are similar to those predicted from 25 µg/L TP and 9 µg/L chl *a* using Equations 3.26–3.28 (3.6 and 1.9 m for o-m and m-e; Table 3.6). These boundary values have recreational and water supply significance. At a mean summer chl *a* greater than 10 µg/L, nuisance algal blooms with maximum concentrations > 30 µg/L begin to occur (Figure 3.7; Walker, 1985, personal communication). That mean chl *a* level is directly related to a summer TP of 25 µg/L. Even beyond 10 µg/L, smaller bloom maxima may occur. That TP threshold for a summer bloom exceeding 30 µg/L chl *a* was also shown for two areas in Lake Okeechobee, Florida (Walker and Havens, 1995).

Although the trophic state *per se* of a lake is determined by the *in situ* concentration of TP, chl *a*, etc., the loading rate that produces that trophic state also has trophic state implications. If improvement of lake quality is desired, the loading (either external or internal) must be reduced.

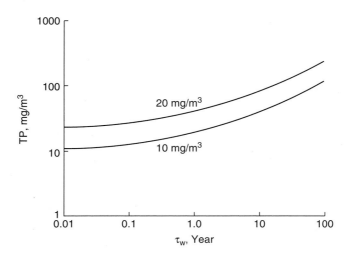

FIGURE 3.8 Relationship of inflow concentration (TP_i) with water retention time (T_w) for two lake concentrations (TP = 10 and 20 mg/m³). (From Cooke et al., 1993. With permission.)

Thus, the "critical" P loading (L_c) to produce a P concentration representing a mesotrophic or eutrophic state is often used as a goal, compared to the current or worsening state of a lake. The critical loading rate for a particular lake and trophic state has been defined as

$$L_c = TP_{e/m;m/o} \overline{z}(\rho + \rho^{0.5})$$

(3.30)

where $TP_{e/m}$ is 20 or $TP_{m/o}$ is 10 mg/m³ for the eutrophic–mesotrophic or mesotrophic–oligotrophic threshold, respectively (Vollenweider, 1976). However, other TP levels can be substituted, e.g., 25 µg/L. The effect of sedimentation is indicated in Figure 3.8, in which the critical inflow concentration, TP_i, is plotted against τ, or $1/\rho$, according to Equation 3.19. If $1/\tau$ is substituted for ρ in Equation 3.19:

$$TP_i = TP(1 + \tau^{0.5})$$

(3.31)

At low τ, the lines for lake concentrations of 10 and 20 mg/m³ become parallel with the abscissa indicating that sedimentation becomes minimal at short residence times (high flushing rates) and the lake concentration equals the inflow concentration. As τ increases, sedimentation becomes increasingly important in permitting higher TP_i without exceeding the critical lake concentration. That is, lakes become more tolerant of increased TP_i as τ increases. See Reckhow and Chapra (1983) for a more detailed discussion of these loading relationships.

Example calculations of P loading, lake P concentration, lake chl a, and Secchi transparency follow.

3.2.3.1 Example 1

(a) Given a lake with a mean depth of 15 m, a flushing rate of 1.5/yr (outflow rate/lake volume), and a mean inflow TP concentration of 80 µg/L, calculate the lake's expected external TP loading in mg/m² per yr.

From the conversion of Equation 3.18 to 3.19 we know that $L / \overline{z}\rho = TP_i$, so

$$L = TP_i z\rho = 80 \text{ mg/m}^3 \times 15 \text{ m} \times 1.5/\text{yr} = 1800 \text{ mg/m}^2 \text{ per yr}$$

(b) Using Equation 3.19, calculate the lake TP:

$$TP = 80/(1 + 1/\rho^{0.5}) = 44 \text{ mg/m}^3$$

If this result differs substantially from the observed TP concentration in the lake, then the model should be calibrated to fit the existing lake data. If this model underestimates the existing lake TP concentration and internal loading has been documented, then an equation similar to 3.21 should be used for the lake.

3.2.3.2 Example 2

Calculate the expected average summer chl a concentration and SD in the lake from Example 1.

Using Equation 3.24, which yields nearly identical results as Equations 3.27 and 3.28 combined, gives

$$\log \text{chl } a = 1.449 \log 44 - 1.136$$

$$= 1.28 \text{ and chl } a = 19.1 \text{ µg/L}$$

and using Equations 3.26 and 3.27 combined gives

$$SD = 7.7/19.1^{0.68} = 1.03 \text{ m}$$

and from Equation 3.26, or 3.27, 3.28, the TSI is 60.

The trend in the 1980s was to develop more specific P loading models for specific lake types. Nurnberg's (1984) separation of anoxic from oxic lakes is an example. Reckhow (1988) developed a set of models for southeastern lakes and reservoirs that included N, P, and τ as predictors of chl a as well as the probability of blue-greens or non-blue-greens representing the dominant algae. Another example is Walker's (1981, 1982, 1985, 1986, 1987, 1996) analysis of USACOE reservoirs, which are typically quite different than lakes due to their higher average flushing rate and hence P loading (see Chapter 2). They also tend to have higher levels of nonalgal turbidity (Lind, 1986).

In analyzing USACOE impoundments, Walker found that the sedimentation rate of P could be appropriately defined as a second order decay rate (rate of decrease dependent on square of the concentration) of the lake TP concentration:

$$P_s = K P^2 \tag{3.32}$$

where P_s is the phosphorus sedimentation rate in mg/m^3 per yr, K is the effective second order decay rate in m^3/mg per yr and P is the reservoir pool phosphorus concentration in mg/m^3. According to Walker, a second order rate gives a more general representation of sedimentation than a first order rate, which is used in the Vollenweider type models for lakes.

An average estimate of decay rate for USACOE reservoirs was 0.1 m^3/mg per yr. However, the rate tended to be lower in reservoirs with low overflow rates (q_s or hydraulic loading) and high inorganic P (i.e., SRP):TP ratios. The overflow rate, q_s, is calculated as the quotient of annual outflow/reservoir area ($or \ \bar{z} / \tau$). This effect of reduced settling with decreased q_s was apparently due to a greater algal assimilation of incoming P. To account for the differences in q_s, Walker (1985) developed two empirical equations:

$$K = \frac{0.17q_s}{q_s + 13.3} \tag{3.33}$$

$$K = \frac{0.056q_s}{Fot(q_s + 13.3)} \tag{3.34}$$

where Fot = tributary inorganic P/TP ratio for reservoirs with high ratios.

Assuming that volume storage does not change and the volume weighted, reservoir pool P concentration equals the outflow concentration, the P mass balance for a reservoir can be represented by:

$$QP_i = QP_o + KVP_o^2 \tag{3.35}$$

where Q is discharge in m^3/yr, P_i and P_o are, respectively, average inflow and outflow P concentrations in mg/m^3, and V is reservoir volume in m^3. Solving that for the average outflow concentration, assuming complete mixing, gives:

$$P_o = \frac{-1 + (1 + 4KP_i\tau)^{0.5}}{2K\tau} \tag{3.36}$$

P_o in reservoirs is usually most sensitive to P_i and least sensitive to the sedimentation term, because residence time in reservoirs is relatively short. P_o becomes more sensitive to the sedimentation term as τ increases and approaches that of P_i as τ decreases below 0.2 yr. This is also apparent from Figure 3.8; if reservoir data are plotted they are usually represented by relatively high P_i and low τ values.

As indicated earlier, reservoirs tend to have higher nonalgal turbidity and shorter residence times (larger ρ) than lakes. Therefore, for a satisfactory prediction of reservoir chl a concentration from predicted P concentration (Equation 3.36), those factors were included (Walker, 1987):

$$\text{chl } a = \frac{\text{chl } a_x}{(1 + 0.025 \text{ chl } a_x G)(1 + Ga)} \tag{3.37}$$

where chl $a_x = (X_{PN})^{1.33}/4.31$, $X_{PN} = \{P^{-2} + [(N\ 150/12)^{-2}]\}^{-0.5}$, $G = Z_{mix}(0.14 + 0.009\ \rho_s)$, a = nonalgal turbidity (as 1/m) = 1/Secchi depth – 0.025 chl a, N is total nitrogen in mg/m^3, Z_{mix} is depth of the mixed layer, and ρ_s is flushing rate during the summer.

If P is assumed or demonstrated to be limiting rather than N, the following simpler model can be used:

$$\text{chl } a + \frac{\text{chl } a_p}{(1 + 0.025 \text{ chl } a_p G)(1 + Ga)} \tag{3.38}$$

where chl $a_p = P^{1.37}/4.88$.

In cases where ρ_s is low (<25/yr), two further simplified models may be used. If either N or P may be limiting and nonalgal turbidity is low, the following may be appropriate:

$$\text{chl } a = 0.2 X_{PN}^{1.25} \tag{3.39}$$

If only P is limiting with low nonalgal turbidity, the following may be used:

$$\text{chl } a = 0.28\,P \tag{40}$$

Walker (1987) developed the computer programs to apply these equations to specific reservoir data. There are three programs; FLUX, which calculates the loadings, PROFILE, which reduces and displays the reservoir constituents, and BATHTUB, which calculates the nutrient balances and predicts response of the whole reservoir or individual segments. This is available from Environ. Lab., USCOE Waterways, 3909 Halls Ferry Road, Vicksburg, MS 39180 (www.Wes.army.mil/el/models/emiiinfo.html).

3.3 SELECTION OF LAKE RESTORATION ALTERNATIVES

The diagnosis and feasibility study is designed to determine the causes of lake problems and to evaluate their current state or severity. The consultant or lake manager will use these data and evaluations to select the appropriate, most cost effective treatment(s) available to achieve the desired lake quality.

Lake restoration techniques are divided into four categories, based on their considered primary objective: (1) to control problems caused by algae, (2) to control excessive macrophyte biomass, (3) to alleviate oxygen problems, and (4) remove sediment. Sediment removal is categorized separately because it may solve several if not all of the above problems. Each of the techniques briefly described below is more thoroughly discussed in the following chapters.

3.3.1 ALGAL PROBLEMS

Because algal biomass is dependent on the concentration of limiting nutrient in the lake's photic zone, the consultant or lake manager must determine, by appropriate evaluation and modeling, the feasibility of controlling the principal sources of the most limiting nutrient. More than one technique may be used at once, but for most in-lake techniques to be effective, important external loading sources should be controlled first.

3.3.1.1 Nutrient Diversion/Advanced Waste Treatment

A reduction in external loading should be the first step. While it is possible to manipulate algal biomass through enhanced grazing or by light limitation (complete mixing), the primary cause of excess algae is high nutrient concentration and if there are important external sources, they should be cost-effectively reduced first.

3.3.3.2 P Inactivation

Internal release of P may be a (or the most) significant source that could delay recovery/improvement of lake quality. Sediment P release can be controlled by adding aluminum salts to the water column resulting in an aluminum hydroxide floc that settles to the sediment surface forming a barrier to further release, even if anoxia persists. This is a powerful, effective, and popular technique. Addition of iron or calcium has also been effective in some circumstances, but frequently repeated treatments are required. Sediment oxidation through enhanced denitrification and resulting improved complexation with iron (Riplox) has been successful and is discussed with P inactivation.

3.3.3.3 Dilution/Flushing

Dilution involves the addition of low-nutrient water to reduce lake nutrient concentration and can be effective where external or internal sources are not controlled. Flushing simply removes algal biomass, although that may require large volumes of water if nutrient concentration is high and not limiting. While effective, employment of these treatments has limited application due to the availability of water, especially low-nutrient water.

3.3.3.4 Lake Protection From Urban Runoff

Land-use modifications can be used to control nutrient loss from the watershed and thus improve lake quality, but they are usually used to protect lake quality from further degradation in areas undergoing development. The effectiveness of these practices at improving/maintaining lake quality has yet to be adequately demonstrated on a whole-lake basis.

3.3.3.5 Hypolimnetic Withdrawal

Nutrient enriched hypolimnetic waters may be preferentially removed through siphoning, pumping, or selective discharge (dams) instead of low-nutrient surface waters. This has been shown to be effective at accelerating P export, reducing surface P concentrations, and improving hypolimnetic oxygen content.

3.3.3.6 Artificial Circulation

This technique is used to prevent or eliminate thermal stratification through the mixing action of a rising column of air bubbles. It will improve DO and reduce iron and manganese, but most importantly it can cause light to limit algal growth in situations where nutrients are uncontrollable and can neutralize the factors favoring dominance by blue-green algae.

3.3.3.7 Food-Web Manipulations

Grazing of algae by large zooplankton (mainly *Daphnia*) can be enhanced by eliminating planktivorous fish through poisoning, physical removal, or increased piscivory (planting piscivores). This technique is relatively inexpensive, and has been successful, but usually for a limited time, and only if lake P has been reduced.

3.3.3.8 Copper Sulfate Treatment

This has been a commonly used treatment for lakes and water supply reservoirs suffering from algal biomass and taste and odor problems for at least a century. It is commonly practiced, although it has significant detrimental aspects.

3.3.4 Macrophyte Problems

Although macrophyte problems often are associated with eutrophication and increased inputs of sediment, control of their growth and biomass cannot be expected to result from reduction of in-lake nutrient concentrations. This is because their nutrient demands are largely supplied through root uptake from the sediment. Therefore, more direct methods are employed to deal with excessive macrophyte biomass.

3.3.4.1 Harvesting

Removing macrophyte biomass from lakes is often an effective, but sometimes cosmetic, treatment to control a nuisance macrophyte problem. Nutrients are removed, which in some lakes can be a significant contribution to internal loading. Thick over story, as well as decomposition of organic matter, contribute to oxygen deficiency and sediment P release, which can be alleviated by plant removal. Quick grow-back sometimes results from cutting plant stems, while grow-back can be curtailed to some extent by removing the roots through the technique of rotovating. However, this is very expensive. Harvesting can have negative effects, such as dispersal of plant fragments to uninfested areas, removal of small fish and encouraging sediment resuspension and sediment P release.

3.3.4.2 Biological Controls

Dispersal of phytophagous insects and fish, especially triploid grass carp, has become more prominent as inexpensive, effective controls on macrophyte biomass. There can be side effects on other fish species and increased sediment nutrient recycling causing algal blooms. There are still uncertainties about effective stocking rates and plant selection.

3.3.4.3 Lake-Level Drawdown

This is truly a multipurpose technique for impoundments. Exposure of rooted plants to freezing or hot conditions eliminates some species, does not affect others, and stimulates a third species group. If the lake is drawn down, other techniques such as sediment removal, placement of screens, or fish management can be employed. Consolidation of sediments and deepening, which are often expected side benefits, usually have not been achieved.

3.3.4.4 Sediment Covers

Screening materials to stop rooted plant growth are expensive, symptomatic, and yet highly effective. Expense prohibits their use over large areas, forcing the consultant or lake manager to choose other means to deal with macrophytes lakewide.

3.3.4.5 Sediment Removal

This technique can be multipurpose, resulting in control of both algae and macrophytes. It is an effective procedure and is frequently recommended for deepening shallow lakes for macrophyte control, curtailing internal nutrient loading by eliminating the enriched sediment layer or eliminating sediments contaminated with toxic substances. It has a significant long-term advantage over nutrient inactivation in that the source is removed rather than bound in place as with P inactivation. The limitations of dredging are its relatively high cost and the requirement for adequate dredged material disposal sites. However, imaginative disposal approaches make dredging more attractive, i.e., forming building blocks from dredged material (USEPA, 2003).

3.3.4.6 Hypolimnetic Aeration

Although a potential exists for control of internal loading from anoxic hypolimnia (especially with iron addition), that potential has not been demonstrated consistently as a technique to control algae. However, it is highly effective at increasing dissolved oxygen in the hypolimnion without destratifying the lake or reservoir. This is usually accomplished with a full airlift device, which brings cold hypolimnetic water to the surface, where gases are exchanged, followed by its return to deep water. Partial airlift devices aerate and distribute the water at depth. Liquid oxygen is also effective. This procedure improves reservoir discharge quality, allows a cold-water fishery to reestablish,

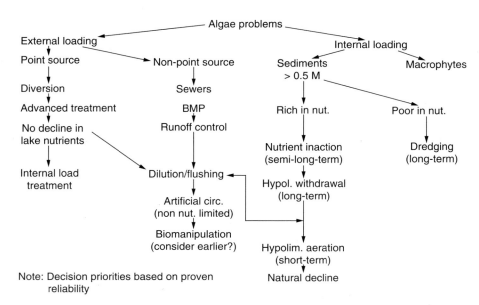

FIGURE 3.9 Decision tree for choice of best restoration procedures for control of algae problems. (From Cooke et al., 1993. With permission.)

provides a daytime refuge for zooplankton grazers (from a warm water fishery), and eliminates problems with iron and manganese in potable water supplies.

3.5   GUIDELINES FOR CHOOSING LAKE RESTORATION ALTERNATIVES

A consultant or lake manager directing a lake project supported with USEPA funds should follow the guidelines of the Clean Lakes Program (CLP) in choosing and defending the chosen methods. While CLP has not been funded since the mid-1990s, lake funds have been available under the non-point program, but mostly for watershed controls. However, there are many State CLPs with similar requirements. Thus, the same guidelines should form the basis for choosing alternatives in lake restoration projects regardless of the funding sources. The reader should consult the USEPA *CLP Guidance Manual* (USEPA, 1980), especially Section 8 and Appendix F, as well as the *Lake and Reservoir Restoration Guidance Manual* (USEPA, 1988), *Managing Lakes and Reservoirs* (NALMS/TI, 2001) and/or *Lake Managers' Handbook* (Vant, 1987).

The diagnostic portion of the feasibility study provides the data to select restoration alternatives. Two fundamental questions are asked, based on the data: How can nutrient diversion (sufficient to protect the lake from further deterioration or sufficient to accomplish a significant change in trophic state or lake quality) be accomplished? What in-lake procedures can be used to accelerate recovery following external load controls or to accomplish further improvement? For each in-lake procedure evaluated, four questions are asked: How effective is it projected to be? How much restoration will be accomplished? How much will it cost? How effective and costly are the alternatives? The alternative of "no action" should also be discussed.

The appropriate technique, or techniques, to apply to a specific lake requires a decision based largely on judgment. Although cost may be the principal criterion, reliability and longevity of the technique(s) will also be important. To make the proper decision, the lake manager will probably go through a decision process in which one or more of the 16 techniques described in this book will be chosen. Such a decision process may take the form of Figure 3.9, for algal problems.

With the aid of at least one year of lake data and a sound nutrient budget, the first consideration should be which nutrient is limiting and what is the principal source of the limiting nutrient. Is the major load coming from external sources or internal recycling? If from external sources, are they point sources or non-point (diffuse) sources? If principally from point sources, diversion of waste-water or stormwater would probably be considered before advanced treatment because it has usually been less costly and operation and maintenance costs are relatively small. If one of those techniques is employed and a slow recovery is predicted or realized, i.e., lake nutrient concentration will not decrease sufficiently to achieve lake quality goals, then dilution/flushing, artificial circulation, and/or biomanipulation could be the next logical consideration. These techniques could provide controls on biomass and nuisance species of algae where controls on the limiting nutrient are not possible. Or, as in the case of dilution, the concentration of limiting nutrient could be controlled without reducing the total load. Dilution/flushing is listed first because there would be control on the causative limiting nutrient concentration. However, costs and scarcity of low-nutrient water may make that technique an unlikely choice.

The problems associated with definition, estimation, and control of non-point nutrient loads, which is often urban stormwater runoff, are discussed in Chapter 5. Stormwater runoff is a principal cause for degradation of urban lakes (and rural lakes in the case of agricultural runoff). The techniques for reduction of non-point loads could be sewers to intercept stormwater and/or septic tank leachate, wet retention ponds, grassy swales ("biofilters"), deep-well injection, chemical treatment in retention ponds, and best management practices (BMPs), such as fertilizer controls (no P) and minimizing impervious surfaces. If any one or a combination of these techniques do not result in improvements (nutrient decline), and the principal source of limiting nutrient is still external, then one of the three techniques previously mentioned could be considered.

The most common reason why lakes or reservoirs do not respond to controls on external inputs of nutrient is due to excessive internal loading or recycling of nutrients from bottom sediments, particularly P. In that case, one can proceed to the right side of Figure 3.9. If sediments are the source of internal loading and the bulk of nutrients are located in the top 0.3 to 0.5 m of a sediment core, then removal of that layer by dredging should provide the most reliable and permanent solution, although it will be the most costly. If sediments are rich in nutrients below that depth, then dredging would result in only exposing more sediment with the same high nutrient content providing little or no expected decrease in internal loading. In that case, there are six techniques that could be considered. These are arranged in sequence of their reliability and expected longevity for control of the nutrient source itself.

Dilution/flushing, artificial circulation, and biomanipulation are, again, aimed at control of algal biomass or nutrient concentration and are not expected to control the source of loading. Riplox, or sediment oxidation, is included with nutrient inactivation and although there has been limited demonstration, its major goal is the restoration of the upper sediment layer and therefore should provide an even longer-term solution than alum. Alum, on the other hand, simply covers the sediment with a floc layer and while its reliability at interrupting sediment P release has been excellent, the layer has been observed to sink through the sediment presumably exposing newly deposited, P-rich sediment that is available for release. Although the record for hypolimnetic withdrawal as a control for internal loading has not been as dramatic as that for alum addition, it has been demonstrated to be reasonably reliable and has the potential to deplete the sediment of nutrients (Chapter 7). Hypolimnetic aeration has not been as effective as alum or sediment oxidation in controlling sediment P release, although it provides direct and effective reaeration and coupled with iron addition has been effective at P control in some cases. Alum addition is the least and dredging the most costly.

If the principal source of internal loading is suspected to be macrophytes, then separate measures for their control must be undertaken. While enclosure and mass balance analyses have indicated the potential significance of macrophyte senescence to internal loading, there are as yet no dem-onstrations of lake water P control through macrophyte control practices. Nevertheless, macrophytes clearly satisfy most of their nutrient demand from the sediment via their roots and, therefore, their control under appropriate circumstances may be a useful approach to reduce lake water P.

A different sequence than that in Figure 3.9 may be needed for a given lake, depending on economic, political, and social demands. Rast and Holland (1988) have suggested a similar organized approach to eutrophication control, although specific restoration techniques are not considered (Figure 3.10). This assessment should very likely precede that in Figure 3.9. There are other benefits and detriments associated with each of these techniques and their success/failure record is more equivocal than implied in this discussion. The reader is thus referred to the individual chapters to gain insight and judgment that will be more pertinent to an individual lake or reservoir.

When an in-lake procedure is chosen, it should be reviewed against this checklist (USEPA, 1980; Table 8.4):

- Will the project displace people?
- Will the project deface existing residences or residential areas?
- Will the project be likely to lead to changes in established land use pattern or an increase in development pressure?
- Will the project adversely affect prime agricultural land or activities?
- Will the project adversely affect parkland, public land, or scenic land?
- Will the project adversely affect lands or structures of historic, architectural, archaeological, or cultural value?
- Will the project lead to a significant long range increase in energy demands?
- Will the project adversely affect short-term ambient air quality?
- Will the project adversely affect short term or long-term noise levels?
- If the project involves physically modifying the lakeshore, its bed, or its watershed, will the project cause any short term or long-term adverse effects?
- Will the project have a significant adverse effect on fish and wildlife or wetlands or other wildlife habitat?
- Will the project adversely affect endangered species?
- Have all feasible alternatives to the project been considered in terms of environmental impacts, resource commitment, public interest, and cost?
- Are there other measures not previously discussed that mitigate adverse impacts resulting from the project?

3.6 THE LAKE IMPROVEMENT RESTORATION PLAN

A technical report is usually required of the consultant or lake manager. The report is a vital part of the diagnosis and feasibility study procedure because it will be used by lake users or homeowners in choosing a course of action and the data will form the benchmark against which future studies will be compared. The report should follow the standard format of a scientific investigation. It should include:

1. Description of the nature of the eutrophication process and a description of the lake's specific problems
2. Listing of the particular questions asked in the diagnosis and feasibility study
3. Description of the area, including maps and a table of morphometric-hydrologic data, and an accurate summary of all measurement methods and sampling locations
4. Compilation of all results in tabular, graphical, and narrative form and an analysis or discussion of the implications of the findings
5. Discussion of the recommendations, including their costs and environmental impacts in relation to goals of lake quality

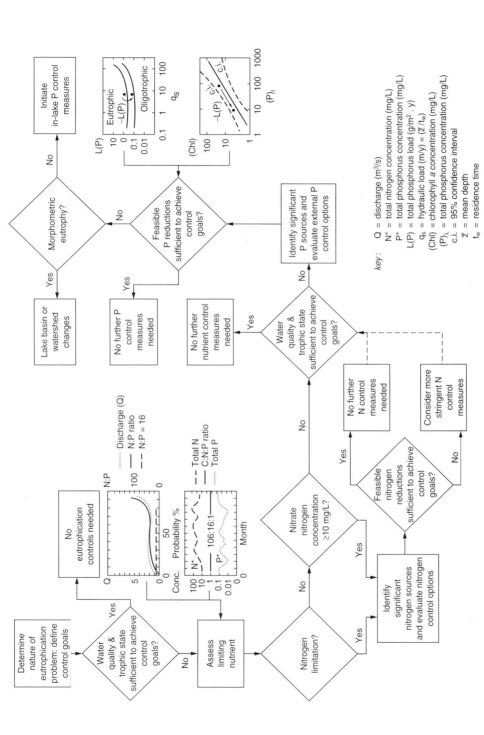

FIGURE 3.10 A typical sequence of events in a diagnostic analysis to determine eutrophication control measures. (From Rast, W. and M. Holland. 1988. *Ambio* 17: 2–12. With permission)

6. Brief summary
7. Citation of literature used in the study

Because few lake users have technical backgrounds, most will not wish to read the technical report. A companion report (or executive summary) that is brief and nontechnical should therefore also be prepared. This second report should include sections on the nature of the problem, questions asked, general findings, recommendations, and costs.

Finally, a public meeting is usually desirable to discuss the results with those who may have to pay for the project and those who will enjoy the benefits. The consultant or lake manager should therefore document the diagnostic field work thoroughly with color slides and should be prepared to make a lucid and brief presentation of the work, the recommendations, and the consequences of the "no action" alternative.

REFERENCES

Ahlgren, I., T. Frisk and L. Kamp-Nielsen. 1988. Empirical and theoretical models of phosphorus loading, retention and concentration vs lake trophic state. *Hydrobiologia* 170: 285–303.

American Public Health Association (APHA). 2003. *Standard Methods for the Examination of Water and Wastewater*, 20th ed., Washington, DC.

Arnell, V. 1982. Estimating runoff volumes from urban areas. *Water Res. Bull.* 18: 383–387.

Auer, M.T., S.M. Doerr, S.W. Effler and E.M. Owens. 1997. A zero degree of freedom total phosphorus model: 1. development for Onondaga Lake, New York. *Lake and Reserv. Manage.* 13: 118–130.

Bachmann, R.W., C.A. Horsburgh, M.V. Hoyer, L.K. Mataraza and D.F. Canfield, Jr. 2002. Relations between trophic state Indicators and plant biomass in Florida lakes. *Hydrobiologia* 470: 219–234.

Barwell, V.K. and D.R. Lee. 1981. Determination of horizontal-to-vertical hydraulic conductivity ratios from seepage measurements on lake beds. *Water Res. Bull.* 17: 565–570.

Belanger, T.V. and D.F. Mikutel. 1985. On the use of seepage meters to estimate groundwater nutrient loading to lakes. *Water Res. Bull.* 21: 265–272.

Boström, B., M. Jannson and C. Forsberg. 1985. Phosphorus release from lake sediments. *Arch. Hydrobiol. Beih. Ergebn. Limnol.* 18: 5–59.

Brater, E.F. and J.D. Sherrill. 1975. Rainfall-Runoff Relations on Urban and Rural Areas. USEPA-670/2-75-046.

Brown, C.D., D.E. Canfield, Jr., R.W. Bachmann and M.V. Hoyer. 1999. Evaluation of surface sampling for estimates of chlorophyll, total phosphorus and total nitrogen concentration in Florida lakes. *Lake and Reserv. Manage.* 15: 121–132.

Burns, N.M. 1995. Using hypolimnetic dissolved oxygen depletion rates for monitoring lakes. New Zealand. *J. Mar. Fresh Water Res.* 29: 1–11.

Canfield, D.E., Jr. and R.W. Bachmann. 1981. Prediction of total phosphorus concentrations, chlorophyll *a*, and Secchi depths in natural and artificial lakes. *Can. J. Fish. Aquatic Sci.* 38: 414– 423.

Canfield, D.E., Jr., K.A. Langeland, M.J. Maceina, W.T. Haller, J.V. Shireman and J.R. Jones. 1983. Trophic state classification of lakes with aquatic macrophytes. *Can. J. Fish. Aquatic Sci.* 40: 1713–1718.

Canfield, D.E., Jr., M.V. Hoyer and C.M. Durarte. 1990. An empirical method for characterizing standing crops of aquatic vegetation. *J. Aquatic Plant Manage.* 28: 64–69.

Carlson, R.E. 1977. A trophic state index for lakes. *Limnol. Oceanogr.* 22: 361–368.

Chapra, S.C. 1975. Comment on an empirical method of estimating retention of phosphorus in lakes. *Water Resour. Res.* 11: 1033–1034.

Chapra, S.C. and K.H. Reckhow. 1979. Expressing the phosphorus loading concept in probabilistic terms. *J. Fish. Res. Bd. Can.* 36: 225–229.

Chapra, S.C. and R.P. Canale, 1991. Long-term phenomenological model of phosphorus and oxygen for stratified lakes. *Water Res.* 25: 707–715.

Christensen, K.K., F.Ø. Andersen and H.S. Jensen. 1997. Comparison of iron, manganese and phosphorus retention in fresh water littoral sediment with growth of littoral uniflora and benthic microalgae. *Biogeochemistry* 38: 149–171.

Clesceri, N.L., S.S. Curran and R.L. Sedlak. 1986. Nutrient loads to Wisconsin lakes: Part I. Nitrogen and phosphorus export coefficients. *Water Res. Bull.* 22: 983–990.

Cooke, G.D., E.B. Welch, S.A. Peterson, and P.P. Newroth. 1993. *Restoration and Management of Lakes and Reservoirs*, 2nd ed. CRC Press, Boca Raton, FL.

Cornett, R.J. and F.H. Rigler. 1979. Hypolimnetic oxygen deficits: Their prediction and interpretation. *Science* 205: 580–581.

Dillon, P.J. and F.H. Rigler. 1974a. A test of simple nutrient budget model predicting the phosphorus concentration in lakewater. *J. Fish. Res. Bd. Can.* 31: 1771–1778.

Dillon, P.J. and F.H. Rigler. 1974b. The phosphorus-chlorophyll relationship in lakes. *Limnol. Oceanogr.* 19: 767–773.

Dillon, P.J. and F.H. Rigler. 1975. A simple method for predicting the capacity of a lake for development based on lake trophic status. *J. Fish. Res. Bd. Can.* 32: 1519–1531.

Downing, J.A. and E. McCauley. 1992. The nitrogen: phosphorus relationship in lakes. *Limnol. Oceanogr.* 37: 936–945.

Duras, J. and J. Hejzlar. 2001. The effect of outflow depth on phosphorus retention in a small hypereutrophic temperate reservoir with short hydraulic residence time. *Int. Rev. ges. Hydrobiol.* 86: 585–601.

Edmondson, W.T. and J.R. Lehman. 1981. The effect of changes in the nutrient income on the condition of Lake Washington. *Limnol. Oceanogr.* 26: 1–28.

Edmondson, W.T. and G.G. Winberg. 1971. *A Manual on Methods for the Assessment of Secondary Productivity in Fresh Waters.* IBP Handbook No. 17. Blackwell Scientific Publ., Oxford, U.K.

Fluck, R.C., C Fonyo and E. Flaig. 1992. Land-use based phosphorus balances for Lake Okeechobee, Florida, drainage basins. *ASAE* 8: 6–13.

Fogle, A.W., J.L. Taraba and J.S. Dinger. 2003. Mass load estimation errors utilizing grab sampling strategies in a karst watershed. *J. Am. Water Res. Assoc.* 39: 1361–1372.

Frodge, J.D., G.L.Thomas. and G.B.Pauley. 1990. Effects of canopy formation by floating and submergent aquatic macrophytes on the water quality of two shallow Pacific Northwest lakes. *Aquatic Bot.* 38: 231–248.

Gächter, R. and J.S. Meyer. 1993. The role of microorganisms in the mobilization and fixation of phosphorus in sediments. *Hydrobiologia* 253: 103–121.

Gächter, R. and B. Müller. 2003. Why the phosphorus retention of lakes does not necessarily depend on the oxygen supply to their sediment surface. *Limnol. Oceanogr.* 48: 929–933.

Goldman, C.R. 1981. Lake Tahoe: Two decades of change in a nitrogen deficient oligotrophic *Lake. Verh. Int. Verein Limnol.* 21: 45–70.

Golterman, H.L. 1969. *Methods for Chemical Analysis of Fresh Waters. IBP Handbook No. 8,* Blackwell Scientific, Oxford, U.K.

Golterman, H.L. 2001. Phosphate release from anoxic sediments or 'What did Mortimer really write?' *Hydrobiologia* 450: 99–106.

Havens, K.E. 2000. Using trophic state index (TSI) values to draw inferences regarding phytoplankton limiting factors and seston composition from routine water quality data. *Korean J. Limnol.* 33: 187–196.

Havens, K.E. and R.T. James. 1997. A critical evaluation of phosphorous management goals for Lake Okeechobee, Florida. *Lake and Reserv. Manage.* 13: 292–301.

Heiskary, S.A. and C.B. Wilson. 1989. The regional nature of lake water quality across Minnesota: An analysis for improving resource management. *J. Minn. Acad. Sci.* 55: 71–77.

Jacoby, J.M., D.D. Lynch, E.B. Welch and M.A. Perkins. 1982. Internal phosphorus loading in a shallow eutrophic lake. *Water Res.* 16: 911–919.

James, W.F. and J.W. Barko. 1997. Net and gross sedimentation in relation to the phosphorus budget of Eau Galle Reservoir, Wisconsin. *Hydrobiologia* 345: 15–20.

Jassby, A.D., J.E. Reuter, R.A. Akler, C.R. Goldman and S.H. Hackley. 1994. Atmospheric deposition of nitrogen and phosphorus in the annual nutrient load of Lake Tahoe (California-Nevada). *Water Resour. Res.* 30: 2207–2216.

Jones, J.R. and R.W. Bachmann. 1976. Prediction of phosphorus and chlorophyll levels in lakes. *J. Water Pollut. Cont. Fed.* 48: 2176–2182.

Jones, C.A. and E.B. Welch. 1990. Internal phosphorus loading related to mixing and dilution in a dendritic, shallow prairie lake. *J. Water Pollut. Cont. Fed.* 62: 847–852.

Jones, J.R., M.F. Knowlton and M.S. Kaiser. 1998. Effects of aggregation on chlorophyll-phosphorus relations in Missouri reservoirs. *Lake and Reservoir Manage.* 14: 1–9.

Jones, J.R. and M.F. Knowlton. Chlorophyll response to nutrients and non-algal seston in Missouri reservoirs and oxbow lakes. *Lake and Reserv. Manage.* In press.

Jones, J.R., M.F. Knowlton and K.G. An. 2003. Trophic state, seasonal patterns and empirical models in South Korean reservoirs. *Lake and Reservoir Manage.* 19: 64–78.

Kann, J. and V.H. Smith. 1999. Estimating the probability of exceeding elevated pH values critical to fish populations in a hypereutrophic lake. *Can. J. Fish. Aquatic Sci.* 56: 1–9.

Kennedy, R.H. 1999. Basin-wide considerations for water quality management: importance of phosphorus retention by reservoirs. *Int. Rev. Hydrobiol.* 84: 557–566.

King County, 2002. Lake Washington existing conditions report. King County Dept. of Nat. Res. and Parks, Seattle, WA.

King County. King County Dept. of Nat. Res. and Parks, Seattle, WA. Personal communication.

Koncsos, L. and L. Somlyódy. 1994. Analysis on parameters of suspended sediment models for a shallow lake. In: *Water Quality International*, '94 IAWQ 17th Biennial International Conference, Budapest, Hungary.

Koski-Vähälä, J. and H. Hartikaine. 2001. Assessment of the risk of phosphorus loading due to resuspended sediment. *J. Environ. Qual.* 30: 960–966.

Kristensen, P., M. Søndergaard and E. Jeppesen. 1992. Resuspension in a shallow eutrophic lake. *Hydrobiologia* 228: 101–109.

La Baugh, J.W. and T.C. Winter. 1984. The impact of uncertainties in a hydrologic measurement on phosphorus budgets and empirical models for two Colorado reservoirs. *Limnol. Oceanogr.* 29: 322–339.

Larsen, D.P. and H.T. Mercier. 1976. Phosphorus retention capacity of lakes. *J. Fish. Res. Bd. Can.* 33: 1742–1750.

Larsen, D.P., J. Van Sickle, K.W. Malueg and P.D. Smith. 1979. The effect of wastewater phosphorus removal on Shagawa Lake, Minnesota: Phosphorus supplies, lake phosphorus, and chlorophyll *a*. *Water Res.* 13: 1259–1272.

Larsen, D.P., D.W. Shults and K.W. Malueg. 1981. Summer internal phosphorus supplies in Shagawa Lake. Minesota. *Limnol. Oceanogr.* 26: 740–753.

Lathrop, R.C., S.R. Carpenter and D.M. Robertson. 1999. Summer water clarity responses to phosphorus, Daphnia grazing, and internal mixing in Lake Mendota. *Limnol. Oceanogr.* 44: 137–146.

Lee, D.R. 1977. A device for measuring seepage flux in lakes and estuaries. *Limnol. Oceanogr.* 22: 140–147.

Lee, D.R. and H.B.N. Hynes. 1978. Identification of groundwater discharge zones in a reach of Hillman Creek in southern Ontario. *Water Pollut. Res. Can.* 13: 121–133.

Lehman, J.T. and T. Naumoski. 1986. Net community production and hypolimnetic nutrient regeneration in a Michigan lake. *Limnol. Oceanogr.* 31: 788–797.

Lind, O.T. 1986. The effect of nonalgal turbidity on the relationship of Secchi depth to chlorophyll *a*. *Hydrobiologia* 140: 27–35.

Lijklema, L. 1980. Interaction of orthophosphate with iron III and aluminum hydroxides. *Environ. Sci. Technol.* 14: 537–541.

Mataraza, L.K. and G.D. Cooke. 1998. Vertical phosphorus transport in lakes of different morphometry. *Lake and Reservoir Manage.* 13: 328–337.

Matson, M.D. and R.A. Isaac. 1999. Calibration of phosphorus export coefficients for total maximum daily loads of Massachusetts lakes. *Lake and Reservoir Manage.* 15: 209–219.

Mitchell, D.F., K.J. Wagner, W.J. Monagle and G.A. Beluzo. 1989. A littoral interstitial porewater (LIP) sampler and its use in studying groundwater quality entering a lake. *Lake and Reservoir Manage.* 5: 121–128.

Moon, C.E. 1973. Nutrient budget following waste diversion from a mesotrophic lake. MS Thesis, University of Washington, Seattle.

North American Lake Management Society (NALMS). 2001. *Management of Lakes and Reservoirs.* USEPA 841-B-01–006. North American Lake Management Society, Madison, WI.

Nichols, S.A. 1999. Floristic quality assessment of Wisconsin lake plant communities with example applications. *Lake and Reservoir Manage.* 15: 133–141.

Nürnberg, G.K. 1984. The prediction of internal phosphorus loads in lakes with anoxic hypolimnia. *Limnol. Oceanogr.* 29: 111–124.

Nürnberg, G.K. 1985. Availability of phosphorus upwelling from iron-rich anoxic hypolimnia. *Arch. Hydrobiol.* 104: 459–476.

Nürnberg, G.K. 1987. A comparison of internal phosphorus loads in lakes with anoxic hypolimnia: Laboratory incubation versus *in situ* hypolimnetic phosphorus accumulation. *Limnol. Oceanogr.* 22: 1160–1164.

Nürnberg, G.K. 1988. Prediction of phosphorus release rates from total and reductant soluble phosphorus in anoxic sediments. *Can. J. Fish. Aquatic Sci.* 45: 453–462.

Nürnberg, G.K. 1995a. The anoxic factor, a quantitative measure of anoxia and fish species richness in central Ontario lakes. *Trans. Am. Fish. Soc.* 124: 677–686.

Nürnberg, G.K. 1995b. Quantifying anoxia in lakes. *Limnol. Oceanogr.* 40: 1100–1111.

Nürnberg, G.K. 1996. Trophic state of clear and colored, soft- and hardwater lakes with special consideration of nutrients, anoxia, phytoplankton and fish. *Lake and Reservoir Manage.* 12: 432–447.

Nürnberg, G.K. and B.D. LaZerte. 2004. Modeling the effect of development on internal phosphorus load in nutrient-poor lakes. *Water Resour. Res.* 40: 1–9.

Organization for Economic Cooperation and Development (OECD). 1982. *Eutrophication of Waters. Monitoring, Assessment and Control.* OECD, Paris.

Omernik, J.M. 1977. Nonpoint Source-stream Nutrient Level Relationships: A Nationwide Study. USEPA 600/3-77-105.

Osgood, R.A. 1988. Lake mixes and internal phosphorus dynamics. *Arch. Hydrobiol.* 113: 629–638.

Penn, M.R., M.T. Auer, S.M. Doerr, C.T. Driscoll, C.M. Brooks and S.W. Effler. 2000. Seasonality in phosphorus release rates from the sediments of a hypereutrophic lake under a matrix of pH and redox conditions. *Can. J. Fish. Aquatic Sci.* 57: 1033–1041.

Perkins, W.W. 1995. *Lake Sammamish Phosphorus Model.* King County Surface Water Manage., Seattle, WA.

Perkins, W.W., E.B. Welch, J. Frodge and T. Hubbard. 1997. A zero degree of freedom total phosphorus model: 2. Application to Lake Sammamish, Washington. *Lake and Reservoir Manage.* 13: 131–141.

Peterson, S.A., R.M. Hughes, D.P. Paulsen and J.M. Omernik. 1995. Regional lake quality patterns: Their relationship to lake conservation and management decisions. *Lakes Reserv. Res. Manage.* 1: 163–167.

Peterson, S.A., N.S. Urquhart and E.B. Welch. 1999. Sample representativeness: A must for reliable regional lake condition estimates. *Environ. Sci. Technol.* 33: 1559–1565.

Pollman, C.D., Tetra Tech., Inc., Gainsville, Florida, personal communication.

Porcella, D.B., S.A. Peterson and D.P. Larsen. 1980. Index to evaluate lake restoration. *J. Environ. Eng. Div. ASCE* 106: 1151–1169.

Prairie, Y.T., C.M. Duarte and J. Kalff. 1989. Unifying nutrient-chlorophyll relationships in lakes. *Can. J. Fish. Aquatic Sci.* 46: 1176–1182.

Prairie, Y.T., R.H. Peters and D.F. Bird. 1995. Natural variability and the estimation of empirical relationships: A reassessment of regression models. *Can. J. Fish. Aquatic Sci.* 52: 7878–7898.

Psenner, R., B. Boström, M. Dinka, K. Pettersson, R. Puckso and M. Sager. 1988. Fractionation of phosphorus in suspended matter and sediment. *Arch. Hydrobiol. Suppl.* 30: 98–103.

Rast, W. and M. Holland. 1988. Eutrophication of lakes and reservoirs: A framework for making management decisions. *Ambio* 17: 2–12.

Rast, W. and G.F. Lee. 1978. Summary analysis of the North American (U.S. Portion) OECD eutrophication project: Nutrient loading-lake response relationships and trophic state. USEPA 600/3-78-008.

Reckhow, K.H. 1983. A method for the reduction of lake model prediction error. *Water Res. Bull.* 17: 911–916.

Reckhow, K.H. 1988. Empirical models for trophic state in southeastern U.S. lakes and reservoirs. *Water Res. Bull.* 24: 723–734.

Reckhow, K.H. and S.C. Chapra. 1983. *Engineering Approaches for Lake Management: Vol. I. Data Analysis and Empirical Modeling.* Butterworths, Boston, MA.

Reckhow, K.H. and J.T. Simpson. 1980. A procedure using modeling and error analysis for the prediction of lake phosphorus concentration from land use information. *Can. J. Fish. Aquatic Sci.* 37: 1439–1448.

Rigler, F.H. 1974. Phosphorus cycling in lakes. In F. Ruttner (Ed.), *Fundamentals of Limnology*, 3rd ed. University of Toronto Press, Toronto, ON, p. 263.

Rochelle, B.P., D.L. Stevens, Jr. and M.R. Church. 1989. Uncertainty analysis of runoff estimates from a runoff contour map. *Water Res. Bull.* 25: 491–498.

Scheffer, M. 1998. *Ecology of Shallow Lakes.* Chapman & Hall, New York, NY.

Scheider, W.A., J.J. Moss and P.J. Dillon. 1979. Measurement and uses of hydraulic and nutrient budgets. In *Lake Restoration*. USEPA 440/5-79-001. p. 77–83.

Seip, K.L. 1994. Phosphorus and nitrogen limitation of algal biomass across trophic gradients. *Aquatic Sci.* 56: 16–28.

Seip, K.L., E. Jeppesen, J.P. Jensen and B. Faafeng. 2000. Is trophic state or regional location the strongest determinant for Chl-a/TP relationships in lakes? *Aquatic Sci.* 62: 195–204.

Shaw, R.D. and E.E. Prepas. 1989. Anomalous short-term influx of water into seepage meters. *Limnol. Oceanogr.* 34: 1343–1351.

Shuster, J.I., E.B. Welch, R.R. Horner and D.E. Spyridakis. 1986. Response of Lake Sammamish to urban runoff control. *Lake and Reservoir Manage.* 2: 229–234.

Siegel, D.I. and T.C. Winter. 1980. Hydrologic Setting of Williams Lake, Hubbard County, Minnesota. U.S. Geol. Survey open-file report 80-403.

Smith, V.H. 1982. The nitrogen and phosphorus dependence of algal biomass in lakes: An empirical and theoretical analysis. *Limnol. Oceanogr.* 27: 1101–1112.

Smith, V.H. and J. Shapiro. 1981. Chlorophyll-phosphorus relations in individual lakes: Their importance to lake restoration strategies. *Environ. Sci. Technol.* 15: 444–451.

Søndergaard, M. 1988. Seasonal variations in the loosely sorbed phosphorus fraction of the sediment of a shallow and hypereutrophic lake. *Environ. Geol. Water Sci.* 11: 115–121.

Søndergaard, M., J.P. Jensen and E. Jeppesen. 1999. Internal phosphorus loading in shallow Danish lakes. *Hydrobiologia* 408/409: 145–152.

Søndergaard M., K.D. Wolter and W. Ripl. 2002. Chemical treatment of water and sediments with special reference to lakes. In M. Perrow and A.J. Davy (Eds.), *Handbook of Ecological Restoration, Vol. 1, Principles of Restoration.* Cambridge University Press, Cambridge, U.K. pp. 184–205.

Sonzogni, W.C. and G.F. Lee. 1974. Nutrient sources for Lake Mendota — 1972. *Trans. Wisc. Acad. Sci.* 62: 133–164.

Soranno, P.A., S.L. Hubler and S.A. Carpenter. 1996. Phosphorus loads to surface waters: A simple model to account for spatial patterns of land use. *Ecol. Appl.* 6: 865–878.

Stauffer, R.E. and D.E. Armstrong. 1984. Lake mixing and its relationship to epilimnetic phosphorus in Shagawa Lake, Minnesota. *Can. J. Fish. Aquatic Sci.* 41: 57–69.

Stauffer, R.E. and G.F. Lee. 1973. The role of thermocline migration in regulation of algal blooms. In E.J. Middlebrooks, D.H. Falkenbor and T.E. Maloney (Eds.), *Modeling the Eutrophication Process.* Utah State University, Water Res. Center, Logan, UT. pp. 73–82.

Stephen, D., B. Moss and G. Phillips. 1997. Do rooted macrophytes increase sediment phosphorus release? *Hydrobiologia* 342: 27–34.

United States Environmental Protection Agency (USEPA). 1974. The relationship of nitrogen and phosphorus to the trophic state of Northeast and North-Central lakes and reservoirs. NES Working Paper 23, Corvallis Environmental Research Laboratory, Corvallis, OR.

USEPA. 1980. *Clean Lakes Program Guidance Manual.* USEPA 440/5-81-003.

USEPA. 1988. *The Lake and Reservoir Restoration Guidance Manual.* USEPA 440/5-88-002.

USEPA. 2003. USEPA and partners demonstrate new technology that turns dredged material into cement. USEPA Region 2 News Release #03138. Nov. 24. New York, NY.

Uttormark, P.D., J.D. Chapin and K.M. Green. 1974. *Estimating Nutrient Loadings of Lakes from non-point Sources.* USEPA 600/3-74-020.

Van Hullebusch, R., F. Auvray, V. Deluchat, P.M. Chazal and M. Baudu. 2003. Phosphorus fractionation and short-term mobility in the surface sediment of a polymictic shallow lake treated with a low dose of alum (Courtille Lake, France). *Water Air Soil Pollut.* 146: 75–91.

Vant, W.N. 1987. *Lake Managers Handbook* — A Guide to Undertaking and Understanding Investigations into Lake Ecosystems, so as to Assess Management Options for Lakes. Water and Soil Misc. Pub. No. 103, Natl. Water and Soil Conserv. Auth., Wellington, New Zealand.

Vollenweider, R.A. 1969a. *A Manual on Methods for Measuring Primary Production in Aquatic Environments.* IBP Handbook, No. 12. Blackwell Scientific Publishers, Oxford.

Vollenweider, R.A. 1969b. Possibilities and limits of elementary models concerning the budget of substances in lakes. *Arch. Hydrobiol.* 66: 1–36.

Vollenweider, R.A. 1976. Advances in defining critical loading levels for phosphorus in lake eutrophication. *Mem. Ist. Ital. Idrobiol.* 33: 53–83.

Vollenweider, R.A. and P.J. Dillon. 1974. The application of the phosphorus-loading concept to eutrophication research. National Research Council, Canada, Tech. Rept. 13690.

Walker, W.W., Jr. 1980. Variability of trophic state indicators in reservoirs. In: *Restoration of Lakes and Reservoirs.* USEPA 440/5-81-010. p. 344.

Walker, W.W., Jr. 1981. Empirical methods for predicting eutrophication in impoundments. Rep. I, Phase I. Tech. Rept. E-81-9, U.S. Army Corps Engineers. Vicksburg, MS.

Walker, W.W., Jr. 1982. Empirical methods for predicting eutrophication in impoundments. Model Testing, Rep. 2, Phase II. Tech. Rept. E-18-9, U.S. Army Corps Engineers, Vicksburg, MS.

Walker, W.W., Jr. 1984. Trophic state indices in reservoirs. In: *Lake and Reservoir Management*. USEPA 440/5-84-001. p. 435–440.

Walker, W.W., Jr. 1985. Empirical methods of predicting eutrophication in impoundments. Applications Manual. EWQOS Program, U.S. Army Corps Engineers, Vicksburg, MS.

Walker, W.W., Jr. 1986. Models and software for reservoir eutrophication assessment. *Lake and Reservoir Manage.* 2: 143–148.

Walker, W.W., Jr. 1987. Empirical methods for predicting eutrophication in impoundments. Rep. 4, Phase III: Application Manual. Tech. Rept. E-81-9, U.S. Army Corps Engineers, Vicksburg, MS.

Walker, W.W. 1996. Simplified procedures for eutrophication assessment and prediction: users manual. Instruction Rep. W-96-2, U.S. Army Corps Engineers, Vicksburg, MS.

Walker, W.W. Jr., personal communication. 1127 Lowell Rd., Concord, MA.

Walker, W.W. Jr. and K.E. Havens. 1995. Relating algal bloom frequencies to phosphorus concentrations in Lake Okeechobee. *Lake and Reservoir Manage.* 11: 77–83.

Welch, E.B. 1989. Alternative criteria for defining lake quality for recreation. In *Proc. Natl. Conf. Enhancing State's Lake Management Programs*, Chicago. p. 7.

Welch, E.B. and G.D. Cooke. 1995. Internal phosphorus loading in shallow lakes: Importance and control. *Lake and Reservoir Manage.* 11: 273–281.

Welch, E.B. and J.M. Jacoby. 2001. On determining the principle source of phosphorus causing summer algal blooms in western Washington lakes. *Lake and Reservoir Manage.* 17: 55–65.

Welch, E.B. and J.M. Jacoby. 2004. *Pollutants in Fresh Water: Applied Limnology.* Spon Press, London and New York.

Welch, E.B. and M.A. Perkins. 1979. Oxygen deficit-phosphorus loading relation in lakes. *J. Water Pollut. Cont. Fed.* 51: 2823–2828.

Welch, E.B., C.A. Rock, R.C. Howe and M.A. Perkins. 1980. Lake Sammamish response to wastewater diversion and increasing urban runoff. *Water Res.* 14: 821–828.

Welch, E.B., D.E. Spyridakis, J.I. Shuster and R.R. Horner. 1986. Declining lake sediment phosphorus release and oxygen deficit following wastewater diversion. *J. Water Pollut. Cont. Fed.* 58: 92–96.

Welch, E.B., C.A. Jones and R.P. Barbiero. 1989. Moses Lake quality: Results of dilution, sewage diversion and BMPs — 1977 through 1988. Dept. Civil Engr., Water Res. Ser. Tech. Rept. 118.

Welch, E.B., E.B. Kvam and R.F. Chase. 1994. The independence of macrophyte harvesting and lake phosphorus. *Verh. Int. Verein. Limnol.* 25: 2301–2314.

Welch, E.B., J. Kann, T.K. Burke and M.E. Loftus. 2004. *Relationships Between Lake Elevation and Water Quality in Upper Klamath Lake, Oregon.* R2 Resources, Inc., Redmond, WA.

Wetzel, R.B. 1983. *Limnology*, 2nd ed. W.B. Saunders, Philadelphia, PA.

Wetzel, R.B. and G.E. Likens. 1991. *Limnological Analysis*. W.B. Saunders, Philadelphia, PA.

Whitmore, T.J. and M.A. Riedinger-Whitmore. 2004. Lake management programs: The importance of sediment assessment studies. *LakeLine* 24: 27–30.

Winter, T.C. 1978. Groundwater component of lake water and nutrient budgets. *Verh. Int. Verein. Limnol.* 20: 438–444.

Winter, T.C. 1980. Survey of errors for estimating water and chemical balances of lakes and reservoirs. In *Symposium on Surface Water Improvements*, sponsored by American Society of Civil Engineering, Minneapolis, MN. p. 224.

Winter, T.C. 1981. Uncertainties in estimating the water balance of lakes. *Water Res. Bull.* 17: 82–115.

Section II

Algal Biomass Control Techniques Directed toward Control of Plankton Algae

The techniques described in Chapters 4 through 10, and Chapter 14, are designed to control nutrients, plankton algae, and other related effects of over production and species composition changes that result from eutrophication, such as, transparency, oxygen, taste/odors, scum formation, toxic algal blooms and trihalomethane (THM) production. Before describing these techniques, some general characteristics of the plankton algae (phytoplankton) should be discussed.

Most of the phytoplankton are in the taxonomic orders Chlorophyta (green algae), Chrysophyta (diatoms, yellow-green, and golden brown algae), Cyanophyta (blue-green algae or cyanobacteria), Pyrrhophyta (dinoflagellates), Euglenophyta (euglenoids), and Cryptophyta (cryptomonads). Species may occur as single cells, or as colonies or filaments composed of numerous cells. They may be grouped into size categories such as ultra-, nano-, and net plankton with separations at about 10 and 50 μm.

The phytoplankton, for the most part, is maintained in the water column by wind-caused turbulence. Their shapes can be irregular, increasing their surface-to-volume ratio, thus decreasing their density, which allows them to resist sinking. Nevertheless, they will settle out of the water column during quiescent conditions, especially as they become senescent and dense. Blue-green algae (cyanobacteria), by virtue of their gas vesicles, and flagellated motile species, can resist sinking. Naturally and artificially caused turbulence and mixing are very important to phytoplankton productivity and species composition as will be discussed in succeeding chapters.

There is a typical seasonal cycle in the phytoplankton that usually begins with a spring diatom bloom in temperate waters. In oligotrophic lakes, the spring bloom consumes much of the available nutrients accumulated during the winter and spring runoff. As sedimentation occurs, nutrient content in the photic zone is depleted and phytoplankton growth slows, which is usually the result of nutrient limitation. With minimal external loading during low summer stream flow and thermal stratification preventing vertical mixing of enriched hypolimnetic waters, nutrient content in the epilimnion reaches low levels causing low algal productivity and biomass. With fall circulation, there may be a bloom of diatoms, but usually in lesser proportion than during the spring. Green algae, desmids, and yellow-green algae are an important part of the summer phytoplankton in oligotrophic lakes, but usually blue-greens (cyanobacteria) are relatively unimportant.

As eutrophication proceeds, the spring diatom bloom is usually succeeded by blue-greens to an increasing degree during the summer. Their abundance depends on nutrient loading from either external or internal sources; in general, the higher the nutrient loading (and concentration), the greater the biomass and extent of dominance by blue-greens, which often extends from early spring through the late fall. Notwithstanding their preference for higher temperature than other groups, the blue-greens may dominate during low-temperature periods with blooms sometimes persisting into winter, under the ice in dimictic lakes or forming surface scums in monomictic lakes. Eutrophication, especially from sewage effluent, results in a reduced ratio of N:P, which favors N-fixing cyanobacteria.

Seasonal patterns vary greatly depending on the availability of light and nutrients. In general, the shallower the lake, the less the control of internal nutrient recycling by thermal stratification and light availability by critical depth as compared to mixing depth. For example, an OI (Osgood Index, Chapter 3; Osgood, 1988) less than 7 will have a strong tendency to mix and thus recycle nutrients from sediments/hypolimnion into the photic zone during summer wind events. Such events have been shown to entrain nutrients and cause summer blooms (Stauffer and Lee, 1973; Larsen et al., 1981). Thus, the pattern of phytyoplankton abundance/species composition varies with wind events in highly dynamic lakes. These are some of the phenomena that must be understood for a given lake in order to estimate the cost effectiveness of respective treatments. Unfortunately, the causes for phytoplankton succession and blue-green (cyanobacteria) dominance are still not well understood and the results from treatments can be predicted only in a general way, which will be apparent in succeeding chapters.

REFERENCES

Larsen, D.P., D.W. Schults and K.W. Malueg. 1981. Summer internal phosphorus supplies in Shagawa Lake, Minnesota. *Limnol. Oceanogr.* 26: 740–753.
Osgood, R.A. 1988. Lake mixis and internal phosphorus dynamics. *Arch. Hydrobiol.* 113: 629–638.
Stauffer, R.E. and G.F. Lee. 1973. The role of thermocline migration in regulating algae blooms. In: D.H. Falkenborg and T.E. Maloney (Eds.), *Modeling the Eutrophication Process*. Utah State University, Logan. pp. 73–82.

4 Lake and Reservoir Response to Diversion and Advanced Wastewater Treatment

4.1 GENERAL

The first step in restoring or improving the quality of eutrophic or hypertrophic lakes and reservoirs is to remove or treat direct inputs of wastewater, stormwater, or both. Such sources usually contain relatively high concentrations of P and N. Unless such external inputs (loading) are reduced, any long term benefits from in-lake treatments will usually not be realized. In some cases, reduction of external inputs is sufficient to restore the water body (e.g., Lake Washington; Edmondson, 1978, 1994), but in others, where internal loading of nutrients is significant, in-lake treatments may be necessary to achieve lake quality improvement (e.g., Lake Trummen; Björk, 1974; see Chapter 20).

Diversion is expected to have similar effects to P removal through advanced wastewater treatment (AWT). Either P is already the most limiting nutrient, or in the case of highly eutrophic or hypertrophic lakes where N is often limiting, P can be made to limit if its concentration is sufficiently reduced. To cause P to limit in enriched N-limited lakes usually requires a substantial reduction in external loading. However, if internal loading from sediments is sufficiently high so that N is still limiting after external load reduction, there could be more benefits from diversion than AWT, since incoming N is also removed. Nevertheless, lakes would probably still remain highly eutrophic even if benefits resulted from N reduction. This was the case in Lake Norrviken, Sweden, as will be discussed later (Ahlgren, 1978). Benefits have also been documented from diluting inflow nitrate concentration (Welch et al., 1984; Chapter 6). Both N and P removal from wastewater may be necessary if background P loading is naturally high (Rutherford et al., 1989).

The important question is usually not whether lakes or reservoirs will recover or improve following external nutrient load reduction, but when and to what extent? Lake P concentrations have decreased in nearly all cases following external load reductions. However, equilibrium P concentrations may still be higher than required to limit algal biomass and N may still be limiting. This is especially likely to happen if internal loading from sediments during summer is substantial.

The rate of recovery or improvement depends on several factors. Lakes usually return to near previous trophic state, or at least improve in quality, after reduction in P loading. Recovery may be slow and incomplete depending on the P retention capacity of the sediments. If the sediment does not retain P (P output > P input, e.g., Søbygaard, Norrviken) then reduction in lake concentration following diversion is due to dilution only and sediment release may continue more or less at the same rate for at least 10 years and maybe longer (Søndergaard et al., 2001).

Deep lakes, and those with smaller wind fetch per unit mean depth, usually respond faster and more completely than shallow lakes. Shallow lakes are more difficult to recover or improve, even though they are "oxic," because of the effectiveness of wind mixing that makes P released from the sediment more available to the photic zone and to algal uptake. Sediment release rates in shallow lakes are as high or higher than in stratified anoxic lakes due to several mechanisms (Welch and Cooke, 1995). Very high release rates of 20–50 mg/m^2 per day have been observed in shallow lakes with low Fe:P ratios, wind mixing and high pH (Søndergaard, 1988; Jensen et al., 1992). Internal P loading is highest during summer due to higher temperature and to biological activity, and the rate increases with trophic state (Søndergaard et al., 2001). Internal loading will decrease eventually, but may remain high for decades. In a few cases internal loading decreased rather soon after input P reduction.

The rate of recovery of P to equilibrium in lakes with post-diversion internal loading can be predicted with mass balance P models that include sediment processes. Internal loading will decline as enriched sediment is buried beneath new, less-rich sediment. Predictions of time to reach 90% of the recovery in lake P to equilibrium concentration following external P reduction were about 80 years for Shagawa Lake (Chapra and Canale, 1991) and 30 years for Lake Okeechobee (Pollman, personal communication).

The long-term response of internal loading to input P reduction is not routinely predicted and where it has been predicted, the response has not been verified. Therefore, at least 10 years of internal loading at the same rate as before treatment must be conservatively assumed. This is based on cases where internal loading declined slowly and even increased to higher rates following treatment (Welch and Cooke, 1995). A relatively few cases actually show a substantial reduction shortly after treatment.

Some general results of lake response to diversion or advanced treatment will be described along with detailed accounts of the recovery of several representative lakes. The role of internal P loading in deep and shallow lakes will be discussed as well as problems in forecasting lake response.

4.2 TECHNIQUES FOR REDUCING EXTERNAL NUTRIENT LOADS

Diversion and AWT are the two techniques most used to reduce external loading. Diversion of treated sewage or industrial wastewater involves installing interceptor lines to convey the waste-waters away from the degraded water body to waters that have greater assimilative capacity (e.g., where light limits). The wastewater may already be collected in a sewer system and represent a "point" source, which requires only a connecting pipe for diversion. Or, where individual household septic tank drainfields or stormwater runoff constitutes non-point sources (Chapter 5), a collection system may be a necessary part of the diversion project. Diversion requires large pipes to transport wastewater long distances at relatively high cost.

AWT reduces the P concentration in wastewater effluents that continue to enter the lake by removal with alum (aluminum sulfate), lime (calcium hydroxide), or iron (ferric chloride). For sewage, the P removal stage follows conventional primary and secondary treatment. Residual TP concentration following AWT is about 1,000 µg/L, which represents a reduction of 80% from 5,000 µg/L, which is typical for secondary treated sewage effluent, but much lower residuals (e.g., 50 µg/L) may be required to reach biomass limiting lake concentrations. Treatment of river inflows has resulted in residual concentrations of only a few µg/L (Bernhardt, 1981; Chapter 5). Treatment costs increase with volume treated and as required residual P concentration decreases.

Once sewage and enriching industrial waste effluents are diverted or treated, the next most important external sources of enrichment may be stormwater runoff, enriched from land-use changes. While P content in stormwater is much lower (2–10%) and less soluble than that in sewage effluent, such non-point sources can represent significant contributions. There are several forms of watershed treatment to reduce P content in runoff water, including P retention in wet detention basins and wetlands, rapid infiltration through soil, and P removal in pre-detention basins (Chapter 5).

Unfortunately, there are few documented cases where stormwater treatment or stormwater diversion has resulted in lake recovery. Although stormwater controls are routinely instituted in watersheds, long-term lake monitoring usually has not been included. Also, stormwater controls are often instituted to protect lakes from increasing development, where there was no history of impairment prior to control measures. Therefore, lake response to external nutrient load reduction will be considered only for wastewater diversion or AWT, and the better-documented cases described.

TABLE 4.1
Results of Diversion and Advanced Treatment of Nutrient Inputs to 42 World Lakes

	TP$_i$	TP$_l$		Chl a	
	% change	% change	conc.	% change	conc.
Type I, n = 15	−74 ± 18	−38 ± 14	28 ± 23	−37 ± 18	5.4 ± 5.4
Type II, n = 9	−76 ± 10	−51 ± 14	118 ± 118	−57 ± 17	26 ± 29
Type III, n = 18	−64 ± 22	−67 ± 14	100 ± 152	+216 ± 394	44 ± 49

Note: See text for type definition; TP$_l$ is equilibrium concentration and TP$_i$ is average inflow concentration in µg/L ± 1 SD).

Source: Data from Cullen, P. and C. Forsberg. 1988. *Hydrobiologia* 170: 321–336. With permission.

4.3 RECOVERY OF WORLD LAKES

There are several reviews of lake/reservoir response to external nutrient load reduction (Uttormark and Hutchins, 1980; Cullen and Forsberg, 1988; Marsden, 1989; Sas et al., 1989; Jeppesen et al., 2002), including accounts of nearly 100 world lakes to which nutrient inputs were reduced. These accounts show that while lakes usually respond to external load reduction, the response may be slow and the degree of improvement less than expected.

Cullen and Forsberg (1988) reviewed the response of 43 lakes to external load reduction. The response varied among "sufficient to change trophic category…" — type I (15), "reduction in lake P and chlorophyll (chl) *a*, insufficient to change trophic category…" — type II (9), and "small or no obvious improvement or reduction in lake P, and with little reduction in chl *a*…" — type III (19). The magnitude of external load (inflow concentration, P$_i$) reduction averaged from about two thirds to three fourths of the pretreatment loading (Table 4.1). Lake P (P$_l$) in the first two categories decreased, but considerably less than the load reduction. Chl *a* averaged a sizable decrease in lakes in the first two categories as well. The reason for trophic state change in only the first category is indicated by the much lower residual concentrations of P$_l$ and chl *a*, compared to the second two categories where residual P$_l$ averaged 100 µg/L or more. The criterion used for trophic state change was 25 µg/L TP for the eutrophic-mesotrophic boundary. Uttormark and Hutchins evaluated 13 additional lakes and found that nine responded with changed trophic state (20 µg/L for the eutrophic-mesotrophic boundary). Seven of those nine lakes are in Austria.

Lakes/reservoirs do respond to external nutrient load reduction, even though the trophic state may not change, meaning that lake P content may not be lowered sufficiently to change trophic state, but improvement in lake quality still occurs. Most lakes do not respond as expected based on flushing and sedimentation rates, especially shallow lakes (Ryding and Forsberg, 1980; Søndergaard et al., 2001). The failure of lakes to recover promptly and as expected is from the recycling of P from sediment, known as internal loading. Internal loading becomes more significant in shallow lakes, because the entire water column can be affected by wind-induced entrainment of both high P bottom water and resuspended particulate P and high pH. However, thermal stratification tends to block availability of hypolimnetic P in deep lakes until the lake destratifies.

Some decline in lake P will occur following external load reduction, even if internal loading is high, so the question is not if recovery will occur, but when and to what extent? The difficulty in forecasting extent of recovery is in predicting equilibrium P concentrations in lakes with substantial internal loading. That is especially a problem in shallow lakes where several mechanisms of internal loading may be operating (Chapter 3). The difference in response between shallow and deep lakes, the difficulty in predicting equilibrium P$_l$ concentrations and the time to equilibrium were well illustrated in a thorough review of nine shallow and nine deep European lakes that experienced external load reduction (Sas et al., 1989). The selected nine shallow lakes, with mean depths, were: Norrviken (Sweden), 5.4 m; Glum Sø (Denmark), 1.8 m; Hylke Sø (Denmark), 7.1 m; Søbygaard (Denmark), 1.0 m; Veluwemeer (Netherlands), 1.3 m; Schlachtensee (Germany), 4.6 m; Cockshoot Broad (UK), 1.0 m; Alderfen Broad (UK), 0.6 m; and Lough Neagh (UK), 8.9 m. The definition of shallow was that most of the lake's epilimnion was in direct contact with bottom sediments. The nine deep lakes were: Gjersjøen (Norway), 23 m; Wahnbach Talsperre (Germany), 18 m; Bodense (Germany, Austria, Switzerland), 100 m; Lac Léman (France, Switzerland), 172 m; Zürichsee-Untersee (Switzerland), 51 m; Walensee (Switzerland), 100 m; Fuschlsee (Austria), 38 m; Ossiachersee (Austria), 20 m; Lago Maggiore (Italy, Switzerland), 177 m.

All lakes, whether shallow or deep, had reduced annual mean lake TP concentration. However, the percent reduction in lake TP was less than the reduction in loading. The mean ratio of pre-diversion inflow TP to post-diversion inflow TP was 5.4 ± 6.8, while the mean ratio of pre-diversion lake TP to post-diversion lake TP was 3.7 ± 5.8 (n = 17). That is, inflow TP decreased 82% [1 − (1/5.4)] on the average while in-lake TP decreased 73%. These means were values based on the highest before and lowest after treatment concentrations.

A net annual release of TP was observed in the shallow lakes for the first few years after external load reduction, but it diminished after about five years, with two exceptions. Continued net release after external load reduction was related to a sediment TP content (top 15 cm) per dry matter in excess of 1 mg/g. Although sediment TP content and P release rate are related, release rate was more closely tied to sediment mobile P content (Nürnberg, 1988). The 1 mg/g level indicates saturation, and sediment P above that level before external load reduction, should produce a slow recovery. However, seasonal (e.g., summer) net release of P continued to occur after loading reduction in many shallow lakes, even though net release on an annual basis ceased. That condition may still result in high summer TP and algal biomass and occur even with sediment P ≤ 1 mg/g (e.g., Long Lake and Green Lake; Chapter 8). On the other hand, net annual release of P never occurred in deep lakes.

With continued net annual release of P from sediments of shallow lakes, a much more gradual reduction in lake TP was observed than for deep lakes, whose annual lake TP concentration responded rather quickly to external load reduction. Moreover, net release of P tended to persist during summer in shallow lakes, although it also tended to decrease as net annual release decreased. Thus, while annual lake TP eventually decreased in both shallow and deep lakes, recovery of lake quality in shallow lakes was slower, because summer algal biomass responds to summer P, which can remain high as long as summer net sediment release occurs (Welch and Jacoby, 2001).

The European lake evaluation suggested that until the summer epilimnion concentration of soluble reactive P (SRP) fell below a mean of 10 µg/L, algae would not be P limited, and even though lake TP declined, algal biomass would not respond (unless due to N reduction). This concept is shown in Figure 4.1, where biomass begins to decline only after P has reached a level low enough to be limiting. Others cited the level of 10 µg/L as critical to initiating algal problem. Sawyer (1947) observed 50 years earlier that Wisconsin lakes with dissolved P exceeding 10 µg/L in the spring would likely have nuisance algal blooms the following summer. The critical concentration is similar in streams, where a nuisance periphytic biomass level of 200 mg/m^2 chl a reached in 30 days accumulation time can be expected at an annual mean SRP ≥ 10 µg/L (Biggs, 2000).

As lower P concentrations are attained, a species composition change may be expected. The percent blue-green algae (cyanobacteria) declined as the ratio of TP:Z_{eu}/Z_{mix} (i.e., P:light) declined. *Oscillatoria* gave way to other blue-greens (*Microcystis, Anabaena,* and *Aphanizomenon*) in shallow lakes before a further decline in the ratio resulted in a decrease in those blue-greens (Figure 4.2). Because *Oscillatoria* does not produce a scum on the lake surface, aesthetic lake quality actually got worse before it got better. The *Oscillatoria* to other blue-green ratio shifted between 50 and 100 µg/L TP. In deep lakes, *Oscillatoria* declined when TP dropped to 10 to 20 µg/L.

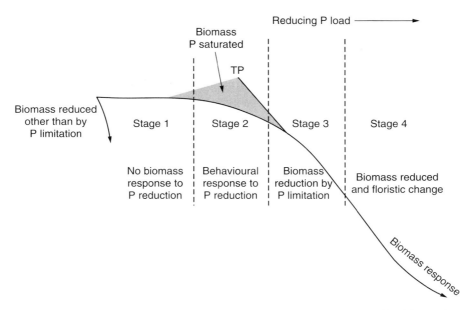

FIGURE 4.1 General expected pattern of algal community response to reduction of in lake nutrient concentration. (From Sas, H. et al. 1989. *Lake Restoration by Reduction of Nutrient Loading: Expectations, Experiences, Extrapolation.* Academia-Verlag, Richarz, St. Augustine, Germany. With permission.)

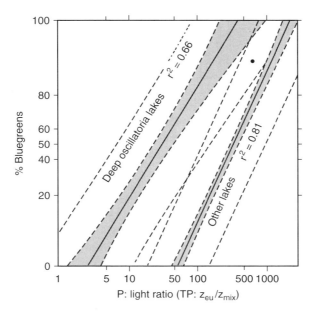

FIGURE 4.2 Log-linear regression relationships for the different categories of blue-green algal responses to restoration. (from Sas, H. et al. 1989. *Lake Restoration by Reduction of Nutrient Loading: Expectations, Experiences, Extrapolation.* Academia-Verlag, Richarz, St. Augustine, Germany. With permission)

TABLE 4.2
Observed and Predicted TP (μg/L) after 20
Years of Equilibration as 5-year Mean Values
following TP Input Reduction in Four Large
Swedish Lakes and Respective Basins

Lake	Basin	TP in-lake	TP predicted
Vättern		6	4
Vänern		8	6
Mälern	Björkfjärden	22	22
	Ekoln	42	63
	Galton	48	53
Hjälmaren	Storhjälmaren	52	28
	Hemfjärden	92	29

Note: See text for method of prediction.

Source: From Wilander, A. and G. Persson. 2001. *Ambio* 30: 475–485. With permission.

European lakes responded to TP reduction according to the following model (Sas et al., 1989):

$$P_l \text{ post} = P_l \text{ pre } (P_i \text{ post}/P_i \text{ pre})^{0.65}$$

where P_l = in-lake mean concentration (May–October) and P_i = annual mean inflow concentration, with pre = pre reduction and post = post reduction equilibrium concentration.

This model was used to evaluate the response of four large Swedish lakes where AWT was installed on all wastewater inputs by the mid 1970s, reducing TP inputs by 50–60% (Wilander and Persson, 2001). The large oligotrophic lakes, Vättern and Vänern, 1,890 and 5,650 km², with mean depths of 39 and 25 m, respectively, were affected only slightly by large reductions in TP input. Equilibrium concentrations were close to values predicted by the Sas model (Table 4.2). The three distinctive basins of Lake Mälern (591 km², 18 m mean depth) responded to input reduction as predicted by the model, although TP in the Ekoln portion is even lower than expected (Table 4.2). Shallow Lake Hjälmaren (402 km², 6.5 m mean depth) did not respond as expected after 20 years following input reduction, due to extensive sediment P internal loading (Table 4.2). In the smaller Hemfjärden (25 km², 1 m mean depth) portion of that lake, TP decreased substantially from pre-treatment concentrations > 150 and even over 500 μg/L during 10 years prior to input reduction, while changes were less in the large basin (Storhjälmaren), more distant from the wastewater source.

A similar, but more complicated response occurred in Lake Balaton, Hungary, where a 45–50% reduction in external TP input resulted in a marked, although delayed, decrease in algal biomass in the small western basin (38 km², 2.3 m mean depth), but a continued high and even increased trophic state was observed in the two larger northeastern basins (600 and 802 km², mean depths 3.2 and 3.7 m; Istvánovics et al., 2002). Net internal loading actually increased by 5–6 fold in the large northwestern basins during the 11-year post input reduction period compared to the pre reduction 8-year period. Internal loading was enhanced by the invasion of a subtropical cyanobacterium (*Cylindrospermopsis raciborskii*), which promoted a positive feedback due to high photo-synthetically-caused pH desorbing P from resuspended sediments.

Analysis of the small western basin showed that the lack of a decrease in TP, despite the algal biomass decrease, was due to reduced P settling caused by upstream processes: (1) reduced loading of TP, relative to Ca, resulted from an upstream reservoir, and (2) increased soluble P, relative to

TP in the summer outflow from an upstream wetland (Istvánovics and Somlyódy, 2001). The decrease in algal biomass was related to increased immobilization of mobile sediment P.

Recovery of 18 lakes in Denmark was recorded over 11 years following P loading reduction (Jeppesen et al., 2002). Four of the 18 lakes were also biomanipulated. TP concentrations declined in all lakes, more in some than others, depending on the magnitude of internal loading. Chl *a* also declined in relation to TP in 10 lakes, even in some over a range of relatively high TP (150–400 µg/L). Taxa composition of the phytoplankton also changed with marked declines in non-heterocystis cyanobacteria with smaller increases in those with heterocysts. Zooplankton biomass did not change significantly with TP reduction, but did in the biomanipulated lakes. However, the zooplankton:phytoplankton ratio increased in all lakes with TP reduction, and the fraction represented by *Daphnia* greatly increased in the biomanipulated lakes. No changes were observed in four untreated lakes.

There are exceptions to poor recovery in shallow lakes with relatively high P content or organic sediments. A chain of three shallow Canadian lakes, Pearce (56 ha, mean depth 2.3 m), June (45 ha, mean depth 2.3 m) and La Cosca (213 ha, mean depth 1.6 m) recovered promptly to an 80% reduction in P loading (Choulik and Moore, 1992). Lake TP in summer decreased 70, 64 and 55% in the three lakes, respectively. Internal loading was not significant despite sediment P levels of 4–20 mg/g. Very high flushing rates, up to 0.5/day during snowmelt, may account for the small internal loading effect.

One exception is the rather quick recovery, in terms of water quality, of a heavily loaded (26.5 g P/m^2 per yr), small (2.8 ha), shallow (0.7 m mean depth) lake following diversion of sewage effluent (98% P load). Little Meer began retaining P annually only three years after diversion (Beklioglu et al., 1999). Fast recovery was attributed to strong planktivory and clear water that continued after diversion despite high residual lake TP (185 µg/L) and high internal loading (38 mg/m^2 per day). The point here is that biotic processes dominated lake quality, rather than P.

Another important point regarding expectations for lake recovery relates to transparency (SD). The improvement in transparency of the water is not linearly related with a reduction in chl *a* and TP concentrations (see equations of Carlson, Chapter 3). The degree of improvement in SD, for an equal amount of P diverted, would become greater as a mesotrophic state (< 25 µg/L) is approached. A graphical display of the Carlson equation in Figure 4.3 illustrates that larger and larger increases in SD occur at each successive decrease in chl *a* content. That is, a given decrease in TP and chl *a* will be more apparent in terms of water clarity improvement following treatment of mesotrophic or lower eutrophic lakes than for higher eutrophic or hypereutrophic lakes.

For additional understanding of the expectations and uncertainty in lake response following external P load reduction, several specific cases will be reviewed in detail. These cases are; Lakes Washington and Sammamish in Washington state, Lakes Norrviken and Vallentuna in central Sweden, Shagawa Lake in Minnesota, the lake chain at Madison, Wisconsin, Lake Zürich, Switzerland, and Søbygaard, Denmark.

4.4   LAKE WASHINGTON, WASHINGTON

The diversion of secondary treated domestic wastewater from Lake Washington from 1964 to 1967 by the Municipality of Metropolitan Seattle, and its subsequent fast recovery, is well known, because its rapid and complete recovery occurred at a time when considerable doubt existed about the prospects for restoring lakes once they had become eutrophic. Lake Washington began recovering before the 3-year construction project, diverting 88% of the lake's external P loading, was completed (Edmondson, 1970, 1978, 1994; Edmondson and Lehman, 1981).

The lake responded precisely as the Vollenweider model (Equation 3.19) predicted. TP declined from a mean annual 64 µg/L prior to diversion to an equilibrium concentration of about 21 µg/L by 1972, 5 years after diversion was complete. However, it had already declined to about 25 µg/L by 1969. The lake should have reached 10% of its total decrease to equilibrium in 2.2 years, based

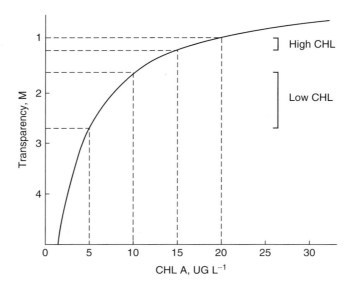

FIGURE 4.3 Chl *a* vs. transparency showing greater absolute benefits to transparency for an incremental change at low vs. high chl *a*. (From Cooke et al., 1993, based on data from Carlson, R.E. 1977. *Limnol. Oceanogr.* 22. With permission.)

on a first order decline [ln $10/(\rho + \rho^{0.5})$, where $\rho = 0.4/yr$]. The predictable response assumed that diversion was completed in 1967 and used an observed retention coefficient that conformed exactly to that of the Vollenweider model, i.e., 0.61 (Edmondson and Lehman, 1981). The post-diversion 1969–1975, 7-year mean was 19 µg/L and the 1976–1979, 4-year mean was 17 µg/L (Table 4.3). TP gradually declined further after 1980, especially in the late 1990s, possibly due to climatic conditions that produced lower flushing rates (Figure 4.4).

Chl *a* decreased from a pre-diversion summer mean of 36 µg/L, in direct proportion to the decrease in TP. Although the lake approached a N-limiting condition prior to diversion, primarily because the ratio of N:P in sewage effluent is rather low (2:1 to 3:1), P was quickly reestablished as the limiting nutrient following diversion (Edmondson, 1970). Chl *a* reached a level of 7 µg/L by 1969 and remained a 7-year mean of 6 µg/L through 1975 (Table 4.3). Secchi transparency increased from a summer mean of 1 to 3.1 m during the same period. This represented some of the first direct evidence of the singular importance of P to algal control.

The lake had another marked improvement after 1975. Transparency more than doubled during the next 4 years to 6.9 m, while chl *a* declined by half to 3 µg/L (Table 4.3). The additional was attributed to *Daphnia* becoming the dominant zooplankter beginning in 1976 (Edmondson and Litt, 1982). *Daphnia* populations increased at that time apparently because *Neomysis mercedis*, a planktivore, decreased in the mid 1960s and blue-green algae (especially *Oscillatoria*) had markedly declined in relative importance by 1976. *Oscillatoria* interfered with the filtering process of *Daphnia* and reduce the efficiency of food consumption (Infante and Abella, 1985). The lake condition in the late 1970s of about 17 µg/L TP, 3 µg/L chl *a*, and nearly a 7 m SD was the result of both chemical and biological recovery. Chl *a* and transparency remained at similar levels during the 1990s, with summer means of 2.7 µg/L and 7.1 m, respectively (King County, 2002).

Lake Washington recovered so promptly and completely because of it's relatively great depth (64 m maximum, 37 m mean), fast renewal rate (0.4/yr), oxic hypolimnion, and relatively short history of enrichment. The large hypolimnetic volume and short period of enrichment (first signs observed in the early 1950s, Edmondson et al., 1956) prevented the hypolimnion from reaching anoxia. Thus, internal loading was insignificant.

TABLE 4.3
Characteristics of Five Lakes, Averaged over Indicated Years before and for Successive Periods following Diversion or Wastewater Treatment (P Removal; AWT)

Lake	$\bar{Z}$	r	L_{int}	Years pre/post		SD pre/post		TP pre/post		Chl *a* pre/post	
Washington[a]	37.0	0.40	No	4	7	1.0	3.1	64	19	36	6
					4		6.9		17		3
Sammamish[b]	18.0	0.55	Yes	2	5	3.2	3.4	33	27	5	7
					4		4.9		19		2.7
Norrviken[c]	5.4	1.2	Yes	2	6	0.7	0.7	NA	236	131	79
					5		1.1		115		45
Shagawa[d]	5.6	1.6	Yes	2	3			51	30	28	24
					18				35		
Søbygaard[e]	1.2	0.08	Yes	2	4		0.41	826	587		617

Note: $\bar{Z}$, mean depth, m; ρ, flushing rate, L/yr; Lint, internal loading; SD, Secchi transparency, m, as summer mean; TP, total phosphorus, μg/L, as annual mean; chl *a*, μg/L as summer mean.

Sources: [a]Edmondson, W.T. and J.R. Lehman. 1981. *Limnol. Oceanogr.* 26: 1–29; King County Dept. Nat. Res., Seattle, WA. [b]Welch, E.B., et al. 1980 *Water Res.* 14: 821–828. Welch, E.B., et al. 1986. In: *Lake Reservoir Management*. USEPA-440/5-84-001. pp. 493–497; King County Dept. Nat. Res., Seattle, WA. [c]Ahlgren, I. 1980. *Arch. Hydrobiol.* 89: 17–32; Sas, H. et al. 1989 *Lake Restoration by Reduction of Nutrient Loading: Expectations, Experiences, Extrapolation*. Academia-Verlag, Richarz, St. Augustine, Germany. [d]Larsen, D.P. et al. 1979. *Water Res.* 13: 1259–1272; Wilson, B. personal communication. [e]Søndergaard, M. et al. 1999. *Hydrobiologia* 408/409: 145–152; Søndergaard, M. et al. 2001 *Sci. World* 1: 427–442. From Cooke et al. 1993. With permission.

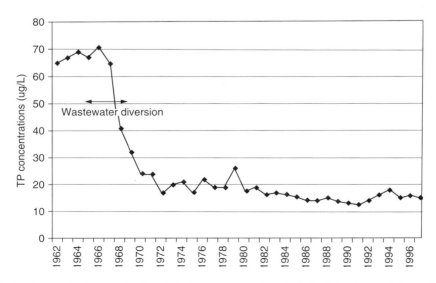

FIGURE 4.4 Changes in January 1 whole-lake TP concentrations in Lake Washington before, during, and after diversion of secondary treated wastewater. (King County, 2002, with early data from Edmondson, W.T. and J.R. Lehman. 1981. *Limnol. Oceanogr.* 26: 1–29.)

The dilution effect of the Cedar River on Lake Washington is another reason for the lake's fast recovery. That is apparent by examining TP inflow and expected lake concentrations from that and other sources. During 1995–2000, the Cedar contributed an average 57% of the water inflow annually, but only 25% of the TP load (Arhonditsis et al., 2003). That represents an annual average TP inflow concentration of 17 µg/L and an expected resulting lake concentration of only about 7 µg/L [TP_{inflow} $(1 - R)$], using the average TP retention coefficient R from Edmondson and Lehman (1981). Thus, if Lake Washington received only Cedar River water, its TP concentration would be only one half the current level. Respective inflow and expected lake concentrations from the remaining inputs averaged 71 and 28 µg/L. The Sammamish River contributes an inflow TP concentration of 82 µg/L with an expected lake concentration of 33 µg/L, more than double the current level. Without the high quality Cedar River inflow, the quality of Lake Washington would be many times poorer, given that 63% of its watershed is urbanized (Arhonditsis and Brett, personal communication).

This case demonstrates the advantage of treating a lake before it reaches an advanced state of eutrophy. Unfortunately, the fast, complete recovery of Lake Washington is atypical.

4.5 LAKE SAMMAMISH, WASHINGTON

The response of nearby Lake Sammamish to sewage and dairy plant effluent diversion in 1968 was slower than that of Lake Washington, but the eventual equilibrium TP concentration was similar in both lakes (Table 4.3). Although the decrease in external loading was not as great as that for Lake Washington (35% vs. 88%), flushing rates for the two lakes are similar, so the rate of TP decline should have been similar as well. While the external load reduction was less in Lake Sammamish, the lake was likewise not as enriched with a pre-diversion mean annual TP concentration of only 33 µg/L, about half that in Lake Washington.

The two principal differences between the two lakes accounting for their dissimilar response are: (1) Lake Sammamish has an anoxic hypolimnion from late summer through mid November when turnover occurs (its mean depth is half that of Lake Washington so its hypolimnetic volume, as well as its initial oxygen supply are much smaller), and (2) Lake Sammamish received its treated-sewage P load via its principal inflow stream, entering the stream about 3 km from the lake, whereas treated effluent was discharged directly into Lake Washington. These differences meant that: (1) Lake Sammamish had a significant internal loading of P, amounting to one third the total post-diversion loading (Welch et al., 1986), and (2) Lake Washington probably received a much greater fraction of its sewage P in a dissolved form, whereas sewage P released to Lake Sammamish had a greater opportunity to be converted to particulate P in the 3 km of stream between discharge and entrance to the lake. Most of the P load to Lake Sammamish entered during winter high flow and two-thirds to three-fourths of the P was particulate, which probably settled before spring and was thus unavailable for spring-summer algal uptake. This would make the lake more responsive to internal than external loading.

During the first 7 years following diversion, Lake Sammamish showed only modest signs of recovery (Welch et al., 1977; 1980). Mean annual whole-lake TP decreased less than 20% from 33 to 27 µg/L in response to a 35% decrease in external loading, while there was no change in summer chl *a* or transparency (Table 4.3). That was much less response than observed in the 18 European lakes where on average, lake TP declined 73% in response to an 82% decrease in external loading (Sas et al., 1989). If internal loading is included (one-third the total), the decrease in total loading (internal + external) was only 19%, about equal to the observed decrease in lake TP (Welch et al., 1986).

The lake's recovery had a subsequent phase, however. There was a delayed TP decline, starting in 1975, to an average of 19 µg/L in the late 1970s and it remained at about 18 µg/L during the early 1980s (1980–1984 mean; Table 4.3). That was followed by a gradual increase through 1997 (Figure

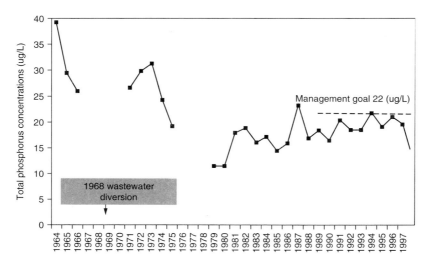

FIGURE 4.5 Mean, annual whole-lake concentration in Lake Sammamish during 1964–2002. Values for 1979 and 1980 were based on only four samples and there were no fall samples in 1981. TSP (total soluble phosphorus) data from 1964 to 1966 were corrected upward to TP by TP/TSP ratio of 1.2. Data for 1964–1966 from Metro; 1971–1975 from University of Washington; 1979–1997 from King County (Dept. Nat. Res., Seattle, WA).

4.5), thought to be caused by land use changes in the watershed, i.e., forest replaced by residences (Perkins, 1995). To preserve lake quality, King County set a whole-lake TP limit of 22 μg/L.

The decline of whole-lake TP from 27 μg/L in the early 1970s to 18 μg/L in the early 1980s was paralleled by a decrease in summer TP in the top 5 m from 20 to 10 μg/L. That accounted for the 50% decrease in summer chl a and an increase in summer transparency to nearly 5 m (Table 4.3). Delay of the TP decrease and recovery of the lake to a near oligotrophic state was apparently due largely to a decrease in anoxic sediment P release rate. The mean release rate was similar between 1964 to 1966 (pre-diversion) and 1971 to 1974 (post-diversion), being 6.1 ± 1.6 and 5.6 ± 3.2 mg/m^2 per day, respectively, but decreased to 2.5 ± 2.1 mg/m^2 per day during 1975, 1979, and 1981 to 1984, Welch et al., 1986). The reduced rate, determined *in situ* as the rate of increase in mean hypolimnetic P during the stratified period, was corroborated with *in vitro* rates for 1973 vs. 1984 and with interstitial P concentrations. While the year-to-year variability in release rate was considerable, rates for the 20 years after 1974 were usually lower, with three exceptions (Figure 4.5).

Oxygen conditions in the hypolimnion were reported to have changed in concert with the reduced sediment P release rate (Welch et al., 1986). However, further analysis shows that wide year-to-year fluctuations occurred in AHOD (± 100 mg/m^2-day) through 2001 with no significant trend since diversion (Chapter 3 for AHOD).

The lower lake TP concentration in the late 1970s and early 1980s may have been partly related to lower external loading resulting from generally lower stream flows. If external loading rates during 1982, 1983, and 1984 are combined with the reduced internal loading, the total loading (internal + external) in the early 1980s represents an ultimate post-diversion decrease of 36%, which is still actually less than the decrease in TP observed in the lake (45%). The subsequent gradual increase in TP, and then the decrease in the late 1990s show the effect of year-to-year inflow variability (Figure 4.5). Nevertheless, TP remained at about one-half the pre- and immediate post-diversion levels.

The principal cause for the later but substantial decrease in lake TP is probably a reduced internal loading. The results from Lake Sammamish, as from other lakes, demonstrate that internal loading decreases following diversion, even though the decrease may be slightly delayed (apparently

7 years in this case). A longer period of increased external loading, prior to diversion, would probably have resulted in greater resistance of internal loading to change.

Results from the 18 European lakes, discussed earlier, indicate that a TP content in surficial sediment of 1 mg/g dry matter may represent an approximate threshold, above which will perpetuate internal loading following diversion. There was no detectable change in TP per unit dry matter in Lake Sammamish surficial sediment; it remained rather uniform through the early 1980s at about 2 mg/g throughout the top 0.5 m. However, the equilibrium concentration of SRP in sediment pore water following anoxic incubation decreased from the early 1970s to 1980s, suggesting that the potential for sediment release declined, which was corroborated by reduced release rates (Welch et al., 1986; Figure 4.5).

4.6 LAKE NORRVIKEN, SWEDEN

The recovery of this lake following diversion in 1969 of 87% of its external loading from sewage and industrial waste was documented by Ahlgren (1977, 1979, 1980, 1988), and poses some contrasts with Lakes Washington and Sammamish. Although Lake Norrviken thermally stratifies, it is much shallower and was hypereutrophic before and after diversion (Table 4.3). Also, internal loading was more significant than in Lake Sammamish, averaging slightly more than external loading on an annual basis during 11 years since diversion, although internal was only about one eighth of external before diversion (data from Ahlgren in Sas et al., 1989).

Lake TP declined as predicted from simple dilution, decreasing from a fall overturn maximum of about 450 µg/L in 1970 to about 175 µg/L in 1975 (Figure 4.6). Apparently, internal loading did not buffer strongly against this "recovery by dilution" because internal was only about one eighth of external before diversion.

Summer TP (June–September) decreased from 260 to 98 µg/L during 1970–1975 and remained at about that level for the next 5 years. Chl a and Secchi transparency improved by 43% and 57%, respectively, during that same 5 years (Table 4.3). Transparency has been as great as 1.2 m and chl a as low as 36 µg/L (both in 1980, the last year of data). Although the lake was still hypereutrophic, its quality improved markedly and the diversion project was considered a success. Moreover, *Oscillatoria agardhii* no longer dominated the phytoplankton as a monoculture during the summer; other blue-greens, e.g., *Aphanizomenon, Anabaena, Microcystis* and *Gomphosphaeria* became important. The change in algal biomass may have resulted from N limitation, because N was diverted as well as P and the correlation of biomass with N was better than with P (Ahlgren, 1978).

Lake Norrviken is an example of a successful diversion even though trophic state may not have changed. Trophic state indices help communicate lake quality, but as this Norrviken example illustrates, the indices should not be used too rigidly to interpret restoration success. Lake Norrviken was classed by Cullen and Forsberg (1988) as a Type II, "reduction in P and chl a but insufficient to change trophic state," while Lake Washington was a Type I, and Lake Sammamish was a Type III (Table 4.1).

As with Lake Sammamish, internal loading in Lake Norrviken decreased after diversion (Ahlgren, 1977). TP in the sediment declined and the sediment release rate, determined by the rate of increase in the hypolimnion (as with Sammamish), decreased from 9.2 to 1.6 mg/m^2 per day, as indicated by the declining amplitude in TP concentration through about 1976 (Figure 4.6). However, the rate subsequently increased again through 1980. That this trend would probably reverse, and the sediments should once again retain P, is suggested from a longer data set for the upstream lake, Vallentunasjön (Figure 4.6). The annual amplitudes in TP in that lake declined and then increased as in Norrviken, but ultimately declined once again (Ahlgren, 1988). Vallentunasjön is shallow and does not thermally stratify; therefore the sediment-water interface is usually oxic. The poor P retention capacity of sediments (continued high internal loading) in that lake is probably due to the relatively low content of iron and large fraction of organic P; there is insufficient iron to complex the soluble P diffusing from the sediment (Lofgren and Boström, 1989).

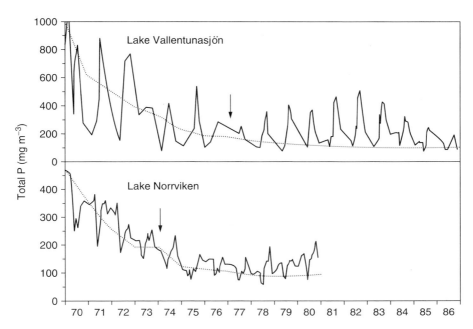

FIGURE 4.6 Response of lake TP concentration in Lakes Norrviken and Vallentunasjön, Sweden to wastewater diversion in 1970. The line with squares was calculated by simple dilution and the arrow represents three detention times. (From Ahlgren, I. 1988. In: G. Blvay (Ed.), *Eutrophication and Lake Restoration — Water Quality and Biological Impacts.* Thoron-les-Bains, France. pp. 79–97. With permission.)

4.7 SHAGAWA LAKE, MINNESOTA

This lake was studied thoroughly from 1971 through 1976 and demonstrates lake response to P-only removal from sewage. AWT was performed by precipitation with calcium carbonate. The lake improved but did not achieve expectations (Larsen et al., 1979).

The average annual, volume weighted TP declined from 51 μg/L before treatment, which began in 1973, to 30 μg/L 3 years following treatment, a 40% reduction (Table 4.3). Soluble P declined 80% to 4.5 μg/L, indicating that P became the more limiting nutrient. Average chl *a* during spring decreased by 50%, but little change was observed during summer (Table 4.3), except that bloom duration decreased.

Average annual TP should have declined to 12 μg/L in 1.5 years according to a first order, steady state model (Larsen et al., 1979), but the lake did not respond as expected due to summer internal loading from anoxic hypolimnetic sediments (Larsen et al., 1981). Net summer release of P averaged 5.3 mg/m^2 per day over the whole lake area. Profundal zone release was nearly double that rate. As much as 6 to 12 mg/m^2 per day could be transported vertically by summer turbulent diffusion, when steep P gradients developed across the thermocline. There is a relatively large area in Shagawa Lake (especially west basin) between 6 and 8 m where anoxia develops. That area is subject to P entrainment by wind mixing (Stauffer and Armstrong, 1986).

Stratified lakes with low ratios of mean depth to area have low resistance to mixing and may experience significant entrainment of hypolimnetic water at the edge of the hypolimnetic-metalimnetic interface through internal seiche activity. For example, Shagawa Lake's west basin has an OI (see Chapter 3) of only 2.28, well below the threshold value where stability begins to limit entrainment (Osgood, 1988). Thus, the internal P supply, available in the photic zone during summer, accounts for the continued high concentration of chl *a*.

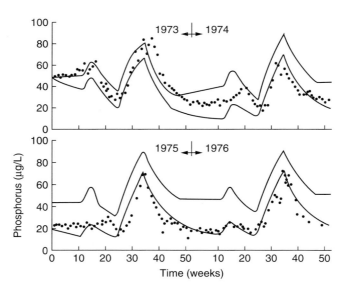

FIGURE 4.7 Comparison of predicted (lines) and observed (dots) total P concentrations in Shagawa Lake, MN as if treatment had not occurred (upper line) and after (lower lines) institution of P removal from sewage effluent. (Reprinted from Larsen, D.P. et al. 1979. *Water Res.* 13: 1259–1272. With permission from Pergamon Press Ltd., Oxford.)

The availability of internally loaded P to the photic zone in Lake Norrviken may be similar to that in Shagawa Lake, considering their similar mean depths (Table 4.3) and OIs (3.30 for Lake Norrviken). As depth increases, and wind fetch and speed are constant, availability of internally loaded P from anoxic hypolimnia (via the edge effect described above) should decrease. However, diffusion may be substantial and the OI may not always indicate availability, as discussed in Chapter 3. Availability of internal P in Lake Sammamish, with three times the mean depth (Table 4.3), but a larger area (OI = 3.98), should also be nearly as great as that in Lakes Norrviken or Shagawa. However, Sammamish is surrounded by rather steep terrain and strong winds and storms usually do not occur in western Washington during summer. A two-layered TP model that includes diffusion and entrainment showed that epilimnetic TP was not highly sensitive to internal loading (Perkins et al., 1997). Nevertheless, Lake Sammamish did show improved clarity and reduced chl *a* in the epilimnion in summer once internal loading declined.

The nonsteady state model for TP (Chapter 3) worked well for describing the early response of Shagawa Lake to reduced loading (Figure 4.7). Larsen et al. (1979) used a constant sedimentation rate (estimated during winter), and determined by calibration the gross internal loading rate that remained constant for the three post-diversion years of analysis. The constant internal loading is evident in Figure 4.7 by the similar rate of TP increase and maximum during summer even though TP overall declined during the 3 years. Internal loading should decline in Shagawa Lake, as in other lakes. However, the rate of decline is not easily predicted. Chapra and Canale's (1991) semi-empirical, long-term model includes hypolimnetic oxygen and sediment P burial, and was calibrated against Shagawa Lake data. By their model, internal loading is expected to decline gradually and require 80 years to reach 90% of the recovery to an equilibrium lake concentration of 12 μg/L.

Shagawa Lake continues to show resistance to further recovery. Annual mean TP content remained at ≥ 30 μg/L through 1994, a level that characterized the initial recovery (Figure 4.8; Wilson personal communication; Table 4.3). Transparency remained relatively unchanged during 1979 through 1996 with a summer mean of 2.06 m (1.68–2.62). Internal loading persisted, as evident by inflow TP exceeding outflow TP, with internal loading accounting for about 30% of the total loading 16 years after input reduction (Wilson and Musick, 1989). These observations are

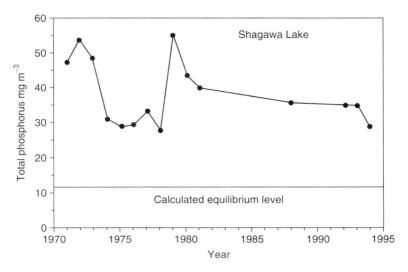

FIGURE 4.8 Average annual TP in Shagawa Lake, MN. (From Wilson, B. personal communication with early data from Larsen, D.P. et al. 1979. *Water Res*. 13: 1259–1272.)

consistent with the prediction of a long-term recovery by Chapra and Canale (1991). The lake has shown signs of further recovery in recent years; transparency averaged 2.95 m (2.68–3.35) during 1997–2003 (Wilson, personal communication).

4.8 MADISON LAKES, WISCONSIN

The Madison chain of lakes includes Lakes Mendota, Monona, Waubesa, and Kegonsa. Sewage effluent from the city of Madison was diverted from the lower two lakes, Waubesa and Kegonsa, in 1958. Prior to that, P from sewage raised the loading to the lower lakes from four- to sevenfold and amounted to 88% of the P loading to Lake Waubesa. Following diversion, winter SRP content in Waubesa decreased from about 500 to about 100 µg/L, which was approximated by a simple hydraulic washout model, i.e., dilution effect (Sonzogni and Lee, 1976). SRP content in Kegonsa also decreased according to simple washout, but not quite so quickly because P washed from Waubesa entered Kegonsa. Although there was little effect on algal biomass due to the continued high P, dominance changed from 99% *Microcystis* to a more mixed assemblage (Fitzgerald, 1964). The continued eutrophy in the lakes may have been responsible for transparency and oxygen content showing no significant changes through the early 1970s (Stewart, 1976).

Sewage effluents entering Lake Mendota from upstream communities were intercepted by the sewer district in 1971, and P loading to Lake Mendota was reduced by about 20% (Sonzogni and Lee, 1976). There was some decrease in SRP in Lake Mendota, with surface SRP dropping from about 100 µg/L in the mid 1970s to about 40 to 80 µg/L in the 1980s, although year-to-year variability was substantial (Lathrop, 1988a, b, 1990). Hypolimnetic SRP decreased about 20% since 1978. Since 1975, spring SRP in the lower lakes (Waubesa and Kegonsa) decreased from between 50 and 90 µg/L to near undetectable levels in the early 1980s (Lathrop, 1988a, b, 1990). This further decrease was attributed to a combination of sewage diversion from upstream of Lake Mendota and less spring runoff since 1977. In response to the reduced SRP, transparency increased and chl *a* decreased. However, SRP may again rise when normal runoff events resume, so that non-point source controls may be necessary to maintain the improved quality of the lakes (Lathrop, 1990).

Attempts to control non-point sources from Lake Mendota's 604 km^2 watershed from the mid 1970s through the 1980s were largely unsuccessful, due primarily to low funding and institutional problems (Lathrop et al., 1998). However, the lake was designated as a "Priority Watershed Project"

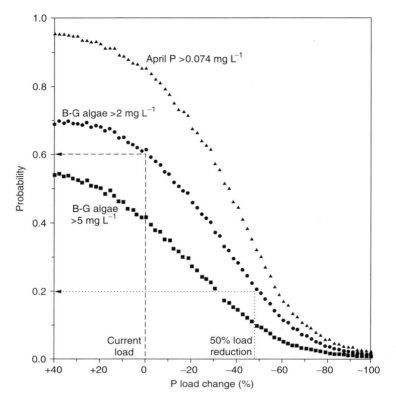

FIGURE 4.9 Probabilities of April P concentrations > 0.074 mg/L (triangles), blue-green algal concentrations > 2 mg/L (circles), and blue-green algal concentrations > 5 mg/L (squares) versus percentage change of the annual surface water P load to Lake Mendota. A scenario is shown for the probability of blue-green algal concentrations > 2 mg/L at the current P loading rate (0% change) and a 50% reduction in P loading. With no loading change, blue-green algae > 2 mg/L are predicted to occur on any given summer day with a probability of 0.6, or 3 out of 5 days over the course of many summers. With a 50% P load reduction, the probability drops to 0.2, or 1 out of 5 days over many summers. (Modified by R. Lathrop from Figure 5 in Lathrop, R.C. et al. 1998. *Can. J. Fish. Aquatic Sci.* 55: 1169–1178. With permission.)

in 1993, which was expected to increase funding and provide more focused control activities. To decide the level of non-point source loading control that was needed, a mass balance TP model was combined with a TP-blue-green algal probability model that produced alternative responses to TP load reduction (Lathrop et al., 1998; Figure 4.9). These model predictions incorporated the variability over a 21-year record to allow the estimates of the probability of lake conditions.

4.8 LAKE ZÜRICH, SWITZERLAND

Conventional wastewater treatment plant construction was initiated in 1955. Over a 10-year period beginning in about 1965, P removal was installed in all treatment plants within the immediate drainage to Lake Zürich. About 54% of the P load entering the lake prior to 1955 was removed by AWT. The inflow TP concentration decreased from 80 μg/L in 1976 to 47 μg/L in 1984. The lake responded positively as lake TP gradually declined from 93 to 54 μg/L over 13 years from 1974 to 1986 (Sas et al., 1989).

A substantial decrease in hypolimnetic oxygen deficit also occurred, with the greatest change observed in the 10 years following conventional wastewater treatment in 1955 (Shantz, 1977). Oxygen deficit averaged 27% less during 1971–1976 compared with 1953–1959.

Annual mean transparency increased about 50% between 1966 and 1975, compared with 1953–1965 (Shantz, 1977). Transparency rarely exceeded 6 m in the years before treatment, but after AWT installation it commonly reached 10 m. The greatest improvement in transparency occurred during winter and autumn. The lake is now clearer since AWT than before the turn of the century. The transparency increase was correlated with the disappearance of *Oscillatoria rubescens* blooms in the mid 1960s, although there was no subsequent decrease in algal biomass or increase in transparency since the mid 1970s (Sas et al., 1989).

4.9 LAKE SØBYGAARD, DENMARK

This lake's recovery was followed closely for 20 years (Søndergaard et al., 1999, 2001). It was heavily loaded with P (~ 30 g/m^2 per yr) for several decades prior to P removal through AWT, beginning in 1982, reducing loading by 80–90%. Phosphorus content in this shallow (1.2 m mean depth, 50 ha) lake was very high following input reduction (Table 4.3) and has not improved over the subsequent 14 years. Mean summer TP concentrations have ranged from 400 to1,000 µg/L and mean chl *a* from 130 to 840 µg/L.

While Lake Washington is an exceptional case for rapid and complete recovery to pre-enrichment conditions, Lake Søbygaard represents the extreme opposite. Recovery, or even improvement in quality, in this lake has not occurred due to the high net rates of internal loading, with maximum rates of 145 mg/m^2 per day and summer averages of 30–50 mg/m^2 per day, maintaining the high summer TP and chl concentrations (Søndergaard et al., 1999, 2001).

Sediment P profiles show that the transient period after input reduction may last for over 30 years before sediments begin retaining P and in-lake P reaches equilibrium with the input. Sediment P content declined over a depth of 25 cm since input reduction, although levels are still high (Figure 4.10). The change in profile concentrations and mass balance analyses show that the sediments lost 57 and 40 mg/m^2, respectively, over the 13-year period, suggesting another 15–20 years are necessary for equilibrium.

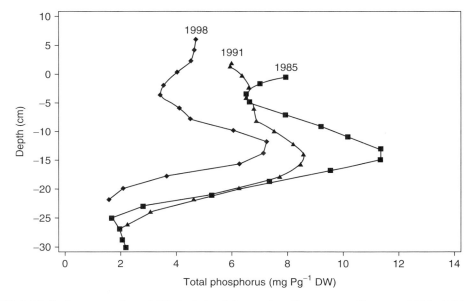

FIGURE 4.10 Sediment profiles of TP in Lake Søbygaard sediment over 13 years (1985, 1991, 1998) following input P reduction, based on sediment cores from a central location that were pooled into one sample. Profiles were adjusted to the 1985 level using a sedimentation rate of 0.6 cm/yr. (From Søndergaard, M. et al. 1999. *Hydrobiologia* 408/409: 145–152. With permission.)

Lake sediments with a P content of only 2 mg/g can have substantial rates of release. Recall that an observation from the 18-lake evaluation by Sas et al. (1989) was that a threshold for recovery might be around 1 mg/g, and this low may not be indicative of low internal loading in some cases (see earlier discussion this chapter). The principal mechanism for the high summer internal loading in this lake is photosynthetically caused high pHs of 10–11 (Søndergaard, 1988). High pH even occurs in pore water and may produce the high loosely sorbed P fraction of 2 mg/g during summer (Søndergaard, 1988).

4.10 COSTS

Effluent diversion costs vary greatly from site to site in relation to transport distance from the lake to the receiving water body or treatment facility. Costs for AWT per capita tend to be more equal from site to site, because cost should depend on wastewater volume, treatment chemicals, and sludge disposal. There should be no additional transport costs unless the wastewater was not treated previously and a new plant and sewer system is required.

Wastewater entered Lake Washington from eleven small secondary treatment plants surrounding the lake and was diverted by the Municipality of Metropolitan Seattle (Metro) during 1963 to 1967 (Edmondson and Lehman, 1981). Diversion involved interception of wastewater from the eleven plants and transporting it to a large primary treatment plant 3 km from the lake, and discharging primary effluent at depth in Puget Sound, a large, usually well-mixed estuary. This has not transferred the eutrophication problem to Puget Sound, because the frequent deep mixing causes light, rather than nutrients, to limit phytoplankton growth in Puget Sound. An upgrade of the treatment plant to secondary was completed, but it has not affected trophic state in the Sound.

Seattle Metro also diverted secondary treated sewage and dairy wastewater from Lake Sammamish in 1968. The collection system transported the wastewater 20 km to a secondary treatment plant that discharged the effluent to the Duwamish River, which after another 20 km enter the surface of Puget Sound. In 1988, that effluent was diverted from the river and piped directly to depth in the Sound.

Costs for diversion vary greatly. The 20-fold cost difference between the Lakes Washington and Sammamish diversions (Table 4.4) was from the marked difference in wastewater volumes. Based on lake area, Lake Sammamish was still fivefold less expensive in capital costs ($11,500 vs. $57,400/ha; 2002 U.S. dollars). Diversion projects involving nine lakes in Florida averaged about $95,000/ha (2002 dollars) in capital costs (Dierberg and Williams, 1989), considerably more than for either Lakes Sammamish or Washington. Diversion was by far the most costly of six techniques employed in the survey of 43 Florida lakes, amounting to 97% of the total expenditures.

The much higher per capita costs for the Lake Sammamish diversion are due to the smaller population served at the time of construction. This population has grown dramatically in the subsequent 35 years so that per capita cost is more in line with that for Lake Washington. Instituting diversion early minimized the non-monetary costs for these projects. Although Lake Washington was eutrophic, it had just reached that state and Lake Sammamish had only reached an upper mesotrophic state. Continued enrichment would probably have lengthened the recovery time (non-monetary cost), especially if Lake Washington's hypolimnion had become anoxic. Thus, the benefits accrued from the recovery of Lake Washington from eutrophy to near oligotrophy and the prevention of Lake Sammamish from becoming eutrophic, and in fact eventually recovering to near oligotrophy, were obtained at even a lower cost than indicated by the dollar value shown in Table 4.4.

Wastewater from domestic and industrial (yeast factory) sources was diverted from Lake Norrviken to Stockholm's treatment system, from which the effluent was discharged to the Baltic Sea. The capital cost of diversion from Lake Norrviken was about half that for Lake Washington (Table 4.4), but on an areal basis (about $760,000/ha; 2002 dollars), it was more expensive than any of the diversion projects mentioned. If all four lakes in the chain (Vallentunasjön, Norrviken,

TABLE 4.4
Estimated Costs for Diversion and Advanced Treatment of Sewage to Restore Five Lakes

Lake	Treatment	Year	Construction			Operation/yr	
			$ × 10^{-6}	$ × 10^{-3}/ha	$/Capita	$ × 10^{-3}	$/Capita
Washington[a]	Diversion	1967	94.9 (505)	41.6	171 (911)	2138	4
Sammamish[a]	Diversion	1968	4.5 (22.9)	8.3	370 (1880)	146	12
Norrviken[b]	Diversion	1969	44.5 (204)	550	106 (486)	6736	16
Shagawa[c]	AWT	1973	1.9 (7.6)	6.0	380 (1520)	389	77
Zürich[d]	AWT	1975	36.0 (119)	13.2	252 (835)	1500	115

Note: Conventional treatment included; PO_4 removal only is $2.5–$4.10 per capita; values in parentheses adjusted to 2002 dollars (USDL, 1987; Economic Indicators, 1986–2002; Lough, personal communication).

Sources: [a]Municipality of Metropolitan Seattle, G. Farris (personal communication); [b]Käppalaforbundet, Arsredovisning 1980, Lidingö, Sweden: Käppalaverket; [c]Vanderboom, S.A. et al. 1976. *Tertiary Treatment for Phosphorus Removal at Ely Minnesota AWT Plant, April 1973 through March 1974*. USEPA-600/2-76-082; R.M. Brice, USEPA (personal communication); [d]Shantz, F. 1977. In: *Lake Pollution Prevention by Eutrophication Control*. Proceedings of a seminar, Killarney, Ireland. pp. 131–139. From Cooke, et al., 1993. with permission.

Edssjön, and Oxundasjön) affected by diversion are considered, the areal cost becomes more in line with the Florida examples above.

AWT was installed at Ely, Minnesota, to remove P from sewage effluent entering Shagawa Lake. The treatment system was paid for by the U.S. Environmental Protection Agency as a test case to determine if AWT would alleviate the effects of eutrophication. Originally there was a two-stage treatment process of lime clarification and dual-media filtration. Effluent TP content was maintained at around 50 µg/L. The research nature of the project accounted in part for the relatively high treatment costs per capita (Vanderboom et al., 1976). The relatively high operating cost has necessitated changes in the process to one that is less burdensome to the local population. Currently, the effluent TP limit is 300 µg/L as an annual mean, and removal is attained with alum (Wilson and Musick, 1989).

The communities around Lake Zürich began constructing 17 wastewater treatment plants in about 1955. Between 1965 and 1975, all treatment plants, serving about 8,400 persons, were equipped with AWT (ferric chloride with activated sludge). The construction costs per capita for Shagawa Lake, serving 5,000, were about 50% higher than for Zürich (Table 4.4), while the per capita operating costs were much higher (by 50%) for Zürich than for the Minnesota plant. Capital costs per lake area were low for the two AWT cases (about $8,300 and 18,200/ha; 2002 dollars) compared to the three diversion projects. Although there are greater operating costs in dealing with chemicals and sludge for AWT, there may also be some hidden costs in exporting pollutants to another receiving water with diversion.

AWT usually involves chemical addition to the existing activated sludge tank, instead of building another (tertiary) unit. That results in greatly reduced capital costs, which would amount to about 20% of the capital plus 40-year operational cost for a 3,785 m^3/day (10 mgd) P removal plant. Operation for P removal is expected to cost about 25% of the total operational costs of a secondary wastewater treatment plant that includes P removal (WPCF, 1983).

4.11   IN-LAKE TREATMENT FOLLOWING DIVERSION

If internal loading of P is expected to retard the recovery following diversion or advanced treatment, then additional in-lake measures may be warranted to hasten recovery. Some recovery occurred in most lakes (even shallow ones) following a substantial reduction in external loading. However, post-treatment expectations of water quality were often not realized during the summer, when internal loading usually occurs. In some cases, in-lake treatment was applied soon after external controls were instituted while in others, additional treatment was instituted after it was clear that recovery was insufficient. Modeling techniques to predict long-term recovery, including sediment dynamics are available (Chapra and Canale, 1991; Pollman, personal communication). Therefore, the time necessary for TP to reach equilibrium may be predictable in many cases. However, use of a long-term response model is not commonplace. Assumptions in such a model regarding sediment P behavior have not been verified over the long term following input reduction. Therefore, in view of results in some lakes discussed above, it may be cost effective to observe lake response for 5 years or so before applying in-lake treatment. This is especially recommended for lakes deep enough to stratify. On the other hand, waiting out the lake's natural recovery may not be desirable and informing the public initially about the possible total costs, including an in-lake application, is an advisable course of action.

Two lakes discussed elsewhere in this book represent examples of applying in-lake treatments because recovery was slow in one case and anticipating a slow recovery in the other case. Sewage effluent was diverted from shallow Lake Trummen, near Vaxjo, Sweden in 1959. Because no improvement in quality was observed during the following 10 years, the top 1 m of sediment was dredged from most of the lake in 1970 to 1971. Removal of the rich layer of sediment, which included the top 30 cm, was sufficient to curtail internal loading and result in a dramatic recovery (Bjork, 1972, 1974). Lake Trummen represents the classic case for dredging in the world and details are presented in Chapter 20.

The recovery of West Twin Lake (reduction in whole-lake P), Ohio, was greatly accelerated by an alum treatment of the lake's hypolimnion (Cooke et al., 1977, 1978). Wastewater that was previously treated through individual on-site septic tank systems was collected and diverted away from both East and West Twin Lakes. Believing that the large internal loading from the anoxic hypolimnion would probably slow recovery, alum was added to upstream West Twin shortly after diversion was complete. Phosphorus content in the treated lake's hypolimnion declined quickly following the alum treatment and that single treatment remained effective for over 15 years. The improvement in lake quality, as evidenced by conditions in the epilimnion, was probably more related to diversion than to the control of internal loading, because of its moderately high mixing resistance (OI = 7.9). In the meantime, the P content and epilimnetic quality in downstream East Twin recovered, possibly from continual dilution with low-P water from West Twin, but more likely from diversion and a long term reduction in its internal loading (see Chapter 8). The details of the treated and untreated lake's response are covered in Chapter 8.

Diversion may not be practical because of cost or feasibility; there may not be a suitable receiving water body. In the case of stormwater, its diffuse nature may require complete sewering of the lake and the fraction of nutrient load removed may not be adequate to significantly improve the lake. In such cases, in-lake treatment without external controls being instituted first, may be deemed more cost effective. This occurred in several lakes in Washington State (Welch and Jacoby, 2001). The relative effect of external vs. internal controls on summer water quality can be estimated from a seasonal P model. For example, as mentioned in Chapter 3, if external loading from stormwater enters the lake primarily during the winter, but internal loading enters during summer and fall, internal loading will have more effect on summer lake P concentration than would otherwise be apparent from an annual P budget. Moreover, if epilimnetic and hypolimnetic P are separated in the model, the significance of internal loading on epilimnetic quality can be determined.

4.12 SUMMARY

The response of lakes and reservoirs to wastewater diversion or AWT has been varied, from lake P concentration being easily predictable from a Vollenweider type model with commensurate change in trophic state to those showing no change in trophic state, as indicated by algal biomass and transparency. The goal in diversion/AWT is to reduce lake P concentration and case histories show that mean concentration can be expected to decrease nearly in proportion to the reduction in inflow concentration on an annual basis. However, lake quality (algal biomass and transparency) is a function of the absolute equilibrium lake P concentration, not the proportional change in concentration. If P does not reach a limiting level, and there is some suggestion that SRP must reach 10 μg/L or less to limit algal biomass, then one cannot expect quality improvements following diversion or AWT.

Annual mean P has less meaning for lake quality in the temperate zone than does the summer mean. If internal loading from sediment release, whether oxic (thermally unstratified) or anoxic (stratified), continues following diversion/AWT, and the equilibrium P content (epilimnetic) in summer is still too high, the expected improvements in lake quality may not occur (e.g., Shagawa Lake). There is some evidence that internal loading eventually declines and quality improves (e.g., Lake Sammamish). However, there is also evidence that internal loading increases before it decreases (e.g., Lake Norrviken). Unfortunately, predicting the long-term behavior of sediment-water exchange is not routine. Current understanding of internal loading and lake response suggests that sediment P release will decline, so one option following diversion/AWT is to wait and see if the trend in P release and lake quality satisfies lake users (i.e., will they wait?). If the choice is to ensure lake quality improvement, and lake P is not expected to reach an algal-limiting level, assuming that the release rate stays constant (e.g., Shagawa Lake model), then an in-lake treatment to control sediment release should be instituted soon after external controls are in place.

REFERENCES

Ahlgren, I. 1977. Role of sediments in the process of recovery of a eutrophicated lake. In: H.L. Golterman (Ed.), *Interactions Between Sediments and Fresh Water.* Dr. W. Junk, The Hague, The Netherlands. pp. 372–377.

Ahlgren, I. 1978. Response of Lake Norrviken to reduced nutrient loading. *Verh. Int. Verein. Limnol.* 20: 846–850.

Ahlgren, I. 1979. Lake metabolism studies and results at the Institute of Limnology in Uppsala. *Arch. Hydrobiol. Beih.* 13: 10–30.

Ahlgren, I. 1980. A dilution model applied to a system of shallow eutrophic lakes after diversion of sewage effluents. *Arch. Hydrobiol.* 89: 17–32.

Ahlgren, I. 1988. Nutrient dynamics and trophic state response of two eutrophicated lakes after reduced nutrient loading. In: G. Blvay (Ed.), *Eutrophication and Lake Restoration — Water Quality and Biological Impacts.* Thoron-les-Bains, France. pp. 79–97.

Arhonditsis, G. Personal communication. Dept. Civil and Environmental Eng., University of Washington, Seattle.

Beklioglu, M., L. Carvalho and B. Moss. 1999. Rapid recovery of a shallow hypereutrophic lake following sewage effluent diversion: Lack of chemical resistance. *Hydrobiologia* 412: 5–15.

Bernhardt, H. 1981. Recent developments in the field of eutrophication prevention. *Z. Wasser Abwasser Forsch.* 17: 14–26.

Biggs, J.F.B. 2000. Eutrophication of streams and rivers: Dissolved nutrient-chlorophyll relationships for benthic algae. *J. North Am. Benthol. Soc.* 19: 17–31.

Björk, S. 1972. Ecosystem studies in connection with the restoration of lakes. *Verh. Int. Verein. Limnol.* 18: 379–387.

Björk, S. 1974. *European Lake Rehabilitation Activities.* Inst. Limnol. Rept. University of Lund, Sweden.

Brett, M.T. Personal communication. Dept. Civil and Environmental Eng., University of Washington, Seattle.

Carlson, R.E. 1977. A trophic state index for lakes. *Limnol. Oceanogr.* 22: 361–368.

Chapra, S.C. and R.P. Canale. 1991. Long-term phenomenological model of phosphorus and oxygen for stratified lakes. *Water Res.* 25: 707–715.

Choulik, O. and T.R. Moore. 1992. Response of a subarctic lake chain to reduced sewage loading. *Can. J. Fish. Aquatic Sci.* 49: 1236–1245.

Cooke, G.D., M.R. McComas, D.W. Waller and R.H. Kennedy. 1977. The occurrence of internal phosphorus loading in two small, eutrophic, glacial lakes in northeastern Ohio. *Hydrobiologia* 56: 129–135.

Cooke, G.D., R.T. Heath, R.H. Kennedy and M.R. McComas. 1978. *The Effect of Sewage Diversion and Aluminum Sulfate Application on Two Eutrophic Lakes.* USEPA-600/3-78-033.

Cooke, G.D., E.B. Welch, S.A. Peterson and P.R. Newroth. 1993. *Restoration and Management of lakes and Reservoirs,* 2nd ed. CRC Press, Boca Raton, FL.

Cullen, P. and C. Forsberg. 1988. Experiences with reducing point sources of phosphorus to lakes. *Hydrobiologia* 170: 321–336.

Dierberg, F.E. and V.P. Williams. 1989. Lake management techniques in Florida, U.S.: Costs and water quality effects. *Environ. Manage.* 13: 729–742.

Edmondson, W.T. 1970. Phosphorus, nitrogen, and algae in Lake Washington after diversion of sewage. *Science* 169: 690–691.

Edmondson, W.T. 1978. *Trophic Equilibrium of Lake Washington.* USEPA-600/3-77-087.

Edmondson, W.T. 1994. Sixty years of Lake Washington: A curriculum vitae. *Lake and Reservoir Manage.* 10: 75–84.

Edmondson, W.T. and J.R. Lehman. 1981. The effect of changes in the nutrient income on the condition of Lake Washington. *Limnol. Oceanogr.* 26: 1–29.

Edmondson, W.T. and A.H. Litt. 1982. *Daphnia* in Lake Washington. *Limnol. Oceanogr.* 27: 272–293.

Edmondson, W.T., G.C. Anderson and D.R. Peterson. 1956. Artificial eutrophication of Lake Washington. *Limnol. Oceanogr.* 1: 47–53.

Fitzgerald, G.P. 1964. In: D.F. Jackson (Ed.), *The Biotic Relationships within Water Blooms, in Algae and Man.* Plenum Press, New York, pp. 300–306.

Infante, A. and S.E.B. Abella. 1985. Inhibition of *Daphnia* by *Oscillatoria* in Lake Washington. *Limnol. Oceanogr.* 30: 1046–1052.

Istvánovics, V. and L. Somlyódy. 2001. Factors influencing lake recovery from eutrophication – the case of Basin 1 of Lake Balaton. *Water Res.* 35: 729–735.

Istvánovics, V., L. Somlyódy and A Clement. 2002. Cyanobacteria-mediated internal eutrophication in shallow Lake Balaton after load reduction. *Water Res.* 36: 3314–3322.

Jensen, H.S., P. Kristensen, E Jeppesen and A. Skytthe. 1992. Iron-phosphorus ratio in surface sediments as an indicator of phosphate release from aerobic sediments in shallow lakes. *Hydrobiologia* 235/236: 731–743.

Jeppesen, E., J.P. Jensen and M. Søndergaard. 2002. Response of phytoplankton, zooplankton and fish to re-oligotrophication: An 11 year study of 23 Danish lakes. *Aquatic Ecosys. Health Manage.* 5: 31–43.

King County. 2002. Lake Washington Existing Conditions Report. King County Dept. Nat. Res. Div. Seattle, WA.

Larsen, D.P., J. Van Sickle, K.W. Malueg and P.D. Smith. 1979. The effect of wastewater phosphorus removal on Shagawa Lake, Minnesota: Phosphorus supplies, lake phosphorus and chlorophyll *a. Water Res.* 13: 1259–1272.

Larsen, D.P., D.W. Schultz and K.W. Malueg. 1981. Summer internal phosphorus supplies in Shagawa Lake, Minnesota. *Limnol. Oceanogr.* 26: 740–753.

Lathrop, R.C. 1988a. Phosphorus Trends in the Yahara Lakes since the mid-1960s. Research Management Findings No. 11. Wisconsin Dept. Nat. Res., Madison.

Lathrop, R.C. 1988b. Trends in Summer Phosphorus, Chlorophyll and Water Clarity in the Yahara Lakes, 1976–1988. Research Management Findings No. 17. Wisconsin Dept. Nat. Res., Madison.

Lathrop, R.C. 1990. Response of Lake Mendota (Wisconsin) to decreased phosphorus loadings and the effect on downstream lakes. *Ver. Int. Verein. Limnol.* 24: 457–463.

Lathrop, R.C., S.R. Carpenter, C.A. Stow, P.A. Soranno and J.C. Panuska. 1998. Phosphorus loading reductions needed to control blue-green algal blooms in Lake Mendota. *Can. J. Fish. Aquatic Sci.* 55: 1169–1178.

Löfgren, S. and B. Boström. 1989. Interstitial water concentrations of phosphorus, iron, and manganese in a shallow, eutrophic Swedish lake — implications for phosphorus cycling. *Water Res.* 23: 1115–1125.

Lough, T. Personal communication. Dept. of Sociology, Sonoma State University.

Marsden, M.W. 1989. Lake restoration by reducing external phosphorus loading: The influence of sediment phosphorus release. *Freshwater Biol.* 21: 139–162.

Nürnberg, G.K. 1988. Prediction of phosphorus release rates from total and reductant-soluble phosphorus in anoxic lake sediments. *Can. J. Fish. Aquatic Sci.* 45: 453–462.

Osgood, R.A. 1988. Lake mixes and internal phosphorus dynamics. *Arch. Hydrobiol.* 113: 629–638.

Perkins, W.W. 1995. Lake Sammamish Total Phosphorus Model. King County Surface Water Management, Seattle, WA.

Perkins, W.W., E.B. Welch, J. Frodge and T. Hubbard. 1997. A zero degree of freedom total phosphorus model: 2. Application to Lake Sammamish, Washington. *Lake and Reservoir Manage.* 13: 131–141.

Pollman, C.D. Personal communication. Tetra Tech, Inc., Gainsville, FL.

Rutherford, J.C., R.D. Pridmore and E. White. 1989. Management of phosphorus and nitrogen inputs to Lake Rotorua, New Zealand. *J. Water Res. Planning Manage.* 115: 431–439.

Ryding, S.O. and C. Forsberg. 1977. Short-term load–response relationships in shallow, polluted lakes. In: J. Barca and L.R. Mur (Eds.), *Hypereutrophic Ecosystems.* W. Junk, The Hague, The Netherlands. pp. 95–103.

Sas, H., I. Ahlgren, H. Bernhardt, B. Boström, J. Clasen, C. Forsberg, D. Imboden, L. Kamp-Nielson, L. Mur, N. de Oude, C. Reynolds, H. Schreurs, K. Seip, U. Sommer and S. Vermij. 1989. *Lake Restoration by Reduction of Nutrient Loading: Expectations, Experiences, Extrapolation.* Academia-Verlag, Richarz, St. Augustine, Germany.

Sawyer, C.N. 1947. Fertilization of lakes by agricultural and urban drainage. *J. N. Engl. Water Works Assoc.* 61: 109–127.

Shantz, F. 1977. Effects of wastewater treatment on Lake Zurich. In: W.K. Downey and G. Ni Vid (Eds.), *Lake Pollution Prevention by Eutrophication Control.* Proceedings of a seminar, Killarney, Ireland. pp. 131–139.

Søndergaard, M. 1988. Seasonal variations in the loosely sorbed phosphorus fraction of the sediment of a shallow and hypereutrophic lake. *Environ. Geol. Water Sci.* 11: 115–121.

Søndergaard, M., J.P. Jensen and E. Jeppesen. 1999. Internal phosphorus loading in shallow Danish lakes. *Hydrobiologia* 408/409: 145–152.

Søndergaard, M., J.P. Jensen and E. Jeppesen. 2001. Retention and internal loading of phosphorus in shallow, eutrophic lakes. *Sci. World* 1: 427–442.

Sonzogni, W.C. and G.F. Lee. 1976. Diversion of wastewaters from Madison lakes. *J. Environ. Eng. Div. ASCE* 100: 153–170.

Stauffer, R.E. and D.E. Armstrong. 1986. Cycling of iron, manganese, silica, phosphorus, calcium and potassium in two stratified basins of Shagawa Lake, Minnesota. *Geochim. Cosmochim. Acta* 50: 215–229.

Stewart, K.M. 1976. Oxygen deficits, clarity and eutrophication in some Madison lakes. *Int. Rev. ges. Hydrobiol.* 61: 563–579.

USDL. 1987 Handbook of Basic Economic Statistics. U.S. Dept. of Labor, Bureau of Labor Statistics, 41, 1. pp. 99–100.

Uttormark, P.D. and M.L. Hutchins. 1980. Input/output models as decision aids for lake restoration. *Water Res. Bull.* 16: 494–500.

Vanderboom, S.A., J.D. Pastika, J.W. Sheehy and F.L. Evans. 1976. Tertiary Treatment for Phosphorus Removal at Ely Minnesota AWT Plant, April 1973 through March 1974. USEPA-600/2-76-082.

WPCF. 1983. Water Pollution Control Federation. Nutrient Control. Manual of Practice FD-7. Facilities Design, Washington, DC.

Welch, E.B. 1977. Nutrient Diversion: Resulting Lake Trophic State and Phosphorus Dynamics. Ecol. Res. Ser. USEPA-600/3-88-003, 91 pp.

Welch, E.B. and G.D. Cooke. 1995. Internal phosphorus loading in shallow lakes: Importance and control. *Lake and Reservoir Manage.* 11: 273–281.

Welch, E.B. and J.M. Jacoby. 2001. On determining the principle source of phosphorus causing summer algal blooms in western Washington lakes. *Lake and Reservoir Manage.* 17: 55–65.

Welch, E.B., C.A. Rock, R.C. Howe and M.A. Perkins. 1980. Lake Sammamish response to wastewater diversion and increasing urban runoff. *Water Res.* 14: 821–828.

Welch, E.B., K.L. Carlson and M.V. Brenner. 1984. Control of algal biomass by inflow nitrogen. In: *Lake Reservoir Management.* USEPA-440/5-84-001. pp. 493–497.

Welch, E.B., D.E. Spyridakis, J.I. Shuster and R.R. Horner. 1986. Declining lake sediment phosphorus release and oxygen deficit following wastewater diversion. *J. Water Pollut. Cont. Fed.* 58: 92–96.

Wilander, A. and G. Persson. 2001. Recovery from eutrophication: Experiences of reduced phosphorus input to the four largest lakes of Sweden. *Ambio* 30: 475–485.

Wilson, B. Personal communication. Minnesota Pollut. Cont. Agency, Minneapolis.

Wilson, B. and T.A. Musick. 1989. Lake assessment program 1988, Shagawa Lake, St. Louis County, Minnesota. Minnesota Poll. Cont. Agency, unpublished manuscript.

5 Lake and Reservoir Protection From Non-Point Pollution

5.1 INTRODUCTION

The major sources of nutrients and organic matter to streams, lakes, and reservoirs in North America and Europe were believed to be "point" sources such as wastewater treatment plant (WWTP) outfalls. These have been greatly upgraded (Welch, 1992), leading to water quality improvement in some lakes (e.g., Lake Washington) because WWTP discharges were their dominant nutrient sources. For many lakes, non-point or diffuse nutrient loading, both internal and external to the lake, is at least as significant as point source loading. This source is difficult to assess and control (Line et al., 1999), and water quality in many lakes has not improved rapidly following diversion or treatment of point sources (Chapter 4). The purposes of this chapter are to describe the origins and nature of non-point loading to streams, lakes, and reservoirs, and to discuss certain methods for managing it.

Urban and agricultural activities are the major non-point sources of silt and nutrients to streams and ultimately to lakes and reservoirs. Loading from these activities is increasing as urban areas expand, food production (especially confined animal operations or CAFOs) increases, and undeveloped land is drained, deforested, tilled, or developed, and stored soil nutrients are released. These land uses in the watershed are good predictors of reservoir and lake productivity. More quantitative indices, such as the drainage ratio (drainage area to lake volume) and the cropland area: livestock density ratio (Pinel-Alloul et al., 2002; Knoll et al., 2003), are being developed and will become more useful with more data.

Agriculture is the primary source of non-point loading through erosion of nutrient-rich soil and from livestock activities, and also is the largest user of fresh water (Novotny, 1999). Demands to increase agricultural yields with fertilizer and manure applications have led to soil nutrient surpluses. For example, the average net gain of phosphorus (P) in U.S. agricultural soils is 26 kg P/ha per year (Carpenter et al., 1998). In Europe, average net gains are higher in some areas (e.g., > 50 kg P/ha per year in The Netherlands), and average 17 kg P/ha per year for general cropping and 24 kg P/ha per year for dairy operations (Haygarth, 1997). Surplus soil P is the basis of non-point runoff, with 3–20% of that applied reaching surface waters (Caraco, 1995).

Soil erosion is a primary mechanism for nutrient transport and for establishing shallow, nutrient-rich littoral zone soils that support macrophyte growth. The average annual soil loss for continuous corn production, for example, has been about 40 metric tons/ha (Brown and Wolf, 1984). CAFO's produce massive quantities of untreated manure that may be discharged directly to water, or added to soils as fertilizer and as a means of waste disposal. Runoff from fields, especially fields treated with manure, is high in biologically available P and may easily reach surface waters. The P load defecated by one cow is equivalent to 18–20 humans, and P concentrations in feedlot runoff may exceed 300 mg P/L (vs < 5 mg P/L in untreated human sewage outfalls) (Novotny, 1999).

Urban runoff, though somewhat less significant than agricultural runoff, is also a large source of nutrients to fresh water. Both urban and agricultural runoff have higher peak discharge and flow volumes than undisturbed areas, although soil type, percent impervious area, climate, and physiography influence these variables. The urban runoff from Madison, Wisconsin may be typical of a U.S. city. In residential areas, the highest runoff P concentrations were from lawns (geometric mean total P (TP) of 2.67 mg P/L). Although lawns produced a relatively low runoff volume, their P loads were relatively large due to the high P concentrations. In residential areas, feeder streets provided the dominant TP and soluble reactive P (SRP) loads, whereas in industrial areas, lawns yielded the highest loads. Streets and parking lots were identified as critical source areas, and lawns were critical areas when runoff volumes became large (Bannerman et al., 1993). Urban runoff also adds bacteria, silt, toxins, and BOD-demanding materials (USEPA, 1993).

Land management procedures, generally known as "best management practices" (BMPs) are the primary methods to protect surface waters from non-point loading, and include conservation tillage, terracing and contour plowing, street sweeping, elimination of combined sewer systems, revised residential development operations, and even vegetarianism (e.g., Novotny and Olem, 1994; Fox, 1999; Sharpley et al., 2000).

Structural and chemical BMPs to protect lakes are effective when correctly designed and maintained. These include stream P precipitation, pond-wetland treatment systems, soil treatments, rain gardens, and riparian repair. This chapter examines their design, effectiveness, and problems, but questions remain about all of them, including long-term cost-effectiveness.

Properly designed and maintained BMPs can be effective, but they are not panaceas and are not substitutes for revised land uses. Humans are becoming more and more urban, producing more and more impermeable areas with associated high runoff volumes, and untreated non-point wastes. In the U.S., the rate of paving is 168,000 ha/yr (Gardner, 1996). Affluent populations are living higher on the food chain, leading to greater production of grain to feed livestock in feedlots, and the seemingly inevitable increased consumption and pollution of fresh water (Brown, 1995: Brown and Kane, 1994). In 1990, the U.S. led the world in meat consumption (12 kg carcass weight/cap per year), and 70% of U.S. and 57% of European Union grain production (often row-crop agriculture that produces high silt and nutrient losses to water) went to livestock (Durning and Brough, 1991). Another continuing trend is the clearing of stream and lakeshore riparian areas for farms and lawns, leading to large transfers of silt and nutrients to fresh water. These trends, linked with the remaining point sources of pollution, suggest that there is a growing issue of attainability regarding fresh water quality.

The following sections provide an introduction to the problems, methods, and results of some procedures used to protect lakes and reservoirs from non-point pollution. Most of these procedures are "ecological engineering," an emerging discipline (Gattie and Mitsch, 2003), and a concept pioneered, in part, by Eugene and Howard Odum (Mitsch, 2003).

We do not consider in this text the very significant and growing problems of non-point pollution from dry and wet deposition of atmospheric materials such as mercury.

5.2 IN-STREAM PHOSPHORUS REMOVAL

Lund (1955) may have been the first to suggest that P removal from streams, or from the lake water column, could lower algal production. Lund stated (p. 93): "It would be interesting to know whether treatment with aluminum sulphate, either of one or more inflows or the reservoir water itself, is a practical proposition." Alum treatments of lake sediments are now common (Chapter 8). Stream treatments are more difficult and expensive because they must be continuous as long as the stream has high nutrient concentrations.

Cooke and Carlson (1986) applied alum directly to the Cuyahoga River, just above a water supply reservoir for Akron, Ohio. Application was continuous, using a manifold spanning the river, with dose flow-proportioned to maintain a river concentration of 1–2 mg Al/L. In 1985, 50–60% of SRP was removed, but TP loading to the reservoir was not lowered significantly. Floc below the manifold built up rapidly, and benthos 60 m below the manifold was eliminated by low pH. In 1986, compressed air was continuously injected at the application site. This prevented floc build-up, pH did not fall, and benthic invertebrate mortality was less (Barbiero et al., 1988). SRP was removed but P loading remained high and algal blooms continued. This crude interception system failed because floc was not produced and contained in a separate structure to protect benthos, and because the dose was too low for sufficient P removal.

Harper et al. (1983) may have been the first to devise a system to treat stormwater inflows with alum. The lower volume and duration of storm flow (versus river flow) allowed treatment of the entire discharge. Harper's early system led to development of a more sophisticated system with sonic flow meters and variable speed pumps that automatically injected alum at a flow-proportioned rate, based on jar tests for dose determination. The floc was discharged to the lake, providing sediment P inactivation, apparently without a significant floc build-up after three years of operation. The system reduced P loading and lake TP fell from > 200 µg P/L to about 25 µg P/L. Algal biomass decreased, and transparency, macrophyte biomass, and dissolved oxygen increased. The USEPA 7 day Chronic Larval Survival Growth Test on fathead minnows (*Pimephales promelas*) demonstrated no chronic toxicity of the alum-treated stormwater as long as pH remained at pH 6.0–6.5. High mortality was evident at pH 7.5 in this low alkalinity system. Floc disposal in the lake was a problem solved by collecting floc in a separate basin, and drying it. The floc is a Grade 1 wastewater sludge that can be disposed of via land application (Harper, 1990).

Ferric iron has been successfully used to remove P, metals, and organics from inflows to drinking water supplies in the U.S., U.K., and The Netherlands. An iron system was established to improve raw water quality of the Amsterdam Rhine Canal and Bethune Polder before their discharge into Lake Loenderveen, part of the water supply of Amsterdam, The Netherlands. The system has been in operation since 1984. Water is treated with $FeCl_3$ (7 mg Fe/L) and detained in a settling basin (mean residence time of 4 h) before it enters the lake. When P content of the raw water is very high, two in-line coagulation and settling systems are used. The basins store floc, which is routinely removed with a hydraulic dredge to drying fields. The Loosdrecht Lakes receive a similar treatment. The process is highly effective, and little final treatment in the potable water supply plant is needed (van der Veen et al., 1987).

Foxcote Reservoir (UK) is a pump-storage water supply. Its nutrient-rich inflow was treated with $Fe_2(SO_4)_3$ to control the algal blooms that had closed the reservoir as a water supply for up to 6 months yearly. Ferric sulfate was injected into the pipeline at an iron-ortho P ratio of 10:1, with a goal of reducing influent P to 10 µg P/L. This was achieved, but internal P loading continued for another two years before it was controlled, apparently by the added iron. Algal blooms were sharply reduced, but macrophytes and mats of filamentous green and blue-green algae appeared as water clarity increased, leading to new taste and odor events. Nevertheless, the treatment was successful because the reservoir is a more reliable water source. The polymictic nature of the reservoir may be a factor in maintaining the sediment iron floc in the oxidized state (Young et al., 1988).

St. Paul, Minnesota withdraws its untreated potable water from Vadnais Lake, a lake that is part of a system of 12 lakes receiving most of their water from the Mississippi River. Cyanobacteria blooms were common, and finished water had severe taste and odor. High silicon source water to the lake (to promote diatom growth), treated with $FeCl_3$, was used (Walker et al., 1989). Laboratory tests demonstrated high ortho-P removal at a dose of 50 µg Fe/L. The added iron also enriched lake sediments, an effect maintained by adding more iron (100 kg Fe/day) through the hypolimnetic aerators in Vadnais Lake. Internal P loading declined because the oxygen-rich hypolimnion maintained iron in an oxidized state. These combined treatments led to improved raw water and lower treatment costs.

Lime ($Ca(OH)_2$) has been suggested as a P precipitant in streams. Diaz et al. (1994) found that P removal was minimal at calcium concentrations less than 50 mg Ca/L and a pH < 8.0. With a dose of 100 mg Ca/L and pH 9.0, up to 76% of P was precipitated. Calcium salts are unlikely to be effective for stream treatments because Ca–P complexes readily solubilize at pH < 8.0, a value often reached during nighttime in many streams. A pH > 9.0 could be toxic.

The most effective P interception system has been the "phosphorus elimination plant" (PEP) concept, first proposed and developed by Bernhardt (1980) for Wahnbach Reservoir, the water

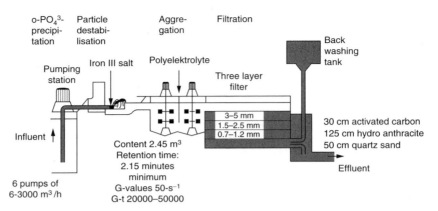

FIGURE 5.1 Principle of the direct-filtration with controlled energy input, "Wahnbach System." (From Bernhardt, H., 1980. *Restoration of Lakes and Inland Waters.* USEPA 440/5-81-010. pp. 272–277.)

supply for Bonn, Germany (Figure 5.1). A pre-reservoir (500,000 m^3) is used as a detention basin and then river water enters the PEP and is treated with 4–10 mg Fe/L (ferric) at pH 6.0–7.0. Treatment with a cationic polyelectrolyte follows and then water is filtered through activated carbon, hydroanthracite, and quartz sand. The Wahnbach PEP has a maximum flow-through rate of 5 m^3/s (79,000 gallons/min), or 5 times the average river flow. The average PEP effluent concentration discharged to Wahnbach Reservoir is 5 µg P/L. Algal blooms and dissolved organic matter (possible trihalomethane precursors) decreased dramatically. The reservoir does not have significant internal P loading (Clasen and Bernhardt, 1987).

At least three other German lakes and water supplies have a PEP (Klein, 1988; Chorus and Wesseler, 1988; Heinzmann and Chorus, 1994; Heinzmann, 1998). These plants are smaller than Wahnbach's, but as effective. The Lake Tegel PEP, the water supply for 100,000 Berliners, has a maximal discharge of 3 m^3/s. It was built for about $333 million (2002 U.S. dollars), with an annual operational cost of about 10% of construction costs. Lake Tegel's TP fell from 750 to 60 µg P/L, and costs to water users for water treatment were lower. Internal P loading in the lake was not a factor (Heinzmann and Chorus, 1994).

Effective chemical interception of P for water supply reservoirs is therefore feasible. There is no technical reason why this procedure could not be applied to recreational lakes and reservoirs.

5.3 NON-POINT NUTRIENT SOURCE CONTROLS: INTRODUCTION

Successful protection of lakes and reservoirs from non-point external loading may appear to be very difficult, especially when drainage area greatly exceeds lake area and there are many sources of potential soil and nutrient loss. Nevertheless, there are several methods with great potential to significantly lower non-point loading of silt and nutrients. These methods all require work in the drainage area itself, meaning that lake managers often have to become land managers and terrestrial ecologists as well.

The Soil Test Phosphorus concentration (STP) (Mehlich, 1984) is a common way to identify a high P source area. Mehlich-3 is one of several methods of extracting and determining P in soil. There is a strong positive relationship between STP and dissolved and TP in runoff water from unfertilized fields. Runoff P concentrations (mostly as dissolved P, the form assimilated by plants) increase greatly in fields receiving fertilizer or manure, and are not related to STP (Sharpley et al., 2001b) (Figure 5.2).

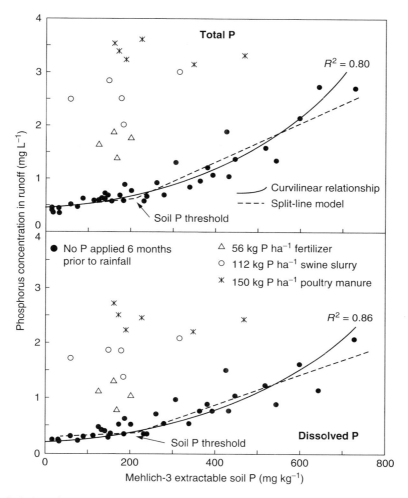

FIGURE 5.2 Relationship between the concentration of dissolved and total P in surface runoff and Mehlich-3 extractable soil P concentration for sites in fields where no P has been applied in the last 6 months and where fertilizer or manure had been applied within 3 weeks of rainfall in FD-36 watershed. Regression equations and corresponding coefficients apply only to plots not having received P in the last 6 months. (From Sharpley, A.N. et al. 2001b. *J. Environ. Qual.* 30: 2026–2036. With permission.)

Not all agricultural or urban areas, even those with apparent intense land use and high STP, are significant P sources to lakes. Gburek et al. (2000), Heathwaite et al. (2000), and Sharpley et al. (2001a, 2003) proposed a modified P index (PI) to identify watershed areas with potential to affect stream P concentrations via runoff. The original PI (Lemunyon and Gilbert, 1993) was developed as a screening tool to evaluate edge-of-the-field P loss, but it did not completely address whether or not the site in question was hydrologically connected to a water body. Most of the P in runoff can come from a relatively small watershed area (Pionke et al., 1997). The modified PI (review by Sharpley et al., 2003) identifies critical P source areas (CSAs), or areas where there is a coincidence of high STP and a high probability that soil and dissolved P will be transported during a runoff event. CSAs should receive the most attention for implementing BMPs.

The relationship between dissolved P, TP, and the PI (Figure 5.3) illustrates the effectiveness of the PI in predicting potential impacts of fertilization or manure application on streams. The PI is far superior to STP alone, as illustrated in Figure 5.2. STP was predictive only when no fertilizer or manure had been applied in the 6 months prior to rainfall (Sharpley et al., 2001b).

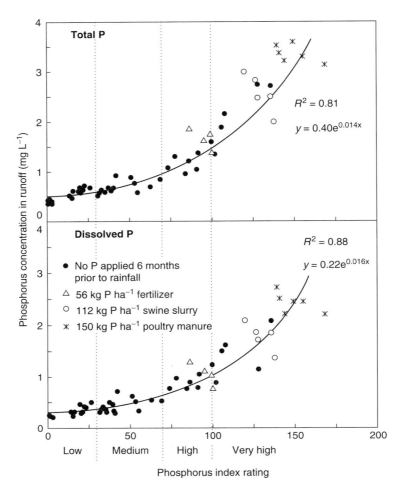

FIGURE 5.3 Relationship between the concentration of dissolved and total P in surface runoff and the P index rating for sites in fields where no P had been applied within the last 6 months and where fertilizer or manure had been applied within 3 weeks of rainfall in FD-36 watershed. (From Sharpley, A.N. et al. 2001b. *J. Environ. Qual.* 30: 2026–2036. With permission.)

The modified PI (Gburek et al., 2000) is useful to lake managers. It provides a watershed-scale evaluation of non-point P sources by first separating source characteristics (e.g., STP, fertilizer application rates), and transport characteristics (e.g., soil erosion, distance to water), weighting their individual importance, and then combining them into an index number that indicates the potential of the site to add P to streams (Figure 5.3). For example, a site with low transport characteristics, but high P source characteristics, might have only medium pollution potential. This approach allows expensive BMPs to be targeted to the most vulnerable sites.

The Pennsylvania modified PI (Sharpley et al., 2001b; Kogelmann et al., 2004) was applied to a small watershed that was 50% soybeans, corn, or wheat, 20% pasture, and 30% woodland (McDowell et al., 2001). Fields were fertilized and/or received poultry or hog manure. Application of the PI demonstrated that only 6% of the watershed (along the stream corridor) had high risk of P transport. These areas had high STP, manure applications, and soil erosion. An additional 17% of the watershed had risk high enough to warrant P management. Other approaches to managing P loss to the stream, such as use of STP only, would have targeted 80–90% of the watershed and may not have produced cost-effective controls of P transport. The PI and lake TP concentrations are correlated ($r^2 = 0.68$) in Minnesota lakes (Birr and Mulla, 2001). The PI approach should be used as part of an ecoregion-based assessment (Chapter 2) to determine strategies to protect a lake, and to provide data on lake quality attainability.

Major nutrient sources to waterways are confined animal feed lots and manure applications to the land. New nutrient management policies, based on P management as well as N, have been established and 47 states have chosen a PI approach. Many of the states have modified the PI to reflect regional ecological differences and state policies. The state strategies and PI modifications are compared in Sharpley et al. (2003). Lake managers should examine their own state's PI (e.g., USDA-NRDC, 2001) before proceeding with this approach to lake and reservoir protection.

There are several BMPs that can reduce the PI value for a watershed and thereby protect lakes and reservoirs (reviews by Robbins et al., 1991; Langdale et al., 1992; Novotny and Olem, 1994; USEPA, 1995; Myers et al., 2000). Only a few can be discussed in this text, including soil amendments, wetland-pond detention systems, buffer strips or zones, and lakescaping. These techniques are meant to intercept or prevent runoff, and do not directly address the land use problem. The total solution to non-point runoff problems involves more complex social, behavioral, political and economic issues beyond the scope of this text. Nevertheless, these broader issues must be addressed for long-term solutions to non-point runoff pollution.

Implementing BMPs is one of the last steps in reducing non-point pollution. Brezonik et al. (1999) listed eight steps when planning and implementing a non-point source pollution control project, emphasizing involvement of all stakeholders throughout the process. Their eight steps begin with problem identification, followed by simultaneous projects to monitor water quality, evaluate pollution sources, and identify relevant physiographic features. These preliminary steps lead to establishing water quality goals, and to identifying cost-effective BMPs and priority drainage areas. This is a "learn as you go process" that may lead to revision of an earlier step.

Cost-effectiveness of BMPs is a central issue. For example, if economic evaluations of several Rural Clean Water Projects (RCWP) had taken place at the project's beginnings, greater economic efficiency would have been possible. In one case, structural BMPs were used to control sediment pollution, but a later analysis showed that it would have been more cost-effective to use crop rotation and conservation tillage. These latter BMPs cost $3,000- $9,000 per percentage drop in sediment load, whereas costs for the structural (e.g., detention basins and animal waste facilities) BMPs exceeded $59,000 per percentage drop (Setia and Magleby, 1988; Magleby, 1992). Many BMPs that reduce sediment loss to the lake are unlikely to be adopted by farmers because of cost (Prato and Dauten, 1991).

Drinking water supply lakes and reservoirs are a critical resource. Many innovative cooperative agreements with farmers have been established to protect them, including federal, state and municipal subsidies to farmers for BMP construction or outright purchase of land and/or livestock. Lake Okeechobee, Florida, the largest lake in the southeastern U.S., was polluted by multiple non-point sources (Gunsalus et al., 1992; Havens et al., 1995). A step-by-step program was developed involving every level of government, expert technical assistance, and all stakeholders. The lake's huge watershed (22,533 km², 13 times lake area) was dominated by cattle ranching. Manure was a major nutrient source, along with backpumping of nutrient-rich irrigation water. In the 1970s, BMPs were initiated including manure management, fencing cattle from streams, and backpumping restrictions. Some dairies were purchased. Although significant declines in non-point loading occurred, non-point internal P loading delayed the lake's improvement (Havens et al., 1995).

The discussions that follow emphasize BMPs to address some of the most significant non-point sources to lakes

5.4 NON-POINT SOURCE CONTROLS: MANURE MANAGEMENT

United States meat consumption is among the world's highest. About 30% of the P input to a livestock farm as feed and fertilizer is exported as crops and meat, leaving a massive surplus in the form of manure (Sharpley et al., 1999). The primary manure disposal method is land application, normally within a few kilometers of production, leading to surplus STP (Carpenter et al., 1998) and high potential for transport to water (Sharpley et al., 1999). The "American Diet" is directly linked to water pollution.

Most P in feed grain is found as phytate-P. Monogastric animals do not digest this molecule, forcing farmers to supplement feed with inorganic P to meet animal P needs. Therefore, poultry and swine manure is very P-rich (Sharpley et al., 2001a). For example, poultry manure typically has an N:P of 3:1, and averages 15.5 g P/kg (Sharpley, et al., undated).

The potential impact of poultry manure is enormous. In Arkansas, for example, poultry farming produces 1 million metric tons of litter and manure annually, or 14,000 metric tons P/yr (Adams et al., 1994; Daniel et al., 1994). Nearly all is land-disposed, and where a PI indicates that transport is possible, there will be runoff, mainly (up to 80% of TP) dissolved P (Shreve et al., 1995).

The potential for P-enriched runoff increases as STP increases (Daniel et al., 1998). The top 5 cm of soil is particularly active as a dissolved P source, but deep tillage reduces surface STP significantly and reduces P and N concentrations in runoff (Sharpley et al., 1996, 1999; Pote et al., 2003), suggesting that plowing-in manure rather than surface disposal could reduce runoff and enhance P uptake into exportable crops. A good measure of the potential of manure-amended soils to yield STP to streams is the water-extractable P concentration of the manure (Kleinman et al., 2002a).

Application of Fe, Al, and Ca salts to manure and poultry litter could reduce the concentration of P in runoff from these materials, though not eliminate it (Moore and Miller, 1994). These salts form compounds with P, removing P from solution. Subsequent solubility of Ca and Fe complexes is pH and redox sensitive, but Al-P salts are redox-insensitive and are insoluble over a wide range of chemical conditions, making them the most effective (Chapter 8).

Adding alum to pig manure at high doses (1:1 molar ratio Al added to P in manure) produced an 84% reduction in SRP in runoff (Smith et al., 2001). Similar results with poultry manure were obtained by Shreve et al. (1995). Application of alum-treated and untreated poultry litter to field test sites produced a 73% SRP reduction in runoff over a 3-year period (Moore et al., 2000) (Figure 5.4). Even with these high percent reductions, SRP concentrations in treated runoff were more than 2.0 mg P/L, or several times greater than P concentrations in tertiary-treated human sewage, and 100 times greater than P concentrations that produce algal blooms.

There are concerns that Al salts used to treat manure will lead to soil contamination. This is unlikely. Al is the third most abundant element on Earth. The amount added to litter and manure is very low relative to soil concentration. As long as soil pH remains in the pH 6–8 range, Al solubility is extremely low.

Al salts are used routinely during potable water treatment, producing an Al-rich water treatment residual (WTR), mainly $Al(OH)_3$. WTRs might be used in controlling P in runoff from manure-treated fields, thereby turning a solid waste into an environmentally useful material (Gallimore et al., 1999; Codling et al., 2000). Preliminary experiments with WTRs (e.g., Haustein et al., 2000) indicated that P was lowered in runoff from WTR-treated manure-rich soils. Some WTR are rich in Cu, a toxic heavy metal, because the supply reservoir has been Cu-treated to kill algae (Hyde and Morris, 2000). This could lead to soil Cu contamination.

Fe and Al salts have greater overall benefits than Ca salts because they reduce litter pH and NH_3 volatilization, leading to fewer poultry diseases, cleaner air, and better fertilizer effect of the litter due to its higher N content (Moore and Miller, 1994). Alum appears to be more effective than coal combustion by-products (e.g., flyash) in controlling SRP released from dairy, swine, and poultry manure in laboratory studies (Dou et al., 2003). Another method to lower the N and P content of manure is to modify poultry diets by reducing protein content and by using phytase supplements to allow digestion of phytate-P compounds, thus eliminating P additions to feed (Nahm, 2002).

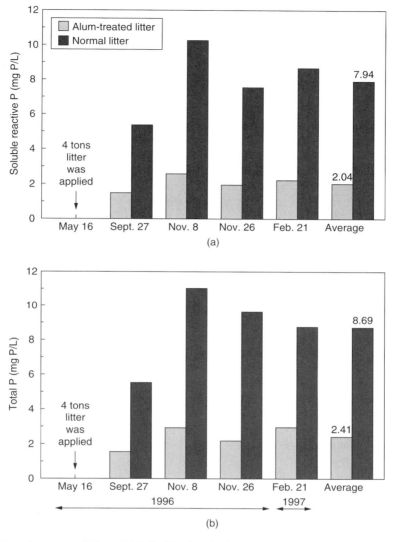

FIGURE 5.4 Phosphorus runoff from fields fertilized with alum-treated and normal litter for first year of the study. (A) Soluble reactive P vs. date; (B) total P vs. date. (From Moore P.A., Jr. et al. 2000. *J. Environ. Qual.* 29: 37–49. With permission.)

Phosphorus transport from soils to water could be lowered by reducing or prohibiting land application of manure to sites with high runoff potential. But even when manure applications are stopped, residual soil P will continue to be transported to streams as subsurface flows for long periods (McDowell and Sharpley, 2001). Treating livestock wastes as human wastes could be the best long-term solution. For example, in just one Arkansas–Oklahoma watershed, the 1996 production of P by confined animals, mostly poultry, was estimated to be 1200 metric tons, the equivalent output of about 3.7 million humans. While only a fraction of this manure reached streams after land disposal, some flowed into a water supply reservoir (Oklahoma Conservation Commission, 1996). Though meat prices might rise, shouldn't manure be transported to a waste treatment plant capable of handling a load of this size? This would transform a non-point nutrient source into a treatable point source, with industry and consumers sharing costs.

5.5 NON-POINT NUTRIENT SOURCE CONTROLS: PONDS AND WETLANDS

5.5.1 INTRODUCTION

Lakes and reservoirs have siltation as well as nutrient problems. Annual suspended solids loading from urban areas can exceed 600 kg/ha, and agricultural sources can be 100 times greater (Weibel, 1969; Piest et al., 1975), leading to turbidity, shallowness, loss of habitat, and creation of plant-choked littoral zones. Modern residential developments often require pre-development placement of structures to detain silt and nutrients, whereas some older developments are being "retrofitted" with these structures. A companion approach is to increase minimum lot sizes, leaving more open spaces and greenbelts, and to restrict developers from clear-cutting vegetation.

Properly designed and maintained constructed ponds and wetlands can protect streams and lakes from non-point runoff, and protect stream banks from erosion. Reviews include Schueler (1987, 1992, 1995. Metropolitan Washington Council of Governments. 202-962-3200. info-center@mwcog.org), Horner et al. (1994), Kadlec and Knight (1996), and Hammer (1997). Wet ponds, wet extended detention basins, pond-wetland systems, buffer zones, and lakescaping are among the most effective BMPs to reduce urban runoff impacts.

5.5.2 DRY AND WET EXTENDED DETENTION (ED) PONDS

Detaining stormwater for more than 24 h, in an otherwise dry basin, reduces the particulate load up to 90%, although minimal soluble nutrients are removed. An additional benefit comes from reducing peak stream velocity, thereby protecting stream banks and riparian zones and reducing the silt load. Nutrient retention, perhaps up to 40–50% of TP, is increased by a two-stage design (Figure 5.5). The top part of the extended detention (ED) pond is dry between storms, and a smaller permanent wet pond remains at the outlet. The pond should be sized to hold the runoff from the mean storm flow, and preferably the volume of a 2.5-cm storm. All ED ponds require regular maintenance and this responsibility should be established prior to construction (Schueler, 1987). Settling of turbidity prior to post-storm release is enhanced with alum (Boyd, 1979) or calcium sulfate (Przepiora et al., 1998).

If properly sized and maintained, wet detention ponds are more effective than dry ponds, and they also lower peak discharge rates. They require a regular water supply to maintain a permanent pool. Their use in drainage basins less than 8 ha (20 acres) is not recommended because of an insufficient water supply (Schueler, 1987).

The principle behind silt retention (and nutrients sorbed to particles) is straightforward. The settling velocity of particles is a function of size and weight, all other factors (temperature, salinity) being equal. Under ideal conditions, particles with a settling velocity greater than the pond overflow rate are retained. In practice, basins are easily built to retain the largest particles, but an incorrectly designed basin does not have sufficient area and volume to detain water long enough to allow finer particles to settle. These are the most nutrient-enriched materials. Design problems become very difficult when the watershed's impervious area is large, leading to a high runoff coefficient (fraction of rainfall existing as runoff) (Wanielista, 1978).

Schueler (1987), Walker (1987), and Panuska and Shilling (1993) reviewed sizing criteria. The most useful pond size indicator is the ratio of pond volume to mean storm runoff volume (VB/VR). A VB/VR of 2.5 is expected to remove 75% of suspended solids and 55% of TP (Schueler, 1987). The National Urban Runoff Program (Athayde et al., 1983) recommended a wet pond with a surface outlet, a mean depth of 1.0 m, and a surface area equal to or greater than 1% of watershed area (with a 0.2 runoff coefficient). Wu et al. (1996) confirmed these criteria, finding that urban wet detention ponds sized at 1% of runoff area had solids removal up to 70% and TP removal of 45%. Deepening the pond is preferable to increasing area for P removal, but very deep ponds could thermally stratify, leading to P recycling. Ponds in series, emphasizing biological removal of

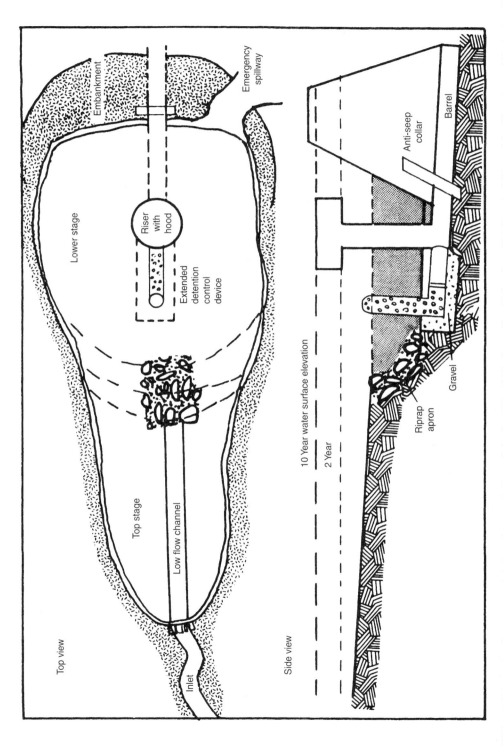

FIGURE 5.5 Schematic of a dry extended detention pond. (From Schueler, T.R. 1987. *Controlling Urban Runoff: A Practical Manual for Planning and Designing Urban BMPs.* Metropolitan Washington Council of Governments, Washington, DC.)

nutrients in the terminal pond, were recommended (Walker, 1987). Figure 5.6 illustrates a wet pond design. Pond (and pond/wetland) construction may require Clean Water Act Section 401 and 404 permits (Schueler, 1995). All ponds should have a dense perimeter of aquatic and bank vegetation to provide protection from shoreline erosion.

Another pond-sizing model calculates sizes for the drainage basin based on desired loading to the lake, land use in each sub-basin, and projected future land uses, using a genetic algorithm (a search technique) to obtain an "optimal decision" about pond sizes, locations, land uses, and costs on a whole basin scale (Harrell and Ranjithan, 2003). This integrated approach needs evaluation because it could be useful for planning purposes for real estate lake developments.

One problem in pond design is the "short-circuiting" that occurs when stormwater passes through the pond with little or no displacement of pond water (Horner, 1995). A minimum length to width of 3:1 may eliminate it (Schueler, 1987), but topography may prevent this design, forcing the use of baffles in the pond to divert inflowing water into all pond areas.

Two wet ponds were used to protect Lake Sammamish (Washington state) from drainage impacts of a 40 ha urban sub-watershed (Comings et al., 2000). Pond C, constructed in a horseshoe shape to minimize short-circuiting, had a detention time of one week and an area that was 5% of its watershed. Pond A was designed with three cells, but allowed short-circuiting through the first two. Its detention time was one day and its area was 1% of its drainage area. Pond performance was evaluated in winter-spring when biological activity was low, but when most of the annual inflows occurred. Pond C removed 81% of total suspended solids (TSS), 46% of TP, 62% of soluble P, and 54% of bioavailable P. Pond A removed 61% of TSS, but only 19% of TP, 3% of soluble P, and 19% of bioavailable P, demonstrating that design and size affect performance.

It is important to establish pond maintenance responsibilities and funds prior to construction because significant sediment removal is required often. Ways to make sediment removal easier are to construct an accessible forebay that retains the largest particles, build a ramp for small dredge access, and to establish a watershed area for sediment disposal (Schueler, 1987).

5.5.3 CONSTRUCTED WETLANDS

Natural wetlands have characteristics of terrestrial and aquatic communities. Among their functions are the capacities to detain water and store materials. However, in some states (e.g., Missouri, Ohio, Illinois, and Iowa) more than 80% of wetlands have been drained or filled (Dahl, 1990), eliminating these important functions.

Wetland rehabilitation has been suggested for returning wetland functions to the landscape, thereby protecting lakes and streams and reducing the volume and frequency of floods (Cairns et al., 1992). Constructing new wetlands is another approach to mitigate losses or to treat urban and agricultural runoff or waste. The use of natural wetlands to treat wastes should be avoided because this will only contribute to their current high rates of destruction unless methods to calculate acceptable P loads are employed (e.g., Keenan and Lowe, 2001).

Reviews and descriptions of constructed wetland designs and effectiveness include Olson (1992), Moshiri (1993), Schueler (1992, 1995), Kadlec and Knight (1996), Hammer (1997), and Kennedy and Mayer (2002).

Surface flow constructed wetlands differ from natural wetlands because they are not dominated by groundwater, their boundaries are defined, there is little internal topographic complexity, they have high inputs of nutrient-enriched suspended solids, and they cannot be maintained without active management (Schueler, 1992). Internal processes, however, are driven by the same ecological processes found in natural wetlands. Sub-surface flow constructed wetlands are described Kadlec and Knight (1996).

The most effective surface constructed wetland, in many cases, is the pond/wetland system (Schueler, 1992) (Figure 5.7). The forebay or pond intercepts suspended solids and protects the wetland (Johnston, 1991; Shutes et al., 1997). There should be access to this forebay for sediment

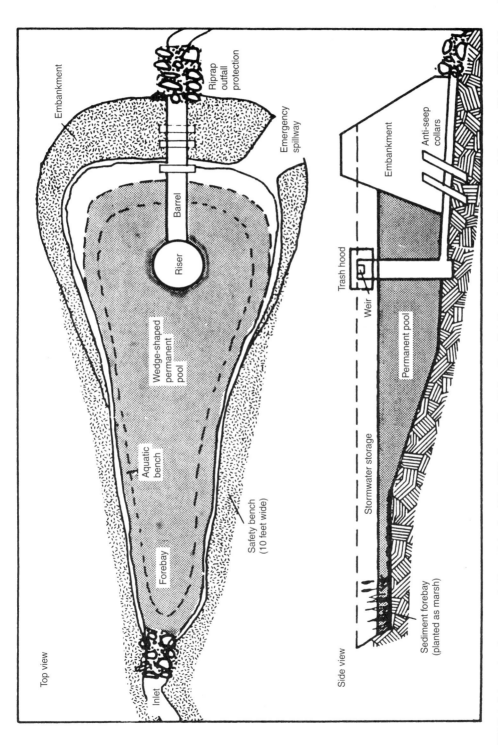

FIGURE 5.6 Schematic of a wet pond. (From Schueler, T.R. 1987. *Controlling Urban Runoff: A Practical Manual for Planning and Designing Urban BMPs.* Metropolitan Washington Council of Governments, Washington, DC.)

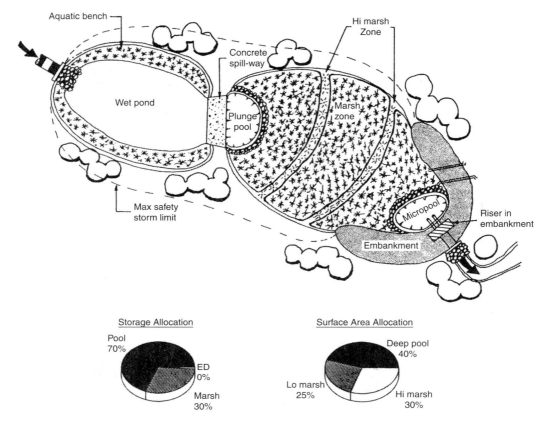

FIGURE 5.7 Design No. 2 — The pond/wetland system. (FromSchueler, T.R. 1992. Design of Stormwater Wetland Systems: Guidelines for Creating Diverse and Effective Stormwater Wetlands in the Mid-Atlantic Region. Metropolitan Washington Council of Governments, Washington, DC.)

removal. Sizing criteria vary with design (Schueler (1992) described 4 basic designs), but the system should have these characteristics: (1) capture and treat at least 90% of the annual runoff volume, (2) have a value of 0.01 as the minimum wetland: watershed area ratio, (3) have about 45% of surface area as a deep pool, 25% as a low marsh, and 30% as a high marsh, (4) have 70% of volume as a deep pool, 30% as marsh, and (5) have a length to width ratio of at least 1.0 (to reduce short-circuiting) (Schueler, 1992). There must be continuous inflow to provide a permanent water body with a depth of 0.5–1.0 m (Shutes et al., 1997). Infiltration from surface to groundwater must be minimal (use clay or other liner). Topsoil is often added to the marsh after construction to allow successful growth of wetland vegetation. A vegetated buffer around the wetland adds wildlife habitat.

Wetlands are effective in nitrate removal. The highest rates were in constructed wetlands dominated by cattails that provided organic carbon for bacterial metabolism, and during periods of highest water temperature (Bachand and Horne, 2000).

Phosphorus retention and storage are among the most important functions of constructed wetlands (reviewed by Richardson and Craft, 1993; Kadlec and Knight, 1996; Reddy et al., 1999). Sediment and peat accumulation are the major mechanisms of long-term P storage. Uptake by plants and their epiphytes, and sorption to soil surfaces are primary processes that change wetland water P concentrations over the short term, but plants and epiphyton release 35–75% of P back to the water column, especially at season's end. Reactions of P with salts of Fe, Al, and Ca are major processes in P storage, and are controlled by initial soil P concentration, pH, and oxidation-reduction potential (Richardson and Craft, 1993). Figure 5.8 summarizes P retention processes in wetlands.

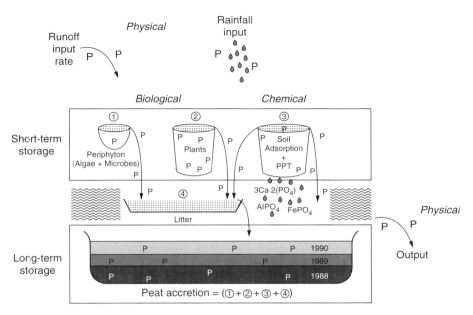

FIGURE 5.8 A conceptual model of phosphorus retention in wetlands. Only the major reservoirs are shown and no attempt was made to show a complete phosphorus cycle among the biotic and abiotic components. Bucket sizes are proportional to storage. (From Richardson, C.J. and C.B. Craft. 1993. In: G.A. Moshiri (Ed.), *Constructed Wetlands for Water Quality Improvement.* Lewis Publishers, Boca Raton, FL. pp. 271–282. With permission.)

If a constructed wetland is to provide sustainable P storage, P input cannot exceed the rate of permanent peat or soil formation. Processes producing soil and peat of the wetland can be impaired by excessive loading. Richardson and Qian (1999) used the North American Wetland Data Base (NAWDB; Knight et al., 1993) to estimate a P "assimilative capacity," based on this concept. They found that when the P load is < 1 g P/m^2 per year, wetland P output to the receiving water remained low and constant. This rate is a North American average representing a conservative and sustainable loading rate. This is called the "One Gram Assimilative Capacity Rule" (Figure 5.9).

Moustafa (1999) used the NAWDB to produce a "Phosphorus Removal Efficiency Diagram," based on surface flow wetland water residence time and P loading rates. Optimal P retention occurred at long residence times and low areal P loading (see also Dierberg et al., 2002). The diagram is useful for both the low water load - high P load application (wastewater) and for high water load - lower P load (stormwater) application.

Rule-of-thumb guidelines are useful in initial feasibility analyses, but other factors must be considered, including seasonality, hydraulic and meteorological constraints, and specific wetland characteristics. Intensive study of wetland ecology (e.g., Mitsch and Gosselink, 2000) by applied limnologists is advisable prior to attempting to design an effective constructed wetland.

Wetlands may be constructed on nutrient-rich agricultural soils, sometimes as part of a mitigation process. P release from flooded soils may be extensive (Pant et al., 2002; Pant and Reddy, 2003), but is controlled with additions of Ca or Al salts (Ann et al., 2000a,b).

5.6 CONSTRUCTED WETLANDS: CASE HISTORIES

Using ponds and wetlands to retain materials and to reduce velocity and amount of water discharged from a watershed is not new. They have been used in China for thousands of years. An example of this cultural-ecological heritage in China is a small (692 ha) agricultural watershed with 193

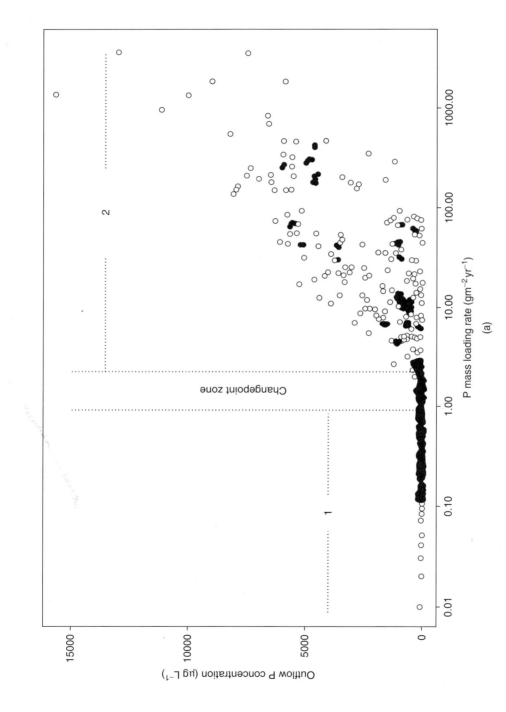

FIGURE 5.9 Input total P loading effects on P output concentrations for the North American Wetland Data Base (total sites = 126, n = 317). In region 1, where loading rate is less than 1 g P/m^2 per year, uniform P output concentrations are found and output P concentration is not a function of loading. In region 2, loading rate > 1 g P/m^2 per year and output P increases significantly as P loading increases. Change point zone is the region where output switches from low and uniform to increasing and non-uniform. (From Richardson, C.J. and S.S. Qian. 1999. *Environ. Sci. Technol.* 33: 1545–1551. With permission.)

constructed ponds ranging in area from 0.01–1.0 ha and with a mean depth of 1.0 m (Yin and Shan, 2001). Runoff from farm fields flows into ditches and then through a pond series. For the whole pond system, solids retention was 86%, TP retention 85%, and SRP retention 51%. Nutrient-rich pond water is pumped back onto the fields, and the ponds are drained and dredged as needed, with dredged materials returned to the land. Because of the ponds, many precipitation events produce no water discharge from the watershed. Plants in the ponds are harvested for livestock feed. Yin and Shan summarized the significance of the ponds by stating (p. 374): (We are) "keeping nine parts of lands free of harm at the cost of converting one part of land into ponds."

The McCarron's pond/wetland (Oberts and Osgood, 1991) was established to protect Lake McCarron (Minneapolis, Minnesota) from the drainage of a 171 ha urban area. Although the pond (1 ha) and 5 in-line wetlands (1.5 ha total) were smaller than recommended, they removed 70% of TP and 51% of dissolved P. The pond was the most effective part of the system because P was largely associated with particulates. When additional drainage area was diverted to the system, short-circuiting reduced performance. Pond/wetland systems of this type in the Minneapolis area have a life span of 5 years or less unless maintained by sediment removal (Oberts et al., 1989).

Despite the effectiveness of the McCarron's pond/wetland, the lake did not improve. Prior to wetland operation, inflowing nutrient and silt-laden stormwater was cooler (and heavier) in the summer than epilimnetic water, and thus plunged below the surface of the lake, reducing its impact on algae growth. The pond/wetland outflows were warmer than epilimnetic waters and tended to float on the lake's surface, contributing nutrients to algae. Also, lake sediments provided significant internal P loading to the epilimnion (Oberts and Osgood, 1991).

Wetland sizing is critical to success. Raisin et al. (1997) described a 0.045 ha wetland used to intercept drainage from a 90 ha pasture. The system's wetland: watershed areal ratio (WWAR) was 0.0005, whereas a ratio of 0.01 is considered a minimum. On an annual basis, this undersized system retained only 11% of N and 17% of P. In contrast, the Clear Lake, Minnesota wetland had a 2-day water retention time and a WWAR of 0.06. It retained 90% of TSS and 70% of TP, but internal P loading in the lake delayed lake recovery, requiring treatment of lake sediments (Barten, 1987).

Wetlands are used to treat the runoff from land disposal of manure (Knight et al., 2000). Despite significant reductions in N and P concentrations, their outflow concentrations often remain in the 10–100 mg P/L range, and therefore can do great damage to streams and lakes. Another type of treatment is required for these concentrated sources.

Agricultural field drainage can be successfully treated with constructed wetlands (Kovacic et al., 2000: Woltemade, 2000). A pond/wetland system with a highly desirable WWAR of 0.09 was used to treat runoff from a potato field. It consisted of a sedimentation basin followed by a level spreader to prevent channelization as water flowed down a 6% slope to the wetland/pond (Figure 5.10). The system handled flows up to the 10-year storm event, and was built for $21,600 (2002 dollars). An access ramp was constructed for sediment removal. During dry summer months, there was 100% total suspended solids (TSS) and P retention. Over 3 years of monitoring, about 48% of TP was retained as pond soil in the sedimentation basin (Higgins et al., 1993). Retention of agricultural runoff P by constructed wetlands is influenced by P loading, season, amount of P attached to solids, and P settling velocity (Braskerud, 2002).

A large constructed wetland (14 km^2) will be used as part of the rehabilitation of Lake Apopka (Florida,), a large (125 km^2), shallow (mean depth 1.6 m) lake that became hypereutrophic from agricultural drainage. Lake water, at a rate twice the lake volume per year, is to be circulated into the wetland to remove algae, resuspended sediments, and other forms of particulate P, and then returned to the lake. A pilot-scale (2.1 km^2) wetland filter was tested over 29 months, and achieved TSS and TP removals of at least 85% and 30%, respectively, indicating that the full-scale implementation of this innovative system will be an integral part of the lake's rehabilitation plan. Costs are estimated at $1.6 million per km^2 for the full-scale project (Coveney et al., 2002).

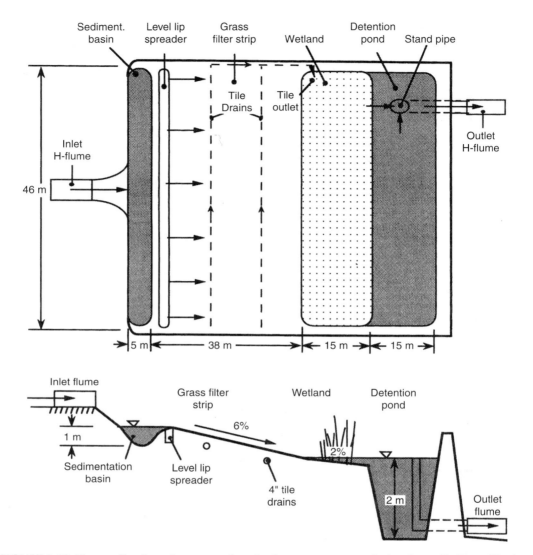

FIGURE 5.10 Plan-profile view of constructed wetland system to treat agricultural runoff. (From Higgins, M.J. et al. 1993. In: *Constructed Wetlands for Water Quality Improvement.* Lewis Publishers, Boca Raton, FL. With permission.)

Constructed wetlands are becoming more common for the treatment of the domestic wastes of small (~ 20 people) communities. The most effective include an aerobic pre-wetland treatment, and they can have up to 95% P removal and retention and be stable for years. They require high treatment area (> 50 m^2/m^3 per day) and Ca- or Fe-rich soils (Luederitz et al., 2001).

In summary, constructed wetlands with a forebay or wet pool have great potential to protect streams and lakes from stormwater solids and nutrients if the system is maintained and is designed to prevent water and P overloads.

5.7 PRE-DAMS

Pre-dams are commonly used in Europe, especially in Germany, to protect downstream reservoirs from nutrients and silt. Pre-dams are less common in the U.S., but one example is Lake Eucha that partially protects Lake Spavinaw, the water supply for Tulsa, Oklahoma from upstream non-point agricultural runoff (Oklahoma Water Resources Board, 2002).

Pre-dams may also protect water supplies from accidental or purposeful spills of toxic or radioactive materials. Most pre-dams have a water residence time of several days, possibly allowing time to close raw water intakes and/or to treat the contaminated water.

Pre-dams normally have a surface overflow plus a deep gate to allow water removal followed by sediment removal. They function primarily through sedimentation of particulates, and nutrient removal occurs by maximizing diatom growth and sedimentation while minimizing blooms of buoyant cyanobacteria and algae grazing *Daphnia* (to prevent remineralization). Effectiveness depends upon retention time, with short times preventing significant settling of particulate P (silt, algal cells). The optimum design of a pre-dam for nutrient retention includes a mean depth that is not significantly greater that euphotic zone depth to prevent internal P loading common in dimictic lakes (Benndorf and Putz, 1987a, b: Putz and Benndorf, 1998).

Three case histories illustrate pre-dam effectiveness and problems. Jesenice Reservoir (Czechoslovakia) has a pre-dam with a 5 day residence time that lowered ortho-P concentration in its effluent to the main reservoir to a range of 10–25 µg P/L, leading to reduced algal biomass (Fiala and Vasata, 1982). Salvia-Castellvi et al. (2001) found that a shallow (mean depth 2.5 m) pre-dam, with a mean residence time of 1.5 day, had a lower P retention than predicted by the Benndorf and Putz (1997a, b) model because green algae with a low sedimentation velocity, and internal P loading, were dominant factors. A deeper, stratified pre-dam (mean depth 7.1 m, mean residence time of 44 days), had a TP retention of up to 90%, agreeing with model predictions. High removal in the deep pre-dam was from cell sedimentation and macrophyte uptake at the pre-dam inflow. Saidenbach Reservoir (Germany) has pre-dams on each of its four tributaries. An average summer SRP elimination approached 50% with a 4-day retention time, similar to Benndorf and Putz (1997a, b) model predictions. Percent TP was less than predicted. Siltation that decreased volume was an important factor in reducing retention time, indicating that frequent sediment removal (Chapter 20) is required for pre-dams (Paul, 2003).

An underwater dam in the riverine zone of the main reservoir may increase sedimentation of particulate P in inflows (Paul, 1995; Paul et al., 1998). This was used in Saidenbach Reservoir. During spring and fall, cold inter- and underflows were trapped behind the submerged dam for a period long enough to promote deposition. The dam also prevented short-circuiting of flows, enhancing deposition.

A primary problem of the pre-dam is that unless it is constructed at the same time as the main reservoir, a retro-fit may be impossible because land may be unavailable or too small in area.

5.8 RIPARIAN ZONE REHABILITATION: INTRODUCTION

The riparian zone is the "gradient-dominated" (Wetzel, 1992, 2001) community between stream or lake and the land, and it has major influences on water quality. The riparian zone has these functions: (1) reduce surface and sub-surface runoff volume, (2) protect banks from erosion, and (3) lower pollutant concentrations in runoff (Dosskey, 2001). In undisturbed temperate and sub-tropical ecosystems, it is characterized by a sharp gradient of submerged, emergent, semi-terrestrial and terrestrial vegetation, high species diversity, very high biomass and productivity, high retention of materials, and periods of significant export of dissolved and particulate organic material that subsidize aquatic food webs (Wall et al., 2001).

Riparian zone destruction, caused by increased stream volume and velocity, increased impervious area, channel straightening, livestock grazing, cultivation to the water's edge, boat traffic, and real estate developments and lawns, is widespread, leading to deposition in stream habitats and downstream reservoirs and lakes. Shoreline erosion from wind and boat-induced waves creates significant losses of materials, producing permanently turbid conditions. For example, shoreline recession at an Ohio real estate reservoir occurred at a rate of 0.12 m/mo (1.4 m/yr), adding 8,300 metric tons of soil and 332 kg P/yr (21% of annual P load). Boat wakes and absence of riparian vegetation were major causes (Wilson, 1979). Similar rates occur on much larger reservoirs (e.g., American Falls Reservoir, Idaho; Hoag et al., 1993).

Game fish biomass may decline when littoral vegetation and coarse woody debris (CWD) are reduced or eliminated. This often leads to increased turbidity and algal biomass (Chapter 9). For example, there was a strong correlation between reduced game fish biomass and reduced emergent and floating plant biomass and diversity in 44 Minnesota lakes, representing a gradient of shoreline development (Radomski and Goerman, 2001). In a survey of 16 north temperate U.S. lakes, shoreline development (houses, lawns) was significantly and negatively correlated with CWD (Christensen et al., 1996). Schindler et al. (2000) found that size-specific growth rates for bluegill and largemouth bass were negatively correlated with the amount of lakeshore development, apparently due to loss of shoreline habitat and CWD.

Riparian zone vegetation removal can lead to increased lake turbidity and/or nutrient release from suspended sediments (Barko and James, 1998). Long Lake (Washington, U.S.) had an inverse relationship between TP content and submersed macrophyte biomass over a 20-year period (Jacoby et al., 2001). Resuspension is strongly related to water depth and shear stresses generated by waves, but submersed vegetation will dampen waves, reducing turbidity and sediment nutrient release, and provide shoreline protection.

Shoreline lawns, and lawns connected to the lake via storm runoff, are major nutrient sources (Shuman, 2001; King et al., 2001). Linde and Watschke (1997) irrigated turf grass, then twelve hours later fertilized it with 4.9 g N/m^2 and 0.3 g P/m^2. Eight hours later, a rainfall event was simulated. They found up to 3.5 mg P/L, averaging 17% of P applied, and up to 6.8 mg N/L, in surface runoff. Leachate to shallow groundwater was similarly enriched. Fertilized Bermuda grass turf yielded five times more P in runoff than manure-treated turf (Gaudreau et al., 2002). Overwatering of urban lawns is common, and leads to nutrient leaching to groundwater and to surface runoff during storms. Kentucky bluegrass plots on sandy loam, fertilized with urea and then overwatered, yielded N in runoff up to 4 mg N/L and an annual loss of 32 kg N/ha (16 times that of overwatered, unfertilized control plots)(Morton et al., 1988). In another example, Bannerman et al. (1993) found a geometric mean TP concentration of 2.67 mg P/L in surface runoff from lawns in Madison, Wisconsin. Fertilized lawns that are hydrologically connected to the lake may be responsible for significant macrophyte and algae growth. Some lake associations have banned the use of P-containing lawn fertilizers.

Nuisance populations of Canada Geese are another symptom of damaged or missing riparian zones and missing natural wetland/pond habitats in lake areas. Geese are attracted to lawns, especially N-fertilized lawns (Owen, 1975), and are significant importers of nutrients to lakes. Waterfowl added 70% of all P entering Wintergreen Lake (Indiana, U.S.), with geese contributing 76% of that load based upon a conservative defecation estimate of 28 times daily. Wild geese may defecate up to 92 times daily (Manny et al., 1994). In other studies, bird contributions to external loading were smaller (e.g., Hoyer and Canfield, 1994; Marion et al., 1994). Birds may also play a role in internal P loading by transforming particulate P (fish, macrophytes) into soluble P, or by enriching littoral sediments that later release P to the water column (Scherer et al., 1995).

Riparian zone rehabilitation may greatly decrease non-point nutrient loading, and is therefore an important part of a lake or reservoir rehabilitation project.

5.9 RIPARIAN ZONE REHABILITATON METHODS

This section serves only as an introduction to these methods. Reviews include Schueler (1987), Herson-Jones et al. (1995), Shields et al. (1995), Henderson et al. (1999), and McComas (2003).

Riparian zone rehabilitation often involves privately owned land and incurs expenses that some landowners cannot or will not meet. Upstream landowners may have no stake in protecting downstream waters. In other cases, the total maximum daily load (TMDL) assessment, or similar efforts, may provide financial assistance or regulatory incentives. Successes, therefore, are likely to come slowly, if at all, especially in North America where monoculture lawns and application of lawn chemicals are common and where real estate or agricultural activities extend to the stream or lake shoreline.

Bank undercut and collapse are significant sources of suspended solids to streams, lakes, and reservoirs. Steep banks continue to erode unless they are regraded, the bank toe protected with stone or other materials, and re-planted. Regrading to a 1:1 slope, and planting ground cover, shrubs, and trees that grow well on stream banks, and construction of flow deflectors or revetments to protect the banks, are common methods (McComas, 2003).

The extensive negative effects of cattle grazing and loafing in riparian zones include reduction or elimination of vegetation, elevated stream temperatures, fish habitat alterations, and stream bank collapse (Armour et al., 1991). Cattle compact the soil, decreasing water infiltration. In the arid U.S. western states, where at least 70% of land is grazed, water loss through runoff and evaporation is increased by the presence of cattle. Cattle impact on riparian zones has been very great in the U.S. west (e.g., 90% of riparian habitat in Arizona is gone) (Fleischner, 1994). Livestock should not have direct access to stream riparian zones and the stream itself.

Even after cattle exclusion, riparian zone recovery may take 10 years (Belsky et al., 1999). Livestock exclusion from stream banks, erosion-prone hillsides, and forests produced an estimated 85% reduction in suspended solids and 25% reduction in TP loads to a New Zealand stream (Williamson et al., 1996). Agriculture contributed 47% of the TP load to Lake Champlain (New York, Vermont, U.S.; Quebec, Canada). Livestock exclusion from streams and reduced numbers of crossing areas, as well as erosion control, produced almost a 20% reduction in TP load to streams flowing to this lake (Meals and Hopkins, 2002).

Creating a vegetated buffer zone between land development and the stream provides stream protection by intercepting nutrients and sediments, and assists in restoring lost biodiversity (Wall et al., 2001; Dosskey, 2001; Brinson et al., 2002; Fiener and Auerswald, 2003). The U.S. Natural Resources Conservation Service (NRCS) of the Department of Agriculture (USDA) issued guidelines for establishment of grass filter strips and forested buffer zones (USDA, 1999). An example of their effectiveness is Bear Creek, Iowa, designated Bear Creek Riparian Buffer National Research and Demonstration Area (Zaimes et al., 2004). Sediment and nutrient transport from three sites were compared. These were: a control site (corn or soybean row crop agriculture to stream's edge), a 7 m wide switchgrass (*Panicum virgatum* L.) filter, and a switchgrass filter (7 m) next to the row crop followed by a forest buffer zone (13 m wide) next to the stream. A pictorial model of the switchgrass and forested buffer zone is illustrated in Figure 5.11. The switchgrass filter alone removed > 90% of sediment and 80% of TP, on average, over the 18 month evaluation. The combined buffer zone (Figure 5.11) reduced average TP loss from 200 g/ha (control) to 19 g/ha, and sediment loss from 587 kg/ha (control) to 16 kg/ha (K.H. Lee et al., 2003). Soil loss through the forested/switchgrass buffer was 65 kg/m of stream per year, whereas pastured fields lost 293 kg/m per year and row crop fields lost 389 kg/m per year (Zaimes et al., 2004).

Buffer zones decrease nutrient and sediment transport by detaining water, allowing particle sedimentation, and by increasing soil infiltration capacity (Lee et al., 2003). It is apparent that lakes and reservoirs can be greatly protected by establishment of buffer zones on streams as well as along the lakeshore.

The effectiveness of a buffer zone can be reduced when there are irregular contours that concentrate runoff area and lower buffer zone area in contact with most of the runoff (Dosskey et al., 2002). Non-cultivation of a 5–7 m wide zone along a stream allows grass growth and good retention of suspended solids, but wider (7–15 m) is better, especially where sloped fields have been cultivated. Wider zones promote more water infiltration and nutrient retention (Schmitt et al., 1999). High nutrient load and hydraulic gradient will reduce buffer zone effectiveness (Sabater et

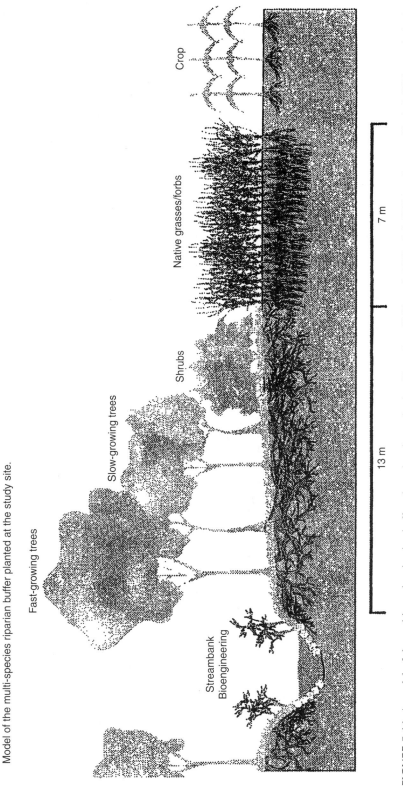

FIGURE 5.11 A model of the multi-species riparian buffer planted at the study site. (From Lee, K.H. et al. 2003. *J. Soil Water Conserv.* 58: 1–8. With permission.)

al., 2003). Suspended solids retention is usually greater than retention of dissolved nutrients, though some nutrients are sorbed to particulates trapped by grass.

While buffer zone width is important, the amount of impervious area, human and livestock impacts, slope, and inspection and maintenance, also determine effectiveness.

Herson-Jones et al. (1995) described how to obtain a "base width" (BW) to protect streams from urban runoff, as follows:

$$BW = 15 \text{ m} + (4 \times \% \text{ slope})$$

BW may be modified by vegetation density and types of soil particles associated with runoff. Nieswand et al. (1990) computed buffer zone width (*W*) for water supply reservoirs, based on overland water flow time of travel (*T*) and slope (*S*) as:

$$W = 2.5 \; T \; S^{-0.5}$$

Application to New Jersey reservoirs yielded a minimum 90 m width for terminal reservoirs and water supply intakes. For perennial streams, a minimum 15 m width or the width calculated from the equation, whichever is widest, was recommended. Buffer zone widths generally range from 15–29 m in North America, with narrower widths more common in the U.S. for similar types of water bodies. Many factors influence buffer zone widths, and in many cases a width is specifically designed for a particular water body rather than from a general formula (Lee et al., 2004).

Shoreline home construction and plowed areas can be major silt sources to lakes and streams, but barriers to retard erosion or silt transport can be effective. Traditional straw mulch appears to reduce runoff volume, but is highly inferior to wood fiber, straw/coconut, and bonded fiber matrix blankets for controlling sediment yields from construction sites (Benik et al., 2003).

Buffer zones are not panaceas and they are not substitutes for better land management. Runoff from areas with high STP may overwhelm a buffer zone. Quantitative data are generally unavailable regarding control of pollutants by buffer zones, and the responses of streams and lakes to them (Dosskey, 2001).

5.10   RESERVOIR SHORELINE REHABILITATION

Reservoir shoreline erosion (bank recession and collapse) is a significant and costly problem. There are 19,000 km of eroded shorelines in U.S. Army Corps of Engineers reservoirs, with half classified as severe (Allen and Tingle, 1993). Surface waves, including boat wakes in smaller reservoirs, and groundwater seepage and runoff, are primary causes (Reid, 1993).

Flexible (stone rip-rap) or rigid armor (retaining walls) are the traditional treatments (Chu, 1993), but these are expensive ($800/km and up). Biological materials are effective and less expensive. An example of the biotechnical approach is Lake Sharpe, South Dakota, a large reservoir on the Missouri River. A 3–5 km fetch on Lake Sharpe produced waves and ice scour and 1 m or higher cut banks. These were regraded, a log breakwater was established 10 m offshore, and bulrush (*Scirpus* sp.) was planted behind it. The slack water behind the logs allowed bulrush establishment that protected against erosion (Figure 5.12). Shoreline slopes greater that 1:1 were difficult to revegetate (Juhle and Allen, 1993).

Bioengineering techniques are more commonplace as solutions to shoreline stabilization (Chapter 12). Some engineers are using dormant brush mats to stabilize soil, in combination with geotextile materials to contain soil (Wendt and Allen, 2002). Shoreline erosion control materials, including interlocking plastic or concrete blocks, fiber blocks, fiber matrix materials, and gabion systems, are available commercially for use in stabilizing and revegetating stream and reservoir shorelines, and in controlling erosion from steep banks while vegetation becomes established. Fiber

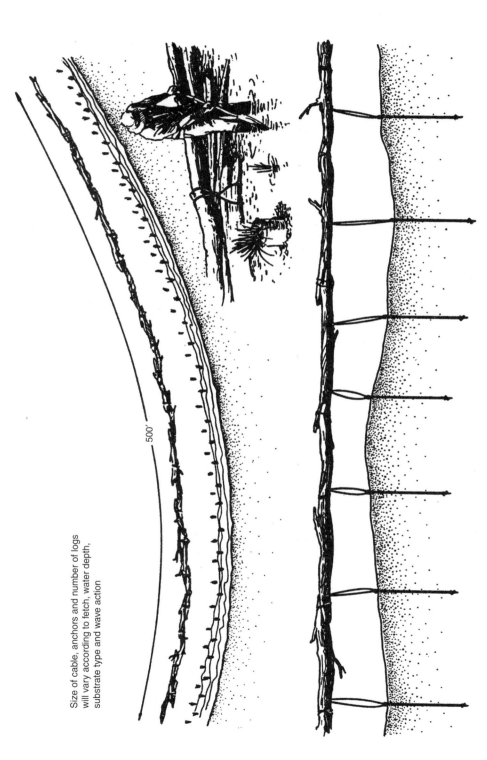

Size of cable, anchors and number of logs will vary according to fetch, water depth, substrate type and wave action

500'

FIGURE 5.12 Plan (top) and profile (bottom) views of log breakwater used at Lake Sharpe, South Dakota. Note vegetation planted shoreward of breakwater. (From Juhle, F.B. and H.H. Allen. 1993. In: Proceedings U.S. Army Corps of Engineers Workshop on Reservoir Shoreline Erosion: A National Problem. Misc. Paper W-93-1. U.S. Army Corps Engineers, Vicksburg, MS. pp. 106–113.).

matrix material can be spray-applied, is biodegradable, and the matrix holds water and reduces raindrop and runoff energy (Spittle, 2002).

Revetments (walls) are another technique to repair and protect shorelines. McComas (2003) developed a model for revetment design, based on wave height and wave runup. Refer to McComas's extensively illustrated book for revetment details.

5.11 LAKESHORE REHABILITATION

Lakeshore property owners play an important role in lake protection and rehabilitation. Fertilized lawns are a source of nutrients and attract nuisance waterfowl. Many owners also have eliminated aquatic vegetation, causing sediment resuspension that adds nutrients to the water column, reduces fish habitat, and contributes to shoreline collapse.

"Lakescaping" is a lawn design that lowers care and maintenance costs, reduces runoff, eliminates the need for fertilizers, discourages or eliminates geese on the lawn, and increases shoreline terrestrial and aquatic biomass and biodiversity (Henderson et al., 1999, available from the Minnesota Department of Natural Resources, 1-888-646-6367). The goal of lakescaping is to return 50–75% of the shoreline to a vegetated state, replacing the monoculture lawn with a diversity of shrubs and trees, and establishing aquatic plants along the shoreline. This is done without obscuring lake views from the house, nor by eliminating use of some lawn and shoreline areas for recreation. In effect, the homeowner creates a customized buffer zone that maintains a selected amount of upland (yard), shoreline (emergent plant zone) and aquatic vegetation. Figures 5.13 and 5.14 illustrate two possible designs and the idea of maintaining sight lines from the house to the lake while retaining a diversity of trees, shrubs, grasses and riparian plants. (See Chapter 12 for further discussion.)

Lakescaping costs are low. Installation for sodded turf grass, seeded turf grass, and lakescaping were (per ha): $46,200, $23,700, and $6,200–25,900. Annual maintenance costs were (per ha): $3,000, $3,100, and $500, respectively (Henderson et al., 1999). Other advantages of lakescaping are the creation of fish habitat, previously extirpated where aquatic plants or sources of CWD had been removed, and the exclusion of geese from the lawn.

A rain garden is an important component of lakescaping. These little gardens receive roof or even street runoff and retain it long enough to allow infiltration into the soil rather than discharge to the lake. A small area at least 5–6 m from the house is dug, forming a flat bottom and square sides to a depth of about 8 cm and to a volume that retains a typical runoff. The basin is planted with species native to the ecoregion that tolerate a wet–dry cycle (see Henderson et al., 1999), and water from the roof is routed to it. This garden attracts birds, butterflies, and possibly amphibians. Water retention times should be short (< 4 days) to prevent mosquito breeding. A brochure entitled *Rain gardens: A household way to improve water quality in your community* is available from Wisconsin–Extension Publications, 45 N. Charter St., Madison, WI 53715.

Lakescaping offers another way to improve lake quality. Declines of Eurasian watermilfoil (*Myriophyllum spicatum*), a major nuisance, have been associated with populations of the native milfoil weevil (Newman and Biesboer, 2000) (Chapter 17). These insects overwinter along the shoreline in the soil-leaf litter interface, and may require undisturbed grasses or forest habitat to survive to re-infest milfoil plants in the following Spring (Newman et al., 2001). Lawns manicured to water's edge may prevent weevil survival, directly contributing to the continued growth of this obnoxious plant by eliminating a natural biocontrol agent.

Like all riparian zone rehabilitations, while understood in practice, lakescaping is often resisted when suggested as a method that property owners might use to improve and protect lake quality. Unlike structural or chemical methods, shoreline rehabilitation is a long-term effort without immediate apparent results. One approach is to encourage (even subsidize) a few property owners to try it, thus building local expertise and advocates (property owners without geese are often happy property owners).

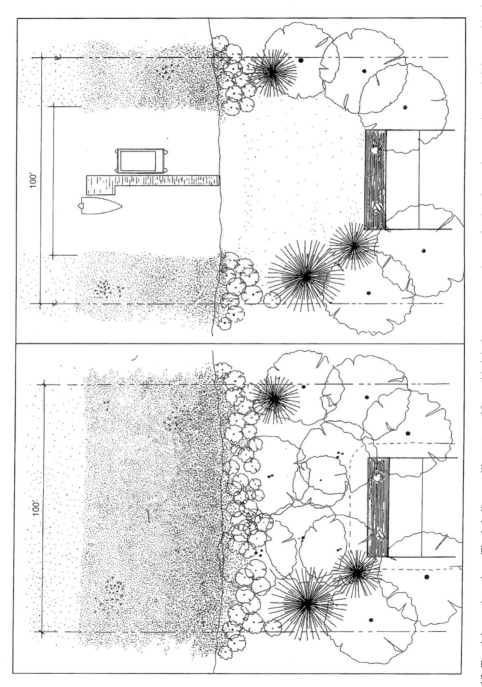

FIGURE 5.13 Two lakescaping plans. The left diagram illustrates a 33 m wide lakefront property where upland and aquatic vegetation are left intact with clearing only for the house. The right diagram illustrates a 33 m wide lakefront property where 60% of upland and aquatic vegetation have been removed, providing some lakeshore protection. (From Henderson, C.L. et al., 1999. *Lakescaping for Wildlife and Water Quality.* Minnesota Department of Natural Resources, St. Paul, MN.)

FIGURE 5.14 Line of sight from house to lake is retained with lakescaping. (From Henderson, C.L. et al., 1999. *Lakescaping for Wildlife and Water Quality*. Minnesota Department of Natural Resources, St. Paul, MN.)

5.12 SUMMARY

Non-point sources of nutrients, silt, organic matter, and toxins are primary threats to fresh water, and their impacts are increasing as human populations increase. The significance of this lies not only with water quality changes, but with the difficulty in controlling it. This chapter provides an introduction to the problem and to some methods of control.

Non-point pollution often stems from personal choices, making effective, long-term correction difficult. The manicured, fertilized, over-watered lawn, for example, is a well-established custom with direct impacts on fresh water. Sometimes personal choices cause indirect impacts, such as those associated with the "American Diet" (e.g., untreated livestock wastes and riparian zone destruction by livestock).

The methods discussed here are not panaceas and serve only to mitigate some of the impacts of land use on fresh water. The issue of water quality attainability remains. Perhaps we will have to temper our expectations of clear, safe water with the realities produced by our desires to live high on the food chain in low habitat diversity ecosystems where erosion, non-point runoff, and successful invasions of exotic plants and animals are dominant forces.

REFERENCES

Adams, R.L., T.C. Daniel, D.R. Edwards, D.J. Nichols, D.H. Pote and H.D. Scott. 1994. Poultry litter and manure contributions to nitrate leaching through the vadose zone. *Soil Sci. Soc. Am. J.* 58: 1206–1211.

Allen, H.H. and J.L. Tingle (Eds.). 1993. *Proceedings, U.S. Army Corps of Engineers Workshop on Reservoir Shoreline Erosion. A National Problem.* Misc. Paper W-93-1. U.S. Army Corps Engineers, Vicksburg, MS.

Ann, Y., K.R. Reddy and J.J. Delfino. 2000a. Influence of chemical amendments on phosphorus immobilization in soils from a constructed wetland. *Ecol. Eng.* 14: 157–167.

Ann, Y., K.R. Reddy and J.J. Delfino. 2000b. Influence of redox potential on phosphorus solubility in chemically amended wetland organic soils. *Ecol. Eng.* 14: 169–180.

Armour, C.L., D.A. Duff and W. Elmore. 1991. The effects of livestock grazing on riparian and stream ecosystems. *Fisheries* 16: 7–11.

Athayde, D.N., P.E. Shelly, E.D. Driscoll, D. Gaboury and G. Boyd. 1983. *Results of the Nationwide Urban Runoff Program. Volume I.* U.S. Environmental Protection Agency, Washington, DC.

Bachand, P.A.M. and A.J. Horne. 2000. Denitrification in constructed free-water surface wetlands: II. Effects of vegetation and temperature. *Ecol. Eng.* 14: 17–32.

Bannerman, R.T., D.W. Owens, R.B. Dodds and N.J. Hornewer. 1993. Sources of pollutants in Wisconsin stormwater. *Water Sci. Technol.* 28: 241–260.

Barbiero, R., R.E. Carlson, G.D. Cooke and A.W. Beals. 1988. The effects of a continuous application of aluminum sulfate on lotic benthic invertebrates. *Lake and Reservoir Manage.* 4 (2): 63–72.

Barko, J.W. and W.F. James. 1998. Effects of submerged aquatic macrophytes on nutrient dynamics, sedimentation, and resuspension. In: E. Jeppesen, M. Søndergaard, M. Søndergaard and Christoffersen (Eds.), *The Structuring Role of Submerged Macrophytes in Lakes.* Chapter 10. Springer-Verlag. New York.

Barten, J.M. 1987. Stormwater runoff treatment in a wetland filter: Effects on the water quality of Clear Lake. *Lake and Reservoir Manage.* 3: 297–305.

Belsky, A.J., A. Matzke, and S. Uselman. 1999. Survey of livestock influences on stream and riparian ecosystems in the Western United States. *J. Soil Water Conserv.* 54: 419–431.

Benik, S.R., B.N. Wilson, D.D. Biesboer, B. Hansen and D. Stenlund. 2003. Evaluation of erosion control products using natural rainfall events. *J. Soil Water Conserv.* 58: 98–105.

Benndorf, J. and K. Putz. 1987a. Control of eutrophication of lakes and reservoirs by means of pre-dams. I. Mode of operation and calculation of nutrient elimination capacity. *Water Res.* 21: 829–838.

Benndorf, J. and K. Putz. 1987b. Control of eutrophication of lakes and reservoirs by means of pre-dams. II. Validation of the phosphate removal model and size optimization. *Water Res.* 21: 839–842.

Bernhardt, H. 1980. Reservoir protection by in-river nutrient reduction. In: *Restoration of Lakes and Inland Waters.* USEPA 440/5-81-010. pp. 272–277.

Birr, A.S. and D.J. Mulla. 2001. Evaluation of the phosphorus index in watersheds at the regional scale. *J. Environ.Qual.* 30: 2018–2025.

Boyd, C.E. 1979. Aluminum sulfate (alum) for precipitating clay turbidity from fish ponds. *Trans. Am. Fish. Soc.* 108: 307–313.

Braskerud, B.C. 2002. Factors affecting phosphorus retention in small constructed wetlands treating agricultural non-point source pollution. *Ecol. Eng.* 19: 41–61.

Brezonik, P.L., K.W. Easter, L. Hatch, D. Mulla, and J. Perry. 1999. Management of diffuse pollution in agricultural watersheds: Lessons from the Minnesota River Basin. *Water Sci. Technol.* 39: 323–330.

Brinson, M.M., L.J. MacDonnell, D.J. Austen, R.L. Beschta, T.A. Dillaha, D.L. Donahue, S.V. Gregory, J.W. Harvey, M.C. Molles, Jr., E.I. Rogers and J.A. Stanford. 2002. *Riparian Areas: Functions and Strategies for Management.* National Academy Press, Washington, DC.

Brown, L.R. 1995. *Who Will Feed China? Wake-up Call for a Small Planet.* W.W. Norton and Co., New York.

Brown, L.R. and H. Kane. 1994. *Full House. Reassessing the Earth's Population Carrying Capacity.* W.W. Norton and Co., New York.

Brown, L.R. and E.C. Wolf. 1984. *Soil Erosion: Quiet Crisis in the World Economy.* WorldWatch Paper 60. WorldWatch Institute, Washington, DC.

Cairns, J., Jr., and Project Committee of the National Research Council. 1992. *Restoration of Aquatic Ecosystems. Science, Technology, and Public Policy.* National Academy Press, Washington, DC.

Caraco, N.F. 1995. Influence of human populations on P transfers to aquatic systems: A regional scale study using large rivers. In: H. Thiessen (Ed.), *Phosphorus in the Global Environment.* John Wiley, New York. pp. 235–244.

Carpenter, S.R., N.F. Caraco, D.L. Correll, R.W. Howarth, A.N. Sharpley and V.H. Smith. 1998. Nonpoint pollution of surface waters with phosphorus and nitrogen. *Ecol. Appl.* 8: 559–568.

Chorus, I. and E. Wesseler. 1988. Response of the phytoplankton community to therapy measures in a highly eutrophic urban lake (Schlachtensee, Berlin). *Verh. Int. Verein. Limnol.* 23: 719–728.

Christensen, D.L., B.R. Herwig, D.E. Schindler and S.R. Carpenter. 1996. Impacts of lakeshore residential development on coarse woody debris in north temperate lakes. *Ecol. Appl.* 6: 1143–1149.

Chu, Y. 1993. Shoreline erosion control — engineering considerations. In: Proceedings, U.S. Army Corps of Engineers Workshop on Reservoir Shoreline Erosion: A National Problem. Misc. Paper W-93-1. U.S. Army Corps Engineers, Vicksburg, MS. pp. 33–40.

Clasen, J. and H. Bernhardt. 1987. Chemical methods of P-elimination in the tributaries of reservoirs and lakes. *Schweiz. Z. Hydrol.* 49: 249–259.

Codling, E.E., R.L. Chaney and C.L. Mulchi. 2000. Use of aluminum- and iron-rich residues to immobilize phosphorus in poultry litter and litter-amended soils. *J. Environ.Qual.* 29: 1924–1930.

Comings, K.J., D.B. Booth and R.R. Horner. 2000. Storm water pollutant removal by two wet ponds in Bellevue, Washington. *J. Environ. Eng. Div. ASCE* 126: 321–330.

Cooke, G.D. and R.E. Carlson. 1986. Water quality management in a drinking water reservoir. *Lake and Reservoir Manage.* 2: 363–371.

Coveney, M.F., D.L. Stites, E.F. Lowe, L.E. Battoe and R. Conrow. 2002. Nutrient removal from eutrophic lake water by wetland filtration. *Ecol. Eng.* 19: 141–160.

Dahl, T.E. 1990. *Wetland Losses in the United States 1780s to 1980s.* U.S. Dept. Interior, Fish and Wildlife Service, Washington, DC.

Daniel, T.C., A.N. Sharpley, D.R. Edwards, R. Wedepohl and J.L. Lemunyon. 1994. Minimizing surface water eutrophication from agriculture by phosphorus management. *J. Soil Water Conserv.* 49: 30–38.

Daniel, T.C., A.N. Sharpley and J.L. Lemunyon. 1998. Agricultural phosphorus and eutrophication: A symposium overview. *J. Environ. Qual.* 27: 251–257.

Dierberg, F.E., T.A. DeBusk, S.D. Jackson, M.J. Chimney and K. Pietro. 2002. Submerged aquatic vegetation-based treatment wetlands for removing phosphorus from agricultural runoff: Response to hydraulic and nutrient loading. *Water Res.* 36: 1409–1422.

Diaz, O.A., K.R. Reddy and P.A. Moore, Jr. 1994. Solubility of inorganic phosphorus in stream water as influenced by pH and calcium concentration. *Water Res.* 28: 1755–1763.

Dosskey, M.G. 2001. Toward quantifying water pollution abatement in response to installing buffers on crop land. *Environ. Manage.* 28: 577–598.

Dossey, M.G., M.J. Helmers, D.E. Eisenahuer, T.G. Franti and K.D. Hoagland. 2002. Assessment of concentrated flow through riparian buffers. *J. Soil Water Conserv.* 57: 336–343.

Dou, Z., G.Y. Zhang, W.L. Stout, J.D.Toth and J.D. Ferguson. 2003. Efficacy of alum and coal combustion by-products in stabilizing manure phosphorus. *J. Environ. Qual.* 32: 1490–1497.

Durning, A.B. and H.B. Brough. 1991. Taking Stock: Animal Farming and the Environment. WorldWatch Paper 103. WorldWatch Institute, Washington, DC.

Fiala, L. and P. Vasata. 1982. Phosphorus reduction in a man-made lake by means of a small reservoir on the inflow. *Arch. Hydrobiol.* 94: 24–37.

Fiener, P. and K. Auerswald. 2003. Effectiveness of grassed waterways in reducing runoff and sediment delivery from agricultural watersheds. *J. Environ. Qual.* 32: 927–936.

Fleischner, T.L. 1994. Ecological costs of livestock grazing in western North America. *Conserv. Biol.* 8: 629–644.

Fox, M.A. 1999. The contribution of vegetarianism to ecosystem health. *Ecosyst. Health* 5: 70–74.

Gallimore, L.E., N.T. Basta, D.E. Storm, M.E. Payton, R.H. Huhnke and M.D. Smolen. 1999. Water treatment residual to reduce nutrients in surface runoff from agricultural land. *J. Environ. Qual.* 28: 1474–1478.

Gardner, G. 1996. Shrinking Fields: Cropland Loss in a World of Eight Billion. WorldWatch Paper 131. WorldWatch Institute, Washington, DC.

Gattie, D.K. and W.J. Mitsch (Eds.). 2003. The Philosophy and Emergence of Ecological Engineering. Special Issue. *Ecol. Eng.* 20(5): 327–454.

Gaudreau, J.E., D.M. Vietor, R.H. White, T.L. Provin and C.L. Munster. 2002. Response of turf and quality of water runoff to manure and fertilizer. *J. Environ. Qual.* 31: 1316–1322.

Gburek, W.J., A.N. Sharpley, L. Heathwaite and G.J. Folmar. 2000. Phosphorus management at the watershed scale: A modification of the phosphorus index. *J. Environ. Qual.* 29: 130–144.

Gunsalus, B., E.G. Flaig and G. Ritter. 1992. Effectiveness of agricultural best management practices implemented in the Taylor Creek/Nubbin Slough watershed and the Lower Kissimmee River Basin. *National Rural Clean Water Symposium.* USEPA/625/R-92/006. pp. 161–171.

Hammer, D.A. 1997. *Creating Freshwater Wetlands.* 2nd ed. Lewis Publishers, Boca Raton, FL.

Harper, H.H. 1990. Long-Term Performance Evaluation of the Alum Stormwater Treatment System at Lake Ella, Florida. Project WM 339. Florida Department of Environmental Regulation, Tallahassee, FL.

Harper, H.H., M.P. Wanielista and Y.A. Yousef. 1983. Restoration of Lake Eola. In: Lake Restoration, Protection, and Management. USEPA 440/5-83-001. pp. 13–22.

Harrell, L.J. and S.R. Ranjithan. 2003. Detention pond design and land use planning for watershed management. J. *Water Resourc. Planning Manage.* ASCE 129: 98–106.

Haustein, G.K., T.C. Daniel, D.M. Miller, P.A. Moore, Jr. and R.W. McNew. 2000. Aluminim-containing residuals influence high phosphorus soils and runoff water quality. *J. Environ. Qual.* 29: 1954–1959.

Havens, K.E., V.J. Bierman, Jr., E.G. Flaig, C. Hanlon, R.T. James, B.L. Jones and V.H. Smith. 1995. Historical trends in the Lake Okeechobee ecosystem. VI. Synthesis. *Arch. Hydrobiol./Suppl.* 107: 101–111.

Haygarth, P. 1997. Agriculture as a source of phosphorus transfer to water: Sources and pathways. *SCOPE Newslett.* No. 21.

Heathwaite, L., A. Sharpley and W. Gburek. 2000. A conceptual approach for integrating phosphorus and nitrogen management at watershed scales. *J. Environ. Qual.* 29: 158–166.

Heinzman, B. 1998. Improvement of the surface water quality in the Berlin region. IAWQ 19th Biennial Int. Conference. Vancouver, B.C., Canada. pp. 187–197.

Heinzman, B. and I. Chorus. 1994. Restoration concept for Lake Tegel, a major drinking and bathing water resource in a densely populated area. *Environ. Sci. Technol.* 28: 1410–1416.

Henderson, C.L., C.J. Dindorf and F.J. Rozumalski. 1999. Lakescaping for Wildlife and Water Quality. Minnesota Department of Natural Resources, St. Paul, MN.

Herson-Jones, L.M., M. Heraty and B. Jordan. 1995. Riparian Buffer Strategies for Urban Watersheds. Metropolitan Washington Council of Governments, Washington, DC.

Higgins, M.J., C.A. Rock, R. Bouchard and R.J. Wengrzinek. 1993. Controlling agricultural runoff by the use of constructed wetlands. In: C.A. Moshiri (Ed.), *Constructed Wetlands for Water Quality Improvement.* Lewis Publishers, Boca Raton, FL.

Hoag, J.C., H. Short and W. Green. 1993. Planting techniques for vegetating shorelines and riparian areas. In: H.H. Allen and J.L. Tingle (Eds.), Proceedings, U.S. Army Corps of Engineers Workshop on Reservoir Shoreline Erosion: A National Problem. Misc. Paper W-93-1. U.S. Army Corps of Engineers, Vicksburg, MS. pp. 114–124.

Horner, R. 1995. Training for construction site erosion control and stormwater facility inspection. In: National Conference on Urban Runoff Management: Enhancing Urban Watershed Management at the Local, County, and State Levels. USEPA.625/R-95/003.

Horner, R.R., J.J. Skupien, E.H. Livingston and H.E. Shaver. 1994. Fundamentals of Urban Runoff Management: Technical and Institutional Issues. Terrene Institute, Washington, DC.

Hoyer, M.V. and D.E. Canfield. 1994. Bird abundance and species richness on Florida lakes: Influence of trophic status, lake morphology, and aquatic macrophytes. *Hydrobiologia* 297/280: 107–119.

Hyde, J.E. and T.F. Morris. 2000. Phosphorus availability in soils amended with dewatered water treatment residual and metal concentrations in residual with time. *J. Environ. Qual.* 29: 1896–1904.

Jacoby, J.M., E.B. Welch and I. Wertz. 2001. Alternate stable states in a shallow lake dominated by *Egeria densa*. *Verh. Int. Verein. Limnol.* 27: 3805–3810.

Johnston, C.A. 1991. Sediment and nutrient retention by fresh water wetlands — effects on surface water quality. *Crit. Rev. Environ. Control* 21: 491–565.

Juhle, F.B. and H.H. Allen. 1993. Corps of Engineers' attempts to solve reservoir shoreline erosion problems using innovative approaches. In: Proceedings U.S. Army Corps of Engineers Workshop on Reservoir Shoreline Erosion: A National Problem. Misc. Paper W-93-1. U.S. Army Corps Engineers, Vicksburg, MS. pp. 106–113.

Kadlec, R.H. and R.L. Knight. 1996. *Treatment Wetlands.* Lewis Publishers, Boca Raton, FL.

Keenan, L.W. and E.F. Lowe. 2001. Determining ecologically acceptable nutrient loads to natural wetlands for water quality improvement. *Water Sci. Technol.* 44: 289–294.

Kennedy, G. and T. Mayer. 2002. Natural and constructed wetlands in Canada: An overview. *Water Qual. Res. J. Canada* 37: 295–326.

King, K.W., R.D. Harmel, H.A. Torbert and J.C. Balogh. 2001. Impact of a turfgrass system on nutrient loadings to surface water. *J. Am. Water Resourc. Assoc.* 37: 629–640.

Klein, G. 1988. Ecodynamic changes in suburban lakes in Berlin (FRG) during the restoration process after phosphate removal. In: *Ecodynamics. Contributions to Theoretical Ecology.* Springer-Verlag, New York. pp. 138–145.

Kleinman, P.J.A., A.N. Sharpley, A.M. Wolf, D.B. Beegle and P.A. Moore, Jr. 2002a. Measuring water-extractable phosphorus in manure as an indicator of phosphorus in runoff. *Soil Sci. Soc. Am. J.* 66: 2009–2015.

Knight, R.L., R.W. Ruble, R.H. Kadlec and S.C. Reed. 1993. Database: North American Wetlands for Water Quality Treatment. Phase II Report. USEPA 600/C-94/002.

Knight, R.L., V.W.E. Payne, Jr., R.E. Borer, R.A. Clarke, Jr. and J.H. Pries. 2000. Constructed wetlands for livestock wastewater management. *Ecol. Eng.* 15: 41–56.

Knoll, L.B., M.J. Vanni and W.H. Renwick. 2003. Phytoplankton primary production and photosynthetic parameters in reservoirs along a gradient of watershed use. *Limnol. Oceanogr.* 48: 608–617.

Kogelmann, W.J., H.S. Lin, R.B. Bryant, D.B. Beegle, A.M. Wolf and G.W. Peterson. 2004. A statewide assessment of the impacts of phosphorus-index implementation in Pennsylvania. *J. Soil Water Conserv.* 59: 9–18.

Kovacic, D.A., M.B. David, L.E. Gentry, K.M. Starks and R.A. Cooke. 2000. Effectiveness of constructed wetlands in reducing nitrogen and phosphorus export from agricultural tile dainage. *J. Environ. Qual.* 29: 1262–1274.

Langdale, G.W., W.C. Mills and A.W. Thomas. 1992. Use of conservation tillage to retard erosive effects of large storms. *J. Soil Water Conserv.* 47: 257–260.

Lee, K.H., T.M. Isenhart and R.C. Schultz. 2003. Sediment and nutrient removal in an established multi-species riparian buffer. *J. Soil Water Conserv.* 58: 1–8.

Lee, P., C. Smith and S. Bouten. 2004. Quantitative review of riparian buffer width guidelines from Canada and the United States. *J. Environ. Manage.* 70: 165–180.

Lemunyon, J.L. and R.G. Gilbert. 1993. The concept and need for a phosphorus assessment tool. *J. Prod. Agric.* 6: 484–496.

Linde, D.T. and T.L. Watschke. 1997. Nutrients and sediments in runoff from creeping bentgrass and perennial ryegrass turfs. *J. Environ. Qual.* 26: 1248–1254.

Line, D.E., G.D. Jennings, R.A. McLaughlin, D.L. Osmond, W.A. Harman, L.A. Lombardo, K.L. Tweedy and J. Spooner. 1999. Nonpoint sources. *Water Environ. Res.* 71: 1054–1069.

Luederitz, V., E. Eckert, M. Lang-Weber, A. Lange and R.M. Gersberg. 2001. Nutrient removal efficiency and resource economics of vertical flow and horizontal flow constructed wetlands. *Ecol. Eng.* 18: 157–172.

Lund, J.W.G. 1955. The ecology of algae and waterworks practice. *Proc. Soc. Water Treatment Exam.* 4: 83–109.

Magleby, R. 1992. Economic evaluation of the Rural Clean Water program. In: The National Rural Clean Water Program Symposium. USEPA/625/R-92/006. pp. 337–346.

Manny, B.A., W.C. Johnson and R.G. Wetzel. 1994. Nutrient additions by waterfowl to lakes and reservoirs — predicting their effects on productivity and water quality. *Hydrobiologia* 280: 121–132.

Marion, L.P. Clergeau, L. Brient and G. Bertu. 1994. The importance of avian-contributed nitrogen (N) and phosphorus (P) to Lake Grandlieu, France. *Hydrobiologia* 279/280: 133–147.

McComas, S. 2003. *Lake and Pond Management Guidebook.* Lewis Publishers, Boca Raton, FL.

McDowell, R.W. and A.N. Sharpley. 2001. Phosphorus losses in subsurface flow before and after manure application to intensively farmed land. *Sci. Rural Environ.* 278: 113–125.

McDowell, R.W., A.N. Sharpley, D.B. Beegle and J.L. Weld. 2001. Comparing phosphorus management strategies at a watershed scale. *J. Soil Water Conserv.* 56: 306–315.

Meals, D.W. and R.B. Hopkins. 2002. Phosphorus reductions following riparian restoration in two agricultural watersheds in Vermont, *Water Sci. Technol.* 45: 51–60.

Mehlich, A. 1984. Mehlich 3 soil test extractant: A modification of Mehlich 2 extractant. *Commun. Soil Sci. Plant Anal.* 15: 1409–1416.

Mitsch, W.J. 2003. Ecology, ecological engineering, and the Odum brothers. *Ecol. Eng.* 20: 331–338.

Mitsch, W.J. and J.G. Gosselink. 2000. *Wetlands.* 3rd ed. John Wiley and Sons, New York.

Moore, P.A., Jr. and D.M. Miller. 1994. Decreasing phosphorus solubility in poultry litter with aluminum, calcium and iron amendments. *J. Environ. Qual.* 23: 325–330.

Moore, P.A., Jr., T.C. Daniel and D.R. Edwards. 2000. Reducing phosphorus runoff and inhibiting ammonia loss from poultry manure with aluminum sulfate. *J. Environ. Qual.* 29: 37–49.

Morton, T.G., A.J. Gold and W.M. Sullivan. 1988. Influence of overwatering and fertilization on nitrogen losses from home lawns. *J. Environ. Qual.* 17: 124–130.

Moshiri, G.A. (Ed.). 1993. *Constructed Wetlands for Water Quality Improvement.* Lewis Publishers, Boca Raton, FL.

Moustafa, M.Z. 1999. Analysis of phosphorus retention in free-water surface treatment wetlands. *Hydrobiologia* 392: 41–54.

Myers, D.N., K.D. Metzger and S. Davis. 2000. Status and Trends in Suspended-Sediment Discharges, Soil Erosion, and Conservation Tillage in the Maumee River Basin — Ohio, Michigan, and Indiana. Water Resources Investigative Report 00-4091. U.S. Department of Interior, U.S. Geological Survey, Denver, CO.

Nahm, K.H. 2002. Efficient feed nutrient utilization to reduce pollutants in poultry and swine manures. *Crit. Rev. Environ. Sci. Technol.* 32: 1–16.

Newman, R.M. and D.D. Biesboer. 2000. A decline in Eurasian Watermilfoil in Minnesota associated with the milfoil weevil *Euhrychiopsis lecontei. J. Aquatic Plant Manage.* 38: 105–111.

Newman, R.M., D.W. Ragsdale, A. Milles and C. Oien. 2001. Overwinter habitat and the relationship of overwinter to in-lake densities of the milfoil weevil *Euhrychiopsis lecontei*, a Eurasian watermilfoil biological control. *J. Aquatic Plant Manage.* 39: 63–67.

Nieswand, G.H., R.M. Hordon, T.B. Shelton, B.B. Chavooshian and S. Blarr. 1990. Buffer strip to protect water supply reservoirs — a model and recommendations. *Water Res. Bull.* 26: 959–966.

Novotny, V. 1999. Diffuse pollution from agriculture — a worldwide outlook. *Water Sci. Technol.* 39: 1–14.

Novotny, V. and H. Olem. 1994. *Water Quality: Prevention, Identification and Management of Diffuse Pollution.* Van Nostrand Reinhard, New York.

Oberts, G.L. and R.A. Osgood. 1991. Water-quality effectiveness of a detention/wetland treatment system and its effect on an urban lake. *Environ. Manage.* 15: 131–138.

Oberts, G.L., P.J. Wotzka and J.A. Hartsoe. 1989. The Water Quality Performance of Select Urban Runoff Treatment Systems. Part One. Metro Council Publication No. 590-89-062a. St. Paul, MN.

Oklahoma Conservation Commission. 1996. Confined Animal Inventory: Lake Eucha Watershed. Final Report. Water Quality Division, Oklahoma City, OK.

Oklahoma Water Resources Board. 2002. Water Quality Evaluation of the Eucha/Spavinaw Lake System. Oklahoma City, OK.

Olson, R.K. (Ed.). 1992. Evaluating the role of created and natural wetlands in controlling non-point source pollution. *Ecol. Eng.* 1(1/2): 1–170.

Owen, M. 1975. Cutting and fertilizing grassland for winter goose management. *J. Wildlife Manage.* 39: 163–167.

Pant, H.K., V.D. Nair, K.R. Reddy, D.A. Graetz and R.R. Villapando. 2002. Influence of flooding on phosphorus mobility in manure-impacted soil. *J. Environ. Qual.* 31: 399–405.

Pant, H.K. and K.R. Reddy. 2003. Potential internal loading of phosphorus in a wetland constructed in agricultural land. *Water Res.* 37: 965–972.

Panuska, J.C. and J.G. Schilling. 1993. Consequences of selecting incorrect hydrologic parameters when using the Walker pond size and P8 urban catchment models. *Lake and Reservoir Manage.* 8: 73–76.

Paul, L. 1995. Nutrient elimination in an underwater pre-dam. *Int. Rev. ges. Hydrobiol.* 80: 579–594.

Paul, L. 2003. Nutrient elimination in pre-dams: Results of long term studies. *Hydrobiologia* 504: 289–295.

Paul, L., K. Schroter and J. Labahn. 1998. Phosphorus elimination by longitudinal subdivision of reservoirs and lakes. *Water Sci. Technol.* 37: 235–244.

Piest, R.F., L.A Kramer and H.G. Heinemann. 1975. Sediment movement from loessial watersheds. In: *Present and Prospective Technology for Predicting Sediment Yields and Sources. Proceedings of a Workshop.* U.S. Department of Interior. Agricultural Research Service. ARS-S-40.

Pinel-Allou, B., E. Prepas, D. Planas and R. Steedman. 2002. Watershed impacts of logging and wildfire: Case studies in Canada. *Lake and Reservoir Manage.* 18: 307–318.

Pionke, H.B., W.J. Gburek, A.N. Sharpley and J.A. Zollweg. 1997. Hydrologic and chemical controls on phosphorus losses from catchments. In: H. Tunney et al. (Eds.), *Phosphorus Loss from Soil to Water.* CAB Int. Press, Cambridge, UK. pp. 225–242.

Pote, D.H., W.L. Kingery, G.E. Aiken, F.X. Han, P.A. Moore, Jr. and K Buddington. 2003. Water-quality effects of incorporating poultry litter into perennial grassland soils. *J. Environ.Qual.* 32: 2392–2398.

Prato, T. and K. Dauten. 1991. Economic feasibility of agricultural management practices for reducing sedimentation in a water supply lake. *Agric. Water Manage.* 19: 361–370.

Przepiora, A., D. Hesterberg, J.E. Parsons, J.W. Gilliam, D.K. Cassel and W. Faircloth. 1998. Field evaluation of calcium sulfate as a chemical flocculent for sedimentation basins. *J. Environ. Qual.* 27: 669–678.

Putz, K. and J Benndorf. 1998. The importance of pre-reservoirs for the control of eutrophication in reservoirs. *Water Sci. Technol.* 37: 317–324.

Radomski, P. and T.J. Goerman. 2001. Consequences of human lakeshore development on emergent and floating-leaf vegetation abundance. *North Am. J. Fish. Manage.* 21: 46–61.

Raisin, G.W., D.S. Mitchell and R.L. Croome. 1997. The effectiveness of a small constructed wetland in ameliorating diffuse nutrient loadings from an Australian rural catchment. *Ecol. Eng.* 9: 19–36.

Reddy, K.R., R.H. Kadlec, E. Flaig, and P.M. Gale. 1999. Phosphorus retention in streams and wetlands. A review. *Crit. Rev. Environ. Sci. Technol.* 29: 83–146.

Reid, J.R. 1993. Mechanisms of shoreline erosion along lakes and reservoirs. In: Proceedings, U.S. Army Corps of Engineers Workshop on Reservoir Shoreline Erosion: A National Problem. Misc. Paper W-93-1. U.S. Army Corps Engineers, Vicksburg, MS. pp. 18–32.

Richardson, C.J. and C.B. Craft. 1993. Effective phosphorus retention in wetlands: Fact or fiction? In: G.A. Moshiri (Ed.), *Constructed Wetlands for Water Quality Improvement.* Lewis Publishers, Boca Raton, FL. pp. 271–282.

Richardson, C.J. and S.S. Qian. 1999. Long-term phosphorus assimilative capacity in fresh water wetlands: A new paradigm for sustaining ecosystem structure and function. *Environ. Sci. Technol.* 33: 1545–1551.

Robbins, R.W., J.L. Glicker, D.M. Bloem and B.M. Niss. 1991. *Effective Watershed Management for Surface Water Supplies.* American Water Works Association Research Foundation. Denver, CO.

Sabater, S., A. Butturini, J.C. Clement, T.P. Burt, D. Dowrick, M. Hefting, V. Maître, G. Pinay, C. Postolache, M. Rzepecki, and F. Sabater. 2003. Nitrogen removal by riparian buffers along a European climatic gradient: Patterns and factors of variation. *Ecosystems* 6: 20–30.

Salvia-Castellvi, M., A. Dohet, P. Vanderborght and L. Hoffman. 2001. Control of the eutrophication of the Reservoir Esch-Sur-Sure (Luxembourg): Evaluation of the phosphorus removal by predams. *Hydrobiologia* 459: 61–72.

Scherer, N.M., H.L. Gibbons, K.B. Stoops and M. Muller. 1995. Phosphorus loading of an urban lake by bird droppings. *Lake and Reservoir Manage.* 11: 317–327.

Schindler, D.E., S.I. Geib and M.R. Williams. 2000. Patterns of fish growth along a residential development gradient in north temperate lakes. *Ecosystems* 3: 229–237.

Schmitt, T.J., M.G. Dosskey and K.D. Hoagland. 1999. Filter strip performance and processes for different vegetation, widths, and contaminants. *J. Environ. Qual.* 28: 1479–1489.

Schueler, T.R. 1987. Controlling Urban Runoff: A Practical Manual for Planning and Designing Urban BMPs. Metropolitan Washington Council of Governments, Washington, DC.

Schueler, T.R. 1992. Design of Stormwater Wetland Systems: Guidelines for Creating Diverse and Effective Stormwater Wetlands in the Mid-Atlantic Region. Metropolitan Washington Council of Governments, Washington, DC.

Schueler, T.R. 1995. Stormwater pond and wetland options for stormwater quality control. In: *National Conference on Urban Runoff Management: Enhancing Urban Watershed Management at the Local, County, and State Levels.* pp. 341–346.

Setia, P. and R. Magleby. Measuring physical and economic impacts of controlling water pollution in a watershed. *Lake and Reservoir Manage.* 4: 63–71.

Sharpley, A.N. 1999. Agricultural phosphorus, water quality, and poultry production: Are they compatible? *Poultry Sci.* 78: 660–673.

Sharpley, A.N., B.J. Carter, B.J. Wagner, S.J. Smith, E.L. Cole and G.A. Sample. Undated. Impact of Long-Term Swine and Poultry Manure Applicaton on Soil and Water Resources in Eastern Oklahoma. Technical Bulletin T-169. Agric. Exper. Station, Stillwater, OK.

Sharpley, A.N., T.C. Daniel, J.T. Sims and D.H. Pote. 1996. Determining environmentally sound soil phosphorus levels. *J. Soil Water Conserv.* 51: 160- 166.

Sharpley, A.N., T. Daniel, T. Sims, J. Lemunyon, R. Stevens and R. Parry. 1999. *Agricultural Phosphorus and Eutrophication.* U.S. Department of Agriculture. ARS-149.

Sharpley, A., B. Foy and P. Withers. 2000. Practical and innovative measures for the control of agricultural phosphorus losses to water: An overview. *J. Environ. Qual.* 29: 1–9.

Sharpley, A.N., P. Kleinman and R. McDowell. 2001a. Innovative management of agricultural phosphorus to protect soil and water resources. *Commun. Soil Sci. Plant Anal.* 32: 1071–1100.

Sharpley, A.N., R.W. McDowell, J.L. Weld and P.J.A. Kleinman. 2001b. Assessing site vulnerability to phosphorus loss in an agricultural watershed. *J. Environ. Qual.* 30: 2026–2036.

Sharpley, A.N., J.L.Weld, D.B. Beegle, P.J.A. Kleinman, W.J. Gburek, P.A. Moore, Jr. and G. Mullins. 2003. Development of phosphorus indices for nutrient management planning strategies in the United States. *J. Soil Water Conserv.* 58: 137–152.

Shields, F.D., Jr., A.J. Bowie and C.M. Cooper. 1995. Control of streambank erosion due to bed degradation with vegetation and structure. *Water Res. Bull.* 31: 475–489.

Shreve, B.R., P.A. Moore, Jr., T.C. Daniel, D.R. Edwards and D.M. Miller. 1995. Reduction of phosphorus in runoff from field-applied poultry litter using chemical amendments. *J. Environ. Qual.* 24: 106–111.

Shuman, L.M. 2001. Phosphate and nitrate movement through simulated golf greens. *Water Air Soil Pollut.* 129: 305–318.

Shutes, R.B.E., D.M. Revitt, A.S. Mungur and L.N.L. Scholes. 1997. The design of wetland systems for the treatment of urban runoff. *Water Sci. Technol.* 35: 19–26.

Smith, D.R., P.A. Moore Jr., C.L. Griffis, T.C. Daniel, D.R. Edwards and D.L. Boothe. 2001. Effects of alum and aluminum chloride on phosphorus runoff from swine manure. *J. Environ. Qual.* 30: 992–998.

Spittle, K.S. 2002. Effectiveness of an hydraulically applied mechanically bonded fiber matrix. *Land and Water* 46(3): 55–60.

U.S. Department of Agriculture (USDA) and U.S. Environmental Protection Agency (USEPA). 1999. *Unified National Strategy for Animal Feeding Operations.* USDA and USEPA, Washington, DC.

U.S. Department of Agriculture-Natural Resources Conservation Service (USDA-NRCS). 2001. Iowa Phosphorus Index. USDA–NRCS Technical Note 25. Des Moines, IA.

U.S. Environmental Protection Agency. 1992. The National Rural Clean Water Program. USEPA/625/R-92/006., Washington, DC.

U.S. Environmental Protection Agency. 1993. Urban Runoff Pollution Prevention and Control Planning. USEPA/625/R-93/004., Washington, DC.

U.S. Environmental Protection Agency. 1995. *National Conference on Urban Runoff Management: Enhancing Urban Watershed Management at the Local, County, and State Levels.* USEPA/625/R-95/003.

van der Veen, C., A. Graveland and W. Kats. 1987. Coagulation of two different kinds of surface water before inlet into lakes to improve the self-purification process. *Water Sci. Technol.* 19: 803–812.

Walker W.W., Jr. 1987. Phosphorus removal by urban runoff detention basins. *Lake and Reservoir Manage.* 3: 314–326.

Walker, W.W.. Jr., C.E. Westerberg, D.J.Schuler and J.A. Bode. 1989. Design and evaluation of eutrophication control measures for the St. Paul water supply. *Lake and Reservoir Manage.* 5: 71–83.

Wall, D.H., M.A. Palmer and P.V.R. Snelgrove. 2001. Biodiversity in critical transition zones between terrestrial, fresh water, and marine soils and sediments: Processes, linkages and management implications. *Ecosystems* 4: 418–420.

Wanielista, M.P. 1978. *Stormwater Management. Quantity and Quality.* Ann Arbor Science Publishers, Ann Arbor, MI.

Weibel, S.R. 1969. Urban drainage as a factor in eutrophication. In: *Eutrophication. Causes, Consequences, and Correctives.* National Academy Press, Washington, DC. pp. 383–403.

Welch, E.B. 1992. *Ecological Effects of Wastewater. Applied Limnology and Pollutant Effects.* Chapman and Hall, London, UK.

Wendt, C.J. and H.H. Allen. 2002. Shoreline stabilization using wetland plants and bioengineering. *Land and Water* 46 (3): 16–22.

Wetzel, R.G. 1992. Gradient-dominated ecosytems — sources and regulatory functions of dissolved organic matter in fresh water ecosystems. *Hydrobiologia* 229: 181–198.

Wetzel, R.G. 2001. *Limnology. Lake and River Ecosystems.* Academic Press. New York.

Wilson, C.B. 1979. A Limnological Investigaton of Aurora Shores Lake: A Study of Eutrophication. M.S. Thesis. Kent State University, Kent, OH.

Williamson, R.B., C.M. Smith and A.B. Cooper. 1996. Watershed riparian management and its benefits to a eutrophic lake. *J. Water Resour. Planning Manage. ASCE* 122: 24–32.

Woltemade, C.J. 2000. Ability of restored wetlands to reduce nitrogen and phosphorus concentrations in agricultural drainage water. *J. Soil Water Conserv.* 55: 303–308.

Wu, J.S., R.E. Holman and J.R. Dorney. 1996. Systematic evaluation of pollutant removal by urban wet detention ponds. *J. Environ. Eng. Div. ASCE* 122: 983–988.

Yin, C. and B. Shan. 2001. Multipond systems: A sustainable way to control diffuse phosphorus pollution. *Ambio* 30: 369–375.

Young, S.N., W.T. Clough, A.J. Thomas and R. Siddall. 1988. Changes in plant community at Foxcote Reservoir following use of ferric sulphate to control nutrient levels. *J. Instt. Water Environ. Manage.* 2: 5–12.

Zaimes, G.N., R.C. Schulz and T.M. Isenhart. 2004. Stream bank erosion adjacent to riparian forest buffers, row-crop fields, and continuously-grazed pastures along Bear Creek in central Iowa. *J. Soil Water Conserv.* 59: 19–27.

6 Dilution and Flushing

6.1 INTRODUCTION

Dilution and flushing can achieve improved quality in eutrophic lakes by reducing the concentration of limiting nutrient (dilution) and by increasing the water exchange (flushing) rate. Both processes can reduce the biomass of plankton algae, by reducing the inflow concentration of limiting nutrient, resulting in a decreased lake concentration, on which maximum biomass depends. By increasing the water input the flushing rate is increased, which in turn increases the loss rate of plankton algae from the lake. Dilution can be effective even when the increase in flushing rate is insufficient to cause a significant loss of algae. On the other hand, flushing rate increase can cause a significant loss without achieving a reduction in the limiting nutrient concentration. Other effects of dilution are also possible, such as increased vertical mixing and a decrease in the concentration of algal excretory products, which can influence the kinds and abundance of algae (Keating, 1977).

Dilution is usually feasible only where large quantities of low-nutrient water are available. Treatment effectiveness is greatest when dilution water is low in limiting nutrient concentration relative to that in the lake and its natural inflow. Lake nutrient concentration can be more effectively lowered if dilution water is the dominant inflow. In some instances, improvements can be achieved by adding water with moderate to high nutrient content, but these results are less certain than with low-nutrient water, largely because lowering the lake concentration is more effective at reducing biomass than washout.

Dilution and flushing have worked successfully in several lakes. Green Lake in Seattle, Washington, was improved markedly by adding city water, beginning in the 1960s (Oglesby, 1969). Moses Lake in eastern Washington has received Columbia River dilution water on a regular basis since 1977, resulting in substantial improvement in lake quality (Welch and Patmont, 1980; Welch and Weir 1987; Welch et al., 1989, 1992). Dilution was instituted in three other lakes in Washington State and was proposed for Clear Lake, California (Goldman, 1968). Lake Bled, Yugoslavia, was flushed intentionally with water from the River Radovna (Sketelj and Rejic, 1966). Lakes Veluwe and Donten in The Netherlands have been diluted with relatively low-P water during winter since 1979 (Hosper, 1985; Hosper and Meyer, 1986). Snake Lake, Wisconsin, was diluted by removing the equivalent of three lake volumes, which allowed low nutrient groundwater to refill the small seepage lake (5 ha, mean depth 2.3 m) (Born et al., 1973).

In what must be one of the world's first lake flushing experiments, water was diverted from Switzerland's Ruess River to Rotsee in 1921 to 1922 to alleviate eutrophic conditions (Stadelman, 1980). The flushing rate of this 460-ha lake was increased from 0.33 to 2.5/yr (or about 0.1 to 0.7%/d) following construction of a canal between the Ruess River and the lake. The lake's state did not improve because the increased flushing rate was insufficient to significantly washout biomass, and because of high concentrations of nutrients in the river water used for flushing. The nutrients originated in sewage effluent from the upstream city of Luzern. There was still no improvement following diversion of direct inputs of sewage effluent to the lake in 1933. There was considerable improvement, however, after nutrient removal from Luzern's wastewater, in the 1970s, resulting in a tenfold P reduction in the Ruess River inflow water.

Following is a discussion of the theoretical basis for the dilution and flushing, reviews of the Moses Lake, Green Lake, and Lake Veluwe cases and general guidelines for application of the technique. The latter includes quantity and quality of water, frequency of application, and project and operating costs.

6.2 THEORY AND PREDICTIONS

Maintenance of low phytoplankton concentrations by high natural rates of dilution and flushing is a commonly observed phenomenon (Dickman, 1969; Dillon, 1975; Welch, 1969). The mechanisms involved in the control of nutrients and/or algal biomass in lakes are in many ways analogous to those in continuous culture systems. When low-nutrient dilution water is added to a laboratory continuous algal culture, the inflow concentration of limiting nutrient is reduced, the maximum biomass concentration possible in the reactor vessel is likewise reduced, and at the same time nutrients and algal biomass are more rapidly washed from the reactor vessel since the water exchange rate is increased. Concentration of limiting nutrient is the critical variable that determines algal biomass in many lakes, as well as in continuous culture systems. The in-lake nutrient concentration may sometimes be more, but is usually less than the inflow concentration, because sedimentation is greater than internal loading. However, increased rates of dilution/flushing will theoretically reduce loss through sedimentation, as demonstrated by predictions using the Vollenweider equation (Figure 3.8), where at short detention times, the lake P concentration equals the inflow concentration. That situation is most typical of reservoirs, which tend to have shorter detention times than lakes and would, therefore, behave more like a continuous culture system with algae (McBride and Pridmore, 1988).

To predict the response of a lake to the addition of low-nutrient water on a day-to-day or year-to-year basis, without consideration of sedimentation or internal loading, the following "dilution only" first order, integrated equation is useful:

$$C_1 = C_i + (C_o - C_i) e^{-\rho t} \tag{6.1}$$

where C_1 is the concentration at time t; C_i is the concentration in the inflow water; C_o is the initial lake concentration; and ρ is the water exchange or flushing rate. This equation assumes the lake is well mixed, that no other nutrient sources exist, and that the limiting nutrient or "percent lake water" can be treated as conservative. Because this equation does not include a sedimentation term, it is normally useful only in the short term as a tracer for nutrient behavior and under conditions of rather large water exchange rates (several percent per day or more). However, in some instances (e.g., Lake Norrviken, Figure 4.6) the response of lake nutrient concentration could follow a simple dilution model if there were no retention of nutrient in sediments, i.e., sediment release equals sediment uptake. In most instances, however, it could be used to estimate the potential for reducing average lake concentrations with a given source of water, and the time necessary for that reduction. Such predictions can be compared with the observed distribution of dilution water — as percent lake water — indicated by a conservative variable, such as sodium or specific conductance. For more realistic predictions, sediment–water interchange of nutrients must be considered. Equations given in Chapter 3 are pertinent in that regard where flushing rate and external loading terms are modified for the added dilution water.

Increased flushing has an indirect effect on lake P concentration, as shown with a Vollenweider steady state, mass balance P model. Adding more water with lower nutrient content also increases nutrient loading, while the resulting increased flushing rate decreases nutrient loss through sedimentation (Uttormark and Hutchins, 1980). These processes could be counteracting the dilution effect in some instances, because, as the authors stated, "a reduction in the influent concentration tends to reduce in-lake concentration, but a reduction in phosphorus retention tends to increase in-lake concentration." They showed that a large increase in the combined flushing rate obtained by adding low-nutrient water (40% of the normal inflow nutrient content) could theoretically increase

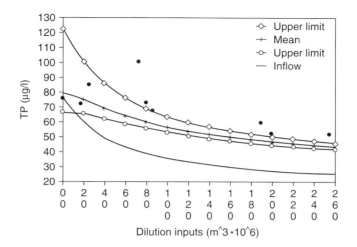

FIGURE 6.1 Predicted and observed (solid circles) column-weighted mean TP as a function of dilution water input in Lower Parker Horn (7) during May–September for the 9-year average TP loading, flushing rate, and range of RTR values observed. (From Jones, C.A. and E.B. Welch. 1990. *J. Water Pollut. Cont. Fed.* 62: 847–852. With permission.)

the lake nutrient concentration if the original flushing rate ρ is low enough, e.g., 0.1/yr. If the flushing rate is relatively large (≥ 1.0/yr) initially, the effect of reduced sedimentation rate is minimized and a reduction in lake concentration will result, but large quantities of water are necessary. Of course, the amount of water needed to achieve a given reduction in inflow concentration is a function of the concentration difference between the normal inflow and dilution water source.

In an actual case, the effect of diminished sedimentation was not significant, but the effect of dilution was reduced because of enhanced sediment release of P (Jones and Welch, 1990). Additions of dilution water (plus normal inflow) to Parker Horn of Moses Lake, Washington resulted in average flushing rates of 8%/d from April to September during 12 years of treatment. Based on a calibrated and verified steady state, mass balance P model, where net internal loading of P was indicated by a negative sedimentation rate coefficient (Equation 3.10), increased dilution water input predicted a much more gradual reduction in lake P concentration than was expected from the diluted inflow concentration alone (Figure 6.1). The minimizing effect of dilution water could not have been caused by reduced sedimentation, because the lower curve of inflow concentration is the highest possible lake concentration with no sedimentation.

6.3 CASE STUDIES

Two lakes where the dilution and flushing technique was implemented can be used as guides for application elsewhere. Moses Lake lies in eastern Washington, and has a surface area of 2,753 ha and mean depth of 5.6 m. Dilution water has been added to one arm of Moses Lake during spring and early summer since 1977. Transport to previously undiluted portions of the lake, by pumping, was begun in 1982. Green Lake, in Seattle, has an area of 104 ha and a mean depth of 3.8 m. The lake received dilution water from the city domestic supply at relatively high rates from 1962 through the mid 1970s, but inputs were subsequently variable resulting in worsening lake quality. The effect of low-nutrient water was to dilute internal loading, the primary cause for summer algal blooms. The cost and effectiveness of other, more reliable dilution-water sources and other controls on internal loading were evaluated and alum (Chapter 8) was implemented in 1991 (URS, 1983, 1987; Jacoby et al., 1994). The suitability of dilution water sources for treating these lakes is apparent from the large ratios of nutrient content in the lakes relative to the dilution inflows, ranging from

5:1 to 10:1. A third case to be discussed is Lake Veluwe in The Netherlands (Hosper, 1985; Hosper and Meyer, 1986), where dilution was used together with wastewater P removal.

6.3.1 MOSES LAKE

Dilution water from the Columbia River has been added to Moses Lake's Parker Horn via the U.S. Bureau of Reclamation's East Low Canal and Rocky Coulee Wasteway (Figure 6.2). The pumping of dilution water from Parker Horn to the previously undiluted Pelican Horn began in 1982, and sewage effluent was diverted from Pelican Horn in 1984. The effects of dilution water only on lake quality can be evaluated for Parker Horn for all years. Because South Lake was affected to some extent by sewage effluent, the added improvement for that area after 1984 was partly due to diversion. The principal effects in Pelican Horn, however, were from sewage diversion. The ashfall from the eruption of Mount St. Helens reduced the lake's internal P loading during 1980 to 1981, so effects of dilution are obscured during those years when dilution water additions were also low. Therefore, periods of evaluation are for 1977 to 1979 and 1986 to 1988 in Parker Horn (dilution only) and 1977 to 1979 (dilution only) and 1986 to 1988 (dilution plus sewage diversion) in South Lake.

The patterns of dilution water addition have been varied but not in a systematic way conducive to determining optimum quantity and seasonal distribution. The average amount of dilution water added from 1977 through 1988 was 169.4×10^6 m^3/yr, which represented a flushing rate in Parker Horn of 17%/d for the 971 d of actual inflow. The average input for April to September (includes days with and without input) was 130×10^6 m^3/yr or 5.8%/d. With dilution water plus the normal input, the flushing rate averaged 7.8%/d for Parker Horn. For the whole lake, these inputs represented a flushing rate of less than 1%/d. Thus, dilution water input created flushing rates in Parker Horn that could have caused some washout of algal cells, but such an effect would not have been significant in the remainder of the lake.

Dilution water addition continued at a slightly higher rate during the 1990s, with input averaging 221×10^6 m^3/yr over the subsequent 13-year period through 2001 — a 30% increase over the previous 12 years. Inputs during the wet years of 1996 and 1997 were only 75 and 32×10^6 m^3/yr. So while dilution has not been constant from year-to-year or evenly distributed in time, water was delivered every year except one during the past 25 years. Nevertheless, low dilution water years (e.g., 1997) still produced large algal blooms and poor lake quality.

Columbia River water was nearly ideal for dilution (Table 6.1). Because the P and N concentrations in Crab Creek (Parker Horn's natural inflow) were so high, apparently due to irrigation and fertilization practices in the watershed, relatively large quantities of Columbia River water were needed to significantly lower the composite inflow concentration, and thus to lower the in-lake concentration. This resulted in larger exchange rates than would otherwise have been necessary without the Crab Creek inflow. Unfortunately, diverting Crab Creek was not economically feasible in this case, but such a manipulation could be considered for other lakes to obtain more efficiency from dilution water quantities.

The addition of dilution water to Moses Lake predictably and rapidly replaced lake water, as judged by tracing a conservative parameter, specific conductance. Values for percent lake water were calculated, assuming that 100% was represented by the conductance of Crab Creek and 0% by the conductance of Columbia River water. For example, using a typical value of 460 μmhos/cm for Crab Creek (CCW), 250 μmhos/cm for lake water (LW), and 120 μmhos/cm for East Low Canal dilution water (ECDW), the percent lake water would be

$$100(LW - ECDW) / (CCW - ECDW) = \%LW$$

$$100(250 - 120) / (460 - 120) = 38$$

(6.2)

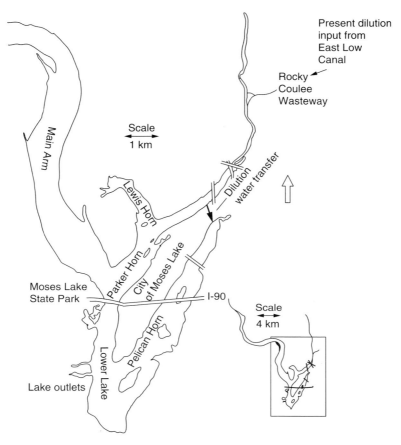

FIGURE 6.2 Moses Lake, Washington, Note source of dilution water from East Low Canal via Rocky Coulee Wasteway and point of pumped transfer of dilution water from Parker Horn to Pelican Horn. (From Cooke, et al. 1993. With permission.)

TABLE 6.1
Nutrient Concentration (µg/L) in Inflow Water to Parker Horn from May to September, 1977 and 1978

	Total P	Total N	SRP	NO$_3$–N
Crab Creek inflow without dilution	148	1331	90	1096
East Low Canal dilution water	25	305	8	19

Source: From Cooke et al., 1993. With permission.

Percent lake water, which decreases with dilution, is used instead of percent dilution water, to represent the behavior of nutrients in the lake. Remaining lake water in Parker Horn, where water enters (Figure 6.2), was reduced to values around 20%, much less than in other parts of the lake (Figure 6.3). This was expected (dashed line, for Equation 6.1), because the average dilution rate during the period from April to June, described here, was 15%/d for Parker Horn, which is a small (8%) portion of the lake volume. The dilution rate decreased, as the water moved through other parts of the lake. As the dilution water input declined in June, the fraction of lake water in Parker

Horn quickly rose to between 50% and 60% (Figure 6.3). Part of that increase was probably caused by wind pushing water from the Main Arm and South Lake into Parker Horn.

Moses Lake is dendritic in shape and most of the lake's volume (63%) is in the Main Arm, out of a direct path from the dilution water inflow. Thus, dilution water was expected to have little effect in the main arm compared to Parker Horn and South Lake, which together represent 29% of the lake volume. However, the lake water residual decreased similarly to that for the lower lake, even if the whole lake volume was used to calculate residual lake water and water exchange rate (solid line and Equation 6.1). Lake water residuals in the whole lake and lower lake reached levels between 50% and 60% in late May and early June and then began a more gradual return to normal as dilution input declined (Figure 6.3).

Improvement in lake quality during the first three years (1977 to 1979), compared to predilution years 1969 to 1970, was near or in excess of 50% for TP, SRP, and chlorophyll (chl) *a*. Secchi transparency increased markedly, not only for Parker Horn but for South Lake as well (Table 6.2). Total N decreased by about the same magnitude, although predilution data for N are incomplete. By 1986 through 1988, lake quality improved even more. The further improvement in Parker Horn was due to reduced inflow P (50%) in Crab Creek during the 12-year study period. The further improvement in most of the lake volume was due partly to sewage diversion, which primarily affected South Lake (Table 6.2). Pelican Horn, which was influenced entirely by groundwater and sewage effluent, showed little effect of dilution water in 1977 to 1979. The extensive improvement there in 1986 to 1988 (Table 6.2) was due primarily to sewage diversion.

While dilution water was distributed throughout the lake, improvement in quality was greatest in Parker Horn, where the fraction of dilution water was highest (Figure 6.3). Wind was probably the main force causing the transport of dilution water from Parker Horn into the Main Arm, with the fraction of dilution water, which reached half the distance through the main arm, being dependent largely upon the fraction existing in Parker Horn (Welch et al., 1982). About half the natural surface inflow and P load entered the Main Arm via Rocky Ford Creek. Phosphorus concentration in that source did not decline during the 12-year study, as it had in Crab Creek. Thus, the continued trend in quality improvement was not as apparent in the Main Arm.

Means for the May to September periods obscure the extreme conditions, such as a maximum Secchi transparency of 3 m in June (4 m was reached in 1982) throughout most of the lake. Chl *a* reached peaks near 50 µg/L in late July and August after dilution water input was curtailed for two to four weeks. Unless dilution water was added continually, blooms returned as the fraction of dilution water in the lake declined. This "boom and bust" phenomenon, promoted by large inputs followed by no input at all, did not produce the optimum effects that would have occurred with a continual input at low rates throughout the summer, while employing similar total amounts of water. The large quantities added over a short period of time — exchanging water in Parker Horn at the rate of about 20%/d and in most of the lake at 2–3%/d — are probably unnecessary, considering that induced flushing rates throughout most of the lake were insufficient to cause significant washout of algal cells and most improvement was due to dilution.

Determining the optimum quantity of dilution water and its distribution over time requires defining the cause(s) for the quality improvement, in this case algal reduction. The reduction of N concentration was the most probable cause for the dilution effect on algal biomass following the initiation of dilution and prior to the Mount St. Helens ashfall. Nitrate, rather than SRP, was the nutrient that most frequently limited algal growth rate (Welch et al., 1972). Nitrate *per se* was not appreciably reduced by dilution, because it was limiting and remained rather low in the lake water during summer both before and after dilution. Although control of biomass was discernable at total N concentrations in lake water below about 600 µg/L (Welch and Tomasek, 1981), the best relationship was found between flow weighted NO_3 concentration in the inflow (Crab Creek) and the average chl *a* concentration in Parker Horn and South Lake ($r = 0.97$; Welch et al., 1984). Controlling algal biomass by reducing the inflow concentration of limiting nutrient, which in its

TABLE 6.2

Average April–September Dilution Rates through Discrete Sections of Moses Lake and Resulting Average May–September Values for TP, SRP, Chl *a* and Secchi Transparency for before (1969–1970) and after (1977–1979) Dilution (except for Pelican Horn) and after Dilution and Sewage Diversion (1986–1988)

Years/Lake Area	Dilution Rate (%/d)	TP (µg/L)	SRP (µg/L)	Chl *a* (µg/L)	Secchi (m)
		Parker Horn			
1969–1970	1.6	152	28	71	0.6
1977–1979	7.8	68	15	26	1.3
1986–1988	8.0	47	6	21	1.5
		South Lake			
1969–1970	1.1	156	48	42	1.0
1977–1979	3.5	86	35	21	1.7
1986–1988	3.6	41	7	12	1.7
		Pelican Horn			
1969–1970	0.0	920	634	48[a]	0.40
1977–1979	0.0	624	441	39[a]	0.45
1986–1988	7.7	77	6	12	0.65

[a] Chl *a*: biolvolume ratios one half rest of lake.

Note: Samples from 0.5 m depth transects.

Source: From Cooke et al., 1993. With permission.

soluble, available form continues to remain at low concentration in the lake (seemingly unrelated to biomass), is analogous to the functioning of a continuous culture system (see Welch, 1992).

Prior to implementation of this project, it was thought that TP would be the most important nutrient to control in order to reduce algal biomass. Although NO_3 limited growth rate during summer, fixation of atmospheric N by blue-green algae should have supplied enough N to complement the available P supply. N fixation is not a rapid process; maximum rates of cell N replacement and growth of about 5% and 10%/d have been reported (Horne and Goldman, 1972; Horne and Viner, 1971). A growth rate by N fixation of only 2.4 ± 1.8%/d was determined for Moses Lake algae (Brenner, 1983). With such slow growth, increasing the flushing rate by about tenfold may well have prevented N uptake by fixation from fully utilizing the available P. In any event, there was a close relationship between inflow NO_3 and chl *a*. Using data before and three years after dilution started (1977 to 1979), that relationship allowed estimates of optimum dilution water input.

The Mount St. Helens ashfall produced a seal over Moses Lake sediments and stopped internal P loading for two years (Welch et al. 1985; Jones and Welch, 1990). That event, coupled with the continued reduction in inflow P concentration to Parker Horn, resulted in a trend toward P limitation (Welch et al., 1989, 1992). Sewage diversion from Pelican Horn also contributed to that trend in the lower lake. For 1986–1988 when P was limiting, the Jones and Bachmann (1976) relationship provided an adequate fit of the data for predicting chl *a* from TP. A steady state model for TP was developed in which internal loading was predicted as a function of flushing rate and relative thermal resistance to mixing (RTRM) (Welch et al., 1989; Jones and Welch, 1990). The resulting relationship between dilution water input and lake TP in Parker Horn shows the moderating effect of internal P loading on the effectiveness of dilution water input (Figure 6.1).

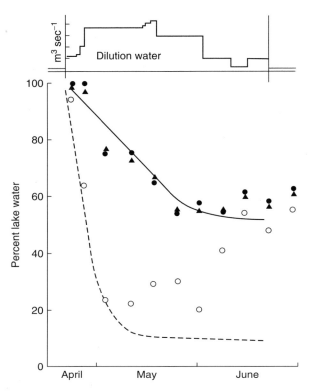

FIGURE 6.3 Residual lake water, in percent, remaining in Parker Horn (open circles), South Lake (closed circles), and the whole lake (triangles) compared with that predicted (based on average inflow from mid-April to mid-June) for the whole lake (solid line) and Parker Horn (dashed line) in response to dilution water input in 1978. Parker Horn, South Lake, and the whole lake represent, respectively, 8%, 21%, and 100% of the lake volume. (Reprinted from Welch, E.B. and C.R. Patmont. 1980. *Water Res.* 14: 1317–1325. With permission from Pergamon Press Ltd., Oxford.)

Goldman (1968) suggested that reducing the N content in Clear Lake, CA (a hypereutrophic, shallow lake in an arid region, like Moses Lake), by adding Eel River water, should reduce algal content (as it did initially in Moses). There was uncertainty about the buffering effect of increased release of N from sediments once the lake water N content decreased and a larger gradient between sediment interstitial N and the overlying water occurred. That effect did occur in Moses Lake, as indicated earlier, although the effect of increasing RTRM was 2.8 times more important, based on the steady state P model, than increasing the flushing rate of dilution water (Jones and Welch, 1990). Reducing the inflow concentration of P, either by adding dilution water or by removing P from the external input, results in an increased gradient in P concentration between sediment and water and, thus, a greater diffusive flux from sediments (Poon, 1977; Sas et al., 1989). A decrease in internal loading as sediments become depleted of P is possible with continued dilution as with wastewater P removal or diversion. The year-to-year variability in internal load was too great to detect any trend in Moses Lake with the 12 years of data through 1988. However, a 2001 P budget shows a negative P internal loading for the April–September period (Carroll, personal communication), compared with internal loading being 53% of the total for that period in 1988.

The physical loss of algal cells by washout has probably contributed to reduced biomass in Parker Horn, where high rates of exchange (20–25%/d) existed for short periods. For example, biomass in upper Pelican Horn decreased from 80 to 10 mm³/L in about 1 month following increased water exchange by pumping Parker Horn water through Pelican Horn (Carlson and Welch, 1983; Welch et al., 1984). The greatest decrease occurred with an exchange rate of 9%/d, although subsequent exchange rates were higher (19%/d). Persson (1981) observed that *Oscillatoria* biomass was greatly influenced by flushing rate during the period (about a month) of maximum growth in a hypereutrophic brackish-water bay. At a flushing rate of 8.1% and 9.4%/d the average biomass was about one half the level at 4.7%/d. At 20.7%/d biomass was only about one third the level at 4.7%/day.

For Parker Horn, therefore, where the mean flushing rate was about 8%/d from April to September (Table 6.2), washout of cells probably represented part of biomass control. In the remainder of the lake, where flushing averaged less than 1%/d, washout was of less significance when compared with a 50%/d maximum growth rate, which was observed for *Aphanizomenon* in the lake. If the N source is fixation, however, even relatively low flushing rates may effectively reduce the biomass of that alga through washout.

Instability of the water column, as indicated by a decreased vertical density gradient (low relative thermal resistance to mixing, RTRM), has contributed to the crash or prevention of blue-green algal blooms in Moses Lake (Welch and Tomasek, 1981). Because the buoyancy capability of blue-greens provides advantages over greens and diatoms when mixing is poor, decreased stability hinders dominance by blue-greens (Knoechel and Kalff, 1975; Paerl and Ustach, 1982; Chapter 19). Daily monitoring showed that blue-green biomass increases and surface accumulations become more pronounced under quiescent conditions, but biomass disperses with increased mixing from wind > 4.9–7.6 m/h (Bouchard, 1989; Welch et al., 1992). Wind has more effect on water column stability in Moses Lake than dilution water input.

While biomass was substantially reduced by dilution, algal composition did not change during the 12 years dilution. Blue-greens dominated the phytoplankton throughout the summer (Welch et al., 1992). The blue-green fraction decreased initially (Welch and Patmont, 1980), but that did not persist. This was unexpected since decreased blue-green dominance has accompanied decreased TP content in most situations (Sas et al., 1989). However, the first intensive monitoring since 1988, conducted during 2001, the fifth highest dilution water input year, showed low mean surface TP (20 µg/L), low chl *a* and a near absence of blue-green algae in Parker Horn and South Lake (Carroll, personal communication).

The optimum use of dilution water for Moses Lake is a moderate, but continuous input from May through August. Water added too early (February to March) would be largely replaced by high-nutrient Crab Creek water by June, when algal blooms begin. Replacement with Crab Creek water creates a problem if dilution water is stopped in June or early July. Unfortunately, the lack of irrigation demand during wet years reduces space available in the downstream impoundment for dilution water routed through Moses Lake. Thus, dilution water transport through the lake is higher in dry years. Nevertheless, based on the earlier relationship between May to August average inflow NO_3 concentration and June to August average chl *a*, a dilution water volume of about 100×10^6 m³ from May through August should control chl *a* to an average of about 20 µg/L, whether that quantity comes as 10 m³/s for the whole period or is divided into 25 m³/s for May and 5 m³/s for June through August. If water input is not maintained through August, blooms with very high chl *a* levels will result. Also as indicated in Figure 6.1, that volume should result in a TP concentration of about 55 µg/L and a chl *a* content of about 24 µg/L if P is limiting. Also as indicated by Figure 6.1, there is a diminishing value to more dilution water and any further improvement in the quality of Moses Lake must involve control of internal P loading. However, if the lack of internal P loading observed in 2001 is representative of conditions in recent years, effectiveness of long-term dilution is greater than expected from Figure 6.1.

Although the input of dilution water has not been distributed over the summer as desired, the water has cost the Moses Lake Irrigation and Rehabilitation District nothing, because water diverted through Moses Lake is used for downstream irrigation. The primary project cost has been the pumping facility for Pelican Horn, at $577,000 (2002 U.S. dollars) plus planning, administrative, and monitoring and research cost. To guarantee a flow of 5 m³/s during July and August, however, it would be necessary for the District to buy water. Although this is not planned and costs and liabilities have not been considered, it is instructive to assume a cost (2002 dollars) similar to that encountered for diluting Green Lake ($0.13/m^3; see Green Lake section). For two months the volume would be about 26×10^6 m^3, with a cost of nearly 3.5×10^6. The cost for dilution water at Moses Lake would probably not be as high as that for the domestic water used in Green Lake. However, it becomes clear that to buy dilution water may be a practical restoration alternative only for relatively small lakes, such as Green Lake, that require much lower rates of dilution water input. Green Lake is not fed naturally by a large input of high-nutrient water that must be diluted in order to dilute the lake nutrient concentration. In that respect, Moses Lake may be a rather unique case of a large lake existing at a location where large amounts of low-nutrient water are available at no cost.

6.3.2 GREEN LAKE

This is another example of the benefits of dilution. The setting is the Seattle metropolitan area; 47,000 people live within 1.6 km of the lake, and as many as 1,162 people per hour use the 4.5-km path around the lake. Perhaps the smaller Green Lake (100 ha) represents a more practical example of dilution than does Moses Lake (2,700 ha).

Dilution was proposed as the primary treatment in 1960 (Sylvester and Anderson, 1964) and was instituted in 1962. In contrast to the high rate of water input to Moses Lake's Parker Horn, the dilution of Green Lake represented a much lower rate of flushing, even less than the whole of Moses Lake (2–3%/day). The average combined flushing rate was increased nearly threefold by adding low-nutrient water from the Seattle domestic supply, which comes from diversions near the source of two Cascade mountain streams. The addition of dilution water to Green Lake from 1965 to 1978 produced a flushing rate, based on dilution water only, ranging from 0.88 to 2.4/yr (0.24–0.65%/d).

A marked improvement in chl a, TP, and Secchi transparency (SD) was noted during the first few years of dilution. Only one year's predilution data existed for comparison with the three years of postdilution monitoring. Water transparency during the summer increased nearly fourfold, to an average of 4 m (because the mean depth is 3.8 m, most of the lake bottom was visible), and chl a decreased more than 90% from 45 to 3 µg/L. The summer mean TP decreased from 65 to 20 µg/L. A substantial decrease in the blue-green algal fraction was observed, particularly during spring and early summer.

Regular monitoring was terminated in 1968, but the lake was again studied intensively for the purpose of proposing a new restoration plan (Perkins, 1983; URS, 1983). Mean chl a and TP had increased to 38 and 55 µg/L, respectively, during the summer of 1981; the lake quality had degraded markedly during the late 1970s, primarily due to declining dilution water inputs. No water was added in 1982, resulting in massive blue-green algal blooms. Dilution water was subsequently added in modest amounts on a regular basis to avoid deterioration in lake quality. Future limitation in the availability of Seattle domestic water necessitated developing a long-term solution.

The percent decrease in TP concentration following initiation of dilution was about what would be expected from Equation 3.10. The expected TP concentration in Green Lake prior to dilution, calculated from estimated external loading, should have been about 80 µg/L, but was only 65 µg/L. Following dilution, the steady state concentration should have been about 35 µg/L; however, it actually declined to 20 µg/L by 1967 (Welch, 1979). The pre- to post-dilution decrease was the same (45 µg/L) for the expected and the actual TP values. The discrepancy is most likely due to overestimating external loading.

Mass balance analysis in the 1980s showed that much of the lake's problem was due to internal P loading, which was unrecognized earlier. Internal loading was high despite nearly the whole lake being unstratified and oxic during summer (Perkins, 1983; URS, 1983). Internal loading accounted for 21% of the total annual P loading during 1981 (determined by difference in the annual mass balance of P). During the three summer months, however, when chl a averaged 38 µg/L, with a maximum of 60 μg/L, the internal source accounted for 88% of the total. By calibrating a non-steady state model (Equation 3.23), Chapter 3), the average areal gross release rate was determined to be 4.0 mg/m² per day (Mesner, 1985). Some of that release can be attributed to migration of P-rich *Gloeotrichia echinulata* from sediments (Barbiero and Welch, 1992). Three storm drains contributed 41% of the annual external loading to the lake, but their inputs are largely confined to the winter, high runoff period.

Dilution remained the principal choice to improve and maintain Green Lake quality, with a goal of achieving a mean summer TP concentration of 28 μg/L. Water from the city supply is most desirable because the delivery facility is in place, and because the city water P concentration is only 10 μg/L. However, city water costs about $0.13/m³ (2002 dollars), and the supply during summer was no longer routinely available (URS, 1983). Other water sources were considered, even though construction of delivery systems would be required. Local groundwater supplies contained too much P. Lake Washington water is sufficiently low in TP concentration (15 μg/L), and its use seemed feasible with construction of a piping system. However, the State's objection to withdrawal, and potential local impacts of the piping system, resulted in favoring an alum alternative (Chapter 8).

If the use of Lake Washington water were feasible, one could calculate a first approximation of the amount of dilution water necessary to reduce the summer lake TP from 55 to 28 μg/L. The annual mean TP of 42 μg/L was 0.76 of the summer mean, so assuming that factor is reasonable, the annual mean should be reduced to 21 μg/L to achieve a summer mean of 28 μg/L. Calibrating a TP steady state model (Equation 3.12) for the 1981 external loading of 203 mg/m² per yr (160 external + 43 internal) gives the following estimate for the sedimentation rate coefficient, which integrates external loading and net internal loading:

$$\sigma = \frac{L}{TP \overline{Z}} - \rho$$

$$\sigma = \frac{203 \text{ mg/m}^2 \cdot yr}{42 \text{ mg/m}^3 \times 3.8 \text{ m}} - 0.92/yr = 0.35/yr$$

To achieve 21 μg/L (mg/m³) as an annual mean, the steady state Equation 3.10 can be solved for the unknown dilution water inflow (X) assuming a Lake Washington TP of 15 μg/L (A_o = area, V = volume):

$$TP = \frac{L + (15X / A_o)}{\overline{Z}(X / V + \sigma)}$$

$$21 \text{ mg/m}^3 = \frac{203 \text{ mg/m} \cdot yr + \dfrac{(15 \text{ mg/m}^3 \cdot X \text{ m}^3/yr)}{1.04 \times 10^6 \text{ m}^2}}{3.8 \text{ m} \left(\dfrac{X \text{ m}^3/yr}{3.95 \times 10^6 \text{ m}^3} + 0.35/yr \right)}$$

Rearranging and solving for *X* gives

$$21.8X + 29 = 203 + 14.4X$$

$$X = 23.5 \times 10^6 \text{ m}^3/yr$$

The addition of this much water would increase the flushing rate from 0.92 to 5.95/yr and reduce the sedimentation rate coefficient, which is not taken into account. However, there are other

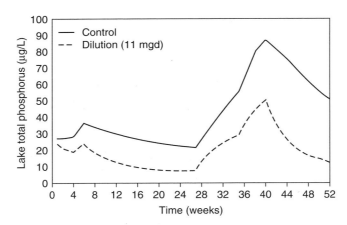

FIGURE 6.4 TP concentrations predicted (calibrated) for calendar year in Green Lake, Washington for 1981, in which the lake received 2 mgd low-nutrient dilution water, compared with predicted TP for 11 mgd dilution water. (From Cooke et al., 1993. With permission.)

errors as well in using a steady state equation, the major one being that a steady state model based on annual means does not separate seasonal variations in loading. In the Green Lake case, summer TP and algal blooms are due primarily to summer internal loading, and the 41% of total annual loading from storm drains is relatively insignificant in determining the summer concentration.

To account for the effect of seasonal variations in loading, a non steady state model (Equation 3.9) was used (URS, 1987; Mesner, 1985). Also with that model, the sedimentation rate coefficient was calibrated as a function of flushing rate ($\sigma = \rho^{0.71}$) so that changes in sedimentation could be accounted for in determining the appropriate dilution volume to achieve a summer average concentration. For Lake Washington water, and based on 1981 data in which about 2.8×10^6 m³/yr of city dilution water was added, only about 15×10^6 m³/yr dilution water was required to achieve a summer mean TP of 28 µg/L. That is only 60% of the previous estimate from a steady state model (23.5 $\times 10^6$ m³/yr). The non steady state model showed that control of storm water inputs would be unnecessary. The effect on summer TP concentration from adding 15×10^6 m³/yr (11 mgd) of Lake Washington water is shown in Figure 6.4.

The estimated present-worth cost for using Lake Washington water, which would be the capital and operating cost for the piping system, was 13.8×10^6 for the annual flow rate of 11 mgd (15 $\times 10^6$ m³/yr) over a 20-year period. The cost for city water at 7.5 mgd (10.4×10^6 m³/yr) was 27.4×10^6 (2002 dollars) over 20 years. Other techniques, such as storm water diversion and artificial circulation, were projected to lower the total project cost. The final decision regarding Green Lake was to abandon the alternative of Lake Washington water due to cost, environmental impacts, and the failure to obtain approval for withdrawal. Instead, an alum treatment of the sediments, along with the construction of a 3 mgd treatment plant (9.3×10^6), to create dilution water by removing P from Green Lake water, was the selected alternative. Ultimately, the treatment plant was deemed unnecessary and the lake was treated with alum in 1991 (Chapter 8). City water is still added occasionally when excessive algal blooms occur.

6.3.3 LAKE VELUWE

This lake in The Netherlands is not a clear-cut case for dilution because dilution water addition was begun coincident with P removal from sewage effluent input in 1979. Internal P loading was high and was expected to minimize the benefits of P removal, so dilution was successfully instituted to mitigate internal loading (Hosper, 1985; Hosper and Meyer, 1986).

Lake Veluwe is large and shallow (3,240 ha, 1.28 m mean depth) with a flushing rate of 1.43/yr. Groundwater, which was low in TP (80 to 100 µg/L) relative to the lake and other sources, was pumped into the lake during winter (November to March) at 19 m³/s, giving a flushing rate of 3.6%/d (5.5/yr). The purpose of dilution was to begin the summer with less algae (*Oscillatoria*) and P, and a lower, more buffered pH (groundwater was hard). The principal cause for summer internal P loading was photosynthetically caused high pH (up to 10) and the resulting dissolution of ironbound P at the sediment–water interface.

The plan apparently worked. TP in the lake should have decreased from 400–600 µg/L to 130–200 µg/L if lake concentration were proportional to loading (i.e., no internal loading). In fact, TP dropped to 100–200 µg/L, indicating that internal loading was substantially curtailed. Internal loading, estimated by calibrating a non steady state model (Chapter 3), decreased from 1.5–5.9 mg/m² per day in 1978 to 1979, before treatment, to 0.0–0.8 mg/m² per day in 1982 to 1983, after treatment.

Chl *a* decreased from 200–400 µg/L before, to 50–150 µg/L after treatment and the transparency range increased from 0.15–0.25 m before to 0.25–0.45 m after treatment. Although the lake was still turbid, green algae and diatoms dominated the plankton for the first time in 20 years, replacing the nearly permanent bloom of *Oscillatoria*. However, the lake did not return to the clear condition that existed prior to the mid 1960s, when nutrient loading increased.

6.4 SUMMARY: EFFECTS, APPLICATIONS, AND PRECAUTIONS

Dilution is frequently used synonymously with flushing. Actually, dilution includes a reduction in the lake concentration of nutrients and a washout of algal cells, while flushing may only cause the latter. For dilution to be cost effective, the inflow water must be substantially lower in concentration than the lake. Effectiveness increases as the difference between inflow and lake concentrations increases. For washout to be an effective control on algal biomass, the flushing rate must be a sizeable fraction of, or preferably approach, the algal growth rate.

In the three lake cases described, facilities for transporting water already existed. While irrigation water added to Moses Lake had a cost, the Bureau of Reclamation delivered the water to users via the treated lake. For Green Lake, a domestic supply was used with little operational cost. Maintenance of water quality in both systems suffered, however, during late summer in Moses Lake, and in Green Lake during some years when low water supplies were encountered. The use of dilution and flushing usually will not be limited as much by costs for facilities or water as by proximity to a supply of low-nutrient water.

Low-P water from the city supply was added to Green Lake in Seattle, beginning in 1962. This raised the water exchange rate from an annual average of 0.83/yr to 2.4/yr. After 5 years of treatment, summer Secchi transparency improved fourfold, chl *a* decreased by over 90%, and TP declined by over 50%. Dilution water inputs declined by the late 1970s and early 1980s, and the lake's quality again deteriorated. While city water was still used on a temporary basis to dilute the lake, that supply was increasingly limited. Other sources of water were considered (including treated lake water), along with stormwater diversion, lake circulation, alum, and biological manipulation. Institutional and supply constraints minimized the use of dilution in favor of an alum application.

Moses Lake receives low-nutrient dilution water from the Columbia River via an irrigation system during spring and summer periods since 1977. Although the lake is highly dendritic in shape, dilution water entering one arm distributed throughout most of the lake, primarily as a result of wind. Prior to dilution, the lake maintained very high nutrient and algal content and very low transparency; TP and chl *a* normally averaged about 150 and 45 µg/L, respectively. After dilution, the respective spring–summer average for Secchi transparency doubled in most of the lake, and TP and chl *a* were reduced by at least 50%. Initially, the biomass of algae was reduced in direct relation to the reduction of NO_3 in the inflow water, although cell washout contributed some to the effect in highly flushed sections. Facilitating the distribution of dilution water to another, previously undiluted arm of the lake demonstrated that algal biomass could be controlled by washout with moderate water exchange rates. With the exception of one year (1984), dilution water addition has continued for 25 years.

The ideal plan would achieve an adequate reduction in limiting nutrient content through a low-rate, annual input of low-nutrient water. Where there is an existing high-nutrient input, it should be diverted, if possible, for the low dilution rate to be most effective. This plan provides for a reduction in biomass primarily through nutrient limitation. If nutrient diversion is not possible, one is faced with higher-rate inputs to sufficiently reduce the inflow nutrient concentration. If only moderate- to high-nutrient water is available, flushing may work well if the loss rate of cells is sufficiently great relative to the growth rate. Flushing rates on the order of 10–15%/d should afford some control through washout. Results from Moses Lake suggest that, until the flushing rate reached 20%/d or more, the zooplankton crop was not reduced and growth rate may even increase.

Internal loading was an important source of P in the three lakes discussed, and dilution achieved substantial control of that source. For Green and Moses Lakes, dilution represented a direct control for internal loading. For Lake Veluwe, dilution indirectly reduced internal loading by decreasing the winter blue-green algal concentration and thereby decreasing summer internal loading as a result of a lower early summer pH.

Costs are highly variable, depending upon the presence of facilities to deliver the water and the quantity and proximity of available water. If the lake is in an urban setting and domestic water is available, then improvement may be possible for less than $150,000 for construction, water costs, and the first year of maintenance and operation. For Wapato Lake in Tacoma, Washington, the cost was less than maintenance and operation of the city's swimming pools, and the lake had more swimmers (Entranco Engineers, personal communication). If the lake is near a free-flowing river and diversion of a portion of the river flow through the lake during summer is feasible, then the costs involve facilities, pumps and pipes, operation, and prevention of side effects (entraining fish).

The advantages of using dilution water include (1) relatively low cost if water is available, (2) an immediate and proven effectiveness if the limiting nutrient can be decreased, and (3) some success even if only moderate- to high-nutrient water is available, through physical limitations to large algal concentrations. However, the principal limitation for use of this technique is the availability of low-nutrient dilution water.

REFERENCES

Barbiero, R.P. and E.B. Welch. 1992. Contribution of benthic blue-green algal recruitment to lake populations and phosphorus. *Freshwater Biol.* 27: 249–260.

Born, S.M., T.L. Wirth, J.O. Peterson, J.P. Wall and D.A. Stephenson. 1973. Dilutional Pumping of Snake Lake, Wisconsin. Wis, Tech. Bull. 66, Dept. Nat. Res., Madison, NJ.

Bouchard, D. 1989. Carbon Dioxide: Its Role in the Succession and Buoyancy of Blue Green Algae at the Onset of a Bloom in Moses Lake. MS Thesis, University of Washington, Seattle.

Brenner, M.V. 1983. The Cause for the Effect of Dilution Water in Moses Lake. MS Thesis, University of Washington, Seattle.

Carlson, K.L. and E.B. Welch. 1983. Evaluation of Moses Lake Dilution: Phase II. Water Res. Tech. Rept. 80. Dept. Civil Eng., Washington Dept. of Ecology, Olympia. Personal communication.

Carroll, J. 2004. Moses Lake Total Maximum Daily Load Phosphorus Study. Pub. No. 04-03-0, WA Dept. of Ecology, Olympia, WA.

Cooke, G.D., E.B. Welch, S.A. Peterson and P.R. Newroth. 1993. *Restoration and Management of Lakes and Reservoirs*, 2nd. ed. CRC Press, Boca Raton, FL.

Dickman, M. 1969. Some effects of lake renewal on phytoplankton productivity and species composition. *Limnol. Oceanogr.* 14: 660–666.

Dillon, P.J. 1975. The phosphorus budget of Cameron Lake, Ontario: the importance of flushing rate relative to the degree of eutrophy of a lake. *Limnol. Oceanogr.* 29: 28–39.

Entranco Engineers, Inc., Bellevue, WA. personal communication.

Goldman, C.R. 1968. Limnological Aspects of Clear Lake, California with Special Reference to the Proposed Diversion of Eel River Water through the Lake. Rept. Fed. Water Pollut. Control Admin.

Horne, A.J. and C.R. Goldman. 1972. Nitrogen fixation in Clear Lake, California. I. Seasonal variation and the role of heterocysts. *Limnol. Oceanogr.* 17: 678–692.

Horne, A.J. and A.D. Viner. 1971. Nitrogen fixation and its significance in tropical Lake George, Uganda. *Nature* 232: 417–418.

Hosper, S.H. 1985. Restoration of Lake Veluwe, The Netherlands, by reduction of phosporus loading and flushing. *Water Sci Technol.* 17: 757–786.

Hosper, H. and M.L. Meyer. 1986. Control of phosphorus loading and flushing as restoration methods for Lake Veluwe, The Netherlands. *Hydrobiol. Bull.* 20: 183–194.

Jacoby, J.M., H.L. Gibbons, K.B, Stoops and D.D. Bouchard. 1994. Response of a shallow, polymictic lake to buffered alum treatment. *Lake and Reservoir Manage.* 10: 103–112.

Keating, K.I. 1977. Blue-green algal inhibition of diatom growth: transition from mesotrophic to eutrophic community structure. *Science* 199: 971–973.

Knoechel, R. and J. Kalff, 1975. Algal sedimentation: The cause of a diatom-blue-green succession. *Verh. Int. Verein. Limnol.* 19: 745–754.

Jones, C.A. and E.B. Welch. 1990. Internal phosphorus loading related to mixing and dilution in a dendritic, shallow prairie lake. *J. Water Pollut. Cont. Fed.* 62: 847–852.

Jones, J.R. and R.W. Bachmann. 1976. Prediction of phosphorus and chlorophyll levels in lakes. *J. Water Pollut. Cont. Fed.* 48: 2176–2182.

McBride, G.B. and R.D. Pridmore. 1988. Prediction of [chorophyll a] in impoundments of short hydraulic retention time: Mixing effects. *Verh. Int. Verein. Limnol.* 23: 832–836.

Mesner, N. 1985. Use of a Seasonal Phosphorus Model to Compare Restoration Strategies in Green Lake. MSE Thesis, University of Washington, Seattle.

Oglesby, R.T. 1969. Effects of controlled nutrient dilution on the eutrophication of a lake. In: *Eutrophication: Causes, Consequences and Correctives.* National Academy of Science, Washington, DC. pp. 483–493.

Paerl H.W. and J.F. Ustach. 1982. Blue-green algal scums: An explanation for their occurrence during freshwater blooms. *Limnol. Oceanogr.* 27: 212–217.

Perkins, M.A. 1983. Limnological Characteristics of Green Lake: Phase I Restoration Analysis. Dept. Civil Eng., University of Washington, Seattle.

Persson, P.E. 1981. Growth of *Oscillatoria agardhii* in a hypertrophic brackish-water bay. *Am. Bot. Finnici.* Vol. 18.

Poon, C.P.C. 1977. Nutrient exchange kinetics in water-sediment interface. *Prog. Water Technol.* 9: 881–895.

Sas, H. et al. 1989. *Lake Restoration by Reduction of Nutrient Loading: Expectations, Experiences and Extrapolations.* Academia-Verlag, Richarz, St. Augustin, Germany.

Sketelj, J. and M. Rejic. 1966. Pollutional phases of Lake Bled. In: *Advances in Water Pollution Research.* Proc. 2nd Int. Conf. Water Pollut. Res. Pergamon, London, pp. 345–362.

Stadelman, P. 1980. *Der zustand des Rotsees bei Luzern.* Kantonales amt fur Gewasserschutz, Luzern.

Sylvester, R.O. and G.C. Anderson. 1964. A lake's response to its environment. *J. Sanit. Eng. Div. ASCE* 90: 1–22.

URS. 1983. Green Lake Restoration Diagnostic Feasibility Study. URS Corp., Seattle, WA.

URS. 1987. Green Lake Water Quality Improvement Plan. URS Corp., Seattle, WA.

Uttormark, P.D. and M.L. Hutchins. 1980. Input–output models as decision aids for lake restoration. *Water Res. Bull.* 16: 494–500.

Welch, E.B. 1969. Factors Initiating Phytoplankton Blooms and Resulting Effects on Dissolved Oxygen in Duwamish River Estuary. U.S. Geol. Surv. Water Suppl. Paper 1873-A. Seattle, WA. p. 62.

Welch, E.B. 1979. Lake restoration by dilution. In: *Lake Restoration.* USEPA-400/5-79-001. pp. 133–139.

Welch, E.B. and C.R. Patmont. 1979. Dilution effects in Moses Lake. In: *Limnological and Socioeconomic Evaluation of Lake Restoration Projects.* USEPA-600/3-79-005. pp. 187–212.

Welch, E.B. and C.R. Patmont. 1980. Lake restoration by dilution: Moses Lake, Washington. *Water Res.* 14: 1317–1325.

Welch, E.B. and M.D. Tomasek. 1981. The continuing dilution of Moses Lake, Washington. In: *Restoration of Lakes and Inland Waters.* USEPA-440/5-81-010. pp. 238–244.

Welch, E.B. and E.R. Weiher. 1987. Improvement in Moses Lake quality by dilution and diversion. *Lake and Reservoir Manage.* 3: 58–65.

Welch, E.B., J.A. Buckley and R.M. Bush. 1972. Dilution as an algal bloom control. *J. Water Pollut. Control Fed.* 44: 2245–2265.

Welch, E.B., K.L. Carlson, R.E. Nece and M.V. Brenner. 1982. Evaluation of Moses Lake Dilution. Water Res. Tech. Rept. 77. Dept. Civil Eng., University of Washington, Seattle.

Welch, E.B., M.V. Brenner, and K.L. Carlson. 1984. Control of algal biomass by inflow nitrogen. In: *Lake and Reservoir Management.* USEPA-440/5-84-001. pp. 493–497.

Welch, E.B., M.D. Tomasek and D.E. Spyridakis. 1985. Instability of Mount St. Helens ash layer in Moses Lake. *J. Fresh Water Ecol.* 3: 103–112.

Welch, E.B., C.A. Jones and R.P. Barbiero. 1989. Moses Lake Quality: Results of Dilution, Sewage Diversion and BMPs — 1977 through 1988. Water Res. Tech. Rept. No. 118. Dept. Civil Eng., University of Washington, Seattle.

Welch, E.B., R.P. Barbiero, D. Bouchard and C.A. Jones. 1992. Lake trophic state change and constant algal composition following dilution and diversion. *Ecol. Eng.* 1: 173–197.

7 Hypolimnetic Withdrawal

7.1 INTRODUCTION

The hypolimnetic withdrawal technique involves changing the depth at which water leaves the lake from the surface to near the maximum depth, so that nutrient-rich, rather than low-nutrient surface water is discharged. Coincidentally, the hypolimnion detention time is shortened, the chance for anaerobic conditions to develop is decreased and the availability of nutrients to the epilimnion, through entrainment and diffusion, is reduced. The technique is accomplished by installing a pipe along the lake bottom from near the deepest point to the outlet, and possibly beyond. The outlet pipe is usually situated below lake level, so the device acts as a siphon. It was named an "Olszewski tube" after its original user (Olszewski, 1961), but the technique itself is more commonly referred to as hypolimnetic withdrawal. This technique is applicable to stratified lakes and small reservoirs in which anaerobic hypolimnia restrict the habitat for fish and promote the release of P, toxic metals, ammonia, and hydrogen sulfide from sediments.

There are two important requirements for treatment success: (1) the lake level must remain relatively constant, and (2) thermal stability should not change. While stratification may be weakened because epilimnetic water tends to be drawn downward, destratification will not occur provided the removal rate of hypolimnetic water is relatively slow. Destratification should be avoided, because it increases the transport of hypolimnetic nutrients and anoxic water to the epilimnion. Polymictic lakes may not be good prospects for withdrawal. To lessen the chances of destratification, directing inlet water to the metalimnion or hypolimnion may be possible. This modification was installed in Lake Ballinger near Seattle (Figure 7.1). While the lake remained stratified with the inflow water directed to depth, the system was not tested without directed flow, for comparison. Destratification has not generally been a problem with the technique. Thermocline depth remained about the same in seven of nine cases examined by Nürnberg (1987). Lowered lake level and thus head loss, will hamper recovery by reducing hypolimnetic water and P export (Livingstone and Schanz, 1994; Dunalska et al., 2001).

Preferentially removing hypolimnetic water, and therefore decreasing the residence time of the hypolimnion, should decrease the period of anoxia and increase the depth of the anoxic boundary resulting in a decrease in internal loading of P. In a large fraction of cases, this has occurred (Nürnberg, 1987). Continued P export should ultimately reduce the sediment P pool. Hypolimnetic withdrawal is an obvious and proven alternative, with relatively low cost, to accelerate recovery in stratified lakes where little improvement has followed wastewater diversion or wastewater P removal because of high internal loading. There have been few new cases reported since the second edition of this book, so most of the following remains relatively unchanged.

The technique is employed inadvertently in reservoirs where hypolimnetic waters are normally discharged for power generation. However, that procedure has not been evaluated for benefits to water quality in the reservoir itself. Low DO content of discharged water has historically been a major problem with deep-discharge impoundments. Multiple and shallower outlets have been incorporated into reservoir design to counteract low DO discharge. Reducing discharge depth minimizes nutrient export. Some combination of deep and shallow outlets may optimize the two goals of sufficient DO and high-nutrient export, but there is little mention of such a practice in the literature.

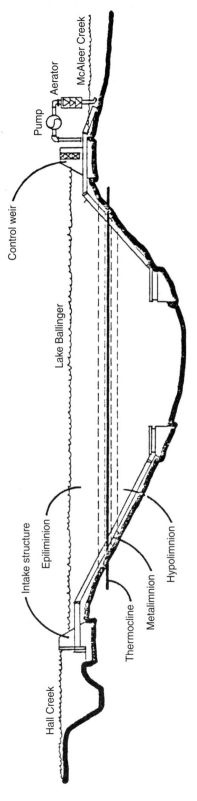

FIGURE 7.1 Inlet and outlet structures designed for hypolimnetic withdrawal in Lake Ballinger. (From KMC, 1981.)

7.2 TEST CASES

7.2.1 GENERAL TRENDS

Hypolimnetic withdrawal installation is documented in 21 lakes and 15 of those are in Europe (Björk, 1974; Nürnberg, 1987). Results are reported from 17 lakes (Nürnberg, 1987; Nürnberg et al., 1987). Four of the 21 lakes, two from the U.S. and two from Canada, are either more recent or not included by Nürnberg. Morphometric and mixing characteristics of these lakes are shown in Table 7.1. There are three other European lakes (Laacher and Lützel in Germany and Rudnickii Wielkie in Poland) with withdrawal systems cited by Dunalska et al., 2001). Ten additional cases were reported from Finland (Keto et al., 2004).

Internal loading from anoxic sediments during summer stratification occurred in all lakes prior to withdrawal and in most cases, external loading was reduced. Prior to withdrawal, Kleiner Montiggler See had been aerated with liquid oxygen and Reithersee was treated with iron chloride to precipitate P followed by dredging (Nürnberg, 1987).

Withdrawal is initiated preferably after stratification, but before anoxic conditions occur. The siphon pipe is located usually 1 to 2 m above the bottom at the greatest depth to maximize P transport (see Tables 7.1 and 7.2). In meromictic lakes however, it may be most effective to position the pipe above the monomolimnion so that it continues to be a sink for P. If there are two basins, withdrawal from the shallowest basin may be more effective at reducing entrainment into the epilimnion (e.g., Lake Wononscopomuc, CT).

Hypolimnetic water withdrawal rates and consequent TP export and duration are shown in Table 7.2. These values varied among the lakes. Years of withdrawal ranged from 1 to 10 for 20 of the listed lakes. Sufficient data were available to estimate TP export and duration in only 11 of the 20 lakes. The longest duration is for Kortowo, Poland, where the first withdrawal pipe was installed in 1956. Recent P budgets show 3.7 and 4.7 times more P exported from the lake than the inputs for 1999 and 2000 (Dunalaska et al., 2001).

Hypolimnetic and epilimnetic data were available on 12 lakes, but data were available for both on only 10 lakes. Maximum hypolimnetic TP concentration decreased in all 11 of 12 lakes and epilimnetic TP decreased in 8 of 12 where data were available. The reduction in hypolimnetic TP is a direct effect, but epilimnetic reduction in TP is an indirect effect demonstrating that entrainment of P from hypolimnion to epilimnion was reduced. The effect of withdrawal on epilimnetic TP was most significant as a function of grand total TP exported over the project life rather than annual export, whether expressed as total mass or per area (Figure 7.2; Nürnberg, 1987). The lakes involved in this analysis were Burgaschi, Hecht, Kleiner Montiggler, Mauen, Meerfelder Maar, Piburger, Waramaug, and Wononscopomuc. Lake Ballinger showed no change in epilimnetic TP due to increased external loading so it was not included.

The longer withdrawal operated the greater was the proportional change in epilimnetic TP (Figure 7.3; Nürnberg, 1987). More data were available for this analysis. The additional lakes besides those listed above for Figure 7.2 are Klopeiner, Kraiger and Wiler, although the latter was eliminated from the regression analysis due to high external P loading (Figure 7.3, open circle; Nürnberg, 1987). While substantial decrease in epilimnetic TP occurred in four lakes, on average, as long as 5 years may be necessary to see a significant decrease in epilimnetic TP (Figures 7.3). Recent data show a reduction of epilimnetic TP from 80 to 18 µg/L during 10 years of withdrawal in Lake Bled (Nürnberg and LaZerte, 2003).

The depth of hypolimnetic anoxia also decreased in 12 of 13 cases with adequate data, but that effect decreased as volume increased, and the days of anoxia decreased in 8 of 10 cases. However, the reduction in anoxia could not be related to withdrawal rate or volume. Thus, the case for lessened anoxia with withdrawal is not strong. Thermocline position remained about the same in 8 of 10 cases and sank 2 to 3 m in 2 cases.

TABLE 7.1
Morphometric Characteristics of Lakes Treated With Hypolimnetic Withdrawal

Lake	Watershed Area (10³ m²)	Lake Area (10³ m²)	Lake Volume (10³ m²)	Water Res. Time (yr)	Mean Depth (m)	Max. Depth (m)	Mixis
Ballinger, Washington[a]	11,720	405	1,838	0.26	4.5	10.0	Monomictic
Bled, Yugoslavia[b]	NA	1438	25,690	3.6	17.9	30.2	Meromictic
Burgäschi, Switzerland[c]	3,190	192	2,483	1.4	12.9	32.0	Meromictic
Chain, British Columbia[d]	—	460	2,760	0.5–3.0	6	9	Polymictic
Devil's, Wisconsin[e]	6,860	1,510	1,390	7.8*	9.2	14.3	Dimictic
Germündener Maar, W. Germany[f]	430	75	1,330	8.0	17.7	39.0	Meromictic
Hecht, Austria[g]	2,221	263	6,428	2.8	24.4	56.5	Meromictic
Kleiner Montiggler, Italy[h]	1,252	52	518	NA	9.9	14.8	Meromictic
Klopeiner, Austria[i]	NA	1106	24,975	1.5	22.6	48.0	NA
Kortowo, Poland[j]	1,020	901	5,293	NA	5.9	17.2	Dimictic
Kraiger, Austria[i]	NA	51	245	2.0	4.8	10.0	Dimictic
Mauen, Switzerland[k]	4,300	510	1,989	0.6	3.9	6.8	Dimictic
Meerfelder Maar, W. Germany[l]	1,270	248	2,270	4.5	9.2	18.0	Dimictic
de Paladru, France[m]	48,000	3900	97,000	4.0	25.0	35.0	Dimictic
Piburger, Austria[g,k]	2640	134	1,835	1.9	13.7	24.6	Meromictic
Pine, Alberta[n]	157,070	4,125	24,088	9.0	5.3	13.2	Dimictic
Reither, Austria[g,o]	NA	15	67	0.3	4.5	8.2	Dimictic
Stubenberg, Austria[i]	NA	450	NA	NA	NA	8.0	Polymictic
Waramaug, Connecticut[p]	37,000	2,866	24,758	0.8	8.6	12.8	Dimictic
Wiler, Switzerland[i,q]	257	31	325	1.0	10.0	20.5	NA
Wononscopomuc, Connecticut[r]	5994	1400	15,500	4.0	11.1	32.9	Dimictic

Source: From Nürnberg, G.K. 1987. *J. Environ. Eng.* 113, with additions. With permission.

Data sources: [a]KCM. 1981. Lake Ballinger Restoration Project Interim Monitoring Study Report; KCM. 1986. Restoration of Lake Ballinger: Phase III Final Report. Kramer, Chin, and Mayo, Seattle, WA. [b]Vrhovsek, D. et al. 1985. *Hydrobiologia* 127; Nürnberg, G.K. and B.D. LaZerte. 2003. *Lake and Reservoir Manage.* 19. [c]Ambühl, H., personal communication. [d]McDonald, R.H. et al. 2004. *Lake and Reservoir Manage.* 20. [e]Lathrop, R.C. personal communication. [f]Scharf, B.W. 1983. *Beitrage Landespflege Reinland-Pfalz* 9. [g]Pechlaner, R. 1978. *Osterreichische Wasserwirtsch.* 30. [h]Thaler, B. and D. Tait. 1981. *Tatigkeitsbericht des Biologischen Landeslabors autonome Provinz Bozen* 2. [i]Hamm, A. and V. Kucklentz. 1981. *Materialien der Bayrischen Landesanstalt fur Wasserforschung, Munchen, FDR,* 15. [j]Olszewski, P. 1961. *Verh. Int. Verein. Limnol.* 14; 1973. *Verh. Int. Verein. Limnol.* 18. [k]Gächter, R. 1976. *Schweiz Z. Hydrol.* 38. [l]Scharf, B.W. 1984. *Natur und Landschaft* 59. [m]Lascombe, C. and J. De Beneditis. 1984. *Verh. Int. Verein. Limnol.* 22. [n]Sosiak, A. 2002. Initial Results of the Pine Lake Restoration Program. Alberta Environment, Edmonton, Alberta. [o]Pechlaner, R. 1975. *Verh. Int. Verein. Limnol.* 19; 1979. *Arch. Hydrobiol. Suppl.* 13. [p]Nürnberg, G.K. 1987. *J. Environ. Eng.* 113. [q]Eschmann, K.H. 1969. *Gesundheitstechnik Zurich* 3. [r]Kortmann, R.W. et al. 1983. In: *Lake Restoration, Protection and Management.* USEPA-440/5-83-001; Nürnberg, G.K. et al. 1987. *Water Res.* 21; NA, not available.

TABLE 7.2
Specific Characteristics of Withdrawal Systems

Lake	Pipe Depth (m)	Withdrawal Volume (10³m³/yr)	Withdrawal Rate (m³/min)	Diameter[b] (cm)	Pipe Outflow[c] (m)	Annual TP Export (kg)	Duration (yr)
Ballinger	9.0	~480	3.4	30.5	NA	NA	3.0
Bled	NA	6307	12.0	NA	NA	NA	10.0
Burgäschi	15.0	1000	3.0	33.0	0.5	147.1	5.0
Chain	6.2	435	4.8	45	1.0	30	9.0
Devil's	14.3	629	9.1	48	2.2	446	1.0
Gemündener Maar	NA	NA	0.1	NA	NA	NA	NA
Hecht	25.0	843	1.5	18.0	2.0	50.8	10.0
Kleiner Montiggler[d]	13.0	16	NA	NA	–0.5	16.0	1.0
Klopeiner	30.0	NA	NA	NA	NA	NA	3.0
Kortowo	13.0	NA	NA	NA	0.5	NA	45.0
Kraiger	NA	NA	NA	20.0	NA	NA	4.0
Mauen	6.5	1000	4.0	30.0	0.5	617.0	6.0
Meerfelder Maar	16.0	190	0.6	30.0	1.2	40.0	1.5
Paladru	31	NA	21.0	NA	NA	416.0	5.0
Piburger	23.0	284	0.6	8.9	11.0	7.8	6.0
Pine	10.2	1140	5.3	53	2.0	153	2
Reither	8.0	126	0.24	10.0	1.0	NA	NA
Stubenberg	NA	NA	NA	NA	NA	NA	NA
Waramaug[a]	8.5	1330	6.3	31.8	0.0	131.9	3.0
Wiler	17.5	NA	0.6	11.0	NA	NA	3.0
Wononscopomuc[a]	15.1	201	0.9	NA	0.0	21.0	5.0

[a] Active pumping.

[b] Inner diameter.

[c] Below lake level — or above for negative values.

[d] Operation only during spring.

Source: From Nürnberg, G.K. 1987. *J. Environ. Eng.* 113, with additions (with permission); NA, not available.

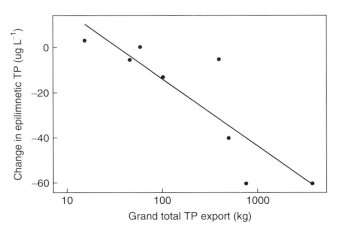

FIGURE 7.2 Changes in epilimnetic TP concentrations (after – before) vs. grand total TP export via hypolimnetic withdrawal (calculated as annual export multiplied by years of operation): Regression line is shown, $y = 46 - 30 \log x$, $n = 8$, $r^2 = 0.75$. (From Nürnberg, G.K. 1987. *J. Environ. Eng.* 113. With permission.)

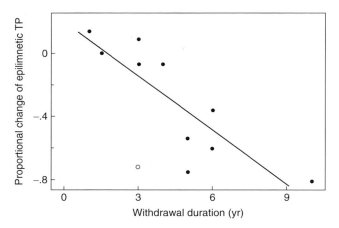

FIGURE 7.3 Proportional change - 0.116 (0.026) x, n = 10, r^2 = 0.72. (From Nürnberg, G.K. 1987. J. Environ. Eng. 113. With permission.)

7.2.1 SPECIFIC CASES

7.2.1.1 Mauen See

This is one of the most successful cases of hypolimnetic withdrawal (Gächter, 1976). An Olszewski tube was installed in 1968 in this Swiss lake at a depth of 6.5 m (Table 7.1). Prior to installation, external P loading was reduced from about 700 to 300 mg/m² per yr (Nürnberg, 1987). The discharge of 4 m³/min provided a hypolimnetic (> 4 m) water residence time of 0.2 years. Marked improvement in lake quality followed installation. Hypolimnetic DO and Secchi visibility increased and hypolimnetic TP decreased by 1,500 µg/L, the most of any lake examined (Nürnberg, 1987). Epilimnetic TP decreased by 60 µg/L. *Oscillatoria* biomass decreased from a before-treatment summer maximum of 152 g/m² to 41 g/m², 7 years after installation.

Before installation, internal P loading from lake sediments during June and July was more than 200 times that of external loading. After installation, internal loading progressively decreased to only four times external loading. Sediment P release progressively declined for the 6 years of observation following installation. During that time, P export exceeded external loading (360 kg/yr) by a total of 3,700 kg, resulting in a decrease in P content of the surficial sediments.

7.2.1.2 Austrian Lakes

Pechlaner (1978) reported on the response of three lakes following installation of Olszewski tubes; Piburger See, Reither See, and Hechtsee. Characteristics of the three lakes are given in Tables 7.1 and 7.2. All three lakes are relatively small but important to local populations and tourists for recreation, especially swimming. The Olszewski tubes were installed to accelerate the restoration process following sewage effluent diversion.

The tube in Piburger See draws water from a depth of 23 m, which is nearly the maximum depth of the lake (24.6 m). Total length of the tube is 639 m, with a diameter of 8.9 cm. Hypolimnetic water is discharged at 0.6 m³/min at a point downstream that is 13.5 m below lake level.

While oxygen content markedly increased, there was no recognizable oligotrophication. In 1970, the same year as the tube was installed, the DO content at the time of ice cover increased by 63% over pre-tube conditions in 1969. DO continued at the improved level or higher for the next 7 years. However, the lake's trophic state did not change because epilimnetic TP declined by 5 µg/L (Pechlaner, 1979).

Piburger See tends to be morphometrically meromictic. The lake mixed completely only twice during the 9 years of observation, even though the monomolimnion was effectively replaced about three times per year by virtue of the tube discharge. Increased circulation of deep water across the bottom sediments, as well as the removal of P-rich water overlying the sediments, resulted in increased internal loading that tended to compensate for the increased losses of P via the tube (see effect of dilution on internal loading, Chapter 6). Phosphorus losses through the tube over 3 years were 79–192% more with the tube than would have occurred without it (Pechlaner, 1979).

In contrast to Piburger See, Reither See improved markedly in quality following installation of a tube (Pechlaner, 1978). The tube was placed near the maximum depth (8.2 m) of the dimictic lake in 1972. Tube diameter was 10 cm and water discharged at 0.24 m^3/min from the hypolimnion of the 1.5-ha lake.

Epilimnetic TP decreased from annual means of 38 and 43 µg/L in 1974 and 1975 to 21 µg/L in 1977. Transparency nearly doubled over the 4-year period following installation. There was some uncertainty about change in phytoplankton biomass, due to interference with detritus. However, there were less blue-green algae after installation.

A larger tube (18 cm) was placed in Hechtsee in 1973. The depth of placement, however was not near the maximum depth as in the other two lakes. Because of meromixis, odors from the monomolimnion were quite strong. Therefore, the tube was placed at 25 m, considerably less than the 56.6 m maximum depth, in order to protect the recreational environment around the lake from nuisance odors. Tube discharge from the 26.3 ha lake varied from 1.2 to 1.8 m^3/min.

Because monomolimnetic water was not withdrawn, DO remained at zero from 25 m to the bottom. DO increased significantly above 25 m after installation of the tube and P transport from the lake increased markedly even though the tube was placed above the monomolimnion. During the first four years following installation, P output (203 kg) exceeded input (93 kg) by 110 kg, which was the actual decrease in lake TP content. TP above 25 m declined by 70–80% from 1973 to 1977, while, as expected, TP below 25 m changed little and actually showed some increase (Pechlaner, 1978).

7.2.1.3   U.S. Lakes

Withdrawal systems were installed in the shallower of two basins of Lake Wononscopomuc, Connecticut, in 1980. Hypolimnetic water was discharged from the shallow basin's maximum depth of 15.1 m at 0.9 m^3/min (Table 7.2), which was sufficient to replace the hypolimnetic volume in 5.6 months (Kortmann et al., 1983; Nürnberg et al., 1987).

Lake quality improved substantially. Hypolimnetic TP decreased from about 400 µg/L before to less than 100 to 50 µg/L over 5 years and epilimnetic TP decreased from 24–30 µg/L to 10–14 µg/L following the start of withdrawal. The decreased TP was apparently due to reductions in internal loading, which was verified by a 79% decrease in measured sediment release in the shallow basin after 2 years of withdrawal (Nürnberg et al., 1987).

DO in the hypolimnion also increased and the anoxic factor (days of anoxia) decreased from 50–65 before to less than 30 after withdrawal. Transparency remained high and unchanged (> 5 m), but metalimnetic blooms of *O. rubescens* were eliminated by the treatment.

Two systems were installed in Lake Waramaug, Connecticut, in 1983. One withdrew water from 8.5 m in one end of the long, S-shaped lake (12.8 m maximum depth) and discharged it at 6.3 m^3/min (Table 7.2). The other system withdrew water from the hypolimnion at the other end of the lake, returning it aerated. There were no significant trends in TP, either in the hypolimnion or epilimnion, during the first 3 years following withdrawal. However, the anoxic factor decreased from 76–89 to 75 days. Reasons for no significant response in TP were: (1) insignificant magnitude of TP removal or duration of removal, or (2) excessive external loading (Nürnberg et al., 1987).

The other U.S. lake treated with withdrawal is Lake Ballinger, north of Seattle. The device was installed in 1982 and allows the lake inlet stream to be directed to the hypolimnion through a 276 m, 30.5 cm diameter pipe. The option also exists to allow all or some fraction to enter the epilimnion if inflow temperature exceeds 16°C and there is a tendency to destratify the water column. A control weir exists at the outlet to adjust the fraction of hypolimnetic and epilimnetic water discharged. The mean flow through the 381 m, 30.5 cm outflow pipe was 3.4 m³/min, which resulted in a replacement time for the hypolimnion of about 3 months (KMC, 1986).

Anoxia occurred for only 2 weeks in 1983, the year after installation, and the hypolimnion remained oxic with at least 3 to 4 mg/L DO during the stratification period in 1984, which was thought to be due to reduced ammonia in the inflow. Hypolimnetic DO remained above 2 mg/L during 1985. Maximum hypolimnetic TP decreased from about 450 to 900 µg/L during 1979–1981, before installation, to about 100 to 150 during 1982 to 1985, after installation. TP at overturn in 1984 was 15 µg/L, the lowest level ever observed.

More recent data are unavailable. Operation of the system has been intermittent in recent years due to odors from the discharge stream that borders a golf course. The lake was treated with alum in 1993.

Internal loading was reduced from a high pre-installation value of 227 kg in 1979 to only 17 kg in 1984. The overall decrease in internal loading was 70%. Unfortunately, a substantial increase in external loading during the late 1970s and early 1980s prevented much reduction in epilimnetic TP and consequent improvement in lake quality (KCM, 1986).

A 1,677-m hypolimnetic withdrawal pipe was installed in Devil's Lake, Wisconsin, in 2002 at the maximum depth, which varies from 13.5 to 15.7 m (Table 7.1,2). The outflow rate is controlled to vary from 6.8 to 10.3 m³/min, depending on lake level. The outflow P concentration averaged 725 µg/L for 48 days of operation in 2002, discharging 446 kg of hypolimnetic TP (Lathrop, personal communication; Lathrop et al., 2004). Data were not yet available to assess lake quality improvement.

The project was initiated because high internal P loading from deep-water sediments caused excessive amounts of planktonic and periphytic algae, even though external inputs from cultural sources were eliminated in prior decades. Potential indirect benefits of reduced productivity included reduction in swimmer's itch by decreasing parasite–host snail densities feeding on periphyton, and reduced fish mercury concentrations by shortening the extent and duration of hypolimnetic anoxia, which is necessary for sulfate-reducing bacteria to convert inorganic Hg to methyl Hg.

Hypolimnetic withdrawal was chosen as the only suitable technique to restore the lake to its original pristine condition because field and laboratory results confirmed that internal P loading could be significantly reduced after multiple withdrawals. Other techniques to reduce internal P loading were rejected due to: (1) high cost, e.g., aeration, hypolimnetic water treatment, (2) opposition to adding chemicals, e.g., alum, to one of the state's high use, "outstanding resource waters," and (3) long-term ineffectiveness of other techniques, e.g., aeration, alum, without continual or periodic retreatment. The siphon withdrawal system has the important advantage of no operation cost. An additional benefit was alleviation of recently recurring flooding problems in the State Park from high lake levels (Lathrop, personal communication; Lathrop et al., 2004).

7.2.1.4 Canada

The restoration of Pine Lake, near Red Deer, Alberta, began in 1991 to improve water quality to a mesotrophic state that existed prior to European settlement (Sosiak, 2002). Epilimnetic TP concentrations reached medians around 100 µg/L during the mid 1990s with chlorophyll (chl) *a* medians of 20–50 µg/L. Most (61%) of the lake TP originated from internal loading. Controls on external loads from surface sources (36%) took place during 1996–1998 and a 1,400-m hypolimnetic withdrawal pipe was installed in 1998 to reduce internal loading (Table 7.1).

The withdrawal system produced high rates of P loss (Table 7.2), and along with external controls, has reduced lake TP and improved water quality (Sosiak, 2002). TP concentration decreased by 44–47% and chl *a* by 76–81% during 1996–2000. Median TP concentrations of 53–61 µg/L equaled those expected from recovery, while chl *a* (7.5–11.1 µg/L) and transparency (2.7–3.4 m) exceeded expectations. Since 2000, TP and chl concentrations have remained relatively low

(Sosiak, personal communication). However, there have been blooms of *Gloeotrichia*, which was apparently absent before treatment.

While external controls probably contributed to the recovery, most of the reduction in lake TP (29%) occurred during, and was attributed to, hypolimnetic withdrawal. Some of the improvement, however, was probably due to lower year-to-year surface runoff, which was positively related to lake TP over the 15 year period of data. Nevertheless, monitoring of other Alberta lakes did not show a widespread decline in lake TP that could be attributed to regional climatic conditions (Sosiak, 2002).

There have been no significant adverse water quality effects in the outlet stream, although temperature and DO were lower immediately downstream. Odors have not been a problem. Such a complete, long-term data set for a project is unusual and is continuing. Such thorough monitoring has allowed a definitive assessment of the recovery and project cost-effectiveness.

Chain Lake is small, shallow and polymictic in British Columbia (Table 7.1). Raising the lake level 1.3 m in 1951 created eutrophic conditions (McDonald et al., 2004). Low nutrient dilution water was diverted to the lake in the 1960s and a small area to 9 m was dredged to enhance stability. Summer TP and chl *a* reached concentrations of 300 and 100 µg/L, respectively, with blue-green algal blooms. Withdrawal has consistently exported water and TP over the 9 years of operation (Table 7.2). Transparency has increased significantly (~ 1 m) over that time period. Downstream adverse effects from degraded water quality were partially mitigated by a fountain aerator.

7.3 COSTS

Installation costs (in 2002 U.S. dollars) for the three systems in the U.S. lakes were as follows: Lake Ballinger (41 ha, 3.4 m^3/min flow) — $420,000; Lake Waramaug (287 ha, 6.3 m^3/min) — $62,000 (Davis, personal communication; KMC, 1981); Devil's Lake (151 ha, 9.1 m^3/min) — $310,000 (Lathrop, personal communication); Pine Lake (412 ha, 5.3 m^3/min) — $282,000, not including contributed labor and equipment. Relatively low cost and low annual maintenance are definite advantages of hypolimnetic withdrawal.

7.4 ADVERSE EFFECTS

Discharge of hypolimnetic water containing high concentrations of P, ammonia, hydrogen sulfide and reduced metals and no oxygen may cause a water quality problem downstream. If the outflow stream contains an important fishery and is otherwise used for recreation or water supply, then special precautions are necessary to minimize adverse effects. Withdrawal water from Lakes Wononscopomuc and Waramaug is aerated and mechanically cleaned before being discharged downstream and the intake pipe end in Lake Waramaug is elevated to avoid high concentrations and fertilization effects downstream (Nürnberg et al., 1987).

Discharge from Lake Ballinger must be interrupted at times due to odors (2 of the 6 months of stratification) and high nutrient content is apparently responsible for extensive periphyton growth downstream from the outlet. Odors are a nuisance to users of the adjacent golf course. The discharges from Hect, Klopeiner, and Kraiger See contained high concentrations of toxic substances so they were stopped during the late summer. Mixing of discharge hypolimnetic water with epilimnetic water would minimize adverse downstream effects.

7.5 SUMMARY

The advantages of hypolimnetic withdrawal are threefold: (1) relatively low capital and operational costs, (2) evidence of effectiveness in a large fraction of cases, and (3) potentially long-term and even permanent effectiveness. In most cases hypolimnetic DO increased, resulting in a decrease in the anoxic volume and the days of anoxia. Internal P loading usually decreased and if there was not an offsetting high external loading, epilimnetic TP also decreased.

The effectiveness of withdrawal apparently depends on magnitude and duration of TP transport from the hypolimnion. Thus, it is important to exchange the hypolimnion volume as frequently as possible. A low rate of replacement may limit the effectiveness of this technique. A desirable exchange rate is severalfold during the stratification period and the desired magnitude can be determined by comparing the oxygen deficit rate with the rate of oxygen transport. For example, the flow directed to the hypolimnion in Lake Ballinger was established based on the desire to add oxygen at double the oxygen demand rate in the hypolimnion. As a result, hypolimnetic water was exchanged about every 3 months. In Mauensee, it was exchanged every 2.4 months. Therefore, an exchange rate of at least once in 2 to 3 months is recommended to assure the effectiveness of withdrawal. In addition, results indicate that at least a 3 and possibly a 5 year duration of TP export may be necessary to see improvement in lake (epilimnetic) quality.

There is a possibility of negative effects on downstream water quality due to low DO, high nutrient content and reduced substances. If the outflow stream contains an important fishery and is otherwise used for recreation or water supply, then special precautions may be necessary to maintain water quality. The extent to which DO in the outflow water will be reduced can be estimated by comparing the existing DO deficit in the lake with the input load of DO (Pechlaner, 1979). If low DO is expected in the outlet, then aeration equipment should be installed. Whether the high P content will cause nuisance attached algal and secondary BOD problems downstream will depend upon the extent to which periphyton growth is limited by nutrients compared with other factors.

REFERENCES

Björk, S. 1974. *European Lake Rehabilitation Activities*. Rep. Inst. Limnol. University Lund, Sweden.

Davis, E.R. 1983. Personal communication. The Hotchkiss School, Lakeville, CT.

Dunalska, J., G. Wisniewski and C. Mientki. 2001. Water balance as a factor determining the Lake Kortowskie restoration. *Limnol. Rev.* 1: 65–72.

Eschmann, K.H. 1969. Die sanierung des wiler Sees durch albeitung des Tiefenwassers. *Gesundheitstechnik Zurich* 3: 125–129.

Gächter, R. 1976. Die Tiefenwasserableitung, ein Weg zur Sanierung von Seen. *Schweiz Z. Hydrol.* 38: 1–28.

Hamm, A., and V. Kucklentz. 1981. Moglichkeiten und Erfolgsaussichten der Seenrestaurierung. *Materialien der Bayrischen Landesanstalt fur Wasserforschung, Munchen, FDR*, 15: 1–221.

Keto, A., A. Lehtinen, A. Mäkelä and I. Sammalkorpi. 2004. Lake Restoration. In: P. Eloranta, Ed., *Inland and Coastal Waters of Finland*, University of Helsinki and Palmina Centre for Continuing Education.

KMC. 1981. Lake Ballinger Restoration Project Interim Monitoring Study Report. Kramer, Chin and Mayo, Seattle, WA.

KMC. 1986. Restoration of Lake Ballinger: Phase III Final Report. Kramer, Chin, and Mayo, Seattle, WA.

Kortmann, R.W., E.R. Davis, C.R. Frink and D.D. Henry. 1983. Hypolimnetic withdrawal: Restoration of Lake Wonoscopomuc, Connecticut. In: Lake Restoration, Protection and Management. USEPA-440/5-83-001. pp. 46–55.

Lascombe, C., and J. De Beneditis. 1984. Une expérience de soutirage des eaux hypolimniques au Lac de Paladru (Isère-France): Bilan des cing premières années de fonctionnement. *Verh. Int. Verein. Limnol.* 22: 1035.

Lathrop, R.C. Personal communication. Wisconsin Dept. Nat. Res., Madison.

Lathrop, R.C., T.J. Astfalk, J.C. Panuska and D.W. Marshall. 2004. Restoring Devil's Lake from the bottom up. *Wisc. Nat. Resour.* 28: 4-9.

Livingstone, D.M. and F. Shanz. 1994. The effects of deep-water siphoning on small, shallow lake. *Arch. Hydrobiol.* 32: 15-44.

McDonald, R.H., G.A. Lawrence and T.P. Murphy. 2004. Operation and evaluation of hypolimnetic withdrawal in a shallow eutrophic lake. *Lake and Reservoir Manage.* 20: 39–53.

Nürnberg, G.K. 1987. Hypolimnetic withdrawal as lake restoration technique. *J. Environ. Eng.* 113: 1006–1016.

Nürnberg, G.K. and B.D. LaZerte. 2003. An artificially induced *Planktothrix rubescens* surface bloom in a small kettle lake in southern Ontario compared to blooms worldwide. *Lake and Reservoir Manage.* 19: 307–322.

Nürnberg, G.K., R. Hartley and E. Davis. 1987. Hypolimnetic withdrawal in two North American lakes with anoxic P release from the sediment. *Water Res.* 21: 923–928.

Olszewski, P. 1961. Versuch einer Ableitung des hypolimnischen Wassers aus einem See. *Verh. Int. Verein. Limnol.* 14: 855–861.

Olszewski, P. 1973. Funfzehn Jahre Experiment auf den kortowo-See. *Verh. Int. Verein. Limnol.* 18: 1792–1797.

Pechlaner, R. 1975. Eutrophication and restoration of lakes receiving nutrients from diffuse sources only. *Verh. Int. Verein. Limnol.* 19: 1272–1278.

Pechlaner, R. 1978. Erfahrungen mit Restaurierungsmassnahmen an eutrophen Badeseen Tirols. *Osterreichische Wasserwirtsch.* 30: 112–119.

Pechlaner, R. 1979. Response to the eutrophied Piburger See to reduced external loading and removal of monomoliminic water. *Arch. Hydrobiol. Suppl.* 13: 293–305.

Scharf, B.W. 1983. Hydrographie und morphometric einiger Eifelmaare. *Beitrage Landespflege Reinland-Pfalz* 9: 54–65.

Scharf, B.W. 1984. Errichtung und Sicherung Schutzwürdige Teile von Natur und Landschaft mit gesamtstaatlich Repräsentativer Bedeutung. *Natur und Landschaft* 59: 21–27.

Sosiak, A. Personal communication. Alberta Environment, Edmonton.

Sosiak, A. 2002. Initial Results of the Pine Lake Restoration Program. Alberta Environment, Edmonton.

Thaler, B. and D. Tait. 1981. Kleiner Montiggler See. Die auswirkungen von Belüftung und Tiefenwasserableitung auf die physikalischen und chemischen Parameter in den Jahren 1979 und 1980. *Tatigkeitsbericht des Biologischen Landeslabors autonome Provinz Bozen* 2: 132–193.

Vrhovsek, D., G. Kosi, M. Karalj, M. Bricelj and M. Zupar. 1985. The effect of lake restoration measures on the physical, chemical, and phytoplankton variables of Lake Bled. *Hydrobiologia* 127: 219–228.

8 Phosphorus Inactivation and Sediment Oxidation

8.1 INTRODUCTION

Nuisance algal blooms can be reduced or eliminated if phosphorus (P) concentrations are lowered to growth-limiting levels by diversion of external loading, by dilution, or a combination of these methods. In cases where loading reduction is significant, where the lake flushing rate is relatively fast, and where recycling from sediments is unimportant, in-lake P can be reduced and trophic state significantly and rapidly improved. The case of Lake Washington (Edmondson, 1970; 1994) is probably the most recognized example of this response (Chapter 4).

For many lakes, however, internal P release prolonged the lake's enriched state and supported continued algal blooms, even though diversion removed a significant fraction of external loading (Cullen and Forsberg, 1988; Sas et al., 1989; Jeppesen et al., 1991; Welch and Cooke, 1995; Scheffer, 1998). Lakes that experience significant internal loading of P to their water columns are the rule rather than the exception. Lakes with extensive littoral and wetland areas (Wetzel, 1990), close proximity between the epilimnion and anoxic sediments (Fee, 1979), or shallow lakes with enriched sediments from a history of high external loading (Jeppesen et al., 1991), will have extensive P recycling. In those lakes, additional in-lake steps may be necessary, following nutrient diversion, to prevent a prolonged eutrophic state. For example, Shagawa Lake, Minnesota (MN), did not respond as rapidly as expected to the reduction of a large fraction of external loading, and it is predicted to require decades to reach equilibrium (Larsen et al., 1981; Chapra and Canale, 1991; Chapter 4). In some lakes, even without reduction of external loading the major input of P, and cause for summer algal blooms, is from sediments (Welch and Jacoby, 2001). In-lake treatments have been effective in such lakes.

Phosphorus inactivation is an in-lake technique, designed to lower the lake's P content by removal of P from the water column (P precipitation) and by retarding release of mobile P from lake sediments (P inactivation). Usually an aluminum salt, either aluminum sulfate (alum), sodium aluminate, or both, is added to the water column to form aluminum phosphate and a colloidal aluminum hydroxide floc to which certain P fractions are bound. The aluminum hydroxide floc settles to the sediment and continues to sorb and retain P within the lattice of the molecule, even under reducing conditions. Alum has been used for coagulation in water treatment for over 200 years and is probably the most commonly used drinking water treatment in the world (Ødegaard et al., 1990). Polyaluminum chloride is another coagulant used in water treatment that has a more favorable floc-forming pH range than alum (Ødegaard et al., 1990), and has been used in lakes (Carlson, personal communication). Iron and calcium salts have also been used to precipitate or sorb P.

Summer lake trophic state is improved when the control of internal P release significantly lowers P concentration in the photic zone, which is the whole water column of polymictic lakes, and the epilimnion and sometimes the metalimnion of eutrophic, dimictic lakes and reservoirs.

This technique has been mistakenly classified as an algicide or herbicide by some agencies. Phosphorus inactivation provides long-term control of algal biomass by significantly reducing the supply of an essential nutrient rather than through poisoning of algal cells. Algicides work by direct toxic action, and are effective only during the brief period when the toxic active ingredient (frequently copper) is present in the water column (Chapter 10). Phosphorus inactivation with aluminum salts is effective for years while algicides are effective for days.

A different method to control internal P loading from anaerobic lake sediments, called sediment oxidation, was developed by Ripl (1976). With this procedure, $Ca(NO_3)_2$ is injected into lake sediments to stimulate denitrification, where nitrate acts as an electron acceptor. This process oxidizes the organic matter. At the same time, ferric chloride is added, if natural levels are low, to remove H_2S and to form $Fe(OH)_3$ to which P is sorbed.

8.2 CHEMICAL BACKGROUND

Aluminum, iron, and calcium salts have been used for centuries for drinking water clarification, and their use today, particularly aluminum, is essential in the treatment of wastewater and drinking water. Lund (1955) appears to be the first to suggest that the addition of aluminum sulfate (alum, $Al_2(SO_4)_3 \cdot 14\ H_2O$) to streams and lakes could be a successful means to control algal blooms. The first published account of such a treatment is Jernelöv (1971), who applied dry alum to the ice of Lake Långsjön, Sweden, in 1968. Iron and calcium are major controllers of the P cycle in lakes, and like aluminum, have been used extensively in wastewater and potable supply treatments, but less frequently than alum in lakes. The first report for iron in lakes to control P was in Dordrecht Reservoir, The Netherlands (Peelen, 1969) and for calcium in a Canadian hard water lake (Murphy et al., 1988).

8.2.1 ALUMINUM

The chemistry of aluminum is complex and incompletely understood (Dentel and Gossett, 1988; Bertsch, 1989). The reactions in water have been reviewed by Burrows (1977), Driscoll and Letterman (1988), and Driscoll and Schecher (1990), among others. The following is drawn from these reports, and from the first detailed lake and laboratory studies of aluminum salts for P inactivation (Browman et al., 1977; Eisenreich et al., 1977).

When aluminum sulfate or other aluminum salts are added to water they dissociate, forming aluminum ions. These are immediately hydrated:

$$Al^{+3} + 6\ H_2O \rightleftharpoons Al\,(H_2O)_6^{3+} \qquad (8.1)$$

A progressive series of hydrolysis (the liberation of hydrogen ions) reactions occurs leading to the formation of aluminum hydroxide, $Al(OH)_3$, a colloidal, amorphous floc with high coagulation and P adsorption properties:

$$Al(H_2O)_6^{3+} + H_2O \rightleftharpoons Al(H_2O)_5OH^{2+} + H_3O \qquad (8.2)$$

$$Al(H_2O)_5OH^{2+} + H_2O \rightleftharpoons Al(H_2O)_4OH_2^+ + H_3O \qquad (8.3)$$

etc.

Omitting coordinating water molecules from the equations, the following occurs:

$$Al^{3+} + H_2O \rightleftharpoons Al(OH)^{2+} + H^+ \qquad (8.4)$$

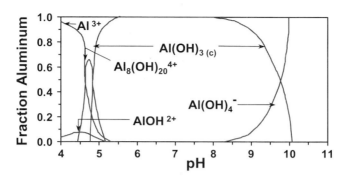

FIGURE 8.1 Fractional distribution of aluminum species as a function of pH (concentration 5.0×10^{-4} M). (Courtesy C. Lind, General Chemical Inc., Parsippany, NJ. With permission.)

$$Al(OH)^{2+} + H_2O \rightleftharpoons Al(OH)_2^+ + H^+ \tag{8.5}$$

$$Al(OH)_2^+ + H_2O \rightleftharpoons Al(OH)_3(s) + H^+ \tag{8.6}$$

where (s) = a solid precipitate.

$Al(OH)_3$ is a visible precipitate or floc that settles through the lake's water column to the sediments. A surface application produces a milky solution, which quickly forms large, visible particles. The floc grows in size and weight as settling occurs and particles within the water column are incorporated. Within hours, water transparency increases dramatically.

The pH of the solution determines which aluminum hydrolysis products dominate and what their solubilities will be (Figure 8.1). At the pH of most lake waters (pH 6 to 8), insoluble polymeric $Al(OH)_3$ dominates and P sorption and inactivation proceeds. At pH 4 to 6, various soluble intermediate forms occur, and at pH less than 4, hydrated and soluble Al^{3+} dominate.

When alum is added to poorly buffered waters, their acid neutralizing capacity (ANC) decreases, pH falls, and soluble aluminum species dominate if ANC is exhausted. At higher pH levels (>8.0), the amphoteric nature (having both acidic and basic properties) of aluminum hydroxide results in the formation of the aluminate ion:

$$Al(OH)_3 + H_2O \leftrightarrow Al(OH)_4^- + H^+ \tag{8.7}$$

At increasing pH levels above 8, as would occur during intense photosynthesis for example, solubility again increases, which could lead to a release of P sorbed to an aluminum salt.

Aluminum salts in water have a time-dependent component to their chemistry (Burrows, 1977). The concentration of monomeric forms (Al^{3+}, $Al(OH)^{2+}$, $Al(OH)_2^+$, and $Al(OH)_4^-$) stabilizes within 24 h. But crystallization takes over a year to complete as larger and larger units of polymeric $Al(OH)_3$ are formed. In lakes, this continued reaction occurs in the sediments, though its consequences to the control of P release are poorly understood. Toxicity studies carried out with a freshly prepared solution of buffered aluminum present a different array of potentially toxic aluminum species than an aged solution with a lower concentration of monomeric species and intermediate polymers (Burrows, 1977). This also may be the reason why continuous exposure to the early hydrolysis products of alum, as would occur in a continuous addition to flowing waters, may be deleterious to biota versus the single treatment to lake sediments (Barbiero et al., 1988). Exposure time to the floc is shorter in a lake treatment due to its relatively quick transport to the bottom.

Properties of $Al(OH)_3$ of greatest interest to lake managers are its apparent low or zero toxicity to lake biota (see later section), its ability to adsorb large amounts of particulate and soluble P, and the binding of P to the floc. In contrast to iron, low or zero dissolved oxygen (DO) concentrations in lake sediments do not solubilize the floc and allow P release, although P may be released from the floc if high pH occurs.

Particulate organic P (cells, detritus) is removed to some extent from the water column by coagulation and entrapment in the $Al(OH)_3$ floc. The settling of the floc through the water column clarifies the water in this manner. That is the reason alum is used extensively in water treatment plants. However, $Al(OH)_3$ may be less effective in removing dissolved organic matter (Browman et al., 1977).

Treatment timing may vary depending on local conditions. Because $Al(OH)_3$ is so sorptive of inorganic P, a seemingly ideal time for treatment is just after ice-out, or in early spring in warmer climates, before the spring bloom of algal cells and corresponding uptake of P occurs. However, temperature alters the rate and extent of reactions of aluminum salts in water (Driscoll and Letterman, 1988). At low temperatures, coagulation and deposition are significantly reduced and high quantities of species such as $Al(OH)_2^+$ that are toxic to some organisms, might occur. This suggests that aluminum solubility is temperature dependent as well as pH dependent. There are other reasons why early spring may not be an ideal time, in spite of high inorganic P. These include (1) sediment P release, not water column P content, is the primary target of P inactivation, (2) early spring months may be windy, making application difficult, (3) wind mixing may distribute the floc to one area of the lake, or scour it from the sediments before the floc consolidates into those sediments, and (4) silicon content, a major complexer of soluble and possibly toxic aluminum species, may be low following a spring diatom bloom. Thus, summer, before blue-green algal blooms appear, or early fall months, may be the most appropriate periods for application. On the other hand, an early spring treatment may avoid the problem of macrophytes, in spite of other risks.

Because hydrogen ions are liberated when an aluminum salt is added to water, H^+ increases in proportion to the decline in alkalinity. In lakes with low or moderate alkalinity (< 30 to 50 mg $CaCO_3$/L), treatment produces a significant decline in pH (increase in H^+) at a low or moderate alum dose, leading to increasing concentrations of toxic, soluble aluminum forms, including $Al(OH)_2^+$ and Al^{3+}. This limits the amount of alum that can be added safely. This problem has been addressed by adding a buffer to the lake or to the alum slurry as it is applied. The work of Dominie (1980) for Lake Annabessacook, Maine, Smeltzer (1990) for Lake Morey, Vermont, and Jacoby et al. (1994) for Green Lake, Washington are examples. Buffering compounds were tested, including sodium hydroxide, calcium hydroxide, and sodium carbonate. The buffer chosen was sodium aluminate ($Na_2Al_2O_4 \cdot N\ H_2O$), a high alkalinity compound with the added benefit of having a high aluminum content (Smeltzer, 1990). Much of this compound's alkalinity comes from the NaOH used in its production (Lind, personal communication). Sodium aluminate and alum should be added to the lake separately to avoid damage to pipes from overheating if mixed together. Sodium carbonate was also successfully used to buffer the treatment of soft water (35 mg/L alkalinity) Long Lake, Washington (Welch, 1996). A mixture of alum and lime has also served the purpose of buffering in soft waters (Babin et al., 1992).

In summary, the primary objective of an in-lake alum treatment is to cover the sediment with $Al(OH)_3$. Mobile P, which otherwise would diffuse into the water column, is sorbed, thereby reducing internal loading. The formation of $Al(OH)_3$ also removes particulate organic and inorganic matter with P from the water column, a secondary objective. The formation of large amounts of $Al(OH)_3$ and negligible amounts of other hydrolysis products depends upon maintaining water column pH between pH 6 and 8. Because lakes differ in alkalinity and sediment mobile-P content, the dose to a lake is lake specific. In some cases, a buffer must be added. Dose determination is discussed in a later section.

8.2.2 IRON AND CALCIUM

Phosphorus forms precipitates and complexes with iron and calcium, and these elements can be used to lower P concentration with less concern for pH shifts and/or the appearance of toxic forms.

The chemistry of these metals with regard to P is probably better understood than that of aluminum (Stumm and Lee, 1960; Stumm and Morgan, 1970).

Inorganic iron exists in solution in lake water and lake sediments in either the oxidized ferric (Fe^{3+}) or reduced ferrous (Fe^{2+}) forms, depending on solution pH and oxidation-reduction potential. Changes in the redox state of iron in lake sediments has an important effect on the P cycle (Mortimer, 1941, 1971). In oxygenated, alkaline conditions, a common state of the entire water column during spring and fall mixing, the redox potential is high and iron is oxidized to the ferric form.

$$Fe^2 + 1/4\ O_2 + 2\ OH^- + 1/2\ H_2O \rightarrow Fe(OH)_3(s) \tag{8.8}$$

$Fe(OH)_3$ sorbs P from the water column, and forms part of an oxidized "microzone" over the sediment surface, providing high sediment P retention. $FePO_4$ also forms, but the primary means of P removal and retention in sediments is sorption to $Fe(OH)_3$, and it is greatest at pH 5 to 7 (Andersen, 1975; Lijklema, 1977).

The generally accepted cycle of Fe and P in lakes is as follows. During periods of thermal stratification in eutrophic lakes, hypolimnetic waters are dark and isolated from mixing for periods of days or weeks (polymictic lakes) to periods of months (dimictic lakes). Without net photosynthesis or aeration by mixing, pH and particularly dissolved oxygen (DO) concentrations decline in the water overlying the sediments. As DO in the overlying water drops below 1.0 mg/L, the oxidized microzone is eliminated, and iron is used by the microbial community as an alternate electron acceptor to oxygen. In the reduced state, ferrous iron (Fe^{2+}) is soluble, and previously iron-bound P (part of mobile P) is free to be released to the water column. This change occurs rapidly so that even brief periods of thermal stability in a shallow eutrophic lake with high sediment oxygen demand can lead to substantial P release. Thermal stability in shallow lakes is largely controlled by wind (see Chapter 4). Also, sediment P release may occur on a diurnal basis in littoral areas of some eutrophic lakes so that P is sorbed to ferric complexes during the day and released during the night (Carlton and Wetzel, 1988). In spite of persistent anoxia in dimictic stratified lakes, the magnitude of sediment P release rates are as great in shallow unstratified lakes and are related more to trophic state than to depth (Nürnberg, 1996).

This generally accepted Fe–P redox cycle may not always hold. Sulfate reduction and the formation of insoluble FeS can remove iron from the cycle, effectively decreasing the Fe:P ratio, producing a greater fraction of P that remains soluble and can be released to overlying water (Smolders and Roelofs, 1993; Søndergaard et al., 2002). However, work by Caraco et al. (1989) seems to contradict the effect of S. In a sample of 23 lakes, only those with an intermediate (100 to 300 μm) sulfate concentration conformed to the iron redox model. Low sulfate (60 μm) systems had low P release under both oxic and anoxic conditions, and high (>3000 μm) sulfate systems had high P release under both conditions. There are many lakes with low sediment P release under anoxic conditions (Caraco et al., 1991a, b).

Phosphorus may be released when OH^- is exchanged for PO_4^{3-} on the iron-hydroxy complex during periods of high pH, even under aerobic conditions (Andersen, 1975; Jacoby et al., 1982; Boers, 1991a; Jensen and Andersen, 1992). This process enhances P release during resuspension events, especially if the suspended particle concentration is relatively low (Koski-Vähälä and Hartikainen, 2001; Van Hullenbusch et al., 2003). That is, the equilibrium shifts from desorption of P from particles to sorption by particles as particle concentration increases.

Iron's reaction to redox and pH conditions means that its addition to lakes as a P inactivant may have to be accompanied by a technique (aeration or artificial circulation) to prevent breakdown of the oxidized microzone or a photosynthetically caused increase in pH. Even aeration may not reduce P release if the sediments have a low Fe:S ratio (Caraco et al., 1991a, b). There is some evidence, however, that iron enrichment of lake sediments may inhibit P release even under anoxic conditions (Quaak et al., 1993; Boers et al., 1994).

Calcium compounds also affect P concentration. Calcium carbonate (calcite) and calcium hydroxide can be added to a lake from allochthonous sources, or produced in hard water lakes during periods of CO_2 uptake during photosynthesis, as follows:

$$Ca(HCO_3)_2 \rightleftharpoons CaCO_3(s) + H_2O + CO_2 \tag{8.9}$$

As plants assimilate CO_2, pH increases and $CaCO_3$ precipitates. Calcite sorbs P, especially when pH exceeds 9.0 (Koschel et al., 1983), and results in significant P removal from the water column (Gardner and Eadie, 1980). At high levels of pH, Ca^{2+}, and P, hydroxyapatite forms, as follows:

$$10\ CaCO_3 + 6\ HPO_4^{2-} + 2H_2O \rightleftharpoons Ca_{10}(PO_4)_6(OH)_2(s) + 10\ HCO_3^- \tag{8.10}$$

Hydroxyapatite, unlike $Fe(OH)_3$ and $Al(OH)_3$, has its lowest solubility at pH >9.5, and P sorbs strongly to it at high pH (Andersen, 1974, 1975). The solubility of calcite and hydroxyapatite increases sharply as CO_2 concentration increases and pH falls, as would be expected in a hypolimnion or dark littoral zone with intense respiration. This will lead to P release. Thus, as with iron, effective P removal and inactivation is possible with calcium, but conditions conducive to continued P sorption can be lost unless an additional management step is taken to maintain an alkaline pH in deep water.

Phosphorus inactivation, by definition, is an attempt to permanently and extensively bind P in lake sediments and thereby lower or essentially eliminate sediments as a P source to the water column. Phosphorus is strongly sorbed to $Al(OH)_3$ and this complex is apparently inert to redox changes, thus providing the possibility of a high degree of treatment permanence. However, the addition of aluminum salts to lakewater produces H^+ ions, and pH falls at a rate dictated by lake water alkalinity and dose of the salt. This can lead to high concentrations of soluble and potentially toxic aluminum species. Thus, unless the lake is well buffered, or buffers are added, the use of aluminum salts may not be appropriate. Sorption of P to iron and calcium complexes can also lead to significant P removal and to P retention in the sediments, but without toxicity problems. However, solubility of these compounds, and hence P sorption, is highly sensitive to pH and redox changes. Anoxia can occur very rapidly in productive lakes, even in shallow water sediments. Aeration or complete mixing would be needed on a continual basis if iron or calcium are employed for P inactivation.

8.3 DOSE DETERMINATION AND APPLICATION TECHNIQUES

8.3.1 ALUMINUM

There have been two approaches to the use of metal salts to control P concentration in lakes. Phosphorus precipitation emphasizes P removal from the water column, while P inactivation emphasizes longer-term control of sediment P release with P removal a secondary objective. During early Al applications, no basis for dose using either procedure existed (see Cooke and Kennedy, 1981 for a review and tabular summary of the first 28 treatments).

Phosphorus removal or precipitation is achieved by adding enough Al to the lake surface to remove the P in the water column at the time of application. Dose is determined by adding increments of $Al_2(SO_4)_3$ to lake water samples until the desired removal of P is achieved. This dose is then used to calculate the amount needed to remove P from the entire lake. Small amounts of Al are usually needed to bring about P removal. However, the goal of nearly all alum treatments in recent years is long-term control of internal loading and that will not be achieved with low doses.

Low dose and continued high external loading are reasons why some of the first treatments were short-lived (e.g., Långsjön). Also, some P fractions, particularly the dissolved organic fraction, may be incompletely removed, leaving a substrate for algal assimilation and growth. Thus, P precipitation to control algae is usually not recommended.

There are three procedures for determining dose to inactivate sediment P. The first is the alkalinity procedure developed by Kennedy (1978) and it has had widespread use. Kennedy believed that P inactivation should provide the greatest control possible. Treatment duration was assumed to be related to $Al(OH)_3$ concentration in lake sediments, because the amount of mobile P bound as Al–P is proportional to Al added. The goal then was to apply as much Al to the sediments as possible, consistent with environmental safety.

As described in an earlier section, the form of Al in water is dictated by pH (Figure 8.1). Between pH 6 and 8, most is in the solid $Al(OH)_3$ form. As pH falls below pH 6.0, other forms, including $Al(OH)_2^+$ and Al^{3+} become increasingly important. These forms are toxic in varying degrees, particularly dissolved Al^{3+} (Burrows, 1977). Everhart and Freeman (1973) found that rainbow trout (*Salmo gairdneri*) could tolerate chronic exposure to 52 µg/L dissolved Al with no obvious changes in behavior or physiological activity. The observation led to adopting 50 µg Al/L as a safe upper limit for post treatment dissolved Al concentration (Kennedy, 1978). Maximum dose was thus defined as the maximum amount of Al that, when added to lake water, would ensure that dissolved Al concentration is less than 50 µg/L (Kennedy and Cooke, 1982). As shown in Figure 8.1, 50 µg Al^{3+}/L should not occur as long as pH remains between pH 6.0 and 8.0. Maintaining pH $\geq$ 6.0 should prevent toxicity from dissolved monomeric Al as judged from experiments with alum applications to lake inflows (Pilgrim and Brezonik, 2004).

There is an added safety factor in that a lake treatment, unlike continuous exposure bioassay, produces a single maximum dose exposure to organisms, followed by a rapid decline in concentration, because the alum floc settles through the water column rather quickly (~ 1 hour). Also, the maximum formation of $Al(OH)_3$ occurs in the 6–8 pH range, leading to maximum deposition and removal of most Al from the water column, and to maximum formation of the P-retaining floc at the sediment-water interface. Dissolved Al concentrations have remained at 100–200 µg/L following treatments of some Washington lakes, without adverse effects, so the Al was probably in an organic, non-toxic form (Welch, 1996). Addition of natural organic matter was shown to reduce toxicity of Al by a factor of two (Roy and Campbell, 1997).

The alkalinity procedure (Kennedy and Cooke, 1980, 1982) provided a toxicological basis for adding enough alum to provide long-term control of sediment P release in lakes with adequate alkalinity (> 35 mg/L $CaCO_3$). There were several problems inherent in developing this procedure. First, few toxicity studies (see later paragraphs) had been conducted on the effects of Al on lake community processes and structure in non-acidified lakes. However, trout are highly sensitive to metals and that may provide a safety factor for the community as a whole. Second, low pH itself can be detrimental. Adverse effects in acidified lakes have been observed to begin at pH $\leq$ 6.0 without Al added (Schindler, 1986). However, those were long-term chronic effects, while alum treatments produce only short-term reductions in pH. Third, some lakes have low alkalinity and only small amounts of alum could be added before pH 6.0 is reached. As noted earlier, this last problem is overcome by the use of buffers.

The following step-by-step procedure, from Kennedy (1978) and Kennedy and Cooke (1982), describes dose determination based on lake water alkalinity.

1. Obtain water samples over the range of lake water alkalinities. Normally this means a series of samples from surface to bottom. Determine alkalinity to a pH 4.5 endpoint.
2. The dose for each stratum is approximated from Figure 8.2, which uses pH 6.0 as the endpoint of alum addition to the lake, rather than 50 µg Al/L. At the determined lake dose, dissolved Al will remain below this limit as long as the pH is between 6.0 and 8.0. Kennedy and Cooke (1982) selected this pH range to provide a safety margin with respect

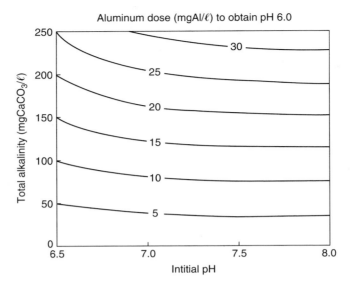

FIGURE 8.2 Estimated aluminum sulfate dose (mg Al/L) required to obtain pH 6 in treated water of varying initial alkalinity and pH. (From Kennedy, R.H. and G.D. Cooke. 1982. *Water Res. Bull.* 18: 389–395. With permission.)

to dissolved Al and excessive hydrogen ion concentration. This allows the addition of sufficient Al to lakes to give long-term control of P release, when lake alkalinities are above about 35 mg $CaCO_3$/L.

A more accurate dose determination is made by titrating water samples with a stock solution of aluminum sulfate of known Al concentration (a solution that contains 1.25 mg Al/ml is made by dissolving 15.4211 g technical grade $Al_2(SO_4)_3 \cdot 18H_2O$ in distilled water and diluting to 1.0 liter). Adding 1.0 ml of this stock to a 500-ml water sample is a dose of 2.5 mg Al/L. As alum is added, the samples are mixed with a stirrer and pH changes are monitored. Optimum dose for each sample is the amount that produces a stable pH of 6.0.

Linear regression is used to determine the relationship between dose and alkalinity. The resulting equation is then used to obtain dose for any alkalinity for this particular lake or reservoir, over the alkalinity range tested. To be cautious, a slightly higher pH (e.g., 6.2–6.3) may be chosen to determine dose (see Jacoby et al., 1994).

The maximum dose for each depth interval from which the alkalinities are obtained is calculated by converting the dose in mg Al/L to lbs (dry) alum/m³, using a formula weight of 666.19 [$Al_2(SO_4)_3 \cdot 18 H_2O$] and a conversion factor of 0.02723 to change mg Al/L to lbs (dry) alum/m³, because English units are used with commercial alum. A conversion factor of 0.02428 is used for $Al_2(SO_4)_3 \cdot 14H_2O$.

3. If liquid rather than granular alum is to be used, as is the usual case, further calculations are necessary to express the dose in gallons of alum/m³. Alum ranges from 8.0 to 8.5% Al_2O_3, which is equivalent to 5.16 to 5.57 pounds dry alum per gallon at 60°F (Lind, personal communication). It is shipped by tank truck at about 100°F and will thus have lower density. The percent Al_2O_3 at 60°F will be stated by the shipper. Convert this to density, expressed as degrees Baumé, using Figure 8.3. Then obtain the shipment temperature and adjust the 60° Baumé number by subtracting the correction factor (using Figure 8.4) from the 60° Baumé number. Pounds per gallon is then obtained from Figure 8.5, using the adjusted Baumé number.

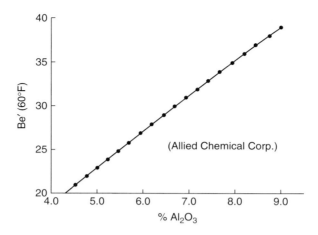

FIGURE 8.3 Relationship of Baumé (60°F) and percent Al$_2$O$_3$. (From Cooke et al. 1978.)

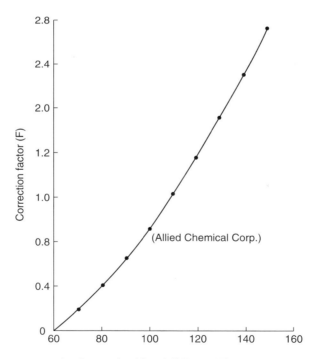

FIGURE 8.4 Temperature correction factors for 32 to 36° Baumé liquors. (From Cooke et al. 1978.)

Maximum dose for each depth interval sampled for alkalinity was calculated earlier as pounds dry alum/m^3. This is converted to gallons/m^3 by dividing pounds (dry)/m^3 by the value in pounds per gallon obtained from Figure 8.5. Total dose to the lake is then the sum of the individual depth interval doses.

4. Accuracy in treating the lake is obtained by dividing the lake into areas marked with buoys or, by using a barge equipped with a satellite guidance system. The volume and alkalinity in each area is measured and the gallons per treatment area determined. This

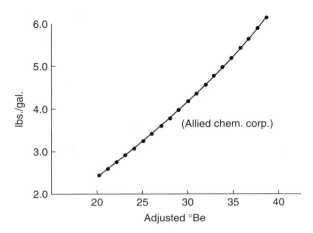

FIGURE 8.5 Curve to determine pounds of alum per gallon, based on adjusted Baumé. (From Cooke et al. 1978.)

approach prevents under-dosing deep areas and overdosing shallow ones, as would occur if an even application were made to the whole lake or reservoir area. Distribution of the alum with respect to lake volume can be accomplished automatically with a barge equipped with an electronic sounding device.

In soft-water lakes, only small amounts of aluminum sulfate can be added before the pH falls below 6.0. Gahler and Powers (personal communication) were perhaps the first to suggest that sodium aluminate, which supplies alkalinity and increases the pH of an aqueous solution, could be used with $Al_2(SO_4)_3$ to maintain a pH between 6.0 and 8.0. Dominie (1980) was apparently the first to successfully use this buffered dose approach on a large scale, when Annabessacook Lake, Maine (alkalinity 20 mg $CaCO_3$/L), was treated with this mixture in an empirically determined ratio of 0.63:1 sodium aluminate to alum. Sodium carbonate was also used successfully to treat Long Lake, Washington (alkalinity 35 mg/L) in 1991 to maintain a pH above 6.2 (Welch, 1996).

By adding a buffer, it is possible to add an amount of alum to lake sediments that is limited only by available funds. While this procedure is based on alkalinity and does not consider the quantity of internal loading, high-alkalinity lakes dosed by the alkalinity method have experienced substantial treatment longevity.

The second procedure for determining dose is based on estimated rates of net internal P loading from the sediments as determined from a mass balance equation. In the first use of this approach, the net internal loading per year in Eau Galle Reservoir, Wisconsin was determined and multiplied by 5, with the goal of controlling P release for 5 years as a desirable target, assuming that the P complexed by Al would ultimately attain a stoichiometric ratio of 1.0 (Kennedy et al., 1987). The Al dose was therefore determined as the quantity of Al equivalent to five times the average summer internal P load. This quantity was then doubled to account for any underestimate of internal loading (release rate can vary year-to-year; Figure 3.5) giving a final dose of 14 g/m^2 Al. Had the alkalinity procedure been used the dose would have been 45 g/m^2, greatly increasing the ratio of Al–P added:Al–P formed. A ratio of 5–10 appears to be appropriate based on observations from core analyses in alum-treated lakes (Rydin et al., 2000). Dose, using the internal loading rate, was expressed as a mass-areal unit (Al/m^2), which is actually more appropriate than is concentration resulting from the alkalinity procedure. The dose expressed as an areal unit is what the sediments should actually receive regardless of water column depth. Eberhardt (1990) has used a similar dose calculation, modified to account for application efficiency of the equipment.

The third procedure to estimate dose is based on a direct determination of mobile inorganic P in the sediments (Rydin and Welch, 1999). The following steps are recommended to apply this procedure:

1. Collect representative 30-cm sediment cores from the areas most actively releasing P. Such an area is seasonally anoxic hypolimnetic sediment in stratified lakes and reservoirs. Maximum depth is probably representative of the active release area, but other hypolimnetic depths may also be appropriate for the most representative estimate. Release can occur from sediments throughout the lake if the water body is unstratified, in which case more cores may be necessary to delimit the active areas.

2. Determine mobile P as Fe–P (or BD-P, bicarbonate dithionate) and loosely sorbed P (according to Psenner et al., 1984) in the top 4 cm of each core. Analyses may be performed at 1-cm intervals in the top 10 cm to increase information, but removing the top 4 cm for one analysis of mobile P is the least expensive, and represents the minimum information needed. A depth of 4 cm was considered appropriate for estimating dose to three Wisconsin lakes, but a greater depth profile may be preferred in some cases to include the majority of mobile P.

3. Convert the volume of sediment to be treated by multiplying the sediment bulk density (g/cm^3) by percent dry matter and then by the mobile P concentration (mg/g) to determine the mass/area to be treated. Values from several sites may be desirable to delimit zones of mobile P content, which would then require differential amounts of alum in much the same way sediment removal is varied with sediment P content in dredging operations. This may be especially important for most cost-effectiveness in unstratified lakes.

4. Determine dose in g Al/m^2 by the product of mobile P content and a ratio of Al added:Al–P formation expected. This ratio was 100:1 as observed in *in vitro* experiments performed with sediments from three Wisconsin lakes (Rydin and Welch, 1999).

Advantages for this procedure are that it (1) measures directly the quantity of mobile P in sediments that should be transformed to Al–P, (2) estimates the Al dose using the 100:1 ratio of Al added:Al–P formed to account for mobile P existing in the 0–4 cm layer and the P that may migrate from greater sediment depths, and (3) optimizes the quantity of alum that should provide the most cost-effective, long-term control of internal loading. Its disadvantage is that an extensive P fractionation analysis of the lake sediments is required.

As the alum dose increases, the loosely-sorbed (labile) P fraction decreases to zero and the Fe–P fraction is proportionately converted to Al–P, according to *in vitro* experiments with sediments from two Swedish lakes and three Wisconsin lakes (Rydin and Welch, 1998, 1999; Figure 8.6). The Al–P formed from Al added in the top 4 cm ultimately reached a plateau that approximated the initial content of mobile P in sediments from the three Wisconsin lakes (Figure 8.7). The line for a ratio of Al added:Al–P formed of 100:1 accounts for most of the mobile P and was the recommended ratio for these three Wisconsin lakes (Figure 8.7).

The Al added:Al–P formed ratio actually observed in Lake Delavan, Wisconsin (Figure 8.6) sediments was only 5 as a result of the 1991 treatment. That ratio was based on the amount of Al that had been added (12 g Al/m^2) as estimated from the Al peak in the sediment profile (Figure 8.7). This 5:1 ratio is due to upward migration from depth of sediment P that saturated the alum floc. The 100:1 ratio observed with an isolated surficial sediment sample *in vitro* represents the response from the deep sediment source. Thus, using the 100:1 ratio and 4 cm sediment depth to calculate dose should provide adequate binding capacity for P in the 4 cm interval, as well as that migrating from depth.

Experiments with surficial sediment (top 5 cm) from Squaw Lake, WI, treated with alum, showed that a ratio of 95:1 (Al added:mobile P) was necessary to bind the mobile P (James and

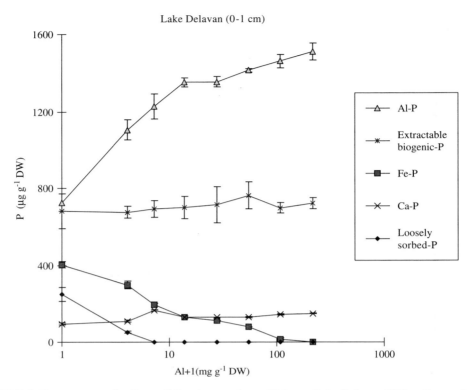

FIGURE 8.6 The response of sediment P fractions to alum addition to Lake Delavan (WI) sediments *in vitro.* (From Rydin, E. and E.B. Welch. 1999. *Lake and Reservoir Manage.* 15: 324–331. With permission.)

Barko, 2003). A sediment depth of 10 cm was used to estimate alum dose, because alum had settled to a similar depth in two other alum-treated Wisconsin lakes. This work corroborated the 100:1 ratio from Rydin and Welch (1999), so using that ratio and the 10 cm depth, a dose of 115 g/m^2 was recommended.

Lower ratios of added Al:Al–P have been observed in other lakes. Short-term experiments in Lake Sonderby, Denmark showed that a ratio of 4:1 was sufficient to greatly reduce sediment P release, which was no less than the release using a ratio of 8:1 (Reitzel et al., 2003). However, extractible organic P was included in their estimate of mobile P, which increased mobile P by ~ 50%. Treatment of sediment in Lake Susan and Lake of the Isles in Minnesota showed ratios of 5.28:1 and 4.68:1 (not including organic P as mobile P) following alum treatment (Huser, personal communication). Even a lower ratio was found in Süsser See, Germany (286 ha, 4.3 m mean depth). Eight years after 16 consecutive annual low-dose (2 mg/L) alum treatments, the added Al layer was found between 10 and 30 cm with an added Al:Al–P formed ratio of 2.1:1 (Lewandowski et al., 2003). The low ratio suggested that dosing a lake over several years at a low rate is more efficient than one large dose; the total 16-year dose in this lake was 138 g/m^2. Evidence was presented that soluble reactive phosphorus (SRP) was still migrating from depth, forming Al–P.

The case of Lake Delavan is useful in evaluating the three dosing procedures (Table 8.1). Calculated doses for the 1991 alum treatment of the lake ranged from 2.3 to 2.8 mg/L (Panuska and Robertson, 1999; Welch and Cooke, 1999; Robertson et al., 2000). Based on a mean depth of 7.6 m, those concentrations result in areal dose rates of 17.5–21.3 g Al/m^2 as averages over the lake. The 1991 dose was based on the observed net internal loading and a 15-year expectation of longevity, which should have yielded a dose of 10 g Al/m^3 or 76 g Al/m^2, 3.6 times what was added (Robertson et al., 2000). According to the *in vitro* results, the dose should have been 150 g Al m^{-2}

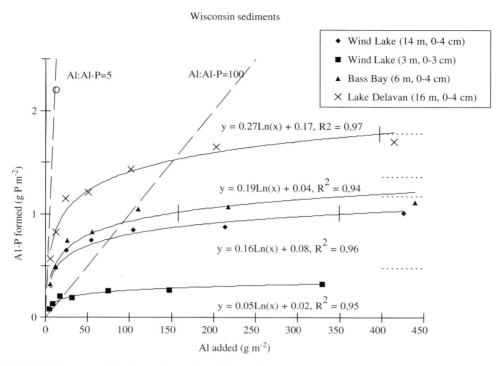

FIGURE 8.7 Response of sediment from three Wisconsin lakes to alum addition *in vitro* showing: (1) Al–P formed from Al added (—), (2) the initial amount of mobile P (-- -), (3) the recommended dose line (dashed) to convert most of the mobile P to Al–P, (4) the ratio of Al added:Al–P formed and approximate dose from the 1991 alum treatment of Delavan Lake (open circle), and (5) doses determined based on alkalinity (vertical bars). (From Rydin, E. and E.B. Welch. 1999. *Lake and Reservoir Manage.* 15: 324–331. With permission.)

TABLE 8.1

Dose Estimates for Lake Delavan, WI Based on the Alkalinity, Internal Loading and Mobile-P Procedures, Compared with the Actual Dose in 1991

Dose method	g Al/m³	g Al/m²	Ref.
1991 treatment	2.3–2.8	17.5–21.3	Robertson et al. (2000)
Internal loading	10	76	Robertson et al. (2000)
Mobile-P (exp. Figure 8.7)	20	150	Rydin and Welch (1999)
Mobile-P (sed. conc.)	25	190	Rydin and Welch (1999)
Alkalinity (Figure 8.7)	51	390	Rydin and Welch (1999)
Alkalinity (jar tests)	33	250	Robertson et al. (2000)

---

to bind all mobile P in the top 4 cm as well as P migrating from depth. Based on the mobile-P content and a 100:1 ratio, the calculated dose should have been about 190 g Al/m². As a result of the under-dose, treatment effectiveness was judged at 50% for 4 years with no effectiveness remaining after 7 years (Robertson et al., 2000). The 2.2 g/m² of mobile P actually inactivated with the treatment represented about 30% of that in the top 10 cm (Rydin and Welch, 1999).

If dosing to Lake Delavan by the internal loading procedure had been fulfilled, the sediments would still have been 50% under-dosed, based on the mobile-P procedure (i.e., 76 vs 150 g Al/m²).

Two maximum allowable doses by the alkalinity procedure were calculated at about 390 g Al/m^2 (Figure 8.7) and 250 g Al/m^2 using jar tests (Robertson et al., 2000). This indicates that for Lake Delavan, the internal loading procedure underestimated the dose compared to that by the alkalinity and mobile P procedures (Table 8.1).

There is also a strong similarity between the under dose of Lake Delavan and that for Eau Galle Reservoir, which was treated in 1986 with a dose estimated from internal loading (Kennedy et al., 1987). The Eau Galle treatment was also short-lived (James et al., 1991), and the final dose was 14 g Al/m^2, even less than added to Lake Delavan (Table 8.1).

This discrepancy may not always occur, however, depending on alkalinity, lake depth and ratio of Al added:Al–P formed. The Al added:Al–P formed ratio (5:1) observed for Lake Delavan sediment, and in other treated lakes (11:1, Rydin et al., 2000), is a result of other substances (e.g., organics) competing with P for binding sites in the alum floc. Thus, the 1:1 ratio (with ×2 correction) used in the internal loading procedure should be increased. From the Lake Delavan experience, use of the internal loading procedure with a ratio of 4:1 or 5:1, without the error correction, should have given effective control for at least 15 years.

The greater the sediment depth considered with the mobile-P procedure, the lower the ratio of Al added:Al–P formed that should be necessary. For example, if a 10 cm depth had been used as the "active layer" to estimate a dose for Lake Delavan instead of 4 cm, and with the mobile-P content of 5 g/m^2, the ratio needed to calculate dose would have been 30:1 to obtain a dose of 150 g/m^2 Al (150/5). That should have provided control for decades (Rydin and Welch, 1999). Nevertheless, James and Barko (2003), using a similar experimental procedure, recommended using both the 100:1 ratio and a 10 cm sediment depth.

The alkalinity procedure has been effective in hard water lakes because a sufficient dose was attained before the critical low pH occurred. These lakes were relatively deep, which contributed to a larger areal dose to the sediments. However, the procedure may not be as effective in shallow lakes. Also, doses by this procedure are apt to be too low in soft water lakes unless mobile-P levels are low. While added buffering capacity with sodium aluminate or sodium carbonate increases the acceptable alum dose, the ultimate stopping point with a buffer is hypothetically unlimited, and thus unknown. Data on the mobile-P content of the sediment defines that limit and thus improve cost-effectiveness. Both the mobile-P and alkalinity procedures are needed for soft water lakes to provide the proper dose and insure adequate buffering.

Dose determination for Green Lake, Washington (alkalinity 35 mg/L) is a case in point. The dose by the alkalinity procedure was about 5 g Al/m^3 (20 g Al/m^2), which was considered inadequate judging from experience in other lakes. Sodium aluminate was added at a ratio of 1.25:1 (sodium aluminate:alum) to increase the dose to 8.7 g Al/m^3 (34 g Al/m^2). That dose was applied in 1991 with an expected pH > 6.75 (Jacoby et al., 1994). The treatment was successful, but effectiveness persisted for only about 4 years. Recent sediment analyses show a relatively uniform concentration of mobile-P with depth (370 mg/g), which amounts to 2.7 mg/g in the top 4 cm or 6.75 g/m^2 in the top 10 cm. To guard against an unacceptable pH, a 10:1 ratio of Al added:Al–P formed and the 10 cm sediment depth was used to calculate a final dose of 72 g Al/m^2 (18.4 g Al/m^3) for a second alum treatment in March 2004. Bench-scale tests showed that an additional 5 g Al/m^3 was needed for the demand for binding sites in the water column for a total of 23.4 g Al/m^3. The minimum pH determined in the lake during the treatment was 6.9.

The dose for the first treatment of Green Lake had no rational basis, other than a desire to avoid low pH by using a buffer in this low alkalinity lake, and still add a reasonable amount of alum judged from other treatments. Had there been high alkalinity and thus adequate buffering in this lake, the alkalinity procedure would have probably yielded an effective long-lasting dose. Using the 100:1 ratio and 4 cm depth, 270 g Al/m^2 in Green Lake would have required much more buffering capacity in these soft waters. While the mobile P in the top 4 cm is higher than that in Lake Delavan (190 g Al/m^2), Green Lake does not go anoxic, except for a very small deep area, so the net internal release rate is only about one-fifth that in Delavan. Such considerations as these may be necessary in determining dose for soft water lakes.

8.3.2 IRON AND CALCIUM

Iron or calcium has been used for years in the wastewater industry to remove P (Jenkins, 1971), but their use to inactivate or precipitate P in lakes has been much less common than the use of aluminum, and there are few guidelines to determine dose. Inactivation with iron is uncommon because low redox potentials in sediments (common in eutrophic lakes) lead to slow solubilization, and high littoral zone pH leads to increased solubility of iron-hydroxide complexes. Some examples of iron doses are available, however. Peelen (1969) added Fe^{3+} to reach 2 mg/L in Dordrecht Reservoir (The Netherlands) to precipitate P from the water column. A dose of 3–5.4 mg Fe^{3+}/L (as ferric sulfate liquor) was applied to the inflow to Foxcote Reservoir (England) to remove P and to inhibit P release from sediments (Hayes et al., 1984; Young et al., 1988). Ferric sulfate and ferric chloride were added at 172–286 kg to the surface of Black Lake, British Columbia during the summers of 1990–1992 to reach whole lake concentrations of 1–2 mg/L Fe (Hall and Ashley, personal communication). Boers (1991b, 1994) added a dose of 100 mg Fe^{+3}/m^2 directly to the sediments of Lake Groot Vogelenzang (The Netherlands). That dose bound 6.6 g/m^2 P throughout a sediment depth of 20 cm (Quaak et al., 1993).

Calcium carbonate and $Ca(OH)_2$ were used by Babin et al. (1989, 1994), Murphy et al. (1990) and Prepas et al. (1990, 2001a, b) in Alberta, to precipitate and inactivate P in storm water detention ponds, water retention basins dug for potable and agricultural supplies, and in lakes. Dose ranges were 13–107 mg/L in lakes, to 5–75 mg/L in stormwater ponds, to as high as 135 mg Ca/L in the dugouts and over 200 mg/L in ponds for macrophyte control. Increases in pH, which occur when lime is added, were kept within the natural range (< 10) of the treated water bodies (Prepas et al., 2001a).

8.3.3 APPLICATION TECHNIQUES FOR ALUM

A P-precipitation treatment of Lake Långsjön, Sweden (Jernelov, 1971), was the first lake treatment to control eutrophication. Granulated (dry) aluminum sulfate was applied directly to the lake surface. While little mention was made of the characteristics of the floc in this treatment, experience with later treatments has shown that floc formation was better with liquid alum. Thus, granular alum was pre-mixed with lake water on-board the delivery barge, prior to its addition to the surface of Horseshoe Lake, WI, the first application in the U.S. (Peterson et al., 1973). Liquid alum has been used almost exclusively for lake treatments since, although a buffered alum mixture is available commercially for small-scale applications (McComas, 2003).

The depth of application in thermally stratified lakes varies depending on treatment objectives, cost, ease of application, and concerns about possible toxicity. Surface applications are easier, faster, and less costly, and provide P precipitation of the entire water column as well as treatment of the pelagic and littoral sediments. Advantages of hypolimnetic-only treatments were noted earlier (Cooke et al., 1993a), but these advantages may no longer be valid. While, some P sorption sites on the floc are lost in surface treatments as the floc falls through the water column and could reduce long-term effectiveness, the quantity of P in the water column is small relative to that in sediments (i.e., often > 100-fold difference). Although alkalinity in surface water is often less than hypolimnetic waters, buffering can resolve this issue and experience has shown an absence of toxicity in properly buffered surface applications. Moreover, surface treatments allow smaller water column Al concentrations that will achieve similar or even greater areal applications than hypolimnetic treatments.

Avoidance of shallow littoral areas may be appropriate, because treatment effectiveness in littoral areas in the summer is hampered if macrophytes are abundant (Welch and Cooke, 1999).

Also, littoral treatments offer no benefit to macrophyte control, because macrophyte growth is unaffected by adding alum to sediments (Mesner and Narf, 1987).

The first hypolimnetic treatment using the alkalinity procedure was Dollar Lake, Ohio (Kennedy, 1978; Cooke et al., 1978). The entire lake's surface also received a light (10% of total dose) application. An advantage of hypolimnetic application is that alum is delivered directly to a primary source of internal P release and the quiescent waters of the hypolimnion allow significant consolidation of the floc and sediment without interference from wind, reducing the possibility of sediment scouring. Hypolimnetic only treatments have been used where extreme precautions were needed to protect organisms in low alkalinity waters, e.g., Ashumet Pond (86 ha), Massachusetts, with < 15 mg/L $CaCO_3$ (ENSR, 2002). Intensive monitoring showed no adverse changes in pH or Al while water quality improved following a buffered alum application at a depth of 10.6 m. Nevertheless, deep applications are slower, require more complicated equipment, are likely to be more costly (e.g., Ashumet Pond), and do not address the problem of P release from oxic littoral sediments.

Alum is usually applied as a one-time dose for reasons of cost as well as effectiveness. Small (~ 2 mg/L Al) annual doses would probably not curtail annual internal loading to the extent of a large dose, designed to inactivate all mobile P in the top several cm. The P remaining unfixed by Al and recycled each year, resettles and re-enriches the surficial sediment layer, continuing to maintain a large concentration of unbound mobile P to provide internal loading. With a large and adequate dose there is no unfixed P available to recycle.

Lewandowski et al. (2003) suggested that the low ratio of added Al:Al–P formed (2.1:1) was due partly to greater efficiency of successive low doses over several years. They argued that successive small doses would offset the reduced effectiveness due to the floc sinking through the sediment. However, there was no evidence that the procedure sufficiently reduced internal loading to actually improve lake quality, the ultimate goal, because external loading was not reduced.

Generally, equipment used to apply alum is similar to that used for the hypolimnetic treatment of Dollar Lake (Kennedy and Cooke, 1982; Figure 8.8). Serediak et al. (2002) reviewed the devices used to apply alum and lime to lakes and ponds in Alberta (Canada). Modern applicators no longer build storage sites on shore. Instead, they use the delivery truck to pump the alum directly to tanks on a barge. A large harvester was effectively used to apply alum and sodium aluminate (Connor and Smith, 1986). Harvesters are designed to carry a heavy payload, are exceptionally maneuverable, and the hydraulically operated front conveyer with the application manifold can be lowered to depths of about 2 meters (see Figure 8.9). A double application manifold with spray nozzle was employed to add the appropriate ratio of alum and aluminate. A fathometer was attached to the aft portion of the hull and the harvester was operated in reverse gear to provide advance notice of bottom contour changes.

More recent application equipment includes a portable, computerized navigational device so that precise swaths are traversed and no areas are missed. This allows applicators to work on windy days when otherwise unknown changes in barge position can occur. Other improvements include the use of an on-board computer to control the output of chemical, based on barge speed and water depth. Such a computer equipped, navigational barge is used by T. Eberhardt, Sweetwater Technology (Figure 8.10).

Pond applications can be simpler. Alum has been added to small ponds by a pump and hose from the shore. Serediak et al. (2002) described a shore-based system for alum or lime that can deliver a slurry directly from shore or pumped to a distribution boat up to 1 km away. Apparently first developed by May (1974), blocks of ferric alum were suspended at mid depth in the pond, allowed to dissolve, and replaced as needed. An application system was described for small lakes and ponds that consisted of mixing dry alum with lake water in a plastic garbage can (McComas, 1989; 2003). A hand operated diaphragm pump was then used to pump alum to a 2-m long manifold pipe, drilled with holes, which was mounted on the stern of a flat-bottom boat or a barge.

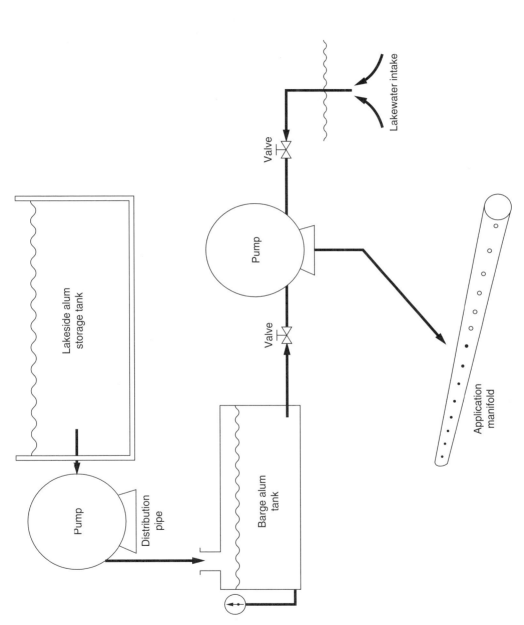

FIGURE 8.8 Basic components of a lake application system. (From Kennedy, R.H. and G.D. Cooke. 1982. *Water Res. Bull.* 18: 389–395. With permission.)

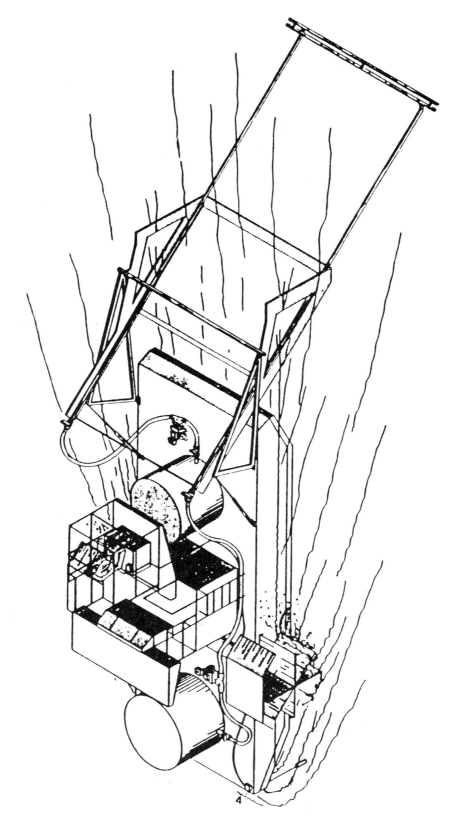

FIGURE 8.9 Modified harvester with alum/aluminate distribution system. (From Connor, J. and M.R. Martin. 1989. NH Dept. Environ. Serv. Staff Rep. 161.)

FIGURE 8.10 Alum application by Sweetwater Inc. to Long Lake, Kitsap County, Washington. (Courtesy of T. Eberhardt. Aiken, Minnesota. With permission.)

Equipment costs were about $190–440/ha (McComas 2003). Two persons can treat a 1.6 ha (4 acre) pond in 1 day.

8.4 EFFECTIVENESS AND LONGEVITY OF P INACTIVATION

8.4.1 Introduction

There have been many (probably hundreds) lake treatments with Al in the past 35 plus years since Långsjön, Sweden, making it one of the more popular lake management tools. While results of only a small fraction of treatments are published, nearly every reported treatment was successful to some degree in reducing sediment P release as well as producing an improvement in trophic state. Treatment areas up to 305 hectares (Irondequoit Bay, Lake Ontario; Spittal and Burton, 1991), doses up to 936 metric tons (12.2 mg/L Al) of alum (Medical Lake, WA; Gasperino et al., 1980), and control of P release for up to 18 years (Garrison and Ihm, 1991; Welch and Cooke, 1999) have occurred. Some treatments have met with limited success due to low doses, focusing of the $Al(OH)_3$ layer by wind mixing, interference from macrophytes, or insufficient reduction of external nutrient loading. However, Al treatments usually have been a reliable lake management technique.

Sediment P inactivation treatments must meet the following criteria to be successful: (1) reduce sediment P release for at least several years, (2) lower the P concentration in the lake's photic zone, and (3) be non-toxic. Determinations of sediment P release either *in situ* or in laboratory cores are used to answer the first question. The second question requires a demonstration that the lake's P-rich hypolimnion was a significant P source to the photic zone (see Chapter 3). In continuously mixed lakes, of course, the second criterion does not apply.

The following paragraphs describe an assessment of the effectiveness and longevity of Al treatments in lakes with adequate data, as judged by the above criteria.

8.4.2 Stratified Lake Cases

Twelve U.S. lakes that received Al treatments (ten hypolimnetic) between 1970 and 1986 were evaluated in the 1990s to determine treatment effectiveness and longevity (Welch and Cooke, 1999). Morphometric characteristics and Al dose are given for those lakes in Table 8.2. Internal loading rate was reduced in seven of the lakes (those with adequate data to determine hypolimnetic P

TABLE 8.2
Characteristics and Alum Doses of Project Lakes

	Lake Name and Location	Treatment Date	Chemicals Used	Dose (gm Al/m^3)	Application Depth (m)	Lake Area (km^2)	Maximum Depth (m)	Mean Depth (m)	Alkalinity (mg/L CaCO$_3$)	Mixis	Ref.
1.	Annabessacook, Winthrop, ME	8/78	AS:SA 1:1.6	25	Hypolimnion	5.75	12.0	5.4	20	Dimictic	Dominie, 1980
2.	Cochnewagon, Winthrop, ME	6/86	AS:SA 2:1	18	Hypolimnion	1.56	9.0	5.7	13–15	Dimictic	Dennis and Gordon, 1991
3.	Kezar, Sutton, NH	6/84	AS:SA 2:1	30	Hypolimnion	0.74	8.2	2.7	3–10	Dimictic	Connor and Martin, 1989
4.	Morey, Fairlee, VT	5–6/86	AS:SA 1.4:1	11.7	Hypolimnion	2.20	13.0	8.4	35–54	Dimictic	Smeltzer, 1990
5.	Irondequoit Bay, Rochester, NY	7–9/86	AS	28.7	Hypolimnion	6.79	23.7	6.9	170	Dimictic	Spittal and Burton, 1991
6.	Dollar, Kent, OH	7/74	AS	20.9	90% Hypolimnion 10% Surface	0.02	7.5	3.9	101–127	Dimictic	Cooke et al., 1978
7.	West Twin, Kent, OH	7/75	AS	26	Hypolimnion	0.34	11.5	4.4	102–149	Dimictic	Cooke et al., 1978
8.	Pickeral, Stevens Point, WI	4/73	AS	7.3	Surface	0.20	4.6	3.0	110	Polymictic	Garrison and Knauer, 1984
9.	Mirror, Waupaca, WI	5/78	AS	6.6	Hypolimnion	0.05	13.1	7.8	222	Dimictic	Garrison and Ihm, 1991
10.	Shadow, Waupaca, WI	5/78	AS	5.7	Hypolimnion	0.17	12.4	5.3	188	Dimictic	Garrison and Ihm, 1991

		AS:SA Ratio	12 (80% of lake v)							
11.	Snake, Woodruff, WI	5/72	unknown	lake v)	Surface	0.05	5.5	2.0	50	Dimictic	Garrison and Knauer, 1984
12.	Horeshoe, Manitowoc, WI	5/70	AS	2.6	Surface	0.09	16.7	4.0	218–278	Dimictic	Garrison and Knauer, 1984
13.	Eau Galle, Spring Valley, WI	5/86	AS	4.5	Surface	0.60	9.0	3.2	144	Dimictic	Barko et al., 1990
14.	Long, Port Orchard, Washington (Kitsap Co.)	9/80	AS	5.5	Hypolimnion	1.40	3.7	2.0	10–40	Polymictic	Welch, et al., 1982
15.	Long, Tumwater, WA (Thurston Co.)	9/83	AS	7.7	Surface	1.30	6.4	3.6	45	Polymictic	Entranco, 1987b
16.	Erie, Mt. Vernon, WA	9/85	AS	10.9	Surface	0.45	3.7	1.8	80–90	Polymictic	Entranco, 1983, 1987a
17.	Campbell, Mt. Vernon, WA	10/85	AS	10.9	Surface	1.50	6.0	2.4	80–90	Polymictic	Entranco, 1983, 1987a
18.	Pattison, Tumwater, WA	9/83	AS	7.7	Surface	1.10	6.7	4.0	45	Polymictic	Entranco, 1987b
19.	Wapato, Parkland WA	7/84	AS	7.8	Surface	0.12	3.5	1.5	NA	Polymictic	Entranco, 1986

Note: AS, aluminum sulfate; SA, sodium aluminate. Dose in $g/m^2 = g/m^3 \times$ mean depth.

Source: From Welch, E.B. and G.D. Cooke. 1999. *Lake and Reservoir Manage.* 15. (With permission.)

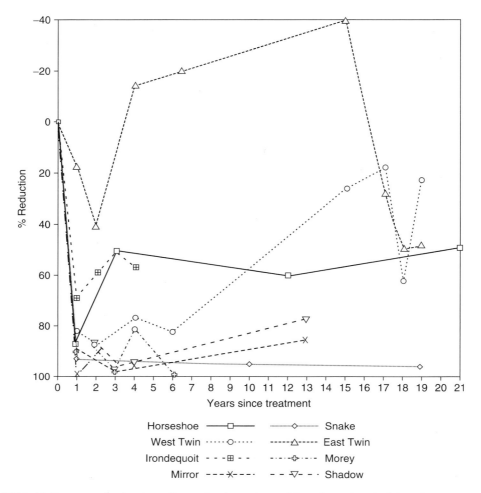

FIGURE 8.11 Percent reduction in sediment P release (rate of hypolimnetic P buildup) for seven treated, stratified lakes and one untreated stratified lake (East Twin). (From Welch, E.B. and G.D. Cooke. 1999. *Lake and Reservoir Manage.* 15. With permission.)

buildup) and remained low for an average of 13 years (4–21 years) after treatment. The treatment was the clear cause for initial control of internal loading immediately following Al addition (Figure 8.11). But the role of Al in improving trophic state is difficult to separate from the effects of diversion in lakes, and some of the longevity of effect ascribed to Al may have been due to sediment recovery (i.e., P burial). A subsequent increasing trend in internal loading following the Al treatment in some lakes (e.g., Mirrow, Shadow, West Twin, Irondequoit Bay) indicated a declining Al effectiveness. One of the treated lakes (West Twin, Ohio) had an experimental control lake (East Twin), and both of those lakes had wastewater (septic drainfield leachate) diversion. The sediment P release rate in treated West Twin Lake was much less than in untreated East Twin for 15 years, and after that, both had less release than initially, apparently an effect of sediment recovery from diversion (Figure 8.11). This was also corroborated by sediment P release rates determined in cores in 1989, 15 years after treatment (Welch and Cooke, 1999).

Seven of the treated lakes with adequate data showed, on average, a substantial improvement in trophic state (Table 8.3). These lakes also showed a long-term average of two-thirds reduction in internal loading (hypolimnetic TP buildup) following treatment and the rate dropped initially by 80% or more in six of these lakes (Figure 8.11). However, the initial decrease (average 39 and

TABLE 8.3
Reductions in Mean Summer Epilimnetic TP and chl *a* in Seven Treated and One Untreated Stratified Lakes

	Pre-treatment (µg/L)		Initial (yr) (% Reduced)		Latest (yr) (% Reduced)	
Lake	TP	Chl *a*	TP	Chl *a*	TP	Chl *a*
E. Twin – (untreated)	48 (4)	57 (4)	51 (1–5)	75 (1–5)	59 (15–18)	81 (17–18)
W. Twin – (treated)	45 (4)	42 (4)	52 (1–5)	66 (1–5)	66 (15–18)	49 (17–18)
Dollar +	82 (1)	41 (1)	65 (1–7)	61 (1–7)	68 (16–18)	29 (17–18)
Annabessacook +	32 (2)	13 (3)	34 (1)	39 (1–2)	41 (9–13)	0 (8–13)
Morey +	13 (1)	13 (6)	30 (1–4)	72 (1–3)	60 (5–8)	93 (5–8)
Kezar +	24 (4)	17 (4)	34 (1–3)	65 (1–3)	37 (4–9)	45 (4–8)
Cochnewagon –	15 (5)	5 (5)	28 (1–3)	67 (1–3)	0 (5–6)	47 (5–6)
Irondoquoit Bay +	47 (4)	23 (4)	13 (1–3)	28 (1–3)	24 (4–5)	30 (4–5)
Mean 7 treated			37	57	42	42

Note: Years of observation in parentheses. Lakes showing availability of hypolimnetic TP to the epilimnion are indicated with +, and those that did not with –.

Source: Modified from Welch, E.B. and Cooke, G.D. 1999. *Lake and Reservoir Manage.* 15. With permission.

57%) in epilimnetic TP and chlorophyll (chl) *a* was less than the hypolimnetic decrease in TP (Table 8.3). Hypolimnetic TP was shown to be available to the epilimnion in five of the seven lakes (Table 8.3). Some of the decrease in epilimnetic TP and chl *a* was probably due to residual effects of wastewater diversion, which occurred 2–3 years earlier. The exception was Lake Morey, which had an Al treatment only (Welch and Cooke, 1999).

West Twin Lake and Lake Morey represent contrasts in hypolimnetic TP availability to the epilimnion. Internal loading rate remained reduced for 15 years in West Twin (Welch and Cooke, 1999) and for at least 12 years in Lake Morey (Smeltzer et al., 1999). Trophic state improved in both lakes, but the primary cause was determined to be wastewater diversion in West Twin, because its improvement was proportional to that of East Twin, the control lake without an Al treatment. Moreover, TP availability to the epilimnion, through entrainment and diffusion, were minimal in West Twin (Mataraza and Cooke, 1997). These processes were obviously important in Lake Morey, which had no wastewater diversion, because epilimnetic TP and chl *a* remained low 12 years after treatment (Smeltzer et al., 1999). Historical aspects of these two treatment cases will be discussed further below. For historical accounts of the other treated, stratified lakes see Welch and Cooke (1999) and references contained therein.

Another interesting case that illustrates hypolimnetic P availability to the epilimnion is the treatment of a 4.6 ha section of 89 ha sandpit Lake Leba, Nebraska (Holz and Hoagland, 1999). The isolated section has mean and maximum depths of 4.2 and 9 m, respectively, and it stratifies strongly (the Osgood Index, or OI = 19.8; Chapter 3). The section was dosed with 10 mg/L Al in 1994. Hypolimnetic SRP and epilimnetic TP remained below pretreatment levels by 97% and 74%, respectively, for 3 years, while chl *a* decreased by 65% and cyanobacteria abundance was 33% less. The hypolimnetic DO 3 mg/L isopleth was 52% deeper compared to the untreated lake. Thus, alum reduced internal loading and improved epilimnetic trophic state. Similarly, massive cyanobacteria blooms were eliminated for at least 7 years in 103 ha Barleber See, Germany, following an alum treatment of only 5.7 mg/L Al to this stratified lake in 1986 (Rönicke et al., 1995). TP decreased from about 120 µg/L to 35–40 µg/L and this persisted for at least that 7 year period. However, in many lakes the hypolimnion is not always a significant source to the epilimnion, so availability of hypolimnetic P should be determined prior to treatment (see Chapter 3 for procedures).

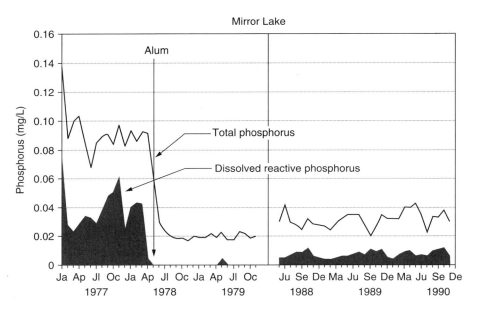

FIGURE 8.12 Volume-weighted mean P concentrations in Mirror Lake before, immediately following, and a decade after completion of restoration work. (From Garrison, P.J. and D.M. Ihm. 1991. *First Annual Report of Long-Term Evaluation of Wisconsin's Clean Lake Projects. Part B. Lake Assessment.* Wisconsin Dept. Nat. Res., Madison.)

8.4.2.1 Mirror and Shadow Lakes, Wisconsin (WI)

Urban storm drainage beginning in 1930 contributed 65% of external P loading to Mirror Lake and 58% to Shadow Lake (Waupaca, WI). Internal P loading from anoxic hypolimnetic sediments also was a major source. Storm drainage was diverted in 1976, decreasing external loading by the above percentages. In 1978, these hardwater lakes received a hypolimnetic alum treatment (Table 8.2). The results were assessed for several years after treatment and then again in 1988, 1989, and 1990 (Garrison and Ihm, 1991) and briefly again in 1991 (Welch and Cooke, 1999). A destratification system was installed in Mirror Lake and operated in spring and fall to ensure full circulation. Its operation could confound the data interpretation.

Figures 8.12 and 8.13 illustrate the volume-weighted mean P concentrations in the two lakes following diversion and again after alum application. Volume-weighted TP and SRP remained well below the pre-diversion concentration for 13 years, but have increased since 1980. The increase appeared to be due to renewed internal P loading to the hypolimnion. Before the alum treatment, but after diversion, the P release rate from Mirror and Shadow Lake sediments under anoxic conditions was 1.3 and 1.27 mg/m^2 per day, respectively. These rates were reduced by the alum treatment to a 1978–1981 average of 0.075 mg/m^2 per day. By 1990, the rate had increased to 0.20 mg/m^2 per day in Mirror and 0.3 mg/m^2 per day in Shadow Lake. The Al(OH)$_3$ layer in 1991 was about 8 to 12 cm below the sediment surface. The new layer of material above the floc contributed to the increased internal P loading. The alum treatment retarded internal P loading for at least 13 years (Figure 8.11).

Although internal P loading increased somewhat, epilimnetic TP levels in 1990 remained low and unchanged from the early post alum years. Part of the reason for low epilimnetic TP concentration may be the alum treatment, because epilimnetic TP in Mirror Lake fell from a post diversion, pre-alum mean of 28 μg P/L to a post-alum (1978) mean of 15 μg P/L and remained at 15 μg P/L in 1990. However, diversion may have contributed to that decrease as well. Notwithstanding the large decrease in hypolimnetic P following Al addition, little of that P may have been available,

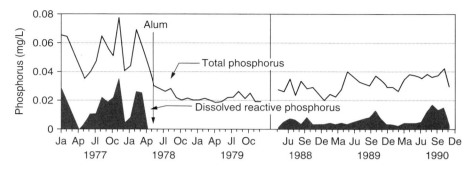

FIGURE 8.13 Volume-weighted mean P concentrations in Shadow Lake before, immediately following, and a decade after completion of restoration work. (From Garrison, P.J. and D.M. Ihm. 1991. *First Annual Report of Long-Term Evaluation of Wisconsin's Clean Lake Projects. Part B. Lake Assessment.* Wisconsin Dept. Nat. Res., Madison.)

especially in a lake like Mirror with a small surface area and relatively large mean depth (OI = 35). Mirror Lake is also surrounded by high hills, further limiting the effects of wind in mixing deeper water with surface water. However, other small lakes with large OIs (Dollar and McDonald Lakes) still showed high availability (Chapter 3).

Water clarity increased and chlorophyll fell after diversion and alum treatment, although the nuisance alga *Oscillatoria agardhii* remained abundant because it was N-limited. Values for SD and chlorophyll in 1988 to 1990 were nearly identical to the 1977 to 1981 years.

This case history is instructive in illustrating the long-term effect (13 years) of alum on anoxic sediment P release. However, the sharply lowered hypolimnetic P concentrations may have been only part of the reason for lower epilimnetic P and chl *a* levels.

8.4.2.2 West Twin Lake (WTL), Ohio

This case history illustrates a highly effective, long-lived P inactivation of hypolimnetic sediments in a dimictic lake. The case is especially important, because an untreated and similar adjacent (200 m), downstream lake, East Twin (ETL), served as a control. This permitted a separation of the effects of diversion of external loading from the hypolimnetic alum treatment. The lakes are small, shallow (Table 8.2), dimictic, and somewhat sheltered from prevailing summer winds by low bluffs and shoreline trees. WTL drains into ETL, though there is little or no flow in summer months.

In 1971 to 1972, septic tank drain-field discharges to both lakes were diverted from the watershed (335 ha), and significant fractions of storm water flows were diverted through shoreline wetlands, which may have further reduced loading. The lakes were very eutrophic (pre-diversion Carlson TP TSI = 62), with intense blue-green algal blooms and high coliform bacteria levels. WTL's hypolimnion was treated with liquid aluminum sulfate (26 mg Al/L) in July 1975, using the alkalinity procedure (Kennedy and Cooke, 1982). The Al treatment was predicted to reduce internal P loading in WTL and increase its post-nutrient diversion rate of recovery over that of ETL. Details of the experiment are reported in Cooke et al. (1978; 1982; 1993b).

The Al treatment was effective in reducing internal loading below that of ETL and that effect persisted for 15 years (Figure 8.14). Anoxic P release, determined from intact cores from both lakes in 1989, showed a rate 2.6 times greater in ETL than WTL. The effect is also apparent in the net rate of change in the P content of the 10 to 11 m contour in both lakes, determined as the difference in content between 1 June and 31 August (Table 8.4). This value is the sum of deposition from upper waters and release from hypolimnetic sediments minus any loss to the sediments or to vertical transport. Although year-to-year rates were variable, as discussed in Chapter 3, rates were much lower in WTL than in the reference lake. There was apparently little difference in release

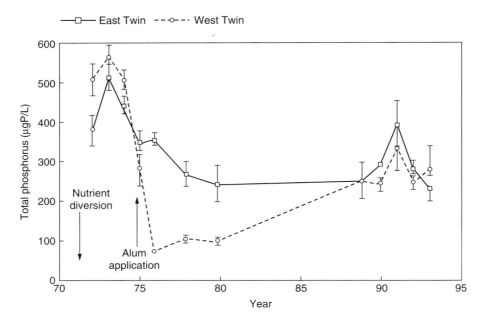

FIGURE 8.14 Mean 10 m total P concentrations in ETL and WTL, Ohio during June to August after nutrient diversion and an alum treatment. (From Welch, E.B. and G.D. Cooke. 1999. *Lake and Reservoir Manage.* 15. With permission.)

TABLE 8.4
Rate of Change in P Content of WTL and ETL Deep Stratum (10 to 11 m) from June to September

Year	WTL (mgP/m² per day)	ETL (mgP/m² per day)
1972	2.55	3.30
1973	4.24	2.83
1974	1.51	2.80
1975[a]	2.68	2.76
1976	0.37	1.76
1978	0.67	3.34
1980	0.00	1.06
1989	2.02	4.05

Note: P content determined as the difference in P content of the latest August sample of the 10 to 11 m contour minus P content of the earliest June sample, divided by contour area and number of days.

[a] Rate up to the alum treatment of West Twin on 26 July 1975.
Source: From Cooke et al., 1993a. With permission.

---

rates between the two lakes by 1989 or possibly earlier (Figure 8.14). This P inactivation treatment therefore met the first criterion (reduce sediment P release) in evaluating treatment success.

The TP concentration in the epilimnion decreased proportionately in both lakes after diversion independent of the hypolimnetic P concentration (Figure 8.15). Therefore, the trophic state improvement observed in WTL was primarily the result of diversion. The trophic state of both lakes in

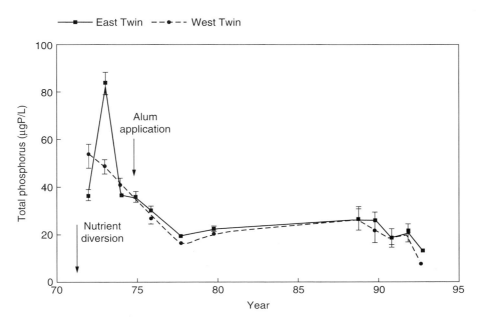

FIGURE 8.15 Mean surface total P concentrations in ETL and WTL, Ohio during June to August after nutrient diversion and an alum treatment. (From Welch, E.B. and G.D. Cooke. 1999. *Lake and Reservoir Manage.* 15. With permission.)

August 1991 was borderline mesotrophic (TP TSI = 46 for WTL, 43 for ETL) and improved slightly with a TP decrease in 1993 (Figure 8.15). TP and chl *a* in the epilimnion of both lakes have remained low for at least 18 years (Table 8.3). Macrophyte abundance and distribution increased greatly after treatment, probably as a response to greatly increased water clarity. Harvesting is now used to manage this problem. Thus, the hypothesis that a nutrient subsidy from ETL's P-rich hypolimnion would maintain its high trophic state was not the case. Vertical P transport was not a major epilimnetic P source in these wind-sheltered, but dimictic lakes (Mataraza and Cooke, 1997). However, the alum treatment of WTL undoubtedly prevented some P loading to the epilimnion. Otherwise it is unlikely that WTL, with its slightly longer water residence time (WTL = 1.28 yr, ETL = 0.58 yr), would have had an epilimnetic P concentration consistently lower than ETL. Nevertheless, the primary cause of the lake improvement was diversion.

Phosphorus input–output data suggested that there were significant internal P sources after treatment, probably from untreated littoral-wetland areas (Cooke and Kennedy, 1978). Foy (1985) reported a similar result in a lake that had received a hypolimnetic alum treatment.

8.4.2.3 Kezar Lake, New Hampshire

This lake is a relatively shallow stratified lake with very low alkalinity (Table 8.2). Wastewater inflows were diverted in 1981, eliminating 71% of the external load. Internal P loading was believed to be a major factor sustaining the blue-green algal blooms after diversion (Connor and Martin, 1989a, b).

The lake's low alkalinity required buffering, so a 2:1 ratio of aluminum sulfate to sodium aluminate was determined through jar tests to provide a high quality floc with maximum P removal while maintaining pH. A modified weed harvester was used to apply alum. A small-scale application to 10 ha occurred in summer 1983 with a dose of 30 g Al/m^3, followed by a full-scale treatment of the lake's hypolimnion (48 ha) in summer 1984 at 40 g/m^3. A heavy dose of copper sulfate was added to the lake's surface prior to the Al treatment.

TP content of the hypolimnion at 6 m decreased from a 4-year mean of 36 µg/L to 16 µg/L, but increased every year through 1987, and then declined again in 1988–1991 to levels close to the post-treatment mean (Welch and Cooke, 1999). Epilimnetic TP also decreased following diversion and again after Al treatment. The pattern of decrease, increase and then decrease following the Al treatment occurred in the epilimnion (2 m) as well as the hypolimnion. That suggests availability of internal loading to algae, which is also suggested by the relatively low OI (3.14). Chl *a* also increased gradually through 1986 following a post-treatment low of about 5 µg/L. However, summer means were usually less than 10 µg/L through 1994, in contrast to pretreatment (USEPA, 1995).

While no sediment P release data are available, the high dose of Al probably controlled hypolimnetic P release. If the hypolimnion had been the only significant P source, water column P should have remained low. Therefore, the observed increased TP after treatment suggests that external P loading was higher than that estimated, there were new sources, and/or that littoral and metalimnetic sediments were a significant P source. As shown in Table 8.3, trophic state was improved and that improvement persisted for at least 8 years. However, this effectiveness was probably only partly due to Al.

Kezar Lake had the softest water of the treated lakes evaluated (Table 8.2). Hypolimnion pH fell to 5.5 and alkalinity was eliminated, though they returned to pretreatment levels in weeks. This was associated with total dissolved Al concentrations at 2 m of up to 400 µg/L at least 1 month after application, and persistent concentrations between 35 and 135 µg/L through 1984. However, these latter values are identical to dissolved Al levels in remote New Hampshire ponds, all with pH above 6 (Connor and Martin, 1989a). There were no reports of mortalities associated with the Kezar Lake treatment, and laboratory bioassays with naturally occurring benthic invertebrates did not reveal detrimental effects on larvae of two insect species, though a decrease in 5-day BOD was observed (Connor and Martin, 1989b).

Kezar Lake represents an important case. Despite great care in the use of a buffer, alkalinity was eliminated causing an excessive decrease in pH, and dissolved Al increased. Lakes with very low alkalinity obviously require more buffering. Nevertheless, sediments were inactivated and no observable adverse effects occurred with the relatively high dose and low alkalinity. This case also indicates the possibility that untreated epilimnetic and metalimnetic sediments were important in internal P loading, with similar mechanisms as occur in shallow lakes (Chapter 3). Perhaps more consideration should be given to understanding the roles of these sediments in P budgets of dimictic lakes.

8.4.2.4 Lake Morey, Vermont

Lake Morey is a relatively large, deep, moderately low alkalinity lake (Table 8.2), located in a mountainous and heavily forested (92%) area. The lake was mildly eutrophic (Spring TP ~ 40 µg/L), in part from external loading, but primarily from internal loading from the hypolimnion as determined by a P budget in 1981–1982. The nutrient-rich hypolimnetic sediments, which account for two thirds of the lake's sediment area, were enriched during an early period of land clearing and poor wastewater disposal practices. Another important factor in the lake's P cycle was the low hypolimnetic Fe:P ratio (about 0.5), which was apparently related to FeS precipitation and low P removal by Fe during turnover (Smeltzer, 1990).

The lake's hypolimnion was dosed with 12 g/m^3 (44 g/m^2) in 1986 at a ratio of aluminum sulfate:sodium aluminate of 1.4:1 (Table 8.2). A modified weed harvester was used for the application. Lake Morey is an excellent case to determine Al effectiveness and longevity due to the importance of internal loading from the relatively large fraction of sediments in the hypolimnion. But that hypolimnetic P may have been unavailable due to the lake's stability (OI = 5.7) and the surrounding forest that provided protection from wind mixing.

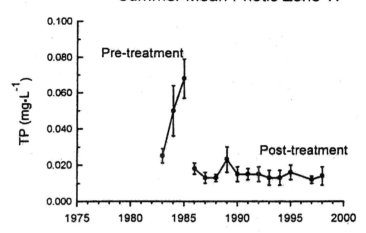

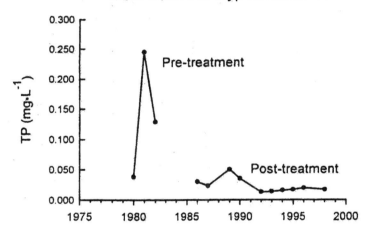

FIGURE 8.16 Long-term water quality monitoring results for Lake Morey. Error bars are 95% confidence intervals for the annual mean values. (From Smeltzer et al. 1999. *Lake and Reservoir Manage.* 15: 173–184. With permission.)

The Al treatment was highly successful in reducing sediment P release by over 90% (Figure 8.11), and late summer mean volume-weighted TP decreased by 83%, from 38–245 μg/L to 13–50 μg/L (Smeltzer et al., 1999). Effectiveness of sediment P control has been consistent through 1998 (Figure 8.16). Trophic state was also improved, demonstrating that internal loading was available prior to treatment. Epilimnetic TP and chl *a* decreased dramatically, with cyanobacteria blooms as high as 31 μg/L chl *a* being eliminated, and that effectiveness persisted for 12 years after treatment (Table 8.3; Smeltzer et al., 1999). Photic zone values reported by Smeltzer et al. (1999) also showed marked reduction in TP and chl *a* of 68 and 61%, respectively. Consistent with the reduction in chl *a*, transparency increased from a summer mean of 4.0 to 7.2 m, and hypolimnetic DO nearly

TABLE 8.5
Effectiveness (% Reduction) and Longevity of Phosphorus Inactivation Based on Mean Summer, Whole-Lake TP Concentrations and Observed P Release Rate

Lake	TP (µg/L)	Reduction in Whole-lake TP Initial%	Latest %	Reduction in Rate Release Initial%	Latest %	Longevity (yr)
Erie[a]	115 (2)	77 (1)	75 (5–8)	79 (1)	82 (5–6)	>8
Campbell[a]	49 (2)	43 (1)	46 (5–8)	57 (1)	64 (5–6)	>8
Long (T)						
North[a]	42 (3)	60 (1–2)	56 (7–8)	84 (1–2)	79 (7–8)	>8
South[a]	31 (3)	32 (1–2)	50 (4–5)	Macrophytes		5
Pattison (T)						
North[a]	28 (3)	43 (1–2)	29 (5–7)	81 (1–2)	73 (5–7)	7
South	30 (3)	–7 (1–2)	–	Macrophytes		<1
Long (K)[a]	63 (3)	48 (1–4)	30 (7–11)	62 (1–4)	40 (7–10)	11
		68 (1)[b]				
Wapato	46 (2)	–24 (1–2)	—	Macrophytes		<1
Pickerel	35 (1)	–26 (1)	—			<1
Mean of 6[a]	55	51	48	73	68	5–11

Note: T = Thurston; K = Kitsap Counties. Years of observation in parentheses.

[a] Six successful treatments.
[b] Second treatment in 1991 at same dose as 1980.

Source: From Welch, E.B. and Cooke, G.D. 1999. *Lake and Reservoir Manage.* 15: 5–27. With permission.

doubled. The Fe:P ratio increased to 3.3 following treatment, which facilitated P precipitation during mixing.

There were temporary increases in dissolved Al in the epilimnion even though pH and alkalinity levels did not change from pre-treatment values. Associated with that was evidence of a decrease in condition factor of yellow perch, and possible changes in species richness of benthic invertebrates (Smeltzer, 1990). Subsequent monitoring shows that despite the temporary adverse effect, there was a long-term benefit to biotic populations (Smeltzer et al., 1999).

8.4.3 Shallow, Unstratified Lake Cases

Internal P loading is important in shallow lakes because P released from sediments is immediately available in the photic zone, and because conditions for P release can be ideal. Microbial activity is enhanced by the higher temperature of sediments in shallow lakes and may lead to sediment anoxia, or a very thin oxidized sediment surface, under temporary water column stability. This condition encourages iron reduction and P release (Jensen and Andersen, 1992; Löfgren and Boström, 1989; Søndergaard et al., 2003). Subsequent wind mixing would entrain the high-P bottom layer; this sequence can provide a series of internal loading events over a summer. Microbial activity may also be important itself in releasing P through mineralization and cell excretion (Søndergaard et al., 2003). High pH in the water column from planktonic and macrophyte photosynthesis may release P through ligand exchange (OH⁻ for PO_4^{-3} on iron–hydroxy complexes (Koski-Vähälä and Hartikainan, 2001; Van Hullebusch et al., 2003). Wind-caused resuspension exposes sediment-bound P for dissolution either via high pH or loosely bound P (Søndergaard, 1988). Some shallow lakes, like some deeper, dimictic lakes, have groundwater inflows in shallow areas that force additional transport of interstitial P from the sediments (Prentki et al., 1979). Internal loading may also involve migration of blue-green algal colonies from the sediment to upper waters, a process that Barbiero and Welch (1992) found to be important in shallow Green Lake, WA. These processes are discussed with respect to simulation modeling in Chapter 3 and elsewhere (Boström et al., 1982; Gaugush, 1984; Carlton and Wetzel, 1988; and Boers, 1991a, b; Søndergaard et al., 2003).

Initially, P inactivation in shallow, continuously mixed lakes was considered to be ineffective because the first three attempts failed. However, Lake Långsjön, Sweden (Jernelöv, 1971) and Lake Lyngby Sø, Denmark (Norup et al., 1975) failed because external loading remained high, and Pickerel Lake, WI (Garrison and Knauer, 1984), failed because a storm after treatment was believed to have redistributed the floc to the lake's center (Table 8.5).

Phosphorus inactivation in shallow, unstratified lakes in Washington state have been largely successful (Welch et al., 1988; Welch and Cooke, 1999). Treatments have been successful because internal loading dominates over external loading during the low-precipitation, low-inflow summers and is thus the main cause for algal blooms during summer in western Washington (Welch and Jacoby, 2001). Alum treatment effectiveness is not confused with lingering effects from wastewater diversion, as is the case with many of the treated dimictic lakes discussed earlier, because with one exception, there has been no point-source diversions from these lakes. Also, the effectiveness was in spite of doses that were probably too low as a result of their low alkalinity and use of the alkalinity dosing procedure.

Of the nine shallow lakes (and basins) receiving alum treatments, six were considered to be successful, averaging about a 50% reduction in TP that lasted for 5–11 years (Table 8.5). Internal loading rate in five of the six lakes with adequate data, determined by the summer increase in lake TP (inflow TP was small), declined by two thirds, which persisted for 7–11 years. Trophic state also improved in the six lakes/basins following the alum treatments and four maintained the improved conditions for 8–11 years (Table 8.6).

Treatment failure in three of the lakes was considered due to extensive coverage by submersed macrophytes. Water from Pattison Lake South, which was completely covered with native macro-

TABLE 8.6
Effectiveness of P Inactivation in Reducing TP and chl *a* in Unstratified Lakes

Lake	% Reduction Initial		% Reduction Latest Years	
	TP	Chl *a*	TP	Chl *a*
Erie (> 8)[a]	77	91	75	83
Campbell (> 8)	43	44	46	28
Long (T)				
North (> 8)	60	89	56	39
South (5)	32	68	50	0
Pattison				
North (7)	43	40	29	—
South (< 1)	–7	6	—	—
Long (K) (> 11)	48	65	30	49
Mean of 6[b]	51	66	48	40

[a] Longevity in years in parentheses.

[b] Unsuccessful treatment of Pattison-South excluded.

Source: From Welch, E.B. and Cooke, G.D. 1999. *Lake and Reservoir Manage.* 15: 5–27. With permission.

phyte species, drains to Long Lake South and provided a source of P from senescing plants. Thick stands of milfoil in Long Lake South itself may have released P as well. The northern basins of these lakes are deeper, with fewer macrophytes and they responded positively. In Wapato Lake, areal coverage by the macrophyte *Ceratophyllum* increased dramatically following treatment, pH increased to 10.1, and TP doubled in concentration between years. This failure may have been due to the increased pH, leading to P release via ligand exchange, and to contributions of P to the water column by *Ceratophyllum.*

8.4.3.1   Long Lake, Kitsap County, Washington

Long Lake, located in Kitsap County, is a shallow (Table 8.2), eutrophic, soft water lake, dominated by *Egeria densa*. Internal P loading was 55% of the total loading during summer 1977 and lake TP concentration was 2–3 times inflow concentration (Jacoby et al., 1982; Welch and Jacoby, 2001). The lake was treated with alum in 1980. TP declined from a three-summer mean of 65 µg/L to a four-summer mean of 32 µg/L, but then returned abruptly to pretreatment levels during 1985–1986, and declined again in 1986–1987, and averaged 41 µg/L from 1987 to 1991 (Figure 8.17). By summer 1990, TP had slowly returned to near pretreatment levels and a second Al application was carried out in October 1991, using Na_2CO_3 buffered alum, which reduced lake TP to an all time low of 20 µg/L (Welch et al., 1994; Welch, 1996).

The TP increase in 1985 was associated with a dramatic and unexplained decline in abundance of macrophytes (*E. densa* from 90% to 10% of plant biomass; Welch et al., 1994). This allowed wind mixing to stir sediment bound or $Al(OH)_3$-bound P into the water column. In 1987, *E. densa*

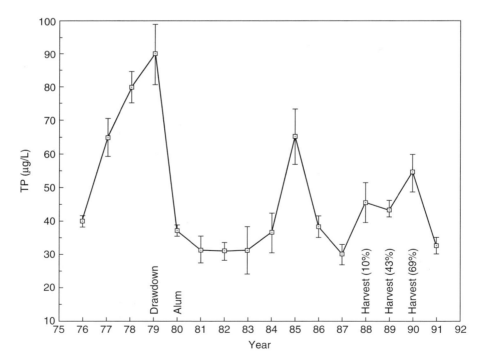

FIGURE 8.17 Whole-lake mean total phosphorus concentration during summer in Long Lake in relation to various lake treatments (From Welch et al., 1994. *Verh. Int. Verein. Limnol.* 25.With permission.)

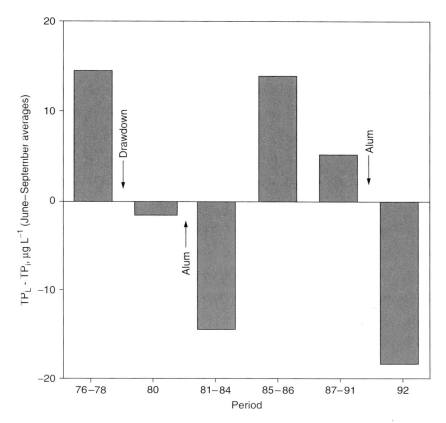

FIGURE 8.18 Mean difference between volume-weighted TP_L (whole-lake) and TP_I (inflow) concentrations during June–September for years before and after the 1991 alum treatment in Long Lake, Kitsap County, Washington. (From Welch, E.B. and G.D. Cooke, 1999. *Lake and Reservoir Manage.* 15. With permission.)

biomass returned to normal levels and P concentration fell again. Thus, the alum treatment of Long Lake was very effective in controlling internal P release for 4 years and partially effective for 11 years (Table 8.5; Figure 8.18). Associated with lower TP in Long Lake was a sharply lowered algal abundance with much less cyanobacteria and increased transparency (Welch et al. 1982). Trophic state improvement was maintained over 11 years in spite of fluctuations in internal loading due to macrophyte declines (Table 8.6). Additional details are in Jacoby et al. (1982), Welch et al. (1986; 1988; 1994), and Welch and Kelly (1990).

8.4.3.2 Campbell and Erie Lakes, Washington

These lakes, located in Skagit County, are connected by a surface water outlet from Erie. Their watersheds are relatively undeveloped (1%) and are over 70% forested. Nevertheless, dense cyanobacteria blooms (*Aphanizomenon*) were common throughout the summer prior to alum treatments. The response of these lakes was the most impressive of all the treated Washington lakes. Part of the reason may be because macrophytes were not abundant prior to treatment (< 10% of surface coverage) due to the dense algal blooms. Also, internal loading was a large part of the total loading during summer (65% and 92%; Welch and Jacoby, 2001).

Summer mean TP concentration had begun to increase during the seventh and eighth years after treatment, but the greatly improved water quality condition continued to prevail (Table 8.6). Moreover, the absence of *Aphanizomenon* and its intense blooms continued for at least eight years

(Welch and Cooke, 1999). Treatment effectiveness may not have lasted much longer, however. Sediment profile determinations in 1997 showed that 3.5 tons of Al–P should have been formed in sediment from 39 tons of Al added (Rydin et al., 2000). Sediment profiles showed that 47% of the added Al was retained. Dividing 3.5 tons by the pre-treatment internal loading rate (0.15 tons/yr) times 0.47 (recovered fraction of added Al) equals 12 years longevity expected. Although monitoring ceased after 8 years, lake quality was still improved over pretreatment levels, but was declining; a 12-year longevity may be realistic.

8.4.3.3 Green Lake, Washington

This lake had been studied off and on for over 30 years prior to the sodium aluminate buffered alum treatment in fall 1991. While the lake may have been "naturally" eutrophic, a permanent lowering of the lake level by about 2.1 m to drain adjoining wetlands in 1911 probably worsened water quality through concentrating the internal P loading flux, which represented 88% of summer total loading in 1981 (Welch and Jacoby, 2001). Except for a single deep hole (1% lake volume) on one side of the lake, 80% of its sediment is above 4 m. Therefore, the lake is largely polymictic and most of the internal loading is readily available. A sizable fraction (up to 60% in 1981) of the internal loading was due to migration of cyanobacteria, mostly *Gloeotrichia*, from the sediments throughout the summer, but other mechanisms in this soft water lake (35 mg/L $CaCO_3$ alkalinity) were probably operating as well (see Chapter 3).

The buffered Al dose was 8.6 mg/L to meet cost constraints and added safety, although jar tests showed that a dose of 15.7 mg/L would have maintained pH above 6.0, with a sodium aluminate (Al):alum (Al) ratio of 1.3 (Jacoby et al., 1994). As treated, pH remained above 6.7 and alkalinity above 28 mg/L. Although TP was reduced by 65% initially, whole lake TP remained at 50% and 35% of the October 1991 pretreatment(40 µg/L) level during the 2 post-treatment years. These 1992–1993 levels were 62% and 50% less than the 1981 summer mean (52 µg/L). Algal blooms were much reduced over 1981 levels (only pretreatment chl *a* data available) by 75% and 55% during 1992–1993. However, the treatment is considered to have been effective for only about 4 years.

The unfortunate aspect of this case is the variability of dilution water inputs. City dilution water was added to the lake during summers beginning in 1962, but inputs declined in the late 1970s (Chapter 6). The bloom problems resulting from low or no dilution, prompted the addition of alum, but there were no dilution-free control data to compare with the near dilution-free (~ 10% of 1960s–1970s rate) post alum years of 1992–1993 — only 1981 when some dilution water was added (~ 40% of 1960s–1970s rate). Because of the unequal dilution, these levels of improvement may underestimate actual effectiveness.

Another complexity involves macrophytes. Eurasian watermilfoil was not present in the lake in 1981, but increased dramatically in the late 1980s. The lake was almost totally covered, with plants reaching the surface in 4 m of water by the time the lake was treated with alum. Given the interferences to Al effectiveness from macrophytes noted in other Washington shallow lakes (Welch and Cooke, 1999), the presence of these dense stands of milfoil, along with the dissolved and particulate carbon they emit to the water column, was probably instrumental in limiting alum effectiveness. Macrophytes may intercept settling alum floc causing an uneven distribution on the sediment surface, and organic carbon would occupy binding sites on the floc (Lind, personal communication). To avoid the potential problems of macrophyte interference, the 2004 treatment of the lake was completed in March, prior to the spring increase in plants. Also, bench-scale tests were conducted to assess the alum demand in the water column and that dose was added to the dose determined from sediment mobile P (see section on dose).

Sediment cores taken in Green Lake in 1998 failed to show a distinctive Al layer that could be ascribed to the 1991 treatment, in contrast to the other five treated Washington lakes (Rydin et al., 2000). The failure detect an Al layer may be due to interference by macrophytes, but the presence of a rather dense population of carp may have disturbed and mixed the floc through the sediment to obscure a marker.

The settling alum floc apparently had no inhibitory effect on migrating cyanobacteria during 1992, 1993 or 1994 (Perakis et al., 1996; Sonnichsen et al., 1997), compared to pre-alum rates determined in 1989–1990 (Barbiero and Welch, 1992). Migration of *Gloeotrichia* and its contained P was variable from year to year, but that variability was related to light and temperature and not the alum treatment. Some other cyanobacteria even increased in migration rate after alum, indicating that these vegetative spores were relatively insensitive to the alum floc. The alum floc was effective in sorbing some of the mobile P fraction and significantly reducing internal loading, albeit short-lived, in spite of the internal loading fraction contributed by cyanobacteria migration.

8.4.4 RESERVOIRS

Phosphorus inactivation treatments of reservoirs have been uncommon. One reason may be that the high hydraulic and thus nutrient and sediment loading to reservoirs dominates water column P concentrations. The resulting high deposition rates of nutrient-rich materials would readily cover over the alum floc and reduce treatment longevity. However, many relatively small reservoirs have water residence times of a year or more with internal loading being substantial.

Eau Galle Reservoir, Wisconsin is a small (0.6 km^2), weakly stratified flood control impoundment, which receives drainage from a 166 km^2, largely agricultural, watershed. External loads of nutrients are high due to the high watershed:lake area ratio, which contributes to excessive algal and macrophyte biomass (Barko et al., 1990). Alum was applied to the hypolimnion in late May 1986 at a rate theoretically sufficient to provide control of sediment P release for 5 years (Kennedy et al., 1987). The treatment was effective in lowering P concentration in the hypolimnion and in controlling treated sediment P release during the summer of 1986. However, external P loading was high during that summer and effectiveness in controlling epilimnetic P was minimal. The alum treatment had no effect on epilimnetic algal biomass in 1986 or in subsequent years (Barko et al., 1990). Recent analysis of dosing procedures suggests that Eau Galle may have been underdosed (see section on dose).

Therefore, the potential for alum effectiveness in reservoirs should not be judged by the Eau Galle case. There were also several other major internal P loading sources besides the hypolimnion, including groundwater and the littoral zone (Barko et al., 1990), as well as the treatment being under-dosed. While alum may not be effective in reservoirs with high watershed:lake area ratios and relatively high external loading, careful analyses may show that sediment P release represents an important fraction of total loading to the epilimnion in some reservoirs, making them appropriate candidates for alum treatment.

8.4.5 PONDS

There are many ponds and small swimming lakes with nuisance blooms of algae caused by internal loading. The most common treatment for such problems is the addition of an algicide, which has short-term effectiveness. Alum addition, or the placement of blocks of ferric alum, $Fe_2(SO_4)_3 \cdot 24H_2O$ are alternative options. A dose of 50 g alum/m^3, as blocks of ferric alum, was sufficient to reduce TP concentrations below 50 µg/L. Blooms of toxic blue-green algae (*Anacystis cyanea* and *Anabaena circinalis*) were suppressed by this procedure in most trials in small farm ponds in New South Wales, Australia (May 1974; May and Baker, 1978).

May and Baker (1978) recommended a single treatment of 100 g alum/m^3 as ferric alum applied just before the onset of warm weather. The blocks were placed in cloth bags and suspended from an anchored float. For larger ponds, several floats were used. The blocks were allowed to dissolve over the season and did not need replacement for 12 months (May and Baker, 1978). Baraclear is a commercially available solid buffered alum product (McComas, 2003).

Continuous addition of alum from dosing stations by means of air diffusers is recommended by McComas (2003). This procedure removes externally loaded P from the water column. Spraying the liquid alum from shore was successful in treating golf-course ponds near Clinton, Washington. The procedure replaced the long-standing use of algaecides. Alum treatments of ponds also have been used to remove clay turbidity (Boyd, 1979).

As with P control, there are cautions. External loading or in-pond mixing are likely to return a treated pond to pretreatment conditions, or, persistent clear water may allow macrophytes to flourish (Chapter 9).

The use of calcium compounds ($CaCO_3$, $Ca(OH)_2$, $CaSO_4$) to remove or inactivate P or to remove turbidity in ponds has had some success. Wu and Boyd (1990) found that gypsum ($CaSO_4 \cdot 2H_2O$) additions to small, fertilized ponds removed SRP and resulted in reduced phytoplankton abundance. The treatments also removed clay particles that produced high pond turbidities. Additions of 250 mg/L of slaked lime ($Ca(OH)_2$) to hardwater ponds in Alberta, Canada successfully reduced P concentrations and algal biomass, and effects of the treatments extended over two summers. Costs to treat these small (1000 to 2000 m^2) ponds, or dugouts, were to $200–400 each. Calcite ($CaCO_3$) treatments reduced algal biomass to a lesser extent and the ponds recovered to pretreatment levels the next summer (Murphy et al., 1990).

Slaked lime treatments of ponds with algal blooms or high nonalgal turbidity are less expensive procedures than alum. Also, there is less concern with toxicity problems when using calcium than with alum, although alum toxicity has not been a problem in well-buffered waters. On the other hand, an excessive increase of pond pH is a risk with additions of $Ca(OH)_2$, and treatment longevity may be less than with alum. Much additional research on the use of calcium salts is needed, including long-term evaluations. These results strongly suggest that lime additions are a more effective, cheaper, and far less toxic alternative to managing hardwater pond problems with algal blooms than the use of copper sulfate. As described in detail in Chapter 9, fish removal may be an even more direct way of managing highly turbid ponds.

8.4.6 Iron Applications

There are only a few cases where iron salts were used to precipitate and/or inactivate P in lakes, largely because low redox values in sediments allow P release from Fe–P complexes to the water column (Cooke et al., 1993b). Nevertheless, iron may be effective in aerobic sediments that have low Fe:P ratios. A study of 15 shallow lakes in Denmark showed that internal loading was controlled by iron if the TFe:TP ratio in surface sediment was ≥ 15 by weight (Jensen et al., 1992). Lakes with ratios < 15 had increasingly higher lake TP concentrations. Therefore, lakes with Fe:P ratios well below 15 would likely benefit from added iron, especially if they had aerobic water columns. For example ferric chloride was injected at 0.5 mg/L into the hypolimnion of Vadnais Lake, MN through hypolimnetic aerators (Walker et al., 1989). The Fe:P ratio in that lake was < 0.5, whereas ratios should be at least > 3.0 for P precipitation from oxygenated water (Stauffer, 1981).

Added iron may also increase P retention in lakes with Fe:S ratios $< 1.2–1.8$ (by weight). Lake Gross-Glienicker, Germany (68 ha, maximum depth 11 m), with a Fe:S ratio in sediment of 1.35, responded positively to an iron treatment of 500 g/m^2, associated with hypolimnetic aeration (Wolter, 1994). Summer TP was lowered from 500 to 12 µg/L and chl to a mean of 10 µg/L from bloom conditions > 100 µg/L.

Iron was applied as a P inactivant to Foxcote Reservoir (England), a small (19 ha), shallow ($\bar{Z}$ = 2.8 m), polymictic reservoir in April 1981, with a dose of 3.5 mg Fe/L as ferric sulfate (Hayes et al., 1984). A 3- to 16-mm thick floc was produced, mean SRP fell from 7 to 3 µg/L, and TP was reduced from a mean of 30 to 16 µg/L. However, the floc layer was eliminated in 30 days, although algal biomass was reduced from April through June. Subsequently, the pumped inflow to this reservoir was dosed with 3–5.4 mg/L Fe that maintained inflow SRP < 10 µg/L (Young et al.,

1988). Aerobic conditions were maintained, phytoplankton declined, while macrophytes and periphyton increased.

Ferric aluminum sulfate (FAS) was added to White Lough (Northern Ireland), a small (7.4 ha) deep ($\bar{Z}$ = 6.2 m, Z_{max} = 10.7 m) dimictic lake (Foy, 1985). Liquid FAS (25 m³) was added to the surface during winter, 1980 at a dose (3.7 mg Fe + Al/L) sufficient to precipitate P. Summer P release to the hypolimnion was reduced by 92% in 1980, but returned to pre-treatment rates by 1982. At fall overturn, iron and P were precipitated, unlike pre-treatment years. However, there was little reduction in phytoplankton (Foy, 1985; Foy and Fitzsimons, 1987). Effectiveness was limited by the low dose of iron and aluminum, coupled with anoxia and continued external loading.

The sediments of shallow ($\bar{Z}$ = 1.8 m) Lake Groot-Vogelenzang (the Netherlands) were treated with a dose of 100 g Fe/m² by mixing FeCl₃ with the upper 15 to 20 cm of sediments (Boers, 1991b, 1994; Quaak et al., 1993). Eighty cubic meters of a 40% solution of FeCl₃ was used, following a 100- to 150-fold dilution of the 40% solution with lake water. Laboratory studies had shown that 100 g Fe/m², but not 50 g Fe/m², would control P release under anoxic conditions, although release was higher than under oxic conditions. The treatment effectiveness was short-lived (3 months). A return to pre-treatment water column P concentrations was attributed to wind storms and high external loading. Although some reduction in external loading was achieved 4 years earlier by wastewater treatment, the residual external load was still sufficient to produce a eutrophic state, in spite of control of internal loading for at least 8 months.

Another water supply reservoir in England (Alton Water), in addition to Foxcote, was treated with iron by dosing the enriched inflow since 1983 (Perkins and Underwood, 2001, 2002). Although the ferric sulfate dose was not stated, the high inflow SRP of 570 µg/L was reduced by 90% with most of the precipitated Fe and P being retained behind a barrier. Nevertheless, low rates of internal loading persisted during summer, even though sediment Fe content was high and the water column (maximum depth 18 m) was artificially and continuously mixed with air. In spite of the treatment, chl a increased in the main body of the 158 ha reservoir throughout the 1980s to early 1990s to bloom peaks of 100 µg/L. Then chl a decreased to summer means of < 10 µg/L in the mid 1990s. Suggested causes of the decrease were a massive kill of roach, a planktivore, and subsequent development of macrophytes.

Biomanipulation in Bautzen Reservoir, Germany (530 ha, mean depth 7.4 m), was enhanced by adding ferrous iron at a total of 113 and 90 tons during May to August 1996 and 1997, respectively (Deppe et al., 1999; Deppe and Benndorf, 2002). The water column was also continuously mixed with air transported from the surface by three underwater pumps through which the iron was injected and dispersed. Preliminary in-lake experiments determined that 5 mg Fe/L was sufficient to remove $\geq$ 90% of SRP. Compared with the reference year 1995, SRP was reduced by 72% and 54% during the treatment period in 1996 and 1997, respectively, while the reduction in TP was 45% both years. *Microcysits* nearly disappeared and was replaced by diatoms after raising the CO₂ concentration by mixing the water column. Fe was preferred for treatment over Al due to the high pH (11) and the potential of toxicity from Al compounds formed at such high pH.

Due to the redox sensitivity of iron, there has been no long-lasting control of sediment P release with iron salts. However, water quality has been improved in the short-term with continuous addition of iron, either at inflows or injected with hypolimnetic or complete circulation devices, so long as oxic conditions exist (Walker et al., 1989; Wolter, 1994; Jaeger, 1994).

8.4.7 Calcium Applications to Hardwater Lakes

Calcite precipitation is a significant regulator of algal productivity in hardwater lakes via sorption of P and labile organic molecules (Otsuki and Wetzel, 1972). Internal loading is also affected by calcite precipitation. For example, sediment P release decreased from > 30 mg/m² per day, associated with the occurrence of natural calcite precipitation in Feldberger-Haussee, Germany

(Kasprzak et al., 2003). Internal loading had remained high for several years despite a 90% reduction in external loding.

Application of this to the management of eutrophic hardwater lakes constitutes a recent lake management technique. The first reported whole-lake treatment was the addition of $Ca(OH)_2$ (slaked lime) to the surface of dimictic eutrophic Frisken Lake, British Columbia (33.8 ha, $\bar{Z}$ = 5.5 m, and Z_{max} = 11 m), during summer 1983 (23 metric tons) and spring 1984 (16 metric tons; Murphy et al., 1988). Phosphorus precipitation from the epilimnion was large; subsequently transparency increased, and *Aphanizomenon flos-aquae* was removed via flocculation. In both summers, all of the precipitate dissolved in the hypolimnion, reducing the treatment's long-term effectiveness. However, the treatment was considered at least as effective as prior uses of copper sulfate, but without the toxicity to biota or accumulation of a metal.

The use of lime has continued in Boreal Plain hardwater eutrophic lakes in Alberta, with multiple treatments of Figure Eight Lake (A = 36.8 ha, Z_{max} = 6 m, $\bar{Z}$ = 3.1 m) and Halfmoon Lake (41 ha, Z_{max} = 8.5 m, Z = 4.7 m) starting in 1986 and 1985, respectively (Prepas et al., 1990, 2001a, b). These lakes stratify weakly and anoxia develops over deepwater sediments with high bottom water TP. Surface water TP and chl a before treatment were 135–266 and 94–113 μg/L, respectively. During *Aphanizomenon* blooms, pH increased to near 10.

Lime was applied as a slurry to lake surfaces during 1986–1992 (five treatments for Figure Eight) and 1988–1993 (four treatments for Half Moon). The quality of both lakes improved markedly over a 7-year period. TP and chl a declined by 91% and 79%, respectively, in Figure Eight and both indicators by 77% in Half Moon (Prepas et al., 2001a, b). Reduced internal P loading was likely due to hydroxyapatite formation in the sediments (Murphy and Prepas, 1990; Prepas et al., 2001b). However, laboratory results showed that alum or lime plus alum were more effective than lime alone plus oxygen in reducing sediment P release (Burley et al., 2001; Prepas et al., 2001b). These results were based on comparisons with reference lakes.

Single doses of lime to lakes and ponds had variable and more temporary effects on TP and chl a and on sediment P release (Prepas et al., 2001a). While such treatments were relatively ineffective in improving water quality, they produced an 80% decline in macrophyte biomass for at least 2 years (Reedyk, 2001). Lime treatments had no effect on cyanobacteria dominance, nor any adverse effect on macroinvertebrates (Yee et al., 1999; Zhang et al., 2001). However, lime treatment of one half a small pond at 250 mg/L $Ca(OH)_2$ reduced macroinvertebrate abundance and diversity to about one half, compared to the untreated side (Miskimmin et al., 1995). Increased pH and failure of macrophytes to colonize were probably the explanation for the adverse effect. Invertebrates and macrophytes were fully recovered 3 years later when the pond was again sampled.

Lime was also used to improve shallow lakes receiving stormwater (Babin et al., 1989, 1992). Even if applications are frequent, as would occur in rapidly flushed systems, costs were not high and a safe, effective, non-toxic material was substituted for an algicide. $Ca(OH)_2$ (slaked lime) was found to be more effective at removal and retention of P than $CaCO_3$, perhaps because it is more soluble and the formation of small, high surface area calcite crystals when $Ca(OH)_2$ is added, furnished more binding sites for P.

Lime has been added to hardwater lakes in northern Germany to improve trophic state. $Ca(OH)_2$ was injected into the hypolimnion of one of two similar basins of Lake Schmaler Luzin (134 ha, maximum depth 34 m) during summers of 1996 and 1997 (Dittrich and Koschel, 2002). The injection was combined with aeration to achieve complete mixing of the suspension. SRP and TP declined during the two summers, respectively, by about one half and one quarter in the whole lake and from 90 to 20 μg/L and 120 to 40μg/L in the hypolimnion. Hypolimnetic injection of calcite also effectively reduced SRP and TP in Lake Dagowsee, Germany (Dittrich et al., 1997). Artificial (mechanical) resuspension of calcareous mud from the bottom of Lake Arendsee, Germany was unsuccessful in reducing P concentration, largely because $CaCO_3$ surfaces were much less effective at sorbing P than iron oxides, which were in low concentration in the mud (Hupfer et al., 2000).

Gypsum [Ca(SO$_4$)], especially ferro-gypsum, effectively reduced sediment P release under anoxic conditions in experiments with lake sediments both *in situ* and in the laboratory (Salonen and Varjo, 2000; Varjo et al., 2003). Methane production, which was considered a principal mechanism for internal P loading by transporting more sediment pore water, was reduced by 96%.

Three forms of calcite were applied as barriers over sediment in laboratory experiments to reduce P release (Hart et al., 2003a, b). Two commercial products of fine particle size (33 and 600 μm) effectively reduced P release, while another was ineffective. The material was mixed with sand at 2% and 5%, amounting to a total mass application rate of 56.3 and 57.4 kg/m^2. The expense and physical magnitude of this barrier material limits it to very small ponds.

These recent experiments suggest that Ca is an effective, but limited, alternative to aluminum salts for P precipitation and inactivation in hard water lakes. The advantages of lime are that it is relatively inexpensive, simple and safe to apply, and without toxic effects (unless pH increases above 10). Where increased pH is a potential problem, alum may be combined with lime as a buffer (Babin et al., 1992). However, more experimentation is needed on questions of dose, application techniques, best seasons for treatment, chemical mechanisms, and treatment longevity. An intriguing possibility would be to treat deep waters of a dimictic, hard water lake with aluminum sulfate to promote long term P inactivation, and to treat epilimnetic and metalimnetic sediments, as needed, with Ca(OH)$_2$.

8.5   PROBLEMS THAT LIMIT EFFECTIVENESS OF P INACTIVATION

As stated earlier, proof of the effectiveness of P inactivation with salts of aluminum, iron, or calcium requires a demonstration that the treatment not only retarded P release, but that it also lowered photic zone P concentration. The effectiveness observed in the shallow, frequently mixed Washington lakes (except those with dense macrophyte cover), where internal loading dominated the P cycle, was clearly due to alum applications. Effective P control should lead directly to water quality improvement in polymictic lakes.

Effectiveness cannot be similarly assumed for well-stratified dimictic lakes, unless there is evidence that vertical P transport added significant amounts of P to the epilimnion prior to treatment. As discussed in Chapter 3, two-layered, dynamic TP models are useful in determining the relative contribution of external and internal loading to the epilimnion. The Osgood Index (OI; chapter 3) may also be useful as well as estimates of diffusion (Mataraza and Cooke, 1997). Small, stratified lakes with high OIs and sheltered from the wind may still have significant transport from the hypolimnion via diffusion. So alum treatment of hypolimnetic sediments would have some effectiveness. Many of the stratified lakes (e.g., Mirror and Shadow Lakes with OIs of 35 and 13) evaluated here also had diversion prior to an alum treatment, so the relative effectiveness of diversion and alum on trophic state improvement was difficult to determine, even though alum was effective in reducing internal loading over 15–20 years. Distinguishing those relative effects was possible with West Twin Lake (OI = 7.9), because East Twin (OI = 9.7) was a control. Not only did the control lake (ETL) improve at the same rate as the alum-treated lake (WTL), their basin morphometry and surrounding shoreline features suggest that their stratification was likely to be stable over the summer period. On the other hand, the trophic states of Morey, Vermont and Kezar, New Hampshire lakes with relatively low OIs (3.13, 5.6) were probably affected by P transport from the deep water. Lake Morey is in a mountainous terrain, which could deflect some of the impact of summer wind mixing. But as shown earlier, the timing of trophic state improvement from the alum treatment, and its persistence for well over 12 years, is evidence of significant vertical P transport. Delavan Lake, Wisconsin, treated with alum in May, 1991, is also a dimictic lake with a low OI (2.8) and probably experiences vertical transport of hypolimnetic P. Unfortunately, that lake was underdosed and treatment effectiveness was short-lived, as discussed earlier (Rydin and Welch, 1999; Robertson et al., 2000).

Effectiveness of P inactivation treatments can thus be assumed only in shallow, continuously circulated and polymictic lakes, and possibly weakly stratified dimictic lakes (with OIs < 6). This is concluded from new information since Cooke et al. (1993a). However, well-calibrated, two-layer TP models may show that even deeper lakes are likely to experience an improvement in surface water quality during mid summer to fall after P inactivation. In those cases where hypolimnetic P availability was pre-determined, treatment would eliminate the introduction of P-rich hypolimnetic waters to the surface via diffusion as the late summer vertical P gradient increases and the lake begins autumn mixis. Treatment could also improve epilimnetic quality during spring mixing. Nevertheless, prospective users of P inactivation in dimictic lakes should determine whether summer algal blooms in their lake are caused by P release from deep lake sediments, which is the traditional target of P inactivants.

As noted earlier, oxic shallow water sediments can be significant P sources, yet they are often not treated when alum is added to dimictic lakes. This omission is at least in part due to concerns by earlier investigators (e.g., Cooke et al., 1978) about exposure of biota to alum. However, the effectiveness of surface applications to softwater shallow lakes in Washington have shown no adverse effects to biota when properly buffered. Total P loading during summer was dominated by internal sources in these shallow, oxic western Washington lakes, analogous to what may occur in littoral areas of dimictic lakes, and alum treatments were effective (Welch and Jacoby, 2001). Much more can be learned about treatment effectiveness. For example, an experimental treatment of only the littoral zone sediments of a eutrophic lake for which there are current water and P input–output data would be instructive. This would allow an assessment of the relative roles of oxic and anoxic sediments in the internal P budget.

Effectiveness of an alum treatment wanes due to sinking of the high-density floc, bioturbation of the floc, and because it becomes covered by new sediment. The P initially inactivated remains complexed. This process is clearly shown in sediment cores from treated lakes where cores were taken (Welch and Cooke, 1999; Rydin et al., 2000; Lewandowski et al., 2003). An Al peak was clearly evident at varying depths below the surface in nearly all treated lakes sampled. An exception was Green Lake, Washington, in which bioturbation by carp may have obscured a peak. The results from WTL are especially interesting, because nearby ETL was untreated. Alum application to the hypolimnion of WTL in 1975 substantially raised the Al content throughout the sediment profile (Figure 8.19). While the Al content of WTL was consistently higher than in ETL throughout the top 20 cm, the peak concentrations were at 18 and 20 cm, 15 years after treatment (Figure 8.19). That translates into a settling rate, by processes mentioned above, of about 1.5 cm/year. On average, the treatment should have increased sediment Al by 30% (~ 6 mg/g) over the 20 cm. The average difference in sediment content between the two lakes over the 20 cm depth was 6.7 mg/g. Concentration of the dose in the top 4 cm would have raised sediment Al content by 165% (Welch and Cooke, 1999).

Effectiveness of alum treatments in lakes with dense macrophyte covers was low, as discussed previously. The reduced effectiveness may be due to the loss of binding sites on the settling floc in water with high dissolved C or entrapment of floc by the plants themselves, causing unequal distribution of floc over the sediment. For example, Green Lake, Washington was covered with Eurasian watermilfoil when treated in October 1991. Sediment cores taken in 1998 showed no evidence of the 1991 treatment; the only one of seven western Washington lakes sampled without a clear Al peak (Rydin et al., 2000). That problem may be overcome by treating early in the spring, prior to development of macrophytes, although that has not been tested.

8.6   NEGATIVE ASPECTS

Aluminum is the among most abundant elements in the Earth's crust, and it is naturally in high concentrations in lake sediments. Thus, an alum treatment only slightly increases sediment Al content. Nevertheless, Al can be toxic, particularly under acidified conditions, and is toxic to fish and macroinvertebrates at concentrations as low as 0.1 to 0.2 mg Al/L at a pH of 4.5 to 5.5 (Baker,

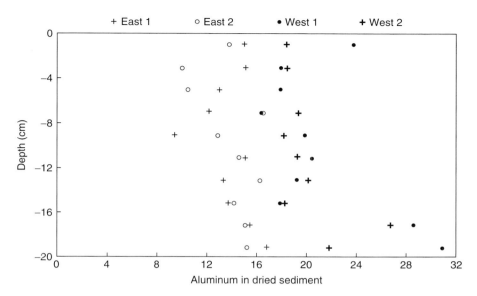

FIGURE 8.19 Distribution of aluminum in sediment cores collected from East and West Twin Lakes, OH, in 1991, 15 years after an alum treatment. (From Welch, E.B. and G.D. Cooke, 1999. *Lake and Reservoir Manage.* 15. With permission.)

1982; Havens, 1993). An understanding of aluminum toxicity is complicated by the varying solubility of several forms, as noted in this chapter. Toxicity from Al under acidified conditions has been ascribed to dissolved monomeric Al, which is composed of the free Al ion as well as simple Al hydroxides, sulfates, fluorides and low-molecular weight humic and fulvic acid complexes (Gensemer and Playle, 1999). The toxic effect of monomeric Al in a pH range of 4.5–5.5 to fish is thought to be due to interactions of cationic Al species at the gill surface causing respiratory and ion regulatory dysfunctions (Witters et al., 1996; Paléo, 1995). However, toxicity of Al to fish has been observed at pHs as high as 6.5 and the cause was ascribed to polymerized Al species, which have maximum formation at pH 6.0 (Paléo, 1995). See reviews of this topic by Burrows (1977), USEPA (1988), Rosseland et al. (1990), and Gensemer and Playle (1999).

While Al toxicity to aquatic animals is well known in permanently acidified lakes and at continuous exposure to pH below circumneutral levels in laboratory experiments, there are safety factors involved under field conditions during alum treatments that do not exist with most of the laboratory experiments described below. Treatment of lakes usually takes several days, even with a sophisticated barge equipped with automatic depth-sensing and location capability. This allows fish to avoid hot spots of low pH in the water column during hydrolysis and settling of the alum floc. Trout have been shown to avoid Al concentrations as low as 27 μg/L and low pH even as high as 5.75 (Exley, 2000). Water column pH and alkalinity usually return to near normal levels after a few days from sediment and atmospheric sources. Formed floc settles through the water column in less than an hour and is stabilized by the time it reaches the sediment. Also, there usually are substances in lakes that complex Al, such as particulate and dissolved organic matter (DOM), especially humic and fulvic acid, as well as hardness-causing ions. As suggested earlier, part of the failure of alum in lakes with dense macrophyte coverage may be high DOM that is sorbed to the alum floc, limiting its capacity to sorb P. Evidence for sorption of organic matter onto the settling floc is the rapid clearing of the water column following application. Therefore, exposure times of several days to weeks to constant Al and pH conditions, as typical in laboratory experiments, are an unrealistic simulation for treatment effects.

There have been few observations of invertebrate and fish populations before and after alum treatments. Every treated lake has exhibited major chemical and physical changes, including sharply increased transparency and reduced algal biomass, but there have been no reported massive biotic changes, such as fish kills, as long as waters were properly buffered. Nevertheless, there are several cases where some component of the lake's biota was examined, in the field or laboratory, where invertebrate populations and/or diversity either increased or decreased, or other effects have been observed, as illustrated in the following reviews.

Lamb and Bailey (1981) conducted acute and chronic laboratory bioassays on alum effects on *Tanytarsus dissimilis*, a chironomid, in which pH was held constant at 7.8 in acute and at 6.8 in chronic test solutions. Test waters were from Liberty Lake, Washington a softwater lake. The acute tests demonstrated no apparent effect on second or third instar larvae at doses ranging from 6.5 to 77.8 mg total Al/L. Dissolved Al remained below 0.1 µg/L at this pH and the larvae used the floc to build tubes. In the chronic tests, using doses ranging from 0.8 to 77.8 mg Al/L, mortalities occurred at all concentrations, even in the control. At a dose of 77.8 mg/L total Al, 50% mortality occurred in 23 days. Dissolved Al remained below 0.1 µg/L at all doses except for the 19 mg/L dose. The 77.8 mg/L dose created a heavy floc that interfered with movement and feeding. No larvae pupated in the 55 days of study at any concentration, even though the normal time to pupation for this species is 23 days. These effects occurred over an alum dose range exceeded in 19 of the 28 alum case histories reviewed by Cooke and Kennedy (1981) and in laboratory test solutions in which dissolved Al remained below 0.1 µg/L. However, long-term effects observed *in situ* following treatments have not been as adverse as these results might portend, and in most cases have been positive.

The most thoroughly investigated response of benthic invertebrate populations to an alum treatment was in softwater (35–45 mg/L CaCO₃) Lake Morey, Vermont. There was a significant decline in benthic invertebrate density and species richness in the upper hypolimnetic sediments during the summer after the 1987 treatment (Smeltzer, 1990). These sediments had been inadvertently exposed to as much as 200 µg Al/L during that summer. In the second summer, the benthic invertebrate community appeared to have recovered and two additional species colonized the lake. Continued monitoring through 1997 showed that invertebrate density had been significantly lower the first summer after treatment (Smeltzer et al., 1999). Moreover, taxa richness at 9 and 12 m and density at 12 m had significantly increased in post treatment years (Figure 8.20). Smeltzer (1990) and Smeltzer et al. (1999) speculated that the recovery may have been related to the improved hypolimnetic DO.

A similar increase in benthic invertebrates following alum treatments occurred in two softwater and three hardwater Wisconsin lakes (Narf, 1990). In contrast to the laboratory studies of Lamb and Bailey (1981), Narf found that benthic insect populations either increased in diversity or density or remained at the same levels after treatment. He suggested that increased insect biomass was related to the post-alum treatment shift of the phytoplankton to green algae, followed by an increase in zooplankton, a main food of chaoborids. Also, greatly reduced oxygen demand of the lake sediments resulted in increased DO concentrations, which increased the habitat for invertebrates. Chaobrid populations doubled after an alum treatment in Newman Lake, and there was no effect on chironomids and oligochaetes (Doke et al., 1995).

There was a significant decrease in species diversity of planktonic microcrustacea, during 2 years of post treatment sampling (1976 and 1978) in West Twin Lake, Ohio (see case history), compared with a downstream reference lake (ETL) and to a detailed series of samples taken in 1968 to 1969, from both lakes before alum treatment (Moffet, 1979). This diversity change could have been due to dissolved Al in the water column, though it was always below 2 µg Al/L, except on treatment day. Alkalinity and pH returned to normal levels (pH 7–9; alkalinity 100 to 150 mg CaCO₃/L) at fall circulation, about 100 days after treatment (Cooke et al., 1978). The decrease in diversity could also be due to effects of the floc on resting stages in treated sediments, to the shift in phytoplankton from blue-greens to dinoflagellates, or to increased transparency and possible increased visual predation by fish. Zooplankton abundance and diversity also declined in Newman Lake, Washington following alum treatment, but only temporarily, recovering within 2 months (Shumaker et al., 1993). There was no lasting adverse effect on zooplankton by the alum treatment

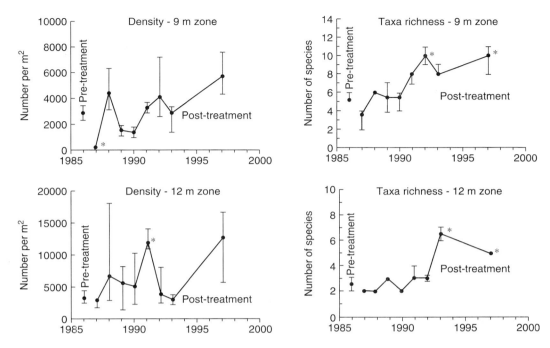

FIGURE 8.20 Benthic invertebrate indicators at 9 and 12 m depth in Lake Morey during 1986–1997 before and after an alum treatment showing median values and interquartile intervals among replicates. Median values that are significantly different ($p < 0.05$) from those in 1986, the year, are indicated by an asterisk. (From Smeltzer, E. et al., 1999. *Lake and Reservoir Manage.* 15. With permission.)

to Liberty Lake, WA. While species diversity was not measured, the abundance of Cladocera, Copepoda, and Rotatoria was reduced only briefly and no unusual losses of species were reported (Gibbons et al., 1984). These varying responses of zooplankton species to alum treatments may be in part due to the wide range of tolerance to Al and pH in these animals (Havens and Heath, 1989; Havens, 1990).

Buergel and Soltero (1983) examined bioaccumulation of Al by rainbow trout (*Salmo gairdneri*) in hardwater Medical Lake, Washington following the addition of 936 metric tons of alum (12.2 mg/L Al). Dissolved Al in the lake ranged from 90 to 420 µg/L following treatment. No mortality, physiological stress, gill hyperplasia or necrosis, or retardation of growth was observed. Gill tissues contained elevated Al, but gill tissues from fish sampled in a nearby hatchery and untreated lake were higher in Al than those of Medical Lake trout. There were no significant (95% level) differences in Al levels in the other tissues of age 2+ trout from Medical Lake or the hatchery, except for liver and kidney. Aluminum levels in composite plankton samples from Medical Lake were ten times greater than in trout tissues, but these were not compared with plankton in untreated lakes.

Lamb and Bailey (1983) conducted an *in situ* study of the effects of the low dose (0.5 mg Al/L) of alum to softwater Liberty Lake, WA. Cages of about 1,000 rainbow trout fry (2.5 cm) were exposed to the application, or placed on treated sediments just after application, or held as controls. All cages were moved to a non-treated site after 1 week. No mortalities occurred during application, but over the ensuing 5 weeks of observation, trout exposed to the application exhibited some increased mortality in excess of control cages. No gill hyperplasia was noted nor was there any change in growth when compared to the other cages.

Exposure to low doses of $Al(OH)_3(s)$, even at pH levels between 7 and 8, can produce some chronic, long term effects. Smallmouth bass (*Micropterus dolomieu*) had significantly reduced activity in 30-day continuous exposures at low doses at pH 7.3 to 7.5 (Kane and Rabeni, 1987).

Reduced feeding and increased mortalities occurred when rainbow trout were continuously exposed to both dissolved and solid forms of Al at concentrations between 52 and 5,200 µg Al/L over a pH range of 7 to 9 (Freeman and Everhart, 1971). Exposure over a period of 6 weeks or more to a concentration of 5,200 µg/L, whether totally soluble at pH 9.0 or nearly insoluble at pH 7.0, produced serious disturbances. Continuous exposure to 52 µg/L produced no acute or chronic physiological or behavioral responses

Mortality to fish at pH levels of 5.5–6.5 are thought to be due to polymerized Al that clogs interlamellar spaces causing hypoxia (Paléo 1995). Trout held continuously in a mixing zone of an acidified tributary, compared with that of the unacidified main stem, died after exposure for 48 hours at pH 6.4 and 75 µg/L total Al. The cause was ascribed to polymerized Al (Witters et al., 1996). Other investigators have also noted that the colloidal-size floc adheres to mucous on fish gill epithelia. If exposure to very fine particulate Al (polymerized Al) is continuous over a long period, hypoxia is possible, as noted earlier (Neville, 1985; Paléo 1995). Normally, this problem should diminish rapidly as larger and larger floc particles form and fall to the sediments after a lake treatment. However, at pH 6.0–6.5, polymerized Al as finely dispersed colloids, could occur in high concentrations and be acutely lethal to rainbow trout (Neville, 1985; Ramamoorthy, 1988). Gensemer and Playle (1999) cite experiments with algae that show that maximum toxicity at 6.0 was largely due to polymeric cations, as well as relaxed competition from H^+.

Continuous exposure experiments do not provide a realistic test of the effect of an alum treatment on fish because in actual practice, fish are exposed only when the $Al(OH)_3$ floc falls through the water column, unless the floc settles on early fish development stages. Continuous exposure of free-swimming fish to Al following a P-inactivation treatment would thus be unusual. Nevertheless, the treatment of the hypolimnion of softwater Lake Morey, Vermont apparently caused a sublethal effect to perch (Smeltzer, 1990; Kirn, 1987). The mixture of alum and sodium aluminate was designed to maintain lake pH above 6.5 and below 8.0. As planned, pH of the hypolimnion remained at about 7.0 and dissolved Al was below 20 µg/L. However, it appears that either a leak in the treatment apparatus, or some vertical transport process, brought Al to the metalimnion-epilimnion boundary, and concentrations of dissolved Al as high as 200 µg/L were observed during brief periods. Levels of 50 to 100 µg/L were found until fall turnover. These high levels were associated with pH values of more than 8.0. The background concentration of 10 to 20 µg/L was restored at turnover. Levels have probably remained low in view of recovered biological indices (Smeltzer et al., 1999). No direct mortalities were observed, but the condition factor of yellow perch (*Perca flavescens*) in an October survey was significantly reduced, and full recovery of the fish had not occurred by the following spring (Kirn, 1987; Smeltzer, 1990). While a direct linkage between fish condition and the treatment was not shown, these changes may be consistent with responses of fish to chronic Al exposures at pH above 6.0.

A fish kill occurred in softwater Long Lake, Wisconsin, but the cause was due to equipment failure. (D. Knauer, personal communication). The delivery of sodium aluminate failed during a mechanical malfunction and unbuffered aluminum sulfate was added to a bay of the lake. The fish kill occurred from low pH, elevated dissolved Al, or both. In Long Lake, Washington, caged hatchery rainbow trout died from low pH in live cages in a section of the lake treated without sodium carbonate buffer. However, caged wild bass and bluegill were unaffected and there were also no adverse effect to any caged fish in the buffered sections (Welch, 1996).

Therefore, Al can be harmful to lake biota, even at pH 6 to 8, especially if exposures are prolonged. While a few short-term negative effects have been observed following alum treatments, in nearly every case where observations have been made there has not been a report of large-scale mortalities nor problems with long-term toxicity or biomagnification. This is likely due to the brief exposure to the large, heavy, preformed floc, and to the very low solubility of $Al(OH)_3$ (s) and absence of Al^{3+} at pH 6 to 8 in the water column and sediment. This is supported by measurements of cyanobacteria recruitment rates from the sediment in Green Lake, Washington, that showed no reduction following a surface alum treatment (Sonnichsen et al., 1997). Also, unless the exposure

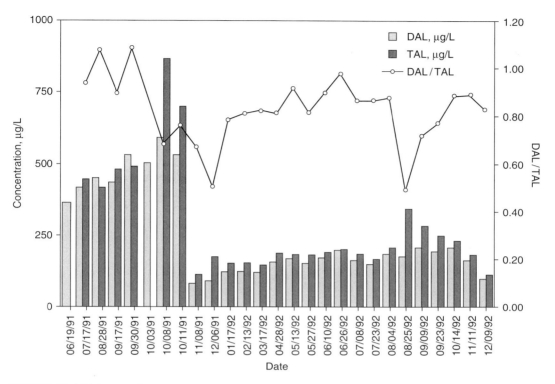

FIGURE 8.21 Volume-weighted mean whole-lake total and dissolved aluminum concentrations in Long Lake, Washington and the ratio of dissolved to total Al (DAL:TAL) before and after the application of alum in October 1991. (From Welch, E.B. 1996. Control of phosphorus by harvesting and alum. Water Res. Series Tech. Rept.152. Univ. of Washington, Dept. of Civil and Environ. Eng.)

to Al has been chronic and/or at high levels, Freeman and Everhart's (1971) data show that fish recovery should be prompt.

The complexing properties of naturally occurring ligands, especially in hard waters, will sharply reduce or eliminate aluminum toxicity (Gensemer and Playle, 1999). Even in soft waters, if the amount of total organic carbon (TOC) is high, aluminum toxicity to brook trout is eliminated when complexed with organic ligands (e.g., humic acid). This observation by Driscoll et al. (1980) led to the conclusion that determination of even dissolved Al in treated waters may greatly overestimate potential toxicity. Many of the experiments described did not include natural organic matter. The low TOC in mesotrophic Lake Morey (Smeltzer, 1990), for example, may be an explanation for the negative effect of Al on perch in this lake. Soft waters like Lake Morey will likely have few inorganic complexing agents as well. However, TOC is high in eutrophic lakes, even when alkalinity is low, so considerable Al complexation will occur in most treated lakes. Dissolved Al actually decreased following treatment of Long Lake, Washington and remained at about one half the pretreatment level for over a year (Figure 8.21; Welch, 1996). That may have been a result of natural levels of Al complexed with humic material by the alum floc in this relatively brown-water lake, as shown elsewhere (Bose and Reckhow, 1998).

In support of this, Ramamoorthy (1988) demonstrated that in a hardwater experimental system, Al was present as a non-exchangeable species (with anion or cation exchange resins) at pH 7 to 9 and was not lethal to rainbow trout. In this experiment, abiotic substances in the water complexed the aluminum released from $Al(OH)_3$ (s), which in turn minimized the amount of Al that could be sorbed to gills.

Silicon has been shown to be at least as significant as circumneutral pH levels in lowering the toxicity of Al. Birchall et al. (1989) demonstrated that at pH 5, with levels of Al acutely toxic to fish fry, a Si:Al ratio of 13 eliminated toxicity and harmful effects to gills. At pH 4 and above, various aluminum-hydroxy species will form hydroxy-alumino-silicate. This solid becomes more stable with increasing pH, and at pH 7 and above is more likely to form than Al complexes with sulfate, phosphate, and strong organic chelators. According to these investigators, increased silicic acid in lakes may be as important as increased pH in alleviating toxicity. The role of Si in alum treatments has not been considered, and should be investigated. In eutrophic lakes, Si levels in epilimnetic waters may fall to very low levels following assimilation by diatoms and then sedimentation. For example, the summer minimum in Lake Michigan is < 0.2 mg SiO_2/L (Schelske, 1988). At the sediment-water interface, however, silicon concentrations are high. A small fraction of the silicon at this interface is normally fixed as an aluminum-silicon-iron complex, but the remainder appears as dissolved silicon (Nriagu, 1978), which may interact with alum added to hypolimnetic waters. Softwater lakes could be low in silicon, depending upon the nature of the drainage basin minerals, which would have in turn an effect on the likelihood of aluminum toxicity.

Phosphorus inactivation with Al salts is a suitable treatment for some but not all lakes. The following treatment categories and characteristics for lakes are suggested:

1. Phosphorus inactivation will be effective and long-lasting, and without significant acute or chronic effects to biota, in lakes with significantly reduced external P loading, an alkalinity above 75 mg/L as $CaCO_3$, and high levels of Si, Ca, SO_4, and TOC. Improvement of trophic state will be most effective in lakes with significant vertical P entrainment and/or diffusion, and continuously mixed or polymictic lakes.

2. Alum treatments of very softwater lakes (< 35 mg/L as $CaCO_3$) require great caution. The approach has been to buffer the alum with sodium aluminate (or sodium carbonate) to maintain pH between 6 and 8. Recent evidence suggests that buffering should be sufficient to hold pH to 6.5 and above. Despite buffering, accidents during application, or leaks, could drive the pH of softwater lakes below pH 6, leading to the appearance of soluble, toxic Al forms.

3. There are lakes that may be unsuitable for P inactivation with alum. These include all lakes with unabated high external P loads. Lakes with very soft, low alkalinity water and/or an acidic pH should not be treated unless special, extreme precautions are taken. Not only is direct toxicity at the time of application possible, but future acidification through acid precipitation could lead to increases in dissolved Al from the treated sediments beyond that released from the sediments from existing aluminum.

4. Lakes with a high pH (pH > 9 to 10) may experience an increase in the dissolved and toxic aluminate ion during application, and are poor candidates for an alum treatment.

There continues to be research needs on the toxicity of Al in basic and circumneutral environments. Careful examination of biotic communities in lakes before and after treatments, with continued long term monitoring, is needed. Such data should provide the documentation to accompany the largely anecdotal reports by treatment operators of an absence of any observable detrimental effects of a properly applied dose of Al salts.

There has been concern that alum treatments to sediments will adversely affect rooted macrophytes. Moss et al. (1996) state that "alum is likely to be toxic to aquatic plant roots and therefore inappropriate in sites of conservation and amenity importance," yet they provided no evidence for the claim. There is the possibility that low pH could occur in non-calcareous sediment if the alum were mixed directly into the sediment and allowed to hydrolyze there. Experiments *in vitro* with *Egeria densa* have shown that alum at 1.5 mg Al/g of sediment (dose to water of

15 g/m^2) mixed into 10 cm of non-calcareous lake sediment, lowered the pH from 6.0 to 4.8 and reduced shoot and root growth by an average of 78% (Jacoby, 1978). There was no such effect when alum was added to the overlying water as is the procedure used in lake treatments. These observations were also supported by experiments with aquatic plants rooted in sandy soil (Maessen et al., 1992). Adverse effects occurred over several weeks when plants were exposed to low sediment pH (< 5). There were no adverse effects from the addition of Al (2.7 mg/L) or increases in Al/Ca ratios. Thus, if hydrolysis has occurred in the water column, prior to the floc reaching the sediment, further acidification should not occur and there should be no adverse effects of alum treatments on rooted aquatic plants. As would be expected, there was no subsequent observable adverse effect on *E. densa* biomass from a treatment of the water surface of Long Lake, Washington, in 1980 and 1991 (Welch et al., 1994). Further evidence is from attempts to control Eurasian watermilfoil by binding sediment P. In that case, Mesner and Narf (1987) mixed alum into sediments in Lake Mendota, Wisconsin, a calcareous sediment, with no effect on plant growth. The problem frequently observed with macrophytes in alum-treated lakes is the opposite of an adverse effect. That is, they increase in biomass and coverage following lake-P reduction and increased transparency (Cooke et al., 1978; Young et al., 1988) and/or retard alum effectiveness (see effectiveness in shallow lakes).

Another possible problem, with no documentation to date, involves the impact of reduced plankton productivity on fish communities. These effects could range from reduced fishing success to changes in predation and food choices (more predation on *Daphnia* or an increased littoral zone fish community) with the clearer water.

TABLE 8.7
Dose, Area Treated, and Worker-Days/ha for P-Inactivation Treatments

Lake	Date	Area (ha)	Dose (g Al/m^3)	Worker-days/ha
Horseshoe, WI[a]	1970	9	2.1[d]	1.33
Welland Canal, NY[a]	1973	74	2.5[d]	1.35
Dollar, OH[a]	1974	1.4	20.9[d]	4.30
West Twin, OH[a]	1975	16	26.1[d]	4.61
Medical, WA[a]	1977	227	12.2[d]	2.03
Annabessacook, ME[a]	1978	121	25.0[d]	1.12
Kezar, NH[b]	1984	48	40.0[e]	0.50
Morey, VT[b]	1986	133	45.0[e]	0.57
Cochnewagon, ME[b]	1986	97	18.0[e]	0.41
Sluice Pond, MA[b]	1987	6	20.0[e]	0.67
3 Mile Pond, ME[c]	1988	266	20.0[e]	0.06

Note: One person working 8 h = 1 worker-day. Most treatments involved 12 to 14 days.

[a] Barge system (Kennedy and Cooke, 1982).

[b] Modified harvester (Connor and Smith, 1986).

[c] Computerized dose and navigation system (T. Eberhardt, 1990).

[d] Aluminum sulfate.

[e] Aluminum sulfate and sodium aluminate.

Source: Data from Cooke, G.D. and R.H. Kennedy. 1981. *Precipitation and Inactivation of Phosphorus as a Lake Restoration Technique.* USEPA-600/3-81-012; Cooke et al. 1993a; Connor, J.N. and M.R. Martin. 1989. *Water Res. Bull.* 25: 845–853. With permission.

8.7 COSTS

Dramatic reductions in P inactivation costs have occurred since Horseshoe Lake was treated in 1970. Table 8.7 lists the area treated, doses, person-days per hectare, and type of application system for 11 lakes (data from Cooke and Kennedy, 1981; Connor and Martin, 1989b; Cooke et al., 1993a). There have been order of magnitude decreases in labor requirements to apply P inactivating chemicals, as application systems have been improved (Conner and Martin (1989b). The first six lakes in Table 8.7 were surface or hypolimnetic treatments and all used a custom-made barge system, following the basic design in Figure 8.8. An order of magnitude decrease in labor per hectare occurred with the introduction of the modified weed harvester design (Connor and Smith, 1986; Figure 8.9). These machines employ paddle wheels on each side of the harvester, making them mobile and easy to turn at the end of a treatment swath. They also can handle heavy loads so that stops to refill alum delivery tanks are less frequent. Most treatments in the U.S. during the 1990s have employed a large, high-speed barge that is capable of holding over 11,250 kg of liquid alum (Sweetwater Technology, Figure 8.10). The barge is guided by a highly accurate LORAN navigation system, and one person can apply about 115 m^3 of liquid alum per day along 15-m wide paths. An efficiency of 0.06 worker-days/ha is typical for this system (Table 8.7). The cost to treat Three Mile Pond (266 ha treated) with the high-speed system was $838/ha (2002 dollars; Connor and Martin, 1989b). In contrast, the cost to treat Cochnewagon Lake, Maine, with the modified harvester was $1,165/ha and the cost to treat Medical Lake, Washington, with the old barge system was $1,520/ha. In most cases, including all of those with the original barge design, costs per hectare have been about twice that of the Three Mile Pond and more recent treatments.

Sediment removal or skimming is the only other operational procedure that could be used to eliminate internal P loading and for which sufficient cost data exist (Chapter 20). Phosphorus inactivation is more economical and effective, although sediment removal does have the long-term benefit of removing the nutrient source.

8.8 SEDIMENT OXIDATION

A restoration technique to oxidize the top 15 to 20 cm of anaerobic lake sediment was developed by Ripl (1976), and has been applied to lakes through equipment development and promotion initially fostered by the Atlas Copco Co. and later by Aquatec, Inc. under the name of Riplox. The objective is to reduce internal P loading in lakes that have anaerobic sediment and high interstitial P concentrations, and in which iron redox reactions control P interchange between sediment and the overlying water. By oxidizing the organic matter through increased denitrification, greater binding of interstitial P with ferric hydroxide complexes should occur, resulting in lower release rates for P (Ripl, 1976; Ripl and Lindmark, 1978). Also, sulfate reduction is prevented, thus decreasing the formation of iron sulfide, leaving iron available to complex P.

Ripl and Lindmark (1978) argued that high interstitial P concentrations, which may lead to high internal P loading, are due largely to metabolic processes. To deplete the organic matter in sediments, and thereby restore an oxidized state, a solution of $Ca(NO_3)_2$ is injected into the sediment to stimulate denitrification. Nitrate is preferred as the electron acceptor and, being a liquid solution, it penetrates into the sediment more readily and is more efficient than adding oxygen to the hypolimnion. Ferric chloride ($FeCl_3$) is added initially to remove hydrogen sulfide (H_2S) and form ferric hydroxide $Fe(OH)_3$, which binds interstitial P. Lime ($Ca(OH)_2$) is added next, to raise the pH to an optimum level and encourage microbial denitrification. Because the redox potential for nitrate reduction is higher than for iron reduction, the latter would be inhibited and P would remain complexed with ferric iron compounds (Foy, 1986).

Ferric chloride and lime additions may be unnecessary in some cases. The pH may be sufficiently high to promote denitrification, and the iron content of the sediments may be adequate (30 to 50 mg/g) for P binding. Willenbring et al. (1984) found that adding iron and lime was unnecessary to substantially decrease the P release from sediment in Long Lake, MN, as long as the calcium nitrate dose was sufficient. The cost savings were significant. The need for all three ingredients and the doses required should be optimized through laboratory experiments for each particular lake sediment. A range of calcium nitrate additions should be studied, with and without added iron. Iron and sulfur content of sediments should also be determined to assess the availability of iron for P complexation.

8.8.1 EQUIPMENT AND APPLICATION RATES

The chemical solutions are applied by direct injection into the sediment with a "harrow" device, which is about 6 to 10 m wide and equipped with flexible tubes that penetrate the sediment (Figure 8.22). The vertical position of the harrow on the lake bottom is adjusted by regulating the injection of compressed air through the tubes. As the harrow is dragged along the lake bottom at a rate of 4 to 5 m/min by an airmotor-driven raft or onshore winch, sediment is disrupted to a depth of about 20 cm and the chemical solutions are injected into the sediment through tubes at the rear of the device. Control of the device's vertical position by air input, as well as other safety features help prevent the device from being obstructed by debris on the lake bottom. The device was described for use in Lake Lillesjön, a small (4.2 ha), shallow (2 m) lake in Sweden (Ripl, 1976). Foy (1986) described an injection system using a propelled barge to drag a harrow weighted with an iron bar and chain to mix the sediment.

The dosages used in Lake Lillesjön were 13 tons of ferric chloride (146 g Fe/m²), five tons of lime (180 g Ca/m²), and 12 tons of calcium nitrate (141 g N/m²). To treat Lake Trekanten, Sweden below the 3-m contour (49 ha) with calcium nitrate required only 160 tons (56 g N/m²) in a 50% solution. The iron naturally present in Lake Trekanten sediments was considered adequate. The solution was further diluted to 10% with lake water, prior to injection into the sediments. The optimum dose of calcium nitrate for Long Lake, Minnesota sediments was found to be the same as for Lake Lillesjön.

The iron content in Lake Lillesjön sediments prior to treatment was considerably less than that in other lakes in the surrounding area: 9 to 23 mg/g dry weight, compared to more typical levels of 30 to 50 mg/g. Sediment P content in Lillesjön was also quite low (1.4 to 3 mg/g) in spite of high interstitial concentrations.

8.8.2 LAKE RESPONSE

Following the treatment of Lake Lillesjön sediments in 1975, interstitial P content in the top 20 cm dropped by 70 to 85% of the 1974 to 1975 levels. The decreased levels persisted through at least 1977 (Figure 8.23). Despite the high loading of NO_3 to the sediments, nitrogen was lost through evolution of N_2 gas, and ammonium actually decreased (Figure 8.23). Denitrification of the added NO_3 was complete after 1.5 months (Ripl, 1981). The oxygen demand of the sediment also decreased by about 30%. Recycling of P and N to the water overlying treated sediment was reduced to between 10% and 20% of the rate observed prior to restoration (Ripl and Lindmark, 1978; Ripl, 1981). Oxygen demand of sediment and sediment P release continued to remain low 10 years after treatment (Ripl, 1986).

The interstitial P content of sediments in Lake Trekanten decreased from 2 to 4 mg/L, prior to the May 1980 treatment, to 0.01 to 0.3 mg/L by July (Atlas Copco, undated). Although P release from sediments decreased, lake P content did not change. Continued P input from external sources and possibly from sediments by other controlling mechanisms than iron redox have been suggested as the reason(s) lake P did not decrease (Ripl, 1986; Pettersson and Böstrom, 1981). According to Ripl (1986), external loading was twice the calculated rate and the continued algal production supplied an energy source for sulfate reduction, leading to iron complexation with sulfide and a recurrence of sediment P release.

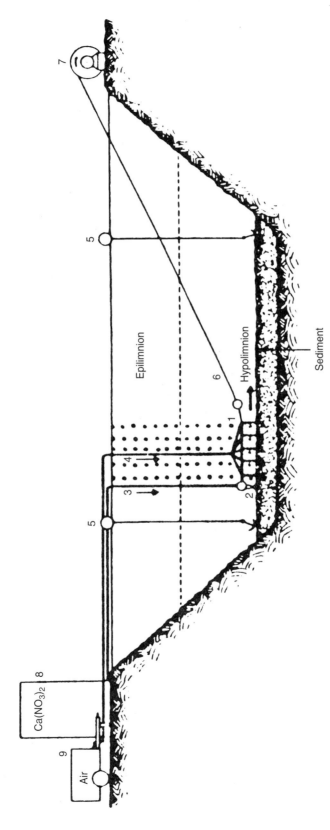

FIGURE 8.22 Riplox treatment system, showing (1) sediment "harrow," (2) injection point of chemicals, (3) supply tube for chemicals, (4) supply tube for compressed air, (5) buoys, (6) towing line, (7) cable winch, (8) tanks for chemicals and (9) air compressor. (From Ripl, W. 1981. In: *Restoration of Lakes and Inland Water.* USEPA 440/5-81-010. With permission.)

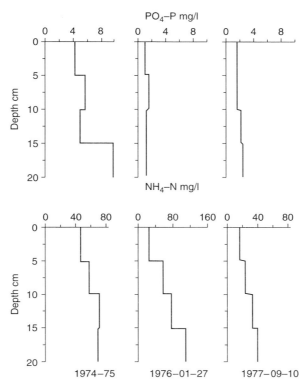

FIGURE 8.23 Phosphate P and ammonium N concentration in interstitial water of Lake Lillesjön sediment before (1974–1975) and after (1976 and 1977) treatment with $Ca(NO_3)_2$. (From Ripl, W. and G. Lindmark. 1978, *Vatten* 34: 135–144. With permission.)

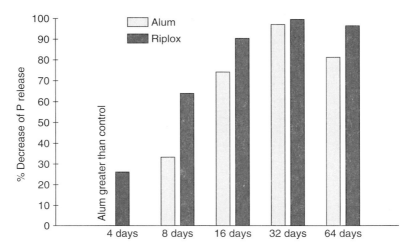

FIGURE 8.24 Relative effectiveness (difference between control and treatment) of alum vs. Riplox in controlling sediment total soluble P release from cores in the laboratory under anaerobic conditions. Doses to the sediment: Alum, 15 g Al/m^2; $Ca(NO_3)_2$, 140 g N/m^2. (From Cooke et al., 1993a. With permission.)

A similar result occurred in Long Lake, Minnesota. Sediment P release decreased by 50 to 80% following the calcium nitrate treatment, but less than expected from laboratory experiments (Willenbring et al., 1984; Noon, 1986). However, there was no change in lake P content. As was probably the case with Lake Trekanten (Ripl, 1986), continued high external loading was the cause for the failure of Riplox to reduce the lake P content. While internal loading from an anoxic hypolimnion represented most of the P entering the lake during summer, lake quality was apparently determined primarily by external loading during the spring runoff (Noon, 1986).

Phosphorus concentration in stratified White Lough, Ireland, decreased following a calcium nitrate treatment of its sediment. Phosphorus release from sediment was delayed (about 1 month) and the maximum hypolimnetic concentration decreased 30% following the May treatment (Foy, 1986). Because lake results conformed to those predicted from laboratory experiments, sufficient control of lake P would have been expected had the dose been the optimum of 30 to 60 g N/m^2 instead of 24 g N/m^2. The release pattern for iron was nearly identical to that for P, but manganese (Mn) and dissolved oxygen were unaffected (Foy, 1986).

Riplox was considered, along with alum, for the inactivation of P release from sediments in unstratified Green Lake, Washington (DeGasperi et al., 1993). The iron (Fe), P, and sulfur (S) contents in Green Lake sediments were, respectively, 38 mg/g, 2 mg/g, and 2.2 μg/g of dry sediment. Sulfur probably had little effect on iron because the S content was so much lower. There was little difference between the effectiveness of Riplox (40 g N/m^2) and alum (15 g Al/m^2); each decreased the sediment P release over 90% in laboratory core experiments (Figure 8.24). Alum was ultimately added to Green Lake in 1991 as described earlier.

8.8.3 COSTS

The total cost of the Lake Lillesjön treatment was about $179,000 (2002 U.S. dollars), of which 44% was for development of the device and preliminary lake investigations. The equipment part of this, of course, would not be necessary in subsequent lake projects. The chemicals applied to the 1.2 ha lake area represented only 6% of the total cost, while 28% of the total went for equipment installation. The latter would not vary greatly with the area treated. The remaining was for equipment rental and labor.

The cost for the 1980 Lake Trekanten treatments, excluding the preliminary investigation, was about $469,000 (2002 dollars), which was only 5% of the cost per area of the Lake Lillesjön project, exclusive of the preliminary investigation. The total area of Lake Trekanten (87 ha) was treated once, and the area greater than 3 m in depth (49 ha) was treated again. A larger device (10-m wide) was used for the treatment of this rather large lake.

The cost of an ideal dose (40 g N/m^2) of calcium nitrate to White Lough was compared with that of an earlier (1980) treatment of the lake with iron/alum. The total costs (2002 dollars) for calcium nitrate and iron/alum were $33,000 and $8,800, respectively. Also, the cost per treated hectare were $7,200 and $1,900, respectively. The much greater cost for Riplox in Lake Lillesjön than in White Lough (5.4×) was largely due to the use of compressed air in Lillesjön (Foy, 1986).

8.8.4 PROSPECTUS

Although greater in cost, this technique appears to be an effective alternative to an alum treatment to inactivate sediment P. Its greatest appeal may be that the chemicals added are normally found in high concentrations even in unpolluted sediments and are introduced directly to and confined largely to sediments. Potential toxicity to animals is probably not an issue with Riplox. Further, the effects may be more permanent than alum, which initially only covers the sediment and then tends to settle and distribute through the sediment column, leaving old as well as newly deposited surficial sediments in contact with overlying water. Ripl (1986) found that sediment oxygen demand remained low for 8 years following treatment of Lake Lillesjön, demonstrating that organic matter was effectively removed.

There is an important precaution in using sediment oxidation. If the internal loading of P is controlled by iron redox reactions, then the P loading should decrease significantly following treatment. If, however, the lake is shallow and internal P loading is controlled to a large extent by high pH and temperature in the near-sediment water during summer, then sediment oxidation may not significantly decrease P loading (Petterson and Böstrom, 1981). High pH may also interfere with alum in shallow lakes. Usually, pH is relatively low (< 7) near the sediment-water interface, in stable water columns, due to microbial decomposition, but high pH in the photosynthetic zone can directly contact surficial sediments or those resuspended during turbulent conditions (Ryding, 1985; Koski-Vähälä and Hartikainen, 2001).

The principal impediment to the selection of Riplox over other inactivation treatments (e.g., alum) is the lack of documented overall successes. Lake Lillesjön is the only case where the experimentally required dose was applied, lake P substantially declined, and the oxidized state of the sediment has persisted. External P loading apparently remained too high for significant lake recovery to occur in Trekanten and Long Lakes, and although expected results were observed in White Lough, the lake was underdosed.

Sediment oxidation (Riplox) is a sound technique and represents an alternative to alum treatment and dredging as a long-term control of internal P loading from anaerobic sediment release. Test cases should be conducted in lakes in which anaerobic P release is the principal contributor to lake P content. Sediment treatment experiments can be conducted *in vitro* (Ripl and Lindmark, 1978; Willenbring et al., 1984), to test the method's potential prior to whole-lake application. Care must be taken, however, to simulate conditions of pH and temperature that occur near the sediment-water interface or during resuspension in the lake, in addition to anaerobic conditions. If these conditions are not duplicated, the potential of the technique may be overestimated. However, similar problems could occur with alum under high pH and resuspension. With either Riplox or alum, if the dose is adequate and P internal loding is substantially reduced (and if external loading is low), high pH should not occur because algae and their photosynthesis are reduced.

Another application of this principle involves the use of effluent from advanced wastewater treatment (AWT) plants (Ripl et al., 1978). Deep-water injection of AWT effluent, with its high NO_3 content, should promote organic matter decomposition and increased sediment P binding, and the oxygen content should increase (Ripl et al., 1978). Furthermore, N-fixing blue-green algae, normally favored by high P and low N, should be discouraged in lakes receiving NO_3-rich waters.

An indication of such changes was observed in three small reservoirs. One received four treatments of NO_3, another received two treatments, and the third received none. A progressive improvement was observed in Secchi transparency, TP, and pH (decreased), and the presence of non blue-green algae increased in response to NO_3 increase (Ripl et al., 1978). Redirecting the effluent discharge in lakes being restored by AWT, but in which internal loading is retarding recovery, offers a complementary method to speed the recovery process.

This technique of hypolimnetic nitrate addition was tested in a small (10 ha), stratified (maximum depth 7.6 m) lake in Denmark (Søndergaard et al., 2000). Nitrate was added at 5-m depth in Lake Lyng five times in dissolved and granulated form during 1995 and 1996 at a rate of 8–10 g N/m^2 per year. There was improved P retention of the sediment, hypolimnetic TP decreased 23–52% and there was less sulfide odor following treatment. Hypolimnetic SRP still reached a maximum of 1 mg/L after treatment. However, the dose of nitrate was much less than added to Lake Lillesjön. The method of addition is less expensive than injection with a harrow and the authors thought the method deserved further investigation.

REFERENCES

Anderson, J.M. 1974. Nitrogen and phosphorus budgets and the role of sediments in six shallow Danish lakes. *Hydrobiologia* 74: 428–550.

Anderson, J.M. 1975. Influence of pH on release of phosphorus from lake sediments. *Arch. Hydrobiol.* 76: 411–419.

Atlas Copco Co. Undated. Wayne, NJ and Wilrijk, Belgium.

Babin, J., E.F. Prepas, T.P. Murphy and H.R. Hamilton. 1989. A test of the effects of lime on algal biomass and total phosphorus concentrations in Edmonton stormwater retention lakes. *Lake and Reservoir Manage.* 5: 129–135.

Babin, J., E.E. Prepas and Y. Zhang. 1992. Application of lime and alum to stormwater retention lakes to improve water quality. *Water Pollut. Res. J. Canada* 27: 365–381.

Babin, J., E.E. Prepas, T.P. Murphy, M. Serediak, P.J. Curtis, Y. Zhang and P.A. Chambers. 1994. Impact of lime on sediment phosphorus release in hardwater lakes: The case of hypereutrophic Halfmoon Lake, Alberta. *Lake and Reservoir Manage.* 8: 131–142.

Baker, J.P. 1982. Effects on fish of metals associated with acidification. In: R.E. Johnson (Ed.), *Acid Rain/Fisheries*. American Fisheries Society. pp. 165–176.

Barbiero, R.P. and E.B. Welch. 1992. Contribution of benthic blue-green algal recruitment to lake populations and phosphorus. *Freshwater Biol.* 27: 249–260.

Barbiero, R., R.E. Carlson, G.D. Cooke and A.W. Beals. 1988. The effects of a continuous application of aluminum sulfate on lotic benthic invertebrates. *Lake and Reservoir Manage.* 4(2): 63–72.

Barko, J.W., W.F. James, W.D. Taylor and D.G. McFarland. 1990. Effects of alum treatment on phosphorus and phytoplankton dynamics in Eau Galle Reservoir: A synopsis. *Lake and Reservoir Manage.* 6: 1–8.

Bertsch, P.M. 1989. Aqueous polynuclear aluminum species. In: G. Sposito (Ed.), *The Environmental Chemistry of Aluminum*. CRC Press, Boca Raton, FL, ch. 4.

Birchall, J.D., C. Exley, J.S. Chappell and M.J. Phillips. 1989. Acute toxicity of aluminum to fish eliminated in silicon-rich acid waters. *Nature* 338: 146–148.

Boers, P.C.M. 1991a. The influence of pH on phosphate release from lake sediments. *Water Res.* 25: 309–311.

Boers, P.C.M. 1991b. The release of dissolved phosphorus from sediments. Ph.D. Dissertation, Limnological Institute (Nieuwersluis) and Institute for Inland Water Management and Waste Water Treatment, Lelystad, The Netherlands.

Boers, P., J. Van der Does, M. Quaak and J. Van der Vlugt. 1994. Phosphorus fixation with iron (III) chloride: a new method to combat internal phosphorus loading in shallow lakes? *Arch. Hydrobiol.* 129: 339–351.

Bose, P. and D.A. Reckhow. 1998. Adsorption of natural organic matter on preformed aluminum hydroxide flocs. *J. Environ. Eng.* 124: 803–811.

Boström, B., M. Jansson and C. Forsberg. 1982. Phosphorus release from lake sediments. *Arch. Hydrobiol. Beih. Ergebn. Limnol.* 18: 5–59.

Boyd, C.E. 1979. Aluminum sulfate (alum) for precipitating clay turbidity from fish ponds. *Trans. Am. Fish. Soc.* 108: 307–313.

Browman, M.G., R.F. Harris and D.E. Armstrong. 1977. Interaction of Soluble Phosphate with Aluminum Hydroxide in Lakes. Tech. Rept.77-05. Water Resources Center, University of Wisconsin, Madison.

Buergel, P.M. and R.A. Soltero. 1983. The distribution and accumulation of aluminum in rainbow trout following a whole-lake alum treatment. *J. Fresh Water Ecol.* 2: 37–44.

Burley, K.L., E.E. Prepas and P.A. Chambers. 2001. Phosphorus release from sediments in hardwater eutrophic lakes: the effects of redox-sensitive and -insensitive chemical treatments. *Freshwater Biol.* 46: 1061–1074.

Burrows, H.D. 1977. Aquatic aluminum chemistry, toxicology, and environmental prevalence. *Crit. Rev. Environ. Control* 7: 167–216.

Caraco, N.F., J.J. Cole and G.E. Likens. 1989. Evidence for sulphate-controlled phosphorus release from sediments of aquatic systems. *Nature* 341: 316–318.

Caraco, N.F., J.J. Cole and G.E. Likens. 1991a. Phosphorus release from anoxic sediments: Lakes that break the rule. *Verh. Int. Verein. Limnol.* 24: 2985–2988.

Caraco, N.F., J.J. Cole and G.E. Likens. 1991b. A cross-system study of phosphorus release from lake sediments. In: J. Cole, G. Lovett and S. Findley (Eds.), *Comparative Analyses of Ecosystems*. Springer-Verlag, New York, NY. pp. 241–288.

Carlton, R.G. and R.G. Wetzel. 1988. Phosphorus flux from lake sediments: effect of epipelic algal oxygen production. *Limnol. Oceanogr.* 33: 562–570.

Carlsson, S.-A. Vattenresurs AB, Tjusta, S-197 93 Sweden.

Chapra, S.C. and R.P. Canale. 1991. Long-term phenomenological model of phosphorus and oxygen for stratified lakes. *Water Res.*, 25: 707–715.

Connor, J.N. and G.N. Smith. 1986. An efficient method of applying aluminum salts for sediment phosphorus inactivation in lakes. *Water Res. Bull.* 22: 661–664.

Connor, J. and M.R. Martin. 1989a. An Assessment of Wetlands Management and Sediment Phosphorus Inactivation, Kezar Lake, New Hampshire. NH Dept. Environ. Serv. Staff Rep. 161.

Connor, J.N. and M.R. Martin. 1989b. An assessment of sediment phosphorus inactivation, Kezar Lake, New Hampshire. *Water Res. Bull.* 25: 845–853.

Cooke, G.D. and Kennedy, R.H. 1981. *Precipitation and Inactivation of Phosphorus as a Lake Restoration Technique.* USEPA-600/3-81-012.

Cooke, G.D., R.T. Heath, R.H. Kennedy and M.R. McComas. 1978. *Effects of Diversion and Alum Application on Two Eutrophic Lakes.* USEPA-600/3-78-033.

Cooke, G.D., R.T. Heath, R.H. Kennedy and M.R. McComas. 1982. Change in lake trophic state and internal phosphorus release after aluminum sulfate application. *Water Res. Bull.* 18: 699–705.

Cooke, G.D., E.B. Welch, S.A. Peterson and P.R. Newroth. 1986. *Lake and Reservoir Restoration.* 1st ed. Butterworth Publishers. Stoneham, MA.

Cooke, G.D., E.B. Welch, S.A. Peterson and P.R. Newroth 1993a. *Restoration and Management of Lakes and Reservoirs*, 2nd ed. CRC Press, Boca Raton, FL.

Cooke, G.D., E.B. Welch, A.B. Martin, D.G. Fulmer, J.B. Hyde and G.D. Schrieve. 1993b. Effectiveness of Al, Ca and Fe salts for control of internal phosphorus loading in shallow and deep lakes. *Hydrobiologia* 253: 323–335.

Cullen, P. and C. Forsberg. 1988. Experiences with reducing point sources of phosphorus to lakes. *Hydrobiologia* 170: 321–336.

DeGasperi, C.L., D.E. Spyridakis and E.B. Welch. 1993. Alum and nitrate as controls of short-term anaerobic sediment phosphorus release: An *in vitro* comparison. *Lake and Reservoir Manage.* 8: 49–59.

Deppe, T. and J. Benndorf. 2002. Phosphorus reduction in a shallow hypereutrophic reservoir by in-lake dosage of ferrous iron. *Water Res.* 36: 4525–4534.

Deppe, T., K. Ockenfeld, A. Meybowm, M. Opitz and J. Benndorf. 1999. Reduction of *Microcystis* blooms in a hypertrophic reservoir by a combined ecotechnological strategy. *Hydrobiologia* 408/409: 31–38.

Dentel, S.K. and J.M. Gossett. 1988. Mechanisms of coagulation with aluminum salts. *J. Am. Water Works Assoc.* 80: 187–198.

Dittrich, M. and R. Koschel. 2002. Interactions between calcite precipitation (natural and artificial) and phosphorus cycle in the hardwater lake. *Hydrobiologia* 469: 49–57.

Dittrich, M., T. Dittrich, I. Sieber and R. Koschel. 1997. A balance analysis of phosphorus elimination by artificial calcite precipitation in a stratified hardwater lake. *Water Res.* 31: 237–248.

Doke, J.L., W.H. Funk, S.T.J. Juul and B.C. Moore. 1995. Habitat availability and benthic invertebrate population changes following alum treatment and hypolimnetic oxygenation in Newman Lake, Washington. *J. Fresh Water Ecol.* 10: 87–102.

Dominie, D.R., II. 1980. Hypolimnetic aluminum treatment of softwater Annabessacook Lake. In: *Restoration of Lakes and Inland Waters.* USEPA-440/5-81-010. pp. 417–423.

Driscoll, C.T., Jr. and R.D. Letterman. 1988. Chemistry and fate of Al(III) in treated drinking water. *J. Environ. Eng. Div. ASCE* 114: 21–37.

Driscoll, C.T., Jr. and W.D. Schecher. 1990. The chemistry of aluminum in the environment. *Environ. Geochem. Health* 12: 28–49.

Driscoll, C.T., Jr., J.P. Baker, J.J. Bisogni and C.L. Schofield. 1980. Effect of aluminum speciation on fish in dilute acidified waters. *Nature* 284: 161–164.

Eberhardt, T. 1990. Alum dose calculations for lake sediment phosphorus inactivation based upon phosphorus release rates and a call for related research. Presentation, North American Lake Management Society Annual Meeting, Springfield, MA and unpublished report, Sweetwater Technology, Aiken, MN.

Eberhardt, T. Personal communication. Sweetwater Technology, Aiken, MN.

Edmondson, W.T. 1970. Phosphorus, nitrogen, and algae in Lake Washington after diversion of sewage. *Science* 169: 690–691.

Edmondson, W.T. 1994. Sixty years of Lake Washington: A curriculum vitae. *Lake and Reservoir Manage.* 10: 75–84.

Eisenreich, S.J., D.E. Armstrong and R.F. Harris. 1977. A Chemical Investigation of Phosphorus Removal in Lakes by Aluminum Hydroxide. Tech. Rept.77-02. Water Resources Center, University of Wisconsin, Madison.

ENSR. 2002. Short-term monitoring report. In: CH2M-Hill, 2002. Ashumet Pond phosphorus inactivation report. ENSR, Willington, CT 06279.

Entranco. 1983. Water Quality Analysis and Restoration Plan: Erie and Campbell Lakes. Report, Entranco Engineering, Inc., Bellevue, WA.

Entranco. 1986. Wapato Lake Restoration: A Discussion of Design Considerations, Construction Techniques and Performance Evaluation. Final Report, Entranco Engineers, Bellevue, WA.

Entranco. 1987a. Final Phase II Report: Erie and Campbell Lakes: Restoration, Implementation and Evaluation. Entranco Engr. Inc., Bellevue, WA.

Entranco. 1987b. Pattison and Long Lakes Restoration Project Final Report. Entranco Eng. Inc., Bellevue, WA.

Everhart, W.H. and R.A. Freeman. 1973. *Effects of Chemical Variations in Aquatic Environments. Vol. II. Toxic Effects of Aqueous Aluminum to Rainbow Trout.* USEPA-R3-73-011b.

Exley, C. 2000. Avoidance of aluminum by rainbow trout. *Environ. Toxicol. Chem.* 19: 933–939.

Fee, E.J. 1979. A relation between lake morphometry and primary productivity and its use in interpreting whole lake eutrophication experiments. *Limnol. Oceanogr.* 24: 401–416.

Foy, R.H. 1985. Phosphorus inactivation in a eutrophic lake by the direct addition of ferric aluminum sulphate: Impact on iron and phosphorus. *Freshwater Biol.* 15: 613–629.

Foy, R.H. 1986. Suppressions of phosphorus release from lake sediments by the addition of nitrate. *Water Res.* 20: 1345–1351.

Foy, R.H. and A.G. Fitzsimmons. 1987. Phosphorus inactivation in a eutrophic lake by the direct addition of ferric aluminum sulphate: Changes in phytoplankton populations. *Freshwater Biol.* 17: 1–13.

Freeman, R.A. and W.H. Everhardt. 1971. Toxicity of aluminum hydroxide complexes in neutral and basic media to rainbow trout. *Trans. Am. Fish. Soc.* 100: 644–658.

Gahler, A.R. and C.F. Powers. 1970. Personal communication. USEPA, Corvallis, OR.

Gardner, W.S. and B.J. Eadie. 1980. Chemical factors controlling phosphorus cycling in lakes. In: D. Scavia and R. Moll (Eds.), Nutrient Cycling in the Great Lakes: A Summarization of Factors Regulating the Cycling of Phosphorus. Special Report No. 83, Great Lakes Research Division, The University of Michigan, Ann Arbor. pp. 13–34.

Garrison, P.J. and D.R. Knauer. 1984. Long term evaluation of three alum treated lakes. In: *Lake and Reservoir Management.* USEPA 440/5-84-001. pp. 513–517.

Garrison, P.J. and D.M. Ihm. 1991. First Annual Report of Long-Term Evaluation of Wisconsin's Clean Lake Projects. Part B. Lake Assessment. Wisconsin Dept. Nat. Res., Madison.

Gasperino, A.F., M.A. Beckwith, G.R. Keizur, R.A. Soltero, D.G. Nichols and J.M. Mires. 1980. Medical Lake improvement project: Success story. In: *Restoration of Lakes and Inland Waters.* USEPA 440/5-81-010. pp 424–428.

Gaugush, R.G. 1984. Mixing events in Eau Galla Lake. In: *Lake and Reservoir Management.* USEPA 440/5-84-001. pp. 286–291.

Gensemer, R.W. and R.C. Playle. 1999. The bioavailability and toxicity of aluminum in aquatic environments. *Crit. Rev. Environ. Sci. Technol.* 29: 315–450.

Gibbons, M.V., F.D. Woodwick, W.H. Funk and H.L. Gibbons. 1984. Effects of a multiphase restoration, particularly aluminum sulfate application, on the zooplankton community of a eutrophic lake in eastern Washington. *J. Fresh Water Ecol.* 2: 393–404.

Hall, K.J. and K.I. Ashley. Personal communication. Ministry of Environment and Dept. of Civil Engineering, Univ. of British Columbia, Vancouver, BC.

Hart, B.T., S. Roberts, R. James, M. O'Donohue, J. Taylor, D. Donnert and R. Furrer. 2003a. Active barriers to reduce phosphorus release from sediments: Effectiveness of three forms of $CaCO_3$. *Aust. J. Chem.* 56: 207–217.

Hart, B., S. Roberts, R. James, J. Taylor, D. Donnert and R. Furrer. 2003b. Use of active barriers to reduce eutrophication problems in urban lakes. *Water Sci. Technol.* 47: 157–163.

Havens, K.E. 1990. Aluminum binding to ion exchange sites in acid-sensitive versus acid-tolerant cladocerans. *Environ. Pollut.* 64: 133–141.

Havens, K.E. 1993. Acid and aluminum effects on the survival of littoral macro-invertebrates during acute bioassays. *Environ. Pollut.* 80: 95–100.

Havens, K. E and R.T. Heath. 1989. Acid and aluminum effects on freshwater zooplankton: An *in situ* mesocosm study. *Environ. Pollut.* 62: 195–211.

Hayes, C.R., R.G. Clark, R.F. Stent and C.J. Redshaw. 1984. The control of algae by chemical treatment in a eutrophic water supply reservoir. *J. Inst. Water Eng. Sci.* 38: 149–162.

Holz, J.C. and K.D. Hoagland. 1999. Effects of phosphorus reduction on water quality: Comparison of alum-treated and untreated portions of a hypereutrophic lake. *Lake and Reservoir Manage.* 15: 70–82.

Huser, B. Personal communication. Dept. of Civil and Environmental Engr., Univ. of Minnesota, Minneapolis, MN.

Hupfer, M., R. Pothig, R. Bruggemann and W. Geller. 2000. Mechanical resuspension of autochthonous calcite (seekreide) failed to control internal phosphorus cycle in a eutrophic lake. *Water Res.* 34: 859–867.

Jacoby, J.M. 1978. Lake phosphorus cycling as influenced by drawdown and alum addition. MS Thesis, Department of Civil Engineering, University of Washington, Seattle, WA.

Jacoby, J.M., D.D. Lynch, E.B. Welch and M.A. Perkins. 1982. Internal phosphorus loading in a shallow eutrophic lake. *Water Res.* 16: 911–919.

Jacoby, J.M., H.L. Gibbons, K.B. Stoops and D.D. Bouchard. 1994. Response of a shallow, polymictic lake to buffered alum treatment. *Lake and Reservoir Manage.* 10: 103–112.

Jaeger. D. 1994. Effects of hypolimnetic water aeration and Fe–P precipitation on the trophic level of Lake Krupunder. *Hydrobiologia* 275/276: 433–444.

James, W.F. and J.W. Barko. 2003. Alum dosage determinations based on redox-sensitive sediment phosphorus concentrations. Water Quality Tech. Notes Coll. (ERDC WQTN-PD-13), US Army Eng. Res. Dev. Center, Vicksburg, MS.

James, W.F., J.W. Barko and W.D. Taylor. 1991. Effects of alum treatment on phosphorus dynamics in a north-temperate reservoir. *Hydrobiologia* 215: 231–241.

Jensen, H.S. and F.Ø. Andersen. 1992. Importance of temperature, nitrate, and pH for P release from aerobic sediments of four shallow, eutrophic lakes. *Limnol. Oceanogr.* 37: 577–589.

Jensen, H.S., P. Kristensen, E. Jeppesen, and A. Skytthe. 1992. Iron-phosphorus ratio in surface sediments as an indicator of phosphate release from aerobic sediments in shallow lakes. *Hydrobiologia* 235/236: 731–743.

Jenkins, L.F., F.Q. Ferguson and A.B. Minar. 1971. Chemical processes for phosphate removal. *Water Res.* 5: 369–389.

Jeppesen, E., P. Kristensen, J.P. Jensen, M. Søndergaard, E. Mortensen and T. Lauridsen. 1991. Recovery resilience following a reduction in external phosphorus loading of shallow, eutrophic Danish lakes: Duration, regulating factors and methods for overcoming resilience. *Mem. Ist. Ital. Idrobiol.* 48: 127–148.

Jernelöv, A. (Ed.) 1971. *Phosphate Reduction in Lakes by Precipitation with Aluminum Sulphate.* 5th International Water Pollution Research Conference. Pergamon Press, New York.

Kane, D.A. and C.F. Rabeni. 1987. Effects of aluminum and pH on the early life stages of smallmouth bass (*Micropterus dolomieui*). *Water Res.* 21: 633–639.

Kasprzak, P., R. Koschel, L. Krienitz, T. Gonsiorczyk, K. Anwand, U. Laude, K. Wysujack, H. Brach and T. Mehner. 2003. Reduction of nutrient loading, planktivore removal and piscivore stocking as tools in water quality management: The Feldberger Haussee biomanipulation project. *Limnologica* 33: 190–204.

Kennedy, R.H. 1978. Nutrient inactivation with aluminum sulfate as a lake reclamation technique. Ph.D. Dissertation, Kent State University, Kent, OH.

Kennedy, R.H. and G.D. Cooke. 1980. Aluminum sulfate dose determination and application techniques. In: *Restoration of Lakes and Inland Waters.* USEPA-400/5-81-010. pp 405–411.

Kennedy, R.H. and G.D. Cooke. 1982. Control of lake phosphorus with aluminum sulfate. Dose determination and application techniques. *Water Res. Bull.* 18: 389–395.

Kennedy, R.H., J.W. Barko, W.F. James, W.D. Taylor and G.L. Godshalk. 1987. Aluminum sulfate treatment of a reservoir: Rationale, application methods, and preliminary results. *Lake and Reservoir Manage.* 3: 85–90.

Kirn, R.A. 1987. Unpublished report on Lake Morey, VT. Vermont Agency of Environmental Conservation, Project F-12-R-22, Job I-1, Roxbury, VT.

Knauer, D. Personal communication. Dept. Nat. Res., Madison, Wisconsin.

Koschel, R., J. Benndorf, G. Proft and F. Recknagel. 1983. Calcite precipitation as a natural control mechanism of eutrophication. *Arch. Hydrobiol.* 98: 380–408.

Koski-Vähälä, J. and H. Hartikainen. 2001. Assessment of the risk of phosphorus loading due to resuspended sediment. *J. Environ. Qual.* 30: 960–6.

Lamb, D.S. and G.C. Bailey. 1981. Acute and chronic effects of alum to midge larva (Diptera: Chironomidae). *Bull. Environ. Contam. Toxicol.* 27: 59–67.

Lamb, D.S. and G.C. Bailey. 1983. Effects of aluminum sulfate to midge larvae (Diptera: Chironomidae) and rainbow trout (*Salmo gairdneri*). In: *Lake Restoration, Protection and Management.* USEPA-440/5-83-001. pp. 307–312.

Larsen, D.P., D.W. Shults and K.W. Malueg. 1981. Summer internal phosphorus supplies in Shagawa Lake, Minesota. *Limnol. Oceanogr.* 26: 740–753.

Lewandowski, J., I. Schauser and M. Hupfer. 2003. Long term effects of phosphorus precipitations with alum in hypereutrophic Lake Süsser See (Germany). *Water Res.* 37: 3194–3204.

Lijklema, L. 1977. The role of iron in the exchange of phosphate between water and sediments. In: H.L. Golterman (Ed.), *Interactions between Sediment and Fresh Water.* W. Junk Publishers, The Hague, The Netherlands. pp. 313–317.

Lind, C. Personal communication. General Chemical Corp., Parsippany, NJ.

Löfgren, S. and B. Boström. 1989. Interstitial water concentrations of phosphorus, iron, and manganese in a shallow, eutrophic Swedish lake: Implications for phosphorus cycling. *Water Res.* 23: 1115–1125.

Lund, J.W.G. 1955. The ecology of algae and waterworks practice. *Proc. Soc. Water Treat. Exam.* 4: 83–109.

Maessen, M., J.G.M. Roelofs, M.J.S. Bellemakers and G.M. Verheggen. 1992. The effects of aluminium, aluminium/calcium ratios and pH on aquatic plants from poorly buffered environments. *Aquatic Bot.* 43: 115–127.

Mataraza, L.K. and G.D. Cooke. 1997. Vertical phosphorus transport in lakes of different morphometry. *Lake and Reservoir Manage.* 13: 328–337.

May, V. 1974. Suppression of blue-green algal blooms in Braidwood Lagoons with alum. *J. Aust. Inst. Agric. Sci.* 40: 54–57.

May, V. and H. Baker. 1978. Reduction of Toxic Algae in Farm Dams by Ferric Alum. Tech. Bull. 19. Dept. Agriculture, New South Wales, Australia.

McComas, S. 1989. Using buffered alum to control algae. *LakeLine* 9: 12–13.

McComas, S. 2003. *Lake and Pond Management Guide Book.* CRC Press, Boca Raton, FL.

Mesner, N. and R. Narf. 1987. Alum injection into sediments for phosphorus inactivation and macrophyte control. *Lake and Reservoir Manage.* 3: 256–265.

Miskimmin, B.M., W.F. Donahue and D. Watson. 1995. Invertebrate community response to experimental lime (Ca(OH)$_3$) treatment of an eutrophic pond. *Aquatic Sci.* 57: 20–30.

Moffett, M.R. 1979. Changes in the microcrustacean communities of East and West Twin Lakes, Ohio, following lake restoration. MS Thesis, Kent State University, Kent, OH.

Mortimer, C.H. 1941. The exchange of dissolved substances between mud and water in lakes (Parts I and II). *J. Ecol.* 29: 280–329.

Mortimer, C.H. 1971. Chemical exchanges between sediments and water in the Great Lakes - speculations on probable regulatory mechanisms. *Limnol.Oceanogr.* 16: 387–404.

Moss, B., J. Madgwick, and G. Phillips. 1996. *A Guide to the Restoration of Nutrient-Enriched Shallow Lakes.* Broads Authority. Norfolk, U.K.

Murphy, T.P. and E.E. Prepas. 1990. Lime treatment of hardwater lakes to reduce eutrophication. *Verh. Int. Verein. Limnol.* 24: 327–334.

Murphy, T.P., K.G. Hall and T.G. Northcote. 1988. Lime treatment of a hardwater lake to reduce eutrophication. *Lake and Reservoir Manage.* 4(2): 51–62.

Murphy, T.P., E.E. Prepas, J.T. Lim, J.M. Crosby and D.T. Walty. 1990. Evaluation of calcium carbonate and calcium hydroxide treatments of prairie water dugouts. *Lake and Reservoir Manage.* 6: 101–108.

Narf, R.P. 1990. Interactions of Chironomidae and Chaoboridae (Diptera) and aluminum sulfate treated lake sediments. *Lake and Reservoir Manage.* 6: 33–42.

Neville, C.M. 1985. Physiological response of juvenile rainbow trout, *Salmo gairdneri*, to acid and aluminum — prediction of field responses from laboratory data. *Can. J. Fish. Aquatic Sci.* 42: 2004–2019.

Noon, T.A. 1986. Water quality in Long Lake, Minnesota, following Riplox sediment treatment. *Lake and Reservoir Manage.* 2: 131.

Norup, B. 1975. *Lyngby Sø Feltundersogelser efter Fosfatfaeldeperiodeni Sommeren 1974: Bundfaunaunder-sogelse Fysisk-kemisk and Planteplankton Production.* Lyngby Taarboek Kommune, Copenhagen, Denmark (cited in Welch et al., 1988).

Nriagu, J.O. 1978. Dissolved silica in pore waters of lakes Ontario, Erie, and Superior sediments. *Limnol. Oceanogr.* 23: 53–67.

Nürnberg, G.K. 1996. Trophic state of clear and colored, soft- and hardwater lakes with special consideration of nutrients, anoxia, phytoplankton and fish. *Lake and Reservoir Manage.* 12: 432–447.

Ødegaard, H., J. Fettig and H.C. Ratnaweera. 1990. Coagulation with prepolymerized metal salts. In: H.H. Hahn and R. Klute (Eds.), *Chemical Water and Wastewater Treatment.* Springer-Verlag, Berlin.

Otsuki, A. and R.G. Wetzel. 1972. Coprecipitation of phosphate with carbonates in a marl lake. *Limnol. Oceanogr.* 17: 763–767.

Paléo, A.B.S. 1995. Aluminium polymerization — a mechanism of acute toxicity of aqueous aluminium to fish. *Aquatic Toxicol.* 31: 347–356.

Panuska, J.C. and D.M. Robertson. 1999. Estimating phosphorus concentrations following alum teatment using apparent settling velocity. *Lake and Reservoir Manage.* 15: 28–38.

Peelen, R. 1969. Possibilities to prevent blue-green algal growth in the delta region of The Netherlands. *Verh. Int. Verein. Limnol.* 17: 763–766.

Perakis, S.S., E.B. Welch and J.M. Jacoby. 1996. Sediment-to-water blue-green algal recruitment in response to alum and environmental factors. *Hydrobiologia* 318: 165–177.

Perkins, R.G. and G.J.C. Underwood. 2001. The potential for phosphorus release across the sediment-water interface in an eutrophic reservoir dosed with ferric sulphate. *Water Res.* 35: 1399–1406.

Perkins, R.G. and G.J.C. Underwood. 2002. Partial recovery of a eutrophic reservoir through managed phosphorus limitation and unmanaged macrophyte growth. *Hydrobiologia* 481: 75–87.

Peterson, J.O., J.T. Wall, T.L. Wirth and S.M. Born. 1973. Eutrophication Control: Nutrient Inactivation by Chemical Precipitation at Horseshoe Lake, WI. Tech. Bull. 62. Wisconsin Dept. Nat. Res., Madison.

Pettersson, K. and B. Boström. 1981. En kritisk granskning av foreslagna metoder for nitratbehandling av sediment. *Vatten* 38: 74.

Pilgrim, K.M. and P.L. Brezonik. 2005. Evaluation of the potential adverse effects of lake inflow treatment with alum. *Lake and Reservoir Manage.* in press.

Prentki, R.T., M.S. Adams, S.R. Carpenter, A. Gasith, C.S. Smith and P.R. Weiler. 1979. The role of submersed weedbeds in internal loading and interception of allocthonous materials in Lake Wingra, Wisconsin. *Arch. Hydrobiol.* 57: 221–250.

Prepas, E.E., R.P. Murphy, J.M. Crosby, D.T. Walty, J.T. Lim, J. Babin and P.A. Chambers. 1990. Reduction of phosphorus and chlorophyll *a* concentrations following $CaCO_3$ and $Ca(OH)_2$ additions to hyper-eutrophic Figure Eight Lake, Alberta. *Environ. Sci. Technol.* 24: 1252–1258.

Prepas, E.E., B. Pinel-Alloul, P.A. Chambers, T.P. Murphy, S. Reedyk, G.J. Sandland and M. Serediak. 2001a. Lime treatment and its effects on the chemistry and biota of hardwater eutrophic lakes. *Freshwater Biol.* 46: 1049–1060.

Prepas, E.E., J. Babin, T.P. Murphy, P.A. Chambers, G.J. Sandland, A. Ghadouani and M. Serediak. 2001b. Long-term effects of successive $Ca(OH)_2$ and $CaCO_3$ treatments on the water quality of two eutrophic hardwater lakes. *Freshwater Biol.* 46: 1089–1103.

Psenner, R., R. Pucsko and M. Sager. 1984. Fractionation of organic and inorganic phosphorus compounds in lake sediments. An attempt to characterize ecologically important fractions. *Arch. Hydrobiol. Suppl.* 70: 111–155.

Quaak, M., J. van derDoes, P. Boers and J. van der Vlugt. 1993. A new technique to reduce internal phosphorus loading by in-lake phosphate fixation in shallow lakes. *Hydrobiologia* 253: 337–344.

Ramamoorthy, S. 1988. Effect of pH on speciation and toxicity of aluminum to rainbow trout (*Salmo gairdneri*). *Can. J. Fish. Aquatic Sci.* 45: 634–642.

Reedyk, S., E.E. Prepas and P.A. Chambers. 2001. Effects of single $Ca(OH)_2$ doses on phosphorus concentration and macrophyte biomass of two boreal eutrophic lakes over 2 years. *Freshwater Biol.* 46: 1075–1087.

Reitzel, K., J. Hansen, H.S. Jensen, F.Ø. Andersen and K.S. Hansen. 2003. Testing aluminum addition as a tool for lake restoration in shallow, eutrophic Lake Sønerby, Denmark. *Hydrobiologia* 506: 781–787.

Ripl, W. 1976. Biochemical oxidation of polluted lake sediment with nitrate — a new restoration method. *Ambio* 5: 132.

Ripl, W. 1980. Lake restoration methods developed and used in Sweden. In: *Restoration of Lakes and Inland Water.* USEPA 440/5-81-010, 495–500.

Ripl, W. 1986. Internal phosphorus recycling mechanisms in shallow lakes. *Lake and Reservoir Manage.* 2: 138.

Ripl, W. and G. Lindmark. 1978. Ecosystem control by nitrogen metabolism in sediment. *Vatten* 34: 135–144.

Ripl, W., L. Leonardson, G. Lindmark, G. Anderson and G. Cronberg. 1978. Optimering av reningsverk/recipient-system. *Vatten* 35: 96.

Robertson, D.M., G.L. Goddard, D.R. Helsel and K.L. MacKinnon. 2000. Rehabilitation of Delavan Lake, Wisconsin. *Lake and Reservoir Manage.* 16: 155–176.

Rönicke, H., M. Bever and J.T. Doretta. 1995. Eutrophierung eines Magdeburger Kiesbaggersees – möglichkeiten zur steuerung des nährstoffhaushaltes und der blaualgenabundanz durch massnahmen zur seenrestaruierung (with English abstract). *Limnologie aktuell, Band* 7: 139–154.

Rosseland, B.O., T.D. Eldhuset and M. Staurnes. 1990. Environmental effects of aluminum. *Environ. Geochem. Health* 12: 17–27.

Roy, R.L. and G.C. Campbell. 1997. Decreased toxicity of Al to juvenile Atlantic salmon (*Salmo salar*) in acidic soft water containing natural organic matter: A text of the free-ion model. *Environ. Chem. Toxicol.* 16: 1962–1967.

Rydin, E. and E.B. Welch. 1998. Dosage of aluminum to absorb mobile phosphate in lake sediments. *Water Res.* 32: 2969–2976.

Rydin, E. and E.B. Welch. 1999. Dosing alum to Wisconsin lake sediments based on possible *in vivo* formation of aluminum bound phosphate. *Lake and Reservoir Manage.* 15: 324–331.

Rydin, E., E.B. Welch and B. Huser. 2000. Amount of phosphorus inactivated by alum treatments in Washington lakes. *Limnol. Oceanogr.* 45: 226–230.

Ryding, S-O, 1985. Chemical and microbiological processes as regulators of the exchange of substances between sediments and water in shallow eutrophic lakes. *Int. Rev. ges. Hydrobiol.* 70: 657.

Salonen, V-P. and E. Varjo. 2000. Gypsum treatment as a restoration method for sediments of eutrophied lakes — experiments from southern Finland. *Environ. Geol.* 39: 353–359.

Sas, H., I. Ahlgren, H. Bernhardt, B. Boström, J. Clasen, C. Forsberg, D. Imboden, L. Kamp-Nielson, L. Mur, N. de Oude, C. Reynolds, H. Schreurs, K. Seip, U. Sommer and S. Vermij. 1989. *Lake Restoration by Reduction of Nutrient Loading: Expectations, Experiences, Extrapolation.* Academia-Verlag, Richarz, St. Augustine, Germany.

Schelske, C.L. 1988. Historical trends in Lake Michigan silica concentrations. *Int. Rev. ges. Hydrobiol.* 73: 559–591.

Schindler, D.W. 1986. The significance of in-lake production of alkalinity. *Water Air Soil Pollut.* 18: 259–271.

Serediak, M.S., E.E. Prepas, T.P. Murphy and J. Babin. 2002. Development, construction and use of lime and alum application systems in Alberta. *Lake and Reservoir Manage.* 18: 66–74.

Scheffer, M. 1998. *Ecology of Shallow Lakes.* Chapman & Hall, New York, NY.

Shumaker, R.J., W.H. Funk and B.C. Moore. 1993. Zooplankton response to aluminum sulfate treatment of Newman Lake, Washington. *J. Fresh Water Ecol.* 8: 375–387.

Smeltzer, E. 1990. A successful alum/aluminate treatment of Lake Morey, Vermont. *Lake and Reservoir Manage.* 6: 9–19.

Smeltzer, E., R.A. Kirn and S. Fiske. 1999. Long-term water quality and biological effects of alum treatment of Lake Morey, Vermont. *Lake and Reservoir Manage.* 15: 173–184.

Smolders, A. and J.G.M. Roelofs. 1993. Sulphate-mediated iron limitation and eutrophication in aquatic ecosystems. *Aquatic Bot.* 46: 247–253.

Sonnichsen, J.D., J.M. Jacoby and E.B. Welch. 1997. Response of cyanobacterial migration to alum treatment in Green Lake. *Arch. Hydrobiol.* 140: 373–392.

Søndergaard, M. 1988. Seasonal variations in the loosely sorbed phosphorus fraction of the sediment of a shallow and hypereutrophic lake. *Environ. Geol. Water Sci.* 11: 115–21.

Søndergaard, M., E. Jeppesen and J.P. Jensen. 2000. Hypolimnetic nitrate treatment to reduce internal phosphorus loading in a stratified lake. *Lake and Reservoir Manage.* 16: 195–204.

Søndergaard, M., K-D. Wolter and W. Ripl. 2002. Chemical treatment of water and sediments with special reference to lakes. In: M. Perrow and A.J. Davy (Eds.), *Handbook of Ecological Restoration, Vol. 1. Principles of Restoration.* Cambridge University Press, Cambridge. pp. 184–205.

Søndergaard, M., J.P. Jensen and E. Jeppesen. 2003. Role of sediment and internal loading of phosphorus in shallow lakes. *Hydrobiologia* 506–509: 135–145.

Spittal, L. and R. Burton. 1991. Irondequoit Bay. Phase II Clean Lakes Project Final Report. Unpublished Report, Monroe County Health Dept., Rochester, NY.

Stauffer, R.E. 1981. *Sampling Strategies for Estimating the Magnitude and Importance of Internal Phosphorus Supplies in Lakes*. USEPA 6001/3-81-015.

Stumm, W. and G.F. Lee. 1960. The chemistry of aqueous iron. *Schweiz Z. Hydrol.* 22: 295–319.

Stumm, W. and J.J. Morgan. 1970. *Aquatic Chemistry. An Introduction Emphasizing Chemical Equilibria in Natural Waters*. John Wiley, New York, pp. xv and 583.

U.S. Environmental Protection Agency (USEPA). 1988. *Ambient Water Quality Criteria for Aluminum — 1988*. USEPA 440/5-88-008. USEPA, Washington, DC.

USEPA. 1995. *Phosphorus Inactivation and Wetland Manipulation Improve Kezar Lake, NH*. USEPA 841-F-95-002.

Van Hullebusch, R., F. Auvray, V. Deluchat, P.M. Chazal and M. Baudu. 2003. Phosphorus fractionation and short-term mobility in the surface sediment of a polymictic shallow lake treated with a low dose of alum (Courtille Lake, France). *Water Air Soil Pollut.* 146: 75–91.

Varjo, E., A. Liikanen, V-P. Salonen and P.J. Martikainen. 2003. A new gypsum-based technique to reduce methane and phosphorus release from sediments of eutrophied lakes (Gypsum treatment to reduce internal loading). *Water Res.* 37: 1–10.

Walker, W.W. Jr., C.E. Westerberg, D.J. Schuler and J.A. Bode. 1989. Design and evaluation of eutrophication control measures for the St. Paul water supply. *Lake and Reservoir Manage.* 5: 71–83.

Welch, E.B. 1996. Control of phosphorus by harvesting and alum. Water Res. Series Tech. Rept.152. Univ. of Washington, Dept. of Civil and Environ. Eng.

Welch, E.B. and G.D. Cooke. 1995. Internal phosphorus loading in shallow lakes: Importance and control. *Lake and Reservoir Manage.* 11: 273–81.

Welch, E.B. and G.D. Cooke. 1999. Effectiveness and longevity of phosphorus inactivation with alum. *Lake and Reservoir Manage.* 15: 5–27.

Welch, E.B. and J.M. Jacoby 2001. On determining the principle source of phosphorus causing summer algal blooms in western Washington lakes. *Lake and Reservoir Manage.* 17: 55–65.

Welch, E.B. and T.S. Kelly. 1990. Internal phosphorus loading and macrophytes: An alternative hypothesis. *Lake and Reservoir Manage.* 6: 43–48.

Welch, E.B., J.P. Michaud and M.A. Perkins. 1982. Alum control of internal loading in a shallow lake. *Water Res. Bull.* 18: 929–936.

Welch, E.B., C.L. DeGasperi and D.E. Spyridakis. 1986. Effectiveness of alum in a weedy, shallow lake. *Water Res. Bull.* 22: 921–926.

Welch, E.B., C.L. DeGasperi, D.E. Spyridakis and T.J. Belnick. 1988. Internal phosphorus loading and alum effectiveness in shallow lakes. *Lake and Reservoir Manage.* 4: 27–33.

Welch, E.B., E.B. Kvam and R.F. Chase. 1994. The independence of macrophyte harvesting and lake phosphorus. *Verh. Int. Verein. Limnol.* 25: 2301–2314.

Wetzel, R.G. 1990. Land-water interfaces: Metabolic and limnological regulators. *Verh. Int. Verein. Limnol.* 24: 6–24.

Willenbring, P.R., M.S. Miller and W.D. Weidenbacher. 1984. Reducing sediment phosphorus release rates in Long Lake through the use of calcium nitrate. In: *Lake and Reservoir Management*. USEPA 440/5-84-002. pp. 118–121.

Witters, H.E., S. Van Puymbroeck, A.J.H.X. Stouthart and S.E.W. Bonga. 1996. Physicochemical changes of aluminium in mixing zones: Mortality and physiological disturbances in brown trout (*Salmo trutta* L.). *Environ. Toxicol. Chem.* 15: 986–996.

Wolter, K-D. 1994. Phosphorus precipitation. In: Restoration of Lake Ecosystems; A Holistic Approach. Publ. 32, Intern. Waterfowl and Wetlands Res. Bureau, Slimbridge, Gloucester, UK.

Wu, R. and C.E. Boyd. 1990. Evaluation of calcium sulfate for use in aquaculture ponds. *Prog. Fish-Cult.* 52: 26–31.

Yee, K.A., E.E. Prepas, P.A. Chambers, J.M. Culp and G. Scrimgeour. 2000. Impact of $Ca(OH)_2$ treatment on macroinvertebrate communities in eutrophic hardwater lakes in the Boreal Plain region of Alberta: *in situ* and laboratory experiments. *Can. J. Fish. Aquatic Sci.* 57: 125–136.

Young, S.N., W.T. Clough, A.J. Thomas and R. Siddall. 1988. Changes in plant community at Foxcote Reservoir following use of ferric sulphate to control nutrient levels. *J. Inst. Water Environ. Manage.* 2: 5–12.

Zhang, Y., A. Ghadouani, E.E. Prepas, B. Pinel-Alloul, S. Reedyk, P.A. Chambers, R.D. Robarts, G. Méthot, A. Raik and M. Holst. 2001. Response of plankton communities to whole-lake Ca(OH)$_2$ and CaCO$_3$ additions in eutrophic hardwater lakes. *Freshwater Biol.* 46: 1105–1119.

9 Biomanipulation

9.1 INTRODUCTION

Intensive research over many decades has developed an understanding of factors regulating distribution, abundance, productivity, and species composition of phytoplankton, especially in deep lakes (reviews by Pick and Lean, 1987; Hecky and Kilham, 1988; Kilham and Hecky, 1988; Seip, 1994, among many). The common approach for long-term control of nuisance algae is to lower nutrient concentrations, an approach supported by controlled experimental laboratory, field enclosure, and whole lake investigations demonstrating that phosphorus (P) and sometimes nitrogen (N) concentrations are causally linked, especially on a long-term basis, to algal production (e.g., Schindler, 1977; Smith and Bennett, 1999). The link between P concentration and algal biomass is frequently illustrated with a log-log TP-chlorophyll regression indicating that most long-term changes in algal biomass are explained by changes in P concentration (Figure 9.1). However, when these data are plotted on a linear scale (Figure 9.2), especially on a short-term basis, variances are apparent, suggesting other factors in addition to nutrients can be important in determining algal biomass. For example, Schindler (1978) found a highly significant correlation ($r = 0.69$) between pelagic productivity and steady state lake P concentrations for 66 P-limited lakes (shallow and deep), ranging in latitude from 38° S to 75° N. The relationship, however, explained only about half of the variance, meaning that P concentration and chlorophyll are highly correlated, but that the relationship might be weak or non-existent for some lakes in some years or parts of years. Grazing, mixing, and/or allelopathic materials, might influence and/or control algal biomass in these lakes.

An example is Square Lake, Minnesota (Osgood, 1984), a lake with a greater Secchi Disc (SD) transparency than expected from its TP concentration of about 20 μg P/L (mesotrophic). SD was more than 7 m during summer months, a depth typical of oligotrophic lakes, and apparently due to *Daphnia* grazing.

One purpose of this chapter is to examine factors other than resources (e.g., nutrients, light) that can control algal biomass in deep lakes, and to discuss how lake managers might use this knowledge to address phytoplankton problems. Since most lakes are shallow and may be dominated by either phytoplankton or macrophytes, this chapter also examines factors determining which producer type dominates, and how this knowledge can be used to manage shallow lakes.

9.2 TROPHIC CASCADE

As illustrated by the Square Lake, Minnesota case history, zooplankton grazing is a source of algal mortality, sometimes leading to lower algal biomass in the water column than expected for a given nutrient level. This might occur when planktivores are suppressed by piscivores. But in some lakes, zooplankton herbivory may be low during some periods of the summer, perhaps due to fish or insect planktivory, allowing phytoplankton to bloom.

A set of hypotheses was developed to explain the roles of resources (nutrients, light) and trophic level interactions. The original organization of these ideas was directed at terrestrial communities (Hairston et al., 1960), and was later (Smith, 1969) proposed for lakes (see Hairston and Hairston, 1993). Hairston et al. (1960) predicted that in systems with three dominant trophic levels (producers, herbivores, primary carnivores), producers would be resource-controlled, whereas in four trophic

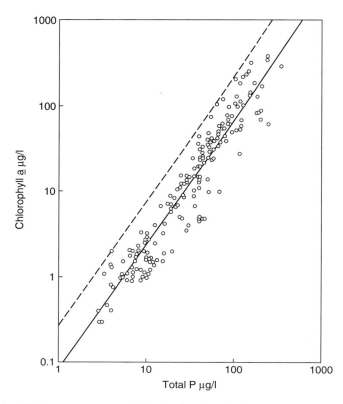

FIGURE 9.1 Relationship between summer chlorophyll and total phosphorus in a number of lakes. (From Shapiro, J. 1979. *U.S. Environmental Protection Agency National Conference on Lake Restoration.* USEPA 440/5-79-001.)

level systems (top predator also), producers would be consumer (herbivore)-controlled and the first carnivore level would be controlled by predation.

Observations of ponds by Hrbacek et al. (1961) supported the above ideas. Fish planktivory (no piscivory) reduced zooplankton grazing, leading to algal blooms. Further early evidence, from Brooks and Dodson (1965), found that planktivory by the alewife (*Alosa pseudoharengus*) in New England lakes led to elimination of the most efficient herbivores (large-bodied *Daphnia*) and to selection for smaller-sized zooplankton (e.g., *Bosmina*) that have lower grazing rates and choose smaller food (algae) particles. They proposed the "size–efficiency" hypothesis to explain grazing impacts of smaller and larger-bodied zooplankton on phytoplankton.

The term "trophic cascade" was introduced by Paine (1980) to describe the roles of species that he termed "strong links" or "strong interactors" in intertidal communities. These are species whose removal (or introduction) produced dramatic changes in prey biomass. If the prey was a competitively superior species, the effects could "cascade" from predator to trophic levels one or two links away. Pace et al. (1999) defined "trophic cascades" as (p. 483): "reciprocal predator–prey effects that alter the abundance, biomass or productivity of a population, community or trophic level across more than one link of the food web." The pelagic trophic cascade, with and without a dominant ("strong link") top carnivore level, is illustrated in Figure 9.3.

Carpenter et al. (1985) argued that trophic cascades could explain the large variances (Figure 9.2) in algal biomass or productivity between lakes with similar nutrient concentrations. They proposed that nutrient levels determined the long-term productivity or trophic state of a lake, but the year-to-year variances from expected trophic state were set by trophic-level interactions. Strong

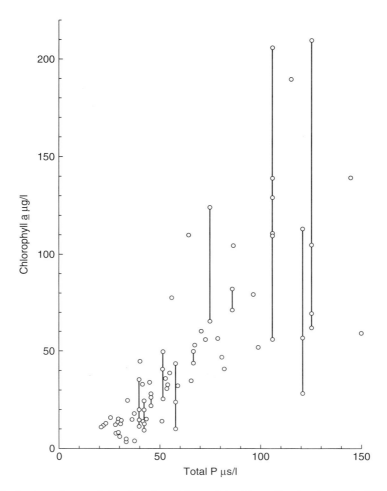

FIGURE 9.2 Replotting of part of the data from Figure 9.1. (From Shapiro, J. 1979. *U.S. Environmental Protection Agency National Conference on Lake Restoration.* USEPA 440/5-79-001, pp. 161–167.)

piscivory suppressed planktivores, allowing zooplankton grazing to reduce algal biomass, a trophic cascade extending from piscivores to phytoplankton.

DeMelo et al. (1992) suggested that evidence for trophic cascades in lakes is weak, except for strongly manipulated lakes, a conclusion contradicted by some experiments, but supported by others. Jeppesen et al. (2000) described trophic cascades in shallow Danish lakes. Brett and Goldman (1996) noted that there have been few whole-lake studies, but in 54 pond and enclosure experiments there was evidence of trophic cascades. Trophic cascades continued over a multi- year period in fertilized, dimictic lakes (Carpenter et al., 2001). In contrast, Drenner and Hambright (2002) found that 10 of 17 experiments, not confounded by other manipulations, failed to support the trophic cascade hypothesis. However, in most lakes dominated by planktivores (three trophic levels) the slope of the chlorophyll: TP regression (Figure 9.1) was three times that of four level (piscivore-dominated) lakes, indicating that piscivore control of planktivory may lead to enhanced zooplankton grazing and lower algal biomass than expected for a given nutrient level. There also may be a "behavioral cascade." Planktivorous fish seek refuge in macrophyte beds in the presence of piscivores caged in open water, allowing longer open water feeding by *Daphnia* than when piscivores are absent (Romare and Hansson, 2003).

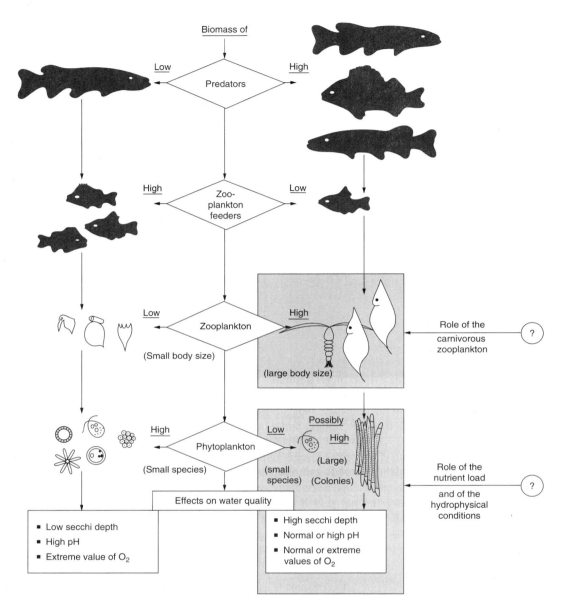

FIGURE 9.3 Hypothetical scheme showing the connections involved in food-chain biomanipulation in lakes. Shaded area represents tentative connections. (From Benndorf, J. et al. 1984. *Int. Rev. ges. Hydrobiol.* 69: 407–428. With permission.)

The trophic cascade hypothesis of Carpenter et al. (1985) is one of the most significant in modern limnology. It stimulated new research and led to a paradigm about control of lake productivity that included both biotic interactions and the role of resources. Readers are urged to examine the many books and review articles describing this concept (e.g., Kerfoot and Sih, 1987; Carpenter, 1988; Gulati et al., 1990; Elser and Goldman, 1991; Carpenter and Kitchell, 1992, 1993; Hansson, 1992; McQueen, 1998).

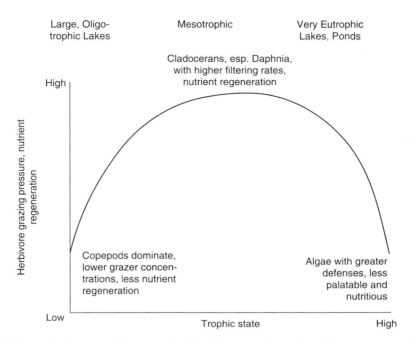

FIGURE 9.4 Schematic diagram of the strength of herbivore grazing and nutrient regeneraton in relation to trophic state. (From Carney, H.J. 1990. *Verh. Int. Verein. Limnol.* 24: 487–492. With permission.)

9.3 BASIC TROPHIC CASCADE RESEARCH

Research on the trophic cascade hypothesis (Carpenter et al., 1985) generated a greater understanding of lakes, and many new questions about lake ecology. Trophic cascades may be more common in mesotrophic lakes than in oligotrophic or hypereutrophic lakes, leading to an "intermediate trophic state" hypothesis (Carney, 1990; Figure 9.4). In mesotrophic lakes, edible, nutritious algae dominate the plankton, and large-bodied zooplankton are abundant. In eutrophic lakes, especially hypereutrophic lakes, the phytoplankton community may be dominated by cyanobacteria that may be toxic, inedible, non-nutritious, and/or produce clogging of the filter-feeding mouthparts of zooplankton (Gliwicz, 1990; Lampert, 1982). These factors may explain the rarity of large-bodied zooplankton in eutrophic lakes, rather than planktivory (deBernardi and Giussani, 1990; Gliwicz, 1990; DeMott et al., 2001).

The "intermediate trophic state" hypothesis was examined by comparing grazing in ultra-oligotrophic Lake Tahoe (California/Nevada), mesotrophic Castle Lake, California, and eutrophic Clear Lake, California, using enclosures with ambient or enhanced ambient zooplankton, and enclosures with added *Daphnia pulex* (Elser and Goldman, 1991). In the eutrophic lake, ambient zooplankton (even at eight times in-lake density) had no impact on phytoplankton biomass. *D. pulex* had a weak effect. It was believed that *Anabaena circinalis*, (42% of the phytoplankton) was either inedible or interfered with filter-feeding. In the mesotrophic lake there were large declines in phytoplankton but increases in primary productivity, suggesting grazing and nutrient recycling by zooplankton. In oligotrophic Lake Tahoe enclosures, copepods dominated with no grazing impact. The trophic cascade was strongest in the mesotrophic lake.

McQueen et al. (1986, 1989, 1992) proposed a "top-down bottom-up" model to explain trophic level interactions, based on enclosure experiments and a multi-year study of eutrophic Lake St. George, Ontario. Regressions between TP, chlorophyll, and fish biomass (piscivore and planktivore) indicated that bottom-up forces (nutrients) were strongest at trophic levels nearest resources, while top-down forces were strongest near the piscivore level. They predicted that bottom-up control becomes increasingly important relative to top-down control in enriched lakes. Therefore, piscivores are less likely to have a cascading effect on algal biomass in the most productive lakes.

A 1982 fish winterkill in Lake George eliminated 72% of largemouth bass, allowing a test of the model, using a 6-year database. In 1983–1984, planktivore biomass increased rapidly, and then began to decrease in 1985 as the bass population recovered. As predicted, a strong top-down cascade from piscivores through planktivores to zooplankton was observed, though negative correlations between planktivore biomass and either total zooplankton or *Daphnia* biomass were not significant. Correlations of zooplankton or *Daphnia* biomass with chlorophyll and transparency also were not significant, but correlations between log chlorophyll and log TP were positive and significant.

McQueen et al. (1989) argued that long-term processes determining trophic level biomass depend on resources and energy flow ("bottom-up"). In accordance with Carpenter et al. (1985), McQueen et al. (1989) concluded that short-term disturbances or cascades set "realized biomass" limits. Lakes are strongly influenced by year to year changes in precipitation and associated water and nutrient loading and by climate (mixing events, fish winterkill), and these stochastic events in turn affect both top and bottom trophic levels, including effects on fish biomass, reproduction, and mortality (Carpenter et al., 1985). These events may not be instantaneous effects of stochastic changes, but instead there will be lags or inertias that produce responses at other times. "Algal production today may depend on yesterday's zooplankton, which depended on zooplanktivores during the past month, which depended on piscivore recruitment the previous year" (Carpenter et al., 1985, p. 637).

The controls of algal biomass in shallow and deep lakes and at various levels of enrichment remain controversial. Empirical data from a large sample ($n = 446$) of shallow and deep lakes, from arctic to temperate zones and ranging from oligotrophic to hypereutrophic, did not support the hypothesis of McQueen et al. (1989) that the cascading effect might be greatest in oligotrophic lakes. Instead, the survey gave partial support to the "intermediate state hypothesis" (Elser and Goldman, 1991), and indicated that at high TP, the most probable condition (even with removal of planktivores) is high algal biomass and turbidity (Jeppesen et al., 2003a).

Trophic cascades are an important determinant of biomass at planktivore, herbivore, and producer trophic levels, and explain some of the variance observed in regressions between resources and biomass. Exciting controversies remain about forces that organize ecosystems, and about the adaptations that species populations make to counter those forces. But lake managers want to know whether manipulations of trophic levels can produce clearer lakes, and if so, how long will the effect last?

9.4 BIOMANIPULATION

Caird (1945) was the first to publish observations about phytoplankton responses to increased piscivorous fish biomass. Caird suspected that largemouth bass addition to a 15 ha lake in Connecticut was associated, through food chain effects, with 4 years of reduced phytoplankton blooms, resulting in a termination of copper sulfate applications

Shapiro et al. (1975) proposed the term "biomanipulation", which he defined (Shapiro, 1990) as "a series of manipulations of the biota of lakes and of their habitats to facilitate certain interactions and results that we as lake users consider beneficial — namely reduction of algal biomass and, in particular, of blue-greens" (p. 13). Shapiro et al. (1975) included effects on algal biomass from "top-down" control of zooplanktivores by piscivores, and "bottom-up" effects on algae such as nutrient cycling by benthivorous fish. Many lake managers apply the term only to top-down control of planktivorous fish (see Drenner and Hambright, 2002). More recently, the term has referred to nearly all ecological manipulations to manage algae and aquatic plants.

There have been many review articles and books about biomanipulation, including Shapiro (1979), Gulati et al. (1990), DeMelo et al. (1992), Carpenter and Kitchell (1992), Moss et al.

(1996a), Hosper (1997), Jeppesen (1998), McQueen (1998), Bergman et al. (1999), and Drenner and Hambright (2002). Lazzaro (1997) is a particularly useful comparative summary.

The following examines biomanipulation case histories, particularly their effectiveness and longevity. Several investigators (e.g., Jeppesen et al., 1997) indicated that resource control of algal biomass is weaker in shallow lakes, suggesting biomanipulation may be more successful in them. Therefore, the examination of shallow and deep lake case histories is separated.

9.5 SHALLOW LAKES

Characteristics of deep and shallow lakes were described in Chapter 2. Briefly, shallow lakes have a mean depth less than 3 m, are usually polymictic, and often have significant nutrient recycling affecting the entire water column. Compared with deep lakes, fish biomass per volume is higher, the impacts of fish on turbidity and sediment nutrient release are greater, and the area colonized by macrophytes may be close to 100% (Cooke et al., 2001). These and other characteristics are summarized in Moss et al. (1996a) and Scheffer (1998).

Shallow lakes are more common than deep ones. Unlike deep lakes, shallow lakes, at moderate (30–100 µg P/L) nutrient levels, appear to exist in alternative states: Either they are clear with rooted plant dominance, or turbid with algae dominance (Scheffer, 1998). At low nutrient concentrations the clear-water vegetated state is most likely, whereas at higher (perhaps > 100 µg P/L) concentrations, the turbid state is more likely (Hosper, 1997). At concentrations between the extremes, either the clear or turbid state can occur. It is the forces determining lake state that are subject to manipulation, leading to the possibility of "switching" from one state to the other. Good examples of clear and turbid shallow lakes existing within a limited geographic area (Alberta) were presented by Jackson (2003).

Macrophyte-dominated, clear-water lakes are resistant to development of algal dominance from increased external nutrient loading because plants reduce wind and boat-generated resuspension of sediments, provide daytime refuge to algae-grazing *Daphnia*, their periphyton may take up significant amounts of nutrients, and some macrophytes release compounds inhibitory to algae. Piscivores may thrive in macrophyte-dominated lakes, controlling fish that prey on zooplankton and on periphyton-consuming snails (*Bronm*ark and Weisner, 1996). This last effect of fish is important because, as discussed later, abundant periphyton may reduce rooted plant growth.

Resistance of the clear water macrophyte-dominated lake to change is reduced by some plant management activities (e.g., harvesting, grass carp; Chapters 14 and 17), by increased fish production (young of the year (YOY) of most fish species are zooplanktivorous), and by introduction of toxins (e.g., copper sulfate, herbicides, insecticides) lethal to *Daphnia* and to plants. There is a nutrient-based stability threshold for the clear water, macrophyte-dominated state of about 50–100 µg P/L. Continued loss of stability as nutrient loading increases and/or plant removal occurs can produce an abrupt switch to the alternative, turbid, macrophyte-free state. Moss et al. (1996a) called these changes leading to the turbid state "forward switches." Figure 9.5 illustrates this model of alternative states and the forces that promote a switch from one state to the other. Note that either the clear or turbid water state can occur without change in overall nutrient concentrations.

Switching a turbid lake to a clear water lake may not occur, even when external loading is significantly reduced. This means that the common advice given to shallow lake owners to reduce external loading as a method to clear up the water may not produce expected results. Sediment resuspension by wind, boat, and fish activity will attenuate light, preventing re-establishment of macrophytes. Extensive internal nutrient recycling may continue to sustain phytoplankton and reduce transparency, preventing re-establishment of clear water and macrophytes. Fish removals, followed by piscivore stocking and enclosures to protect plants from birds, are among the biomanipulation procedures that trigger the switch to a clear water state. Reduction of nutrient concentrations through diversion (Chapter 4) or P inactivation (Chapter 8) increases the probability of the

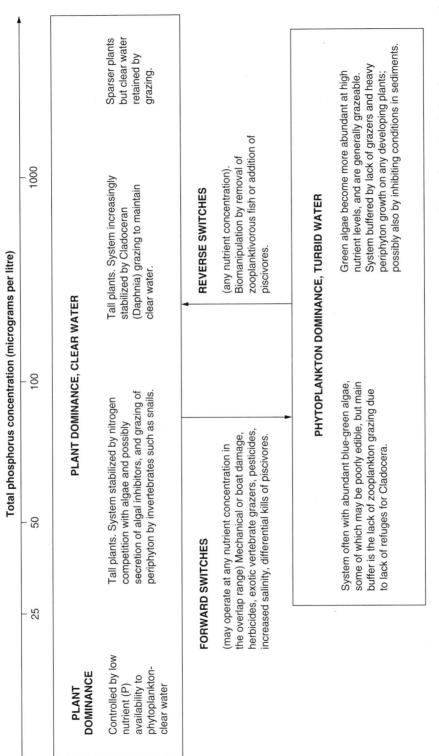

FIGURE 9.5 The alternative stable states model for dominance by aquatic plants or phytoplankton in shallow lakes, over the gradient of total phosphorus concentrations that includes both pristine values and those encountered in polluted conditions. (From Moss, B. et al. 1996. *A Guide to the Restoration of Nutrient-Enriched Lakes.* Broads Authority, Norwich, Norfolk, UK. With permission.)

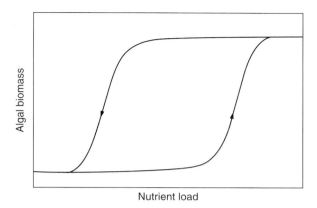

FIGURE 9.6 Eutrophication and oligotrophication in relation to algal biomass, showing a typical hysteresis curve. (From Hosper, H. 1997. *Clearing Lakes. An Ecosystem Approach to the Restoration and Management of Shallow Lakes in The Netherlands.* RIZA, Lelystad, The Netherlands. With permission.)

switch. Figure 9.6 illustrates the pattern of resistance of clear and turbid shallow lakes to increasing or decreasing external nutrient loading.

Biomanipulation, especially top-down procedures, is more likely to be successful in shallow lakes, and in turn, shallow lakes are easier to biomanipulate because nearly all fish can be removed. The following case histories illustrate the outcomes of some efforts to restore turbid lakes to the clear, macrophyte-dominated state. Most examples are European because they prefer rooted plant-dominated, clear water conditions in their shallow lakes. In North America, with a high density of shallow lakes, ponds, and reservoirs, lake users appear to want an algae-free, non-turbid, macrophyte-free lake, regardless of factors preventing this condition from being stable or even possible (e.g., high internal nutrient recycling, high external loading, and/or stocking of exotic herbivorous fish). This unrealistic goal is possible only with continual reliance on expensive mechanical and/or chemical controls.

More realistic expectations about the trophic state of shallow lakes might be of value to North American lake users. For example, a common tactic is to attempt to manage a macrophyte-dominated lake toward an intermediate biomass of plants that may satisfy the lake users who want a macrophyte-free lake. This condition is unlikely where external and internal nutrient loads are high or increasing, and where plant removal (e.g., harvesting, stocking of grass carp) is extensive. These conditions may drive the lake to the turbid state. It could be more realistic, in some cases, to manage some shallow lakes in an area toward macrophyte-free water that may be compatible with boating and swimming, while managing other nearby lakes toward the clear water condition (Van Nes et al., 1999, 2002). For clear water lakes with high resilience (low probability of switching to the turbid state because nutrient loading and benthivorous fish biomass are low), some macrophyte removal could occur in high use areas. Management goals should be consistent with reality (Welch, 1992a).

The following case histories were chosen from situations where the lake was intentionally manipulated rather than from instances of unplanned and drastic biological changes such as a winter fish kill or a drought.

9.6 BIOMANIPULATION: SHALLOW LAKES

9.6.1 Cockshoot Broad (UK)

Cockshoot Broad (3.3 ha, mean depth 1.0 m) is one of several small, riverine, shallow lakes in eastern England. Originally they were macrophyte-dominated, but recently dense phytoplankton replaced macrophytes (Moss et al., 1996b). Aquatic plants in the UK are considered an asset, providing high biodiversity, and efforts were made to rehabilitate some of these lakes, including Cockshoot Broad.

Consistent with conventional wisdom of the time (early 1980s), P removal at municipal waste-water facilities was believed sufficient to rehabilitate Cockshoot Broad. However, high internal P recycling was a major P source, and the Broad was therefore isolated from the adjacent nutrient-rich river, and about one meter of P-rich sediment removed. TP fell, and aquatic plants returned. However, by 1984–1985, submersed plants declined and the phytoplankton-dominated state returned because *Daphnia pulex,* abundant following isolation and dredging, declined to small numbers by 1984 as planktivorous fish populations recovered (Moss et al., 1986b).

Biomanipulation via nearly complete fish removal occurred in winter 1989 and 1990. Fish removal in winter is easier because fish tend to aggregate at this time, making electro-fishing and seining easier. Maintenance fish removal continued in subsequent winters. *Daphnia* returned, chlorophyll concentration declined, and submersed macrophytes recolonized the broad. High nutrient concentrations occurred in the macrophyte and phytoplankton-dominated conditions, indicating that these alternative states were influenced by biological interactions. When *Daphnia* were absent, chlorophyll concentrations were highly correlated with TP. There was no correlation in years of high *Daphnia*–low planktivore densities.

Grazer control of phytoplankton appeared to be linked to macrophytes that served as physical refuges from planktivory (Timms and Moss, 1984; Moss et al., 1986b, 1994). Unlike deep lakes, where vertical migration can provide a daytime refuge for zooplankton from fish predation (Gliwicz, 1986), shallow lake zooplankton may employ diel horizontal migration (DHM) to and from the littoral zone to provide daytime refuge from planktivory. While *Daphnia* appear to be chemically repelled by some macrophytes (e.g., *Myriophyllum exalbescens*), they use other macrophytes to avoid fish (Lauridsen and Lodge, 1996; Burks et al., 2001). However, if littoral zones are dominated by planktivores (including YOY piscivores), *Daphnia* mortality may be high (Perrow et al., 1999). These authors suggested that if macrophytes comprise 30–40% of lake volume that is sufficient refuge from fish for zooplankton to maintain clear water. The clear-water stabilizing effect of DHM appears to be high when macrophytes are abundant and littoral-associated piscivores control planktivory (Burks et al., 2002). The clear water state may be possible, even at elevated nutrient concentrations, when *Daphnia* grazing is extensive. Additional studies on DHM are needed.

An assessment of zooplankton grazing in increasing transparency may be difficult to determine using traditional sampling methods. Only nighttime sampling reveals the actual density of zooplankton in macrophyte-dominated shallow lakes. Daytime open water sampling fails to capture zooplankton in refuges (Meijer et al., 1999). Artificial refugia for large-bodied zooplankton in the English broads, including bundles of brush, strands of polypropylene rope, and mesh cages did not enhance zooplankton survival (Moss, 1990; Irvine et al., 1990).

9.6.2 LAKE ZWEMLUST (AND OTHER DUTCH LAKES)

The Lake Zwemlust (1.5 ha, mean depth 1.5 m) case history is instructive because of its long-term data set, and because of problems in maintaining the clear water state after biomanipulation. In 1968, a broad-spectrum herbicide (diuron) was applied, eliminating macrophytes. There was a rapid shift to the turbid, algae-dominated state. A minimum transparency of 1.0 m in swimming lakes is required in The Netherlands, but blooms of *Microcystis aeruginosa* reduced transparency below this criterion. In winter, 1987, Lake Zwemlust was seined, electro-fished, and drained to eliminate planktivorous and benthivorous fish. It was then stocked with pike (*Esox lucius*) and rudd (*Scardinius erythrophthalmus*), willow twigs were added as shelter for pike fingerlings, yellow water lily (*Nluphar lutea*) and *Chara* were planted, and *Daphnia magna* and *D. hyalina* (1 kg wet weight) were introduced (Gulati, 1990; van Donk et al., 1990).

Though external nutrient loading remained high (2.4 g P/m^2 per year; van Donk et al., 1993), the water became clear. In 1988–1989 *Elodea nuttalli* dominated, and phytoplankton became N-

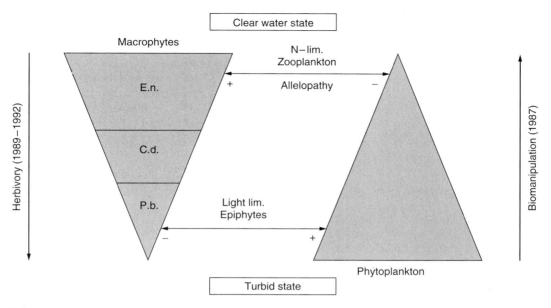

FIGURE 9.7 Schematic presentation of mechanisms buffering the stability (horizontal arrows) and inducing the transition (vertical arrows) of the clear water state and turbid state in Lake Zwemlust. (From van Donk, E. and R.D. Gulati. 1995. *Water Sci. Technol.* 32: 197–206. With permission.)

limited. Summer chlorophyll concentrations were low in 1987, the first summer of the treatment, partly from intense *Daphnia* grazing. By 1991, small-bodied *Daphnia* became dominant after planktivory resumed, and grazing on algae declined. *Ceratophyllum demersum* was the dominant macrophyte in 1990–1991, but was nearly absent in 1992–1994. Late summer algal blooms resumed in 1992–1994. *Potamogeton berchtholdii* appeared in the spring of 1992–1994, but became covered with epiphytes and was eliminated (van Donk and Gulati, 1995). Another planktivore removal took place in 1999, followed by a clear water period and a slow return of macrophytes.

What caused the shift back to the turbid water condition? Herbivorous birds (coot, *Fulica atra*) invaded the lake in 1989, removing *Elodea* at rates up to 7 kg dry weight/day during late autumn. Rudd are also herbivorous, and the two grazers (fish and birds) shifted plant dominance to coontail (*Ceratophyllum demersum*) and *Potamogeton*. Coot then grazed on *Ceratophyllum* in 1991–1992. Because *Daphnia* grazing was effective only in 1987, it may have been macrophytes that maintained the clear water state from 1988–1991 (perhaps through alleleopathy, by inducing N-limitation of phytoplankton, and/or by preventing sediment resuspension). After macrophyte elimination by fish, birds, and epiphytes, algal blooms returned to Lake Zwemlust (van Donk et al., 1994; van Donk and Gulati, 1995; Figure 9.7). Ultimately high nutrient loading prevented establishment of a permanent clear water state. Only the turbid state appeared to be stable (van de Bund and van Donk, 2002).

One factor causing loss of macrophytes is light limitation due to epiphyte coverage (Sand-Jensen and Søndergaard, 1981). This apparently occurred in Lake Zwemlust (van de Bund and van Donk, 2002). Bronmark and Weisner (1992) proposed that biomanipulation of the littoral zone food web to eliminate molluscivores (often Centrarchidae or sunfish), thereby maintaining biomass of epiphyton-grazing snails, could play a role in stabilizing the clear water state. This hypothesis has gained support. In nutrient-enriched aquaria, *C. demersum* exhibited extensive production of new biomass at high snail (*Physa, Helisoma*) densities (coontail obtains its nutrients directly from the water). High nutrient conditions with low or no snails permitted heavy epiphyton growth and apparent light limitation of *C. demersum* (Lombardo, 2001). In shallow U.K. lakes, macrophyte biomass was negatively correlated with periphyton biomass, which, in turn, was negatively correlated with invertebrate biomass ($r^2 = 0.714$). Invertebrate density on plant surfaces was lower in lakes with high fish biomass (Jones and Sayer, 2003). Thus the trophic cascade in shallow lakes that involves fish-invertebrates-periphyton may be as significant as the fish-zooplankton-phytoplankton cascade in determining whether macrophytes or phytoplankton are the dominant producers. The Bronmark–Weisner hypothesis should receive more study.

The Lake Zwemlust case history illustrates complexity in lakes, and also the most significant issue with biomanipulation — how to stabilize the new and presumably more desirable state (Shapiro, 1990). Resilience describes how fast a system returns to equilibrium after disturbance (e.g., piscivore addition or planktivore removal). The three-level system (planktivore, herbivore, producer) apparently is very resilient, possibly due to high planktivore reproductive rate, slow growth of piscivores when they are added (Carpenter et al., 1992), and to the effects of high nutrient concentrations (Jeppesen et al., 2003a). Large and continuous efforts may be needed to maintain the four level, non-equilibrium system, especially when the lake is nutrient-enriched (as was the case with Lake Zwemlust).

There have been many biomanipulations in The Netherlands. The most successful in maintaining the clear water state were those with extensive fish removal (at least 75%) and lower nutrient levels (Meijer et al., 1999). Biomanipulation success could be enhanced by excluding herbivorous birds from selected lake areas with nets over plant beds, thus allowing plant establishment (Moss, 1990).

Alternative states of clear or turbid water exist in shallow lakes, and at the extremes of nutrient concentrations, apparently are stable. Maintenance of biomanipulated states between the extremes requires continual attention in the face of continuing ecological succession.

9.6.3 LAKE VAENG (AND OTHER DANISH LAKES)

Extensive biomanipulation experiments have been conducted in Denmark. The data from hundreds of lakes, plus enclosure and whole lake manipulations, indicate that this procedure is more likely to be effective in shallow lakes. High effectiveness is associated with a large increase in macrophyte coverage. Macrophytes in turn stabilize sediments, sequester nutrients, provide a zooplankton refuge, and release allelochemics. Also, there must be significant removal of planktivorous and benthivorous fish. Otherwise, the lake may revert to the turbid state soon after biomanipulation. Reduction of TP to concentrations below the 50–100 µg P/L range is also important (Jeppesen et al., 1997, 2000).

An example is Lake Vaeng (15 ha, mean depth = 1.2 m), where high internal P recycling, and benthivorous and planktivorous fish, prevented recovery after a large fraction (65%) of wastewater was diverted. Five years later, 50% of the fish were removed. Large-bodied Cladocera replaced rotifers, phytoplankton biomass declined sharply, diatoms replaced Cyanophyta, and macrophytes recolonized the lake. Internal P recycling was reduced, perhaps from fish removal, improved redox conditions, and/or lower pH. Predatory fish control of planktivores and the associated high transparency were stable for at least 8 years after fish removal (Søndergaard et al., 1990; Lauridsen et al., 1994; Jeppesen, 1998).

9.6.4 LAKE CHRISTINA, MINNESOTA

Shallow lakes in midwestern North America are important feeding and staging areas for migratory waterfowl, and provide hunting, fishing, and aesthetic attractions. Their quality as waterfowl and game fish habitats declines dramatically when benthivorous/planktivorous fish become dominant, leading to algal blooms, loss of macrophytes, and high turbidity. The goal of the Lake Christina project was restoration of its migratory bird habitat (Hanson and Butler, 1994a, b; Hansel-Welch et al., 2003).

Lake Christina is a large (1600 ha), shallow (mean depth = 1.5 m) lake in west-central Minnesota. Very high chlorophyll and turbidity combined to light-limit submersed plants. By 1977–1980, plants and waterfowl were largely absent. The fish community, dominated by bullhead (*Ictalurus nebulosus*), bigmouth buffalo (*Ictiobus cyprinellus*), yellow perch (*Perca flavescens*) and northern pike (*Esox lucius*), was eliminated with rotenone (3.0 mg/L) in autumn, 1987. Largemouth

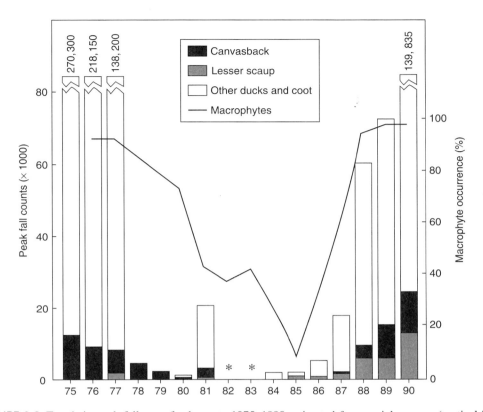

FIGURE 9.8 Trends in peak fall waterfowl counts 1975–1990 estimated from aerial surveys (vertical bars) and occurrence of any submerged macrophytes (solid line). Asterisks indicate that only canvasback peak density was recorded. (From Hanson, M.A. and M.G. Butler. 1994. *Can. J. Fish. Aquatic Sci.* 51: 1180–1188. With permission.)

bass (*Micropterus salmoides*) and walleye (*Stizostedion vitreum vitreum*) were then stocked to suppress re-invading rough fish. The lake was monitored from 1985–1998, making it one of the most well documented restoration projects.

The treatment was successful and long-lasting (at least through 1998), and switched Lake Christina to a stable, clear-water, macrophyte-dominated habitat. Small Cladocera (*Bosmina, Chydorus*) were replaced with large *Daphnia* (*D. pulex, D. galeata*) in 1988, leading to clear water phases from grazing and a return of submersed plants. A "pioneer" plant community (1988–1992) of *Najas* spp., *Myriophyllum sibiricum,* and *Ruppia maritima* was replaced (1992–1998) with an understory of *Chara* and a canopy of *Potamogeton pusillus* and *P. pectinatus*. Apparently pioneer plants stabilized the sediments, allowing *Chara* and *Potamogeton* to succeed. The appearance of *Chara* may be particularly important because charophytes have a strong effect in clearing water, possibly through the release of allelopathic materials that inhibit phytoplankton (van Donk and van de Bund, 2002). Macroinvertebrates, especially *Hyallela azetca*, became abundant and migratory waterfowl returned (Hanson and Butler, 1994b; Figure 9.8).

Plant biomass varied with changes in light attenuation, not with waterfowl grazing. Waterfowl had negative effects on submersed vegetation only at high bird densities (Marklund et al., 2002). Fish bioturbation was not associated with turbidity prior to the rotenone treatment. Rather, wind resuspension of sediments, before the reestablishment of plants, may have been a major factor in maintaining the turbid state, along with phytoplankton production in the nutrient-rich (mean TP = 76 µg/L) lake (Hanson and Butler, 1994a). After fish removal, light was initially increased by zooplankton grazing, leading to macrophyte establishment, but grazing declined in later years. In some years, mats of filamentous algae, especially *Cladophora*, played a role in light attenuation to macrophytes (Hansel-Welch et al., 2003). Phytoplankton blooms did not recur, despite the high P concentration and reduced grazing, suggesting that factors such as allelopathy (Chapter 17) plus sediment stabilization may have played more important roles.

The alternative stable state theory was supported, and the lasting effects of a powerful shallow lake biomanipulation were demonstrated. The Lake Christina project suggests that re-suspension of soft, organic sediments of a previously turbid, eutrophic, shallow lake will not prevent re-establishment of macrophytes, and that these sediments are not too "loose" for rooted plants. This issue continues to be debated (e.g., Bachmann et al., 1999; Meijer et al., 1999; Schelske and Kenney, 2001). A long-term study of 15 shallow lakes found that recovery after biomanipulation and reduction of nutrient loading was not affected by sediment re-suspension (Jeppesen et al., 2003b). The potential for sediment re-suspension appears to be at least partially dependent on lake morphometry (Bachmann et al., 2000), and on effects of fish disturbance of sediments (Scheffer et al., 2003). The near absence of benthivorous fish in Lake Christina may have been the most important factor in plant re-establishment.

An assessment procedure has been developed to estimate the probability of success with shallow lake biomanipulation (Hosper and Jagtman, 1990; Hosper and Meijer, 1993), and these sources should be reviewed before initiating a project. The following criteria should be met: (1) Removal of planktivore and benthivore biomass should exceed 75% in 2 years or less, (2) Piscivores should be stocked, especially where YOY planktivory is extensive, and (3) Immigration of new fish should be pr*evented* (Hosper and Meijer, 1993; Perrow et al., 1997; Hansson et al., 1998). Lakes exposed to strong wind may remain turbid and plants will not become established, regardless of biomanipulation efforts.

9.7 BIOMANIPULATION: DEEP LAKES

Shallow lakes respond to extensive biomanipulation, primarily because a large fraction of the fish community can be removed, and because macrophytes can develop and contribute to maintaining clear water. Many dimictic lakes have small littoral zones and macrophytes probably have limited effects in maintaining clear water. However, the following case histories indicate that biomanipulation of pelagic communities can produce clearer water through the trophic cascade, although nearly continuous maintenance may be required.

9.7.1 Lake Mendota, Wisconsin

Lake Mendota (area = 40 km^2, maximum depth = 25 m) experienced algal blooms for at least a century, and these became common after 1945 as external nutrient loading increased. Wastewater diversion occurred in 1971. Non-point loading, which is strongly tied to climate and runoff at this eutrophic lake, continued and even increased due to urban development in its large (600 km^2) watershed. Phosphorus concentrations in the lake were strongly related to external loading and to blue-green algal biomass (Lathrop et al., 1998). Examination of Secchi transparency data from 1900–1993 revealed that changes in summer water clarity were linked to algal blooms. The lowest transparency values were during high nutrient, low herbivory years, and greater summer transparency was found during high herbivory years, suggesting that this lake would respond to biomanipulation (Lathrop et al., 1996).

The lake was stocked with 2.7 million fingerling walleye (*Stizostedion vitreum*) and 170,000 fingerling northern pike (*Esox lucius*) between 1987 and 1999. Angler restrictions (increased minimum size limit, reduction in catch limit) were imposed in 1988, and further restrictions added in 1991 for walleye, and in 1996 for pike (Lathrop et al., 2002).

Walleye biomass increased rapidly to a peak in 1998 whereas northern pike increased initially, then declined. Planktivore biomass and predation on *Daphnia* declined sharply, in part from a large-scale natural die-off of cisco (*Coregonus artedi*) in 1987. Low planktivore levels were maintained until 1999 when yellow perch (*Perca flavescens*) increased, even with the high piscivory. Large-bodied *D. pulicaria* dominated during high piscivory years, and high transparency occurred in some of those years. Lake P concentration due to runoff increased to high levels in 1993 and remained high during the biomanipulation. Even with higher P loading (average annual load = 0.85 g P/m ^{2} per year), the effect of reduced planktivory cascaded to the phytoplankton, causing increased transparency. In view of the die-off of cisco in 1987, controlled planktivory could not be ascribed entirely to enhanced piscivory (Lathrop et al., 2002). Decreased P loading from watershed management will continue to improve water clarity.

9.7.2 Bautzen Reservoir And Grafenheim Experimental Lakes (Germany)

Bautzen Reservoir is large (553 ha), moderately deep (mean depth = 7.4 m), polymictic, and hypereutrophic, with blue-green algal blooms (*Microcystis aeruginosa*) and a potentially effective algae grazer (*Daphnia galeata*). The reservoir was created in 1973 and the northern pike population (*Esox lucius*) developed rapidly. Sport fishing began in 1976 and the pike population was decimated in two years, allowing development of small, planktivorous perch (*Perca fluviatilis*). Biomanipulation began in 1977 with the stocking of pike-perch (*Stizostedion lucioperca*) and the imposition of fish catch restrictions. Stocking, at rates of 20,000 to 80,000 pike-perch per year continued from 1980–1982 and 1984–1988. Changes in lake condition during the pre-biomanipulation years (1977–1980) were compared with the biomanipulation period (1980–1988) (Benndorf, 1987, 1988, 1989, 1990; Benndorf and Miersch, 1991; Benndorf et al., 1984, 1988, 1989, 2002).

Planktivores (perch) were controlled but not eliminated by pike-perch. But low level planktivory prevented development of a large population of invertebrate planktivores (*Chaoborus*). This pelagic food web (a piscivore and reduced populations of vertebrate and invertebrate planktivores), may have allowed an increase in *Daphnia* biomass and the stabilization of zooplankton community structure. In 1976–1980, *Bosmina* and *Ceriodaphnia* were dominant. After 1980, *D. galeata* increased, planktivorous invertebrates initially increased, but then decreased to low density, extended early summer clear water phases occurred, and inedible algae (*Microcystis, Pediastrum, Hydrodictyon*) became dominant (Figure 9.9). Algal biomass was less than in pre-biomanipulation years in only 1 of the 8 years of biomanipulation, but the dominance of colonial blue-greens permitted more light penetration.

Biomanipulation in Bautzen Reservoir was enhanced in 1996 and 1997 with the addition of ferrous iron while the water column was continuously mixed with a circulation system (Chapter 8; Deppe and Benndorf, 2000).

Multiple feedbacks apparently occur in eutrophic lakes, whereby short term positive effects of trophic cascades can enhance longer term processes, ultimately leading to reduced TP concentrations. For example, biomanipulation can produce fewer bottom-feeding fish, more macrophyte coverage when the water becomes clearer, and reduced yield of algae per unit of TP. Phosphorus release from epilimnetic sediments may also decrease as a consequence of lower algal productivity and hence lower pH (Benndorf, 1989). A net export of TP from the epilimnion to the hypolimnion may occur during the daytime *Daphnia* migration to deep water to avoid fish predation (Wright and Shapiro, 1984). This occurred in Lake Haugatjern, Norway in the years following a rotenone treatment to eliminate fish (Reinertsen et al., 1990).

Positive changes associated with food web manipulation also may be countered by longer term feedback processes leading to increased TP concentrations and decreased water clarity, including dominance by large, slow growing, inedible blue-green colonies. Long-term adaptations of the *Daphnia* population to reduced epilimnetic predation may lead to the establishment of a population that does not migrate and remains in the epilimnion continuously. This might decrease export of TP to the hypolimnion (Benndorf, 1989; Benndorf and Miersch, 1991). There is observational and experimental evidence that food quantity and predation affect the selection of *D. magna* and *D.*

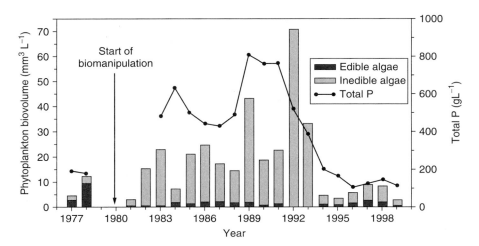

FIGURE 9.9 Biovolume of edible and inedible phytoplankton (summer averages, May–October) and concentration of total phosphorus (annual averages) in Bautzen Reservoir (Germany) before and after biomanipulation. See Benndorf, 1995 for experimental details; external P-loading was dramatically reduced after 1990 (extended from Benndorf, 1995). (From Benndorf, J. et al. 2002. *Freshwater Biol.* 47: 2282–2295. With permission.)

galeata mendotae genotypes found in the epilimnion and hypolimnion (Dumont et al., 1985; King and Miracle, 1995). This supports Benndorf's prediction of a longer-term selection on grazers leading to additional changes in the food web, including a reduction in, or cessation of, vertical migration. Longer-term studies are needed to determine if biomanipulation has an impact on TP concentrations.

Planktivorous fish were nearly eliminated by adding piscivores to the Grafenheim Lakes, demonstrating the possibility of "over-biomanipulation." (Benndorf et al., 2000). Large invertebrate planktivores (*Chaoborus*) emerged in the absence of planktivorous fish, leading to *Daphnia* elimination by the chaoborids and no grazer control of phytoplankton in either the piscivore-dominated lake or the planktivore-dominated control lake. The linear food chain model of the trophic cascade hypothesis did not predict that other species (e.g., *Chaoborus*) could play major roles and offset cascades (McQueen et al., 2001). A moderate density of planktivores (e.g., Bautzen Reservoir) may be needed to control invertebrate planktivores, though at a level permitting high *Daphnia* biomass (Wissel et al., 2000). This could require intensive management.

Many southern eutrophic reservoirs in the U.S. have a three-level food web dominated by gizzard shad (*Dorosoma cepedianum*), and piscivores may not be found as a functionally distinct trophic level. Shad have high fecundity, are omnivorous, and are apparently not controlled by top-down or bottom-up forces. They spawn in early spring and their larvae exhibit strong planktivory, depriving planktivorous bluegill larvae (*Lepomis macrochirus*) of food. This reduces the largemouth bass food base. Further, larvae from early shad spawning will reach sizes unavailable to YOY bass. Both of these factors lead to poor growth of each bass year class, reduced YOY bass survival over the winter, and ultimately to smaller bass populations. Shad productivity is not controlled by zooplankton availability because shad also consume large phytoplankton. Omnivorous shad are thus a "keystone species," affecting trophic levels above (piscivores) and below (herbivores and phytoplankton). Where gizzard shad are abundant, piscivores decline, and the chain-like food web that produces a trophic cascade in some natural lakes is absent in these reservoirs (Stein et al., 1995). In Swedish lakes, roach (*Rutilus rutilus*), an omnivore, can become abundant and depress piscivorous perch populations (Persson et al., 1988). Though not yet demonstrated, some natural U.S. lakes dominated by omnivorous European carp (*Cyprinus carpio*), may have fewer piscivores, possibly through a mechanism similar to that proposed by Stein et al. (1995) for shad-dominated reservoirs.

The biomanipulation experiences at Lake Mendota and Bautzen Reservoir indicate that a strong top-down manipulation by the addition of large numbers of piscivores may cascade to phytoplankton. But not all piscivore additions are successful and none appears to be stable. An experimental addition of largemouth bass to a Texas water supply reservoir, for example, did not lead to lower algal biomass, nor to reduced nutrient concentrations (Drenner et al., 2002). Piscivore additions apparently must be large and continuous from year to year, and success can be thwarted by anglers.

Biomanipulation is not a substitute for significant external nutrient loading reduction, especially in deep lakes where macrophytes have a small role. Benndorf et al. (2002), in support of the "bottom-up top-down" hypothesis of McQueen et al. (1989), argued that reduced P concentration is critical to long-term success. Benndorf and Miersch (1991) proposed a "Biomanipulation Efficiency Threshold of P Loading," estimated as 0.6–0.8 g P/m^2 per year, above which biomanipulation-induced P losses from the epilimnion (e.g., increased sedimentation, *Daphnia* vertical migration) would be overcome by external loading. Elser and Goldman (1991) also supported the idea that mesotrophic lakes are more responsive to top-down control.

There is disagreement about nutrient concentrations and biomanipulation success. Carpenter et al. (2001) observed changes in grazers and phytoplankton over a 7-year period in small, deep Wisconsin lakes that were either planktivore or piscivore-dominated. The lakes were enriched over a 5-year period with N and P, but remained P-limited. Even though nutrient concentrations were similar, the planktivore lake had more algal biomass and primary productivity. *Daphnia* grazing was apparent in the piscivore lake, even with high pelagic P concentrations. This experiment does not support the P-loading threshold hypothesis. Based on analyses of 17 case histories that were not confounded by other manipulations, Drenner and Hambright (2002) concluded that piscivore-dominated lakes can have lower chlorophyll for a given P concentration than planktivore-dominated lakes.

9.8 COSTS

An analysis of biomanipulation costs has not been developed, but there is little question that fish removal or poisoning (and fish disposal) is expensive (see Welch (1992b) for preliminary cost estimates of these procedures), as are follow-up procedures such as piscivore re-stocking or macrophyte planting (needed in some cases). As with most other in-lake treatments, there is little assurance of long-term biomanipulation success if external nutrient loading remains high, making diversion costs part of biomanipulation costs. There are some lower cost biomanipulation activities that may change lake trophic state. For example, stable water level in shallow reservoirs may be important in the switch from clear to turbid conditions, and for the decline in emergent wetland plants (Chapter 13). If water levels are allowed to fluctuate, sediment exposure at low levels enhances sediment compaction, leading to lower re-suspension and turbidity, and perhaps to spring development of submersed plants by exposure of sediments to light (Coops and Hosper, 2002). Other lower cost techniques for shallow lakes include construction of artificial refuges for zooplankton, herbivorous bird exclosures (e.g., Moss, 1990, 1998), and attempts to sharply reduce Canada Geese populations by addling eggs and by lakescaping (Chapter 5). Other low cost alternatives are described by McComas (2003).

9.9 SUMMARY AND CONCLUSIONS

Drenner and Hambright (1999, 2000) reviewed methods and successes of biomanipulation experiments. Most (80%) experiments were in Europe and on small (< 25 ha), shallow (< 3 m mean depth) lakes. While success was high (61%), the least successful procedure (29%) was piscivore stocking. Partial fish removal in shallow lakes was most successful, but this conclusion was confounded by nutrient diversion, which occurred in 60% of cases. One reason for the relatively low success of piscivore stocking is that planktivorous and benthivorous fish may grow to sizes beyond the piscivore mouth gape (Hambright et al., 1991).

There are additional lake management strategies that enhance biomanipulation projects. Reduction of external loading is a prerequisite. If a biomanipulated lake remains eutrophic, the potential for algal blooms, including blooms of inedible species, remains high. But internal P recycling may also be a significant source, especially in shallow lakes where all P released is in the photic zone. These lakes could be alum-treated (Chapter 8). Phosphorus inactivation is effective in shallow lakes, and there is no evidence that alum applications are detrimental to macrophytes. An alum treatment may enhance macrophytes by making water clearer. In deep lakes, artificial circulation could enhance and stabilize top-down effects because deep mixing can light-limit phytoplankton productivity, provide an aerobic but dark refuge from planktivory for grazers, and possibly limit P release from deepwater sediments by producing aerobic conditions at the sediment-water interface (Chapter 19).

There is much to be learned about biomanipulation, especially in deep lakes. High effectiveness can be expected in shallow lakes with P concentrations < 100 µg P/L, although in both deep and shallow lakes, stable effects may require continuous maintenance.

The field of limnology is indebted to the pioneers (e.g., Shapiro, Carpenter, Moss, Gulati, Søndergaard, Benndorf, McQueen, Jeppesen, Hosper) in the study of manipulation of lake food webs. Whether or not biomanipulation is a major lake management tool, these workers, and those following them, have added very greatly to our basic understanding of lake ecology. This is clearly a case where *applied* limnology led the way in developing new and basic understanding of biological processes in aquatic habitats.

While biomanipulation has been successful, it is not a panacea for eutrophication problems. The most enriched lakes are apparently the most resistant to change following a manipulation. Simply put, biomanipulation can be an effective lake management procedure, but it is not a long-term substitute for controlling the forces that made the lake eutrophic, namely high loading of dissolved and particulate organic and inorganic substances.

REFERENCES

Bachmann, R.W., M.V. Hoyer and D.E. Canfield, Jr. 1999. The restoration of Lake Apopka in relation to alternative stable states. *Hydrobiologia* 394: 219–232.

Bachmann, R.W., M.V. Hoyer and D.E. Canfield, Jr. 2000. The potential for wave disturbance in shallow Florida lakes. *Lake and Reservoir Manage.* 16: 281–291.

Benndorf, J. 1987. Food web manipulation without nutrient control: A useful strategy in lake restoration? *Schweiz. Z. Hydrol.* 49: 237–248.

Benndorf J. 1988. Objectives and unsolved problems in ecotechnology and biomanipulation: A preface. *Limnologica* 19: 5–8.

Benndorf, J. 1989. Food-web manipulation as a tool in water-quality management. *JWSRT Aqua* 38: 296–304.

Benndorf, J. 1990. Conditions for effective biomanipulation — conclusions derived from whole-lake experiments in Europe. *Hydrobiologia* 200: 187–203.

Benndorf, J. 1995. Possibilities and limits for controlling eutrophication by biomanipulation. *Int. Rev. Hydrobiol.* 80: 519–534.

Benndorf, J. and U. Miersch. 1991. Phosphorus loading and efficiency of biomanipulation. *Verh. Int. Verein. Limnol.* 24: 2482–2488.

Benndorf, J., H. Kneschke, K. Kossatz and E. Penz. 1984. Manipulation of the pelagic food web by stocking with predacious fishes. *Int. Rev. ges. Hydrobiol.* 69: 407–428.

Benndorf, J., H. Schultz, A. Benndorf, R. Unger, E. Penz, H. Kneschke, K. Kossatz, R. Dumke, U. Horning, R. Kruspe and S. Reichel. 1988. Food-web manipulation by enhancement of piscivorous fish stocks: Long-term effects in the hypertrophic Bautzen Reservoir. *Limnologica* 19: 97–110.

Benndorf, J., H. Schultz, A. Benndorf, R. Unger, E. Penz, H. Kneschke, K. Kossatz, R. Dumke, U. Hornig, R. Kruspe, S. Reichel and A. Köhler. 1989. Food web manipulation by enhancement of piscivorous stocks: Long-term effects in the hypertrophic Bautzen Reservoir. *Arch. Hydrobiol. Beih. Ergebn. Limnol.* 33: 567–569.

Benndorf, J., B.Wissel, A.F. Sell, U. Hornig, P. Ritter and W. Boing. 2000. Food web manipulation by extreme enhancement of piscivory: An invertebrate predator compensates for the effects of planktivorous fish on a plankton community. *Limnologica* 30: 235–245.

Benndorf, J., W. Boing, J. Koop and I. Neubauer. 2002. Top-down control of phytoplankton: The role of time scale, lake depth and trophic state. *Freshwater Biol.* 47: 2282–2295.

Bergman, E., L.-A. Hansson and G. Andersson. 1999. Biomanipulation in a theoretical and historical perspective. *Hydrobiologia* 404: 53–58.

Brett, M.T. and C.R. Goldman. 1996. A meta-analysis of the fresh water trophic cascade. *Proc. Natl. Acad. Sci USA* 93: 7723–7726.

Bronmark, C. and S.E.B. Weisner. 1992. Indirect effects of fish community structure on submerged vegetation in shallow eutrophic lakes — an alternative mechanism. *Hydrobiologia* 243/244: 293–301.

Bronmark, C. and S.E.B. Weisner. 1996. Decoupling of cascading trophic interactions in a fresh water, benthic food chain. *Oecologia* 108: 534–541.

Brooks, J.L. and S.J. Dodson. 1965. Predation, body size, and composition of plankton. *Science* 150: 28–35.

Burks, R.L., E. Jeppesen and D.M. Lodge. 2001. Littoral zone structures as refugia against fish predators. *Limnol. Oceanogr.* 46: 230–237.

Burks, R.L., D.M. Lodge, E. Jeppesen and T.L. Lauridson. 2002. Diel horizontal migration of zooplankton: Costs and benefits of inhabiting the littoral. *Freshwater Biol.* 47: 343–366.

Caird, J.M. 1945. Algae growth greatly reduced after stocking pond with fish. *Water Works Eng.* 98: 240.

Carney, H.J. 1990. A general hypothesis for the strength of food web interactions in relation to trophic state. *Verh. Int. Verein. Limnol.* 24: 487–492.

Carpenter, S.R. (Ed.) 1988. *Complex Interactions in Lake Communities.* Springer-Verlag, New York, NY.

Carpenter, S.R. and J.F. Kitchell. 1992. Trophic cascade and biomanipulation interface of research and management — a reply to the comment by DeMelo et al. *Limnol. Oceanogr.* 37: 208–213.

Carpenter, S.R. and J.F. Kitchell. 1993. *The Trophic Cascade in Lakes.* Cambridge University Press, Cambridge, UK.

Carpenter, S.R., J.F. Kitchell and J.R. Hodgson. 1985. Cascading trophic interactions and lake productivity. *BioScience* 35: 634–639.

Carpenter, S.R., C.E. Kraft, R. Wright, H. Xi, P.A. Soranno and J.R. Hodgson. 1992. Resilience and resistance of a lake phosphorus cycle before and after food web manipulation. *Am. Nat.* 140: 781–798.

Carpenter, S.R., J.J. Cole, J.R. Hodgson, J.F. Kitchell, M.L. Pace, D. Bade, K.L. Cottingham, T.E. Essington, J.N. House and D.E. Schindler. 2001. Trophic cascades: Nutrients and lake productivity: Whole-lake experiments. *Ecol. Monogr.* 71: 163–186.

Cooke, G.D., P. Lombardo and C. Brant. 2001. Shallow and deep lakes: Determining successful management options. *LakeLine* 21: 42–46.

Coops, H. and S.H. Hosper. 2002. Water-level management as a tool for the restoration of shallow lakes in The Netherlands. *Lake and Reservoir Manage.* 18: 293–298.

De Bernardi, R. and G. Giussanig. 1990. Are blue-green algae a suitable food for zooplankton — an overview. *Hydrobiologia* 200: 29–41.

De Melo, R., R. France and D.J. McQueen. 1992. Biomanipulation: Hit or myth? *Limnol. Oceanogr.* 37: 192–207.

Demott, W.R., R.D. Gulati and E. Van Donk. 2001. *Daphnia* food limitation in three hypereutrophic Dutch lakes: Evidence for exclusion of large-bodied species by interfering filaments of Cyanobacteria. *Limnol. Oceanogr.* 46: 2054–2060.

Deppe, T. and J. Benndorf. 2002. Phosphorus reduction in a shallow hypereutrophic reservoir by in-lake dosage of ferrous iron. *Water Res.* 36: 4525–4534.

Drenner, R.W. and K.D. Hambright. 1999. Biomanipulation of fish assemblages as a lake restoration technique. *Arch. Hydrobiol.* 146: 129–166.

Drenner, R.W. and K.D. Hambright. 2002. Piscivores, trophic cascades, and lake management. *Sci. World* 2: 284–307.

Drenner, R.W., R.M. Baca, J.S. Gilroy, M.R. Ernst, D.J. Jensen and D.H. Marshall. 2002. Community responses to piscivorous largemouth bass: A biomanipulation experiment. *Lake and Reservoir Manage.* 18: 44–51.

Dumont, H.J., Y. Guisez, I. Carels and H.M. Verheye. 1985. Experimental isolation of positively and negatively phototactic phenotypes from a natural population of *Daphnia magna* Strauss: A contribution to the genetics of vertical migration. *Hydrobiologia* 126: 121–127.

Elser, J.J. and C.R.Goldman. 1991. Zooplankton effects on phytoplankton in lakes of contrasting trophic status. *Limnol. Oceanogr.* 36: 64–90.

Gliwicz, M. 1986. Predaton and the evolution of vertical migration. *Nature* 320: 746–748.

Gliwicz, M. 1990. *Daphnia* growth at different concentrations of blue-green filaments. *Arch. Hydrobiol.* 120: 51–65.

Gulati, R.D. 1990. Structural and grazing responses of zooplankton community to biomanipulation of some Dutch water bodies. *Hydrobiologia* 200/201: 99–118.

Gulati, R.D., E.H.R.R. Lammens, M.-L. Meijer and E.Van Donk. 1990. *Biomanipulation: Tool for Water Management.* Kluwer Academic, Dordrecht. The Netherlands.

Hairston, N.G., Jr. and N.G. Hairston., Sr. 1993. Cause–effect relationships in energy flow, trophic structure, and interspecific interactions. *Am. Nat.* 142: 379–411.

Hairston, H.G., F.E. Smith and L.R. Slobodkin. 1960. Community structure, population control, and competition. *Am. Nat.* 94: 421–425.

Hambright, K.D., R.W. Drenner, S.R. McComas and N.G. Hairston. 1991. Gape-limited piscivores, planktovore size refuges, and the trophic cascade hypothesis. *Arch. Hydrobiol.* 121: 389–404.

Hanson, M.A. and M.G. Butler. 1994a. Responses of plankton, turbidity, and macrophytes to biomanipulation in a shallow prairie lake. *Can. J. Fish. Aquatic Sci.* 51: 1180–1188.

Hanson, M.A. and M.G. Butler. 1994b. Responses to food web manipulation in a shallow waterfowl lake. *Hydrobiologia* 280: 457–466.

Hansel-Welch, N., M.G. Butler, T.J. Carlson and M.A. Hanson. 2003. Changes in macrophyte community structure in Lake Christina (Minnesota), a large shallow lake, following biomanipulation. *Aquatic Bot.* 75: 323–338.

Hansson, L.A. 1992. The role of food chain composition and nutrient availability in shaping algal biomass development. *Ecology* 73: 241–247.

Hansson, L.A. H. Annadotter, E. Bergman, S.F. Hamrin, E. Jeppesen, T. Kairosalo, E. Luokkanen, P.-A., Nilsson, M. Søndergaard and J. Strand. 1998. Biomanipulation as an application of food chain theory: Constraints, synthesis, and recommendations for temperate lakes. *Ecosystems* 1: 558–574.

Hecky, R.E. and P. Kilham. 1988. Nutrient limitation of phytoplankton in fresh water and marine environments: A review of recent evidence on the effects of enrichment. *Limnol. Oceanogr.* 33: 796–822.

Hosper, H. 1997. *Clearing Lakes. An Ecosystem Approach to the Restoration and Management of Shallow Lakes in The Netherlands.* RIZA, Lelystad, The Netherlands.

Hosper, S.H. and E. Jagtman. 1990. Biomanipulation additional to nutrient control for restoration of shallow lakes in The Netherlands. *Hydrobiologia* 200: 523–534.

Hosper, S.H. and M.-L. Meijer. 1993. Biomanipulation, will it work for your lake? A simple test for the assessment of chances for clear water, following drastic fish-stock reduction in shallow, eutrophic lakes. *Ecol. Eng.* 2: 63–72.

Hrbacek, J., M. Dvorakova, V. Korinek and L. Prochazkova. 1961. Demonstration of the effect of the fish stock on the species composition of zooplankton and the intensity of metabolism of the whole plankton assemblage. *Verh. Int. Verein. Limnol.* 14: 192–195.

Irvine, K., B. Moss, and J. Stansfield. 1990. The potential of artificial refugia for maintaining a community of large-bodied Cladocera against fish predation in a shallow eutrophic lake. *Hydrobiologia* 200: 379–389.

Jackson, L.J. 2003. Macrophyte-dominated and turbid states of shallow lakes: Evidence from Alberta lakes. *Ecosystems* 6: 213–223.

Jeppesen, E. 1998. *The Ecology of Shallow Lakes.* National Environmental Research Institute. Technical Report 247. Copenhagen, Denmark.

Jeppesen, E., J.P. Jensen, M. Søndergaard, T. Lauridsen, L.J. Pearson and L. Jensen. 1997. Top-down control in fresh water lakes: The role of nutrient state, submerged macrophytes and water depth. *Hydrobiologia* 342/343: 151–164.

Jeppesen, E., J.P. Jensen, M. Søndergaard, T. Lauridsen and F. Landkildehus. 2000. Trophic structure, species richness and biodiversity in Danish lakes: Changes along a phosphorus gradient. *Freshwater Biol.* 45: 201–218.

Jeppesen, E., J.P. Jensen, C. Jensen, B. Faafeng, D.O. Hessen, M. Søndergaard, T. Lauridsen, P. Brettum and K. Christoffersen. 2003a. The impact of nutrient state and lake depth on top-down control in the pelagic zone of lakes: A study of 466 lakes from the temperate zone to the Arctic. *Ecosystems* 6: 313–325.

Jeppesen, E. J.P. Jensen, M. Søndergaard, K.S. Hansen, P.H. Moller, H.V. Rasmussen, V. Norby and S.E. Larsen. 2003b. Does resuspension prevent a shift to a clear water state in shallow lakes during reoligotrohication? *Limnol. Oceanogr.* 48: 1913–1919.

Jones, J.I. and C.D. Sayer. 2003. Does the fish-invertebrate-periphyton cascade precipitate plant loss in shallow lakes? *Ecology* 84: 2155–2167.

Kerfoot, W.C. and A. Sih. (Eds.). 1987. *Predation. Direct and Indirect Impacts on Aquatic Communities.* University Press of New England. Hanover, NH.

Kilham, P. and R.E. Hecky. 1988. Comparative ecology of marine and freshwater phytoplankton. *Limnol. Oceanogr.* 33(4 part 2): 776–795.

King, C.E. and M.R. Miracle. 1995. Diel vertical migration by *Daphnia longispina* in a Spanish lake: Genetic sources of distributional variation. *Limnol. Oceanogr.* 40: 226–231.

Lampert, W. 1982. Further studies on the inhibitory effect of toxic blue-green *Microcystis aeruginosa* on the filtering rate of zooplankton. *Arch. Hydrobiol.* 95: 207–220.

Lathrop, R.C., S.R. Carpenter, and L.G. Rudstam. 1996. Water clarity in Lake Mendota since 1900: Responses to differing levels of nutrients and herbivory. *Can. J. Fish. Aquatic Sci.* 53: 2250–2261.

Lathrop, R.C., S.R. Carpenter, C.A. Stow, P.A. Soranno and J.C. Panuska. 1998. Phosphorus loading reductions needed to control blue-green algae blooms in Lake Mendota. *Can. J. Fish. Aquatic Sci.* 55: 1169–1178.

Lathrop, R.C., B.M. Johnson, T.B. Johnson, M.T. Vogelsang, S.R. Carpenter, T.R. Hrabik, J.F. Kitchell, J.J. Magnuson, L.G. Rudstam and R.S. Stewart. 2002. Stocking piscivores to improve fishing and water clarity: A synthesis of the Lake Mendota biomanipulation project. *Freshwater Biol.* 47: 2410–2424.

Lauridsen, T.L. and D.M. Lodge. 1996. Avoidance by *Daphnia magna* of fish and macrophytes: Chemical cues and predator-mediated use of macrophyte habitat. *Limnol. Oceanogr.* 41: 794–798.

Lauridsen, T.L., E. Jeppesen and M. Søndergaard. 1994. Colonization and succession of submerged macrophytes in shallow Lake Vaeng during the first five years following fish manipulation. *Hydrobiologia* 275/276: 233–242.

Lazzaro, X. 1997. Do the trophic cascade hypothesis and classical biomanipulation approaches apply to tropical lakes and reservoirs? *Verh. Int. Verein. Limnol.* 26: 719–730.

Lombardo, P. 2001. Effects of fresh water gastropods on epiphyton, macrophytes, and water transparency under meso- to eutrophic conditions. Ph.D. Dissertation. Kent State University, Kent, OH.

Marklund, O., H. Sandsten, L.-A. Hansson and I. Blindow. 2002. Effects of waterfowl and fish on submerged vegetation and macroinvertebrates. *Freshwater Biol.* 47: 2049–2059.

McComas, S. 2003. *Lake and Pond Management Guidebook.* Lewis Publishers, Boca Raton, FL.

McQueen, D.J. 1998. Fresh Water food web biomanipulation: A powerful tool for water quality improvement, but maintenance is required. *Lakes Reservoirs Res. Manage.* 3: 83–94.

McQueen, D.J., J.R. Post and E.L. Mills. 1986. Trophic relationships in fresh water pelagic ecosystems. *Can. J Fish. Aquatic Sci.* 43: 1571–1581.

McQueen, D.J., M.R.S. Johannes, J.R. Post, T.J. Stewart and D.R.S. Lean. 1989. Bottom-up and top-down impacts on fresh water pelagic community structure. *Ecol. Monogr.* 59: 289–309.

McQueen, D.J., R. France and C. Kraft. 1992. Confounded impacts of planktivorous fish on fresh water biomanipulations. *Arch. Hydrobiol.* 125: 1–24.

McQueen, D.J., C.W. Ramcharan and N.D. Yan. 2001. Summary and emergent properties. *Arch. Hydrobiol. Spec. Iss. Adv. Limnol.* 56: 257–288.

Meijer, M.-L., I. deBoois, M. Scheffer, R. Portielje and H. Hosper. 1999. Biomanipulation in shallow lakes in The Netherlands: An evaluation of 18 case studies. *Hydrobiologia* 409: 13–30.

Moss, B. 1990. Engineering and biological approaches to the restoration from eutrophication of shallow lakes in which plant communities are important components. *Hydrobiologia* 200: 367–377.

Moss, B. 1998. Shallow lakes biomanipulation and eutrophication. SCOPE Newsletter No. 29. October, 1998.

Moss, B., H.R. Balls, K. Irvine and J. Stansfield. 1986. Restoration of two lowland lakes by isolation from nutrient-rich water sources with and without removal of sediment. *J. Appl. Ecol.* 23: 391–414.

Moss, B., S. McGowan and L. Carvalho. 1994. Determination of phytoplankton crops by top-down and bottom-up mechanisms in a group of English lakes, the West Midland Meres. *Limnol. Oceanogr.* 39: 1020–1029.

Moss, B., J. Madgwick and G. Phillips. 1996a. *A Guide to the Restoration of Nutrient-Enriched Lakes.* Broads Authority, Norwich, Norfolk, UK.

Moss, B., J. Stansfield, K. Irvine, M. Perrows and G. Phillips. 1996b. Progressive restoration of a shallow lake: A 12-year experiment in isolation, sediment removal and biomanipulation. *J. Appl. Ecol.* 33: 71–86.

Osgood, R.A. 1984. Long term grazing control of algal abundance — a case history. In: *Lake and Reservoir Management.* USEPA 440/5-84-001. pp. 144–150.

Pace, M.L., J.J. Cole, S.R. Carpenter and J.F. Kitchell. 1999. Trophic cascades revealed in diverse ecosystems. *Trends Ecol. Evol.* 14: 483–490.

Paine, R.T. 1980. Food webs — linkage, interaction strength and community infrastructure — The Third Tansley Lecture. *J. Anim. Ecol.* 49: 667–685.

Perrow, M.R., M.-L. Meijer, P. Dawidowicz and H. Coops. 1997. Biomanipulation in shallow lakes: State of the art. *Hydrobiologia* 342/343: 355–365.

Perrow, M.R., A.J.D. Jowitt, J.H. Stansfield and G.L. Phillips. 1999. The practical importance of the interactions between fish, zooplankton, and macrophytes in shallow lake restoration. *Hydrobiologia* 396: 199–210.

Persson, L., G. Andersson, S.F. Hamrin and L. Johansson. 1988. Predator regulation and primary production along the productivity gradient of temperate lake ecosystems. In: S.R. Carpenter (Ed.), *Complex Interactions in Lakes.* Springer-Verlag, New York, NY. pp. 45–65.

Pick, F.R. and D.R.S. Lean. 1987. The role of macronutrients (C,N,P) in controlling cyanobacterial dominance in temperate lakes. *N.Z.J. Mar. Fresh Water Res.* 21: 425–434.

Reinertsen, H., A. Jensen, J.I. Koksvik, A. Langeland and Y. Olsen. 1990. Effects of fish removal on the limnetic ecosystem of a eutrophic lake. *Can. J. Fish. Aquatic Sci.* 47: 166–173.

Romare, P. and L.-A. Hansson. 2003. A behavioral cascade: Top-predator induced behavioral shifts in plank- tivorous fish and zooplankton. *Limnol. Oceanogr.* 48: 1956–1964.

Sand-Jensen, K. and M. Søndergaard. 1981. Phytoplankton and epiphyte development and their shading effect on submerged macrophytes in lakes of different nutrient status. *Int. Rev. ges. Hydrobiol.* 66: 529–552.

Schelske, C.L. and W.F. Kenney. 2001. Model erroneously predicts failure for restoration of Lake Apopka, a hypereutrophic substropical lake. *Hydrobiologia* 448: 1–5.

Scheffer, M. 1990. Multiplicity of stable states in fresh water systems. *Hydrobiologia* 200/201: 475–486.

Scheffer, M. 1998. *Ecology of Shallow Lakes.* Kluwer Academic Publishers, Dordrecht, The Netherlands.

Scheffer, M., R. Portielje and L. Zambrano. 2003. Fish facilitate wave resuspension of sediment. *Limnol. Oceanogr.* 48: 1920–1926.

Schindler, D.W. 1977. Evolution of phosphorus limitation in lakes. *Science* 195: 260–262.

Schindler, D.W. 1978. Factors regulating phytoplankton production and standing crop in the world's lakes. *Limnol. Oceanogr.* 23: 478–486.

Seip, K.L. 1994. Phosphorus and nitrogen limitation of algal biomass across trophic gradients. *Aquatic Sci.* 56: 16–28.

Shapiro, J. 1979. The need for more biology in lake restoration. In: *U.S. Environmental Protection Agency National Conference on Lake Restoration.* USEPA 440/5-79-001. pp. 161–167.

Shapiro, J. 1990. Biomanipulation: The next phase — making it stable. *Hydrobiologia* 200: 13–27.

Shapiro, J., V. LaMarra and M. Lynch. 1975. Biomanipulation: An ecosystem approach to lake restoration. In: P.L. Brezonik and J.L. Fox (Eds.), *Symposium on Water Quality Managemnt and Biological Control.* University of Florida, Gainesville, FL. pp. 85–96.

Smith, F.E. 1969. Effects of enrichment in mathematical models. In: *Eutrophication: Causes, Consequences, Correctives.* National Academy of Sciences. Washington, DC.

Smith, V.H. and S.J. Bennett. 1999. Nitrogen:phosphorus supply ratios and phytoplankton community structure in lakes. *Arch. Hydrobiol.* 146: 37–53.

Søndergaard, M., E. Jeppesen, E. Mortensen, E. Dall, P. Kristensen and G. Sortkjaer. 1990. Phytoplankton biomass reduction after planktivorous fish reduction in a shallow, eutrophic lake — a combined effect of reduced internal P-loading and increased zooplankton grazing. *Hydrobiologia* 200: 229–240.

Stein, R.A., D.R. DeVries and J.M. Dettmers. 1995. Food-web regulation by a planktivore: Exploring the generality of the trophic cascade hypothesis. *Can. J. Fish. Aquatic Sci.* 52: 2518–2526.

Timms, R.M. and B. Moss. 1984. Prevention of growth of potentially dense phytoplankton populations by zooplankton grazing, in the presence of zooplanktivorous fish, in a shallow wetland ecosystem. *Limnol. Oceanogr.* 29: 472–486.

van de Bund, W. and E. van Donk. 2002. Short-term and long-term effects of zooplanktivorous fish removal in a shallow lake: A synthesis of 15 years of data from Lake Zwemlust. *Freshwater Biol.* 47: 2380–2387.

van Donk, E. and R.D. Gulati. 1995. Transition of a lake to turbid state six years after biomanipulation: Mechanisms and pathways. *Water Sci. Technol.* 32: 197–206.

van Donk, E. and W.J. van de Bund. 2002. Impact of submerged macrophytes including charophytes on phyto- and zooplankton communities: Allelopathy versus other mechanisms. *Aquatic Bot.* 72: 261–274.

van Donk, E., R.D. Gulati and M.P. Grimm. 1990. Restoration by biomanipulation in a small hypertrophic lake: First year results. *Hydrobiologia* 191: 285–295.

van Donk, E., R.D. Gulati, A. Ledema and J.T. Meulemans. 1993. Macrophyte-related shifts in the nitrogen and phosphorus contents of the different trophic levels in a biomanipulated lake. *Hydrobiologia* 251: 19–26.

van Donk, E., E. DeDeckere, J.G.P. Klein Breteler and J.T. Meulemans. 1994. Herbivory by waterfowl and fish on macrophytes in a biomanipulated lake: Effects on long term recovery. *Verh. Int. Verein. Limnol.* 25: 2139–2143.

van Nes, E.H., M.S. van den Berg, J.S. Clayton, H. Coops, M. Scheffer and E. van Ierland. 1999. A simple model for evaluating the costs and benefits of aquatic macrophytes. *Hydrobiologia* 415: 335–339.

van Nes, E.H., M.Scheffer, M.S. van den Berg and H. Coops. 2002. Aquatic macrophytes: Restore, eradicate or is there a compromise? *Aquatic Bot.* 72: 387.

Welch, E.B. 1992a. Reexamining management goals for shallow waters. *WALPA News.* December 1992: 1–2.

Welch, E.B. 1992b. *Ecological Effects of Wastewater. Applied Limnology and Pollutant Effects.* Chapman and Hall, New York.

Wissel, B., K. Freier, B. Muller, J. Koop and J. Benndorf. 2000. Moderate planktivorous fish biomass stabilizes biomanipulation by suppressing large invertebrate predators of *Daphnia. Arch. Hydrobiol.* 149: 177–192.

Wright, D.J. and J. Shapiro. 1984. Nutrient reduction by biomanipulation: An unexpected phenomenon and its possible cause. *Verh. Int. Verein. Limnol.* 22: 518–524.

10 Copper Sulfate

10.1 INTRODUCTION

Copper, an effective algicide, is registered for use in potable water supplies. Its effects are temporary (days), annual treatment costs can be high, there are major negative impacts to non-target organisms, and significant copper contamination of sediments is possible. Several U.S. states have started to restrict or phase out copper use or to lower the permissible dose. The search for an alternative algicide with fewer negative effects has been unsuccessful. Copper sulfate is also used in tank mixes of herbicides to enhance macrophyte control (Chapter 16).

The purposes of this chapter are to describe copper sulfate's dose and application procedures, and to discuss its positive and negative effects. There are several reviews of copper use for algae control (i.e., AWWARF, 1987; Cooke and Carlson, 1989; Demayo et al., 1982; McKnight et al., 1981, 1983; Raman and Cook, 1988).

10.2 PRINCIPLE OF COPPER SULFATE APPLICATIONS

The primary toxic form of copper to algae is the cupric ion (Cu^{2+}) (McKnight et al., 1981), although other forms such as copper-hydroxy complexes may also be toxic (Erickson et al., 1996). Effects on algae include inhibitions of photosynthesis, phosphorus (P) uptake, and nitrogen fixation (Havens, 1994), but effects vary with algal species. Cyanobacteria are particularly sensitive, with concentrations as low as 5–10 µg Cu/L suppressing activity (Demayo et al., 1982; Horne and Goldman, 1974). Copper treatments are likely to be most effective in controlling blooms of nitrogen fixing cyanobacteria, possibly through frequent low doses (Elder and Horne, 1978).

The activity of the cupric ion is affected by: (1) inorganic complexation, (2) precipitation ($Cu(OH)_2CO_3$, CuO, CuS), (3) complexation with compounds such as humic and fulvic acids, (4) adsorption on materials such as clays, and (5) biological uptake (McKnight, 1981; McKnight et al., 1981, 1983; Fitzgerald, 1981). Effective doses therefore may vary among lakes.

pH has a significant effect on the appearance of the cupric form (Cu^{2+}), requiring higher $CuSO_4$ doses in lakes with high alkalinity and pH (Figures 10.1 and 10.2). Copper is less toxic in hard water, in part due to the precipitation of malachite ($Cu(OH)_2CO_3$) and to competition with calcium and magnesium for binding sites on the algal cell membrane.

The experiments by Button et al. (1977) in Hoover Reservoir (alkalinity = 96 mg/L as $CaCO_3$, pH = 7.8), a water supply for Columbus, Ohio, illustrate the brief period of high Cu^{2+} that can be expected in a water body of this alkalinity. Cu^{2+} concentration in the water column fell rapidly after the application of 1.56 g $CuSO_4 \cdot 5H_2O/m^2$. About 95% of the total $CuSO_4$ dissolved in the top 1.75 m of the water column. At the end of 2 h, soluble Cu^{2+} fell to pre-treatment levels (Figure 10.3), perhaps through precipitation, dilution by incoming water, or by washout. An algae bloom, consisting of taste and odor causing diatoms *Melosira* sp., *Asterionella* sp. and *Stephanodiscus* sp., was controlled. Formation of insoluble malachite may have been responsible for a substantial fraction of the loss of Cu^{2+} because conditions for its formation were ideal (Button et al., 1977).

Complexation by dissolved humic substances in Mill Pond Reservoir, a Massachusetts water supply with a high humic content, apparently prevented the rapid loss of copper to the lake's bottom, making the treatment more effective. Biomass of the taste and odor causing dinoflagellate, *Ceratium*

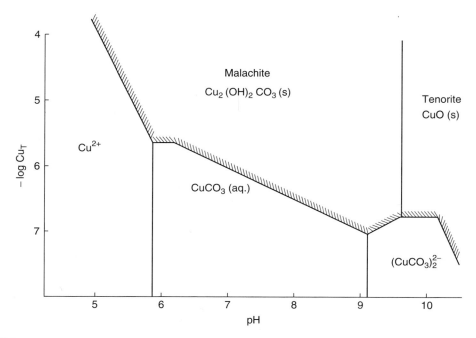

FIGURE 10.1 Relationship between pH and concentration and forms of copper in high alkalinity water. (From McKnight, D.M. et al. 1983. *Environ. Manage.* 7: 311–320. With permission.)

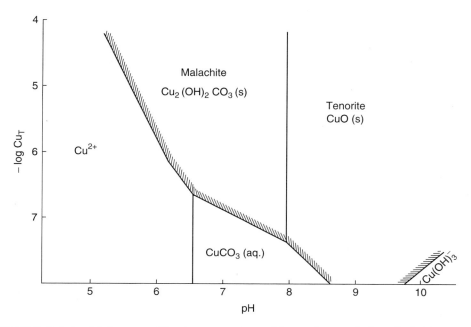

FIGURE 10.2 Relationship between pH and concentration and forms of copper in low alkalinity water. (From McKnight, D.M. et al. 1983. *Environ. Manage.* 7: 311–320. With permission.)

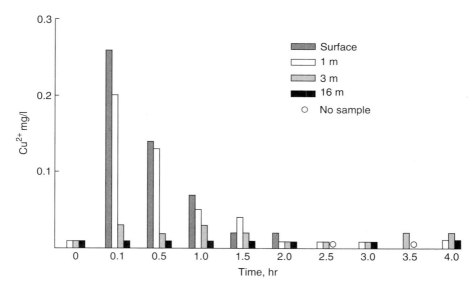

FIGURE 10.3 Depth of soluble copper penetration after application to Hoover Reservoir, OH. (From Button, K.S. et al. 1977. *Water Res.* 11: 539–544. With permission.)

hirundinella, was reduced by 90%, although the green algae *Nanochloris* and *Ourococcus* were unaffected by the complexed copper and appeared to be copper tolerant (McKnight, 1981). In this case, the dose of $CuSO_4$ saturated the organic complexing agent and still provided enough Cu^{2+} to control the dinoflagellate. Presumably, much higher doses would have been needed to control the other species.

The effectiveness of copper has been enhanced by either complexing copper with a carrier molecule, or by chelating it to non-metal ions, to keep copper in solution (DuBose et al., 1997). These formulations allow effective treatment at lower doses.

Mat-forming filamentous algae can be pond and littoral zone nuisances, and doses to control them vary widely. Using Cutrine-Plus (Applied Biochemists, Milwaukee, WI 53218, U.S.), an ethanolamine–copper complex, *Oedogonium* and *Spirogyra* had a very low EC_{50} of 3 μg Cu/L (dose producing a 50% biomass reduction), but *Hydrodictyon, Pithophora,* and *Rhizoclonium* were 15 times more tolerant, and *Oscillatoria* was six times more tolerant than *Pithophora* (Lembi, 2000). Field dose ranges could be wider than these laboratory doses, indicating the need for correct identification of algae and for recognition that *Pithophora, Oscillatoria, and Lyngbya* form thick mats or "scums" that may resist copper penetration.

10.3 APPLICATION GUIDELINES

Guidelines for $CuSO_4$ treatments for planktonic algae were developed by Mackenthun (1961). However, reservoirs and lakes are sufficiently unique to require experience and judgment of the applicator for the dose most likely to produce control. These are Mackenthun's guidelines: For lakes with a methyl orange alkalinity > 40 mg/L as $CaCO_3$, the dose for planktonic algae is 1.0 mg $CuSO_4 \cdot 5H_2O$/L, as copper sulfate crystals, for the upper 0.3 m depth regardless of actual depth. In water with this alkalinity, 0.3 m is considered the maximum effective depth range, after which copper is rapidly lost to complexation. If alkalinity is < 40 mg/L, the dose is 0.3 mg $CuSO_4$ $5H_2O$/L. Copper sulfate is more effective at water temperatures > 15°C. Doses at these concentrations will be toxic to many species of algae and to some non-target organisms (Nor, 1987). Control of *Chara* and *Nitella* requires a dose of 1.5 mg/L, or higher, and must be applied early in the season before these algae become encrusted with marl.

Applicators may increase doses to compensate for copper sulfate ineffectiveness when water column conditions promote complexation and precipitation. At a water column pH of 8.0, less than 10% of the added copper is in the dissolved form. Photosynthesis can drive pH up to 9 or above, and copper effectiveness will be minimal. A chelated or complexed form may be needed in high alkalinity waters (Raman and Cook, 1988).

Planktonic and filamentous algae are rarely controlled with a single application. The "guideline" dose of 1.0 mg/L as $CuSO_4 \cdot 5H_2O$ for waters with alkalinity > 40 mg/L as $CaCO_3$ (Mackenthun, 1961: Fitzgerald, 1967) is often followed, but it appears that lower doses (e.g., 0.15 mg/L) at daily intervals for 3–5 days could be more effective (DuBose et al., 1997). The problems with low doses, however, are algal tolerance (McKnight et al., 1981; Twiss et al., 1993), the rapid loss of copper via complexation, precipitation, or washout to concentrations that are too low, and the costs associated with re-applications.

Treatment methods range from the traditional burlap bag of $CuSO_4$ towed behind a boat, to mechanical spreaders, sprayers, and helicopters. Large quantities (e.g., 4,500 to 7,000 kg per day) have been applied to water supply reservoirs, using barges and chemical spreaders (McGuire et al., 1984) to treat taste and odor causing periphytic species of *Oscillatoria*. Copper can be added to a reservoir's inflow (Bean, 1957), or introduced near an artificial circulation device. Recreational lake users may wait until an algal bloom develops before making application, an approach that could be effective, although severe dissolved oxygen (DO) depletions are possible. Water supply managers face the problem of preventing episodes of unacceptable tastes and odors, or the appearance of a bloom of a toxic algal species. Some potable water supply operators monitor the algal community on a frequent and regular basis during the summer and fall and treat the reservoir to prevent a "bloom." This may require several treatments. This approach emphasizes the need for continuous and detailed monitoring of the water body.

10.4 EFFECTIVENESS OF COPPER SULFATE

The chemical and hydrological features of the treated water determine how rapidly copper will be lost through precipitation, adsorption, washout, or dilution. There have been suggestions that some algae species population have become resistant to low doses of copper, thus requiring either the chelated or complexed forms, or a greater concentration, for effectiveness. These and other factors are significant in the few published case histories about algae responses to copper.

The experimental treatments to periphytic blue-green algae in the highly buffered (alkalinity 150 mg/L as $CaCO_3$) Casitas Reservoir, California are among the few published case histories about control of these taste and odor producing algae (AWWARF, 1987). Several chelated and non-chelated copper compounds were studied for effects on *Oscillatoria limosa* and other species of this genus. Dry $CuSO_4$ crystals in chelated (ethanolamine) and non-chelated forms were applied to surface waters over the periphyton mats, at doses from 0.2 to 0.3 mg Cu/L (chelated) and 0.4 to 1.7 mg Cu/L (non-chelated). Liquid copper citrate (chelated) and $CuSO_4$ solutions were applied directly on the periphyton via a submerged hose, at 0.2 to 2.2 mg Cu/L and 0.2 to 1.0 mg Cu/L, respectively. Divers were used to monitor results and water samples were obtained to determine changes in taste and odor causing compounds.

The submerged applications of $CuSO_4$ and $CuSO_4$ citrate solutions at Casitas Reservoir had little effect on periphyton. Applications of $CuSO_4$ at 1.7 mg Cu/L to the lake's surface, based on an estimated volume of water near the periphyton growths, had some effect but produced significant benthic invertebrate mortality. Application of chelated granular copper to the surface at doses of 0.2 to 0.4 mg Cu/L was effective at periphyton control, but regrowth was apparent in 4 weeks. This formulation was toxic to benthic invertebrates and was the most costly treatment (Table 10.1). Copper use was stopped at Casitas Reservoir due to environmental concerns.

Copper sulfate treatments of nuisance phytoplankton "blooms" frequently are successful for brief periods. Species other than the target algae may become dominant, or algal biomass may

TABLE 10.1
Costs of Copper Sulfate Treatments at Casitas Reservoir, California

Treatment	Cost (2002$)
CuSO$_4$ solution	\$169–499 ha^{-1} (\$19–202 acre^{-1})
CuSO$_4$ crystals	\$152–913 ha^{-1} (\$72–370 acre^{-1})
CuSO$_4$–citric acid solution	\$98–1106 ha^{-1} (\$40–446 acre^{-1})
Copper-ethanolamine granular	\$547–2263ha^{-1} (\$221–916 acre^{-1})

Source: Modified from AWWARF. 1987. Current Methodology for the Control of Algae in Surface Waters. Research Report. AWWA, Denver, CO. With permission.

"rebound" to levels similar to or higher than the original bloom condition. Copper sulfate is unquestionably effective as long as the cupric ion concentration remains high, but water masses are hydraulically dynamic, leading to washout, dilution, and reinoculation with algae, and chemical and physical conditions may lead to loss of copper. In situations where eutrophication continues, increasingly frequent and heavier doses may be needed (Hanson and Stefan, 1984).

Copper sulfate has been used to kill snails in bathing beach areas to limit the release of immature (cercaria) forms of the blood flukes (Trematoda, Schistosomatidae) that penetrate human skin, causing "swimmer's itch." Humans are not the normal host and the cercaria die in the skin, producing severe itching. The Minnesota Department of Natural Resources (undated pamphlet) recommended treating with 1.5 kg/100 m^2 out to the edge of the littoral zone. This dose is lethal to most invertebrates and could produce sediment contamination.

10.5 NEGATIVE EFFECTS OF COPPER SULFATE

The benefits of copper sulfate treatments for algae control in recreational lakes should be weighed against the exposure of non-target organisms to concentrations of a heavy metal greater than the median lethal dose from laboratory studies. Copper sulfate negatively impacts aquatic communities, and could create human health problems. Resistance may develop in target algae, and algae grazing by zooplankton may be eliminated. Dissolved oxygen depletions can occur when large volumes of dead algal cells decompose, creating conditions causing increases in iron, P, manganese, hydrogen sulfide, and ammonia concentrations.

Laboratory test procedures for copper toxicity often involve exposure of the test organism for 96 h, a test period that may obscure effects. Copper is rapidly lost from solution, even with highly simplified, possibly soft-water, experimental conditions, suggesting that 48-h exposures may be more realistic in determining an LC$_{50}$ (concentration lethal to 50% of test organisms) (Mastin and Rodgers, 2000).

Laboratory toxicity tests have demonstrated lethal and sublethal effects on bluegills (*Lepomis macrochirus*). The 96-h LC$_{50}$ ranged from 1.0–3.0 mg Cu/L (Blaylock et al., 1985), to as high as 16.0 mg. Cu/L (Ellgaard and Guillot, 1988) in test waters of moderate alkalinity (46–82 mg/L as CaCO$_3$). However, locomotor activity was impaired at much lower concentrations (e.g., 40 µg Cu/L; Ellgaard and Guillot, 1988). Hatchability and survival of 4-day old larvae were affected by concentrations above 77 µg Cu/L (Benoit, 1975). The risk of direct bluegill mortality apparently is low, but sublethal effects on behavior and reproduction, and on feeding behavior, could lead to reduced growth, and occur at concentrations more than an order of magnitude less than recommended for algae treatment (Sandheinrich and Atchison, 1989). Other species (e.g., trout) may be even more copper sensitive.

Does copper accumulation in lake sediments pose a bioaccumulation or toxicity risk? Anderson et al. (2001) compared the hepatic concentrations of copper in largemouth bass (*Micropterus salmo-*

ides) and common carp (*Cyprinus carpio*) in Lake Mathews and Copper Basin Reservoir, both in California. Lake Mathews, a water supply reservoir, received more than 2000 tons of granular copper sulfate over a 20-year period. The lake retained 80% of the applied copper, mainly associated with oxidizable and carbonate-bound phases that could release copper under some chemical conditions (Haughey et al., 2000). Copper Basin Reservoir was untreated. Sediment copper in Lake Mathews averaged 290 mg Cu/kg dry weight; Copper Basin's was 8 mg Cu/kg dry weight. Hepatic accumulation of copper was found in smaller bass (< 41 cm length) and in all carp in the treated lake, but there were no apparent effects of copper on these species, as estimated by condition factors. Copper in treated lake sediments was found in organic, carbonate, and iron-oxide forms, with a small amount in bioavailable form. Toxicity bioassays, using amphipods (*Hyallela azecta*) and cattails (*Typha latifolia*), did not reveal impaired survival or growth when these species were exposed to re-wetted pond soils that had an average concentration of 173 mg Cu/kg dry weight (vs. 36 mg Cu/kg dry weight in untreated sediments) (Han et al., 2001). Accumulation in fish may be through food web transfer, or through direct exposure during applications.

Copper may be highly toxic to benthic invertebrates (Giudici et al., 1988; Harrison et al., 1984; Mastin and Rodgers, 2000; Nor, 1987), but it does not appear to continue to interact with the water column after its deposition, at least in sediments with high carbonate content (Sanchez and Lee, 1978). Copper accumulation in sediments could produce a sufficiently high concentration to delay or greatly increase the costs of a sediment removal project, but sediment contamination has not been shown to impair certain fish, invertebrate, or vascular plant species.

Copper could become a problem in low alkalinity lakes and reservoirs, with low carbonate-containing sediments, if acidification of the system occurred, perhaps through acid precipitation. For example, in laboratory tests, copper was toxic to fathead minnows (*Pimephelas promelas*) at concentrations as low as 2 µg Cu/L at pH 5.6 and dissolved organic carbon (DOC) of 20 µg/L. A multiple regression model found that pH and DOC explained 93% of the variance in toxicity in test systems (Welsh et al., 1993). Similar results occurred with *Ceriodaphnia dubia* (Cladocera) where the copper LC_{50} increased (toxicity decreased) in direct proportion to pH and DOC increases (Kim et al., 2001). Prolonged use of copper to control algae could create a situation where an acidified lake or reservoir was rendered unusable. Copper algicides should not be used in low pH, low DOC, poorly buffered waters.

The potential for copper toxicity in contaminated sediments can be predicted by pore water concentration, or by acid-volatile sulfide (AVS) concentration. AVS binds with metals, mole for mole, to form an insoluble metal complex. Thus, if AVS concentration in sediments exceeds the concentration of a simultaneously extracted metal (SEM), all of the metal exists as a sulfide (e.g., CuS) and cannot be directly toxic to benthos (Ankley et al., 1996). However, as these authors note, resuspended sediments, or contamination of food webs via ingestion of contaminated benthos, detritus, or sediments, may produce toxicity that cannot be predicted from the AVS/SEM analysis. This predictive analytical tool should be useful where there are concerns about copper toxicity of lake sediments following extensive $CuSO_4$ applications.

The "rebound" of algal biomass after $CuSO_4$ treatment may be from copper toxicity to algae-grazing zooplankton (McKnight, 1981; Cooke and Carlson, 1989). $CuSO_4$ is highly toxic to species of *Daphnia*, a common and effective grazer of planktonic algae, and an important item in fish diets (Chapter 9). Copper concentrations 100 times less than needed for algae control inhibit reproduction or are lethal to zooplankton (Blaylock et al., 1985; Naqvi et al., 1985; Winner et al., 1990). *Daphnia magna, D. pulex, D. parvula,* and *D. ambigua*, tested in waters with an alkalinity of 100–119 mg/L as $CaCO_3$, exhibited reductions in survival and reproduction when copper concentrations exceeded 8 µg Cu/L (Winner and Farrell, 1976). The 48-h LC_{50} for *D. magna* exposed to the complexed products Clearigate and Cutrine-Plus (Applied Biochemists Inc., Milwaukee, Wisconsin), and to granular $CuSO_4$, were 29, 11, and 19 µg Cu/L, respectively. Alkalinity in these test systems ranged from 55–95 mg/L as $CaCO_3$ at pH 7–8 (Mastin and Rodgers, 2000). These concentrations are more than an order of magnitude lower than recommended doses for lakes with moderate or high alkalinity. In many copper-treated waters, natural mortality of algae through grazing may be reduced or eliminated and a brief chemical-based mortality substituted, perhaps creating a "chemical dependency" on the part of lake users.

The responses of lake communities to copper, or presumably to any toxicant, may be poorly estimated from single species laboratory studies. Taub et al. (1990) treated species-rich laboratory ecosystems with copper during different periods in ecological succession. Copper was an effective algicide early in succession but became less effective as pH and dissolved organic carbon increased over time from community metabolism. This study suggested that copper should be applied during the initial stages of an algal bloom before cells have altered the water's chemical content sufficiently to limit copper toxicity, and when cells are actively dividing.

Copper stress impairs food web functions. When planktonic communities in *in situ* mesocosms were exposed to 140 µg Cu/L for 14 days, not only were *Daphnia*, phytoplankton, and Protozoa (ciliates, flagellates) greatly reduced in abundance, but carbon flow through the food web was impaired. Bacteria increased significantly, but there was little energy transfer via the microbial loop to higher trophic levels (Havens, 1994).

Fifty-eight years of granular $CuSO_4$ treatments of four Minnesota recreational lakes and a water supply reservoir may have produced significant deterioration of their quality. The deposition of dead organic matter in deeper water after a $CuSO_4$ application was large enough to stimulate microbial metabolism and eliminate DO. Low or zero DO conditions apparently stimulated P release from enriched sediments, which in turn stimulated algal blooms, requiring yet another algicide application. A state regulatory agency terminated $CuSO_4$ use in all of these systems due to copper contamination of sediment. Phytoplankton problems did not become worse (Hanson and Stefan, 1984).

Copper does not appear to be directly teratogenic, mutagenic, or carcinogenic to humans. Unlike aquatic organisms, humans tolerate moderately high concentrations (< 1.5 mg Cu/L) (Nor, 1987). However, the use of $CuSO_4$ to control cyanobacteria blooms in potable water supply lakes and reservoirs poses a potential human health risk. Cyanobacteria, especially species of *Microcystis, Anabaena, and Anabaenopsis* (*Cylindrospermopsis*) may produce powerful hepatotoxins and neurotoxins. Consumption of raw water (prior to appropriate potable water treatment) has been associated with livestock and human illnesses and deaths (Carmichael et al., 1985, 2001). When copper is used to treat the reservoir, cell lysis occurs, releasing toxins (Kenefick et al., 1992). In northern Queensland, Australia, 148 people, mostly children, were affected with hepatoenteritis. Most were hospitalized. An epidemiologic study found that only people who had consumed water from Soloman Dam, which had been copper-treated several days earlier, had become ill. The source of the toxin was *Cylindrospermopsis raciborskii* (Bourke et al., 1983; Hawkins et al., 1985). Most modern water supply treatment plants that treat eutrophic raw water use granular activated carbon (GAC) to remove dissolved organic compounds. GAC may remove algal toxins as well. However, some plants process copper-treated eutrophic raw water without GAC. Unless the operators are aware of a potentially toxic cyanobacteria bloom, and take appropriate steps, toxin-laden water could be sent into the distribution system. Drinking water supply managers should monitor algal species composition and density on a daily basis, at sites along the reservoir's length (Cooke and Carlson, 1989) in order to anticipate an algal bloom. Cyanobacteria blooms may originate in the riverine zone, or be inoculated from sediments and develop in the water column (Barbiero and Kann, 1994). In either case, early and regular algicide treatment may prevent the bloom from materializing. But even this "early warning system" (Means and McGuire, 1986) can fail to prevent the bloom.

10.6 COSTS OF COPPER SULFATE

The costs for $CuSO_4$ use in algae management are dictated by dose, frequency of reapplication, area to be treated, type of algal nuisance, and other lake-specific factors. The more costly chelated or complexed forms may be needed in hardwater situations, but may be longer lasting and more effective.

In four Minnesota recreational lakes and a water supply lake, with over 58 years of $CuSO_4$ treatment, 1.5 million kg of $CuSO_4$ were applied at an estimated cost of $4.04 million (2002 U.S. dollars), including labor and operating costs. During summer months, 35% of the chemical costs at the water treatment plant were for $CuSO_4$. Costs for chemicals to operate the plant have not increased since terminating $CuSO_4$ applications. The treatments were not sufficiently cost effective, given that benefits were temporary and there were long-term environmental changes (Hanson and Stefan, 1984).

The variation of single treatment costs with copper formulation is illustrated by the treatments at Casitas Reservoir, California (Table 10.1). Granular copper sulfate costs about $2.00 per kilogram, and liquid Cutrine Plus costs about $10.00 per liter (McComas, 2003). Application costs vary greatly.

Copper sulfate application, the standard treatment for algal problems for many decades, is often effective for brief periods and may be the only short-term solution to a current algae problem, particularly in water supply reservoirs. However, there is substantial evidence against the continued use of this compound, in part from the low or non-existent margin of safety for non-target organisms. There are other longer term and more permanent options, including control of external and internal nutrient loading, to manage algae. Water supply operators should exercise caution in using copper sulfate, particularly during algal blooms, and should develop a diagnosis-feasibility and management plan to address causes of algal blooms (Cooke and Carlson, 1989; Chapter 3).

REFERENCES

American Water Works Association Research Foundation (AWWARF). 1987. Current Methodology for the Control of Algae in Surface Waters. Research Report. AWWA, Denver, CO.

Anderson, M.A., M.S. Giusti and W.D. Taylor. 2001. Hepatic copper concentrations and condition factors of largemouth bass (*Micropterus salmoides*) and common carp (*Cyprinus carpio*) from copper sulfate-treated and untreated reservoirs. *Lake and Reservoir Manage.* 17: 97–104

Ankley, G.T., D.M. DiToro, D.J. Hansen and W.J. Berry. 1996. Technical basis and proposal for deriving sediment quality criteria for metals. *Environ. Toxicol. Chem.* 15: 2056–2066.

Barbiero, R.P. and J. Kann. 1994. The importance of benthic recruitment to the population development of *Aphanizomenon flos-aquae* and internal loading in a shallow lake. *J. Plankton Res.* 16: 1581–1588.

Bean, E.L. 1957. Taste and odor control at Philadelphia, *J. Am. Water Works Assoc.* 49:205–216.

Benoit, R.A. 1975. Chronic effects of copper on survival, growth, and reproduction of the bluegill (*Lepomis macrochirus*). *Trans. Am. Fish. Soc.* 104: 353–358.

Blaylock, B.G., M.L. Frank and J.F. McCarthy. 1985. Comparative toxicology of copper and acridine to fish, *Daphnia,* and algae. *Environ. Toxicol. Chem.* 4: 63–71.

Bourke, A.T.C., R.B. Hawes, A. Neilson and N.D. Stallman. 1983. An outbreak of hepato enteritis (the Palm Island mystery disease) possibly caused by algal intoxication. *Toxicon* 3 (suppl.): 45–48.

Button, K.S., H.P. Hostetter and D.M. Mair. 1977. Copper dispersal in a water supply reservoir. *Water Res.* 11: 539–544.

Carmichael, W.W., C.L.A. Jones, N.A. Mahmood and W.C. Theiss. 1985. Algal toxins and water-based diseases. *CRC Rev. Environ. Control* 15: 275–313.

Carmichael, W.W., S.M.F.O. Azevedo, J.S. An, R.J.R. Molica, E.M. Jochimsen, S. Lau, K.L. Rinehart, G.R. Shaw and G.K. Eaglesham. 2001. Human fatalities from Cyanobacteria: Chemical and biological evidence for cyanotoxins. *Environ. Health Perspect.* 109: 663–668.

Cooke, G.D. and R.E. Carlson. 1989. *Reservoir Management for Water Quality and THM Precursor Control.* American Water Works Association Research Foundation (AWWARF). Denver, CO.

Demayo, A., M.C. Taylor and K.W. Taylor. 1982. Effects of copper on humans, laboratory and farm animals, terrestrial plants, and aquatic life. *CRC Rev. Environ. Control* 12: 183–255.

DuBose, C.K. Langeland and E. Philips. 1997. Problem fresh water algae and their control in Florida. *Aquatics* 19: 4–11.

Elder, J.F. and A.J. Horne. 1978. Copper cycles and CuSO₄ algicidal activity in two California lakes. *Environ. Manage.* 2: 17–30.

Ellgaard, E.G. and J.L. Guillot. 1988. Kinetic analysis of the swimming behavior of bluegill sunfish, *Lepomis macrochirus* Rafinesque, exposed to copper: Hypoactivity induced by sublethal concentrations. *J. Fish. Biol.* 33: 601–608.

Erickson, R.J., D.A. Benoit, V.R. Mattson, H.P. Nelson Jr. and E.N. Leonard. 1996. The effects of water chemistry on the toxicity of copper to fathead minnows. *Environ. Toxicol. Chem.* 15: 181–193,

Fitzgerald, G.P. 1967. Current methods for algae control. IN: Proceedings Fourth Annual Water Quality Research Symposium. New York State Department of Health, Albany, NY. pp. 72–81.

Fitzgerald, G.P. 1981. Selective algicides. In: *Proceedings of Workshop on Algal Management and Control.* Tech. Rept. E-81-7. U.S. Army Corps of Engineers, Vicksburg, MS. pp. 15–31.

Giudici, M.D., L. Migliore, C. Gambardella and A. Marotta. 1988. Effect of chronic exposure to cadmium and copper on *Asellus aquaticus* (L.) (Crustacea, Isoposa). *Hydrobiologia* 157: 265–269.

Han, F.X., J.A. Hargreaves, W.L. Kingery, D.B. Huggett and D.K. Schlenk. 2001. Accumulation, distribution, and toxicity of copper in sediments of catfish ponds receiving periodic copper sulfate applications. *J. Environ. Qual.* 30: 912–919.

Hanson, M.J. and H.G. Stefan. 1984. Side effects of 58 years of copper sulfate treatment of the Fairmont Lakes, Minnesota. *Water Res. Bull.* 20: 889–900.

Harrison, F.L., J.P. Knazovich and D.W. Rice. 1984. The toxicity of copper to the adult and early life stages of the fresh water clam, *Corbicula manilensis*. *Arch. Environ. Toxicol. Chem.* 13: 85–92.

Haughey, M.A., M.A. Anderson, R.D. Whitney, W.D. Taylor and R.F.Losee. 2000. Forms and fate of Cu in a source drinking water reservoir following CuSO₄ treatment. *Water Res.* 34: 3440–3452.

Havens, K.E. 1994. Structural and functional responses of a fresh water plankton community to acute copper stress. *Environ. Pollut.* 86: 259–266.

Hawkins, P.R., M.T.C. Runnegar, A.R.B. Jackson and I.R. Falconer. 1985. Severe hepatotoxicity caused by the tropical cyanobacterium (blue-green alga) *Cylindrospermopsis raciborskii* (Woloszynska) Seenaya and Subba isolated from a domestic water supply reservoir. *Appl. Environ. Microbiol.* 50: 1292–1295.

Horne, A.J. and C.R. Goldman. 1974. Suppression of nitrogen fixation by blue-green algae in a eutrophic lake with trace additions of copper. *Science* 83: 409–411.

Kenefick, S.L., S.E. Hrudey, H.G. Peterson and E.E. Prepas. 1992. Toxin release from *Microcystis aeruginosa* after chemical treatment. *Water Sci. Technol.* 27: 433–440.

Kim, S.D., M.B. Gu, H.E. Allen and D.K. Cha. 2001. Physicochemical factors affecting the sensitivity of *Ceriodaphnia dubia* to copper. *Environ. Monitor. Assess.* 70: 105–116.

Lembi, C.A. 2000. Relative tolerance of mat-forming algae to copper. *J. Aquatic Plant Manage.* 38: 68–70.

Mackenthun, K.M. 1961. The practical use of present algicides and modern trends toward new ones. In: *Algae and Metropolitan Wastes.* Trans. of 1960 Seminar, U.S. Dept. Health, Education, and Welfare, U.S. Public Health Service. PB-199-296. Cincinnati, OH. pp. 148–154. (*Note*: this article contains the original dose chart and design specifications for copper sulfate applications.)

Mastin, B.J. and J.H. Rodgers, Jr. 2000. Toxicity and bioavailability of copper herbicides (Clearigate, Cutrine-Plus, and copper sulfate) to fresh water animals. *Arch. Environ. Contam. Toxicol.* 39: 445–451.

McComas, S. 2003. *Lake and Pond Management.* Lewis Publishers and CRC Press, Boca Raton, FL.

McGuire, M.J., R.M. Jones, E.G. Means, G. Izaguirre and A.E. Preston. 1984. Controlling attached blue-green algae with copper sulfate. *J. Am. Water Works Assoc.* 76: 60–65.

McKnight, D. 1981. Chemical and biological processes controlling the response of a fresh water ecosystem to copper stress: a field study of the CuSO₄ treatment of Mill Pond Reservoir, Burlington, Massachusetts. *Limnol. Oceanogr.* 26: 518–531.

McKnight, D.M., S.W. Chisholm and F.M.M. Morel. 1981. Copper Sulfate Treatment of Lakes and Reservoirs: Chemical and Biological Considerations. Tech. Note No. 24. Dept. Civil Eng., Massachusetts Institute of Technology, Cambridge, MA.

McKnight, D.M., S.W. Chisholm and D.R.F. Harleman. 1983. CuSO₄ treatment of nuisance algal blooms in drinking water reservoirs. *Environ. Manage.* 7: 311–320.

Means, E.G. III and M.J. McGuire 1986. An early warning system for taste and odor control. *J Am. Water Works Assoc.* 78(3): 77–83.

Minnesota Department of Natural Resources. *Control of Swimmers' Itch and Leeches.* Undated informational leaflet #8. Ecological Services Division, Division of Fish and Wildlife, Minneapolis.

Naqvi, S.N., V.D. Davis and R.M. Hawkins. 1985. Percent mortalities and LC$_{50}$ values for selected micro-crustaceans exposed to Treflan, Cutrine-Plus, and MSMA herbicides. *Bull. Environ. Contam. Toxicol.* 35: 127–132.

Nor, Y.M. 1987. Ecotoxicity of copper to biota: a review. *Environ. Res.* 43: 274–282.

Raman, R.K. and Cook, B.C. 1988. Guidelines for Applying Copper Sulfate as an Algicide: Lake Loami Field Study. ILENR/RD-WR-88/19. Illinois Dept. Energy Natural Resources, Springfield, 9 pp.

Sanchez, I. and Lee, G.F. 1978. Environmental chemistry of copper in Lake Monona, Wisconsin. *Water Res.* 12: 899–903.

Sandheinrich, M.B. and G.J. Atchison. 1989. Sublethal copper effects on bluegill, *Lepomis macrochirus*, foraging behavior. *Can. J. Fish. Aquatic Sci.* 46:1977–1985.

Taub, F.B., A.C. Kindig, J.P. Meador and G.L. Swartzman. 1990. Effects of "seasonal succession" and grazing on copper toxicity in aquatic microcosms. *Verh. Int. Verein. Limnol.* 24: 2205–2214.

Twiss, M.R., P.M. Welbourn and E. Schwartzel. 1993. Laboratory selection for copper tolerance in *Scenedesmus acutus* (Chlorophyceae). *Can. J. Bot.* 71: 333–338.

Welsh, P.G., J.F. Skidmore, D.J. Spry, D.G.Dixon, P.V. Hodson, N.J. Hutchinson and B.E. Hickie. 1993. Effect of pH and dissolved organic carbon on the toxicity of copper to larval fathead minnow (*Pimelphales promelas*) in natural lake waters of low alkalinity. *Can. J. Fish. Aquatic Sci.* 50: 1356–1362.

Winner, R.W. and M.P. Farrell. 1976. Acute and chronic toxicity of copper to four species of *Daphnia*. *J. Fish. Res. Bd. Canada* 33: 1685–1691.

Winner, R.W., H.A. Owen and M.V. Moore. 1990. Seasonal variability in the sensitivity of fresh water lentic communities to a chronic copper stress. *Aquatic Toxicol.* 17: 75–92.

Section III

Macrophyte Biomass Control

11 Macrophyte Ecology and Lake Management

11.1 INTRODUCTION

"Macrophyte" refers to all macroscopic aquatic vegetation (vs. microscopic plants like phytoplankton), including macroalgae such as the stoneworts *Chara* and *Nitella*; aquatic liverworts, mosses, and ferns; as well as flowering vascular plants. Understanding aquatic plant biology is important to the immediate problems of managing aquatic plants and aquatic ecosystems. A thorough knowledge of macrophyte biology makes the development of new management techniques, the efficacy of present techniques, and the assessment of environmental impacts more efficient. Understanding macrophyte biology also makes management results more predictable, especially when considered in a long-term ecosystem context.

Aquatic plant management refers to controlling nuisance species, to maximizing the beneficial aspects of plants in water bodies, and to restructuring plant communities. As a natural part of the littoral zone and of the entire lake, producing stable, diverse, aquatic plant communities containing high percentages of desirable species is a primary management goal.

A single chapter cannot review all macrophyte biology that might be relevant to management. Potential topics range from subcellular biology as it relates to genetic engineering; to the physiology of resource gain, allocation, and transport; and to plant relationships with their habitat and other organisms in the ecosystem. This chapter discusses aquatic plant biology as it relates to other chapters in this book; that is, types of aquatic plants, nutrient relationships, reproduction, phenology, the physiology of growth, and community and environmental relationships. It briefly discusses the importance of planning for aquatic plant management. For more detailed information on topics relating to aquatic plant biology refer to Hutchinson (1975), Sculthorpe (1985), Barko et al. (1986), Pieterse and Murphy (1990), Wetzel (1990, 2001), Adams and Sand-Jensen (1991), Hoyer and Canfield (1997), Jeppesen et al. (1998), and the references contained within these publications. Two excellent resources for retrieving aquatic plant information, either "on-line" or by traditional methods are the Aquatic, Wetland, and Invasive Plant Information Retrieval System (APIRS) at Center for Aquatic and Invasive Plants, University of Florida (http://plants.ifas.ufl.edu/) and the U.S. Army Corps of Engineers, Aquatic Plant Control Research Program (www.wes.army.mil/el/aqua/) at Vicksburg, Mississippi.

11.2 PLANNING AND MONITORING FOR AQUATIC PLANT MANAGEMENT

Without a plan, aquatic plant management is haphazard. Objectives remain undefined, leaving no way to gauge progress. Ineffective treatments are discarded without knowing why they failed. In short, the same failures are repeated every year. A successful aquatic plant management plan uses basic planning principles: (1) the problem is defined; (2) an assessment discovers the underlying cause of the problem; (3) plant ecology and the plant community relationships form the scientific basis for the plan; (4) the efficacy; cost; health, safety, and environmental impacts; regulatory appropriateness; and public acceptability of all management options are considered and compared; (5) results are monitored to evaluate the effectiveness of management and to detect impacts to the lake ecosystem; and (6) a strong educational component keeps team members, opinion leaders, lake users, governmental officials, and others in the general public well informed. When comparing control techniques, a method should be discarded if it does not work or if it causes unacceptable environmental harm. It may be discarded if it is more expensive than other suitable techniques.

Aquatic plant management plans need not be complex and there is a variety of good advice on how to develop a management plan (Mitchell, 1979; Nichols et al., 1988; Washinton Department of Ecology, 1994; Hoyer and Canfield, 1997; Korth et al., 1997). Computer technology helps develop and evaluate more complex aquatic plant management plans (Grodowitz et al., 2001a).

Assessing the situation and evaluating and monitoring management practices are key components of an aquatic plant management strategy where aquatic plant sampling is needed. Sampling schemes are many and a sampling method should be designed to answer specific management questions. A number of references are available to help design a sampling program for assessment, evaluation, and monitoring (Dennis and Isom, 1984; NALMS, 1993; Clesceri et al., 1998).

11.2.1 CASE STUDY: WHITE RIVER LAKE AQUATIC PLANT MANAGEMENT PLAN

The Wisconsin Department of Natural Resources gives grants for lake management planning. Small-scale lake planning grants of up to $3,000 are available for obtaining and disseminating basic lake information, conducting education projects, and developing management goals. Large-scale lake planning grants up to $10,000 per project are available for bigger projects that conduct technical studies for developing elements of, or completing comprehensive management plans. In addition to monies supplied by the state, the grantee must supply 25% of the cost as cash or in-kind services. The grants are funded by a motorboat fuel tax.

The White River Lake Management District, with the aid of a consultant, used lake planning grant money to prepare an aquatic plant management plan in the year 2000 (Aron & Associates, 2000). White River Lake has a surface area of 25.9 ha, a maximum depth of 8.8 m, and is located in central Wisconsin. The White River Lake Management District was created approximately 20 years ago in response to growing water quality concerns. The district acquired an aquatic plant harvester approximately 15 years ago to control *Chara* sp. They are also concerned about the invasions of the exotic species Eurasian watermilfoil (*Myriophyllum spicatum*) and curly-leaf pond-weed (*Potamogeton crispus*). The district desires to: (1) preserve native plants, (2) protect sensitive areas, (3) control exotic and nuisance plants, (4) provide improved navigation, and (5) educate district members on the value of aquatic plants and the threats to a balanced plant population. The Table of Contents (Table 11.1) shows the topics considered in the plan including goals and objectives, background and problem definition, and plant management alternatives. From this and sampling information a plant management plan was developed that included a strong educational component.

Macrophytes were sampled along 15 transects placed at approximately equal intervals around the lake (Figure 11.1). Sampling points were randomly selected at approximately 0.5, 1.5, 3, and 4 m depths along each transect. At each sampling location, the species present were noted and the density of each species was estimated on a 1–5 basis, with 5 representing the heaviest growth. The survey showed that *Chara* sp. was dominant (Table 11.2) and that Eurasian watermilfoil occurred in the lake. Water star grass (*Zosterella dubia*), white water lily (*Nymphaea* sp.), and curly-leaf pondweed were found in the lake but not at the sampling locations.

The aquatic plant management plan recommendations are as follows (Aron & Associates, 2000):

RECOMMENDATIONS

White River Lake continues to have an excellent aquatic plant community with a wide range of diversity. Eurasian watermilfoil was only found in isolated patches. Management efforts should be directed toward

TABLE 11.1
Table of Contents for the White River Lake Aquatic Plant Management Plan

Chapter I 2
    Introduction... 2
    Goals & Objectives.. 2
Chapter II — Background **3**
    Shoreline Development 3
    Recreational Uses 3
    Value of Aquatic Plants 5
    Current Conditions 12
    Sensitive Areas.. 13
    Fish and Wildlife 14
Chapter III — Problems... **15**
Chapter IV — Historical Plant Management **16**
Chapter V — Plant Management Alternatives **17**
    Drawdown... 17
    Nutrient Inactivation 17
    Dredging for Aquatic Plant Control.. 18
    Aeration 18
    Screens 18
    Chemical Treatment.. 19
    Native Species Reintroduction 21
    Harvesting... 21
    Hand Controls.. 22
    Biomanipulation... 23
Chapter VI — Plant Management Plan **24**
    Recommendations... 24
    Other Recommendations 24
    Education and Information 24
    Chemical Treatment.. 24
    Riparian Controls... 24
    Harvesting... 25
    Plan Reassessment.. 26
    Finding of Feasibility 26
Chapter VII — Summary.. **27**

Source: From Aron & Associates. 2000. White River Lake — Aquatic Plant Management Plan. Unpublished report. Wind Lake, WI. With permission.

protection and maintenance of the resource with a focus on controlling Eurasian watermilfoil. Small patches of Eurasian watermilfoil should be eradicated using hand-raking, pulling, or chemical treatment. Additionally, signs should be placed at all access locations that describe this species and ask boaters to remove all plant material from their boats and trailers prior to and after using White River Lake.

OTHER RECOMMENDATIONS

Education and Information

The District should take steps to educate property owners regarding their activities and how they may affect the plant community in White River Lake. Informational material should be distributed regularly to residents, landowners, and lake users and local government officials. A newsletter, biannually or quarterly, distributed to landowners and residents should be part of the plant management budget. Topics

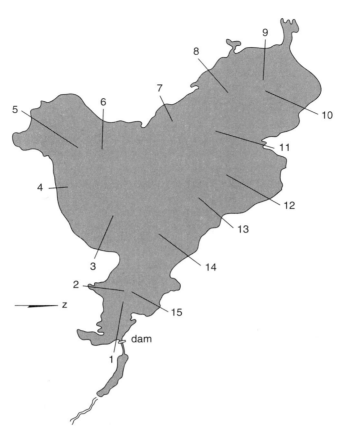

FIGURE 11.1 Sampling transect locations in White River Lake, Wisconsin. (From Aron & Associates. 2000. White River Lake — Aquatic Plant Management Plan. Unpublished report. Wind Lake, WI. With permission.)

should include information relating to lake use impacts, importance and value of aquatic plants, land use impacts, etc. Other issues that should be addressed may include landscape practices, fertilizer use, and erosion control. Existing materials are available through the Wisconsin Department of Natural Resources (WDNR) and the University of Wisconsin Extension (UWEX). Other materials should be developed as needed. The District should also enlist the participation of the local schools. The schools could use White River Lake as the base for their environmental education programs. Regular communications with residents will improve their understanding of the lake ecosystem and should lead to long-term protection.

Chemical Treatment

If there is local public acceptance, the District may continue selective chemical treatment to control Eurasian watermilfoil. If conducted, a WDNR permit must be obtained and selective herbicides should be used to protect native aquatic plant species.

Riparian Controls

Riparians should be encouraged to use the least intensive method to remove nuisance vegetation. This could include minimal raking and pulling. If screens are considered by individuals, a WDNR permit will be required. Riparians should be encouraged to allow native plants to remain. This will help prevent

TABLE 11.2
Aquatic Vegetation of White River Lake, Wisconsin for 2000

Species	Frequency (%)	Relative Frequency (%)	Average Density[a]
Chara sp.	92	35.5	3.8
Myriophyllum spicatum	7	2.7	1.3
Potamogeton zosteriformis	42	16.2	1.9
Vallisneria americana	10	3.9	2.2
Potamogeton richardsonii	5	1.9	3.3
Najas flexilis	12	4.6	2.6
Potamogeton pectinatus	33	12.7	1.6
Ceratophyllum demersum	17	6.6	2.1
Ranunculus longirostris	8	3.1	1.2
Myriophyllum heterophyllum	20	7.7	1.3
Elodea canadensis	2	0.8	1.3
Potamogeton amplifolius	7	2.7	2.0
Polygonum amphibium	2	0.8	1.3
Utricularia vulgaris	2	0.8	2.0

[a] Average density of species rated on a 1–5 basis in sampling units where the species occurred.

Source: From Aron & Associates. 2000. White River Lake — Aquatic Plant Management Plan. Unpublished report. Wind Lake, WI. With permission.

infestation of the areas by Eurasian watermilfoil and curly-leaf pondweed. The native plants will also help stabilize the sediments and minimize shoreline erosion.

Harvesting

The District may continue to harvest as needed to control the nuisances. The equipment should be maintained regularly. Operators should be trained in aquatic plant identification to help protect native non-target plants.

Plant management should be avoided in areas with species of special interest such as wild celery. Operators need to make sure that cutter bars and paddle wheels are kept out of the sediments or to cut one foot above the plant beds when possible.

Operators should operate equipment at speeds only sufficient to harvest the plant material. Excessive speeds will increase the inefficiency of the harvester, causing plants to lay over rather than be cut, and it will increase the numbers of fish trapped.

Operators should work to aggressively control the number of "floaters" and if they do occur, should be removed immediately. Equipment should be operated so that cut plant material does not fall off the harvester.

Plan Reassessment

The District should review or contract to review, the plant populations of White River Lake every 3–5 years. Eurasian watermilfoil removal efforts should be reviewed for effectiveness. The management plan should also be reviewed, and if necessary modified, every 3–5 years. This will be especially important to determine the continued health of the aquatic plant population.

Finding of Feasibility

The harvesting program is necessary to maintain minimal recreational access to White River Lake. It is necessary to maintain a stable clear-water condition for the lake.

The District has shown the ability to maintain and operate an effective harvesting program. The District harvests approximately 50% (30 acres) of White River Lake. Approximately 60 acres (94%) of the lake are available for aquatic plant growth.

In this plan the problem was defined, there was an assessment made of the underlying problem, management options were considered, and there is a strong educational component. There are recommendations for periodic monitoring of the plant community in the future. Additional recommendations could include some periodic testing, even simple Secchi depth readings that monitor water quality, to determine if habitat conditions in the lake are changing, or if plant management might be causing some unforeseen circumstance.

11.3 SPECIES AND LIFE-FORM CONSIDERATIONS

Control tactics are often species-specific. When devising a management plan it is important to know each species' identity, location, and abundance. Each species has unique physiological, habitat, and ecological requirements. The more known about the species of interest, the more successful management will be. The first step is identifying species. Refer to Cleseri et al. (1998) to find taxonomic keys that are regionally appropriate. There are computer programs that help identify aquatic plants (Grodowitz et al., 2001a, b) and The Center for Aquatic and Invasive Plants' website is an excellent place to find species-specific information, lists of taxonomic keys, and "on-line" help identifying plants.

Depending on the definition of "aquatic" and "weed," fewer than 20 of approximately 700 aquatic species are major weeds (Spencer and Bowes, 1990). Because of their prolific growth and reproduction, they often interfere with utilization of fresh waters and may displace indigenous vegetation. Much macrophyte research has been stimulated by the need to control nuisance plants so there is a wealth of information about a limited number of species.

Aquatic plants form four distinct groups based on life form: (1) submergent, (2) free-floating, (3) floating-leaved, and (4) emergent, that differ in habitat, structure and morphology, and the means they obtain resources. Plants in the same life-form group often have similar adaptations to their environment. By grouping species according to life-form, species that are well known may be used as models for species that are less well known but have similar life-forms.

Emergent macrophytes such as reeds (*Phragmites* spp.), bulrushes (*Scirpus* spp.), cattails (*Typha* spp.) and spikerushes (*Eleocharis* spp.) are rooted in the bottom, have their basal portion submersed in water, and have their tops elevated into air. This is ideal for plant growth. Nutrients are available from the sediment, water is available from the sediment and overlying water, atmospheric carbon dioxide and sunlight are available to emergent portions of the plant.

Floating-leaved macrophytes, such as waterlilies (*Nymphaea* spp.), spatterdock (*Nuphar* spp.), and watershield (*Brasenia* sp.), are rooted in the bottom with leaves that float on the water surface. Floating leaves live in two different habitats, water on the bottom, air on top. A thick, waxy coating protects the upper leaf surface from the aerial environment. Floating leaves do not have the structural support of emergents so they can be ravaged by wind and waves. Floating-leaved species are usually found in protected areas.

Submergent species include such varied groups as quillworts (*Isoetes* spp.), mosses (*Fontinalis* spp.), stoneworts, and numerous vascular plants like the many pondweeds (*Potamogeton* spp.), wild celery (*Vallisneria americana*.), and watermilfoils (*Myriophyllum* spp.). They face special problems obtaining light for photosynthesis and they must obtain carbon dioxide from the water where it is much less available than it is in air. They invest little energy in structural support because they are supported by water and water accounts for about 95% of their weight.

Free-floating macrophytes float on or just under the water surface. Their roots are in water, not in sediment. Small free-floating plants include duckweeds (*Lemna* spp.), mosquito fern (*Azolla caroliniana*), and water fern (*Salvinia* sp.). Water hyacinth (*Eichornia crassipes*), and frog's bit (*Limnobium spongia*) are examples of larger free-floating plants. They depend on the water for nutrients and their leaves have many characteristics of floating-leaved species. Their location is at the whims of wind, waves, and current so they are usually found in quiet embayments.

11.4 AQUATIC PLANT GROWTH AND PRODUCTIVITY

The aquatic habitat moderates extremes of temperature and water stress that commonly limits terrestrial plant productivity. Water, however, exerts a high resistance to solute diffusion and selectively attenuates the quality and quantity of light, which can limit aquatic productivity. Species of a similar life-form, although taxonomically diverse, encounter the same habitat limitations. Some species have traits that allow them to exploit conditions in an opportunistic and competitive manner. These species are more productive and thus more likely to become aquatic nuisances.

11.4.1 LIGHT

The quality and quantity of light in aquatic systems have important influences on the growth and development of submergent species. The quality and quantity of light depend upon dissolved materials and suspended particulate matter in the water, and upon water depth. Light becomes more limited and the quality changes with increasing depth and with turbidity from algae, silt, and resuspended bottom sediments. Zonation of macrophytes along depth gradients can be caused by the light regime (Spence, 1967) and increased turbidity can decrease the maximum depth of plant growth (Spence, 1967; Nichols, 1992). Light may also play an important role in seasonal changes in macrophyte dominance and interspecific competition.

Emergent, free-floating, and floating-leaved plants grow in atmospheric sunlight. They are sun plants. Each leaf can potentially utilize all the solar energy it receives for growth (Spencer and Bowes, 1990). Their productivity, at least for emergents, is similar or even greater than terrestrial sun plants.

Submergent species are shade plants. Leaf photosynthesis is saturated by a fraction of full sunlight. The light compensation point (i.e., where the photosynthetic rate equals the respiration rate) for some species is as low as 0.5 percent of full sun (Spencer and Bowes, 1990). Some of the most important nuisances have the lowest compensation points. This may give them a slight but decided advantage over other species for accumulating energy resources.

Light generally limits the lakeward edge of the littoral zone and there is evidence that increased turbidity decreases maximum plant biomass (Robel, 1961). Clear water lakes usually have deeper littoral zones. Nichols (1992) found a 1.2–7.8 m range of maximum plant growth depths for a suite of Wisconsin lakes. This depth range is similar to those reported by Hutchinson (1975), is broader than the 1.0–4.5 m range reported by Lind (1976) for eutrophic lakes in southeastern Minnesota, and is more shallow than the 12 m maximum depth for Lake George, New York (Sheldon and Boylen, 1977) and the 11 m for Long Lake, Minnesota (Schmid 1965). All these depths are considerably more shallow than the 18 m maximum depth for *Utricularia geminiscapa* in Silver Lake, New York (Singer et al., 1983), the 20 m maximum depth for bryophytes in Crystal Lake, Wisconsin (Fassett, 1930), and the approximately 150 m maximum depth for charophytes and bryophytes in Lake Tahoe, California (Frantz and Cordone, 1967). Even shallow lakes, if they are turbid enough, will have sparse aquatic plant growth (Engel and Nichols, 1994; Nichols and Rogers, 1997).

Hutchinson (1975), Dunst (1982), Canfield et al. (1985), Chambers and Kalff (1985), Duarte and Kalff (1990), and Nichols (1992) found a significant regression between Secchi depth and the

TABLE 11.3
Regression Equations of Secchi Depth versus Maximum Depth of Plant Growth

Equation	Region	Reference
MD = 0.83 + 1.22 SD	Wisconsin	Dunst, 1982
MD$^{0.5}$ = 1.51 + 0.53 ln SD	Wide variety	Duarte and Kalf, 1987
MD = 0.61 log SD + 0.26	Finland; Florida; Wisconsin,	Canfield et al., 1985
MD = 2.12 + 0.62 SD	Wisconsin	Nichols, 1992
MD$^{0.5}$ = 1.33 log SD + 1.40	Quebec and the World	Chambers and Kalf, 1985

Note: MD = maximum depth of plant growth in meters; SD = Secchi depth in meters.

maximum depth of plant growth (Table 11.3). In many cases these regressions are similar (Duarte and Kalff, 1987) and are used as models to predict the maximum depth of plant growth for management such as dredging depth to eliminate plant growth (see Chapter 20).

Light also affects a number of morphogenetic processes in submerged aquatic plants including the germination of fruits, anthocyanin production in stems and leaves, the positioning of chloroplasts, leaf area, branching, and stem elongation (Spence, 1975). The most important for management purposes may be stem elongation. For some of the worst nuisance species like *Hydrilla verticillata*, *Egeria densa*, and *M. spicatum,* low light stimulates substantial increases in shoot length (Spencer and Bowes, 1990). These species quickly form a surface canopy so they are no longer light limited, they can shade out slower growing competitors, and they greatly restrict water use by forming a tangled mass of stems and leaves on the water surface.

11.4.2 NUTRIENTS

Submergent macrophytes use both aqueous and sedimentary nutrient sources, and sites of uptake (roots vs. shoot) are related at least in part to nutrient availability in sediment versus the overlying water. In other words, submergent plants operate like good opportunistic species should operate; they take nutrients from the most available source.

Rooted macrophytes usually fulfill their phosphorus (P) and nitrogen (N) requirements directly from sediments (Barko et al., 1986). The role of sediment as a source of P and N for submergent macrophytes is ecologically significant because available forms of these elements are normally low in the open water during the growing season. This is important knowledge because there is a common misconception that excessive external nutrient loading directly to the water column causes macrophyte problems. External nutrient loading usually produces algal blooms, shading and reducing macrophyte biomass. The availability of micronutrients in open water is usually very low, but relatively available in sediments. However, the preferred source of potassium (K), calcium (Ca), magnesium (Mg), sulfate (SO$_4$), and chloride (Cl) appears to be the open water (Barko et al. 1986). Free-floating species obtain their nutrients from the water column and may compete directly with algae for available nutrients.

There are few substantiated reports of nutrient related growth limitation for aquatic plants (Barko et al., 1986). Nutrients supplied from sediments, combined with those in solution are generally adequate to meet nutritional demands of rooted aquatic plants, even in oligotrophic systems. There are exceptions to this statement so there is not a clear consensus on the relationship of nutrient supplies to plant productivity under natural conditions. In Lake Memphremagog (Quebec-Vermont border), Duarte and Kalff (1988) demonstrated that biomass increases averaged 2.1 times greater for fertilized plants (fertilized with 3:1:1, N to P to K ratio) than paired controls. The biomass increase was greatest in shallow water (1 m depth) and with perennial plants. In Lawrence

Lake, Michigan *Scirpus subterminalis* and *Potamogeton illinoensis* biomasses increased with nitrogen and phosphorus fertilization (Moeller et al., 1998). Nutrient limitation also reduced productivity in plants such as wild rice (*Zizania* spp.) that annually produce high biomasses (Dore, 1969; Carson, 2001). There is evidence that nitrogen needs to be replenished to sustain annual macrophyte growth in infertile sediments (Rogers et al., 1995). The nitrogen can be supplied by non-point sources such as sedimentation from shoreline erosion and silt loading, or from lawn fertilization. Multiple nutrient deficiencies appear to diminish growth on extremely low density and extremely high density (usually meaning highly organic or highly sandy) substrates (Barko and Smart, 1986). Plant tissue analysis suggested to Gerloff (1973) that the elements most likely to limit macrophyte growth differed by lake and that nitrogen, phosphorus, calcium and copper were growth limiting or close to growth limiting in different Wisconsin lakes. When available, plants take up nutrients well above their physiological needs (e.g., luxury consumption), which confounds the analysis of the direct relationship between nutrients and growth (Gerloff, 1973; Moeller et al., 1998).

Attempts to control plant growth by limiting sediment nutrients through dredging or covering nutrient rich sediments, or chemically making nutrients unavailable with alum have been unsuccessful (Engel and Nichols, 1984; Messner and Narf, 1987). Attempts to control macrophytes by controlling nutrients in the water column are counter-productive. Phytoplankton obtain their nutrients exclusively from the water column so the first response to nutrient limitation (primarily P) is improved water clarity that improves macrophyte growth.

Although this information suggests that nutrients do not limit aquatic plant growth, oligotrophic lakes generally maintain less total plant biomass and usually contain different species than more nutrient rich lakes. Many species found in oligotrophic lakes have the ability to seasonally conserve both biomass and nutrients.

11.4.3 Dissolved Inorganic Carbon (DIC), pH, and Oxygen (O_2)

Dissolved inorganic carbon (DIC) most likely limits submergent macrophyte photosynthesis (Barko et al., 1986; Spencer and Bowes, 1990). Photosynthesis in terrestrial plants is limited by CO_2 transport and it is even more critical in submersed species. Carbon dioxide diffusion is much slower in water than in air. Free CO_2 is the most readily used carbon form for photosynthesis. Some species can utilize bicarbonate, but they do so less efficiently and they expend more energy doing so. The ability to use bicarbonate has adaptive significance in many fresh water systems because the largest fraction of inorganic carbon may exist as bicarbonate. Eurasian watermilfoil (*M. spicatum*), a notorious nuisance species, has a substantial capacity to use bicarbonate for photosynthesis. The ratio of CO_2 to bicarbonate to carbonate is determined by the alkalinity and pH of the water, and by CO_2 uptake by plants.

In dense plant beds free CO_2 and bicarbonate can be depleted in a few hours of photosynthesis. This shifts the carbon equilibrium toward carbonates that are not used for photosynthesis and increases O_2 concentration and pH. These water conditions cause O_2 inhibition of photosynthesis and photorespiratory CO_2 loss (Spencer and Bowes, 1990). All three conditions lower net photosynthesis. In addition to the utilization of bicarbonate, submergent macrophytes have a number of anatomical, morphological, and physiological mechanisms to enhance carbon gain (Spencer and Bowes, 1990; Wetzel, 1990).

Emergent, free-floating, and floating-leaved plants use atmospheric CO_2 so photosynthesis is not hampered by the slow diffusion rates of gases in water. In addition, lack of water stress allows their stomata to remain open so photosynthesis proceeds unhindered during daylight hours.

Oxygen concentrations determine redox conditions and thus nutrient release from sediments. The underground biomass of rooted species may be living in an anaerobic environment. Lack of oxygen hinders nutrient acquisition. Some species, especially emergents, produce aerenchyma that allows oxygen diffusion from the aerial environment to submerged organs (Wetzel, 1990). Even dead stems are capable of conducting oxygen to rhizomes (Linde et al., 1976). Cutting off emergent plant stems (including dead stems) so they remain below the water surface, thus depriving rhizomes and roots of oxygen for a long period of time is an effective technique for controlling cattails (Beule, 1979) and possibly other emergent species

Increased levels of a single nutrient are likely to increase plant growth only to the point where another nutrient becomes growth limiting. Smart (1990) described laboratory experiments where a reciprocal relationship was found between inorganic C supply and sediment N availability. High levels of both factors stimulated plant growth, increasing the demand on the other factor until one of them limited growth. High levels of aquatic plant production required both an abundance of inorganic C and high sediment N availability (see section above on nutrients).

11.4.4 SUBSTRATE

Substrates provide an anchoring point for rooted plants and, as explained above, are the nutrient source for critical nutrients like N and P. Some sediments (e.g., rocks or cobble) are so hard that plant roots cannot penetrate them; others are so soft, flocculent, and unstable that plants cannot anchor in them. Coarse textured sediments can be nutritionally poor for macrophyte growth. Small accumulations of organic matter stimulate plant growth on these sediments.

Low sediment oxygen concentrations, or high concentrations of soluble reduced iron and manganese or soluble sulfides, can be toxic to plants. High soluble iron concentrations interfere with sulfur metabolism. Sediments containing excessive organic matter may contain high concentrations of organic acids, methane, ethylene, phenols, and alcohols that can be toxic to vegetation (Barko et al., 1986).

The above conditions are most common in eutrophic lakes. To some degree, aquatic plants protect themselves from these toxins with oxygen release from their roots. This eliminates the anaerobic conditions that create toxic substances in the rhizosphere surrounding the root.

Also, as explained above, sediment density has important impacts on nutrient acquisition by plants. Consolidating flocculent sediments using drawdown is one method of improving the habitat for aquatic plant restoration (see Chapter 12).

11.4.5 TEMPERATURE

Water buffers temperature extremes for plant growth but submerged plants can be exposed to temperature extremes from near zero to as much as 40 C (Spencer and Bowes, 1990). Some submerged plants can grow at temperatures as low as 2°C (Boylen and Sheldon, 1976) and it is not unusual to find some species in a green condition living under ice cover. Weed problems are generally most severe in the 20–35°C range.

Water temperature interacts with light to affect plant growth, morphology, photosynthesis, respiration, chlorophyll composition, and reproduction (Barko et al., 1986). High temperatures, within the thermal tolerance range, promote greater chlorophyll concentration and productivity, with a concomitant increase in both shoot length and shoot number. Increasing temperature and light appear to cause opposing response in shoot length (Barko et al., 1986). Different metabolic processes show differing responses to temperature so growth represents an integration of temperature responses. In thermally stratified lakes, depth related temperature decreases could reduce the length of the growing season if plant growth reaches the thermocline or below (Moeller, 1980).

Eurasian watermilfoil and curly-leaf pondweed, two aquatic nuisances, are examples of cool water strategists. Although optimum photosynthetic temperatures for both species appear to be between 30 and 35°C, which is high when compared to terrestrial plants and suggests a preference for warm climates, their photosynthetic rate at low temperatures is a higher percentage of their maximum rate and higher than some other species (Nichols and Shaw, 1986). For milfoil, the responsiveness of dark respiration to temperature, as indicated by a high Q_{10} (2.28), likely results in this lower optimum growth temperature. In other words, because dark respiration rises quickly with temperature, milfoil growth is more efficient at a lower temperature than is suggested by its optimum photosynthetic temperature. Titus and Adams (1979) compared milfoil to the native wild celery and found that milfoil is much better able to photosynthesize at low temperatures. Curly-leaf pondweed thrives in cool water. The active part of its life cycle occurs during cool water conditions and it is dormant during warm water conditions (Stuckey et al., 1978). Hydrilla (*H. verticillata*), another notorious nuisance species, appears to grow better at elevated temperatures than most other submerged plants (Spencer and Bowes, 1990). The ability to photosynthesize at temperature extremes influences the competitive relationships among coexisting species.

Temperature is important when using herbicides and water level drawdown. Herbicides are most effective when target plants are actively growing so herbicide applications at cold water temperatures are generally not effective. However, herbicide applications in cool water could be a selective means of treating nuisance species like curly-leaf pondweed and Eurasian watermilfoil. Drawdown is most effective when troublesome plants are aerially exposed to sub-freezing temperatures and cold desiccation, conditions more extreme than are normally found in the aquatic environment.

11.5 PLANT DISTRIBUTION WITHIN LAKES

Turbidity, nutrient concentration, sediment texture, sediment organic matter, siltation rates, and wind and wave action determine plant distribution and abundance. These parameters are interrelated and interact with lake morphology — basin depth, bottom slope, surface area, and shape to determine the lake's littoral zone. Lake basins are extremely variable and reflect their mode of origin. Lake morphology is constantly modified by water movements within the basin, by accumulations of plant detritus, and by sediment inputs from silt loading and bank erosion.

The steepness of the littoral slope is inversely related to the maximum submerged macrophyte biomass. In Lake Memphramagog, Duarte and Kalff (1986) found that about 87% of the variance in maximum submerged macrophyte biomass was explained by littoral zone slope and sediment organic matter. This is probably due to the difference in sediment stability on gentle and steep slopes. A gentle slope allows the deposition of fine sediments that promote plant growth. Steep slopes are areas of erosion and sediment transport; areas not suitable for plant growth.

Surface area and shape significantly influence the effect wind has on wave size and current strength. Large lakes have large fetches and thus have greater wave and current energy than small lakes. Wave action and current erode shorelines. The directions and strength of the wind, slope, and lake shape determine sediment movement. Points and shallows are swept clean by wind and waves; bays and deep spots fill with sediment. Thus basin size, shape and depth determine the distribution of sediments in a lake and therefore the distribution of plants. In shallow water the direct physical forces of wind, waves, and ice also determine plant distribution (Duarte and Kalff, 1988, 1990).

For management purposes, macrophytes are likely to establish and proliferate in lakes with large areas of shallow, warm water; rich, fine-textured, moderately organic sediments; and moderately clear water. This means that homeowner expectations for living on a weed-free lake can be unrealistic if they are located on lakes with the above characteristics. The manipulation of lake depth and littoral bottom slope, while not always easy or inexpensive, is a powerful management tool for encouraging or discouraging aquatic plant growth in specific areas of a lake.

11.6 RESOURCE ALLOCATION AND PHENOLOGY

Understanding resource allocation is critical to understanding a plant's life history. Applying control tactics when carbohydrate reserves are low in storage organs is one means of optimizing management. Times of low energy reserves are called control points because plants are least likely to recover from management stress without adequate energy. To be useful to managers, low energy reserves need to be correlated with observable phenological events such as the onset of flowering. This provides the manager with a rapid means for timing management practices such as harvesting or herbicide applications.

Linde et al. (1976) found that total non-structural carbohydrates (TNC) were lowest in cattail (*Typha glauca)* rhizomes when the spathe leaf around the pistillate flower is shed (i.e., the cattail flower is emerging from the surrounding leaf). They suggested that this is an excellent time to control cattails. Later, carbohydrates are produced in excess of the plant's immediate needs and are stored in rhizomes where they are available to help the plant recover from severe injuries such as cutting. Titus (1977) found that TNC in Eurasian watermilfoil dropped to about 5% of dry weight in early summer and in late autumn. The early summer depression corresponds to the spring growth flush; the late autumn drop may correspond to carbohydrate allocation to reproductive fragments that are abscised. Minimum TNC storage for hydrilla occurs at the end of July, indicating a primary physiological weak point that can be utilized for management (Madsen and Owens, 1996). Unfortunately, hydrilla shows no visual phenological indicators of low TNC. Potential control points for water hyacinth appear to be shortly before blooming, when flowers are actively developing, and around mid-October when the plants are actively translocating carbohydrates to the stem bases (Luu and Getsinger, 1988).

The ability to stress plants during times of low TNC levels is open to question. Perkins and Sytsma (1987) interrupted carbohydrate accumulation in Eurasian water-milfoil roots by fall harvesting. However, TNC stores were rapidly replenished after harvesting and increased over winter. Growth was not reduced the following year.

11.7 REPRODUCTION AND SURVIVAL STRATEGIES

Reproductive ability usually determines whether or not a plant becomes a major weed problem. Macrophytes range from obligate sexually reproducing annuals to those that persist only by vegetative reproduction. Madsen (1991) divided reproductive strategies into three types: (1) annuals, where overwintering (or survival of other adverse conditions such as seasonal drought) is strictly by seeds; (2) perennial herbaceous, where vegetative propagules such as turions, tubers, or winter buds are used for overwintering; and (3) perennial evergreen, where vegetative, non-reproductive biomass is used for overwintering. Simulation modeling suggests that there is an optimum biomass allocation with respect to investment in overwintering structures (Van Nes et al., 2002). Too little investment reduces the chances to regain dominance in the subsequent year, while too much investment in dormant structures reduces photosynthesis. Vegetative reproduction usually predominates in most species because vegetative propagules are probably sufficient for overwintering without the high energy investment needed for sexual reproduction. Vegetative propagules have large energy reserves so they are usually more successful at initial establishment and rapid initial growth than are seeds. Two distinct advantages to sexual reproduction are that sexual propagules are more resistant to environmental stress and sexual propagation allows for a recombination of genetic traits that might fit the environment better as conditions change. Also, seeds are likely to be dispersed more widely than vegetative propagules.

The vegetative spread of plants falls into two categories. The production of tubers, turions, stem fragments, or other specialized reproductive structures that detach from the parent plant is one method of vegetative propagation. These propagules disperse for varying distances by wind, waves, current, and by humans or other animals. For plant dispersal, these propagules are functionally similar to seeds. A second method is the elongation of rhizomes or the production of stolons or runners where the new plant is attached to the parent plant for a period of time. This method allows young plants to grow quickly but limits the area of spread. Some emergent species colonize deeper water through spreading rhizomes, stolons, or runners.

Many species are intermediate and reproduce both sexually and vegetatively. Wild celery for example takes advantage of all forms of reproduction. It produces winter buds that may be dislodged from the sediment and spread; during the growing season plants form runners with new rosettes attached; and the plant produces flowers, fruits, and seeds. The habitat conditions that favor one mode of reproduction over the other are undetermined (Titus and Hoover, 1991).

The ability of aquatic plants to recolonize areas from seed banks may be crucial to the recovery of aquatic plant communities following severe or prolonged habitat disturbance. Kimber et al. (1995) germinated 12 macrophyte species from the sediment seed bank of Lake Onalaska, Wisconsin and they found the seed bank did not reflect the composition of the vegetation within the lake. If Lake Onalaska was recolonized from the seed bank, the plant community could be considerably different than they found. Many aspects of reproduction and survival have management implications. Knowing about seed banks and the types of plant propagules that survive best under different habitat conditions is critical for plant restoration. Species that become aquatic nuisances are usually prolific vegetative reproducers. Harvesting can spread many nuisance species because they propagate rapidly from plant fragments. Hydrilla, for example, can regrow from a single node (Langeland and Sutton, 1980) and when chopped into small pieces (Sabol, 1987). Using harvesting or herbicides in mixed plant communities can change the community structure from species that grow, spread, and reproduce slowly to those that do so aggressively (Nicholson, 1981). Timing of the formation of reproductive structures is also important. Properly timed harvests, for instance, can greatly reduce the number of vegetative buds produced by curly-leaf pondweed. Damage to reproductive structures can enhance management effectiveness. Cooke et al. (1986) reported that harvesting Eurasian watermilfoil deep enough to remove or disrupt root crowns was more successful than merely cutting stems.

11.8 RELATIONSHIPS WITH OTHER ORGANISMS

Aquatic plants are food and habitat for a wide variety of organisms ranging from epiphyton to manatees; making them an important and desirable part of the lake ecosystem (Figure 11.2). Macrophytes are colonized by a rich array of microbes, especially in hard water. Many invertebrates feed on this epiphytic flora that grows directly on macrophytes or that grows on macrophyte detritus. Stem boring and case building invertebrates use macrophytes for habitat. Large zooplankton use macrophytes as protection from fish predation. Large zooplankton are needed to moderate algae blooms in lakes (see Chapter 9). In North America, few fish feed directly on macrophytes but they feed on the invertebrates that are associated with macrophytes. Macrophytes are important fish habitat and the relationship with fish differs depending on whether the fish is a predator or prey species. Sometimes small areas of the littoral habitat are critical for fish spawning. Seeds, tubers, and foliage of submersed species are food for a variety of wildlife, especially waterfowl. Invertebrates living in macrophyte beds are important for wildlife production. They produce protein that is vital to laying hens and chicks of many waterfowl and related water birds. Higher up the food chain eagles, osprey, loons, mergansers, cormorants, mink, otter, raccoons, and herons, to name a few, feed on fish, shellfish, and invertebrates that live in aquatic plant beds. Nesting sites in, or nesting materials from the emergent zone are important to muskrats and birds like red-winged and yellow-headed blackbirds, marshwrens, grebes, bitterns, and Canada geese. Basically aquatic plants (along with phytoplankton) form the base of the aquatic food web. They are the primary producers of energy that powers the aquatic ecosystem.

Three groups of relationships are important for management: (1) herbivory, (2) intra- and interspecific competition, and (3) pathogenic relationships. Some relationships have developed into management techniques. Others have potential for developing new management strategies.

Herbivorous control of macrophytes has been widely studied (see Chapter 17 for a review). Undesirable plants have been converted into a variety of desirable, or at least innocuous organisms at a higher trophic level. Herbivores (or grazers) include snails, crayfish, turtles, waterfowl, fish, insects and other arthropods, and aquatic mammals. Some organisms are native to the management region; others were imported specifically for aquatic nuisance control. The impacts native grazers

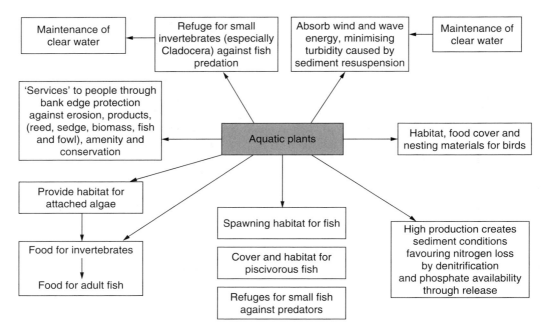

FIGURE 11.2 Links between aquatic plants and other organisms including humans. (From Moss, B. et al. 1996. *A Guide to the Restoration of Nutrient-Enriched Shallow Lakes*. Broads Authority, Norwich, UK. With permission.)

have on macrophyte biomass are under-appreciated (Lodge et al., 1998; Mitchell and Perrow, 1998; Søndergaard et al., 1998). Pelikan et al. (1971) reported that 9–14% of the net annual cattail production was consumed or used as lodge construction by muskrats. Smith and Kadlec (1985) reported that waterfowl and mammalian grazers reduced cattail production by 48% in the Great Salt Lake marsh, Utah, and Anderson and Low (1976) estimated that waterfowl consumed 40% of the peak standing crop of sago pondweed (*Potamogeton pectinatus*) in Delta Marsh, Manitoba. Non-consumptive destruction by grazers also reduces plant biomass. Submerged macrophyte shoots are clipped off by crayfish near the sediment and float away (Lodge and Lorman, 1987). Stem boring insects destroy much more plant tissue than they consume and some insects bore into seeds, rendering them infertile, while consuming little plant tissue. Herbivory can be a serious problem when trying to re-establish aquatic plant communities.

The antagonistic relationships between certain algae and macrophytes have been known for a long time (Hasler and Jones, 1949; Fitzgerald, 1969; Nichols, 1973). This is partially explained by competition for nutrients and light and may be partially explained by the production of allelopathetic substances (Wetzel and Hough, 1973; Phillips et al., 1978; Kufel and Kufel, 2002; van Donk and van de Bund, 2002;). In some shallow, eutrophic lakes there are alternate stable states (Scheffer et al., 1993) where, at different times, a lake is dominated by algae or macrophytes. Biomanipulation management techniques (see Chapter 9) are developing from an understanding of the competition between macrophytes and algae (and a variety of other factors including nutrient status and fish and zooplankton populations), and the alternate stable state, that attempt to change algae dominated lakes into macrophyte dominated lakes (Moss et al., 1996).

Interspecific competition between macrophytes is difficult to study in natural settings (Elakovich and Wooten, 1989; McCreary, 1991) but there is great interest in the area. Competition for resources or allelopathy change plant community structure by allowing one species to replace another or by reducing the productivity or fecundity of a species. There are field reports of dwarf arrowhead (*Sagittaria subulata*) and spikerushes (*Eleocharis acicularis* and *E. coloradoensis*) crowding out pondweeds in irrigation canals (Elakovich and Wooten, 1989). Sutton and Portier (1991) found that the spikerushes *E. cellulosa* and *E. interstincta* contain substances that are phytotoxic to hydrilla. It also appears common for pondweeds and other macrophytes to replace *Chara* and *Najas flexilis* in a successional scheme after major habitat disturbances like dredging or bottom covering (Engel and Nichols, 1984; Nichols, 1984). A desirable outcome of this knowledge would be to plant, or in other ways encourage the growth of desirable but highly competitive or allelopathic plants like the spikerushes, to discourage or eliminate the growth of noxious plants, but this is an area that needs further study (see Chapter 17). An annotated bibliography developed by Elakovich and Wooten (1989) provides more information and resources about aquatic plant allelopathy.

Plant pathogens include fungi, bacteria, and viruses. They have the potential to make desirable biocontrol agents because they are (1) numerous and diverse, (2) often host specific, (3) easily disseminated and self-maintaining, (4) capable of limiting populations without eliminating the species, and (5) non-pathogenic to animals. Aquatic plant management or the potential for aquatic plant management using plant pathogens is discussed in Chapter 17.

11.9 THE EFFECTS OF MACROPHYTES ON THEIR ENVIRONMENT

Habitat and environment influence macrophyte distribution and productivity. Macrophytes also impact the lake ecosystem. How? — The effects are physical, chemical, and biological.

Dense stands of aquatic plants form heavy shade that significantly alters the photosynthetically available light under the canopy (Adams et al., 1974). Shading and reduced water circulation allows vertical temperature gradients as steep as 10° C/m to develop under macrophyte canopies (Dale and Gillespie, 1977).

Reduced water flow through macrophyte beds enhances deposition of fine sediment that would otherwise be eroded (James and Barko, 1990, 1994). Macrophyte beds act as a sieve, retaining coarse particulate organic detritus (Prentki et al., 1979). Both mechanisms increase sediment accumulation.

Daily dissolved oxygen (DO) changes as large as 8 mg/L occur in dense submersed macrophyte beds (Engel, 1990). When plants are photosynthesizing, water can become supersaturated with oxygen. Dark respiration can deplete dissolved oxygen in dense plant beds with little water circulation. Dense growths of floating or matted submersed species decrease oxygenation by inhibiting atmospheric oxygen exchange. Locations with low or widely fluctuating DO concentration provide poor habitat for fish or zooplankton.

Submersed aquatic plant metabolism strongly influence DIC and pH. In dense plant beds, pH can change by three pH units (increasing during rapid photosynthesis then decreasing with respiration and atmospheric CO_2 exchange) during a 24-hour period (Barko and James, 1998). Macrophytes remove inorganic carbon from the water by both assimilation and marl production. Marl production increases sedimentation and can precipitate phosphorus. Macrophytes release dissolved organic compounds into the water that contribute to bacterial metabolism and elevate biological oxygen demand in the littoral zone (Carpenter et al., 1979).

Macrophytes influence nutrient cycles. Phosphorus, for example, is removed from sediments via plant roots and is incorporated into plant biomass (Figure 11.3). When plant tissue dies and decays, phosphorus is circulated, at least briefly, back into the water column. The extent and timing of this cycling greatly influences phytoplankton growth. If nutrients are sequestered in macrophyte biomass during the growing season, little is available for phytoplankton. In northern lakes, if the nutrients are released in the fall, water temperatures are cool so noxious phytoplankton blooms do not occur. Many native species, like wild celery, do not die and decay until fall. Eurasian watermilfoil and hydrilla, however, slough off leaves during the warm season; curly-leaf pondweed typically dies in early summer and Eurasian watermilfoil autofragments during the summer, making nutrients available to phytoplankton during the height of the growing season (Barko and Smart, 1980; Nichols and Shaw, 1986). In Lake Wingra, Wisconsin for example, Adams and Prentki (1982) reported that

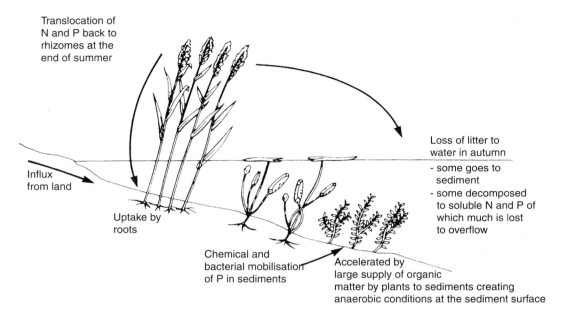

Translocation of N and P back to rhizomes at the end of summer

Influx from land

Uptake by roots

Chemical and bacterial mobilisation of P in sediments

Loss of litter to water in autumn
- some goes to sediment
- some decomposed to soluble N and P of which much is lost to overflow

Accelerated by large supply of organic matter by plants to sediments creating anaerobic conditions at the sediment surface

FIGURE 11.3 Nutrient transfers that occur between plant beds and open water in lakes. (From Moss, B. et al. 1996. *A Guide to the Restoration of Nutrient-Enriched Shallow Lakes.* Broads Authority, Norwich, UK. With permission.)

about two thirds of the seasonal biomass accumulation of milfoil decomposed during the year of production, 50–75% of this is lost in the first 3 weeks of decomposition, and plant biomass was decaying all summer long. Macrophyte decay accounted for about half of the internal phosphorus loading in Lake Wingra (Carpenter, 1983). Nutrient cycles probably differ in southern waters where macrophytes grow year around or where the cold-water season is short or non-existent.

Water chemistry changes caused by abundant macrophyte growth also affect nutrient dynamics. Phosphorus release from littoral sediments is enhanced at high pH. Increases in pH from 8.0 to 9.0 at least doubles the rate of P release from oxic littoral sediments (Barko and James, 1998). As stated above this pH change easily occurs in actively growing macrophyte beds. Anoxic conditions caused by night respiration also enhance sediment P release. For the inlet of Lake Delevan, Wisconsin, Barko and James (1998) reported that 600 kg P were indirectly mobilized from littoral sediments during the summer by altering pH and dissolved oxygen conditions and an additional 600 kg were mobilized directly from sediments by root uptake. Together, the P release from sediments and macrophyte tissue was twice the external P load contributed to Lake Delevan from the watershed.

Macrophyte death and decay also adds organic matter to the sediments. Dissolved oxygen concentrations are influenced by when and how much organic matter is added to the sediments. If large amounts of dead organic matter are added to a lake under warm conditions, DO depletion and the associated impacts on aquatic organisms and nutrient cycling are a concern. In northern climates DO depletion under ice can be critical to fish survival if decaying vegetation is extremely abundant.

Over the short term, organic matter addition is food for benthic organisms. Over the long term, accretion of organic sediments along with sediment trapping and marl precipitation cause expansion of the littoral zone and fill the lake. In general, macrophyte beds are sinks for particulate matter and sources of dissolved phosphorus and inorganic carbon (Carpenter and Lodge, 1986).

Managers need to understand the many effects of macrophytes on their environment and the relationship of macrophytes with the surrounding watershed. They relate directly to the environmental impacts of management. For a more detailed description of these relationships Carpenter and Lodge (1986), Engel (1990), Wetzel (1990), and Jeppesen et al. (1998) are recommended.

In general, aquatic plants are a natural and desirable part of the aquatic ecosystem and the shallow waters of lakes and reservoirs are ideal habitats for plant growth. The desire to have a "weed-free" lake is both naïve and unreasonable. Aquatic plants will become more abundant, at least up to the point of extreme eutrophy or high turbidities, as lakes and reservoirs lose depth to internal processes and to additions of allochthonous material from runoff; and when exotic plants invade. A thorough understanding of macrophyte biology is the basis for developing innovative management approaches. Continued research and development will improve our understanding of the relationship of aquatic plants to overall lake and reservoir quality and our ability to manage aquatic plant communities to maintain or enhance that quality.

REFERENCES

Adams, M.S. and R.T. Prentki. 1982. Biology, metabolism and functions of littoral submersed weedbeds of Lake Wingra, Wisconsin, U.S.: A summary and review. *Arch. Hydrobiol. Suppl.* 62 3/4: 333–409.

Adams, M.S. and K. Sand-Jensen. 1991. Ecology of Submersed Macrophytes. *Aquatic Bot.* 41(1–3), Special Issue.

Adams, M.S., J.E. Titus and M.D. McCracken. 1974. Depth distribution of photosythetic activity in a *Myriophyllum spicatum* community in Lake Wingra. *Limnol. Oceanogr.* 19: 377–389.

Anderson, M.G. and J.B. Low. 1976. Use of sago pondweed by waterfowl on the Delta Marsh, Manitoba. *J. Wildlife Manage.* 40: 233–242.

Aron & Associates. 2000. White River Lake — Aquatic Plant Management Plan. Unpublished report. Wind Lake, WI.

Barko, J.W., M.S. Adams and N.L. Clesceri. 1986. Environmental factors and their consideration in the management of submersed aquatic vegetation: A review. *J. Aquatic Plant Manage.* 24: 1–10.

Barko, J.W. and W.F. James. 1998. Effects of submerged aquatic macrophytes on nutrient dynamics, sedimentation, and resuspension. In: E. Jeppesen, M. Sondergaard, M. Sondergaard and K. Christoffersen (Eds.), *The Structuring Role of Submerged Macrophytes in Lakes*. Ecol. Studies 131. Springer-Verlag, New York, NY. pp. 197–214.

Barko, J.W. and R.M. Smart. 1980. Mobilization of sediment phosphorus by submergent fresh water macrophytes. *Freshwater Biol.* 10: 229–239.

Barko, J.W. and R.M. Smart. 1986. Sediment related mechanisms of growth limitation in submersed macrophytes. *Ecology* 67: 1328–1340.

Beule, J.D. 1979. Control and Management of Cattails in Southeastern Wisconsin Wetlands. Tech. Bull. 112. Wisconsin Dept. Nat. Res., Madison.

Boylen, C.W. and R.B. Sheldon. 1976. Submergent macrophytes: Growth under winter ice cover. *Science* 194: 841–842.

Canfield, D.E., K.A. Langeland, S.B. Linda and W.T. Haller. 1985. Relations between water transparency and maximum depth of macrophyte colonization in lakes. *J. Aquatic Plant Manage.* 23: 25–28.

Carpenter, S.R. 1983. Submersed macrophyte community structure and internal loading: relationship to lake ecosystem productivity and succession. In: J. Taggart and L. Moore (Eds.), *Lake Restoration, Protection and Management*, Proc. Second Ann. Conf. NALMS, Vancouver, BC. pp. 105–111.

Carpenter, S.R., A. Gurevitch and M.S. Adams. 1979. Factors causing elevated biological oxygen demand in the littoral zone of Lake Wingra, Wisconsin. *Hydrobiologia* 67: 3–9.

Carpenter, S.R. and D.M. Lodge. 1986. Effects of submersed macrophytes on ecosystem processes. *Aquatic Bot.* 26:341–370.

Carson, T.L. 2001. Assessing wild rice (*Zizania palustris*) productivity and the factors responsible for decreased yields on Rice Lake, Rice Lake National Wildlife Refuge. In: McGregor, M.N. (Ed.), *Abstracts 41st Annual Meeting, Aquatic Plant Manage. Soc.*, Minneapolis, MN.

Chambers, P.A. and J. Kalff. 1985. Depth distribution and biomass of submersed aquatic macrophyte communities in relation to Secchi depth. *Can. J. Fish. Aquatic Sci.* 42: 701–709.

Clesceri, L.S., A.E. Greenberg and A.D. Eaton. 1998. *Standards Methods for the Examination of Water and Wastewater.* Amer. Publ. Health Assoc., Amer. Water Works Assoc., Water Environ. Fed., Washington, DC.

Cooke, G.D., E.B. Welch, S.A. Peterson and P.R. Newroth. 1986. *Lake and Reservoir Restoration*, 1st Edition. Butterworths, Boston, MA.

Dale, H.M. and T.J. Gillespie. 1977. The influence of submersed aquatic plants on temperature gradients in shallow water bodies. *Can. J. Bot.* 55: 2216–2225.

Dennis, W.M. and B.G. Isom. 1984. *Ecological Assessment of Macrophyton.* ASTM, Philadephia, PA.

Dore, W.G. 1969. *Wild-Rice.* Canada Department of Agriculture, Queens Printer, Ottawa.

Duarte, C.M. and J. Kalff. 1986. Littoral slope as a predictor of maximum biomass of submerged macrophyte communities. *Limnol. Oceanogr.* 31: 1072–1080.

Duarte, C.M. and J. Kalff. 1987. Latitudinal influences on the depths of maximum colonization and maximum biomass of submerged angiosperms in lakes. *Can. J. Fish. Aquatic Sci.* 44: 1759–1764.

Duarte, C.M. and J. Kalff. 1988. Influence of lake morphometry on the response of submerged macrophytes to sediment fertilization. *Can. J. Fish. Aquatic Sci.* 45: 216–221.

Duarte, C.M. and J. Kalff. 1990. Patterns in the submerged macrophyte biomass of lakes and the importance of the scale of analysis in the interpretation. *Can. J. Fish. Aquatic Sci.* 47: 357–363.

Dunst, R.C. 1982. Sediment problems and lake restoration in Wisconsin. *Environ. Int.* 7: 87–92.

Elakovich, S.D. and J.W. Wooten. 1989. Allelopathic Aquatic Plants for Aquatic Plant Management; A Feasibility Study. Tech. Rept. A-89-2. U.S. Army Corps of Engineers, Vicksburg, MS.

Engel, S. 1990. Ecosystem Responses to Growth and Control of Submerged Ma rophytes: A Literature Review. Tech. Bull. 170. Wisconsin Dept. Nat. Res., Madison.

Engel, S. and S.A. Nichols. 1984. Lake sediment alteration for macrophyte control. *J. Aquatic Plant Manage.* 22: 38–41.

Engel, S. and S.A. Nichols. 1994. Aquatic macrophyte growth in a turbid windswept lake. *J. Fresh Water Ecol.* 9: 91–109.

Fassett, N.C. 1930. Plants of some northeastern Wisconsin lakes. *Trans. Wis. Acad. Sci. Arts Lett.* 25: 157–168.

Fitzgerald, G.P. 1969. Some factors in the competition or antagonism among bacteria, algae, and aquatic weeds. *J. Phycol.* 5: 351–359.

Frantz, T.C. and A.J. Cordone. 1967. Observations on deepwater plants in Lake Tahoe, California and Nevada. *Ecology* 48: 709–714.

Gerloff, G.C. 1973. Plant Analysis for Nutrient Assay of Natural Waters. Rept. USEPA-R1-73-001. Washington, DC.

Grodowitz, M.J., S.G. Whitaker and L. Jeffers. 2001a. *Aquatic Plant Information System.* U.S. Army Eng., Waterways Exp. Sta., Vicksburg, MS.

Grodowitz, M.J., S.G. Whitaker and L. Jeffers. 2001b. Noxious and Nuisance Plant Management Information System. Eng. Res. Dev. Center, U.S. Army Corps of Engineers, Vicksburg, MS.

Hasler, A.D. and E. Jones. 1949. Demonstration of the antagonistic action of large aquatic plants on algae and rotifers. *Ecology* 30: 359–364.

Hoyer, M.V. and D.E. Canfield. 1997. *Aquatic Plant Management in Lakes and Reservoirs.* NALMS, Madison, WI and Lehigh, FL.

Hutchinson, G.E. 1975. *A Treatise on Limnology-Limnological Botany.* John Wiley, New York.

James, W.F. and J.W. Barko. 1990. Macrophyte influences of the zonation of sediment accretion and composition in a north-temperate reservoir. *Verh. Int. Verein. Limnol.* 120: 129–142.

James, W.F. and J.W. Barko. 1994. Macrophyte influences on sediment resuspension and export in a shallow impoundment. *Lake and Reservoir Manage.* 10: 95–102.

Jeppesen, E., M. Sondergaard, M. Sondergaard and K. Christoffersen. 1998. *The Structuring Role of Submerged Macrophytes in Lakes.* Ecol. Studies 131. Springer-Verlag, New York.

Kimber, A., C.E. Korschgen and A.G. van der Valk. 1995. The distribution of *Vallisneria americana* seeds and seedlings light requirements in the Upper Mississippi River. *Can. J. Bot.* 73: 1966–1973.

Korth, R., S. Engel and D.R. Helsel. 1997. *Your Aquatic Plant Harvesting Program.* Wisconsin Lakes Partnership, Madison.

Kufel, L. and I. Kufel. 2002. *Chara* beds acting as nutrient sinks in shallow lakes - a review. *Aquatic Bot.* 72: 249–260.

Langeland, K.A. and D.L. Sutton. 1980. Regrowth of hydrilla from axillary buds. *J. Aquatic Plant Manage.* 18: 27–29.

Lind, C.T. 1976. The phytosociology of submerged macrophytes in eutrophic lakes of southeastern Minnesota. Ph.D. Thesis, Univ. Wisconsin, Madison.

Linde, A.F., T. Janisch and D. Smith. 1976. *Cattail — The Significance of its Growth, Phenology and Carbohydrate Storage to its Control and Management.* Tech. Bull. 94. Wisconsin Dept. Nat. Res., Madison.

Lodge, D.M., G. Cronin, E. Van Donk and A.J. Froelich. 1998. Impact of herbivory on plant standing crop: comparison among biomes, between vascular and nonvascular plants, and among fresh water herbivore taxa. In: E. Jeppesen, M. Sondergaard, M. Sondergaard and K. Christoffersen (Eds.), *The Structuring Role of Submerged Macrophytes in Lakes.* Ecol. Studies 131. Springer-Verlag, New York, NY. pp. 149–174.

Lodge, D.M. and J.G. Lorman. 1987. Reduction of submersed macrophyte biomass and species richness by the crayfish *Orconectes rusticus. Can. J. Fish. Aquatic Sci.* 44: 591–597.

Luu, K.T. and K.D. Getsinger. 1988. *Control Points in the Growth Cycle of Waterhyacinth.* Aquatic Plant Cont. Res. Prog. Rept., Vol. A-88-2. U.S. Army Eng., Waterways Exp. Sta., Vicksburg, MS.

Madsen, J.D. 1991. Resource allocation at the individual plant level. *Aquatic Bot.* 41: 67–86.

Madsen, J.D. and C.S. Owens. 1996. Phenological Studies to Improve Hydrilla Management. Aquatic Plant Cont. Res. Prog. Rept., Vol. A-96-2. U.S. Army Eng., Waterways Exp. Sta., Vicksburg, MS.

McCreary, N.J. 1991. Competition as a mechanism of submersed macrophyte structure. *Aquatic Bot.* 41: 177–193.

Messner, N. and R. Narf. 1987. Alum injection into sediments for phosporus inactivation and macrophyte control. *Lake and Reservoir Manage.* 3: 256–265.

Mitchell, D.S. 1979. Formulating aquatic weed management programs. *J. Aquatic Plant Manage.* 17: 22–24.

Mitchell, S.F. and M.R. Perrow. 1998. Interactions between grazing birds and macrophytes. In E. Jeppesen, M. Sondergaard, M. Sondergaard and K. Christoffersen (Eds.), *The Structuring Role of Submerged Macrophytes in Lakes.* Ecol. Studies 131. Springer-Verlag, New York, NY. pp. 175–196.

Moeller, R.E. 1980. The temperature-determined growing season of a submerged hydrophyte: tissue chemistry and biomass turnover of *Utricullaria purpurea. Freshwater Biol.* 10: 391–400.

Moeller, R.E., R.G. Wetzel and C.W. Osenberg. 1998. Concordance of phosphorus limitations in Lakes: bacterioplankton, phytoplankton, epiphytes-snail consumers, and rooted macrophytes. In E. Jeppesen, M. Sondergaard, M. Sondergaard and K. Christoffersen (Eds.), *The Structuring Role of Submerged Macrophytes in Lakes.* Ecol. Studies 131, Springer-Verlag, New York, NY. pp. 318–325.

Moss, B., J. Madgwick and G.L. Phillips. 1996. *A Guide to the Restoration of Nutrient-Enriched Shallow Lakes.* Broads Authority, Norwich, UK.

NALMS. 1993. Aquatic Vegetation Quantification Symposium. *Lake and Reservoir Manage.* 7: 137–196.

Nichols, S.A. 1973. The effects of harvesting macrophytes on algae. *Trans. Wis. Acad. Sci. Arts Lett.* 61: 165–172.

Nichols, S.A. 1984. Macrophyte community dynamics in a dredged Wisconsin lake. *Water Res. Bull.* 20: 573–576.

Nichols, S.A. 1992. Depth, substrate and turbidity relationships of some Wisconsin lake plants. *Trans. Wis. Acad. Sci. Arts Lett.* 80: 97–119.

Nichols, S.A., S. Engel and T. McNabb. 1988. Developing a plan to manage lake vegetation. *Aquatics* 10(3): 10–19.

Nichols, S.A. and S.J. Rogers. 1997. Within-bed distribution of *Myriophyllum spicatum* L. in Lake Onalaska, upper Mississippi River. *J. Fresh Water Ecol.* 12: 183–191.

Nichols, S.A. and B.H. Shaw. 1986. Ecological life histories of the three aquatic nuisance plants, *Myriophyllum spicatum, Potamogeton crispus,* and *Elodea canadensis. Hydrobiologia* 131: 3–21.

Nicholson, S.A. 1981. Changes in submersed macrophytes in Chautauqua Lake, 1937–1975. *Freshwater Biol.* 11: 523–530.

Pelikan, J., J. Svoboda and J. Kvet. 1971. Relationship between the population of muskrats (*Ondatra zibethica*) and primary production of cattail (*Typha latifolia*). *Hydrobiologia* 12: 177–180.

Perkins, M.A. and M.D. Sytsma. 1987. Harvesting and carbohydrate accumulation in Eurasian watermilfoil. *J. Aquatic Plant Manage.* 25: 57–62.

Phillips, G.L., D. Eminson and B. Moss. 1978. A mechanism to account for macrophyte decline in progressively eutrophicated fresh waters. *Aquatic Bot.* 4: 103–126.

Pieterse, A.H. and K.J. Murphy. 1990. *Aquatic Weeds, The Ecology and Management of Nuisance Aquatic Vegetation.* Oxford Univ. Press, Oxford, UK.

Prentki, R.T., M.S. Adams, S.R. Carpenter, A. Gasith, C.S. Smith and P.R. Weiler. 1979. The role of submersed weedbeds in internal loading and interception of allochthonous materials to Lake Wingra, Wisconsin. *Arch. Hydrobiol. Suppl.* 57. 2: 221–250.

Robel, R.J. 1961. Water depth and turbidity in relation to growth of sago pondweed. *J. Wildlife Manage.* 25: 436–438.

Rogers, S.J., D.G. McFarland and J.W. Barko. 1995. Evaluation of the growth of *Vallisneria americana* Michx. in relation to sediment nutrient availability. *Lake and Reservoir Manage.* 11: 57–66.

Sabol, B.M. 1987. Environmental effect of aquatic disposal of chopped hydrilla. *J. Aquatic Plant Manage.* 25: 19–23.

Scheffer, M., S.H. Hosper, M.-L. Meijer, B. Moss and E. Jeppesen. 1993. Alternative equilibria in shallow lakes. *Trends Ecol. Evol.* 8: 275–279.

Schmid, W.D. 1965. Distribution of aquatic vegetation as measured by line intercept. *Ecology* 46: 816–823.

Sculthorpe, C.D. 1985. *The Biology of Aquatic Vascular Plants.* 2nd ed. Koeltz Scientific Books, Konigstein, Germany.

Sheldon, R.B. and C.W. Boylen. 1977. Maximum depth inhabited by aquatic vascular plants. *Am. Mid. Nat.* 97: 248–254.

Singer, R., D.A. Roberts and C.W. Boylen. 1983. The macrophyte community of an acidic lake in Adirondack (New York, U.S.A.): A new depth record for aquatic angiosperms. *Aquatic Bot.* 16: 49–57.

Smart, R.M. 1990. Effects of Water Chemistry on Submersed Aquatic Plants: A Synthesis. Misc. Paper A-90-4, U.S. Army Corps of Engineers, Vicksburg, MS.

Smith, L. and J. Kadlec. 1985. Fire and herbivory in a Great Salt Lake marsh. *Ecology* 66: 259–265.

Sondergaard, M., T.L. Lauridsen, E. Jeppesen and L. Bruun. 1998. Macrophyte-waterfowl interactions: tracking a variable resource and the impact of herbivory on plant growth. In: E. Jeppesen, M. Sondergaard, M. Sondergaard and K. Christoffersen (Eds.), *The Structuring Role of Submerged Macrophytes in Lakes.* Ecol. Studies 131. Springer-Verlag, New York, NY. pp. 298–306.

Spence, D.H.L. 1967. Factors controlling the distribution of fresh water macrophytes with particular reference to the lochs of Scotland. *J. Ecol.* 55: 147–170.

Spence, D.H.L. 1975. Light and plant response in fresh water. In: G. Evans, O. Rackham and C. Bainbridge (Eds.), *Light as an Ecological Factor: II.* Blackwell Sci., Oxford, UK. pp. 93–133.

Spencer, W.E. and G. Bowes. 1990. Ecophysiology of the world's most troublesome weeds. In: A. Pieterse and K. Murphy (Eds.), *Aquatic Weeds, The Ecology and Management of Nuisance Aquatic Vegetation.* Oxford Univ. Press, Oxford, UK. pp. 39–73.

Stuckey, R.L., J.R. Wehrmeister and R.J. Bartolotta. 1978. Submersed aquatic vascular plants in ice-covered ponds of central Ohio. *Rhodora* 80: 575–580.

Sutton, D.L. and K.M. Portier. 1991. Influence of spikerush plants on growth and nutrient control of hydrilla. *J. Aquatic Plant Manage.* 29: 6–11.

Titus, J.E. 1977. The comparative physiological ecology of three submerged macrophytes. Ph.D. thesis. Univ. Wisconsin, Madison.

Titus, J.E. and M.S. Adams. 1979. Coexistence and the comparative light relations of the submersed macrophytes *Myriophyllum spicatum* and *Vallisneria americana. Oecologia* 40: 273–286.

Titus, J.E. and D.T. Hoover. 1991. Toward predicting reproductive success in submersed fresh water angiosperms. *Aquatic Bot.* 41: 111–136.

van Donk, E. and W.J. van de Bund. 2002. Impact of submerged macrophytes including charophytes on phyto- and zooplankton communities: allelopathy versus other mechanisms. *Aquatic Bot.* 72: 261–274.

van Nes, E.H., M. Scheffer, M.S. van den Berg and H. Coops. 2002. Dominance of charophytes in eutrophic shallow lakes - when should we expect it to be an alternative stable state? *Aquatic Bot.* 72: 275–296.

Washington Department of Ecology. 1994. A Citizen's Manual for Developing Integrated Aquatic Vegetation Managment Plans. Olympia, WA.

Wetzel, R.G. 1990. Land-water interfaces: Metabolic and limnological regulators. *Verh. Int. Verein. Limnol.* 24: 6–24.

Wetzel, R.G. 2001. *Limnology.* 3rd ed. Academic Press, London, UK.

Wetzel, R.G. and R.A. Hough. 1973. Productivity and the role of aquatic macrophytes in lakes-an assessment. *Pol. Arch. Hydrobiol.* 20: 9–19.

12 Plant Community Restoration

12.1 INTRODUCTION

Because of the vital role plants play in the aquatic ecosystem there is a growing interest in restoring aquatic plant communities. Aquatic plant restoration may: (1) improve fish and wildlife habitat; (2) reduce shoreline erosion and bottom turbulence; (3) buffer nutrient fluxes; (4) shade shorelines; (5) reduce nuisance macrophyte and algae growth; (6) treat stormwater and wastewater effluent; (7) replace exotic invaders with native species; (8) improve aesthetics; and (9) generally moderate environmental disturbance. Although there is some debate about the proper term(s) — enhance, restore, rehabilitate, develop, restructure — for these efforts (Haslam, 1996; Moss et al., 1996; Munrow, 1999) the essence is to return aquatic plants to areas where they were previously found, to develop areas where they should be found, or to restructure present plant populations to provide the ecological assets of a healthy macrophyte community. For purposes of this discussion, restoration is broadly and loosely defined. It can mean planting a single species where plants were previously extirpated. It can mean changing habitat conditions so revegetation occurs naturally. It can mean restoring diversity to a monotypic, exotic plant community. It can mean doing nothing and letting nature take its course. In few, if any, cases is an aquatic plant community restored or rehabilitated in the strictest, ecological definition of the terms (Haslam, 1996; Moss et al., 1996; Munrow, 1999; Chapter 1).

Various techniques have been used to restore saline and fresh water marshes, swamps, seagrasses, and fresh water plants in lakes and streams (Kadlec and Wentz, 1974; Johnston et al., 1983; Orth and Moore, 1983; Marshall, 1986; Storch et al., 1986; Moss et al., 1996). The technology for aquatic plant community restoration is quickly developing but presently it is still as much of an art as it is a science. Much more is known about restoring wetlands, which includes shoreline emergent plants, than is known about restoring submergent communities.

Table 12.1 lists decision items for estimating the potential for success or the amount of work involved in a plant restoration. If the habitat for restoration has most of the items in the right or "increase success" side of the table, little or nothing other than patience may be needed to restore plants. If most items are in the "decrease success" column, anticipate more work, expense, and potential for failure. The suggested remedies are broad categories. They may not be suitable because of cost, physical limitations, environmental impact, or regulatory or political realities at any specific location. For instance, drawdown may be physically impossible, prohibitively expensive, or not approved by regulatory agencies on a natural lake without a control structure. Some techniques are untested. Would algicide treatments, a selective plant management technique, temporarily increase water clarity for macrophyte establishment? Most restoration areas need some remediation or a desirable plant community would be present. Remediation and restoration should not be viewed as a single effort. For example, after macrophytes are planted, they may need protection from predators and waves before they become successfully established and spread. Careful selection of plant material can overcome some habitat limitations. Some species are more tolerant of turbidity or fluctuating water levels, or are able to grow in deeper water than are other species. Many of the suggested remedies are discussed in other sections of this book (e.g., nutrient limitation and inactivation to increase water clarity, drawdown, dredging) or are discussed more thoroughly in the following sections and in the case histories.

TABLE 12.1
Decision Items for Assessing Plant Restoration Potential and Suggested Remediation Techniques

Factors for Assessing Plant Restoration Potential	Decrease Success	Increase Success	Remedies[a]
Water clarity	Turbid water	Clear water during most of growing season	1, 2, 3, 4, 12
Sediment characteristics			
Density	Low density	Moderate to high density	2, 6, 7
Organic matter content	High	Moderate to low	2, 6, 7
Toxicity	Toxic	Non-toxic	5, 6, 7
Predator population	High	Low	3, 4, 8
Environmental energy (current, waves, etc.)	High	Low	4, 9
Water			
Depth	Deep	Shallow	2
Stability	Fluctuating	Stable water level	4, 10
Plant population			
Residual plants	Few or none	Abundant	11
Sediment seed bank	Few or none	Abundant	11
Plant population in area	Few or none	Abundant	11
Non-desirable species	Abundant	Few or none	11, 12, 13

[a] Types of remedies: (1) nutrient limitation, (2) drawdown, (3) fish population manipulation, (4) physical barriers, (5) aeration, (6) shallow dredging, (7) sand blanket, (8) predator population control, (9) slow-no-wake or no-motor regulations, (10) stabilize water level, (11) macrophyte planting, (12) selective plant management, (13) do nothing.

Some tests needed to enhance restoration success are simple. Secchi depths explain a lot about water clarity and whether algal blooms, benthivorous fish, wind and waves, or heavy powerboat use causes turbidity. Aquarium tests determine sediment seed banks, sediment suitability for plant growth, and propagule viability. Wind and wave impacts are estimated from local weather summaries, a lake map, and observation of plant distribution. Simple observation is used to determine animal and human use (e.g., carp (*Cyprinus carpio*) spawning, powerboating, bank fishing). Plant collections determine species occurrence and distribution. These tests may not be all that are needed but they will answer some of the basic questions needed for a successful restoration.

Aquatic plant restoration is discussed based on the level of effort needed to complete a project. The least effort method is "doing nothing," followed by habitat protection and alteration, and finally by active establishment. In reality all three might be needed in a single project. Habitat may need to be altered before any plants will grow. After alteration, doing nothing for a growing season or two determines if natural revegetation will occur. If natural revegetation occurs, the additional cost and time-consuming effort of planting may not be needed. If this is not the case, planting is needed to increase desirable species, diversity, or to revegetate difficult areas. Even after successful plant establishment, further efforts are usually needed to protect the plant community. For instance, herbivory may be a problem or other aquatic plant management techniques may be needed to control nuisance macrophytes.

12.2 THE "DO NOTHING" APPROACH

There is evidence that aquatic plant management techniques such as harvesting and herbicidal treatment favor rapidly reproducing, aggressively growing species — the weeds (Cottam and Nichols, 1970; Nicholson, 1981; Bowman and Mantai, 1993; Doyle and Smart, 1993; Nichols and

Lathrop, 1994). Plant succession is continually "set back." So what can be done? — "do nothing" and hope that natural successional trends will re-establish a diverse community of non-weedy, native species. The advantages of doing nothing are that the developing plants are from local sources and they are adapted to local conditions so they may have the best chance for survival. The technique is inexpensive and plant succession is not continually "set back" so the community that develops may be the most stable for existing conditions. The disadvantages are that it may take a long time for a plant community to develop or change, especially if the area was not previously vegetated and/or if there is no natural source of propagules in the area (Smart and Dick, 1999; Nichols, 2001). Little is known about the dynamics of aquatic plant community change so the results are unpredictable and doing nothing may be politically unpalatable. There is also evidence that plant communities can change from a diverse native community to one dominated by exotics without cause or manipulation. For example, the plant community in the Cassadaga Lakes, New York, with little or no management, changed from one dominated primarily by native pondweeds to one dominated by curly-leaf pondweed and Eurasian watermilfoil — an obvious case where doing nothing did not work (Bowman and Mantai, 1993). In some locations the "do nothing" approach is codified by designating areas as critical habitat, which is a regulatory approach to protect areas so restoration is not needed or can occur naturally.

12.2.1 CASE HISTORY: LAKE WINGRA, "DOING NOTHING"

Lake Wingra is a 137-ha, shallow (mean depth of 2.4 m), urban lake located in Madison, Wisconsin. The University of Wisconsin Arboretum and city parkland surrounds it so, unlike many urban lakes, the shoreline is not heavily developed. Around 1900, *Equisetum* spp., *Zizania* sp., *Typha latifolia*, *T. angustifolia,* and *Scirpus validus* were common species of the broad marshes surrounding the lake. Dense growths of *Chara* spp. were interspersed between the emergents. Wild celery (*Vallisneria americana*) was particularly abundant. There were at least 34 species of aquatic plants in Lake Wingra at this time and the lake bottom was completely vegetated (Bauman et al., 1974). During the first half of the 20th century dredging, filling, water-level fluctuation, and the introduction of carp decimated the aquatic vegetation. Macrophytes were sparse from the late 1920s through 1955 (Bauman et al., 1974). Eurasian watermilfoil (*Myriophyllum spicatum*) invaded Lake Wingra in the early 1960s and by 1966 it was dominant and replaced the remaining native species. From the mid-1960s to the early 1970s *M. spicatum* was present in dense stands in shallow areas of the lake. The milfoil stands declined in 1977 (Carpenter, 1980). Except for some minor plant harvesting around a public boat livery and a swimming beach, there was little or no management on Lake Wingra after the early 1950s when carp were seined to low levels.

The reason for the milfoil decline was never adequately determined. Between 1969 and 1996 species number increased slightly, Simpson's (1949) diversity increased dramatically from 0.52 to 0.88, the relative frequency of exotic species (*M. spicatum* and *Potamogeton crispus*) dropped from 68.9% to 35.9%, and the relative frequency of species sensitive to disturbance (Nichols et al., 2000) increased from 0.1% to 19.1% (Table 12.2). The maximum depth of plant growth increased from 2.7 m to 3.5 m. Wild celery and *Potamogeton illinoensis* returned — they were last reported in the lake in 1929. The vegetation recovery in Lake Wingra was more dramatic than in the other Madison, Wisconsin area lakes that had a similar history of an Eurasian watermilfoil invasion (Nichols and Lathrop, 1994) but are more heavily managed.

The vegetation recovery in Lake Wingra was neither planned nor predicted so why did the vegetation recover? No reason can be given with absolute certainty because the results are observational and were not part of an experimental program. Historically Lake Wingra had a rich aquatic flora and even at the height of the milfoil invasion there were more than 15 species of plants in the lake. Dane County, Wisconsin also has 24 lakes greater than 30 ha in size so there is an abundant supply of aquatic plant propagules in the vicinity for invasion and there is probably a seed (propagule) bank in the sediment, although this was never tested. After the abundant carp population

TABLE 12.2
Comparison of Species Relative Frequencies in Lake Wingra,
Wisconsin between 1969 and 1996[a]

Plant Species	Rel. Freq. (%) 1969[b]	Rel. Freq. (%) 1996
Myriophyllum spicatum	68.4	27.4
Potamogeton pectinatus	8.1	6.6
Potamogeton natans	6.2	1.3
Nuphar variegatum	4.8	0.4
Potamogeton nodosus	3.0	—
Ceratophyllum demersum	2.9	8.4
Nymphaea tuberosa	2.6	3.5
Chara sp.	—	7.1
Najas flexilis	0.3	2.2
Potamogeton crispus	0.5	8.4
Potamogeton foliosus	0.1	5.8
Potamogeton richardsonii	0.2	6.6
Potamogeton zosteriformis	0.5	9.3
Vallisneria americana	—	5.3
Potamogeton sp.[c]	—	4.0
Other species[d]	2.4	3.7
Simpson diversity[e]	0.52	0.88

[a] Does not include emergent species.

[b] After Nichols, S.A. and S. Mori. 1971. *Trans. Wis. Acad. Sci. Art Lett.* 59: 107–119.

[c] Probably *Potamogeton illinoensis.*

[d] Species with less than 1.0% relative frequency in either or both sampling periods; includes *Elodea canadensis, Zosterella dubia,* and *Ranunculus longirostris.*

[e] A modification of Simpson, W. 1949. *Nature* 163: 688.

was seined to low levels in the early 1950s they never regained their former abundance. The lake is shallow, with fine, moderately organic, and moderately nutrient rich sediments. There has been no major disturbance of the plant beds due to management activities and there is a "slow-no-wake" boating ordinance on the lake. In total, Lake Wingra is an ideal location for aquatic plant growth and given the chance, they returned. Eurasian watermilfoil declines occurred in other lakes and native species are returning (Smith and Barko, 1992; Nichols, 1994; Helsel et al., 1999;) so the Lake Wingra experience is not unusual.

12.3 THE HABITAT ALTERATION APPROACH

The degradation or decimation of aquatic plant communities often resulted from major habitat alterations. Plant communities were lost because of water level increases; wind and wave erosion; actions of benthivorous fish or plant predators; and cultural eutrophication, aquatic plant management, or other human activities. Often a combination of these factors led to the demise of macrophyte communities (Nichols and Lathrop, 1994). The end result is turbid water and/or high-energy environments that are unsuitable for aquatic plant growth. Reversing unsuitable habitat conditions allows vegetation to return. Both regulatory and more active approaches involving engineering or biomanipulation are used to alter habitat. The disadvantages to these approaches are that there is no way of predicting the results and they may be politically unpalatable, especially regulatory approaches. Restoration may take a long time but experience indicates that revegetation occurs rapidly once limiting habitat factors are removed. An advantage is the plant community that develops is from local sources so it should be adapted to local conditions. Costs and environmental impacts are highly variable, depending on the technique. Regulatory approaches like establishing no-motor or slow-no-wake zones are inexpensive and environmentally benign or beneficial. Fencing "founder" colonies of remaining plants to protect them from predation is of moderate cost. Constructing islands and breakwalls to protect plants from wind and waves, large-scale fish removal projects, and nutrient reduction techniques are expensive, some costing in the millions of dollars; and they may have moderate to severe environmental impacts, at least over the short-term.

12.3.1 Case History: No-Motor, Slow-No-Wake Regulations

12.3.1.1 Long and Big Green Lakes: Heavily Used Recreational Lakes in Southeastern Wisconsin

12.3.1.1.1 Long Lake

Long Lake has a surface area of 169 ha and a maximum depth of 14.3 m. A dam installed in 1855 raised the natural level of this glacial lake by 2 m. This created an extensive littoral zone extending from shore by as much as 120 m before dropping sharply into deep water. The Long Lake State Recreation Area occupies the east shore of the lake and the west shore is developed with permanent and seasonal homes. A 1989 survey found that Long Lake had 7,088 boating days of annual use, which corresponds to 41 boating days per hectare per year (Asplund and Cook, 1999). Peak boating activity occurs in July, with as many as 60 boats present on some weekends. The lake is long and narrow in a north-south direction, which makes it ideal for water-skiing and inner tubing. The lake has at least 22 species of floating and submerged plants with *Chara* sp. being the most abundant species.

Local property owners wanted to protect aquatic vegetation for fish habitat. They were also worried about water quality problems from the exposed sediments and that disturbed areas might be colonized by Eurasian watermilfoil. Aerial photos showed major areas of the shallow littoral zone that were devoid of plant growth. The worst area was along the eastern shore.

In May 1997 the Long Lake Fishing Club placed slow-no-wake buoys for approximately 1,500 m along the east shore of the lake. The buoys were placed approximately 120 m out from the shore so the slow-no-wake zone extended from the buoys into the shoreline. Two no-motor zones of about 125 m each were placed within the slow-no-wake zone. Although the restrictions were technically voluntary, a concerted effort was made to educate lake users about the importance of respecting the special boat-use zones.

Asplund and Cook (1999) assessed the submerged macrophyte community in late August of 1997. They found that the large scour (non-vegetated) areas seen in 1995 were almost completely covered with *Chara*. Scour areas were reduced to as little as 1.5% of the area (Table 12.3). Boat tracks were still evident in the no-wake area, but at a much lower frequency. This suggests that boats still uproot plants or cut off stems at no-wake speeds. Alternative explanations are that boaters occasionally traveled through the area at faster speeds or anchors were dragged along the bottom. No boat tracks were observed in the no-motor plots. Sampling found very little vegetational difference between management areas in terms of overall stand density and canopy height. One can only speculate on the reason, but unprotected comparison areas may have historically received less boat use and were in fact protected because boaters avoided the east side of the lake that was largely a no-wake zone. In 1998 the local town board permanently established a no-wake zone along the eastern side of the lake, but the buoys were placed closer to the shoreline so that about one-third to one-half the area protected in 1997 was outside the no-wake zone. Aerial photography revealed that boat scour and tracks eliminated much of the *Chara* that grew in 1997 in this newly unprotected area.

TABLE 12.3
Comparison of the Percentage of Unvegetated Area in Protected and Unprotected Areas of Long Lake, Wisconsin between 1995 and 1997[a]

Area	1995 (before protection) (%)	1997 (during protection) (%)
No-motor	2.7	1.5
No-wake	17.4	2.0
Unprotected	12.0	2.2

[a] Vegetation consisted primarily of *Chara* sp. and native milfoils.

Source: After Asplund, T. and C.E. Cook. 1999. *LakeLine* 19(1): 16.

TABLE 12.4
Bulrush Density and Bed Size in Big Green Lake, Wisconsin

Bed location	Bed Area (m²)			Bulrush stem density (stems/m²)		
	1997	1998	2002	1997	1998	2002
Southwest	114	94	77	8	45.3	12.5
Center	1223	1268	1157	20.9	26.3	19.3
Northeast	505	662	432	24.3	25.3	23.2
Total	1842	2024	1666			

Source: Data supplied by Chad Cook, Wis. Dept. Nat. Res. Personal communications, 2002.

12.3.1.1.2 Big Green Lake

Big Green Lake is large (surface area 2,974 ha) and deep (maximum depth 72 m). However, it has shallow bays where hardstem bulrush (*Scirpus acutus*) was an important part of the emergent vegetation. Historical accounts identified five bulrush stands in the lake ranging in size from 3,500 to 255,000 m² (Asplund and Cook, 1999). The largest remaining stand was about 1,840 m² in size in 1997 and appeared to be shrinking.

Motorboat activity was thought to be a major factor in the decline of this remaining stand. The sandbar area adjacent to the stand is a popular place for mooring boats and wading. To address this concern the local town board enacted an ordinance in 1997 to place no-motor buoys around the stand. The extent of the stand was divided into three sections and mapped in 1997, 1998, and 2002 using GPS. Stem densities were also determined for those years. In 1998 the stand size and stem densities appeared to be somewhat greater or at least not shrinking (Table 12.4; Asplund and Cook, 1999). By 2002 stand size and stem densities have not increased and may have decreased slightly (Table 12.4). After 5 years of protection, it appears, at best, that restricting motorboat traffic has slowed the decline of the bulrush bed.

12.3.1.2 Active Habitat Manipulation: Engineering and Biomanipulation Case Studies

12.3.1.2.1 Lake Ripley, Wisconsin: Boat Exclosures

Lake Ripley has a surface area of 169 ha, a maximum depth of 13.4 m, and an extensive littoral area less than 2 m deep. Littoral sediments are very flocculent and easily resuspended due to a high percentage of marl. Homes ring the lake, there are more than 300 boats docked around the

TABLE 12.5
Average Plant Growth in Protected and Unprotected Areas in Lake Ripley, Wisconsin

Location	Percent Cover (%)	Max. Plant Height (cm)	Biomass (g/m²)
Unprotected area	58	46	434
Protected area — mesh fencing	84	82	823
Protected area — solid fencing	82	61	1063

Source: After Asplund, T. and C.E. Cook. 1997. *Lake and Reservoir Manage.* 13: 1–12.

lake, and on weekends boat use approaches 50 boats on the lake at one time (Asplund and Cook, 1997). Historically, Lake Ripley had a diverse plant community dominated by wild celery, pond-weeds (primarily *P. illinoensis* and *P. pectinatus*), and water lilies (*Nuphar variegata* and *Nymphaea odorata*). Eurasian watermilfoil dominated the vegetation in the 1980s but has since declined. Native species were slow to recolonize areas suitable for plant growth.

Motorboats had a major impact on aquatic plants. Boat "tracks" or scour lines through the remaining plant beds were visible from aerial photos in areas of high boat traffic (Asplund and Cook, 1997). It was not known whether the impact was due to increased turbidity caused by resuspension of bottom sediments, turbulence from boat wakes and prop wash, direct scouring of the sediment, direct cutting by motor propellers, or breakage from contact with boat hulls.

Asplund and Cook (1997) examined the impact of motorboating on the aquatic plant community by constructing two solid plastic and two mesh fencing exclosures in the lake that excluded boat access. After a single growing season, species composition was similar between the plots with *Chara* sp. and *Najas marina* being the predominant species. However, plant growth between the areas varied considerably. The plant growth in the protected areas was not significantly different between those areas protected with mesh or solid fencing. The protected areas had about one and one-half times as much area covered, about one and one-half to two times the maximum plant height, and two to two and one-half times the biomass as the unprotected areas (Table 12.5). Through additional water chemistry testing they concluded that motorboats reduced plant biomass by sediment scouring and direct cutting of the plants, but not by turbidity generation.

Similar exclosures of varying sizes and designs were needed to protect remaining plants from herbivorous fish, wading or aquatic mammals, and waterfowl in other restoration efforts (Moss et al, 1996; van Donk and Otte, 1996).

12.3.1.2.2 Big Muskego and Delavan Lakes, Wisconsin: Drawdown, Benthivorous Fish Removal, and Nutrient Reduction

Big Muskego Lake has a surface area of 840 ha and is very shallow (mean depth is 0.75 m). A dam built in the 1800s flooded this former deep-water marsh. The lake is eutrophic and drains a predominantly agricultural watershed of 7,600 ha. Before the treatment the submersed plant community was dominated by Eurasian watermilfoil and common carp dominated the fishery. Although the lake was well vegetated, with approximately 95% of the area containing vegetation; soft, flocculent and highly organic sediments, carp, wind and wave action, and turbidity limited the growth of desirable, native aquatic plants. Besides increasing plant diversity, wildlife managers were interested in increasing the extent of the emergent zone. Before treatment cattails were found at 9.9% of the sampling points and all other emergent species were found at less than 1% of the sampling points.

Drawdown (Chapter 13) of Big Muskego Lake started in October 1995 and the water level was lowered by about 0.5 m between December 1995 and July 1996. A channel was excavated to promote further drawdown in mid-July 1996. This caused an additional 0.5 to 0.6 m drawdown

TABLE 12.6
The Vegetation of Big Muskego Lake before and after Drawdown and Carp Removal

Species	Rel. Freq. (%) Pre-treatment (1995)	Rel. Freq. (%) Post-treatment (1997)
Ceratophyllum demersum	2.9	0.6
Chara sp.	—	15.5
Lemna minor	—	12.2
Lythrum salicaria	6.5	2.8
Myriophyllum sibiricum	3.8	1.0
Myriophyllum spicatum	61.6	8.3
Najas marina	1.3	4.2
Nuphar variegata	4.2	0.1
Nymphaea odorata	3.6	2.8
Potamogeton amplifolius	2.3	0.1
Potamogeton crispus	0.6	1.7
Potamogeton illinoensis	—	1.3
Potamogeton pectinatus	1.6	11.9
Ranunculus longirostris	0.6	3.8
Scirpus spp.	0.3	14.2
Typha latifolia	6.8	18.1
Other species[a]	3.9	1.4

[a] Other species include: *Carex* spp., *Ceratophyllum echinatum, Elodea canadensis, Zosterella dubia, Najas flexilis, Potamogeton nodosus, P. pusillus, Sagittaria latifolia,* and *Zizania aquatica.* They had a relative frequency of less than 1% in both sampling periods.

Source: Data from John Madsen, Department of Biology, Minnesota State University, Mankato. Personal communications, 2002.

between July 1996 and January 1997. The lake was allowed to refill during late winter and early spring 1997. Normal water levels returned by April 1997. Overall, about 13% of the sediment area was exposed for approximately 1 year, while over 80% of the sediment area was exposed for about 6 months (James et al., 2001a, b). In addition, drawdown concentrated undesirable fish so they were more easily removed with a piscicide.

Prior to drawdown Muskego Lake sediments were very fluid. Surface sediments were over 90% water, sediment density was low, and organic content of the sediment was very high (more than 40%) (James et al., 2001a, b). Hopefully, desiccation would consolidate sediments, reducing resuspension potential and turbidity. A concern was the effect oxidation of aerially exposed sediments might have on mobilizing sediment organic nitrogen and phosphorus. Internal nutrient loading after reflooding exposed sediments could stimulate excessive algal blooms that would be counterproductive to macrophyte growth.

Lake drawdown effectively consolidated sediments (e.g., increased sediment density) and decreased organic matter content. Mean porewater concentrations of soluble reactive phosphorus and NH_4–N initially increased after reflooding but declined markedly 1 year later (James et al., 2001a, b). Macrophyte growth responded to the new habitat conditions. Mean macrophyte biomass increased from pretreatment levels of 150 g/m^2 in 1995 to post-treatment levels of 1,400 g/m^2 in 1998. This high biomass may have played a role in depleting sediment phosphorus reserves (James et al., 2001a, b).

The plant community also changed dramatically (Table 12.6). Species number increased from 18 to 25 taxa. The relative frequency of emergent species increased from 14.2% to 35.3%. The relative frequency of exotic species decreased from 70% to 17%. Simpson's (1949) diversity increased from 0.61 to 0.88.

The areal extent of the plant community changed very little between pre- and post-treatment. This was not surprising since there was little room for plant community expansion. Before treatment only about 1% of the area was not vegetated. After treatment only about 0.5% was not vegetated. From a wildlife management perspective the treatment was very successful. The desired increase in emergent coverage was achieved and there was a substantial increase in sago pondweed (*Potamogeton pectinatus*), a prime waterfowl food.

Success needs to be carefully defined when planning restorations since results are unpredictable and riparian property owners may not appreciate increased aquatic vegetation. An example is Delavan Lake in southeastern Wisconsin. It has a surface area of 725 ha, a maximum depth of 16.5 m, and a mean depth of 7.6 m. A major rehabilitation in the late 1980s and early 1990s included efforts to reduce internal and external phosphorus loading; eradicate benthivorous fish, primarily carp and buffalo (*Ictiobus cyprinellus*); restock predatory game fish; and temporarily draw down lake levels. Historically Delavan Lake had a rich aquatic flora. Surveys done between 1948 and 1975 identified 25 macrophyte taxa (not all in the same survey), but vegetation was declining by the 1950s. In 1955, the Izaak Walton League planted a number of desirable species in the southwest end of the lake because of a concern over the loss of aquatic vegetation. In the early 1960s only seven species were reported and by 1968 only four species remained. The aquatic vegetation for several years before rehabilitation consisted of a single pondweed species (*Potamogeton* sp.) and white water lily (*N. odorata*). As expected the diversity and abundance of aquatic plants increased because of rehabilitation. The number of species increased to six in 1990 and 20 in 1993. By 1998, however, species number decreased to 13 and Eurasian watermilfoil, curly-leaf pondweed (*P. crispus*), and coontail (*Ceratophyllum demersum*) reached nuisance levels in parts of the lake, especially in areas less than 3 m deep in the northern and southern ends of the lake and near the inlet and outlet (Robertson et al., 2000). Additional plant management was anticipated and macrophyte harvesting and chemical treatment were part of the original rehabilitation plan; but macrophyte growth was much greater than expected. A total of 5,376 m³ of plant material was harvested during the 1997, 1998, and 1999 growing season. Heavy macrophyte growth near the inlet was partially blamed for remobilizing sediment phosphorus that reduced the success of phosphorus limitation efforts (Robertson et al., 2000). By 2001, 12 submergent or free floating species were found but the relative frequency of exotic species was 36.7% (Table 12.7).

12.3.1.2.3 Breakwaters of All Sizes

Breakwaters are used to reduce the impact of wind and wave erosion on aquatic plants (Chapter 5). They are used to protect established plants, new plantings, or to make suitable habitat for plant invasion or community expansion.

The simplest breakwaters are wave breaks; V-shaped wave deflectors constructed out of two half-sheets of plywood or other suitably sturdy material (approximately 1.2 m by 1.2 m in size). They are joined at an approximately 90° angle and staked to the bottom on the lakeward side of remaining plants or new plantings (Bartodziej, 1999).

Sandbags containing sediment and rhizomes of reeds (*Phragmites australis*) and burlap bags containing sediment and rhizomes of reeds placed inside old tires filled with sand were used to try to stabilize sediments on Lake Poygan, Wisconsin. These plantings eventually failed (Kahl, 1993).

Coir (coconut fiber) geotextile rolls, plant rolls, geotextile mats, branch box breakwaters, brush mattresses, and wattling bundles were used as wave breaks and erosions control devises in some Missouri impoundments (Fischer et al., 1999). Coir rolls were 0.4 m in diameter and placed in shallow trenches. Emergent species were planted on 0.5-m centers on the shoreward side of the roll. A plant roll is similar to a coir roll. It is a cylinder of plant clumps and soil wrapped in burlap and placed in a trench. The ones used in Missouri were 3 m long. Coir geotextile non-woven mats, placed flat and anchored on the reservoir bottom, with emergents planted on 0.3-m centers through-

TABLE 12.7
Relative Frequency (%) of Aquatic Plants in Delavan Lake, Wisconsin for 2001

Species	Rel. Freq. (%)
Ceratophyllum demersum	9.3
Elodea canadensis	2.2
Myriophyllum spicatum	29.0
Potamogeton crispus	7.7
P. foliosus	3.3
P. pectinatus	35.5
P. zosteriformis	2.2
Vallisneria americana	2.2
Zannichellia palustris	1.6
Zosterella dubia	6.0
Other species[a]	1.0

[a] Other species were *Lemna minor* and *Chara* sp.

Source: Data from Kevin MacKinnon, District Administrator, Delavan Lake Sanitary District, Delavan, WI. Personal communications, 2002.

out the mat was another technique used. Brush mattresses and wattling bundles consisted of young willow (*Salix* sp.) shoots tied in bundles or in a long roll and staked to the bottom. Brush boxes were similar except the willow shoots were woven between and wired to posts driven into the bottom. These techniques are most easily installed in reservoirs under drawdown conditions.

The Missouri impoundments project is very recent so the results are inconclusive (Fischer et al., 1999). The Missouri researchers learned that patience is the key. Do not expect lush aquatic vegetation covering the entire littoral zone after one year unless you have a small pond and plenty of time and money. Wave action appeared to be the primary limiting factor to initial plant survival and dispersion. The growth of thick algal mats in the protected areas; fluctuating water levels, especially during cold weather; herbivory; and drifting logs and debris that knocked down protective devises were also problems.

Floating booms of logs or old tires have been used to dampen wave action (see Chapter 5). Probably the most interesting floating devises are the Schwimmkampen (Germany) or Ukishima (Japan) — artificially constructed floating wetlands (Hoeger, 1988; Mueller et al., 1996). They are constructed on floating platforms that support wetland vegetation. They move up and down with fluctuating water levels and improve water quality primarily by dissipating wave action thus reducing shoreline and bottom erosion. They provide nursery areas for small fish and crustaceans and in urban areas they have been used for aesthetics by enhancing privacy and dissipating noise. Depending on size, they can be towed to different areas of the lake as needed.

One of the larger projects was the Terrell's Island breakwall constructed on Lake Butte des Morts, Wisconsin. Lake Butte des Morts is a 3,587-ha lake with a mean depth of 1.8 m and a maximum depth of 2.7 m. Originally Lake Butte des Morts and other upriver lakes of the Winnebago Pool were large riverine marshes. Dams constructed in the 1850s raised water levels by about 1 m. Initially they were rich in aquatic vegetation but vegetation decline accelerated from the 1930s through the present because of high water levels, extreme flooding, erosion of shorelines and bottom sediments, lake shore development, plant removal, carp, and accelerated nutrient inputs (Wisconsin Department of Natural Resources (WDNR), 1991). Between 1994 and 1998, 3,245 m of breakwall was constructed connecting the mainland to a series of small islands and enclosing around 243 ha

FIGURE 12.1 Carp exclusion gate at Terrell's Island restoration area. The center of the gate is spring loaded so that a boat pushes it down when entering or leaving the area.

of water (Arthur Techlow, WDNR, Oshkosh, Wisconsin, personal communication, 2002). The breakwall was constructed of limestone with a planned 3.7 m wide top, 3 to 1 side slopes, and a height of 0.9 m above the ordinary summer water level. It was constructed with only one gap to allow boater access. This gap was gated to prevent access by large carp (Figure 12.1) but to allow other fish access for spawning. The gap was also built with sufficient overlap to reduce waves. Small islands were also constructed to reduce wind fetch inside the breakwater. Waterfowl food and habitat and a fish spawning area were the main reasons for restoring vegetation to the area. The total cost for everything including feasibility studies, engineering, administration, and construction was approximately 1.7 million (not adjusted to current prices) U.S. dollars (Arthur Techlow, WDNR, Oshkosh, Wisconsin, personal communications, 2002).

Vegetation in the area was sampled pre-construction from 1988 to 1994 and after construction from 1999 to 2001(Tim Asplund, Mark Sessing, and Chad Cook, WDNR, Madison, Wisconsin, personal communications, 2002). In 1999 and 2000 other water quality parameters were compared between the area enclosed by the breakwall and open water areas. The percent frequency of vegetated sampling points increased from 15% before construction to 39%, then 55%, then 99% from 1999 to 2001 (Figure 12.2); species numbers also increased. The plant community composition changed dramatically (Table 12.8). Sago pondweed and wild celery were the dominant pre-construction species making up over 60% of the relative frequency. After construction elodea (*Elodea canadensis*) was the dominant species with a relative frequency varying from 34% to 52%, while the combined relative frequency of sago pondweed and wild celery varied from 11.6% to 17%. For plant restoration the breakwall is working. Although the importance of sago pondweed and wild celery, very desirable waterfowl food plants, declined in the plant community their abundance increased on an absolute basis. During 1999, the overall water clarity improved inside the breakwall but it was lower than mid-lake sites during the end of July and into early August (Tim Asplund, WDNR, personal communication, 2002). In the fall, turbidity and suspended solids declined, greatly improving water clarity and light penetration inside the breakwall. It appeared that the quiescent environment created by the breakwall resulted in greater algae blooms during the summer compared to the rest of the lake, but lower inorganic sediment resuspension throughout the year (Tim Asplund, WDNR, personal communication, 2002). Water quality data for 2000 show higher water clarity,

FIGURE 12.2 Aquatic plants growing behind the breakwater at the Terrell's Island restoration area, August 2002.

TABLE 12.8
Comparison of Species Relative Frequencies at Terrell's Island Area before and after Breakwall Construction

Species	Average Rel. Freq. (%), 1988–1994, Before Construction	Rel. Freq. (%), After Construction		
		1999	2000	2001
Potamogeton pectinatus	45	15.4	5.8	1.5
Vallisneria americana	15.8	1.6	5.8	12.8
Ceratophyllum demersum	4.0	4.1	0.7	0.2
Elodea canadensis	0.3	33.7	51.7	35.0
Zosterella dubia	6.9	8.2	10.9	23.4
Najas sp.	12.7	11.0	0	0
Chara sp.	4.1	0.9	2.7	1.5
Myriophyllum spicatum	10.9	14.6	6.5	5.5
Potamogeton pusillus	0.5	10.6	2.0	0
P. crispus	—	—	6.2	17.5
Myriophyllum sp.	—	—	1.6	0.8
Other species[a]	—	—	6.2	0.9

[a] Other species were *Nitella* sp., *P. natans, P. nodosus,* and unidentified species.

Source: Data from Tim Asplund, Mark Sessing, and Chad Cook, WDNR, personal communications, 2002.

lower fertility, and fewer algae at stations within the breakwater when compared to stations in the open lake. The suspended solids within the breakwater come primarily from algae while those in the open lake are mostly non-organic (Mark Sessing, WDNR, personal communication, 2002). Wind and wave action inside the breakwater are still sufficient that erosion of the constructed islands is a problem.

Artificial islands were constructed in pools of the upper Mississippi River between Wisconsin and Minnesota. The objective for island construction was to improve habitat conditions for aquatic plants by reducing wave resuspension of fine materials, thereby improving light penetration in localized areas. Aquatic vegetation, mainly wild celery, became established on the down-stream side of the islands but detailed results are not available (Janvrin and Langreher, 1999).

12.4 AQUASCAPING

Aquascaping — a term describing the planting of aquatic and wetland plants — is landscaping in and around water (see Chapter 5 for additional discussion). The vision of landscaping may not seem appropriately applied to lakes or reservoirs but natural landscaping is a term that has been used for many years in terrestrial systems, which is applied to planning, restoring, and managing extensive areas such as prairies, savannahs, and woodlands. Admittedly, aquascaping is often applied to small projects like water gardens and sedimentation ponds but good planning, cultural, and management principles are needed regardless of the size of the project.

The advantage of aquascaping is that with success, you get what you want where you want it. In many areas, such as the southeastern and western United States, active revegetation may be the only option. Reservoirs in these regions are often constructed in areas that lack natural lakes and they may be remote from any aquatic plant populations that could serve as a propagule source. As a result these reservoirs have no aquatic plant seed bank and receive only limited inputs of seed or other plant propagules. If they are colonized it is often with nuisance species that are adapted for exploiting disturbed conditions (Smart and Dick, 1999). The disadvantages of active revegetation are that it is expensive, labor intensive, and can easily fail. Plant restoration makes a good volunteer project so some of the expense for plant material and labor can be minimized (Figure 12.3).

Aquascaping matches plant material with the habitat at specific locations within the restoration area. Water chemistry, water depth, substrate, turbidity, wave action, and human or animal uses are important considerations when developing an aquascaping plan and selecting plant material. Also important is selecting plants that provide the desired function. Is the restoration being done to prevent shoreline erosion, to provide fish or wildlife habitat, to intercept nutrients, to be aesthetically pleasing, or for some other reason? A final plan should include a map(s) showing habitat characteristics, species locations, plant densities, planting methods, and protective measures. Specific questions to ask include:

1. What are the habitat limitations of the site(s) within the restoration area?
2. What plant species have the desired properties needed for the restoration? Will they thrive in the physical and chemical habitat? Are they able to withstand wind, waves, turbidity, human and animal traffic, or drawdown? Are the plants good waterfowl food, fish or waterfowl habitat, aesthetically pleasing, or whatever functional characteristics are wanted?
3. Are the desired species readily and locally available at a reasonable price?
4. What is the best way to propagate the plants for existing conditions?
5. Do the selected species have good reproductive potential? Once established can the plants grow and reproduce well enough to maintain and increase the population?
6. Are the selected species native to the region?

FIGURE 12.3 Volunteer planting emergent plants around constructed islands behind the Terrell's Island breakwater. Twine will be strung between stakes to prevent predation of new planting by large waterfowl like geese and swans.

7. Do the selected species have weedy tendencies? Will any become a nuisance in the future?
8. How large a population is needed to ensure a viable stand despite losses from herbivores, pathogens, poor reproductive success, wind, waves, turbidity, competition with other species, and climatic conditions?
9. Where will each species be located and in what densities?
10. What habitat remediation techniques are needed?
11. What cultural and protective measures are needed?

The answers to these questions are not simple. Many introductions, reintroductions, and transplants of species fail because innate species characteristics, interactions between species, or habitat characteristics are not considered (Botkin, 1975).

Even if a species formerly grew in an area, the habitat might be altered so that it is no longer suitable for that species or for plants in general. As discussed earlier (Table 12.1), habitat remediation may be needed for any plants to grow. However, careful selection of plant material can overcome some habitat limitations. As an aid to plant selection, Table 12.9 provides information on median depth, substrate preference, and turbidity tolerance for a number of species. Table 13.1 (Chapter 13) of this volume shows how selected aquatic plants respond to drawdown. Consult this table if water level fluctuation is a concern in the rehabilitation area.

Although many aquatic plants are broadly tolerant of water chemistry conditions, this is not always the case. The more information that is known about the plant material, the better the potential for a successful restoration. There are regional studies for North America, Europe, Japan, and probably other areas that provide aquatic plant distribution with regard to a variety of chemical parameters (Moyle, 1945; Seddon, 1972 Hutchinson, 1975; Beal, 1977; Pip, 1979, 1988; Hellquist

TABLE 12.9
Habitat Preferences of Selected Lake Plants[a]

Species	Median Growth Depth[b]	Substrate Preference[c]	Turbidity Tolerance[d]
Bidens beckii	3	S	N
Brasenia schreberi	2	O	Y
Ceratophyllum demersum	3	S	Y
Ceratophyllum echinatum	2	S	—
Decodon verticillatus	E	S	E
Dulichium arundinaceum	1	S	Y
Elatine minima	1	—	—
Eleocharis acicularis	2	H	Y
E. palustris	2	H	E
E. robbinsii	3	S	E
Elodea canadensis	3	S	Y
Eriocaulon aquaticum	3	H	N
Gratiola aurea	3	—	—
Isoete echinospora	3	O	—
I. lacustris	3	H	—
Lobelia dortmanna	3	H	—
Myriophyllum farwellii	3	S	—
M. heterophyllum	4	O	—
M. sibiricum	3	S	N
M. tenellum	3	H	—
M. verticillatum	3	O	N
Najas flexilis	3	H	N
N. gracillima	3	—	—
Nuphar variegata	2	S	O
Nymphaea odorata	2	O	O
Polygonum amphibium	–	O	O
Pontederia cordata	2	O	N
Potamogeton amplifolius	3	S	O
P. diversifolius	3	S	—
P. epihydrus	3	O	O
P. filiformis	3	O	—
P. foliosus	2	S	Y
P. gramineus	3	H	O
P. illinoensis	3	O	N
P. natans	2	O	O
P. nodosus	2	O	Y
P. obtusifolius	3	S	—
P. pectinatus	3	O	O
P. praelongus	3	S	N
P. pusillus	3	S	Y
P. richardsonii	3	O	O
P. robinsii	3	O	O
P. strictifolius	3	H	—
P. zosteriformis	3	S	N
Ranunculus flammula	3	H	—
R. longirostris	2	O	O
R. trichophyllus	2	O	O
Sagittaria graminea	2	O	E
S. latifolia	1	O	E
S. rigida	2	S	E

TABLE 12.9 (Continued)
Habitat Preferences of Selected Lake Plants[a]

Species	Median Growth Depth[b]	Substrate Preference[c]	Turbidity Tolerance[d]
Scirpus americanus	2	O	N
S. validus	2	O	N
Sparganium chlorocarpum	2	S	—
S. eurycarpum	1	O	O
Typha latifolia	1	O	N
Utricularia geminiscapa	3	S	—
U. gibba	3	S	—
U. intermedia	3	S	—
U. vulgaris	3	S	Y
Vallisneria americana	3	H	Y
Zannichellia palustris	2	H	Y
Zizania aquatica	2	S	O
Zosterella dubia	3	O	Y

[a] Many emergent species are not included. Emergent species are usually found in shallow water and turbidity tolerance does not apply unless otherwise noted.

[b] 1, less than 0.5 m; 2, 0.5–1 m; 3, 1–2 m; 4, greater than 2 m; E, emergent species, usually found in shallow water; —, unknown or unreported.

[c] S, prefers soft substrate; H, prefers hard substrate; O, no substrate preference; —, unknown or unreported.

[d] Y, turbidity tolerant; N, not turbidity tolerant; O, no turbidity preference, probably tolerant; E, emergent species, probably turbidity tolerant; —, unknown or unreported.

Source: After Nichols, S.A. 1999. Distribution and Habitat Descriptions of Wisconsin Lake Plants. Bull. 96. Wisconsin Geol. Nat. Hist. Surv., Madison. Conditions primarily for Wisconsin lake plants.

and Crow, 1980, 1981, 1982, 1984; Crow and Hellquist, 1981, 1982, 1983, 1985; Kadono, 1982a, b; Nichols, 1999). These studies should be consulted if there are questions about the suitability of selected species for the chemical conditions found at a restoration site.

Plants have different functional values and should be selected accordingly, depending on the objectives of the rehabilitation. Table 12.10 lists the wildlife and environmental values of selected lake plants. Remember, when interpreting this table that fish and fowl are not taxonomists. In cases like value as cover, structure is more important than the plant species involved. For instance, the structure of large-leaf pondweeds makes good fish cover. It probably makes little difference to the fish whether the species is *Potamogeton illinoensis*, *P. praelongus*, or *P. richardsonii*. Likewise, most strongly rooted, emergent plants stabilize substrate whether they are *Carex* spp., *Scirpus* spp., or *Typha* spp. Aesthetics are subjective and not included in this list, but emergent and floating-leaf species like *Pontederia cordata*, *Sagittaria spp*, *Sparganium* spp., *Nymphaea* spp., *Nelumbo lutea*, and *Nuphar* spp. have showy flowers and are often used for water gardening (Chapter 5). *Acorus calamus*, *Carex* spp., *Cyperus* spp., *Juncus* spp., *Scirpus* spp., *Phragmites australis*, *Typha* spp., *Brasenia schreberi* and *Zizania* spp. have life forms that people may find aesthetically pleasing. Even species with small flowers like elodea, *Ranunculus longirostris*, *R. trichophyllus*, *Zosterella dubia*, or *Polygonum amphibium;* that have an interesting natural history like *Utricularia* spp., wild celery, and *Brasenia schreberi*, or that were used as food by indigenous people like *Zizania* spp., *Sagittaria* spp. and *Nelumbo lutea*, can add interest or educational value to a project.

When possible, use native plant material from a locality near the restoration. These species and ecotypes will likely be the most successful for local conditions. Using native plant material avoids the problem of introducing exotic species (be careful that exotics aren't introduced with the natives). Most people are familiar with aquatic nuisance problems caused by non-North American exotics

TABLE 12.10
Wildlife and Environmental Values of Selected Lake Plants

Species	Waterfowl Food Part[a]	Waterfowl Food Value[b]	Waterfowl Cover[c]	Other birds Food Part[a]	Other birds Cover[c]	Muskrat Food[c]	Substrate Stabilize[c]	Nuisance Potential[c]	Fish Value[d]
Acorus calamus	—	P	X	—	—	—	X	—	—
Brasenia schreberi	S	G	—	—	—	—	—	X	C
Carex spp.	S	F	X	—	—	—	X	—	S
Ceratophyllum demersum	S,F	F	X	—	—	—	—	X	F,S
Chara spp.	F	G	—	—	—	—	—	X	F
Cyperus spp.	S	F	—	S	—	—	—	X	—
Decodon verticillatus	S	P	—	—	—	—	—	—	—
Eleocharis spp.	T	G	—	S	X	X	X	—	F,S,C
E. acicularis	—	F	—	—	—	—	X	X	S
E. palustris	—	F	X	—	—	—	—	—	F
Elodea canadensis	F	F	—	—	—	—	—	X	F
Equisetum spp.	F	P	—	—	—	X	X	X	—
Juncus spp.	—	—	—	—	—	—	X	X	S
Lemna minor	F	G	—	—	—	—	—	X	F
Lemna trisulca	F	G	—	—	—	—	—	—	—
Myriophyllum spp.	S,F	P	—	S	—	—	—	X	F,C
M. exalbescens	—	F	—	—	—	—	—	X	—
Najas flexilis	S,F	E	—	S	—	—	—	X	F,C
N. guadalupensis	S,F	E	—	—	—	—	—	X	—
Nelumbo lutea	—	—	—	—	—	—	X	X	F,C
Nuphar variegata	—	F	—	—	—	—	—	—	F,C
Nymphaea spp.	S	P	—	S,T,F	—	—	X	X	F,C
Phragmites australis	—	—	X	—	X	—	X	X	F
Polygonum amphibium	S	E	—	—	—	—	X	X	—
Pontederia cordata	S	P	X	—	—	X	—	X	C
Potamogeton amplifolius	S	F	—	—	—	—	—	—	F
P. capillaceus	S	F	—	—	—	—	—	—	—
P. diversifolius	S	F	—	—	—	—	—	—	—
P. epihydrus	S,T,F	G	—	—	—	—	—	—	F,C
P. foliosus	S,T,F	G	—	—	—	—	—	—	—
P. friesii	S,F	G	—	—	—	—	—	—	—
P. gramineus	S,T	G	—	—	—	—	—	—	—
P. illinoensis	S	F	—	—	—	—	—	X	C
P. natans	S,T	G	—	—	—	—	X	—	—
P. nodosus	S	G	—	—	—	—	—	—	F,C
P. obtusifolius	—	—	—	—	—	—	—	—	F,C
P. pectinatus	S,T	E	—	—	—	—	—	X	F,C
P. praelongus	S,T,F	F	—	—	—	—	—	—	F,C
P. pusillus	S,T,F	G	—	—	—	—	—	—	—
P. richardsonii	S,T,F	G	—	—	—	—	—	—	F,C
P. robinsii	—	—	—	—	—	—	—	X	F,C
P. spirillus	S	F	—	—	—	—	—	—	—
P. strictifolius	S	G	—	—	—	—	—	—	—
P. zosteriformis	S	F	—	—	—	—	—	—	—
Ranunculus spp.	S,F	P	—	—	—	—	—	—	F
Ruppia maritima	S,T,F	E	—	S	—	—	—	—	—
Sagittaria spp.	—	—	X	S	X	X	—	—	—

TABLE 12.10 (Continued)
Wildlife and Environmental Values of Selected Lake Plants

Species	Waterfowl Food Part[a]	Waterfowl Food Value[b]	Waterfowl Cover[c]	Other birds Food Part[a]	Other birds Cover[c]	Muskrat Food[c]	Substrate Stabilize[c]	Nuisance Potential[c]	Fish Value[d]
S. cuneata	S,T	F	—	—	—	—	X	—	—
S. latifolia	—	F	—	—	—	—	X	—	—
Scirpus spp.	—	—	X	S,T	X	X	X	X	F,C
S. acutus	S	E	—	—	—	—	X	—	F,C
S. americanus	S	G	—	—	—	—	X	—	F,C
S. fluviatilis	S	P	—	—	—	—	X	—	—
S. validus	—	—	—	—	—	—	X	—	F,C
Sparganium chlorocarpum	—	F	—	—	—	—	—	—	—
S. eurycarpum	S	F	—	—	—	—	—	—	—
Spirodela polyrhiza	F	G	—	—	—	—	—	X	F
Typha spp.	T,F	P	X	S	X	X	X	X	F
Utricularia purpurea	—	—	—	—	—	—	—	—	F,C
Vallisneria americana	S,T,F	E	—	—	—	—	—	X	F,C
Wolffia columbiana	F	F	—	—	—	—	—	—	—
Zannichellia palustris	S,F	G	—	S	—	—	X	—	F
Zizania aquatica	S	E	X	S	X	—	X	—	—
Zosterella dubia	S	P	—	—	—	—	—	X	F,S

[a] S = seeds or comparable structure; T = tubers or roots; F = foliage and stems; —, information unknown or unreported.

[b] E, excellent; G, good; F, fair; P, poor; —, information unknown or unreported.

[c] X, plant is functional in specified category; —, information unknown or unreported.

[d] F, direct food or supports fish food fauna; C, cover; S, spawning habitat, —, information unknown or unreported.

Source: After Nichols, S.A. and J.G. Vennie. 1991. Attributes of Wisconsin Lake Plants. Inf. Cir. 73. Wis. Geol. Nat. Hist. Surv., Madison.

Compiled from Kadlec, J.A. and W.A. Wentz. 1974. State-of-the Art Survey and Evaluation of Marsh Plant Establishment Techniques: Induced and Natural. Contract Rep. D-74-9. Dredged Materials Research Program, U.S. Army Coastal Eng. Res. Ctr., Fort Belvoir, VI; Trudeau, P.N. 1982. Nuisance Aquatic Plants and Aquatic Plant Management Programs in the United States: Vol. 3: Northeastern and North-Central Region. Rep. MTR-82W47-03, Mitre Corp., McLean, VI; Carlson, R.A. and J.B. Moyle. 1968. Key to the Common Aquatic Plants of Minnesota. Spec. Publ. 53. Minn. Dept. Cons., St. Paul, MN; U.S. Army Corps of Engineers. 1978. Wetland Habitat Development with Dredged Material: Engineering and Plant Propagation. Tech. Rept. DS-78-16, Office, Chief of Engineers, Washington, DC; Fassett, N.C. 1969. *A Manual of Aquatic Plants.* University of Wisconsin Press, Madison.

such as water hyacinth (*Eichornia crassipes*), hydrilla (*Hydrilla verticillata*), egeria (*Egeria densa*), and Eurasian watermilfoil. Tables 12.9–12.11 list native North American species. Remember, these species are not indigenous to all locations in North America and they could cause serious aquatic nuisance problems if they are transported out of their native range. A similar warning is given for non-North America restorations. Les and Mehrhoff (1999) reported that 76% of the non-indigenous aquatic plant introductions in southern New England were the result of escapes from cultivation. Their list of introductions includes egeria, hydrilla, Eurasian watermilfoil, curly-leaf pondweed (*Potamogeton crispus*), and *Trapa natans*.

Wildlife managers learned to propagate wetland and aquatic species for waterfowl food and habitat. A wealth of information about collecting, storing, culturing, planting, and managing aquatic species is found in the older wildlife management literature. Kadlec and Wentz (1974) summarized much of this literature and is recommended reading for anyone considering an aquatic plant restoration project. Plants are established from seed by direct broadcasting or by packing the seed in mud balls before sowing. Some seeds need a special treatment of stratification or scarification before they will germinate. Whole plants or vegetative propagules can be placed directly in bottom sediment or weighted with rubber bands and nails, or mesh bags and gravel, and sown from the water surface (Brege, 1988). Some species can be spread by cutting with a harvesting machine and not collecting the cut parts or the cut parts can be collected, gathered in bunches, and weighted with a rubber band and nail so they sink into bottom sediments. Emergent species can be propagated by drawing down the water and spreading "marsh hay" with ripened seeds on the mud flats, or by spreading marsh soil from established stands on the desired area.

Table 12.11 reviews methods to propagate many common aquatic species. The method chosen depends on the size of the area; the type of propagules, labor and funds available; habitat conditions; and the time available to wait for results. For some emergent species, as an example, seeds can be used on exposed mudflats but plants or rootstock are needed for planting under water. Transplanting mature plants is recommended for areas with severe habitat limitations or for areas where quick results are needed (Smart and Dick, 1999).

Plant material can be purchased from aquatic nurseries, started in your own nursery, or collected from the wild. Seed, especially from emergent species, can be gathered in abundance with little harm to wild populations. Some species can be cut as hay and threshed using conventional agricultural equipment. Plant cuttings also do little harm to many species. Seeds, tubers, rhizomes, and rootstock are often found in flotsam, washed up on shorelines after violent storms. Sometimes plant material can be gathered in abundance from this source. Aquatic plant nurseries advertise in landscaping and nursery publications, outdoors magazines, and trade publications like *Land and Water.*

Using plant material that grows and reproduces well enough to be successful posses a dilemma. Some species are so aggressive that they cause aquatic nuisance problems in the future. In severely degraded habitats only weedy species may survive; their aggressive growth may be desirable for habitat restoration. Weedy species may modify the habitat enough so other species can invade. However, established plants can ward off invaders. There is little evidence that a new planting can displace an established stand of nuisance species (Lathrop et al., 1991; Storch et al., 1986).

Table 12.10 lists the nuisance potential of selected species. This is a generalized statement that probably indicates plant growth potential. Aquatic nuisance is very subjective and nuisance potential varies within the plant's range. American lotus (*N. lutea*), for example, is protected in the northern part of its range (e.g., Michigan), unobtrusive in the central part of its range, and actively managed as an aquatic nuisance in the southern part of its range.

Planting density depends on the species and the plant material used. For seeding, some recommendations are 23 kg/ha for bulrushes (*Scirpus* spp.), 46 kg/ha for wild rice (*Zizania* spp.), 1,240 seeds/ha for American lotus, and 28 kg/ha for wild celery (Lemberger, undated). For vegetative parts, 1,850–2,470 tubers, rhizomes, or rootstocks per hectare are recommended for submersed, larger floating-leaf, and small to medium sized emergent species (Lemberber, undated). Butts et al. (1991) recommended spacing plants of small emergent species on 1-m centers, large emergent species and large floating-leaf species (water lilies and American lotus) on 2-m centers. The advice of Moss et al. (1996) is, "For submerged species, as large a quantity as possible is desirable." They recommend planting four rhizomes, each containing one node per square meter for emergent species and ten fragments, each about 10 cm long or turions per square meter for submerged species. This is a higher density than other recommendations but they are allowing for predation by waterfowl and aquatic mammals. When establishing founder colonies (see next section), Smart and Dick (1999) recommended using 0.4 to 0.8 mature plants per square meter in fenced exclosures.

Planting densities are based on viable propagules. Testing for propagule germination (e.g., germination percentage to calculate pure live seed) is always wise and may cause an upward adjustment in the amount planted.

TABLE 12.11
Plant Propagation Methods for Selected Aquatic Plants

Species[a]	Life Cycle[b]	Transplant (seedlings or whole plants)[c]	Roots,[c] Tubers or Rhizomes	Cuttings[c]	Winter Buds[c]	Seeds[c]
Acorus calamus	P	X	X	—	—	—
Brasenia schreberi	P	X	—	—	X	X
Carex spp.	B	X	X	—	—	X
Ceratophyllum spp.	P	X	—	—	—	—
Chara spp.	—	X	—	—	—	—
Cyperus spp.	B	X	X	—	—	X
Eleocharis spp.	B	X	X	—	—	X
Elodea spp.	P	X	—	X	—	—
Juncus spp.	P	X	—	—	—	—
Lemna spp.	—	X	—	—	—	—
Myriophyllum spp.	P	X	—	X	—	X
Najas spp.	A	—	—	—	—	X
Nelumbo lutea	P	X	X	—	—	X
Nuphar spp.	P	X	X	—	—	X
Nymphaea spp.	P	X	X	—	—	X
Phragmites australis	P	X	X	—	—	X
Polygonum spp.	P	X	X	—	—	X
Pontederia cordata	P	X	X	—	—	X
Potamogeton spp.	—	X	—	—	—	X
P. amplifolius	P	X	X	X	—	X
P. foliosus	P	X	—	—	X	X
P. gramineus	P	X	X	X	—	X
P. natans	P	X	X	—	—	X
P. nodosus	P	X	X	X	—	X
P. pectinatus	P	X	X	X	—	X
P. pusillus	P	X	—	—	X	X
P. richardsonii	P	X	X	—	—	—
P. spirillus	—	X	—	—	—	X
P. zosteriformis	P	X	—	—	X	X
Ranunculus spp.	P	X	—	—	—	X
Ruppia maritima	P	X	X	X	X	—
Sagittaria latifolia	P	X	X	—	—	X
Sagittaria rigida	P	X	X	—	—	—
Scirpus	P	X	X	—	—	X
Sparganium spp.	P	X	X	—	—	X
Spirodela polyrhiza	P	X	—	—	—	—
Typha spp.	P	X	X	—	—	X
Vallisneria americana	P	X	X	—	—	X
Wolffia spp.	—	X	—	—	—	—
Zannichellia palustris	A	—	—	—	—	X
Zizania spp.	A	—	—	—	—	X
Zosterella dubia	P	X				X

[a] When the propagation methods noted for all species in a genus were the same it is listed as the generic name and spp. When only one species is listed for a genera, closely related species can probably be propagated with similar methods. Most species can be transplanted but for annual species the effort is of questionable value.

[b] A = annual; P, perennial; B = both: — = information unknown or unreported.

[c] X, can be propagated by the designated method; — = information unknown or unreported.

Source: Adapted from Nichols, S.A. and J.G. Vennie. 1991. Attributes of Wisconsin Lake Plants. Inf. Cir. 73. Wis. Geol. Nat. Hist. Surv., Madison.

Original data from: Kadlec, J.A. and W.A. Wentz. 1974. State-of-the Art Survey and Evaluation of Marsh Plant Establishment Techniques: Induced and Natural. Contract Rep. D-74-9. Dredged Materials Research Program, U.S. Army Coastal Eng. Res. Ctr., Fort Belvoir, VI; Lemberger, J.J. Undated. Wildlife Nurseries Catalog. Wildlife Nurseries, Oshkosh, WI; Kester, D. 1989. Kester's Wild Game Food Nurseries Catalog. Omro, WI; U.S. Army Corps of Engineers. 1978. Wetland Habitat Development with Dredged Material: Engineering and Plant Propagation. Tech. Rept. DS-78-16, Office, Chief of Engineers, Washington, DC.

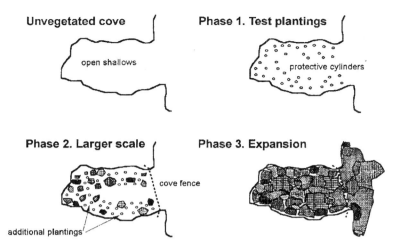

FIGURE 12.4 Diagrammatic representation of the founder colony approach. Phase I involves planting of test plants within small protective exclosures. During the second growing season (Phase 2), a larger fenced areas is constructed, if necessary, and additional plantings of the most suitable species are made. During the third and subsequent growing seasons (Phase 3), the founder colonies vegetate the rest of the reservoir. (From Smart, R.M. et al. 1998. *J. Aquatic Plant Manage.* 36: 44–49. With permission.)

12.5 THE FOUNDER COLONY: A REASONABLE RESTORATION APPROACH

Founder colony is a term used by Smart et al. (1998) and Smart and Dick (1999) to describe establishing small plant colonies at strategic locations in the rehabilitation area and allowing them to spread. Moss et al. (1996) and van Donk and Otte (1996) recommended a similar approach. Plant establishment is done in three phases (Figure 12.4). During Phase 1 test species are planted within small protective exclosures. Well-protected (from wind and waves), shallow (less than 2 m deep) coves or bays with a gradual slope are the best planting locations. They generally have the clearest water available. A fine textured substrate is preferred and generally indicates a favorable, low-energy environment. Use mature transplants with well-developed shoots and leaves. They are robust so they have the highest probability of success. Mature transplants can be planted over a relatively long time period. This is advantageous in habitats with seasonal limitations such as spring flooding or winter drawdown. Planting can be delayed until normal conditions return. Recommended water depths for planting submerged species range from 0.5 to 1.0 m, floating-leaved plants from 25 to 75 cm, and emergent plants from 0 to 25 cm. Planting submerged species in water depths that are about or slightly less than the height of the plant ensures that at least some leaves will receive adequate photosynthetic light, even in turbid water. Protection from herbivores is vital (Moss et al., 1996; van Donk and Otte, 1996; Smart and Dick, 1999) and cannot be overemphasized. Protective devices can be constructed for individual plants, for multiple plants, or for larger areas such as fenced coves. Several designs for protective exclosures are provided in Smart and Dick (1999). The top of large exclosures may need covering to prevent waterfowl or other birds from entering (Moss et al., 1996; van Donk and Otte, 1996; Figure 12.3).

Assuming some success in Phase 1, additional protected transplants of successful species are added in Phase 2, usually the second growing season. Depending on the level of herbivory in Phase 1, larger fenced areas may be required to allow plants to expand beyond their protective cages. Additional species can be tested to add diversity to the restoration. During subsequent growing seasons, the founder colonies expand through vegetative and sexual reproduction into adjacent unvegetated areas (Phase 3). They serve as a propagule source for natural colonization throughout the lake.

TABLE 12.12
Results of Founder Colony Trials

Lake	Species Planted	Phase[a]	Results
North	*Vallisneria americana*	1	75–100% coverage within exclosures
			Growth outside exclosures in cool months
			Heavy herbivory outside exclosure in warm months
			With time may outgrow herbivory
Lewisville	*Potamogeton nodosus*	1	Plants within exclosures survived and spread
	V. americana		No survival of unprotected plantings outside of exclosures or in damaged exclosures
			No radial spread beyond exclosures
Guntersville	*V. americana*	1	Were able to get plants to grow within exclosures only with heavy protection from herbivores
	P. nodosus		
Conroe	*V. americana*	2	Greater than 95% survival of all three species
	P. nodosus		Single plants formed colonies up to 3 m in diameter
	Zosterella dubia		*Chara* and *Najas guadalupensis* invaded protected areas
			Z. dubia formed new colonies in protected areas

[a] Phase described in founder colony section.

Although untested over the long term, the founder colony idea is a very reasonable approach. It does not require a large initial investment in plant material. It allows species survival and expansion potential to be tested before expanding the project area. It provides an assessment of herbivory problems. In short, it tests a number of site-specific questions about the restoration potential before expending a lot of resources on a potential failure. If the founder colonies are successful that may be all that is needed to revegetate large areas.

12.5.1 CASE STUDIES

12.5.1.1 Founder Colonies in North Lake, Lake Lewisville, and Lake Conroe, Texas and Guntersville Reservoir, Alabama

North Lake, Lakes Lewisville, and Conroe and Guntersville Reservoir are all impoundments in the southeastern United States. They vary in size from 330 ha (North Lake) to 27,490 ha (Guntersville Reservoir). All trials reported are Phase 1 or Phase 2 of the founder colony process (Doyle and Smart, 1993; Doyle et al., 1997; Smart et al.; 1998). The number of species planted was limited in all cases and herbivory was a serious problem for establishing plants or getting beyond Phase 1 at all locations (Table 12.12). One serious herbivore not previously mentioned was the red-ear pond slider turtle (*Trachemys scripta elegans*) that could climb over the exclosure fencing. The aquatic mammals, nutria (*Myocastor coypus*), beaver (*Castor canadensis*), and muskrats (*Ondatra zibethica*) were blamed for destroying some exclosures and allowing turtles to enter. Lake Conroe was the most successful effort and reached Phase 2 (Smart et al., 1998). Grass carp (*Ctenopharyngodon idella*) were previously used to eliminate a hydrilla infestation but the residual grass carp population prevented native plant establishment. They also prevented expansion from small-scale, Phase 1 plantings. Six cove sites were fenced off for a larger protected area in Phase 2. Within the larger fenced area, single mature transplants were planted in individual plant protection cylinders. All plants spread, forming colonies up to 3 m in diameter. New colonies of *Zosterella dubia* that likely resulted from shoot fragments of the original plants formed within the fenced coves. *Chara* and *Najas guadalupensis* found their own way into the protected area.

The results from North Lake also looked promising. Wild celery spread throughout the exclosures and grew outside the exclosures during cool months when herbivory was low. In warm months it was grazed back to the exclosure boundary but with enough colonies and the proper conditions plant growth could exceed the rate of herbivory (Doyle et al., 1997). The authors cite literature that suggests that increased patch size sharply increases patch survival.

The ability of plants to colonize and spread in Lewisville Lake and Guntersville Reservoir did not look promising (Doyle and Smart, 1993; Doyle et al., 1997). Plants in these two impoundments survived only with strong protection from herbivory.

12.5.1.2   Cootes Paradise Marsh: Volunteers in Action

Cootes Paradise Marsh is a 250 ha, drowned river mouth, deep-water marsh, on the shoreline of Lake Ontario, near the city of Hamilton, Ontario, Canada. It once contained abundant emergent and submergent aquatic vegetation. By 1990 there was only a fringe plant community remaining on the western end of the marsh (Chow-Fraser, 1999). Loss of emergent cover was accompanied by a reduction in species diversity and replacement of native species by exotic invaders such as *Lythrum salicaria*, *P. australis*, and *Typha angustifolia*. Increased water levels, disturbance by common carp, and cultural eutrophication caused the demise of the aquatic plant community.

Seedling emergent stock of *T. angustifolia*, *T. latifolia*, *Sagittaria latifolia*, *Cephalanthus occidentalis*, *Acorus calamus*, *Decodon verticillatus*, and *Calla palustris* was planted between mid-July and late August, 1994. They were planted in 2.4 m square exclosures, in water depths ranging from mudflats to 45 cm. Depending on plant size and the available plant material, between 100 and 385 seedlings were planted in each exclosure. The submergents, sago pondweed, *P. foliosus*, and elodea were planted in August 1995 into 12, 7.3-m square silt-screen exclosures in water depths that varied from 45 to 60 cm.

Water depth had a tremendous effect on the ability of emergent species to survive. None of the *C. occidentalis*, *A. calamus*, or *C. palustris* planted in water depths greater than 10 cm survived to 1995. *Typha* spp. would not grow in deeper areas unless they were placed in exclosures. Screened exclosures allowed them to grow in water depths exceeding 40 cm. This was also true for *Sagittaria* and *Rumex*. None of the *Scirpus* and *Decodon* grew at depths greater than 30 cm, whether they were in screened or unscreened exclosures (Chow-Fraser, 1999).

Growth of submerged plants was related to the structural integrity of the exclosures, which failed to overwinter intact. The four exclosures that received the least damage contained varying amounts of sago pondweed. Since sago was not planted in all the exclosures, it appears that it successfully colonized all the moderately intact exclosures because it is turbidity tolerant (Chow-Fraser, 1999).

The volunteer Classroom Aquatic Plant Nursery Program raised seedling plants. This program began because commercial plant material used in experimental 1993 plantings were not ecotypically adapted to conditions at Cootes Paradise and their high cost was not sustainable in the long run. A lot of volunteer help was also used constructing exclosures and planting macrophytes.

Other things learned include: (1) plastic construction type fencing does not make suitable exclosures where muskrats are a problem, (2) exclosures are easily destroyed during winter ice conditions in northern climates, (3) using silt-screen as part of the exclosure increased water clarity and was deemed essential for submerged plant survival, and (4) resurgence of cattail beds outside the planting site was attributed to the successful exclusion of carp.

12.5.1.3   Rice Lake at Milltown, Wisconsin: Lessons Learned

"Mistakes seen in hindsight are the fodder of increased understanding" (Moss et al., 1996) summarizes the Rice Lake experience. Rice Lake is a 52 ha, shallow lake in northwestern Wisconsin. About 87% of the surface area is less than 1.5 m deep and the maximum depth is 1.8 m deep. It

TABLE 12.13
Macrophytes Planted in Rice Lake, Wisconsin during 1988–1989

Species	Propagules planted	No. of plots planted	Planting date
	Emergent Plants		
Sparganium eurycarpum	500 root stocks	2	5/24/89
Polygonum amphibium	500 root stocks	2	5/24/89
Zizania palustris	12 kg seed	5	9/9/88
Zizania palustris	45 kg seed	2	9/18/89
	Submergent Plants		
Ceratophyllum demersum	300 shoot bundles	2	6/14/89
Elodea canadensis	500 shoot bundles	2	6/14/89
Potamogeton richardsonii	500 root stocks	2	5/24/89
P. pectinatus	1000 tubers	4	5/24/89
Vallisneria americana	1000 tubers	4	5/24/89

Source: After Engel, S. and S.A. Nichols. 1994. Restoring Rice Lake at Milltown. Wisconsin Tech. Bull. 186. Wis. Dept. Nat. Res., Madison, WI.

is a classic case for shallow lake restoration. Historically it had good water clarity, abundant wild rice (*Zizania palustris*), waterfowl, and a desirable fishery for bass (*Micropterus salmoides*) and bluegills (*Lepomis macrochirus*). Damming of the outlet stream by beaver and increased groundwater inputs raised the water level. This combined with wind, erosion, ice heaving, runoff from farms, and nutrients from a municipal wastewater treatment plant converted this formerly clear water lake into a turbid one, dominated by phytoplankton, with a depauperate macrophyte flora (Engel and Nichols, 1994). Black bullheads (*Ameiurus melas*), a bottom feeding fish, were 76% of the electrofishing catch. Although 25 species of aquatic plants were found, most were in the marsh fringe surrounding the lake. Water lilies (*N. variegata* and *N. odorata*), sago pondweed, and *P. natans* were the dominant offshore species.

Increasing water clarity by restoring emergent and submergent species to stabilize bottom sediment, reduce wind fetch, and sequester plant nutrients so they were unavailable for algae growth was the plan for Rice Lake. The species used and the numbers planted are listed in Table 12.13. Wild rice seed, the species of primary importance, was sown in the fall of 1988 and 1989. Seed for the 1988 planting was collected from a nearby lake and sown at the rate of 2.3 kg per plot in five, 19×22-m plots. The 1989 planting used commercially purchased seed that was planted in two plots, one 180×45 m near the lake outlet and the other 150×54 m near the lake inlet. Each plot received 27.7 kg of seed. There were about 16, 348 wild rice seeds/kg. The other species were planted in monotypic, 50×20-m plots in various areas of the lake. Emergent species were planted from the shoreline to 1/2 m deep; submergent species were planted offshore in water 0.5 to 1.0 m deep. The emergent species, and *Potamogeton richardsonii*, sago pondweed, and wild celery were purchased from a commercial nursery. Shoots of *C. demersum* and elodea were collected from a southern Wisconsin lake. Wild rice was hand sown, emergent species were hand planted in slits, and the submergent species tubers and root stocks were attached to nails with rubber bands and hand sown. Shoots were bundled together with rubber bands, weighted with a nail and hand sown. All plant material was aquarium tested for viability in Rice Lake sediment.

None of the macrophytes planted from tubers, root stocks, or shoot bundles in 1989 grew well. *Sparganium eurycarpum, Polygonum amphibium,* and *P. richardsonii* failed to sprout in aquaria, indicating poor nursery stock. Other plants grew well in aquaria but not in the lake. Elodea and *C. demersum* were unable to reach the surface before water clarity deteriorated. Sago pondweed disappeared or the sprouted shoots became indistinguishable from wild plants invading the plots.

FIGURE 12.5 Muskrat "nipping" of wild rice plants in Rice Lake, Wisconsin. (Photo courtesy of Sandy Engel, WDNR.)

Wild celery tubers showed promise, sending leaves to the water surface by late summer, but they failed to grow the next summer.

Wild rice from both plantings sprouted in all lake plots by June, formed emergent leaves by July, and set seed by September of the year following planting. In August 1989 rice densities averaged 47 stems/m², about 60% of the average density of rice plants found in the lake where the seed was collected. In August 1990 stem density averaged 30 stems/m². Wind, waves, and ice affected rice production in both years. An island, torn loose during ice-out in 1989, carried seed and sediment away from the initial plots. A storm in August 1990 uprooted plants and thinned the rice bed by the inlet, reducing its density to less than half that at the outlet. Muskrats caused the most damage. Muskrats nipped off more than 90% of the emerging shoots in July 1990 (Figure 12.5). A local trapper caught 21 adult muskrats in the inlet and outlet areas of the lake in June 1991. Beaver also dammed the outlet, causing water levels to fluctuate, so beaver control was needed.

What were the lessons learned from hindsight? Many have already been discussed. Use good plant material. Effort and money were wasted at Rice Lake because of poor nursery stock. Plants need protection. Large patch size and muskrat reduction were not sufficient to protect wild rice. Exclosures would probably have worked better. The Rice Lake lesson that has not been previously discussed is that there was a "window of opportunity" from about mid-April when ice left the lake until the first week of June, when dense algal blooms occurred, or approximately 50 days, when seeds, tubers, or rhizomes needed to sprout and reach the water surface. Subtract from this time, days when water temperatures were too cold for plant growth. Using simulation modeling van Nes et al. (2002) suggest that a short clear-water phase enhances the probability of vegetation survival. The optimal timing for a clear-water phase is in the end of May and into June. In Rice Lake, plants that did not grow and form a surface canopy by early June were destined to fail. With dense algae blooms by the first week of June, plantings in late May were probably too late and planting in mid-June were certainly too late for successful plant establishment. Other techniques such as bullhead removal, nutrient inactivation, herbicide treatment of algae, or wave barriers to reduce turbidity may have increased the chance for success but they were not tried at Rice Lake.

TABLE 12.14
Macrophyte Community Change in Rice Lake, Wisconsin

Species	Relative Frequency (%)			
	August 1987[a]	August 1989[a]	August 1999[b]	July 2002[c]
Ceratophyllum demersum	35.4	19	—	8.1
Chara sp.	—	—	0.2	8.9
Elodea canadensis	—	—	37.1	28.2
Lemna spp.	—	—	—	0.2
Najas flexilis	—	—	—	9.6
Nuphar variegata	—	—	—	0.6
Nymphaea odorata	—	—	—	2.2
Potamogeton natans	—	6.9	—	7.7
P. pectinatus	64.5	72.4	25.4	5.1
P. zosteriformis	—	—	37.1	24.3
Ranunculus longirostris	—	1.7	0.2	5.1
Vallisneria americana	—	—	—	5.5

Sources: [a]After Engel, S. and S.A. Nichols. 1994. Restoring Rice Lake at Milltown. Wisconsin. Tech. Bull. 186. Wis. Dept. Nat. Res., Madison; [b]after Roesler, C.P. 2000. The Recovery of Rice Lake, a Water Quality Success Story. Unpublished report, Wis. Dept. Nat. Res., Spooner; [c]Data provided by Brook Waalen and Eric Wojchik, Polk County Land and Water Conservation Department, Balsam Lake, WI. Personal communications, 2002.

Although macrophyte planting was not successful, and there were lessons learned, the Rice Lake saga continued to a successful ending. A hint of water quality improvement appeared in 1995. A spring total phosphorus concentration of only 34 µg/L was found along with a Secchi depth of 1.1 m. Both values were spring records for the period beginning in 1988 (Roesler, 2000). June and July was a return to more "normal" values, but a record August Secchi depth of 0.5 m was measured.

In 1996, a dramatic change in water quality occurred. The summer mean Secchi depth was 1 m. The summer mean total phosphorus concentration was 44 µg/L, a 71% decrease over the 1988–1989 value and the summer mean chlorophyll *a* concentration was 26 µg/L, a 72% decrease over the 1988–1989 value (Roesler, 2000).

Water quality improvements continued and the plant community responded accordingly. A steady increase in submergent macrophyte coverage and density was observed each year from 1996 to 1999. An aquatic macrophyte survey conducted in 1999 found submergent macrophytes present in 100% of the sampling sites (Roesler, 2000). This compares to 31% vegetated plots in 1987 and 51% in 1989 (Engel and Nichols, 1994). Vegetation coverage continued in 2002 with 99% of the sampled plots being vegetated and the plant community was much more diverse (Table 12.14). Wild rice is still sparse in the lake despite several years of additional planting by the Great Lake Indian Fish and Wildlife Commission. Muskrats continued to be a problem for growing rice.

The reasons for this dramatic change are uncertain. It may be partially explained by upgrading the Village of Milltown wastewater treatment plant, up stream from the lake, from primary treatment to secondary treatment in 1978. If so, this illustrates the length of time it may take to see in-lake results of such changes (Chapter 4). Improvements in agricultural practices in the direct watershed may also have contributed to water quality improvements in the lake. In the past 10–15 years the number of cattle present declined by 75% or more. Significant areas of erosive cropland were enrolled in the Conservation Reserve Program and conventional tillage has been largely replaced by conservation tillage (Roesler, 2000). A tertiary wastewater treatment plant was constructed for Milltown in 1997 that should contribute to the future long-term health of Rice Lake.

12.6 CONCLUDING THOUGHTS

Aquatic plant restoration lags behind terrestrial and wetland restoration. Since it is relatively new, lessons learned from successes and failures are limited.

The need for aquatic plant restoration falls into three general categories: (1) low species diversity, (2) adverse abiotic or biotic conditions including infestations by exotic species, and (3) no plants and a paucity of propagules in the region. In the case of low species diversity the "do nothing" approach may be appropriate if there is a historical record of good species diversity in the restoration area or in adjacent areas and habitat conditions support plant growth. The "do nothing" approach may also work where the invasion with some exotic species is a problem. There are many cases where an exotic species like Eurasian watermilfoil declined and native species returned to the area. Active management such as harvesting and herbicide treatment may favor aggressive exotics. Here again, doing nothing may be a viable alternative management strategy. There is no guarantee, however, that native species will replace exotics by doing nothing and trying to replace exotics by planting native species has been unsuccessful.

Often efforts to correct adverse biotic conditions resulted in lush growth of *Chara*. Most studies have not lasted long enough to see if the *Chara* community evolves into a more diverse community of higher vascular plants as happened in cases of shallow dredging or other bottom treatments (Engel and Nichols, 1984). *Chara* can be very desirable for stabilizing the clear water state in shallow, eutrophic lakes (Kufel and Kufel, 2002; van Donk and van de Bund, 2002). Increased algal growth that is counter productive to macrophyte growth was a problem with some habitat restoration efforts where quiescent water formed behind breakwaters.

Lack of plants requires active planting. Besides correcting adverse abiotic conditions, a big threat to success is herbivory or other destruction by a variety of animals. Although one of the reasons for establishing plants is for fish and wildlife habitat, many animals do not have the patience to wait, and ravage the area before plantings are successful. Founder colonies using well protected mature plants appear to form a reasonable approach but time has not allowed its success to be demonstrated over the long term and over large areas. Protective exclosures are difficult to maintain. They can be breached by a variety of animals, especially medium-sized and large mammals. Floating debris and ice heaving during northern winters can also destroy them. Timing of plantings can be critical where algal turbidity is a concern.

Volunteer labor should be used where appropriate. It is an under-exploited resource for low-tech, low-budget, labor-intensive restoration activities. It is also a successful way to raise community awareness and instill a sense of stewardship and collective responsibility for the water resource.

REFERENCES

Asplund, T. and C.E. Cook. 1997. Effects of motor boats on submerged aquatic macrophytes. *Lake and Reservoir Manage.* 13: 1–12.

Asplund, T. and C.E. Cook. 1999. Can no-wake zones effectively protect littoral zone habitat from boating disturbance. *LakeLine* 19(1): 16.

Bartodziej, W. 1999. A bid to revegetate an urban lake shoreland. *LakeLine* 19(1): 10.

Bauman, P.C., J.F. Kitchell, J.J. Magnuson and T.B. Kayes. 1974. Lake Wingra, 1837–1973: A case history of human impact. *Trans. Wis. Acad. Sci. Arts Lett.* 62: 57–94.

Beal, E.O. 1977. A Manual of Marsh and Aquatic Vascular Plants of North Carolina with Habitat Data. Tech. Bull. 247. North Carolina Agric. Exp. Sta., Raleigh.

Botkin, D.B. 1975. Strategies for the reintroduction of species to damaged ecosystems. In: J. Cairns, K. Dickson and E. Herricks (Eds.), *Recovery and Restoration of Damaged Ecosystems*. University of Virginia Press, Charlottesville. pp. 241–261.

Bowman, J.A. and K.E. Mantai. 1993. Submersed aquatic plant communities in western New York: 50 years of change. *J. Aquatic Plant Manage.* 31: 81–84.

Brege, D. 1988. Fresh greens. *Wisconsin Nat. Res. Mag.* 12(2): 9.

Butts, D., J. Hinton, C. Watson, K. Langeland, D. Hall and M. Kane. 1991. *Aquascaping: Planting and Maintenance.* Circ. 912. Florida Coop. Ext. Service, University Florida, Gainesville.

Carlson, R.A. and J.B. Moyle. 1968. Key to the Common Aquatic Plants of Minnesota. Spec. Publ. 53. Minn. Dept. Cons., St. Paul.

Carpenter, S.R. 1980. The decline of *Myriophyllum spicatum* in a eutrophic Wisconsin lake. *Can. J. Bot.* 58: 527–535.

Chow-Fraser, P. 1999. Volunteer-based experimental planting program to restore Cootes Paradise Marsh, an urban coastal wetland of Lake Ontario. *LakeLine* 19(1): 12.

Cottam, G. and S.A. Nichols. 1970. *Changes in Water Environment Resulting from Aquatic Plant Control.* Water Res. Ctr. Tech. Rept. OWRR B-019-Wis. University Wisconsin, Madison.

Crow, G.E. and C.B. Hellquist. 1981. Aquatic vascular plants of New England: Part 2. *Typhaceae* and *Sparganiaceae. Station Bull.* 517. University New Hampshire Agric. Exp. Sta., Durham.

Crow, G.E. and C.B. Hellquist. 1982. Aquatic vascular plants of New England: Part 4. *Juncaginaceae, Scheuchzeriaceae, Butomaceae, Hydrocharitaceae. Station Bull.* 520. University New Hampshire Agric. Exp. Sta., Durham.

Crow, G.E. and C.B. Hellquist. 1983. Aquatic vascular plants of New England: Part 6. *Trapaceae, Haloragaceae, Hippuridaceae. Station Bull.* 524. University New Hampshire Agric. Exp. Sta., Durham.

Crow, G.E. and C.B. Hellquist. 1985. Aquatic vascular plants of New England: Part 8. *Lentibulariaceae. Station Bull.* 528. University New Hampshire Agric. Exp. Sta., Durham, NH.

Doyle, R.D. and R.M. Smart. 1993. Potential Use of Native Aquatic Plants for Long-Term Control of Problem Aquatic Plants in Guntersville Reservoir, Alabama. Report 1, Establishing Native Plants. Tech. Rept. A-93-6. U.S. Army Corps of Engineers, Vicksburg, MS.

Doyle, R.D., R.M. Smart, C. Guest and K. Bickel. 1997. Establishment of native aquatic plants for fish habitat: test planting in two north Texas reservoirs. *Lake and Reservoir Manage.* 13: 259–269.

Engel, S. and S.A. Nichols. 1984. Lake sediment alteration for macrophyte control. *J. Aquatic Plant Manage.* 22: 38–41.

Engel, S. and S.A. Nichols. 1994. Restoring Rice Lake at Milltown. Wisconsin. Tech. Bull. 186. Wis. Dept. Nat. Res., Madison, WI.

Fassett, N.C. 1969. *A Manual of Aquatic Plants.* University of Wisconsin Press, Madison.

Fischer, S., P. Cieslewicz and D. Seibl. 1999. Aquatic vegetation reintroduction efforts in Missouri impoundments. *LakeLine* 19(1): 14.

Haslam, S.R. 1996. Enhancing river vegetation: conservation, development, and restoration. *Hydrobiologia* 340: 345–348.

Hellquist, C.B. and G.E. Crow. 1980. Aquatic vascular plants of New England: Part I. *Zosteraceae, Potamogetonaceae, Zannichelliaceae, Najadaceae. Station Bull.* 515. University New Hampshire Agric. Exp. Sta., Durham, NH.

Hellquist, C.B. and G.E. Crow. 1981. Aquatic vascular plants of New England: Part 3. *Alismataceae. Station Bull.* 518. University New Hampshire Agric. Exp. Sta., Durham.

Hellquist, C.B. and G.E. Crow. 1982. Aquatic vascular plants of New England: Part 5. *Araceae, Lemnaceae, Xyridaceae, Eriocaulaceae,* and *Pontederiaceae. Station Bull.* 523. University New Hampshire Agric. Exp. Sta., Durham, NH.

Hellquist, C.B. and G.E. Crow. 1984. Aquatic vascular plants of New England: Part 7. *Cabombaceae, Nymphaeaceae, Nelumbonaceae, Ceratophyllaceae. Station Bull.* 527. University New Hampshire Agric. Exp. Sta., Durham.

Helsel, D.R., S.A. Nichols and R.W. Wakeman. 1999. Impacts of aquatic plant management methodologies on Eurasian watermilfoil populations in Southeastern Wisconsin. *Lake and Reservoir Manage.* 15: 159–167.

Hoeger, S. 1988. Schwimmkampen, Germany's artificial floating islands. *J. Soil Water Cons.* 43(4): 304–306.

Hutchinson, G.E. 1975. *A Treatise on Limnology — Limnological Botany.* John Wiley, New York.

James, W.F., J.W. Barko, H.L. Eakin and D.R. Helsel. 2001a. Changes in sediment characteristics following drawdown of Big Muskego Lake, Wisconsin. *Arch. Hydrobiol.* 151: 459–474.

James, W.F., H.L. Eakin and J.W. Barko. 2001b. Rehabilitation of a Shallow Lake (Big Muskego Lake, Wisconsin) via Drawdown: Sediment Responses. Aquatic Plant Cont. Res. Prog., Tech. Note EA-04. U.S. Army Corps of Engineers, Vicksburg, MS.

Janvrin, J.A. and H. Langrehr. 1999. Aquatic vegetation response to islands constructed in the Mississippi River (abstract). Presented at: A Symposium of the Upper Mississippi River Conservation Committee, Lacrosse, WI.

Johnston, D.L., D.L. Sutton, V.V. Vandiver and K.A. Langeland. 1983. Replacement of *Hydrilla* by other plants in a pond with emphasis on growth of American lotus. *J. Aquatic Plant Manage.* 21: 41–43.

Kadlec, J.A. and W.A. Wentz. 1974. State-of-the Art Survey and Evaluation of Marsh Plant Establishment Techniques: Induced and Natural. Contract Rep. D-74-9. Dredged Materials Research Program, U.S. Army Coastal Eng. Res. Ctr., Fort Belvoir, VA.

Kadono, Y. 1982a. Distribution and habitat of Japanese *Potamogeton. Bot. Mag. Tokyo* 95: 63–76.

Kadono, Y. 1982b. Occurrence of aquatic macrophytes in relation to pH, alkalinity, Ca^{++}, Cl^-, and conductivity. *Jpn. J. Ecol.* 32: 39–44.

Kahl, R. 1993. Aquatic Macrophyte Ecology in the Upper Winnebago Pool Lakes, Wisconsin. Tech. Bull. 182. Wis. Dept. Nat. Res., Madison.

Kester, D. 1989. *Kester's Wild Game Food Nurseries Catalog.* Omro, WI.

Kufel, L. and I. Kufel. 2002. *Chara* beds acting as nutrient sinks in shallow lakes — a review. *Aquatic Bot.* 72: 249–260.

Lathrop, R.C., E.R. Deppe, W.T. Seybold and P.W. Rasmussen. 1991. Attempts at reestablishing a native pondweed (*Potamamogeton amplifolius*) in Lake Mendota, Wisconsin. In: *Abstracts of the 11th Annual International Symposium of Lake, Reservoir, and Watershed Management*, NALMS, Madison, WI.

Lemberger, J.J. Undated. *Wildlife Nurseries Catalog.* Wildlife Nurseries, Oshkosh, WI.

Les, D.H. and L.J. Mehrhoff. 1999. Introduction of nonindigenous aquatic vascular plants in southern New England: a historical perspective. *Biol. Invas.* 1: 281–300.

Marshall, S. 1986. Transplanting bulrush to enhance fisheries and aquatic habitat. *Aquatics* 8(4): 16–17.

Moss, B., J. Madgwick and G.L. Phillips. 1996. *A Guide to the Restoration of Nutrient-enriched Shallow Lakes.* Broads Authority, Norwich, U.K.

Moyle, J.B. 1945. Some chemical factors influencing the distribution of aquatic plants in Minnesota. *Am. Mid. Nat.* 34: 402–421.

Mueller, G., J. Sartoris, K. Nakamura and J. Boutwell. 1996. Ukishima, floating islands, or schwimmkampen? *LakeLine* 16(3): 18.

Munrow, J. 1999. Ecological resotration: Rebuilding nature. *Vol. Monitor* 11(1): 1–6.

Nichols, S.A. 1994. Evaluation of invasion and declines of submersed macrophytes for the Upper Great Lakes region. *Lake and Reservoir Manage.* 10: 29–33.

Nichols, S.A. 1999. Distribution and Habitat Descriptions of Wisconsin Lake Plants. Bull. 96. Wis. Geol. Nat. Hist. Surv., Madison.

Nichols, S.A. 2001. Long-term change in Wisconsin lake plant communities. *J. Fresh Water Ecol.* 16: 1–13.

Nichols, S.A. and S. Mori. 1971. The littoral macrophyte vegetation of Lake Wingra. *Trans. Wis. Acad. Sci. Art Lett.* 59: 107–119.

Nichols, S.A. and R.C. Lathrop. 1994. Cultural impacts on macrophytes in the Yahara lakes since the late 1800s. *Aquatic Bot.* 47: 225–247.

Nichols, S.A. and J.G. Vennie. 1991. Attributes of Wisconsin Lake Plants. Inf. Circ. 73. Wis. Geol. Nat. Hist. Surv., Madison, WI.

Nichols, S.A., S.P. Weber and B.H. Shaw. 2000. A proposed aquatic plant community biotic index for Wisconsin lakes. *Environ. Manage.* 26: 491–502.

Nicholson, S.A. 1981. Changes in submersed macrophytes in Chautauqua Lake, 1937–1975. *Freshwater Biol.* 11: 523–530.

Orth, R.J. and K.A. Moore. 1983. The biology and propagation of eelgrass, *Zostera marina,* in Chesapeake Bay — project summary. USEPA-600/3-82-090. Annapolis, MD.

Pip, E. 1979. Survey of the ecology of submerged aquatic macrophytes in central Canada. *Aquatic Bot.* 7: 339–357.

Pip, E. 1988. Niche congruency of aquatic macrophytes in central North America with respect to 5 water chemistry parameters. *Hydrobiologia* 162: 173–182.

Robertson, D.M., G.L. Goddard, D.R. Helsel and K.L. MacKinnon. 2000. Rehabilitation of Delavan Lake, Wisconsin. *Lake and Reservoir Manage.* 16: 155–176.

Roesler, C.P. 2000. The Recovery of Rice Lake, a water quality success story. Unpublished report, Wis. Dept. Nat. Res., Spooner.

Seddon, B. 1972. Aquatic macrophytes as limnological indicators. *Freshwater Biol.* 2: 107–130.

Simpson, W. 1949. Measurement of diversity. *Nature* 163: 688.

Smart, R.M. and G. Dick. 1999. Propagation and Establishment of Aquatic Plants: A Handbook for Ecosystem Restoration Projects. Tech. Rept. A-99-4. U.S. Army Corps of Engineers, Vicksburg, MS.

Smart, R.M., G.O. Dick and R.D. Doyle. 1998. Techniques for establishing native aquatic plants. *J. Aquatic Plant Manage.* 36: 44–49.

Smith, C.S. and J.W. Barko. 1992. Submersed Macrophyte Invasions and Declines. Aquatic Plant Cont. Res. Prog. Rept., Vol. A-92-1. U.S. Army Corps of Engineers, Vicksburg, MS.

Storch, T.A., J.D. Winter and C. Neff. 1986. The employment of macrophyte transplanting techniques to establish *Potamogeton amplifolius* beds in Chatauqua Lake, New York. *Lake and Reservoir Manage.* 2: 263–266.

Trudeau, P.N. 1982. Nuisance Aquatic Plants and Aquatic Plant Management Programs in the United States: Vol. 3: Northeastern and North-Central Region. Rep. MTR-82W47-03, Mitre Corp., McLean, VI.

U.S. Army Corps of Engineers. 1978. Wetland Habitat Development with Dredged Material: Engineering and Plant Propagation. Tech. Rept. DS-78-16, Office, Chief of Engineers, Washington, DC.

van Donk, E. and A. Otte. 1996. Effects of grazing by fish and waterfowl on the biomass and species composition of submerged macrophytes. *Hydrobiologia* 340: 285–290.

van Donk, E. and W.J. van de Bund. 2002. Impact of submerged macrophytes including charophytes on phyto- and zooplankton communities: allelopathy versus other mechanisms. *Aquatic Bot.* 72: 261–274.

van Nes, E.H., M. Scheffer, M.S. van den Berg and H. Coops. 2002. Dominance of charophytes in eutrophic shallow lakes - when should we expect it to be an alternative stable state? *Aquatic Bot.* 72: 275–296.

Wisconsin Department of Natural Resources. 1991. Upriver Lakes Habitat Restoration Project, Environmental Impact Statement. Wis. Dept. Nat. Res., Madison.

13 Water Level Drawdown

13.1 INTRODUCTION

Water level drawdown is an established, multipurpose reservoir and pond management procedure to control certain aquatic plants and fish populations, and possibly to produce a switch in alternative stable states (Chapter 9). It is less commonly used in lakes without an outlet control because siphoning or pumping (Chapter 7) is needed. It provides opportunities to repair structures such as dams or docks, to remove or consolidate flocculent sediments, and to carry out dredging or sediment cover installation.

This chapter emphasizes drawdown to reduce macrophyte biomass, and describes case studies from several North American climates. Responses of 74 plant species to whole-year, winter, or summer drawdown are presented as a user guideline. A discussion of its use in fish management is included, and positive and negative factors of the procedure are summarized. Drawdown can also be used to encourage regrowth of emergent species (Chapter 12). Reviews include Cooke (1980), Culver et al. (1980), Ploskey (1983), and Leslie (1988).

13.2 METHODS

The primary mode of action of water level drawdown for macrophyte biomass management is exposure of plants, especially root systems, to dry and freezing, or dry and hot conditions for a period sufficient to kill the plants and their reproductive structures. Winter drawdowns are more successful than summer, although the number of reported summer drawdowns is too small for adequate evaluation. The advantages of winter drawdown, in addition to effectiveness on some target plants, are: (1) there will be no invasion of moist lake soils by semi-terrestrial plants, (2) there will be no proliferation of aquatic emergents, and (3) there will be less interference with recreation. Also, runoff is often highest in the spring, so refill should occur. The decision to employ a summer or a winter drawdown to control plants depends upon target species susceptibility, uses of the reservoir, and other management objectives.

Aquatic plants do not respond uniformly to drawdown. Table 13.1 is a list of the responses of 74 species. Some are unaffected or increase in biomass, while others are very susceptible. Because of this, accurate plant identification is required.

Table 13.2 is a summary of responses of 19 common plants to drawdown. Cutgrass and smartweed are among those that grow well in moist soils and shallow water, and will proliferate in some drawdown situations. This may be desirable when attempting to enhance a fishery, as explained in later paragraphs. Alligator weed and hydrilla are serious nuisances in southern U.S. waters and are rarely controlled by this procedure. Milfoil and water hyacinth have been controlled by winter drawdown, particularly *Myriophyllum spicatum* (Eurasian watermilfoil). This plant, however, as shown by experience in Tennessee Valley Authority (TVA) reservoirs and in Oregon, withstands low temperatures if the plant remains moist or if the exposed hydrosoil is not frozen for several weeks. Milfoil is also well adapted to rapid vegetative spread. It may recolonize areas dominated by native plants prior to drawdown.

In lakes with a mixture of species, exposure of littoral communities to dry and hot or to dry and cold conditions may eliminate or curtail one plant species and favor the development of a

TABLE 13.1
Responses of 74 Aquatic Plants to Water Level Drawdown

Species	Increased			Decreased			No Change		
	A	W	S	A	W	S	A	W	S
Alternanthera philoxeroides	10	9	15						
			31						
Bidens sp.			13						
Brasenia schreberi					1	13			
					11	14			
					22	15			
						26			
Cabomba caroliniana			15		11	17			
					23	26			
Carex spp.			13						
Cephalanthus occidentalis			15						
Ceratophyllum demersum	28	20		14	1	13		21	15
					2	17			
					9				
					11				
					16				
					32				
Chara vulgaris		16	17			15		30	14
								35	
Cyperus spp.	10								
Eichhornia crassipes		9	15	10	11				
			31		23				
					35				
Eleocharis baldwinii			15			17			
Eleocharis acicularis			13		1				
			17		22				
Elodea canadensis		21			1			2	
					6			30	
					20				
					33				
Elodea densa					9	12			
					16	17			
Elodea sp.					11				
Glyceria borealis		21							
Hydrilla verticillata		3		18				36	
(see section on Florida)		9							
Hydrochloa caroliniensis				10					
Hydrotrida caroliniana					21				
Jussiaea diffusa					7				
Leersia oryzoides		21	13						
Lemna minor	28								
Lemna sp.					1				
Limnobium spongia						26			
Myriophyllum brasiliense						15			14
Myriophyllum exalbescens					2			30	
Myriophyllum heterophyllum						26			15
Myriophyllum spicatum		4			5			30	
					24				

TABLE 13.1 (Continued)
Responses of 74 Aquatic Plants to Water Level Drawdown

Species	Increased			Decreased			No Change		
	A	W	S	A	W	S	A	W	S
					25				
					33				
					35				
Myriophyllum sp.					1				
					11				
Megalodonta beckii		1							
Najas flexilis	28	1	13					27	15
		6							
		21							
		24							
		33							
Najas guadalupensis				10	9			17	
					14				
					16				
Nelumbo lutea						15		23	7
Nuphar advena					22				
Nuphar luteum						26			
Nuphar macrophyllum		9							
Nuphar polysepalum									12
Nuphar variegatum					20				13
					21				
Nuphar sp.					1				
Nymphaea odorata			26			14			12
						15			
Nymphaea tuberosa					19				
Panicum sp.	10								
Polygonum coccineum		21	8		1				
Polygonum natans								21	
Pontederia cordata							10		
Potamogeton americanus		21							
Potamogeton amplifolius		20			1				
					2				
Potamogeton crispus					33		6		
					35				
Potamogeton diversifolius		1					15		
		19							
Potamogeton epihydrus		19						1	
		21							
Potamogeton foliosus		19						6	
Potamogeton gramineus		19						6	
Potamogeton natans		1				13			
Potamogeton nodosus								32	
Potamogeton pectinatus		28		34		6			
		34							
Potamogeton Richardsonii		21						1	
Potamogeton Robbinsii					1				
					2				
					20				

TABLE 13.1 (Continued)
Responses of 74 Aquatic Plants to Water Level Drawdown

Species	Increased			Decreased			No Change		
	A	W	S	A	W	S	A	W	S
					33				
Potamogeton zosteriformis		19						1	
								2	
Potamogeton spp.	8								14
Ranunculus tricophyllus								1	
Sagittaria graminea								10	
Sagittaria latifolia		20						1	
Salix interior		21							
Scirpus americanus		1							
Scirpus californicus				10					
Scirpus validus		21	29						
Sium suave		21							
Sparganium chlorocarpum						29			
Spirodela polyrhiza				1					
Typha latifolia	18	21				29	10	1	
Utricularia purpurea						26			
Utricularia vulgaris				1					
Utricularia sp.					22				17
Vallisneria Americana		1		10			2		

Note: A, whole-year drawdown; W, winter drawdown; S, summer drawdown. Numbers refer to references given as sources below.

[a] Summer-fall drawdown.

Sources are from the Reference list as follows: 1. Beard, 1973; 2. Dunst and Nichols, 1979; 3. Fox et al., 1977; 4. Geiger, 1983; 5. Goldsby et al., 1978; 6. Gorman, 1979; 7. Hall et al., 1946; 8. Harris and Marshall, 1963; 9. Hestand and Carter, 1975; 10. Holcomb and Wegener, 1971; 11. Hulsey, 1958; 12. Jacoby et al., 1983; 13. Kadlec, 1962; 14. Lantz et al., 1964; 15. Lantz, 1974; 16. Manning and Johnson, 1975; 17. Manning and Sanders, 1975 (summer–fall drawdown); 18. Massarelli, 1984; 19. Nichols, 1974; 20. Nichols, 1975a; 21. Nichols, 1975b; 22. Pierce et al., 1963; 23. Richardson, 1975; 24. Siver et al., 1986; 25. Smith, 1971; 26. Tarver, 1980; 27. Tazik et al., 1982; 28. van der Valk and Davis, 1978; 29. van der Valk and Davis, 1980; 30. Wile and Hitchin, 1977; 31. Williams et al., 1982; 32. Godshalk and Barko, 1988; 33. Crosson, 1990; 34. Van Wijck and DeGroot, 1993; 35. Wagner and Falter, 2002; 36. Poovey and Kay, 1996.

resistant one. Some susceptible plants such as milfoil, as noted above, are normally so successful that few other species coexist. In these cases several years of winter water level drawdown, followed by no drawdown for 1 to 2 years, may prevent establishment of resistant species by allowing other species to reestablish. The drawdown cycle can then be repeated.

Management of macrophyte biomass, and fishery enhancement, through systematic changes in water level, are not possible with every water body where water level can be regulated. Hydropower storage and flood control reservoirs are most amenable to water level management. The strong influence of flow and the limited storage capacity of main stem reservoirs limit their water level manipulations for management purposes (Ploskey et al., 1984). Other factors prevent or limit the use of water level drawdown for management, including water supply use, summer or winter

TABLE 13.2
Summary of Responses of 19 Aquatic Plants to Water Level Drawdown

Species That Usually Increase

1. *Alternanthera philoxeroides* (alligator weed): Annual (Holcomb and Wegener, 1971), winter (Hestand and Carter, 1975), summer (Lantz, 1974)
2. *Hydrilla verticillata* (hydrilla): Winter (Fox et al., 1977; Hestand and Carter, 1975)
3. *Leersia oryzoides* (cutgrass): Winter (Nichols, 1975b), summer (Kadlec, 1962)
4. *Najas flexilis* (bushy pondweed): Annual (van der Valk and Davis, 1978), winter (Beard, 1973; Crosson, 1990; Gorman, 1979; Nichols, 1975b; Siver et al. 1986), summer (Kadlec, 1962)
5. *Polygonum coccineum* (smartweed); Winter (Nichols, 1975b), summer (Harris and Marshall, 1963. Beard, (1973) reported a decrease in this species in a winter drawdown
6. *Potamogeton epihydrus* (leafy pondweed): Winter (Nichols, 1974, 1975b). Beard (1973) reported no change in this species in a winter drawdown
7. *Scirpus validus* (softstem bulrush): Winter (Nichols, 1975b), summer (van der Valk and Davis, 1980)

Species That Usually Decrease

1. *Brasenia schreberi* (water shield): Winter (Beard, 1973; Hulsey, 1958; Richardson, 1975), summer Kadlec, 1962; Lantz et al., 1964; Lantz, 1974; Tarver, 1980)
2. *Cabomba caroliniana* (fanwort): Winter (Hulsey, 1958; Richardson, 1975), summer (Manning and Sanders, 1975; Tarver, 1980)
3. *Ceratophyllum demersum* (coontail): Annual (Lantz et al., 1964), winter (Beard, 1973; Dunst and Nichols, 1979; Godshalk and Barko, 1988; Hestand and Carter, 1975; Hulsey, 1958; Manning and Johnson, 1975), summer (Kadlec, 1962; Manning and Sanders, 1975). Increases or no change in this species were reported by Lantz, 1974; Nichols, 1975a, b; and van der Valk and Davis, 1978
4. *Egeria densa* (Brazilian elodea): Winter (Hestand and Carter, 1975; Manning and Johnson, 1975), summer (Jacoby et al., 1983; Manning and Sanders, 1975)
5. *Myriophyllum* spp. (milfoil): Winter (Beard, 1973; Crosson, 1990; Dunst and Nichols, 1979; Goldsby et al., 1978; Hulsey, 1958; Smith, 1971; Siver et al. 1986), summer (Lantz, 1974; Tarver, 1980; Van Wijck and DeGroot, 1993). Increases and no change in milfoil have occasionally been reported; see Table 13.1 for species and references
6. *Najas guadalupensis* (southern naiad): Annual (Holcomb and Wegener, 1971), winter (Hestand and Carter, 1975; Lantz et al., 1964; Manning and Johnson, 1975). Manning and Sanders (1975) reported no change in this species in a summer–fall drawdown
7. *Nuphar* spp. (yellow water lily): Winter (Beard, 1973; Nichols, 1975a, b; Pierce et al., 1963), summer (Tarver, 1980). Increases and no change in Nuphar have occasionally been reported; see Table 13.1 for species and references
8. *Nymphaea odorata* (water lily): Summer (Lantz et al., 1964; Lantz, 1974). Jacoby et al. (1983) reported no change in this species in a summer drawdown; Tarver (1980) reported an increase in a summer drawdown
9. *Potamogeton robbinsii* (Robbins's pondweed): Winter (Beard, 1973; Crosson, 1990; Dunst and Nichols, 1979; Nichols, 1975a)

Species That Do Not Change, Or Whose Response Is Variable

1. *Eichhornia crassipes* (water hyacinth): Hestand and Carter (1975), Holcomb and Wegener (1971), Hulsey (1958), Lantz (1974), Richardson (1975)
2. *Elodea canadensis* (elodea): Beard (1973), Dunst and Nichols (1979), Gorman (1979), Nichols (1975a, b), Wile and Hitchin (1977)
3. *Typha latifolia* (cattail): Beard (1973), Holcomb and Wegener (1971), Nichols (1975b), van der Valk and Davis (1980)

recreation, shoreline development such as parks or homes, the need to maintain water levels for downstream low-flow augmentation, and dam design that will not allow sufficient water release (Culver et al., 1980). Also, undesirable effects on non-target littoral zone or wetland species could prevent the use of this technique. A permit to discharge enough water to expose the littoral area could be needed where wetland alteration or destruction could occur, or where discharge may affect downstream uses.

13.3 POSITIVE AND NEGATIVE FACTORS OF WATER LEVEL DRAWDOWN

Control of susceptible nuisance plants and fish management are two of the several ways that drawdown can be used to improve or restore lakes. Ideally, if this procedure is to be implemented for plant control, the possibility of carrying out every other lake improvement procedure that drawdown makes possible should be considered.

Grass carp and herbicide applications are effective for managing nuisance macrophytes in certain circumstances (Chapters 16 and 17). Water level drawdown can reduce the amount of grass carp needed, or improve their effectiveness (Stocker and Hagstrom, 1986), and provides opportunity for pelletized herbicide applications (Westerdahl and Getsinger, 1988).

Loose, flocculent sediments are common in eutrophic systems and represent a significant source of turbidity, discomfort to swimmers, and a source of nutrients to the water column. Drawdown is effective in consolidating some types of lake sediments. The effects of drying on muck-type (organic and nutrient-rich, high in water content), flocculent (poorly defined sediment-water interface), and peat-type (fibrous, organic, low water content) sediments from Lake Apopka, Florida were examined in the laboratory. Muck-type sediments consolidated 40–50% after exposure to rain and sun for 170 days. Peat consolidated about 7% under identical conditions (Fox et al., 1977). The 40–50% water loss may be sufficient to make the sediments firm to walk on, and consolidated sediments appear to remain firm after reflooding (Kadlec, 1962), although groundwater seepage might prevent these changes. In some lakes, there is only a slight consolidation of sediments after a summer drawdown (e.g., Long Lake, Washington ; Jacoby et al., 1982).

Sediment removal could be combined with sediment consolidation to bring about deepening of selected areas. A bulldozer could be used for sediment removal instead of expensive hydraulic dredges, assuming sediments can support heavy equipment (Chapter 20). Since consolidated sediments have lower water content, and little water is removed with them during bulldozer operation, runoff from disposal sites is minimal and land reuse at the disposal site could be immediate. An extreme drawdown of Lake Tohopekaliga, Florida was followed by a sediment removal of 165,000 m^3 in 1987, 340,000 m^3 in 1991, and 3,000,000 m^3 in 2002. The 2002 dredging was projected to remove 120 t of P and 2,500 t of N (Williams, 2001). When sediments are exposed, debris can be removed and artificial reefs for anglers can be constructed.

Loose flocculent sediments can inhibit growth of desirable macrophytes and prevent fish spawning. Some of these sediments can be removed via lake outflow during a drawdown, although there may be downstream impacts. At Newnan's Lake, Florida, summer drawdown scoured the lake's bottom, removing 270 kg P and 59,000 kg of flocculent sediments, and produced some sediment compaction (Gottgens and Crisman, 1991).

Water level management is an important part of the restoration of lake or reservoir "fringe" wetlands (Levine and Willard, 1989), and fluctuating water levels are essential to maintaining the vegetation supporting a waterfowl community (Kadlec, 1962) A Michigan waterfowl reservoir was drawn down in summer to stimulate the growth of plants attractive to ducks. Emergent plants such as *Typha* (cattail) and *Scirpus* (bulrush) prefer bare mudflats as a seedbed, a condition not met in stable water level systems; drawdown provided conditions for the germination of their seeds (Kadlec, 1962). Up to 20,000 seeds per square meter were found in the upper 5 cm of exposed sediments in an Iowa marsh, a seed bank that should allow establishment of a community of emergent and annual species (van der Valk and Davis, 1978).

Drawdown presents other possibilities for lake improvement. Sediment covers are more easily and cheaply installed on dry, consolidated sediments than by the use of SCUBA (Chapter 15). Repair or construction of docks, placement of riprap on banks, maintenance of dams, and removal of litter can be carried out effectively after drawdown. Finally, this procedure has the lowest cost of any macrophyte management method unless pumps are required to lower the water level (Dierberg and Williams, 1989).

Partial water level drawdown could be used to re-establish rooted macrophytes in order to stabilize sediments. This has been proposed for Lake Okeechobee, Florida where transparency has fallen sharply in some areas, possibly from migration of bottom mud toward the shore when water levels exceed 4.6 m. Lowering water level by 1.0 m may reduce sediment transport, clarify the water, and promote macrophyte establishment (Havens and James, 1999).

Algal blooms have occurred after reflooding of dried and/or frozen lake sediments (Hulsey, 1958; Beard, 1973), suggesting that drawdown may be a factor in switching a lake from a clear water, macrophyte-dominated condition to an algae-dominated, turbid condition (Chapter 9). Factors causing blooms may be P release from reflooded sediments, along with fish control of algae grazers.

Total P concentration in the top layer of drying, highly organic marsh sediments increased, and decreased in bottom layers (as deep as 40 cm), as dessication proceeded by an upward flux of water to the sediment surface (DeGroot and Van Wijck, 1993). In Big Muskego Lake, Wisconsin, pore-water ammonium N and SRP, and laboratory-based P release, increased following a drying/freezing drawdown (James et al., 2001). Dried sediments from a eutrophic reservoir had significantly lower affinity for P than continuously wet sediments, perhaps because redox cycling of Fe–P species stops when sediments are air-dried, leading to the formation of crystalline Fe molecules with low P sorption (Baldwin, 1996). These data suggest there may be significant P flux to the water column at reflooding, especially from hydrosoils with high organic content (Watts, 2000). Phosphorus released from dried and/or frozen lake sediments may be P associated with bacteria cells (Sparling et al., 1985; Qui and McComb, 1995). Relatively brief periods of drying or freezing may be all that is needed to produce P release at reflooding (Klotz and Linn, 2001).

Field observations of P release following reflooding are uncommon and conflicting. A summer 1979 drawdown (June–October) to control *Egeria densa* in Long Lake, Washington, a U.S. region of low summer precipitation, was successful in lowering the 1980 standing crop by 84%. *Nuphar polysephalum* and *Nymphaea odorata* were unaffected, and macrophyte biomass recovered by 1981. Water column total P and pH were lower, and dense cyanobacteria blooms were absent in summer 1980. In the months of reflooding following the 1979 drawdown, there was no increase in water column P (Jacoby et al., 1982). In contrast, P increased after reflooding of Backus Lake, Michigan (Kadlec, 1962). Algal blooms are not always a consequence of drawdowns. Drawdown of a hypereutrophic, and previously regulated lake (Zeekoevlei, South Africa) led to large *Daphnia* and clear water, even though nutrients increased significantly following reflooding. Fish biomass was apparently lowered by washout, allowing large-bodied zooplankton and the clear water state to occur (Harding and Wright, 1999). More field observations about immediate water chemistry changes following reflooding are needed.

Drawdown exposes wetlands adjacent to the lake and may have impacts to wetland biota. Drawdown at Lake Bomoseen, Vermont, produced major effects on a wetland containing several threatened or endangered plant species. Effects on invertebrates were also severe. The elimination of native plant species from exposed littoral areas may allow nuisance species from deep water, such as Eurasian watermilfoil, to invade the exposed areas (Crosson, 1990).

Failure to refill following drawdown is a potentially serious problem. This may be from a failure to close the dam at the proper time, or to drought. Reflooding should begin in late winter so that lake users can be assured of access to the lake during recreation season, and other uses of the reservoir can occur.

There is a potential for low DO and an associated fish kill during drawdown, particularly if incoming water is rich in nutrients and organic matter and the remaining pool is small in volume. Once water level is down, there are few possibilities for aerating it. Fish kills due to low DO have been a concern, but reports are contradictory. Beard (1973) found no fish mortality despite a 70% winter drawdown in a eutrophic reservoir, and E.B. Welch (personal communication) found that DO in Long Lake, Washington, did not fall below 5 mg/L during a summer drawdown of 2 m (Z_{max} = 3.5 m). Low DO (but no fish kill) occurred in Mondeaux Flowage, Wisconsin, during a winter drawdown (Nichols, 1975a). In contrast, a fish kill in Chicot Lake, Louisiana occurred during a summer drawdown (Geagan, 1960), and Gaboury and Patalas (1984) observed a fish kill in a Manitoba lake following winter drawdown and a loss of DO. DO problems can occur when summer drawdown causes turnover of a thermally stratified lake so that water low in oxygen is suddenly introduced into surface waters (Richardson, 1975). During a winter drawdown of an enriched Wisconsin reservoir, sediments became resuspended in the river-like areas of the upper reservoir. These sediments were high in organic matter, were anaerobic, and contained significant H_2S and reduced iron. High chemical and biological oxygen demand extracted any remaining DO from the water. This condition moved downstream, removing DO from the lower reaches of the reservoir. A delay of drawdown until mid-January and a release limit of 25% of reservoir volume were recommended to prevent future DO problems (Shaw, 1983). However, this might not provide sufficient exposure to cold for macrophyte control. The possibility that drawdown will produce an oxygen depletion in the remaining pool should be assessed, and aeration or artificial circulation devices (Chapters 18 and 19) may be necessary.

Drawdown may have severe consequences to the invertebrate community, which in turn could reduce fish productivity as well as species diversity of the benthic community. Also, the release of large volumes of water can create flooding conditions downstream. In addition, release of nutrient-rich and/or anaerobic water will be deleterious to stream biota. A late fall water release would most likely be oxygenated with lower nutrient concentrations (assuming water release at fall overturn), and thus have lower impact on downstream biota.

There can be safety concerns with a winter drawdown if inflows (e.g., a winter rain/snow melt) cause the ice cover on the remaining water to float. This could create open water or thin ice near shore.

13.4 CASE STUDIES

The object of water level drawdown for nuisance plant control is to expose plants to freezing-desiccation or to heat-desiccation, destroying the plant body and the rhizomes or roots. Exposure to heat or cold may also be detrimental to seeds, turions, and tubers. In some regions (e.g., Louisiana), water level fluctuations have been a principal plant control method (Richardson, 1975), whereas in others (e.g., the U.S. Pacific Northwest) climate extremes are usually too narrow to provide the necessary harsh conditions. The following U.S. case histories illustrate responses to drawdown in several climates.

13.4.1 Tennessee Valley Authority (TVA) Reservoirs

Hall et al. (1946) were among the first to describe flooding and dewatering effects on aquatic plants. Several woody species require dewatering for establishment, including black willow (*Salix nigra*), buttonball (*Cephalanthus occidentalis*), green ash, (*Fraxinus lanceolata*), tupelo gum (*Nyssa aquatica*), and bald cypress (*Taxodium distichum*). Each of these must be dewatered to become established. Similarly, herbaceous weeds will not develop on sites that remain inundated until June. Reflooding of woody and herbaceous plants, as discussed later, is an important technique to enhance development of populations of fish food organisms.

Alligator weed (*Alternanthera philoxeroides*) is a nuisance in some TVA reservoirs. Subfreezing temperatures are lethal to above-ground parts in this mid-latitude, milder climate region of the U.S., but below-ground roots show little or no injury and overwintering fragments recolonize sites after spring reflooding. The water primrose (*Jussiaea diffusa*) forms floating mats and is destroyed by dewatering and freezing. In the case of the primrose, as well as two other nuisance plants (bladderwort, *Utricularia biffa*, and milfoil, *Myriophyllum scabratum*), plants may survive if the soil remains moist during a winter drawdown (Hall et al., 1946).

The TVA reservoirs have been infested with Eurasian watermilfoil (*M. spicatum*). It is particularly troublesome in reservoirs with differences of only 0.6 to 1.0 m between minimum and maximum water levels (Goldsby et al., 1978). Although the herbicide 2,4-D was used extensively, drawdown was the most effective control method along shorelines where herbicide dilution occurred (Chapter 16). A 1.8-m (6-ft) winter drawdown at Watts Bar and Chickamauga Reservoirs killed all milfoil plants on well-drained shorelines. In some areas, landforms of milfoil developed that later reverted to the aquatic form when inundated (Smith, 1971).

Eurasian watermilfoil in Melton Hill Reservoir, Tennessee was managed with 2,4-D and winter drawdown from December to mid-February (1971 to 1972). The area colonized in 1972 was less than 1971, especially in shallow water, and deeper water plants did not increase in biomass. From 1973 to 1976, 2,4-D was used, but costs increased steadily. The herbicide brought about a 68% reduction in areal coverage in 1973 compared to the 1972 coverage after drawdown, but reinfestation was rapid unless herbicides were reapplied or drawdown was used. Semi-monthly winter drawdowns were effective in destroying root crowns of plants exposed to freezing, but even harsh winters did not reduce infestations unless the hydrosoil was completely dewatered. A combination of maintenance 2,4-D applications and high frequency, short duration winter drawdowns was most effective and economical for control of *M. spicatum* in Melton Hill Reservoir (Goldsby et al., 1978).

13.4.2 LOUISIANA RESERVOIRS

Water level manipulation is an important method of reservoir management in Louisiana, a southern region of the U.S. that experiences some periods of freezing weather in most winters. Chemical controls were costly, and harvesting (Chapter 14) was expensive and promoted plant spreading through fragmentation. Water hyacinth biomass (*Eichhornia crassipes*) can be controlled by drying and freezing. As with milfoil, plants left in a few centimeters of water survive. Unfortunately dewatering promotes seed germination, but after 1 or 2 years of drying and freezing there is a significant reduction in viable seeds (Richardson, 1975).

Anacoco Reservoir, Louisiana was drawn down 1.5 m from midsummer to mid-October, reducing reservoir area from 1,052 ha to 526 ha. It was refilled by mid-February. About 40% of the reservoir was closed to fishing due to *Potamogeton* sp. and *Najas guadalupensis*, but after drawdown and refill, only about 5% of the area was closed. The drawdown eliminated water shield (*Brasenia schreberi*), restricted the spread of parrot feather (*Myriophyllum brasiliense*) and water lily (*Nymphaea. odorata*), and enhanced *Chara vulgaris* (Lantz et al., 1964; Lantz, 1974).

Bussey Reservoir (northeastern Louisiana) was drawn down in October and refilled in May. In the summer prior to drawdown, 280 ha were infested with *Potamogeton* sp. and *N. guadalupensis*. In the two summers following refill, only 16 ha (40 acres) were infested. The treatment was considered 90% effective. Lafourche Reservoir, also in northeastern Louisiana, had a partial drawdown in winter and further water removal in the summer to determine effects on *Ceratophyllum demersum* (coontail), which infested 80% of the reservoir. In the summer following refill, over 60% of the reservoir was clear of coontail (Lantz et al., 1964; Lantz, 1974).

Drawdown in Louisiana is a successful control method for many plants, but cannot be used for eradication. However, lake managers could stagger fluctuation years to prevent plants from adapting to it. The recommended schedule is 2 to 3 years of drawdown followed by 2 years without water level fluctuation (Lantz, 1974). Nichols (1975b) also recommended staggered drawdowns for Wisconsin reservoirs. Presumably intervals without water fluctuation allow susceptible species to regain dominance over drawdown-resistant species. Subsequent drawdown then frees the reservoir once again from susceptible nuisance plants.

13.4.3 FLORIDA

A fall-winter drawdown (September 1972 to February 1973) was used to control nuisance vegetation in a central Florida reservoir, Lake Ocklawaha (Rodman Reservoir). The dominant plants before drawdown were *Ceratophyllum demersum, Egeria densa, Hydrilla verticillata, Eichhornia cras-*

sipes (water hyacinth), and *Pistia stratiotes* (waterlettuce). Water level was lowered 1.5 m and the study sites were dry or had very shallow water. By the end of the second growing season following reflooding, *Ceratophyllum* coverage was reduced by 47% and *Egeria* coverage by 56%. Hydrilla and water hyacinth had a lake-wide increase of 64 times and 33 times, respectively, after reflooding. Failure to control these species was due in part to a winter without frost, allowing spread to new areas (Hestand and Carter, 1975). Water hyacinth can be controled by drying and freezing in northern Louisiana reservoirs, although drying appears to enhance seed germination (Lantz, 1974). In the milder climate of central Florida, conditions appropriate for control of these species through winter drawdown seldom occur.

Drawdown to control hydrilla should be based on its life cycle (Massarelli, 1984; Leslie, 1988; Poovey and Kay, 1998). Hydrilla produces dessication-resistant subterranean tubers in early fall. Drawdown at this time gives hydrilla an additional competitive advantage and might ensure a monoculture following refill. Release of the weevil *Bagous affinis* could reduce tuber density (Buckingham and Bennett, 1994) (Chapter 17). A spring drawdown might be used to kill the standing crop, followed by another drawdown before tuber formation to kill newly sprouted plants (Haller et al., 1976). A third drawdown in the following spring may eliminate remaining plants, a successful approach in managing hydrilla in Fox Lake, Florida, though cattails (*Typha* sp.) invaded dewatered soils making overall success questionable (Massarelli, 1984).

A lock and spillway were built in 1964, reducing the natural water level fluctuations by 71% in Lake Tohopekaliga, one of the lakes in the Kissimmee chain of lakes in central Florida. Shoreline agricultural and housing development, and increased wastewater discharges followed. Organic deposits from algae and weeds, particularly water hyacinth, degraded the littoral habitat and fishing success declined. A drawdown of 2.1 m from March through September 1971, with final refill by March 1972, exposed 50% of the lake's bottom. The sediments dried and consolidated, and desirable (for fisheries) submerged plants returned. Drawdowns were needed again in 1979, 1987, and 1990, along with organic sediment removal with frontend loaders and bulldozers, to maintain the improved lake condition (Wegener and Williams, 1974a; Williams et al., 1979; Williams, 2001).

13.4.4 WISCONSIN

Winter water level drawdown of the 172 ha Murphy Flowage, Wisconsin was successful in opening the flowage (reservoir) to recreation. Between mid-October and mid-November, 1967 and 1968, water level was lowered 1.5 m and maintained at that level until March, and then brought to full volume. In 1967, 30 ha were closed to fishing from late spring through summer by *Potamogeton robbinsii, P. amplifolius, Ceratophyllum demersum, Myriophyllum* spp. *and Nuphar* spp. The first winter drawdown opened 26 of the 30 ha for fishing, and none of the above species returned as dominants in 1969. *Megalondonta beckii* (water marigold), *Najas flexilis,* and *P. diversifolius* all increased following drawdown, and *P. natans* was unchanged. Even with these resistant species, there were still 24 of the original 30 ha open to fishing in 1969 (Figures 13.1 and 13.2). Success was attributed to freezing and drying of vegetative reproductive structures in this cold climate region of the U.S. The reduction of the *Nuphar* population was thought to be due to deep frost and sediment upheaval. The three resistant species were beginning to come back, but the flowage was destroyed by a flood in 1970, preventing a longer term evaluation. Fishing success for largemouth bass increased in summer 1968 (Snow, 1971; Beard, 1973).

The primary negative effect was the appearance of a phytoplankton bloom in August 1968 (Beard, 1973). This response is not uncommon when plants are controlled, perhaps in part due to P release from reflooded soils and to the absence of the clear water stabilizing effects of macrophytes (Chapter 9). The suggestion that drawdown could be used to switch the lake to the macrophyte-dominated state (Coops and Hosper, 2002) is not supported in this case.

Legend:

Potamogeton robbinsii
Nuphar spp.
Ceratophyllum demersum
Potamogeton robbinsii and Nuphar spp.
Potamogeton amplifolius
Myriophyllum spp.
Potamogeton natans
Megalodonta beckii
Potamogeton diversifolius
Najas flexilis

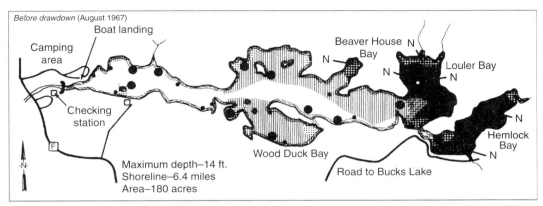

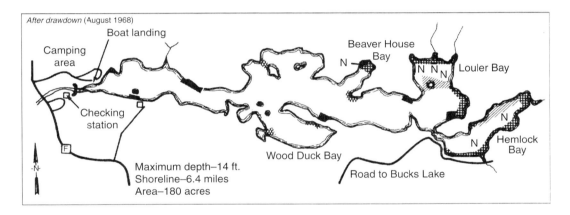

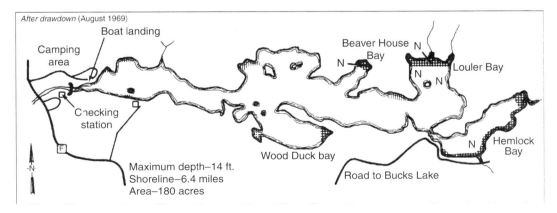

FIGURE 13.1 Abundance of aquatic plants before and 2 years after an overwinter drawdown at Murphy Flowage, Wisconsin. Ranking was based on the percentage within the 210 quadrants, covering the entire flowage. (From Beard, T.D. 1973. Overwinter Drawdown. Impact on the Aquatic Vegetation in Murphy Flowage, Wisconsin. Tech. Bull. No. 61. Wisconsin Department of Natural Resources, Madison.)

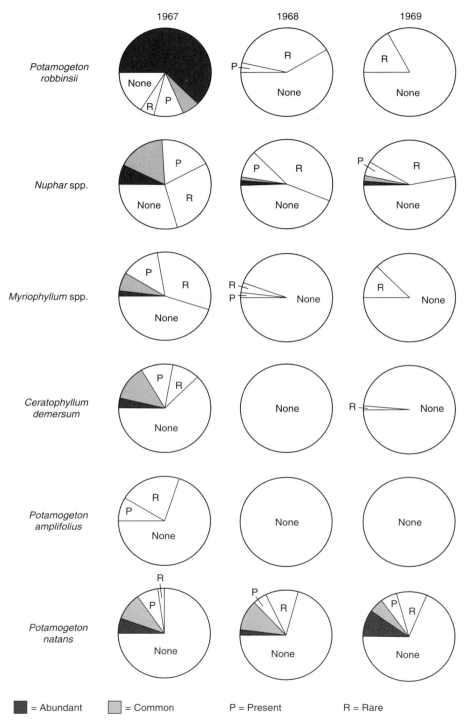

FIGURE 13.2 Distribution of the major species of aquatic plants in Murphy Flowage, Wisconsin, before and after overwinter drawdowns. The distribution includes only the areas in which the species were abundant, common, and present. (From Beard, T.D. 1973. Overwinter Drawdown. Impact on the Aquatic Vegetation in Murphy Flowage, Wisconsin. Tech. Bull. No. 61. Wisconsin Department of Natural Resources, Madison.)

13.4.5 CONNECTICUT

Candlewood Lake, Connecticut, a pump-storage reservoir, supported a monoculture of *M. spicatum* (Eurasian watermilfoil) in 1983, severely limiting recreation. Water level was lowered 2 m in winter 1983 to 1984 and 2.7 m in winter 1984 to 1985 in this cold climate region of the U.S. Where sediments were dry and frozen, milfoil was essentially eliminated. Moist areas of the exposed lake sediments supported milfoil growth in the summer. The exposed frozen soil areas supported an infestation of *Najas minor* and *N. flexilis*, but these low-growing plants did not interfere with most water uses (Siver et al., 1986).

Among the eastern U.S. state lake management programs, Connecticut, Pennsylvania, Delaware, New Jersey, Maryland, and Virginia have successfully used winter drawdowns for macrophyte control (Culver et al., 1980).

13.4.6 OREGON

An attempt to control *M. spicatum* in a Portland, Oregon reservoir through winter drawdown was unsuccessful. Water level was dropped from mid-December to mid-February, 1981 to 1982, to the base of the milfoil beds. Subsurface seepage, high water retention by the sediments, and high rainfall kept the roots moist throughout the drawdown. The roots were exposed to about 32 h of air temperatures between 1 and 4°C. Above ground milfoil biomass was eliminated, but root crowns were unaffected and regrowth began in March. The reservoir did not refill to previous levels, and live plants were common along exposed areas in July, demonstrating milfoil resistance to these conditions. An application of 2,4-D was required to obtain control (Geiger, 1983). The U.S. Pacific rainforest climate is probably not suitable for this procedure because winters are mild and wet, and dewatering and freezing may not occur.

13.5 FISH MANAGEMENT WITH WATER LEVEL DRAWDOWN

Water level drawdown is an effective, inexpensive, and widely recognized reservoir fishery management method. A detailed discussion of fish management is not within the scope of this text, but lake managers should be aware of its use to enhance fish habitat. Drawdowns appear to stimulate fish productivity by reestablishing conditions similar to the first years of a newly filled reservoir. Monocultures of submersed aquatic plants can be eliminated, terrestrial vegetation that is established on exposed hydrosoils is flooded and colonized with invertebrates, and extreme densities of forage fish (relative to predators) are reduced through predation. The result may last for several years, with a sharply increased biomass and individual sizes of game fish, and a reduction in biomass or abundance of rough fish and stunted panfish or other planktivores. These fishery changes can mean a better sports fishery, clearer water, and fewer algal blooms (Chapter 9). Reviews of drawdown in fish management include Bennett (1954), Pierce et al. (1963), Culver et al. (1980), Ploskey (1983), Ploskey et al. (1984), and Randtke et al. (1985).

13.6 CASE HISTORIES

A major water level fluctuation where water levels are lowered for several months every 3 to 5 years, can create new reservoir conditions on a small scale (Ploskey, 1983; Randtke, et al., 1985). The effect of the drawdown is, in part, related to its timing during the year. In Kansas, where midsummer air temperature regularly exceeds 35°C, summer drawdown allows seeding of exposed soils with rapidly growing vegetation, such as millet, and promotes invertebrate growth after reflooding. The smaller volume of the remaining pool allows intense predation on smaller fish and rapid predator growth (Randtke et al., 1985). Summer drawdown in a central Missouri reservoir not only contributed to largemouth bass growth, but many small fish were eliminated when stranded in pools and vegetation. The reservoir was partially refilled in autumn to increase waterfowl habitat. Drawdown occurred again in the winter and rising levels in the spring inundated planted areas and terrestrial vegetation, adding terrestrial fauna to fish diets and providing a substrate for littoral invertebrates (Heman et al., 1969). Flooded vegetation is essential for the spring spawning of northern pike (Hassler, 1970).

A fall drawdown is recommended to increase predator foraging on smaller fish, provided drawdown is extensive and water temperature exceeds 13°C. Prey fish are presumably more vulnerable when they are forced to abandon littoral refuges. Two months of drawdown appears to be a minimum for predators to have the desired impact on forage fish. An early autumn drawdown of 2 or more months can allow time for grasses or other terrestrial plants to become established or to be seeded. Also, there is less chance of a severe DO depletion during this season, assuming algal photosynthesis in the remaining pool (Ploskey, 1983).

Forage size bluegill decreased from 850/ha to 163/ha in an Arkansas reservoir during partial autumn drawdown. Biomass of gizzard shad was sharply reduced. In the following year, there should be larger game fish, greater fishing success, and reduced predation on large sized zooplankton (Hulsey, 1958). This latter effect, in turn, could mean more effective grazing on algae by zooplankton, clearer water, and possibly a return of macrophytes (Chapter 9). If this scenario were realized, it would support the hypothesis that drawdown can switch a lake to the macrophyte-dominated state (Coops and Hosper, 2002).

Drawdown can be used to manage specific fish populations. For example, common carp (*Cyprinus carpio*) produce turbid conditions and add significant loads of nutrients to the water column. Carp can be controlled by lowering the water level at spawning time. Water temperature, gonad conditions, and presence and water depth of eggs were monitored in a South Dakota reservoir. Water was withdrawn, exposing the eggs and stranding fry in pools (Shields, 1958). Also they can be removed by seining following drawdown (Hulsey, 1958; Lantz, 1974).

Drawdown is a convenient method of adding fish attractors or structure. Standing crops of channel catfish, bluegill, largemouth bass, and white crappie were 16–20 times higher in areas of brush shelters than in control areas (Pierce and Hooper, 1979).

A drawdown enhanced the Lake Tohopekaliga (Florida,) fishery. Natural water level fluctuations in the Kissimmee chain of lakes were sharply reduced following channelization and damming, leading to organic sediment build-up and loss of submersed aquatic vegetation. Fish-food organisms declined in abundance and diversity, and the sports fishery, an important component of the local economy, was affected. After reflooding, the terrestrial and semi-aquatic plants that flourished during sediment exposure, and invertebrate populations, increased sharply. Fishing success increased and the worth of the added fish alone was estimated to be over $6 million. The beneficial effects of the drawdowns lasted for several years, during which algal blooms and water hyacinth decay gradually produced fish habitat deterioration (Wegener and Williams, 1974b; Williams et al., 1979, 1982; Williams, 2001; Moyer, 1987).

Ecological succession occurs when stable water levels resume. The rate of change can be rapid, as expected in most ecosystems that have been perturbed to the point of having characteristics of an early successional stage. As long as the regulatory agency and the public agree, drawdowns can be used at regular intervals to maintain the "newness" of the system.

Drawdown for fishery management can have negative consequences. Exposure of the littoral zone constitutes loss, and sometimes destruction, of habitat for benthic invertebrates, and they may exhibit great changes in density and diversity following drawdown. In the year following a summer drawdown in which loose, flocculent sediments became solid enough to walk on, insect populations were greatly reduced, molluscs were absent, and hardened sediments might have retarded recolonization by certain species (Kadlec, 1962). The sharp decrease in fishing in an Illinois lake following an autumn water level drawdown was believed to be due to a decline in abundance of invertebrates (Bennett, 1954). Paterson and Fernando (1969) found that 150 days of exposure (southern Ontario, Canada), during which sediments froze to a depth of 20 cm, destroyed a large portion of the benthic fauna. Where drawdown and exposure are not severe, or where newly planted vegetation or terrestrial vegetation is flooded, the density of invertebrates may increase rapidly, along with improved fishing (Wegener et al., 1974). Short-term drawdowns during cool periods may protect burrowing invertebrates (McAfee, 1980), but may not control nuisance macrophytes or produce desired changes in fish species composition.

Winter drawdowns of Lake Pend Oreille (Idaho) were severe in order to provide hydropower and capacity to store snow pack runoff in the spring. Drawdown was reduced from 3.0–3.7 m to 2.1 m during two winters to enhance salmon spawning area and survival, and to increase warmwater fish habitat (bass, bluegill). Increased water levels led to increased macrophyte biomass and possibly to increased overwintering fish habitat and fish survival (Wagner and Falter, 2002).

The summer drawdown of Cross Lake, Manitoba, reduced fish habitat and standing crops of lake whitefish (*Coregonus clupeaformis*), walleye (*Stizostedion vitreum vitreum*), northern pike (*Esox lucius*), and cisco (*C. artedii*) became lower. A severe fish kill occurred after a winter drawdown. Prolonged winter drawdowns reduced white fish and cisco hatching success, whereas low spring water levels denied walleye and pike access to spawning areas (Gaboury and Patalas, 1984).

13.7 SUMMARY

Water level drawdown has been used to produce at least short-term control of some aquatic plant species, but it is species specific and some plants are unaffected or may even thrive, particularly if competitive species are eliminated. Winter drawdowns appear to be most effective in plant control, assuming the littoral area can be exposed to several weeks of dry, freezing conditions. Drawdown at this time produces the least impact on downstream biota because fall destratification of the reservoir will assure release of aerated, lower nutrient waters. Drawdown is ineffective in moist, mild climates and where seepage in winter keeps lake sediments moist.

Drawdown is among the least expensive lake management techniques. In Florida, drawdown costs were estimated to be $7.25/ha per year (2002 U.S. dollars) (Dierberg and Williams, 1989). Its use reduces the cost of other procedures such as sediment removal or application of sediment covers. Dam construction for new ponds and reservoirs should allow deep-water release.

Water level drawdown is an effective and well-established fish management technique. It is used to enhance the growth of predator species, to control the density of forage fish, and to assist in management of nuisance species such as common carp.

Additional research is needed about species responses to drawdown, the release of nutrients from reflooded sediments, and the comparative merits of dry-hot vs. dry-cold exposure. There is also a need for research on the impacts on fish and other animal populations and the use of this technique as part of food web manipulation. There can be major negative impacts to non-target littoral species, including the invertebrate community and wetlands.

REFERENCES

Baldwin, D.S. 1996. Effects of exposure to air and subsequent drying on the phosphate sorption characteristics of sediments from a eutrophic reservoir. *Limnol. Oceanogr.* 41: 1725–1732.

Beard, T.D. 1973. Overwinter Drawdown. Impact on the Aquatic Vegetation in Murphy Flowage, Wisconsin. Tech. Bull. No. 61. Wisconsin Department of Natural Resources, Madison.

Bennett, G.W. 1954. The effects of a late summer drawdown on the fish population of Ridge Lake, Coles County, Illinois. In: *Trans. 19th North American Wildlife Conf.*, pp. 259–270.

Buckingham, G.R. and C.A. Bennett. 1994. Biological and Host Range Studies with *Bagous affinis*, an Indian Weevil that Destroys Hydrilla Tubers. Tech. Rept. A-94-8. U.S. Army Corps Eng., Vicksburg, MS.

Cooke, G.D. 1980. Lake level drawdown as a macrophyte control technique. *Water Res. Bull.* 16: 317–322.

Coops, J. and S.H. Hosper. 2002. Water-level management as a tool for the restoration of shallow lakes in the Netherlands. *Lake and Reservoir Manage.* 18: 293–298.

Crosson, H. 1990. Impact Evaluation of a Lake Level Drawdown on the Aquatic Plants of Lake Bomoseen, Vermont. Vermont Department of Environmental Conservation, Waterbury.

Culver, D.A., J.R. Triplett, and G.G. Waterfield. 1980. The Evaluation of Reservoir Waterlevel Manipulations as a Fisheries Management Tool in Ohio. Final Report to Ohio Department of Natural Resources, Division of Wildlife, Project F-57-R, Study 8, Columbus, p. 67.

DeGroot, C.-J. and C. Van Wijck. 1993. The impact of dessication of a freshwater marsh (Garcines Nord, Camargue, France) on sediment-water-vegetation interactions. Part 1. The sediment chemistry. *Hydrobiologia* 252: 83–94.

Dierberg, F.E. and V.P. Williams. 1989. Lake management techniques in Florida, USA: Costs and water quality effects. *Environ. Manage.* 13: 729–742.

Dunst, R. and S.A. Nichols. 1979. Macrophyte control in a lake management program. In: J.E. Breck, R.T. Prentki, and O.L. Loucks. (Eds.), *Aquatic Plants, Lake Management, and Ecosystem Consequences of Lake Harvesting.* Center for Biotic Systems and Institute for Environmental Studies, University of Wisconsin, Madison. pp. 411–418.

Fox, J.L., P.L. Brezonick and M.A. Keirn. 1977. *Lake Drawdown as a Method of Improving Water Quality.* USEPA-600/3-77-005.

Gaboury, M.N. and J.W. Patalas. 1984. Influence of water level drawdown on the fish populations of Cross Lake, Manitoba. *Can. J. Fish Aquatic Sci.* 41: 118–125.

Geagan, D. 1960. A report of a fish kill in Chicot Lake, Louisiana during a water level drawdown. *Proc. La. Acad. Sci.* 23: 39–44.

Geiger, N.S. 1983. Winter drawdown for the control of Eurasian watermilfoil in an Oregon oxbow lake (Blue Lake, Multnomah County). In: *Lake Restoration, Protection and Management.* USEPA 440/5-83-001. pp. 193–197.

Godshalk, G.L. and J.W. Barko. 1988. Effects of winter drawdown on submersed aquatic plants in Eau Galle Reservoir, Wisconsin. In: Proc. 22nd Annual Meeting, Aquatic Plant Control Research Program. Misc. Paper A-88-5. U.S. Army Corps Eng., Vicksburg, MS.

Goldsby, T.L., A.L. Bates and R.A. Stanley. 1978. Effect of water level fluctuation and herbicide on Eurasian watermilfoil in Melton Hill Reservoir. *J. Aquatic Plant Manage.* 16: 34–38.

Gorman, M.E. 1979. Effects of an overwinter drawdown and incomplete refill on autotroph distribution and water chemistry in a permanent recreational pond. MS Thesis. Kent State University, Kent, OH.

Gottgens, J.F. and T.L. Crisman. 1991. Newnan's Lake, Florida: removal of particulate organic matter and nutrients using a short-term partial drawdown. *Lake and Reservoir Manage.* 7: 53–60.

Hall, T.F., W.T. Penfound and A.D. Hess. 1946. Water level relationships of plants in the Tennessee Valley with particular reference to malaria control. *J. Tenn. Acad. Sci.* 21: 18–59.

Haller, W.T., J.L. Miller, and L.A. Garrard. 1976. Seasonal production and germination of hydrilla vegetative propagules. *J. Aquatic Plant Manage.* 14: 26–29.

Harding, W.R. and S. Wright. 1999. Initial findings regarding changes in phyto- and zooplankton composition and abundance following the temporary drawdown and refilling of a shallow, hypertrophic South African coastal lake. *Lake and Reservoir Manage.* 15: 47–53.

Harris, S.W. and W.H. Marshall. 1963. Ecology of water-level manipulations on a northern marsh. *Ecology* 44: 331–343

Hassler, T.J. 1970. Environmental influences on early development and year-class strength of northern pike in lakes Oake and Sharpe, North Dakota. *Trans. Am. Fish. Soc.* 99: 369–375.

Havens, K.E. and R.T. James. 1999. Localized changes in transparency linked to mud sediment expansion in Lake Okeechobee, Florida: Ecological and management implications. *Lake and Reservoir Manage.* 15: 54–69.

Heman, M.L., R.S. Campbell and L.C. Redmond. 1969. Manipulation of fish populations through reservoir drawdown. *Trans. Am. Fish. Soc.* 98: 293–304.

Hestand, R.S. and C.C. Carter. 1975. Succession of aquatic vegetation in Lake Ocklawaha two growing seasons following a winter drawdown. *Hyacinth Control J.* 13: 43–47.

Holcomb, D. and W. Wegener. 1971. Hydrophytic changes related to lake fluctuation as measured by point transects. *Proc. Southeast Assoc. Game Fish Commun.* 25: 570–583.

Hulsey, A.H. 1958. A proposal for the management of reservoirs for fisheries. *Proc. Southeast Assoc. Game Fish Commun.* 12: 132–143.

Jacoby, J.M., D.D. Lynch, E.B. Welch and M.A. Perkins. 1982. Internal phosphorus loading in a shallow eutrophic lake. *Water Res.* 16: 911–919.

Jacoby, J.M., E.B. Welch and J.T. Michaud. 1983. Control of internal phosphorus loading in a shallow lake by drawdown and alum. In: *Lake Restoration, Protection and Management.* USEPA-440/5-83-001. pp. 112–118.

James, W.F., J.W. Barko, H.L. Eakin and D.R. Helsel. 2001. Changes in sediment characteristics following drawdown of Big Muskego Lake, Wisconsin. *Arch. Hydrobiol.* 151: 459–474.

Kadlec, J.A. 1962. Effects of a drawdown on a waterfowl impoundment. *Ecology* 43: 267–281.

Klotz, R.L. and S.A. Linn. 2001. Influence of factors associated with water level drawdown on phosphorus release from sediments. *Lake and Reservoir Manage.* 17: 48–54.

Lantz, K.E. 1974. Natural and Controlled Water Level Fluctuation in a Backwater Lake and Three Louisiana Impoundments. Report to Louisiana Wildlife and Fisheries Comm., Baton Rouge.

Lantz, K.E., J.T. Davis, J.S. Hughes and H.E. Schafer. 1964. Water level fluctuation — its effects on vegetation control and fish population management. *Proc. Southeast Assoc. Game Fish Comm.* 18: 483–494.

Leslie, A.J., Jr. 1988. Literature review of drawdown for aquatic plant control. *Aquatics* 10: 12–18.

Levine, D.A. and D.E. Willard. 1989. Regional analysis of fringe wetlands in the Midwest: creation and restoration. In: J.A. Kusler and M.E. Kentula, (Eds.), *Wetland Creation and Restoration: the Status of the Science. Vol. I. Regional Reviews.* USEPA 600/3-89-038a. pp. 305–332.

Manning, J.H. and R.E. Johnson. 1975. Water level fluctuation and herbicide application: an integrated control method for hydrilla in a Louisiana reservoir. *Hyacinth Control J.* 13: 11–17.

Manning, J.H. and D.R. Sanders, 1975. Effects of water fluctuation on vegetation in Black Lake, Louisiana. *Hyacinth Control J.* 13: 17–21.

Massarelli, R.J. 1984. Methods and techniques of multiple phase drawdown — Fox Lake, Brevard County, Florida. In: *Lake and Reservoir Management.* USEPA 440/5-84-001. pp. 498–501.

McAfee, M. 1980. Effects of a water drawdown on the fauna in small cold water reservoirs. *Water Res. Bull.* 16: 690–696.

Moyer, E.J. 1987. *Kissimmee Chain of Lakes Studies. Study I. Lake Tohopekaliga Investigations.* Florida Game and Fresh Water Fish Commission. Kissimmee.

Nichols, S.A. 1974. *Mechanical and Habitat Manipulation for Aquatic Plant Management. A Review of Techniques.* Tech. Bull. No. 77. Wisconsin Department of Natural Resources, Madison, WI.

Nichols, S.A. 1975a. The use of overwinter drawdown for aquatic vegetation management. *Water Res. Bull.* 11: 1137–1148.

Nichols, S.A. 1975b. The impact of overwinter drawdown on the aquatic vegetation of the Chippewa Flowage, Wisconsin *Trans. Wisc. Acad. Sci.* 63: 176–186.

Paterson, C.G. and C.H. Fernando. 1969. The effect of winter drainage on reservoir benthic fauna. *Can. J. Zool.* 47: 589–595.

Pierce, B.E. and G.R. Hooper. 1979. Fish standing crop comparisons of tire and brush fish attracters in Barkley Lake, Kentucky. *Proc. Ann. Conf. Southeast Assoc. Fish Wildlife Agencies* 33: 688–691.

Pierce, P.C., J.E. Frey and H.M. Yawn. 1963. An evaluation of fishery management techniques utilizing winter drawdowns. *Proc. Southeast Assoc. Game Fish Comm.* 17: 347–363.

Ploskey, G.R. 1983. A Review of the Effects of Water-Level Changes on Reservoir Fisheries and Recommendations for Improved Management. Tech. Rept. E-83-3. U.S. Army Corps Eng., Vicksburg, MS.

Ploskey, G.R., L.R. Aggus and J.M. Nestler. 1984. Effects of Water Levels and Hydrology on Fisheries in Hydropower Storage, Hydropower Mainstream, and Flood Control Reservoirs. Tech. Rept. E-84-8. U.S. Army Corps Eng. Vicksburg, MS.

Poovey, A.G. and S.H. Kay. 1998. The potential of a summer drawdown to manage monoecious hydrilla. *J. Aquatic Plant Manage.* 36: 127–130.

Qiu, S. and A.J. McComb. 1994. Effects of oxygen concentration on phosphorus release from reflooded air-dried wetland sediments. *Aust. J. Mar. Fresh Water Res.* 45: 1319–1328.

Randtke, S.J., F. de Noyelles, D.P. Young, P.E. Heck and R.R. Tedlock. 1985. A Critical Assessment of the Influence of Management Practices on Water Quality, Water Treatment, and Sport Fishing in Multipurpose Reservoirs in Kansas. Kansas Water Resources Research Institute, Lawrence.

Richardson, L.V. 1975. Water level manipulation: a tool for aquatic weed control. *Hyacinth Control J.* 13: 8–11.

Shaw, B.H. 1983. *Agricultural Runoff and Reservoir Drawdown Effects on a 2760-Hectare Reservoir.* USEPA-600/S3-82-003.

Shields, J.T. 1958. Experimental control of carp reproduction through water drawdowns in Fort Randall Reservoir, South Dakota. *Trans. Am. Fish. Soc.* 87: 23–33.

Siver, P.A., A.M. Coleman, G.A. Benson and J.T. Simpson. 1986. The effects of winter drawdown on macrophytes in Candlewood Lake, Connecticut. *Lake Reservoir Manage.* 2: 69–73.

Smith, G.E. 1971. Resumé of studies and control of Eurasian watermilfoil (*Myriophyllum spicatum* L.) in the Tennessee Valley from 1960–1969. *Hyacinth Control J.* 9: 23–25.

Snow, H.E. 1971. Harvest and Feeding Habits of Largemouth Bass in Murphy Flowage, Wisconsin. Tech. Bull. No. 50. Wisconsin Department of Natural Resources, Madison.

Sparling, G.P., K.N. Whale and A.J. Ramsay. 1985. Quantifying the contribution from the soil microbial biomass to the extractable P levels of fresh and air-dried soils. *Aust. J. Soil Res.* 23: 613–621.

Stocker, R.K. and N.T. Hagstrom. 1986. Control of submerged aquatic plants with triploid grass carp in southern California irrigation canals. *Lake and Reservoir Manage.* 2: 41–45.

Tarver, D.P. 1980. Water fluctuation and the aquatic flora of Lake Miccosukee. *J. Aquatic Plant Manage.* 18: 19–23.

Tazik, P.P., W.R. Kodrich and J.R. Moore. 1982. Effects of overwinter drawdown on bushy pondweed. *J. Aquatic Plant Manage.* 20: 19–21.

van der Valk, A.G. and C.B. Davis. 1978. The role of seed banks in the vegetation dynamics of prairie glacial marshes. *Ecology* 59: 322–335.

van der Valk, A.G. and C.B. Davis. 1980. The impact of a natural drawdown on the growth of four emergent species in a prairie glacial marsh. *Aquatic Bot.* 9: 301–322.

Van Wijck, C. and C.J. De Groot. 1993. The impact of dessication of a fresh water marsh (Garcines Nord, Camargue, France) on sediment-water-vegetation interactions. 2. The submerged macrophyte vegetation. *Hydrobiologia* 252: 95–103.

Wagner, T. and C.M. Falter. 2002. Response of an aquatic macrophyte community to fluctuating water levels in an oligotrophic lake. *Lake and Reservoir Manage.* 18: 52–65.

Watts, C.J. 2000. Seasonal phosphorus release from exposed, reinundated littoral sediments in an Australian reservoir. *Hydrobiologia* 431: 27–40.

Wegener, W. and V. Williams. 1974a. Extreme Lake Drawdown: A Working Fish Management Technique. Florida Game and Fresh Water Fish Commission, Kissimmee.

Wegener, W. and V. Williams. 1974b. Fish population responses to improved lake habitat utilizing an extreme drawdown. *Proc. Southeast Assoc. Game Fish Comm.* 28: 144–161.

Wegener, W., V. Williams and T.D. McCall. 1974. Aquatic macroinvertebrate responses to extreme drawdown. *Proc. Southeast Assoc. Game Fish Comm.* 28: 126–144.

Westerdahl, H.E. and K.D. Getsinger. 1988. Efficacy of Sediment-Applied Herbicides Following Drawdown in Lake Ocklawaha, Florida. Info. Exch. Bull. A-88-1. U.S. Army Corps Eng., Vicksburg, MS.

Wile, I. and G. Hitchin. 1977. An Evaluation Of Overwinter Drawdown as an Aquatic Plant Control Method for the Kawartha Lakes. Ontario Ministry of the Environment, Toronto.

Williams, V.P. 2001. Effects of point-source removal on lake water quality: A case history of Lake Tohopekaliga, Florida. *Lake and Reservoir Manage.* 17: 315–329.

Williams, V.P., E.J. Moyer and M.W. Halen. 1979. *Water Level Manipulation Project.* Report F-29-8. Florida Game and Fresh Water Fish Commission, Kissimmee.

Williams, V.P., E.J. Moyer and M.W. Halen. 1982. *Water Level Manipulation Project. Study IV. Lake Tohapekaliga Drawdown.* Report F-29-11. Florida Game and Fresh Water Fish Commission, Kissimmee.

14 Preventive, Manual, and Mechanical Methods

14.1 INTRODUCTION

Preventive, manual, and mechanical methods form a continuum of plant management options. Avoiding aquatic nuisance problems is the most desirable so preventive measures are needed. If new infestations of nuisance plants are found or if only small areas of aquatic plants need to be managed, manual methods may be appropriate. If a nuisance is already large and can't be managed manually, then mechanized plant removal is an option or can become part of an integrated aquatic plant management program.

Contingency planning cannot be overemphasized. To paraphrase Benjamin Franklin — a gram of foresight prevents a metric ton of milfoil. Typically, aquatic plant invasions have been unnoticed or overlooked until they become problematic. Contingency planning for exotic invasions is similar to planning for other natural disasters. The threat is identified and the resources for dealing with it including people, equipment, and finances are known and can be deployed quickly and easily. Barriers to rapid action, such as the need for permits or legislative approval, are taken care of ahead of time. Preventive, manual, and mechanical approaches form part of the armory of techniques available to manage aquatic plants.

14.2 PREVENTIVE APPROACHES

Many aquatic plants have large ranges and are spread naturally by birds, wind, and water current (Johnstone et al., 1985). Many exotic and nuisance aquatic plants spread vegetatively. Natural dispersal of whole plants or long-stemmed fragments long distances is unlikely (Johnstone et al., 1985). As examples, whole plants of water hyacinth (*Eichornia crassipes)* were found in a wastewater treatment pond and waterlettuce (*Pistia stratiotes)* was found in a stream in northern Wisconsin, during the summer of 2002 (Frank Koshere, Wisconsin Department of Natural Resources (WDNR), personal communication, 2002). It is unlikely that birds, wind, or water current carried these plants all the way from the southern United States where they are common. Human transport, either knowingly or by accident, is the probable explanation. Human activities that transport plants can be grouped into: (1) equipment related dispersal such as attachment of plant fragments onto boats, boat trailers, float-planes, and fishing gear such as nets; (2) plant- or animal-related dispersal where exotic plants are introduced from aquarium discards, fish stocking, or use of aquatic plants as packaging material for fishing bait or packing in nursery stock of ornamental plants such as water lilies; and (3) deliberate dispersal as a means of habitat enhancement or water gardening (see aquascaping in Chapters 5 and 12), scientific transplant experiments, agriculture (e.g., rice seeds), or anti-social behavior (Johnstone et al., 1985).

The magnitude of this problem should not be underestimated. Schmitz (1990) reported that at least 22 species of exotic aquatic and wetland plants have been introduced into Florida. Of the 17 species of aquatic plants that Les and Mehrhoff (1999) identified as non-indigenous to southern New England, 13 escaped from cultivation, two were natural dispersal or accidental introductions, and the mode of introduction for two species was uncertain. Even a location as remote as New

Zealand is plagued with aquatic nuisances caused by the introduction of the exotics coontail (*Ceratophyllum demersum*), egeria (*Egeria densa*), elodea *(Elodea canadensis)*, hydrilla *(Hydrilla verticillata)*, and *Lagarosiphon major* (Johnstone et al., 1985).

14.2.1 THE PROBABILITIES OF INVASION

Johnstone et al. (1985) found that exotic plant distribution was significantly associated with boating and fishing activities in New Zealand. They expressed the probability of a species dispersing from an infested lake to an uncolonized lake in given time period as the product of the frequency of lakes uncolonized by the species, the frequency of the species being transported by interlake boat traffic, the frequency of interlake boat traffic traveling a defined range of interlake distances; and the number of fragments arriving (all species) at all lakes per unit of time. In addition the propagule must be viable when it reaches the lake, it must find suitable habitat for growth, and it must compete with other species to become successfully established. It must then propagate and spread to become invasive. Waters most at risk for invasion are those with suitable habitat found along the pathway of expansion. Lack of success at dispersal, survival, or reproduction prevents a species from expanding its range.

Johnstone et al. (1985) also found that the probability of interlake plant dispersal by boats decreased rapidly as the distance between lakes increased and in New Zealand it was extremely small beyond distances of 125 km. Dispersal distances by boats vary by region and are likely longer in North America (although for Wisconsin, Buchan and Padilla (2000) reported that the average distance traveled by recreational boaters was 45 km) but these distances are usually short and are probably unintentional. This type of dispersal is considerably different than the dispersal that concerns Les and Mehrhoff (1999) where plants are intentionally introduced into an area. Intentional introductions can spread plants long distances because of the care given to insure survival.

For purposes of unintentional invasions, lakes can be viewed as islands in a sea of unfavorable aquatic plant habitat (i.e., land). To successfully invade a new lake, aquatic plant viability depends upon surviving desiccation as it crosses the land barrier. The degree of desiccation depends on the time out of water and the desiccation rate. For coontail, hydrilla, elodea, egeria, and *L. major*, survivorship dropped off dramatically with a 75% or greater weight loss (Johnstone et al., 1985). Viability of desiccated fragments was not obvious from visual inspection. After about 50% weight loss, all the leaves on plant fragments die, but the fragment retained the ability to grow from lateral buds (Johnstone et al., 1985). Coontail was the most desiccation resistant followed by *L. major*, egeria, elodea, and finally hydrilla. Under laboratory conditions, coontail remained viable for up to 35 hours when dried at 20°C and 50% relative humidity (Johnstone et al., 1985). Studies from British Columbia indicate that Eurasian watermilfoil (*Myriophyllum spicatum*) lost viability in 7 to 9 hours when dried in the shade in still air (Anonymous, 1981).

Desiccation rate depends on the time of day; weather conditions; degree of protections from drying factors such as wind, sun, and vehicle speed; and the species. While laboratory studies of survival rates are informative, they may bear little reality to conditions where invasive plants are found in live wells, bilge water, minnow buckets, the bottom of leaky boats, or in moist gobs wrapped around trailer axles (Figure 14.1).

Johnstone et al. (1985) suggest that dispersal, rather than habitat type, are responsible for the distribution patterns of exotic aquatic plants, and Cook (1985) concluded that the establishment of introduced aquatic plants was more dependent on human disturbance of the environment than on plant mobility. The species Johnstone et al. (1985) studied are able to occupy a wide range of habitats and they found that lake trophic status and species distribution patterns were unrelated. This may also be typical of other invasive species. However, if resources are limited, it is prudent to first search for invasives in habitats where they are most likely to occur or become a problem. Knowing preferred habitats informs riparian property owners, lake managers, and government officials of the potential for future lake invasions.

FIGURE 14.1 Boat and trailer leaving a boat launching area showing exotic plants (mainly *Myriophyllum spicatum*) "hitch hiking" on trailer parts. All plant material should be removed before launching in a different lake.

Using limnological data from over 300 lakes in the United States and southern Canada, Madsen (1998) found that total phosphorus (TP) and Carlson's Trophic State Index (TSI) were the best predictors of Eurasian water-milfoil dominance in a lake. Lakes with a TP of 20–60 µg/L or a TSI of 45–65 were most at risk of *M. spicatum* dominance. Crowell et al. (1994) compared total plant biomass and Eurasian water-milfoil biomass to water clarity and sediment characteristics in Lake Minnetonka, Minnesota, as a means of identifying habitat conditions conducive to producing nuisance biomass conditions. Using habitat information as a tool, monitoring and management resources can first be allocated to the lakes or areas of a lake most likely to develop substantial nuisances.

Buchan and Padilla (2000) also developed models to predict the likelihood of Eurasian water-milfoil presence in lakes. They found that the most important factors affecting the presence or absence of *M. spicatum* were those that influenced water quality factors known to impact milfoil growth, rather than factors associated with human activity and dispersal potential. Their models do not consider dispersal probability to the lake so their concluding remark is, "Lakes with the greatest risk of being invaded will be those with the highest likelihood of both providing suitable milfoil habitat and being recipients of the greatest frequency of recreational boat traffic." An advantage of some of their models is they are based on data that usually exists in publicly available databases so it is inexpensive to collect and use.

Using bioindicators as a quick and inexpensive way of determining habitat suitability, Nichols and Buchan (1997) found that *Potamogeton illinoensis, P. pectinatus, P. gramineus,* and *Najas flexilis* were native Wisconsin species that commonly occurred with Eurasian watermilfoil. Their presence should indicate lakes with good milfoil habitat. The preferred depth, pH, alkalinity, and conductivity ranges for *P. illinoensis* and *P. pectinatus* are very similar to milfoil. *Sparganium angustifolium* was negatively associated with milfoil and its preferred water chemistries were quite different. It is a good indicator of lakes where Eurasian watermilfoil is not likely to flourish.

The U.S. Army Corps of Engineers (USCAOE) is developing a simulation model (CLIMEX) for analyzing species ranges to determine climate compatibility of potential invasion locations with those of the species home range or known distribution (Madsen, 2000a). It is a promising tool to identify potentially problematic plants for prevention efforts and regulatory exclusion. It requires more information on species life histories, growth potential, distributions, and habitat requirements to become fully useable (Madsen, 2000a). However, using preliminary information, Madsen (2000a) assessed the potential for *Cabomba caroliniana, E. densa, H. verticillata* (monoecious and dioecious biotypes), *Hydrocharis morsus-ranae, Ludwigia uruguayensis, Marsilea quadrifolia, Myriophyllum heterophyllum, Najas marina, N. minor, Nymphoides peltata,* and *Trapa natans* to pose realistic nuisance threats to ecosystems in Minnesota, *C. caroliniana, H. verticillata* (monoecious biotype), *N. peltata, M. heterophyllum, H. morsus-ranae,* and *T. natans* showed the highest probability for success in Minnesota. *T. natans, M. heterophyllum, H. verticillata,* and *C. caroliniana* were likely to cause the most severe problems if they successfully invaded.

Habitat, the time of year a viable plant propagule arrives at a lake, and stored energy in the propagule determine colonization success. Kimbel (1982) found for Eurasian watermilfoil that low propagule (stem fragments in this case) mortality occurred during late summer, in shallow water. Mortality increased during early autumn, in deep water. Substrate type did not affect mortality. Low total nonstructural carbohydrate (TNC) content was linked to increased mortality.

14.2.2   EDUCATION, ENFORCEMENT, AND MONITORING AS PREVENTIVE APPROACHES

Preventive approaches delay or negate nuisance species introductions into uninfested lakes. They depend primarily on regulation, education, monitoring, and mechanical barriers. They are not fail-safe. Public cooperation and the full support of lakeshore residents at uninfested locations are essential. Education, monitoring, and enforcement is most cost effective and practical where there are limited access points to uninfested waters because they are most easily monitored. Education usually involves public information campaigns involving pamphlet distribution, use of news media, and warnings posted at infested locations (Figure 14.2).

Minnesota state statutes prohibit a person from possessing, importing, purchasing, selling, propagating, transporting, or introducing a prohibited exotic species and prohibit transporting any aquatic macrophyte on a highway (MDNR, 1998). Other states, Canadian provinces, New Zealand, Australia, and probably others have developed or are developing similar legislation (Clayton, 1996). Citations, usually issued by conservation officers, can result from violating regulations. Often, citations are a very effective educational tool. Whether state regulations are enough to tackle a national or global issue of exotic species is questionable. A review of the broader aspects of non-indigenous species, aquatics included, and suggested technologies for preventing and managing problems on a nationwide basis are provided by USOTA (1993).

Lake monitoring by trained volunteers, especially at boat launches is another effective prevention tool. The Volunteer Monitor (Smagula et al., 2002) reported locations in New Hampshire, Wisconsin, Massachusetts, and Vermont where volunteers discovered exotic aquatic plant invasions in time for swift management action.

The web site http://www.invasivespecies.gov provides a lot of information about the vectors and pathways of aquatic plant species invasions. Also included are a variety of educational and monitoring resources.

14.2.3   BARRIERS AND SANITATION

Physical barriers can be used to reduce or eliminate free-floating species or floating plant fragments from spreading to downstream locations (Deutsch, 1974; Cooke et al., 1993). The barriers must be

FIGURE 14.2 Sign at a boat-launching area warning users that those waters contain exotic species and that it is illegal to place a boat or trailer in navigable water with exotics attached.

constantly maintained and they are usually not 100% effective. With some species, like water hyacinth, the shear mass of plants makes using barriers problematic (Deutsch, 1974).

In British Columbia barriers of welded mesh were placed at selected lake outlets and cleaned regularly to prevent the downstream spread of Eurasian watermilfoil. Generally, barriers were effective in reducing the volume of fragments moving downstream, but some fragments were not retained and milfoil became established downstream (Cooke et al., 1993).

Removing floating plant rafts at the water intake was the most cost effective means of plant control at New Zealand hydropower stations (Clayton, 1996). Barriers and nets were an efficient means of removing cut aquatic plants that were concentrated by wind and current in Weyauwega and Buffalo Lakes, Wisconsin (Livermore and Koegel, 1979). Log booms were used in Lake Cidra, Puerto Rico to contain floating mats of water hyacinth after they were broken apart and pushed to a take-out point (Smith, 1998). Once captured, the mats were removed with a bucket excavator.

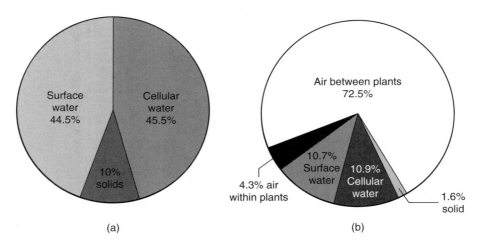

FIGURE 14.3 Percentage of constituents by weight (a) and volume (b) of harvested *Myriophyllum spicatum*. (After Livermore, D.F. and R.G. Koegel. 1979. In: J. Breck, R. Prentki and O. Loucks (Eds.), *Aquatic Plants, Lake Management, and Ecosystem Consequences of Lake Harvesting*. Inst. Environ. Stud., University Wisconsin, Madison. pp. 307–328.)

Removing nuisance plants at boat launch sites is important for preventing species spread from lake to lake. In New Zealand, Johnstone et al. (1985) found that if the area near the boat ramp was plant-free, even if the lake contained nuisance exotics, no plants were found on boats or trailers.

14.3 MANUAL METHODS AND SOFT TECHNOLOGIES

Manually pulling or using hand tools such as cutters, rakes, forks, and hooks are the most common mechanical type of aquatic plant management in the world (Madsen, 2000b). It is the method most widely used by lakeshore property owners in the United States.

Inexpensive equipment, very selective methods, rapidly deployed techniques, few use restrictions, no foreign substances added to the water, and immediately useable areas are the advantages of manual methods (i.e., soft technologies). However, the methods are labor intensive and hard work. Fatigue often results before management is complete. The areas treated are small and productivity is limited. The methods are usually inexpensive unless labor costs are high. Therefore, manual treatments make good volunteer projects. A local SCUBA club, for example, annually removes Eurasian watermilfoil from Devils Lake, Wisconsin as a service project (Jeff Bode, WDNR personal communication, 2002). The techniques do little environmental harm; mainly because treatment areas are small. There are safety issues while wading or swimming in dense plant beds and when wielding sharp tools, underwater, with limited visibility.

Many tools used in manual techniques are available from local hardware or farm supply stores. Some can be found in "junk" piles of outdated farm equipment (McComas 1989). To increase efficacy and efficiency it is important to match the tool to the task (Table 14.1, McComas, 1993).

Manual uprooting was used to reduce Eurasian watermilfoil biomass and change plant community structure in high use areas (e.g., swimming beaches) of Chautauqua Lake, New York (Nicholson, 1981a). Two treatments were tested; one where only Eurasian watermilfoil was removed and another where all plants were removed. One year after treatment, milfoil biomass was between 25% and 29% less in the treated areas than in untreated areas. Total plant biomass was between 21% and 29% less (Nicholson, 1981a). Even in the complete removal areas, revegetation was noticeable within a few weeks after treatment.

In University Bay of Lake Mendota, Wisconsin, Eurasian watermilfoil was cut as close to the bottom as possible using SCUBA and a sickle or divers knife (Nichols and Cottam, 1972). One

TABLE 14.1
Recommended Manual Methods for Removing Aquatic Plants Based on Rooting Strength[a]

Method	Non-Rooted, Free Floating[b]	Weakly Rooted	Strongly Rooted	Very Strongly Rooted
Cutters				
Straight-edge weed cutter		X	X	X
Electric weed cutter		X	X	
Scythe, machete, corn knife, diver's knife, sickle[c]			X- emergent species only	X-emergent species only
Rakes				
Garden rake	X			
Modified silage fork	X	X		
Landscape rake	X	X		
Hand pulling	X	X	X	
Hay or pulp hook				X
Drag		X	X	
Garden cultivator		X		
Skimmers				
Modified fish net or seine	X			

[a] X-rated by McComas (1993) as an excellent or good technique; assumes the user is wading or working from shore, a pier, or boat.

[b] Non-rooted, free floating include free-floating species, plant fragments, and species like *Ceratophyllum demersum* and *Chara* sp.; weakly rooted species are plants that can be easily pulled out by the roots like some *Potamogeton* spp., *Elodea* spp., and *Najas* spp.; strongly rooted species are hard to pull by hand, the stems often break before the roots are pulled out, an example is *Myriophyllum spicatum;* strongly rooted plants are very difficult to uprooted by hand, they are often floating-leaf species like *Nymphaea* spp. and *Nuphar* spp. and emergents like *Typha* spp. and *Scirpus* spp. Sometimes rooting strength depends on bottom sediments. If in doubt, give a "pull" test.

[c] Recommended for emergents only for safety reasons. Divers knives and sickles are safer when used in conjunction with SCUBA.

harvest reduced regrowth by at least 50%, two harvests by 75%, and three harvests virtually eliminated plant material during the year of treatment. Harvesting one year reduced the biomass the following year, especially in deep water. Three harvests during the previous year were most effective in controlling biomass the second year. Root removal significantly reduced milfoil biomass in Cayuga Lake, New York 1 year after treatment (Peverly et al., 1974).

14.4 MECHANICAL METHODS

14.4.1 THE MATERIALS HANDLING PROBLEM

Mechanical control of aquatic plants is both a biological and a materials handling problem. Somewhat depressing is the fact that a pile of harvested plants (Eurasian watermilfoil in this case) is approximately 90% water by weight and 75% air by volume (Livermore and Koegel, 1979/ Figure 14.3). A great deal of effort and money is spent on removing and transporting water and air. There are a variety of ways to mechanically remove aquatic plants and every step involves materials handling (Figure 14.4). Understanding and enhancing materials handling increases harvesting efficiency. It is wise to enlist someone with materials-handling experience (engineer, public works department director) to work with a lake consultant or biologist on a harvesting program.

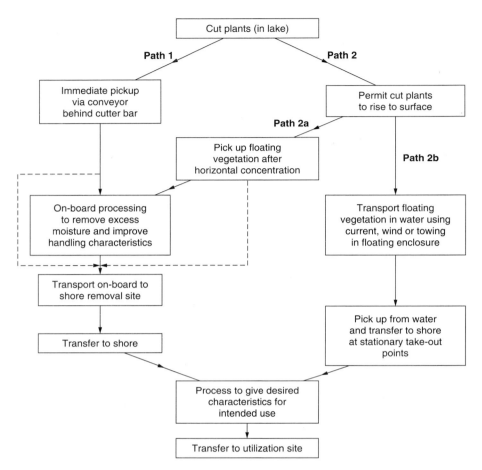

FIGURE 14.4 Flow chart of alternative harvesting options. (From Livermore, D.F. and R.G. Koegel. 1979. In: J. Breck, R. Prentki and O. Loucks (Eds.), *Aquatic Plants, Lake Management, and Ecosystem Consequences of Lake Harvesting.* Inst. Environ. Stud., University of Wisconsin, Madison, WI. pp. 307–328.)

14.4.2 MACHINERY AND EQUIPMENT

"The diversity of machines devised to cut, shred, crush, suck, or roll aquatic plants would be large enough to fill a museum" (Wade, 1990). Aquatic cutters and harvesters evolved from agricultural equipment. Over the years there have been numerous designs to make machinery more efficient, less costly, safer, more reliable, or to use in special circumstances (Deutsch, 1974; Dauffenbach, 1998). The two basic designs are those with a bow reciprocating cutter or a bow rotary cutter (Livermore and Koegel, 1979). "Sawfish," "Waterbug," "Chub," "Cookie Cutter," "Sawboat" and "Swamp Devil" were some colorful names given to these machines.

Bow rotary cutting machines are used primarily on emergent or floating-leaved plants. They chop plants into small pieces and return them to the water, "blow" them on to the bank, or "blow" them into transport equipment.

Bow reciprocating cutters are the industry standard (Figure 14.5). Some machines only cut plants, others are harvesters that elevate cut plant from the water and load them for transport. Sizes range from small, boat mounted cutters to large harvesters with up to 3 m wide cutters that can cut to a 2-m depth, and can transport 30 m^3 of harvested material. A transport barge, shoreline conveyor, a trailer or wheels to transport the harvester on land, and dump trucks are additional equipment often used in a harvesting operation (Figure 14.5). Diver-operated suction dredges,

(a)

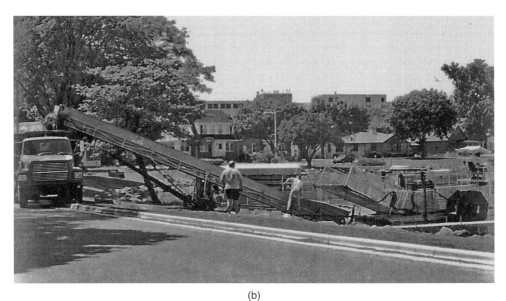

(b)

FIGURE 14.5 Mechanical harvester (a) and shoreline unloading equipment (b) operating in Lake Monona, Wisconsin.

machines that use water pressure to "wash" plants out of the bottom, and cultivating and rototilling machines are also used for aquatic plant management.

Harvesters are somewhat awkward to maneuver, have a limited cutting depth, and, because of the large conveyor, have a limited forward speed (Figure 14.5). Efforts to overcome these limitations have led to numerous innovations including two stage harvesting where plants are cut in one stage and removed in a secondary operation (Livermore and Koegel, 1979). Therefore, the distinction cannot always be made between a cutter and a harvester based solely on the machinery used. Harvesting means that the plants are removed from the water but it may not be done in a single operation. There is a continuum of options between cutting and harvesting.

14.4.3 CUTTING

Cutting is more rapid than harvesting, the machinery is usually less costly, it may be the most appropriate method for managing annual and emergent species in shallow water, it can be done in deeper water than harvesting, small cutters can operate in areas harvesters can not, and efficiency might be increased by cutting and removing plants in separate operations. However, cutting may spread the aquatic plant nuisance, a secondary operation may be needed to remove plants, and floating plants may become a health, safety, or environmental problem.

14.4.3.1   Case Study: Water chestnut (*Trapa natans*) Management in New York, Maryland, and Vermont

Water chestnut is a floating-leaf aquatic plant introduced into the United States from Eurasia by at least the late 1800s. It is found in the northeastern United States as far south as northern Virginia. Water chestnut is a true annual that over winters entirely by seeds that germinate in late May. By early June a dense canopy of rosettes form on the water surface. Flowering occurs in early July, the first fruits reach maturity in August, and seed production continues until the plant dies in the fall. The seeds sink when released. Water chestnut grows aggressively, lacks food or shelter value to most fish and waterfowl, impedes boat traffic, and its spiny fruits cause painful wounds to swimmers. However, because it is an annual, populations can be controlled if the plant is eliminated before seed set. Because some seeds may remain viable in sediments for at least 12 years (Elser, 1966) a plant infestation will not be eliminated in a single year.

The USACOE started cutting water chestnut in the Potomac River in the 1920s and 10 years of annual cutting reduced infestations to very low levels. Tidal currents carried cut plants to salt water where they were apparently killed. Water chestnut was not eliminated but could be maintained by annual hand pulling of plants (Elser, 1966).

In 1955 large patches of water chestnut were found in the Bird River, Maryland. After seven seasons of cutting and the use of chemicals (2,4,-D, see Chapter 16) the species appeared to be exterminated and the project was terminated (Elser, 1966). This assessment proved to be premature and several large patches were discovered in 1964 along with patches in the Sassafras River system, Maryland. These areas were harvested but the infestations grew so rapidly they could not be managed by harvesting alone in 1964. In 1965 about 73 ha were harvested and rosettes on the remaining plants turned brown and fell off — possibly from saltwater intrusion (Elser, 1966). No results were reported after 1965 but it is obvious that continued vigilance is needed to manage water chestnut by cutting or chemicals but management efforts can be reduced to low levels once plants are under control (see section on maintenance management in Chapter 16).

In Watervliet Reservoir, New York (175 ha, 3.5 m mean depth) water chestnuts were cut 10 cm below the water surface with a sharp, V-shaped metal blade mounted on the front of an air boat (Methe et al., 1993). In an uncut area of the reservoir water-chestnut seeds were recruited to the seed bank while in the cut areas the seed bank declined (Madsen, 1993). Rosettes were not removed after cutting and Methe et al. (1993) found that rosette fragments containing buds or flowers at the time of cutting were capable of producing mature seeds. The cutting experiment at Watervliet Reservoir apparently was not continued long enough to determine whether cutting could eliminate the water-chestnut problem. However, the lesson learned is that cutting early and often is needed to eliminate water chestnut and vigilance is needed for a number of years so an area is not reinfested from a seed bank.

Water chestnut has been an aquatic nuisance problem in Lake Champlain, on the New York–Vermont border for decades. It occupies approximately 121 ha of the southern portion of the lake. Mechanical shredding is one alternative for controlling large expanses of water chestnut where conventional harvesting or herbicides are impractical or cost-prohibitive. A concern with shredding plants and returning the biomass to the system is the impact on water quality. In July 1999 a 10,000-m^2 area was shredded to study water quality changes (James et al., 2000). Results showed that shredding resulted in improved dissolved oxygen conditions, increased turbidity, and a buildup of N and P in the water column.

14.4.3.2 Case Study: Pre-Emptive Cutting to Manage Curly-Leaf Pondweed (*Potamogeton crispus*) in Minnesota

Curly-leaf pondweed in Minnesota acts like a winter annual; most plants sprout from turions. By controlling plants before turions production, turion density and thus stem density should decrease.

Laboratory trials showed and field observations confirmed that curly-leaf did not grow back if it was cut after growth reached 15 nodes but turions were not produced until growth reached 20–22 nodes (McComas and Stuckert, 2000). There is a "window of opportunity" to manage curly-leaf during the year of cutting and to prevent additional turions being recruited to the propagule bank by cutting it between the 15- and 20-node stage. Volunteers were organized to cut curly-leaf on French, Alimagnet, Diamond, and Weaver lakes, Minnesota in May or early June of 1996, 1997, and 1998 (McComas and Stuckert, 2000). Volunteers targeted the worst infestations first and about 50% of the total coverage, but 70–80% of the nuisance coverage, in each lake was cut (McComas and Stuckert, 2000).

After 3 years of cutting, the nuisance plant coverage in French Lake was reduced from 36 ha to 10 ha, in Alimagnet Lake from 18 to 4 ha, in Diamond Lake from 8 to 0 ha, and in Weaver Lake from 10 to 2 ha. Stem densities in cut areas of French and Alimagnet lakes were reduced by about 65 to 80% in the year after 2 years of cutting. It is uncertain whether all decreases in coverage and stem density could be attributed to cutting. Reference areas in Diamond Lake, Alimagnet Lake and an uncut reference lake showed some natural decline in curly-leaf (McComas and Stuckert, 2000). Stem density may not tell the entire story because a single turion can produce runners that grow numerous stems.

McComas and Stuckert (2000) concluded that the degree of nuisance control was a direct function of the intensity of cutting prior to turion formation on an annual basis. Although cutting is likely to be an annual event, as stem densities decline, maintenance cutting should be easier. Nuisance conditions are likely to return if cutting is neglected for a year or two.

14.4.3.3 Case Study: Deep Cutting, Fish Lake, Wisconsin

Fish Lake is a 101-ha seepage lake, in south-central Wisconsin, with a maximum depth of 19.5 m and an average depth of 6.6 m. Eurasian watermilfoil formed a continuous ring around the lake's perimeter at depths ranging from 1.5 to 4.5 m. Milfoil comprised 90% of the plant biomass and covered approximately 40% of the lake bottom (Unmuth et al., 1998).

The ultimate objective of deep cutting in Fish Lake was to create persistent edge for fish habitat within dense plant beds by establishing narrow, open, channels (Unmuth et al., 1998). To accomplish the deep cutting, a conventional harvester was retrofitted with a cutting bar (Figure 14.6) that allowed plants to be cut near the sediment surface in water depths ranging from 1 to 6.5 m. It cost approximately $10,000 to replace the cutter bar, add a hydraulic boom, and install a depth finder to the harvester.

During August 1994, 262 1.8-m wide channels, ranging in length from 30 to 1200 m, were cut in a radial pattern, perpendicular to the shoreline. A total of 36,200 m of channel were cut at depths ranging from 1.5 m near the shoreline to 4.5 m at the outer edge of the plant beds. The deep-cutter

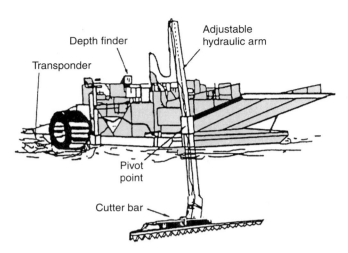

FIGURE 14.6 A modified close-cut harvester. (From Unmuth, J.M.L. et al. 1998. *J. Aquatic Plant Manage.* 36: 93–100. With permission.)

required two people to operate. One person drove the machine and a second person monitored the depth finder and adjusted the cutting bar to maintain a target cutting height of no more than 0.6 m above the bottom. The machine cut about 854 m of channel per hour. The total cut was about 6.4 ha, which represented 19% of the milfoil by area and 18% of the original milfoil biomass (Unmuth et al., 1998). A conventional harvester followed the deep-cutter to pick up plant material as it floated to the surface.

Surveying 16% of the channels, Unmuth et al. (1998) assessed the immediate success of close cutting. At each sampling point divers classified the height of the remaining stubble as short (< 0.3 m), medium (0.3–0.6 m), and tall (> 0.6 m). The 0.3- and 0.6-m criteria were selected because research found that over-wintering shoots of milfoil generally exceeded 0.6 m in height by late summer in Fish Lake and they produced side branches from the main stem at heights between 0.3 and 0.6 m above the root crown (Unmuth et al., 1998). Cutting plants below these heights may hinder regrowth by interfering with carbohydrate resource allocation and root mass. This assessment showed that 83% of the sites were cut within 0.6 m and 45% were within 0.3 m of the sediment surface.

The persistence of close-cut channels was analyzed by using vertical aerial photographs and by using divers to measure regrowth in the channels. Divers compared plant regrowth in the center of the channel to plant height of the surrounding bed. Categories used were no regrowth, minimal regrowth (< 50% height of adjacent bed), and moderate regrowth (> 50% height of adjacent bed). Early assessment of channel persistence (1995) showed that only 50 channels, representing 2,300 m of channel length, about 7% of the original, were visible (Unmuth et al., 1998). In addition, 72% of the sites within the visible channels had plant regrowth of over 50% of the surrounding channel, and the majority of the visible channels were less than 3 m deep.

The longer term response to close-cutting was more pronounced. In 1996, remnants of 170 channels, totaling 7700 m (about 21% of the total channel length) were clearly visible from the air. About half of all sites surveyed in visible channels had regrowth less than 50% of the surrounding bed. About 50% of the channel length cut in the 3- to 4.5-m zone was visible (Unmuth et al., 1998). By 1997, remnants of 123 channels, totaling 3500 m of channel length (10% of the original) remained detectable. Of the channels cut in 3 to 4.5 m, 46% remained visible. The remnant channel length in the shallow zone declined to 4% of the original cut (Unmuth et al., 1998).

It is uncertain why the persistence of the channels in 1995 appeared to be less than in 1996 and 1997 but a possible explanation was collapse of the surrounding beds in 1995 due to an invasion of the milfoil weevil (*Euhrychiopsis lecontei*) making detection more difficult (Unmuth et al., 1998).

Remember, as a registered agent / landlord, you can access The DPS website and:

View all the deposits you have submitted to The DPS

Submit new deposits

Request the repayment of deposits

Make general enquiries

Thank you for choosing The DPS as your deposit protection provider.

Yours sincerely,

The DPS Support Team

www.depositprotection.com

Follow The DPS on **Twitter and visit our** **blog to get the latest on deposit protection and related topics.**

The long-term persistence of deep-water channels varied considerably among different regions of the lake for no apparent reason. Unmuth et al. (1998) also found no significant relationship between the success rate of the original cut (e.g., stubble height) and long-term channel persistence.

Close cutting was slower than conventional harvesting, needed a larger crew to operate, and required secondary pick-up of cut plants. However, a single cut was successful at creating persistent channels for fish habitat that lasted for at least three years in water deeper than 3 m.

14.4.3.4 Case Study: Cutting the Emergents, Cattails (*Typha* spp.) and Reeds (*Phragmites* spp.)

Cutting cattails and reeds is a common practice, especially in Europe (Wade, 1990). For best results they are cut twice during the growing season and are cut below the water level. The cut shoots become flooded with water, die, and rot. For cattails and probably for other emergents a rapid decline in oxygen to submersed parts probably causes death (Sale and Wetzel, 1983). A fall cutting was less effective at controlling reeds and a winter cutting when the reeds were hardened and carbohydrates were in the rhizome was not damaging. Winter cutting may enhance reed growth by removing dead culms that harbor pathogenic fungi and insect larvae. Winter-cut reeds were more productive the following year than uncut stands (Wade, 1990). Likewise, in the European climate, there was no difference between winter-cut and uncut *T. angustifolia* stands relative to regrowth the following year (Wade, 1990).

As reported in Chapter 11, Linde et al. (1976) found that total nonstructural carbohydrates (TNC) were lowest in cattail rhizomes just before flowering and they suggested this was an excellent time to control cattails. However, recommendations for cattail control in the fall appear to be different in the northern United States than in Europe. Cutting cattail stems, including dead stems, below the water line in the fall prevented cattail rhizomes from getting oxygen for respiration under winter ice conditions and plant death resulted (Beule, 1979).

Because of shallow water, large harvesters can not operate in emergent stands. The water quality impacts of not removing cut plants in the emergent zone may not be as great as in deeper water. Most emergents decay slowly when compared to submergent species and often the water and bottom sediments in this zone are already nutrient rich and anoxic.

14.4.4 HARVESTING

14.4.4.1 Efficacy, Regrowth, and Change in Community Structure

There is little doubt that harvesting reduces aquatic nuisances — at least temporarily. If a species is soft enough to cut, grows in a location that can be reached by a harvester, and floats to the water surface, it can be removed by harvesting. Long-term management is enhanced when the recovery of nuisance species is slow or when the replacement community is less of a nuisance than the original community. The questions are: (1) how rapid is regrowth, (2) are there techniques that extend harvesting efficacy, (3) does harvesting change the plant community structure, and (4) what harvesting techniques, if any, enhance community structure? Most information on regrowth and community change was developed from studies of undifferentiated biomasses of a variety of plants or from populations of plants strongly dominated by Eurasian watermilfoil. Long-term studies are few in number.

The longevity of harvesting depends on initial plant biomass, regrowth rates, and reproduction methods; the depth, frequency, completeness, and seasonal timing of cuts; and ecosystem factors such as the productivity of the area being harvested. There is general agreement (Nichols, 1974; Peverly et al., 1974; Wile, 1978; Johnson and Bagwell, 1979; Newroth, 1980; Kimbel and Carpenter, 1981; Mikol, 1984; Cooke et al., 1990, 1993; Engel, 1990a) that more than one harvest is needed to control the regrowth of a variety of plants in a variety of geographic areas over the growing season. Even more harvests are likely needed in areas with longer growing seasons such as the southeastern United States. As examples, Johnson and Bagwell (1979) reported that egeria reached the surface in Lake Bistineau, Louisiana 3 months after cutting. Trials to control *Nymphaea odorata* in Mill Lake, British Columbia, showed that harvesting provided only 3 to 4 weeks of control (Cooke et al., 1993). Six weeks after harvesting Eurasian watermilfoil in Lake Wingra, Wisconsin, biomass in the harvested plot was similar to that in the unharvested plots (Kimbel and Carpenter, 1981). Pre-harvesting levels of Eurasian watermilfoil returned to Saratoga Lake, New York 1 month after harvesting (Mikol, 1984). Biomass of macrophytes in LaDue Reservoir, Ohio returned to pre-harvest quantities within 23 days (Cooke et al., 1990). It took about 6 weeks for the biomass in harvested areas of Lake Minnetonka, Minnesota to reach that of unharvested areas (Crowell et al., 1994). Macrophytes quickly regrew to pre-harvest levels in Halverson Lake, Wisconsin (Engel, 1990a). Hydrilla biomass at harvested sites exceeded those at undisturbed sites within 23 days in the Potomac River (Serafy et al., 1994).

Engel (1990a) reported that at least 30% of the total standing crop of macrophytes in Halverson Lake remained after "complete" harvesting. Some plants grew in water too shallow or too deep for operating the harvester. Paddle wheels stirred the sediments, creating turbidity that hid plants below the water surface. Occasional stumps and boulders forced the harvester operator to raise the cutter bar and cut plants well above the bottom.

Regrowth varied with the timing of the first harvest and multiple harvests were more effective than a single harvest. The recovery from a single harvest declined as the date of harvesting became progressively later, at least for milfoil growth and some other species (Kimbel and Carpenter, 1981; Engel, 1990a). The effectiveness of harvesting in Chemung Lake, Ontario depended upon the time of year of harvesting and the number of harvests per season. Harvests in June and July were least effective in lowering the regrowth rate and plant density. Two harvests and three harvests per season were most effective in reducing stem number and height (Cooke et al., 1986). The results of multiple hand cuttings of milfoil in Lake Mendota (Nichols and Cottam, 1972) were discussed earlier in this chapter.

Regrowth also varied with the habitat and the type of cut. For example, Howard-Williams et al. (1996) found markedly different regrowth patterns in Lake Aratiatia, as compared to Lake Ohakuri, New Zealand. Both lakes contained mixed species but *L. major* was the primary species of concern. In harvested areas of Lake Aratiatia the remaining plant beds were patchy and regrowth was highly variable. In some areas there was no plant regrowth. They attributed the patchy regrowth to water flow. Where current velocity regularly exceeded 0.15 m/s there was little or no regrowth of *Lagarosiphon*. In Lake Ohakuri, with negligible water flow, regrowth was not patchy. Plant height increased at a relatively uniform rate.

Regrowth was slower in deep-water areas or where cutting was close to the bottom (Nichols and Cottam, 1972; Cooke et al., 1986, 1990). Cutting milfoil close enough to the bottom to injure the root crown significantly slowed regrowth in LaDue Reservoir and East Twin Lake, Ohio (Conyers and Cooke, 1982; Cooke et al., 1990). After 7 weeks the biomass in the harvested plot of East Twin Lake was only 12% of the unharvested plot biomass. Nearly summer long control was achieved following a "touch-up" harvest on day 42 in LaDue Reservoir. Non-harvested area biomasses averaged at least 100 g/m^2 compared to root-crown harvested area biomasses of less than 20 g/m^2. Below sediment harvesting was used to control *Chara* in Paul Lake, British Columbia. A shearing blade replaced the horizontal cutter bar assembly at the bottom of the front conveyor. The harvester operator lowered the conveyor to the lake bottom, moved slowly forward, pushed the blade into the soft substrate, and collected *Chara* along with the soft surface sediment (Cooke et al., 1993). These projects illustrate the importance to efficient harvesting management of knowing the location(s) of meristematic tissue in the target plant species.

Intensive harvesting for one or more years can reduce plant biomass in subsequent years (Neel et al., 1973; Nichols and Cottam, 1972; Wile et al., 1979 Kimbel and Carpenter, 1981; Painter and Waltho, 1985; Cooke et al., 1986). However, some reductions would not impress aquatic plant managers. Although results were statistically significant, the biomass reduction was only 20 g/m^2 in areas harvested the previous year in Lake Wingra when compared to unharvested areas (Kimbel and Carpenter, 1981). In Chemung Lake the milfoil biomass decline after years of intense harvesting was more dramatic but it was uncertain whether the result could be attributed to harvesting or an unexplained decline in milfoil seen in many lakes (Wile et al., 1979; Smith and Barko, 1992). Aquatic plants were about one-quarter as dense the year following intensive harvesting in Lake Sallie, Minnesota (Neel et al., 1973). Painter and Waltho (1985) experimented with the timing and number of harvests of Eurasian watermilfoil in Buckhorn Lake, Ontario. They concluded that a June/August or June/September double cut was the most desirable management option and that milfoil biomass was significantly affected the year following an October cut. Two to three cuts a season, including a late season harvest appear to be most effective in reducing stem density and plant regrowth (Cooke et al., 1986). Surveying 27 lakes in Wisconsin, Michigan, and Minnesota with harvesting programs, Nichols (1974) reported that people on 17 lakes thought harvesting improved lake conditions over the short-term, six thought there was a long-term benefit, and four thought conditions worsened.

A likely explanation for limited growth after intensive harvesting is the reduction of energy reserves (often measured as total nonstructural carbohydrates — TNC (Kimbel and Carpenter, 1981). Harvesting at times when TNC levels are low in storage organs or when TNC are being transported to storage organs to support the next year's growth may have the greatest impact (see the section on Resource allocation and phenology in Chapter 11). Kimbel and Carpenter (1981) reported that TNC levels, both per plant and per unit area were lower in plots harvested 11 months earlier in Lake Wingra, Wisconsin. They concluded, however, that Eurasian water-milfoil was resilient to harvesting stress despite lower TNC values during the summer following treatment. In Washington state, Perkins and Systma (1987) were able to interrupt carbohydrate accumulation in milfoil roots with a fall harvest. However, TNC stores were rapidly replenished after harvest and increased over winter. Milfoil growth was not reduced the following year. Late season harvesting may be more effective in regions with a severe winter climate or if more stress were placed on the plant. Although a likely explanation, reduced growth from harvesting caused by reduced energy reserves has not been conclusively demonstrated in operational harvesting situations.

The longer-term impacts of harvesting are even less definitive. Nichols and Lathrop (1994) compared an area in Lake Wingra, Wisconsin with a history of mechanical harvesting with other areas of the lake with no known harvesting. Species diversity and taxa richness in three out of four unharvested areas were greater than in the harvested area but differences appeared to be more related to an increase in *Ceratophyllum demersum* after the Eurasian water-milfoil decline of the mid-1970s. In assessing the long-term impact of plant management methodologies on Eurasian watermilfoil in southeast Wisconsin, Helsel et al. (1999) found that in seven out of nine lakes studied, native aquatic plant species increased or remained the same and in eight out of nine lakes, Eurasian watermilfoil remained the same or declined regardless of the aquatic plant management methods used. Management methods included mechanical harvesting, chemical treatment, a combination of the two, and no management.

After a short regrowth period some studies concluded that harvesting had little impact on plant biomass or it increased. There was a lack of long-term effect on *E. densa* biomass in Long Lake, Washington in spite of years of heavy harvesting (Welch et al., 1994). No significant reduction of stem biomass or plant vigor was seen in Eurasian water-milfoil growth in Okanagan Valley, British Columbia lakes after repeated harvesting, and growth may have been stimulated in some cases (Anonymous, 1981; Cooke et al., 1993). Plant growth rates in harvested plots were greater than those in adjacent non-harvested areas of Lake Minnetonka, Minnesota (Crowell et al., 1994) and plants became denser in Halverson Lake after harvesting (Engel, 1990a). As stated above, hydrilla biomass in the Potomac River was greater 23 days after harvesting than in non-harvested areas (Serafy et al., 1994).

Harvesting removes the shading plant canopy. This might increase plant biomass by allowing plants deeper in the water column to receive sufficient light for growth. Harvesting also removes terminal plant growth, which allows more energy for lateral growth, i.e., the "pruning effect"; plants become more "bushy." Another possibility is that the harvested area became severely reinfested with cut plant parts from harvesting, which ultimately grew into new plants.

Harvesters cut all species in the managed area so using harvesting to selectively manage a plant community is difficult. Harvesting can be selective by altering the depth and time of cut and by having harvest and no harvest areas. The latter case is applicable where there are monotypic stands of a nuisance species in some areas of a lake and diverse native plant communities in other areas (Nichols and Mori, 1971; Unmuth et al., 1998). The results of harvesting on community structure are similar to those reported for chemical control (see Chapter 16) and somewhat unpredictable. That is, the resulting community can be (1) dominated by species not present immediately prior to harvesting, (2) dominated by species that were dominant immediately prior to harvesting, or (3) dominated by species that were present before management but not dominant (Wade, 1990). Management examples illustrate these changes.

Harvesting a dense canopy of narrow-leaved pondweeds (*Potamogeton* spp.) in Halverson Lake allowed *Zosterella dubia* to flourish and dominate the plant community for 7 years after the last harvest (Engel, 1990a). Engel (1987) also reported cases where years of harvesting a canopy of *Myriophyllum sibiricum* allowed *Vallisneria americana* to dominate, and where wild rice (*Zizania aquatica*) greatly expanded its range when competing submergents were removed by harvesting. *Nitella* spp. showed a marked increase in dominance after harvesting *L. major* in Lake Aratiatia, New Zealand and coontail followed by elodea, egeria, and *Potamogeton crispus* became more dominant in Lake Ohakuri, New Zealand after harvesting (Howard-Williams et al., 1996).

Nichols and Cottam (1972), Johnson and Bagwell (1979) and Welch et al. (1994) reported no change in plant community structure after harvesting. The harvested plant communities replaced themselves. In Chatauqua Lake, New York, harvesting appeared to promote the growth of Eurasian watermilfoil at the expense of *Potamogeton* spp. (Nicholson, 1981b). Species that reproduce sexually, regenerate poorly from fragments, and heal and grow slowly after cutting are at a competitive disadvantage under a harvesting regime. Conversely, species like Eurasian watermilfoil that grow rapidly after cutting and regenerate from fragments are likely to replace themselves, become more dominant, or easily invade areas managed by harvesting.

14.4.4.2 The Nutrient Removal Question

Nutrient removal is a frequently cited advantage of harvesting (Carpenter and Adams, 1978). Calculating the potential for removing nutrients is straightforward. By knowing the area of the lake covered with macrophytes (m^2), the average biomass of the plants in the area (g dry wt/m^2 per year) and the nutrient concentration of the plants (g nutrient/g dry weight of plants) an estimate of the total nutrient available for removal can be calculated (Burton et al., 1979). This number is reduced by the percentage of the total area harvested and the efficiency of the harvest (e.g., even in harvested areas all plant biomass is not removed). This number is often compared with nutrient loading to the lake to determine the percent of the net annual loading that might have been or was removed by harvesting. These numbers varied widely (Table 14.2) but obviously more nutrients will be removed if macrophyte biomass is high, the nutrient concentration within the biomass is high, the lake areas covered with macrophytes is high, and the percentage of the macrophyte biomass harvested is high. Harvesting has the greatest impact on the nutrient budget if nutrient removal is high and nutrient loading is low. In eutrophic lakes, even where nutrient loading is controlled, it still may take several years for harvesting to have an impact on nutrient concentrations (Carpenter and Adams, 1977; Burton et al., 1979).

Simple calculations of nutrients removed by plant harvesting may be misleading. The nutrient content of plant tissue varies by season, waterbody, and species (Wile, 1974; Hutchinson, 1975;

TABLE 14.2
Phosphorus Removal by Macrophyte Harvesting

	Lower Chemung[a]	Sallie[b]	Wingra[c]	East Twin[d]
Surface area covered by macrophytes	430 ha	34%	34%	11.7 ha
Macrophytes harvested	18.7%	100%	100%	50%
Dry weight removed (kg)	3,020 metric tons wet weight	30,400	130,100	18,720
Mean tissue phosphorus concentration (% dry weight)	0.25%	0.27%	0.39%	0.15%
Phosphorus removed by harvesting (kg)	560 kg	100 kg	580 kg	23.1 kg
Net annual phosphorus load (kg)	610	10360	1592	8.1–62
Percentage of net annual load removed by harvesting	92%	0.96%	36.4%	46%–100%

[a] Based on data for 1975. From Wile, I. et al. 1979. In: J. Breck, R. Prentki and O. Loucks (Eds.), *Aquatic Plants, Lake Management, and Ecosystem Consequences of Lake Harvesting.* Inst. Environ. Stud., University Wisconsin, Madison. pp. 145–159.

[b] From Neel, J.K. et al. 1973. *Weed Harvest and Lake Nutrient Dynamics.* Ecol. Res. Series, USEPA-660/3-73-001. Peterson, S.A. et al. 1974. *J. Water Pollut. Cont. Fed.* 46: 697–707.

[c] Based on estimates of the nutrient pool. Full-scale harvesting did not occur. From Carpenter, S.R. and M.S. Adams. 1978. *J. Aquatic Plant Manage.* 16: 20–23.

[d] Phosphorus budget based on 1972–1976 sampling. Phosphorus content of plants, plant density, and areal coverage was based on 1981 data. Only limited harvesting was done in 1981. Removal was based on a realistic estimate of 50% plant removal by harvesting. From Conyers, D.L. and G.D. Cooke. 1982. In: J. Taggart and L. Moore (Eds.), *Lake Restoration, Protection and Management*, Proc. Second Annu. Conf. North American Lake Management Society. USEPA, Vancouver, BC. pp. 317–321.

Zimba et al., 1993). Rooted macrophytes extract nutrients from both the sediment and the water column so removing nutrients in plant biomass may have a different effect on lake nutrient budgets than preventing nutrients from entering the lake (Carpenter and Adams, 1977).

The plant community may not be able to maintain the high biomass production needed for extensive nutrient removal over the long term. In Lake Sallie, Minnesota, harvesting took place each summer from 1970 through 1972. A single operator harvested in the same manner, using the same harvester, each year. Figure 14.7 shows data that were normalized to a rate function for the same areas based on daily harvest records (Peterson, 1971). All things being equal, operator proficiency should have improved with each successive year's experience, thus increasing the harvest yield rate. However, the yield (kg/h) decreased with each successive year. Harvesting was started in July 1973, but was halted almost immediately because the macrophyte yield was very poor. This suggests that successive harvests reduced plant biomass from year to year. Unfortunately, there was no control lake, to help determine if the plant decline in Lake Sallie was due to harvesting or just a general, regional phenomenon. However, other findings, already discussed in this chapter support the idea that repeated harvesting reduces plant biomass from year to year.

Many lake renewal efforts failed because the role of internal nutrient loading wasn't appreciated (for instance Shagawa Lake, Minnesota; Larsen et al., 1979; Wile et al., 1979). Internal nutrient loading in many eutrophic lakes is greater than external loading (see Chapters 4 and 8) particularly as external loading is reduced. The role aquatic plants play in internal nutrient loading is being increasingly appreciated and macrophyte harvesting may be a way of reducing internal nutrient cycling (see Chapter 11, The effects of macrophytes on their environment). As stated in Chapter 11, Barko and James (1998) calculated that abundant plant growth at the inlet contributed about 1200 kg of P to the nutrient budget of Delevan Lake, Wisconsin. Water chemistry changes caused

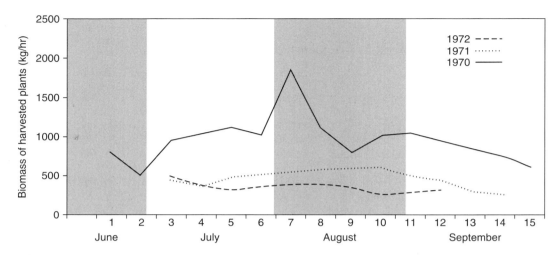

FIGURE 14.7 Yield of harvested plants from Lake Sallie, Minnesota showing the decline in biomass with successive years of harvesting. (After Peterson, S.A. 1971. Nutrient dynamics, nutrient budgets, and weed harvest as related to the limnology of an artificially enriched lake. Ph.D. Thesis, University North Dakota, Grand Forks. Figure courtesy of Spencer Peterson.)

by an abundant macrophyte growth accounted for one half of the P and nutrient mobilization by macrophytes from littoral sediments accounted for the other half. Macrophyte decay accounted for about half the internal P loading in Lake Wingra (Carpenter, 1983). Through modeling, Asaeda et al. (2000) estimated that phosphorus released from decaying *P. pectinatus* could be reduced by at least 75% by harvesting above ground biomass at the end of the growing season. Aquatic plant removal by harvesting could change water chemistry conditions, remove nutrients in plant biomass that would otherwise be recycled, and reduce sedimentation of macrophyte biomass. Sediment nutrients might also be depleted by harvesting rooted plants that obtain N and P from sediments (Carpenter and Adams, 1977). The impact of sediment nutrient depletion is difficult to calculate because plants obtain an unknown fraction of nutrients from the water; nutrients in the sediments, at least available P, are continually replenished by equilibration with insoluble forms; and sedimentation continually adds nutrients to sediments (Carpenter and Adams, 1978).

Although there have been numerous measurements, models, and speculation about the role harvesting plays in nutrient budgets, there are few if any examples where harvesting reduced nutrient concentration (at least for P) in the water column. Most studies found P concentrations were unchanged or increased under a harvesting regime; or secondary indicators of higher nutrient levels like algal growth increased or were unchanged. Welch et al. (1994) found that summer lake total P concentrations were higher during harvest years than non-harvest years in Long Lake. Root crown harvesting in LaDue Reservoir was associated with elevated levels of total P, chlorophyll, blue-green algae, and seston (Cooke et al., 1990).

In the Lake Sallie example the phytoplankton productivity changed with harvesting. Phytoplankton productivity in 1969, a year prior to plant harvesting, was relatively high and typical of eutrophic conditions (Smith, 1972; Figure 14.8). However, phytoplankton productivity increased noticeably in 1970, the first year of harvesting. It peaked in 1971, and in 1972 and 1973 productivity was above the 1969 pre-harvest levels (Brakke, 1974). Figures 14.7 and 14.8 show that increased phytoplankton productivity was probably related to reduced plant biomass caused by harvesting. Thus the gain from one management effort was offset by a response from another segment of the ecological community likely because of a change in nutrient pathways.

Harvesting had little effect on phytoplankton in Halverson Lake (Engel, 1990a). No significant changes in ambient nutrient levels or phytoplankton species composition were seen in Chemung

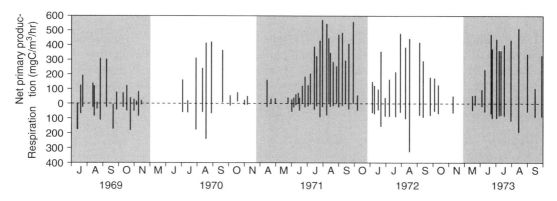

FIGURE 14.8 Phytoplankton productivity in the surface waters of Lake Sallie, Minnesota for 1 year prior to the beginning of harvesting and for 4 years after harvesting began. (After Smith, W.L. 1972. Plankton, weed growth, primary productivity, and their relationship to weed harvest in an artifically enriched lake. Ph.D. Thesis, University North Dakota, Grand Forks; Brakke, D.F. 1974. Weed harvest effects on algal nutrients and primary productivity in a culturally enriched lake. M.S. Thesis, University North Dakota, Grand Forks. Figure courtesy of Spencer Peterson.)

Lake during the harvesting period (Wile et al., 1979). Painter and Waltho (1985) found no observable change in sediment total P or N in the rooting depth of Buckhorn Lake after 2 years of harvesting. They concluded that the sediment nutrient pool was much greater than that needed for milfoil growth. There are a number of possible explanations why nutrient levels increased or remained unchanged including: (1) denuding the littoral zone by harvesting may have allowed a greater percentage of dissolved and particulate allochthonous inputs to reach the pelagic zone; (2) the intact littoral zone may have acted as a nutrient sink and the harvested one did not; (3) the relationship between macrophyte nutrient removal and the nutrient budget may not be direct or immediate so sampling did not detect change; (4) the nutrients removed by harvesting were a minor part of the nutrient budget; (5) nutrient cycles and other ecological processes are complex enough that nutrient removal may have been compensated for in some other manner; (6) the growth limiting nutrient for algae may not have been removed; (7) some other mechanism such as allelopathy between macrophytes and algae may have limited phytoplankton growth before harvesting; and (8) harvesting may not have continued long enough or the area harvested may not have been large enough to impact the nutrient budget.

Does all this mean that harvesting to remove nutrients should be discounted? — Certainly not. It does mean that harvesting alone is not likely to solve an excess nutrient problem, at least in the short term. When money for nutrient removal is limited (which it often is) the cost of removing nutrients by harvesting needs to be compared to the cost of removing, sequestering, or preventing nutrient inputs by other means. Nutrient removal by harvesting is generally expensive (Neel et al., 1973) relative to the return. Harvesting can be used as part of an integrated nutrient management plan that includes reducing nutrient inputs, sequestering in-lake nutrient sources, and nutrient removal. If harvesting is used, or is planned to be used, to manage aquatic plant nuisances, consider nutrient removal as an additional benefit. However, the timing of harvests for maximum nutrient removal and for maximum reduction of long-term growth is likely to be different than timing for the maximum seasonal reduction of an aquatic nuisance. Carpenter and Adams (1978) calculated that harvesting Eurasian watermilfoil in late August from Lake Wingra would remove the maximum amount of P. The above discussions indicate that a mid-fall harvest may have the greatest long-term impact on regrowth. From a management perspective, in northern states, this may be unpalatable because the water recreation season is ending. Managers are unlikely to spend money on harvesting at that time of year. Users want "weed" free lakes in June, July, and August.

14.4.4.3 Environmental Effects

The environmental impacts of harvesting include: (1) immediate and protracted physical and chemical effects, (2) effects on the biota, and (3) effects on ecosystem processes. Removal of dense plant canopies should result in physical and chemical water quality more like open water areas. Aquatic plant harvesting is usually limited to relatively small areas of a water body so environmental effects are likely to be minor but could be more profound in small shallow lakes or in localized areas of large lakes with dense macrophyte growth.

14.4.4.3.1 Physical and Chemical Effects

Some immediate physical and chemical effects of harvesting could include water temperature changes; increases in suspended material from machinery disturbance of sediments; dissolved oxygen changes caused by reduced photosynthesis, by allowing better atmosphere-water contact, or by decomposition of cut plants; and P concentration changes due to leakage from cut macrophyte stumps or from sediment disturbance. Carpenter and Gasith (1978) studied the immediate effects on littoral water chemistry and metabolism of small mechanically harvested plots in Lake Wingra. They found no significant water temperature, seston concentration, dissolved organic carbon, or conductivity differences between harvested and unharvested plots. Dissolved reactive P concentrations were variable, usually at the detection limit, and not significantly different between harvested and unharvested areas. The leakage of P from cut stems was insignificant if it occurred. Community photosynthesis was depressed in shallow areas where stem removal was complete, but in deep areas, where removal was not complete, photosynthesis by remaining macrophyte stumps and phytoplankton approximated that of the undisturbed littoral zone. They also found that material suspended by harvester operation settled from the water column in less than one hour. Their conclusion was that mechanical harvesting of limited areas caused little immediate detrimental physical or chemical impacts to the littoral environment. Madsen et al. (1988) found that harvesting dense macrophyte beds reduced the diel DO variations without increasing the average oxygen concentration.

Over the longer term, harvesting could effect nutrient cycling between the water column and lake sediments, depress photosynthesis (with a decrease in pH), and change oxygen levels. Many changes are speculative because few if any studies have followed harvesting long enough, over large enough areas, and monitored the environmental effects of their impacts. Littoral zone erosion has been demonstrated in areas where plants were removed by harvesting or other means (Howard-Williams et al., 1996; James and Barko, 1994). Welch et al. (1994) speculated that increased total P levels in Long Lake resulted from increased, wind-driven, sediment resuspension after harvesting removed the macrophyte cover. Mechanical harvesting may reduce sediment accumulation or enrichment by removing organic matter that partially decomposes in the littoral zone. Particulate materials can be trapped in unharvested vegetation and, in water courses with moderate flows, aquatic plants may remove and accumulate significant amounts of dissolved and particulate nutrients. Again, changes in these parameters, other than short-term flow through of particulate and dissolved materials in devegetated areas, have not been measured in areas under harvesting management.

14.4.4.3.2 Biotic Effects

The biotic effects of harvesting interest lake managers because they include features that are most conspicuous to lake users such as macrophyte density, water clarity, phytoplankton concentration, and fish stocks. The main biotic effect is the removal of non-target plant species. This impact is covered above under the heading of "Efficacy, regrowth, and change in community structure."

Plant harvesting directly removes fish, invertebrates, and other creatures, including a variety of microbes, which live in or on aquatic plants. In shear numbers, the organisms removed by harvesting are impressive but the magnitude of the impact is variable. Engel (1990a) estimated that between 11% and 22% of all plant-dwelling macroinvertebrates and over 50,000 fish were removed from 4 ha Halverson Lake with 2 years of harvesting. Monahan and Caffrey (1996) reported that, in Irish canals, harvesting reduced macroinvertebrate numbers by 60–85% and about one million macroinvertebrates were removed with each ton of *Ranunculus* sp. harvested. Mikol (1984) estimated that 2,220 to 7,410 fish were removed per hectare harvested in Saratoga Lake, New York. Hydrilla harvesting in Florida removed 85 kg/ha of fish (Haller et al., 1980). About 50 to 100 fish were collected in each load of harvested plants in the Okanagan Lakes of British Columbia (Cooke et al., 1993). Unmuth et al. (1998) estimated a removal rate of 2,254 fish/ha using conventional harvesting in Fish Lake. An estimate of 700 turtles/year along with an unknown number of mud puppies *(Necturus maculosus)* and adult and immature bullfrogs *(Rana catesbeiana)* were removed from 96 ha Lake Keesus, Wisconsin by harvesting (Booms, 1999). Weed cutting removed no mussels from British canals (Aldridge, 2000).

The impact of direct removal on the fish population is questionable. In all cases the fish removed were small, generally slow moving, panfish or forage fish species. The most common size class of fish removed from Lake Keesus was 2–4 cm long and no fish over 12 cm long was removed (Booms, 1999). Engel (1990a) was the only one to report removal of large numbers of large-mouth bass *(Micropterus salmoides)*. Mikol (1984) estimated fish removal to be about 2.4–2.6% of the standing fish crop in Saratoga Lake. But, Haller et al. (1980) estimated that 32% of the fish numbers and 18% of the fish biomass was removed by harvesting in Florida. Wile (1978) reported fish population (except for yellow perch, *Perca flavescens*) in Chemung Lake remained stable throughout the extensive harvesting operation and she did not believe the change in perch population was related to harvesting. Opinions vary on the impact of fish removal. Haller et al. (1980) valued the fish removed at $410,000 but most authors thought the impact of fish removal was insignificant.

Harvesting also removes a food source and covering habitat for a variety of organisms. Macrophytes provide microheterotrophs a substrate for colonization and a reduced carbon source through extracellular secretion and decay of macrophyte tissue (Carpenter and Adams, 1977). Loss of this labile organic matter through harvesting could decrease both mineralization rates and microbial production in lakes.

Macroconsumers of macrophytes include waterfowl, mammals, and invertebrates such as crayfish and insects. Direct grazing losses in fresh water are often negligible relative to macrophyte production (Carpenter and Adams, 1977) but they can be significant, especially on emergents and with mammals and waterfowl (see Chapters 11 and 12). Much macrophyte tissue enters the food web as detritus after plant death (Fisher and Carpenter, 1976). Many consumers inhabiting macrophyte shoots graze the complex of algae and detritus on the macrophyte surface rather than the macrophyte tissue. All the above consumers are directly vulnerable to loss of food and habitat when macrophytes are harvested.

The effects of macrophyte removal have food chain impacts. The relationships between macrophyte cover, zooplankton distribution, and the diet and growth of roach *(Rutilus rutilus)* were studied in the River Great Ouse, U.K. before, directly after, and over several weeks following weed cutting. Fish and zooplankton were significantly associated with the macrophyte zone that provided high food densities and refuge during high flow periods. Removal of all but a 2-m marginal macrophyte zone led to a rapid decline of mean cladoceran densities, probably the result of increased washout, fish predation, and starvation (Garner et al., 1996). This was accompanied by a rapid decline in growth rates that was attributed to roach being forced to feed on less nutritious aufwuchs.

Harvesting may benefit fish growth. Stunted growth is common in many lakes with high macrophyte densities. Removal of stunted fish via harvesting can increase the size structure of the fish population by making limited food energy available to a smaller number of fish. In the deep cutting experiment on Fish Lake, Unmuth and Hanson (1999) found the mean abundance of largemouth bass and bluegills *(Lepomis macrochirus)* did not change significantly, but growth increased for age 2–4 largemouth bass and declined for age 5 largemouth bass and age 4–6 bluegill. Population size structure increased for both species. Although deep cutting channels only gave short-term macrophyte control in other Wisconsin lakes, there were strong positive growth responses for some age classes of bluegills and largemouth bass that will persist for the lifetime of the affected age classes (Olson et al., 1998). The increased growth in some age classes may be related to increased predation efficiencies resulting from more edge created by cutting channels in dense plant beds. Removal of fish could also allow larger zooplankton to survive, which is desirable for biomanipulation efforts (see Chapter 9).

A problem that has not been studied is the impact of harvesting on spawning fish. Bluegills and largemouth bass, at least in the Upper Midwest, U.S., typically spawn in early to mid-June, a prime time for harvesting. Their preferred spawning habitat is openings in macrophyte beds. What happens to the nests, eggs, and guarding parental fish when a harvester goes "plowing" through these areas? Fry of these species also use dense plant beds for cover later in the season.

Recovery of the biota after harvesting is variable and there are techniques to mitigate harvesting impacts. In Irish canals it took 8 to 10 months for macroinvertebrate numbers to return to pre-harvest levels (Monahan and Caffrey, 1996). Leaving unharvested refuges for fish and invertebrates is one suggestion. Monahan and Caffrey (1996), Garner et al. (1996) and Aldridge (2000) suggested harvesting only one side or only the center of a river or canal during a growing season. Close cutting techniques reduced fish removal rates from 2,254 fish/ha, typical of conventional harvesting, to 36 fish/ha (Unmuth et al., 1998). Fish had the opportunity to escape from the weed mass between cutting and removal. The optimum amount of macrophyte cover required for optimum fish habitat is ill defined. The relationship is parabolic so that fish foraging and growth is optimized at an intermediate level of plant density (Trebitz, 1995; Olson et al., 1998). In Wisconsin, deep cutting removed about 20% of the macrophyte cover in the lakes where littoral zones macrophyte coverage was over 90%. Wholesale plant removal negatively impacts phytophilic fish species but may benefit other species (Bettoli et al., 1993). Harvesting to an intermediate level of plant density probably has few long-term negative impacts on fish populations and it is probably beneficial.

Harvesting could spread nuisances because many species are able to propagate rapidly from plant fragments. Hydrilla, for instance, can regrow from a single node (Langeland and Sutton, 1980) and when chopped into small pieces (Sabol, 1987). Even well designed harvesters can lose between 7% and 15% of the cut plants (Engel, 1985). Two-stage harvesting purposely leaves fragments in the water for later collection. The magnitude of the problem is area and species specific. In areas of severe plant infestations there may be little additional habitat for plant growth. In other areas, wind, water current, and boat traffic can spread plant fragments that will cause aquatic nuisances in uninfested areas.

The fragment problem may not be as severe as it first appears. Naturally produced Eurasian water-milfoil fragments grow better than artificially cut stems and they have a higher TNC content, which suggests they could better survive the winter (Kimbel, 1982). Milfoil fragments generated by harvesting may be less problematic than naturally produced fragments and harvesting may reduce the parent stock producing autofragments.

14.4.4.3.3 Ecosystem Effects

The impact of harvesting on ecosystem processes could take a long time to develop and the repercussions could be complex. Therefore, predicting or measuring the ecosystem harvesting impacts is difficult. For managers, Engel (1990b) provides a concise yet easy to read literature review of short-term, long-term, and ecosystem effects that are likely to occur from harvesting.

An area where there is some management experience with ecosystem impacts is the change in stable state in shallow, eutrophic lakes (see Chapter 9). Any treatment that removes large areas of plants in shallow, eutrophic lakes may shift the stable state from a macrophyte dominated lake to an algae dominated lake (see a similar statement in Chapter 16 regarding chemical controls) (Scheffer et al., 1993; Moss et al., 1996). Even with experience, it is very difficult to calculate how much management will cause a shift (van Nes et al., 2002) and once the shift occurs it can be difficult to return to a macrophyte dominated state (Scheffer, 1998). Jacoby et al. (2001) reported large year-to-year shifts between high total P and algal biomass, low transparency and low macrophyte biomass on one hand; and low total P and algal biomass, high transparency and high macrophyte biomass on the other that they attributed to plant harvesting in Long Lake.

14.4.4.4 Operational Challenges

There are many organizational tasks that help make harvesting operationally successful. There are multiple pieces of equipment to deploy; crew members, whether paid or volunteer, to find, train, and schedule; launching and unloading sites to coordinate; disposal areas to secure; salaries and insurance to provide; and permits to obtain. It does not take long for the operational aspects of harvesting to become a full-time job — at least during the harvesting season.

Harvesting is like, "trying to mow a yard full of rocks at night without a moon" (Helsel, 1998). Safety is a big concern while operating all the machinery involved with a harvesting. There are cutter bars, aprons, motors, paddle wheels, boats, trucks, barges, fuel, and lubricants to deal with. There are many areas where poor judgment can lead to accidents.

Working on the water dictates that when high winds, large waves, and/or lightening occurs, operations must be shut down for safety's sake. Improper loading of transport barges has caused them to overturn. Recreational use on some lakes is so high that harvesters do not operate on holidays and weekends. What does one do with tons of wet, smelly aquatic plants? At worst, if done improperly, nutrients and oxygen demanding organic matter are returned to the lake. On shore they can become odiferous and attract nuisances. However, proper disposal can turn aquatic plants into compost or green manure.

Public relations are another important operational challenge. Some people's idea of a peaceful sunrise over the lake does not include a harvester lumbering back and forth in front of their dock. Others want every weed in the lake trimmed "yesterday."

14.4.5 SHREDDING AND CRUSHING

Shredding and crushing reduce the bulk of harvested material, thus reducing transport and disposal costs. Shredding and crushing machines are of two basic designs and date back to the early 1900s. One design uses a front conveyor, with or without a cutting blade that lifts the plants onto a barge where they are crushed or chopped (Wunderlich, 1938; Livermore and Koegel, 1979; Sabol, 1982). This design is most commonly used for free-floating or submergent species. The remains are returned to the waterway, or conveyed or blown onto a transport barge or onto shore. The other design uses a bow mounted rotary cutter to shred plants (Dauffenbach, 1998). This design is most commonly used on free-floating and emergent species and can often work in very shallow water. The efficacy of shredding and crushing is not reported but it should be similar to conventional harvesting.

The major concerns with shredding and crushing are returning viable plant fragments and nutrient-containing and oxygen-demanding materials to the water. The changes in water chemistry resulting from shredding water chestnut was mentioned earlier in this chapter (James et al., 2000) as was the result of the regeneration of hydrilla from cut parts (Sabol, 1987). Harvesting and onboard chopping of hydrilla in Orange Lake, Florida showed that the daily minimum oxygen content in the water column was not affected by in-water disposal. Removal of aquatic plants reduced oxygen accrual during the day but in-water disposal did not reduce further accrual. Chlorophyll *a* concentration increased, thermal stratification decreased, and a small amount (0.6%) of hydrilla fragments remained viable (Sabol, 1982). Most stem fragments sank to the bottom within 2 hours of disposal. The longest fragments, those with the greatest regrowth potential, remained floating for three days or more. Leaving an unharvested buffer zone around the harvested area to catch floating fragments is one way suggested by Sabol (1982) to reduce fragment spread. Alligator weed (*Alternanthera philoxeroides*) fed once through a conventional brush chipper had a regrowth rate of 5%. When the vegetation was fed through the chipper a second time it was reduced to a sludge with no regrowth (Livermore and Wunderlich, 1969). A realistic concern in southern waters is the attraction of large carnivores (e.g., alligators) to the "chum" of chopped fish and other organisms that are a "by-catch" of shredding (Madsen, 2000b).

A combination of shredding and conventional harvesting was used in Lake Istokpoga, Florida to remove floating tussock plant communities. A rotary cutting "cookie cutter" was used to chop the tussocks into small pieces. A harvester picked up the pieces and transported them to disposal sites on shore. Minor, but statistically detectable water chemistry differences occurred at the harvest sites. Chlorophyll *a*, total N, and total P concentrations decreased and turbidity and dissolved solids increased during harvest (Alam et al., 1996). Dissolved oxygen differences between harvested and unharvested sites were minor.

14.4.6 DIVER-OPERATED SUCTION DREDGES

Divers operating small suction dredges (Figure 14.9) mechanizes hand removal of plant roots and stems. A pump on a barge provides a vacuum through a hose. A diver takes one end of the hose (approximately 10 cm diameter) to the lake bottom. There the diver selectively removes the target vegetation using a sharp implement to dig roots from the substrate. Plant material moves up the hose by suction and collects at the surface in mesh baskets.

Advantages of diver dredging include: (1) selective removal of target vegetation, (2) plant parts are collected minimizing the risk of further spread, (3) only localized turbidity results because substrates are minimally disrupted, and (4) operations are site specific and suctioning may be used in places where other methods are impossible. Limitations of diver dredging include: (1) low rates and high unit operational costs, (2) health risks to the diver, and (3) in dense plant populations removal of target species is probably too slow to be practical. Diver-operated dredging is probably most useful for removing initial infestations of nuisance species.

Diver-operated dredges were extensively tested in British Columbia to control sparse colonies of Eurasian watermilfoil. The objectives were to: (1) provide long-term milfoil control, (2) remove root systems, and (3) permit treatment where no other methods were practical. Depending on local

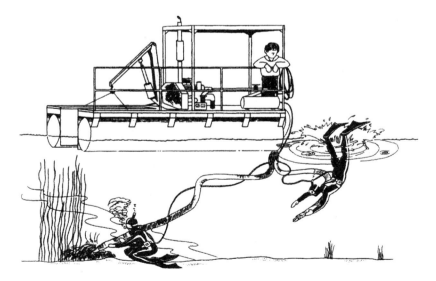

FIGURE 14.9 A diver operated dredge designed by the British Columbia Ministry of Environment, Lands, and Parks. (From Cooke, G.D. et al. 1993. *Restoration and Management of Lakes and Reservoirs*, 2nd ed. Lewis Publishers and CRC Press, Boca Raton, FL. With permission.)

conditions, 85–97% root removal could be achieved by diver dredging (Cooke et al., 1993). Because of high operational costs and changes in control strategy, extensive use of this method was discontinued in favor of using bottom barriers.

Extensive diver dredging removed hydrilla from a marina on the Potomac River in the U.S. Effectiveness was 100% for biomass removal and 91% for tuber removal based on measurements before and after dredging (Cooke et al., 1993). Hydrilla regrew rapidly in the test plots because of reinfestation from adjacent untreated areas.

Suction harvesting reduced both the biomass and percent cover of Eurasian watermilfoil in Lake George, New York. Milfoil was the most abundant pre-harvest species in localized areas of the lake. It declined to the fifth most abundant species after suction harvesting — from a 30% pre-harvest cover to less than 5% after harvesting (Boylen et al., 1996). One year latermilfoil averaged about 7% cover. Native species showed a variable response to suction harvesting. *Potamogeton amplifolius* and *Vallisneria americana* coverage declined; *P. robbinsii*, *Zosterella dubia*, *Elodea canadensis*, and *P. gramineus* coverage remained about the same; and *Najas flexilis* coverage increased substantially (Boylen et al., 1996). Depending on the site, harvesting the regrowth required between 64% and 89% less effort than the initial harvest, and hand removal of regrowth the following year required only about 20% of the effort for initial harvesting (Boylen et al., 1996).

Clayton (1996) reported that diver operated suction harvesting was especially useful in areas of irregular bottom contours and obstacles, and in areas too deep for conventional harvesting. However, he also reported that suction harvesting had a small and short-lived impact on extensive *Lagarosiphon major* beds in Lake Whakamarino, New Zealand.

14.4.7 HYDRAULIC WASHING

Hydraulic washing uses a water pump and high pressure nozzles to "wash" plants out of bottom substrates. Pressure washing machines had interesting names like Aqua-Beach Comber and Water Witch but they were used in very limited areas and were considered prototypes (Deutsch, 1974; Nichols, 1974). McComas (1993) mentions the hand-held "water rake" that can be used around piers and for beach cleaning. In British Columbia, hydraulic washing achieved a high degree of root removal and was most successful in soft substrates or following tillage where root systems were sheared or dislodged (Cooke et al., 1993). Hydraulic washing was unsatisfactory as a primary control measure for Eurasian watermilfoil since large root masses were not broken up or dislodged from the sediments despite repeated passes (Cooke et al., 1993). Generally, hydraulic washing is useful only in small areas.

14.4.8 WEED ROLLERS: AUTOMATED, UNTENDED AQUATIC PLANT CONTROL DEVICES

The commercially available weed roller consists of a horizontal roller arm that attaches to a vertical pivot arm. The pivot arm reaches above the water where it is anchored to a dock or a free-standing tripod. An electric power head, attached to the top of the pivot arm, drives the roller. The whole mechanism looks like a giant L, lying on its back. The standard model has a 6.4-m roller arm and a 1.4-m pivot arm. Additional 2.1-m sections can be added to the roller and two additional 0.6-m sections can be added to the pivot arm.

The roller arm, with attached fins, rolls slowly back and forth across the lake bottom in a 270° arc. The roller arm turns at about 5 rpm and can cover the total arc in about 1/2 to 1 hour. The fins and the continuous rolling motion stir up the lakebed, uprooting plants. Plants are "worn down," float away, or become wrapped in the roller. According to the manufacturer's instructions, the roller should be operated continuously until all aquatic plants are cleared. Then, operating it once every week or two maintains a plant free area. Stirring causes light sediments to float away, resulting in a sandier bottom that may be less conducive to plant growth.

Minnesota Department of Natural Resources calls these units "automated untended aquatic plant control devices." They feel weed rollers have the potential to remove a larger swath of vegetation, displace more sediment, and eliminate plants for a longer period of time than many other devices used by riparians to control aquatic plants. They, along with some other states, require a permit to ensure weed rollers are used appropriately. Their effectiveness probably depends on the plant type, plant density, bottom type and the amount of time the unit is operated. However, no studies were found that confirm this speculation.

14.4.9 MECHANICAL DEROOTING

Derooting methods were developed to remove or kill plant root crowns, which would, hopefully, give longer term plant control than conventional harvesting. Rototilling, using angled derooting bars or tines to dig out roots, and cultivating, using agricultural equipment towed behind amphibious tracked, wheeled, or floating vehicles to shear root crowns, are the two main derooting approaches. McComas (1989, 1993) describes many agricultural tools that can be used for small scale derooting. Derooting has been used primarily for Eurasian water-milfoil control on the west coast of North America, especially in British Columbia. Eurasian watermilfoil's buoyant root crowns make it vulnerable to derooting.

Advantages of mechanical derooting include: (1) a high proportion of root crowns are removed in most areas depending on water depth, substrate type, and operator skill; (2) derooting equipment can be used over a longer period of time, generally the ice-free season, which allows management to take place during times of low public use and may provide more continuous employment for management crews; (3) effectiveness may persist for 2 or 3 years; and 4) no fish are removed during derooting. The disadvantages of derooting are: (1) derooting is not effective on all substrate types because of bottom irregularities or hard-packed sands or clays; (2) obstacles may prevent efficient operation or damage machinery; (3) fragmentation and dispersal of root crowns or stems may not be desirable and collection of floating material is difficult and time consuming; (4) the treatment area is disrupted to the tilling depth; (5) effective treatment requires a conscientious operator; (6) guidance of equipment can be difficult so many overlapping passes may be needed to prevent missed areas; (7) derooting is not recommended in dense plant stands as stems clog tilling equipment and reduce effectiveness; and (8) mechanical derooting should not be used in newly infested areas where spread of plant fragments is undesirable.

Treatment depth varies depending on the equipment used. Cultivating equipment works best in shallow water or on shorelines under drawdown conditions. Rototillers can work in a variety of depths. The maximum depth that can be reached by the rototiller arms limits their operational depth. Rototilling is relatively slow when compared to cultivating. Derooting treatments are usually limited to high priority recreation areas such as public beaches and heavily used boating facilities.

Derooting treatments require two passes through the treatment area, normally at right angles to each other, to provide acceptable control. In a number of British Columbia trials, plant density was reduced immediately after derooting treatment by cultivation from 49–98% and 80–97% for rototilling (Cooke et al., 1993). A single pass with a rototiller provided an 83% reduction in stem density and a 94% root biomass reduction in sandy substrates (Cooke et al., 1993). This was improved to 98% and 100%, respectively, after many passes. In silty substrates one pass resulted in a 55% stem density reduction and a 70% root biomass reduction that was raised to 88% and 84%, respectively, after many passes (Cooke et al., 1993). Sand or gravel substrates consistently permitted easier dislodgement of Eurasian water-milfoil roots than did silt or clay substrates. The latter substrates bound to roots so that root crowns were only partially dislodged after tilling. Spring treatment gave the best results for milfoil root crown removal and in Pend Oreille Lake, Washington conventional harvesting beforehand enhanced root removal (Cooke et al., 1993).

Reinfestation of treated areas from roots crowns that were missed or only partially dislodged by tilling and from stem fragments drifting in from untreated areas began early in the following growing season (assuming a fall, winter, or early spring treatment). British Columbia experiences found that Eurasian watermilfoil regrew about 50% during the first growing season after treatment and there was almost complete reinfestation during the second growing season (Cooke et al., 1993). There was rapid regrowth unless 90% to 95% of root crowns were removed. A pattern of post-tilling reduction of Eurasian watermilfoil, followed by varying growth of native plants from turions, tubers, and seed, followed by a reinfestation by Eurasian watermilfoil, was observed in many British Columbia locations (Cooke et al., 1993).

No persistent changes in 27 water quality variables were documented during tests of different tilling machinery (Cooke et al., 1993). Turbidity increased dramatically immediately following tillage but then rapidly declined. More research is needed to evaluate the impacts of derooting on the benthic invertebrate community and derooting may stimulate the growth of some plant species that reproduce by seeds, tubers, and turions by removing nuisance target plants.

14.4.10 Costs and Productivity

Variables that affect the productivity and thus the cost of harvesting include: (1) width of cut, (2) forward velocity of the harvester, (3) harvester maneuverability, (4) plant species and density, (5) weather conditions (i.e., wind and waves), (6) water depth and bottom obstructions, (7) skill and incentive of the operator, (8) matching accessory equipment (transport barges, conveyors, trucks) to the harvester and overall conditions, (9) mechanical design and condition of the harvester as related to breakdown time, and (10) percentage of time spent on non-harvesting operations such as transporting harvested plants or moving equipment from location to location. With this number of variables, cost and productivity comparisons between harvesting machines and programs are difficult to make. Making comparisons between widely different mechanical methods such as cutting, harvesting, shredding and crushing, suction harvesting, hydraulic washing, weed rolling, and mechanical derooting, is even more difficult. Besides these difficulties, most cost estimates are dated and most cost and productivity estimates are based on "once through" costs. They do not consider the cost of achieving a certain level of customer satisfaction (for instance some areas may have to be harvested three times a growing season for the area to be useable for recreation). Perhaps the best way of discussing costs is to provide a worksheet to calculate costs, compare costs of other methods to that of standard harvesting, and compare productivities of the methods. Techniques for increasing efficiencies will also be discussed.

Table 14.3 provides cost categories that should be considered for a harvesting operation. Any one category may or may not be appropriate to a specific case. For instance, if harvesting is done on a single lake and the harvester is used to transport cut material to shore where it is directly unloaded onto the bank, the transport barge, shore conveyor, and trailer may not be needed. However, a loader of some sort may be needed to pick up the piles of harvested plants. Utilizing volunteers obviously reduces labor costs. Capital costs are often amortized over a 10-year period, the estimated life of the equipment, hence capital costs were divided by 10. In practice, different age machines are likely used so each should be amortized separately. Often harvesting machines never wear out — they get totally rebuilt after numerous repairs. If some other form of mechanical removal is used, its capital costs can be substituted for harvester costs. Contingency costs are added to cover unexpected expenses. Equipment can be leased or rented and there are harvesting contractors so that capital costs could be transferred to annual operating costs. Leasing, renting, or contracting is a good way to try mechanical removal before committing to a long-term program. Studies of the Dane County, Wisconsin annual harvesting expenses indicated that about 26% of the expenses was for operating costs (fuel, oil, wages); 33% was for repair, overhaul, and modification of equipment; 26% was for investment in equipment; 11% was for supervision; and 4% was for shop facilities (Koegel et al., 1977). In a detailed study of four harvesting operations in North America, Smith (1979) reported that, on average, 55.4% of the budget was spent on labor, 12.3% on fuel and maintenance, 1.4% on repairs, 9.9% on disposal, 21.2% on depreciation and financing, and 3% on miscellany.

TABLE 14.3
Budget Calculation Sheet

I. Capital Investment for Equipment

Harvester(s)	_____	
Shore conveyor(s)	_____	
Transport barge(s)	_____	
Trailer or wheel assemblies	_____	
Truck(s)	_____	**Annual**
Other	_____	**capital cost**
Total capital cost	_____ + Interest	_____
	/(10)	

II. Operation and Maintenance Costs

Salaries	_____
Fringe benefits	_____
Taxes (income and social security)	_____
Fuel	_____
Lubricants	_____
Insurance	_____
Maintenance and repair (parts and labor)	_____
Disposal costs	_____
Supervisory costs	_____
Permits	_____
Public relations	_____
Monitoring and reporting	_____
Other	_____
Total Annual Operation and Maintenance Costs	_____

III. Contingency Costs
(Total Annual Capital + Operation and Maintenance Costs) × 10%   _____

Total annual cost   _____

Often results of harvesting are disappointing because too small or too few pieces of equipment are used. Browns Lake, Wisconsin provides an example of the equipment used in a harvesting program on a medium sized lake (162 ha, average depth 2.3 m) (Korth et al., 1997). Although the area harvested is not given, only about one third of the lake is over 3 m deep so there are probably about 110 harvestable hectares in the lake. The vegetation is mixed; Eurasian watermilfoil, *Chara* sp., *Najas flexilis, N. marina, Nitella* sp., *Nymphaea odorata,* and eight species of pondweeds are the predominant species. Harvesting occurs 5 days a week from the first week in June to the last week in August. Plants that homeowners pile on their shorelines are picked up 7 days a week. This operation uses a 3-m and a 1.8-m cut harvester, a 1.5-m cut skimmer, an 3.3- by 8.5-m transport barge, a 10.7-m shoreline conveyor, a dump truck, a pickup truck, and a knuckle-boom loader.

Proper sizing of equipment is also important for operational and cost efficiencies. Korth et al. (1997) calculated that, in a hypothetical case of a 243 ha lake with 20 ha of aquatic plants to harvest once a month for 3 months, buying a 6.4 metric ton capacity harvester was more economical than buying a 4.6 or a 3.6 metric ton capacity harvester, even though all machines could accomplish the harvesting task. The 6.4-ton harvester minimized staff overtime and increased efficiency through reduction of idle time spent waiting for a full dump truck load of plants. The 6.4-ton harvester saved 7% over the 4.6-ton harvester, 18% over the 3.6-ton harvester, and 80% over the small harvester with a transport barge. It would cost about $28,864 per year or almost $500/ha to harvest the 60 ha with the 6.4-ton harvester in the above scenario.

TABLE 14.4
Rate and Relative Cost Estimates for Various Mechanical Removal Methods

Method	Rate	Reference	Cost[a]
Cutting (3 m cut)	0.45 ha/h	Koegel and Livermore, 1979	20%
Deep Cutting (2.4 m cut)	0.15 ha/h	Unmuth et al., 1998	—
Harvesting (2.4 m cut?)	1.25 ha/day	Anonymous, 1978	100%
Shredding (2.4 m cut)	0.3 ha/h	Canellos, 1981	40%
Suction harvesting	0.35 ha/day/person	Anonymous, 1978	425%
Hydraulic washing	0.8 ha/day	Anonymous, 1978	30%
Weed rolling	129 m²/h	Manufacturer	—
Cultivating[b]	0.7–1.6 ha/day	Anonymous, 1978	7–22%
Rototilling	0.2 ha/day	Anonymous, 1978	140%

[a] Costs are compared on a relative basis to the cost of harvesting at 100%. Estimate for the "once through" cost of harvesting is about $500/ha. From Korth, R. et al. 1997. *Your Aquatic Plant Harvesting Program.* Wisconsin Lakes Partnership, Stevens Point, WI.

[b] Costs varied by the type of cultivating and towing equipment.

FIGURE 14.10 Pile of decomposed aquatic plants 1 year after being harvested and deposited in a wetland adjacent to the Yahara River, Wisconsin.

Costs vary widely. Smith (1979) found a tenfold difference, based on per hectare or per load, harvesting cost between the least expensive and the most expensive of four harvesting operations he surveyed. Table 14.4 provides a "rough" estimate of plant removal rates and the "once through" cost compared to standard harvesting cost. This table is most valuable for displaying the relative expense of each technique.

The obvious way of increasing harvesting efficiencies is to keep machines harvesting. This has many operational aspects. Harvesting in long, straight passes eliminates time spent turning the

TABLE 14.5
Characteristics of Mechanical Management Techniques[a]

Method	Description	Advantages	Disadvantages	Where used Effectively	Plant Response
Hand cutting/pulling	Hand pulling or use of hand tools	Low-technology, affordable, selective	Labor-intensive, cost is labor based	Initial infestations, volunteer or cheap labor	Very effective in localized areas
Cutting	Cut with mechanical device, with or without secondary plant removal	More rapid and less expensive than harvesting	Cut plants can be health and environmental problem, may spread infestations, may be illegal without secondary collection	Heavily infested systems	Non-selective, length of treatment dependent on number of times cut
Harvesting	Mechanical cutting with plant removal	Removes plant biomass, immediate relief in harvested areas	Slower and more expensive than cutting, may spread fragments, may stimulate growth, no roots removed, disposal of harvested plants	Widespread use in areas of chronic plant problems	Non-selective, length of treatment dependent on number of time cut
Shredding	Mechnical cutting and shredding of plant material, may occur on-board	Immediate relief of nuisance, can be used for volume reduction for easier transport, effective for free-floating and emergent plants, can operate in shallow water	Decomposition and spread of vegetative parts if in-lake disposal is used	Heavy mats of floating or emergent plants. Volume reduction for easier transport of submergent species	Non-selective, length of treatment dependent on the timing and number of cuts
Diver-operated suction harvesting	Vacuum lift used to remove entire plants including roots from sediment	Moderately selective, longer term, removes roots to any depth, can be successfully used in areas with obstacles	Slow and expensive, labor intensive	Best suited for small areas of moderate density nuisance plants	Minimal regrowth of Eurasian water-milfoil, not as effective against tuber producing plants
Hydraulic washing	Uses water pressure to wash plants from sediment	Washes sediment off roots and dislodged fragments float and can be collected	Machines mainly prototypes, water jets can't dislodge old root crowns or turf, need to collect floating fragments	Best suited for small areas of moderate density plants	Effective for removing reproductive parts from soft sediment
Weed rolling	Roller with fins stirs bottom sediments to dislodge plants	Good for small areas adjacent to docks, can be moved to different locations, highly automated	Allows sediment and plants to drift away, may clog or stop in heavy plant growth or other obstructions, unknown impact to benthos, may be navigation hazard	Moderate plant infestations close to docks	After initial infestation cleared, running once every week or two maintains area
Cultivating	Tills bottom sediments to remove rooted plants	Removes roots and root crowns, may operate throughout the ice-free season	Depth of operation limited, must contain floating fragments, avoid obstacles; requires more than "one pass" for good control, doesn't work well in heavy plant growth	Most effective under drawdown conditions and moderate plant density	About 90% effective in removing rooted plants but regrowth can be rapid from missed plants

TABLE 14.5 (Continued)
Characteristics of Mechanical Management Techniques[a]

Method	Description	Advantages	Disadvantages	Where used Effectively	Plant Response
Rototilling	Tills bottom sediments to remove rooted plants	Removes roots and root crowns, may operate throughout the ice-free season, will operate in deeper water than other mechnical methods	If operated in the summer, fragments must be collected and tops removed first, requires more than "one pass" for good control, doesn't work well in heavy plant growth, works better in sand and gravel than in silt or clay	Most effective in high-use recreation areas	Most effective at disrupting plants with floating root crowns, not selective, intermediate length of results

[a] After Anonymous, 1978 and Madsen, 200b.

machine or trimming between piers. Close and convenient unloading sites reduce time spent transporting plants to shore. Using a machine on a single lake reduces the time lost transporting equipment from lake to lake. Double cutting shifts allow more machine use during daylight hours and reduce overtime wages. Separate transport barges allow a harvester to keep cutting while the barge carries plants to shore. Two-stage harvesting allows the cutting machine to travel faster. The "pick-up" machine can also transport plants while the cutter keeps cutting. Methods of reducing plant volume and weight on the lake allow machines to harvest more before unloading. Contingency plans for areas to harvest during marginal weather conditions reduce lost time due to weather. Good maintenance is critical to efficient harvesting.

As some "real-life" examples of the above ideas, the Dane County harvesting operation was able to increase its operating machine hours on one harvester from 48.9% of total machine hours to 61.2% by increasing preventive maintenance from 6.9% to 8.5%, and by keeping the machine on one lake thus reducing moving hours from 23.2% to 4.1% (Koegel et al., 1977). On a second machine increasing preventive maintenance by 0.8% and reducing moving time by 6.2% increased operating time by 24% of total machine hours (Koegel et al., 1977).

Finding convenient unloading sites is often an operational challenge. In Dane County there are stretches of the Yahara River that are harvested to maintain water flow. Some areas are more than 3.2 km from an unloading site and the river is shallow and rocky. In these stretches, aquatic plants are dumped in the adjacent wetlands. By the end of the second growing season plant piles that were originally over 2 m high were reduced to almost nothing (Figure 14.10). In the meantime, waterfowl used the plant piles as nesting sites. In a somewhat similar situation, Elser (1966) describes "bottomless pits" designed for water-chestnut disposal in the Chesapeake Bay region. A 3.7 m square frame was constructed, covered over the outside with snow fence, and anchored to the bottom. Harvested plants were dumped into the pit. They dried on the top, rotted on the bottom, and in about a day the mass was reduced to a fraction of the original volume. The pit could be filled every day — almost an infinite capacity. After about 2 weeks of drying and rotting the plants became so tightly matted, the pit could be removed, and a stake put through the center of the mat to anchor it to the bottom. A better idea would have been to remove the mat from the water thereby avoiding nutrient recycling into the water.

To summarize, the ways to minimize costs and maximize harvesting productivity include: (1) maximizing equipment deployment through the season; (2) minimizing crew and equipment mobilization time and costs; (3) preplanning harvesting strategies; (4) finding appropriate financing if needed; (5) implementing a responsive institutional systems; and (6) selecting appropriate equipment (Smith, 1979). Another interesting point made by Smith (1979) is that many harvesting machines don't depreciate and some actually appreciate in value so the 10-year, straight-line depreciation that is commonly used in cost calculations (Table 14.3) may not be realistic. Avoiding depreciation drastically reduces harvesting costs.

14.5 CONCLUDING REMARKS

There is no quantitative definition of an aquatic plant management problem. The assessment is generally made by water body users, whether they are riparian land owners, an informal or formal lake organization (or sportsman's group, conservation society, etc.), or some level of government. Determine if there is a management problem. This may take some consensus building. It should be borne in mind that managing a natural lake system is different from managing a lawn, park, garden, or cornfield. Our society is becoming increasingly conditioned to trimmed lawns, exotic flora in parks, and weed-free cornfields. Lakes vary tremendously in their productivity and many lakes are naturally "messy" looking. Many lakes never were nor will they ever be the pristine looking mountain or Canadian Shield lakes seen in advertisements. Learning to appreciate a lake for what it is, not what you think you can make saves a lot of time, effort, and expense.

Not all lakes have aquatic plant management problems. This is the ideal situation. However, it does not mean aquatic plant management should not be considered. Most aquatic plant nuisances are caused by invasive, exotic species. An aquatic plant management plan based on education and monitoring can pay dividends in the long run for lakes with no existing aquatic plant problems.

If invasive exotics are quickly detected they can often be eradicated and prevented from spreading by hand removal, suction dredging, herbicides, fragment barriers, quarantine areas, boat launch cleaning stations, and other methods. Diligence is needed. One treatment or 1 year of treatments is not likely to eradicate the problem and reinfestation is likely if it occurred once.

If the infestation has already spread to the point that eradication is not likely, then management is needed. Management will be long-term and there are many options that can be used (see chapters on biomanipulation, drawdown and sediment covering, biological controls, aquatic plant community rehabilitation, chemical controls, and sediment removal). Mechanical methods provide a variety of options that can be integrated with other methods. The strengths and weaknesses of the mechanical methods are summarized in Table 14.5.

Harvesting is often compared to chemical control. Certainly there are differences in cost, efficacy, and environmental impacts to consider. An item not often considered is the differences in the level of organization and commitment needed to carry out each type of program. Herbicide treatments can cover large areas, quickly. The major cost is for materials. Unless a manager is responsible for a large number of lakes it is questionable whether it is worthwhile owning the equipment and the contingency items (training, insurance, etc.) needed for chemical applications. Equipment costs are a major item in harvesting. To make harvesting economical the machinery has to be used continually during the harvesting season and for a number of years. This takes a long-term view of and commitment to management, and a strong and stable management organization. On-the-job experience is invaluable.

Harvesting is often considered to be environmentally benign when compared to herbicide treatments. Any management that removes large areas of aquatic plants from a system has environmental impacts. Some impacts are beneficial and some are not. Some impacts may be so long term that they are not recognized or are not measurable in a realistic management timeframe. Harvesting is non-selective in the cut areas and conventional harvesting removes small fish and invertebrates. Conventional harvesting removes nutrients and organic matter from the water-body. Harvesting should be considered in an overall nutrient management program but there is no evidence that harvesting alone will reverse eutrophication.

REFERENCES

Alam, S.K., L.A. Ager, T.M. Rosegger and T.R. Lange. 1996. The effects of mechanical harvesting of floating plant tussock communities on water quality in Lake Istokpoga, Florida. *Lake and Reservoir Manage.* 12(4): 455–461.

Aldridge, D.C. 2000. The impact of dredging and weed cutting on a population of fresh water mussels (Bivalva: Unionidae). *Biol. Cons.* 95: 247–257.

Anonymous. 1978. Aquatic Plant Management Program Vol. IV. A Review of Mechanical Devices Used in the Control of Eurasian Watermilfoil in British Columbia. Brit. Col. Min. Environ.,Victoria.

Anonymous. 1981. A Summary of Biological Research on Eurasian Watermilfoil in British Columbia, Vol. XI. Brit. Col. Min. Environ., Victoria.

Asaeda, T., V.K. Trung and J. Manatunge. 2000. Modeling the effects of macrophyte growth and decomposition on the nutrient budget in shallow lakes. *Aquatic Bot.* 68:217–237.

Barko, J.W. and W.F. James. 1998. Effects of submerged aquatic macrophytes on nutrient dynamics, sedimentation, and resuspension, In: E. Jeppesen, M. Sondergaard, M. Sondergaard and K. Christoffersen (Eds.), *The Structuring Role of Submerged Macrophytes in Lakes*. Springer-Verlag, New York. pp. 197–214.

Bettoli, P.W., M.J. Maceina, R.L. Noble and R.K. Betsill. 1993. Response of a reservoir fish community to aquatic vegetation removal. *North Am. J. Fish. Manage.* 13: 110–124.

Beule, J.D. 1979. Control and Management of Cattails in Southeastern Wisconsin Wetlands. Tech. Bull. 112. Wis. Dept. Nat. Res., Madison, WI.

Booms, T.L. 1999. Vertebrates removed by mechanical harvesting in Lake Keesus, Wisconsin. *J. Aquatic Plant Manage.* 37: 34–36.

Boylen, C.W., L.W. Eichler and J.W. Sutherland. 1996. Physical control of Eurasian watermilfoil in an oligotrophic lake. *Hydrobiologia* 340: 213–218.

Brakke, D.F. 1974. Weed harvest effects on algal nutrients and primary productivity in a culturally enriched lake. M.S. Thesis, Univ. North Dakota, Grand Forks.

Buchan, L.A. and D.K. Padilla. 2000. Predicting the likelihood of Eurasian watermilfoil presence in lakes, a macrophyte monitoring tool. *Ecol. Appl.* 10: 1442–1455.

Burton, T.M., D.L. King and J.L. Ervin. 1979. Aquatic plant harvesting as a lake restoration technique. In: Anonymous (Ed.), *Lake Restoration, Proceedings of a National Conference*. USEPA 440/5-79-001. pp. 177–185.

Canellos, G. 1981. Aquatic Plants and Mechanical Methods of Their Control. Mitre Corp., McClean, VI.

Carpenter, S.R. 1983. Submersed macrophyte community structure and internal loading: relationship to lake ecosystem productivity and succession, In: J. Taggart and L. Moore (Eds.), *Lake Restoration, Protection and Management*, Proc. Second Annu. Conf. NALMS. Vancouver, BC. pp. 105–111.

Carpenter, S.R. and M.S. Adams. 1977. Environmental Impacts of Mechanical Harvesting on Submersed Vascular Plants. Rep. 77. Inst. Environ. Stud., Univ. Wisconsin, Madison.

Carpenter, S.R. and M.S. Adams. 1978. Macrophyte control by harvesting and herbicides: Implications for phosphorus cycling in Lake Wingra, Wisconsin. *J. Aquatic Plant Manage.* 16: 20–23.

Carpenter, S.R. and A. Gasith. 1978. Mechanical cutting of submerged macrophytes: Immediate effects of littoral water chemistry and metabolism. *Water Res.* 12: 55–57.

Clayton, J.S. 1996. Aquatic weeds and their control in New Zealand Lakes. *Lake and Reservoir Manage.* 12(4): 477–486.

Conyers, D.L. and G.D. Cooke. 1982. A comparison of the costs of harvesting and herbicides and their effectiveness in nutrient removal and control of macrophyte biomass. In: J. Taggart and L. Moore (Eds.), *Lake Restoration, Protection and Management*, Proc. Second Annu. Conf. NALMS. Vancouver, BC. pp. 317–321

Cook, C.D.K. 1985. Range extensions of aquatic vascular plant species. *J. Aquatic Plant Manage.* 23: 1–6.

Cooke, G.D., E.B. Welch, S.A. Peterson and P.R. Newroth. 1986. *Lake and Reservoir Restoration*, 1st ed. Butterworths, Boston, MA.

Cooke, G.D., A.B. Martin and R.E. Carlson. 1990. The effects of harvesting on macrophyte regrowth and water quality in LaDue Reservoir, Ohio. *J. Iowa Acad. Sci.* 97(4): 27–32.

Cooke, G.D., E.B. Welch, S.A. Peterson and P.R. Newroth. 1993. *Restoration and Management of Lakes and Reservoirs*, 2nd ed. Lewis Publishers and CRC Press, Boca Raton, FL.

Crowell, W., N. Troelstrup, L. Queen and J. Perry. 1994. Effects of harvesting on plant communities dominated by Eurasian watermilfoil in Lake Minnetonka, MN. *J. Aquatic Plant Manage.* 32: 56–60.

Dauffenbach, G. 1998. Part I: Past, present, and future of mechanical harvesting. *LakeLine* 18(1): p. 16.

Deutsch, A. 1974. *Some Equipment for Mechanical Control of Aquatic Weeds.* International Plant Protection Center, Oregon State University, Corvallis, OR.

Elser, H.J. 1966. Control of water chestnut by machine in Maryland, 1964–1965. *Proc. Northeastern Weed Cont. Conf.* 20: 682–687.

Engel, S. 1985. Aquatic Community Interactions of Submerged Macrophytes. Tech. Bull 156. Wisconsin Dept. Nat. Res., Madison, WI.

Engel, S. 1987. The impact of submerged macrophytes on largemouth bass and bluegills. *Lake and Reservoir Manage.* 3: 227–234.

Engel, S. 1990a. Ecological impacts of harvesting macrophytes in Halverson Lake, Wisconsin. *J. Aquatic Plant Manage.* 28: 41–45.

Engel, S. 1990b. Ecosystem Responses to Growth and Control of Submerged Macrophytes: A Literature Review. Tech. Bull. 170. Wis. Dept. Nat. Res., Madison.

Fisher, S.G. and S.R. Carpenter. 1976. Ecosystem and macrophyte primary production of Fort River, Massachusetts. *Hydrobiologia* 47: 175–187.

Garner, P., J.A.B. Bass and G.D. Collett. 1996. The effect of weed cutting upon the biota of a large regulated river. *Aquatic Cons. Mar. Fresh Water Ecosyst.* 6: 21–29.

Haller, W.T., J.V. Shireman and D.F. DuRant. 1980. Fish harvest resulting from mechanical control of hydrilla. *Trans. Am. Fish. Soc.* 109: 517–520.

Helsel, D.R. 1998. Part II: Surface water management through aquatic plant harvesting. *LakeLine* 18(1): 26.

Helsel, D.R., S.A. Nichols and R.W. Wakeman. 1999. Impacts of aquatic plant management methodologies on Eurasian watermilfoil populations in Southeastern Wisconsin. *Lake and Reservoir Manage.* 15: 159–167.

Howard-Williams, C., A.M. Schwarz and V. Reid. 1996. Patterns of aquatic weed regrowth following mechanical harvesting in New Zealand hydro-lakes. *Hydrobiologia* 340: 229–234.

Hutchinson, G.E. 1975. *A Treatise on Limnology-Limnological Botany.* John Wiley, New York.

Jacoby, J.M., E.B. Welch and I. Wertz. 2001. Alternate stable states in a shallow lake dominated by *Egeria densa. Verh. Int. Verein. Limnol.* 27: 3805–3810.

James, W.F. and J.W. Barko. 1994. Macrophyte influences on sediment resuspension and export in a shallow impoundment. *Lake and Reservoir Manage.* 10: 95–102.

James, W.F., J.W. Barko and H.L. Eakin. 2000. Macrophyte Management via Mechanical Shredding: Effects on Water Quality in Lake Champlain (Vermont-New York). Aquatic Plant Cont. Res. Prog., Tech. Notes, ERDC TN-APCRP-MI-05. U.S. Army Corps of Engineers, Vicksburg, MS.

Johnson, R.E. and M.R. Bagwell. 1979. Effects of mechanical cutting on submersed vegetation in a Louisana lake. *J. Aquatic Plant Manage.* 17: 54–57.

Johnstone, I.M., B.T. Coffey and C. Howard-Williams. 1985. The role of recreational boat traffic in the interlake dispersal of macrophytes: A New Zealand case study. *J. Environ. Manage.* 20: 263–279.

Kimbel, J.C. 1982. Factors influencing potential intralake colonization by *Myriophyllum spicatum* L. *Aquatic Bot.* 14: 295–307.

Kimbel, J.C. and S.R. Carpenter. 1981. Effects of mechanical harvesting on *Myriophyllum spicatum* L. regrowth and carbohydrate allocation to roots and shoots. *Aquatic Bot.* 11 (2): 121–127.

Koegel, R.G. and D.F. Livermore. 1979. Reducing capital investment in aquatic plant harvesting systems. pp. 329–338. In: J. Breck, R. Prentki and O. Loucks (Eds.), *Aquatic Plants, Lake Management, and Ecosystem Consequences of Lake Harvesting.* Inst. Environ. Stud., University Wisconsin, Madison. pp. 329–338.

Koegel, R.G., D.F. Livermore and H.D. Bruhn. 1977. Cost and productivity in harvesting of aquatic plants. *J. Aquatic Plant Manage.* 15: 12–17.

Korth, R., S. Engel and D.R. Helsel. 1997. *Your Aquatic Plant Harvesting Program.* Wisconsin Lakes Partnership, Stevens Point.

Langeland, K.A. and D.L. Sutton. 1980. Regrowth of hydrilla from axillary buds. *J. Aquatic Plant Manage.* 18: 27–29.

Larsen, D.P., J. Van Sickle, K.W. Malueg and P.D. Smith. 1979. The effect of wastewater phosphorus removal on Shagawa Lake, Minnesota: phosphorus supplies, lake phosphorus and chlorophyll *a. Water Res.* 13: 1259–1272.

Les, D.H. and L.J. Mehrhoff. 1999. Introduction of nonindigenous aquatic vascular plants in southern New England: a historical perspective. *Biol. Invas.* 1: 281–300.

Linde, A.F., T. Janisch and D. Smith. 1976. Cattail — The Significance of Its Growth, Phenology and Carbohydrate Storage to its Control and Management. Tech. Bull. 94. Wisconsin Dept. Nat. Res., Madison.

Livermore, D.F. and R.G. Koegel. 1979. Mechanical harvesting of aquatic plants: an assessment of the state of the art. In: J. Breck, R. Prentki and O. Loucks (Eds.), Aquatic Plants, Lake Management, and Ecosystem Consequences of Lake Harvesting. Inst. Environ. Stud., University Wisconsin, Madison, WI. pp. 307–328.

Livermore, D.F. and W.E. Wunderlich. 1969. Mechanical removal of organic production from waterways. In: Anonymous (Ed.), *Eutrophication: Causes, Consequences, Correctives.* Natl. Acad. Sci., Washington, DC. pp. 494–519.

Madsen, J.D. 1993. Waterchestnut seed production and management in Watervliet Reservoir, New York. *J. Aquatic Plant Manage.* 31: 271–272.

Madsen, J.D. 1998. Predicting invasion success of Eurasian watermilfoil. *J. Aquatic Plant Manage.* 36: 28–32.

Madsen, J.D. 2000a. A Quantitative Approach to Predict Potential Nonindigenous Aquatic Plant Species Problems. Aquatic Res. Brief. U.S. Army Corps of Engineers, Vicksburg, MS.

Madsen, J.D. 2000b. Advantages and Disadvantages of Aquatic Plant Management Techniques. Rep. ERDC/EL MP-00-01. U.S. Army Corps of Engineers, Vicksburg, MS.

Madsen, J.D., M.S. Adams and P. Ruffier. 1988. Harvesting as a control for sago pondweed (*Potamogeton pectinatus* L.) in Badfish Creek, Wisconsin: frequency, efficiency and its impact on the stream community oxygen metabolism. *J. Aquatic Plant Manage.* 26: 20–25.

McComas, S. 1989. Small-scale macrophyte control: what we can learn from the farmer. *LakeLine* 9 (2): p. 2.

McComas, S. 1993. *Lake Smarts.* Terrene Institute, Washington, DC.

McComas, S. and J. Stuckert. 2000. Pre-emptive cutting as a control technique for nuisance growth of curly-leaf pondweed, *Potamogeton crispus. Verh. Int. Verein. Limnol.* 27: 2048–2051.

MDNR. 1998. Harmful Exotic Species of Aquatic Plants and Wild Animals in Minnesota: Annual Report for 1997. Minn. Dept. Nat. Res., St. Paul.

Methe, B.A., R.J. Soracco, J.D. Madsen and C.W. Boylen. 1993. Seed production and growth of waterchestnut as influenced by cutting. *J. Aquatic Plant Manage.* 31: 154–157.

Mikol, G.F. 1984. Effects of mechanical control of aquatic vegetation on biomass, regrowth rates, and juvenile fish populations at Saratoga Lake, New York. In: Anonymous (Ed.), *Lake and Reservoir Management,* Proc. Third Annu. Conf. NALMS. pp. 456–462.

Monahan, C. and J.M. Caffrey. 1996. The effect of weed control practices on macroinvertebrate communities in Irish canals. *Hydrobiologia* 304: 205–211.

Moss, B., J. Madgwick and G.L. Phillips. 1996. *A Guide to the Restoration of Nutrient-enriched Shallow Lakes.* Broads Authority, Norwich, Norfolk, U.K.

Neel, J.K., S.A. Peterson and W.L. Smith. 1973. *Weed Harvest and Lake Nutrient Dynamics.* Ecol. Res. Series, USEPA-660/3-73-001.

Newroth, P.R. 1980. Case study of aquatic plant management for lake restoration and preservation in British Columbia,. In: Anonymous (Ed.), *Proceedings of an International Symposium on Restoration of Lakes and Inland Waters.* USEPA-440/5-81-010. pp. 146–152.

Nichols, S.A. 1974. Mechanical and Habitat Manipulation for Aquatic Plant Management. Tech. Bull. 77. Wisconsin Dept. Nat. Res., Madison, WI.

Nichols, S.A. and L. Buchan. 1997. Use of native macrophytes as indicators of suitable Eurasian watermilfoil habitat in Wisconsin lakes. *J. Aquatic Plant Manage.* 35: 21–24.

Nichols, S.A. and G. Cottam. 1972. Harvesting as a control for aquatic plants. *Water Res. Bull.* 8: 1205–1210.

Nichols, S.A. and R.C. Lathrop. 1994. Impact of harvesting on aquatic plant communities in Lake Wingra, Wisconsin. *J. Aquatic Plant Manage.* 32: 33–36.

Nichols, S.A. and S. Mori. 1971. The littoral macrophyte vegetation of Lake Wingra. *Trans. Wis. Acad. Sci. Arts Lett.* 59: 107–119.

Nicholson, S.A. 1981a. Effects of uprooting on Eurasian watermilfoil. *J. Aquatic Plant Manage.* 19: 57–58.

Nicholson, S.A.1981b. Changes in submersed macrophytes in Chautauqua Lake, 1937–1975. *Freshwater Biol.* 11: 523–530.

Olson, M.H., S.R. Carpenter, P. Cunningham, S. Gafny, B.R. Herwig, N.P. Nibbelink, T. Pellet, C. Storlie, A.S. Trebitz and K.A. Wilson. 1998. Managing macrophytes to improve fish growth: A multi-lake experiment. *Fish. Manage.* 23(2): 6–12.

Painter, D.S. and J.I. Waltho. 1985. Short-term impact of harvesting of Eurasian watermilfoil. In: L.W.J. Anderson (Ed.), *Proc. First Int. Symp. on Watermilfoil (Myriophyllum spicatum) and Related Haloragaceae Species*. Aquatic Plant Manage. Soc., Vancouver, BC. pp. 187–201.

Perkins, M.A. and M.D. Sytsma. 1987. Harvesting and carbohydrate accumulation in Eurasian watermilfoil. *J. Aquatic Plant Manage.* 25: 57–62.

Peterson, S.A. 1971. Nutrient dynamics, nutrient budgets, and weed harvest as related to the limnology of an artificially enriched lake. Ph.D. Thesis, University North Dakota, Grand Forks.

Peterson, S.A., W.L. Smith and K. Maleug. 1974. Full scale harvest of aquatic plants: nutrient removal from a eutrophic lake. *J. Wat. Pollut. Cont. Fed.* 46: 697–707.

Peverly, J.H., G.M. Miller, W.H. Brown and R.L. Johnson. 1974. Aquatic Weed Management in the Finger Lakes. Tech. Rept. 90. Water Res. Marine Sci. Cent., Cornell University, Ithaca, NY.

Sabol, B.M. 1982. Improved Aquatic Plant Material Disposal Techniques for Mechanical Control Operations. Aquatic Plant Cont. Res. Prog., Inf. Exchange Bull. A-82-1. U.S. Army Corps of Engineers, Vicksburg, MS. pp. 1–4.

Sabol, B.M. 1987. Environmental effect of aquatic disposal of chopped hydrilla. *J. Aquatic Plant Manage.* 25: 19–23.

Sale, P.J.M. and R.G. Wetzel. 1983. Growth and metabolism of *Typha* species in relation to cutting treatments. *Aquatic Bot.* 15: 321–324.

Scheffer, M. 1998. *Ecology of Shallow Lakes*. Chapman Hall, London.

Scheffer, M., S.H. Hosper, M.-L. Meijer, B. Moss and E. Jeppesen. 1993. Alternative equilibria in shallow lakes. *Trends Ecol. Evol.* 8: 275–279.

Schmitz, D. 1990. The invasion of exotic aquatic and wetland plants into Florida. In: *Proceedings, National Conference on Enhancing the States' Lake and Wetland Management Programs*. USEPA and NALMS, Chicago, IL. pp. 87–92.

Serafy, J.E., R.M. Harrell and L.M. Hurley. 1994. Mechanical removal of *Hydrilla* in the Potomac River, Maryland: local impacts on vegetation and associated fishes. *J. Fresh Water Ecol.* 9: 135–143.

Smagula, A.P., L. Herman, M. Robinson and A. Bove. 2002. Vigilant volunteers fight invasives. *Vol. Monitor* 14(2): 26.

Smith, W.L. 1972. Plankton, weed growth, primary productivity, and their relationship to weed harvest in an artifically enriched lake. Ph.D. Thesis, University North Dakota, Grand Forks.

Smith, G.N. 1979. Recent case studies of macrophyte harvesting costs: options by which to lower costs. In: J. Breck, R. Prentki and O. Loucks (Eds.), *Aquatic Plants, Lake Management, and Ecosystem Consequences of Lake Harvesting*. Inst. Environ. Stud., University Wisconsin, Madison. pp. 345–356.

Smith, G.N. 1998. Water hyacinth harvesting at Lake Cidra, Puerto Rico-A big operation. *LakeLine* 18(1): p. 20.

Smith, C.S. and J.W. Barko. 1992. Submersed Macrophyte Invasions and Declines. U.S. Army Corps of Engineers, Aquatic Plant Cont. Res. Prog. Rep., Vol. A-92-1. U.S. Army Corps of Engineers, Vicksburg, MS.

Trebitz, A. 1995. *Predicting bluegill and largemouth bass response to harvest of aquatic vegetation*. Ph.D. Thesis, University Wisconsin, Madison.

Unmuth, J.M.L. and M.J. Hanson. 1999. Effects of mechanical harvesting of Eurasian watermilfoil on largemouth bass and bluegill populations in Fish Lake, Wisconsin. *North Am. J. Fish. Manage.* 19: 1089–1098.

Unmuth, J.M.L., D.J. Sloey and R.A. Lillie. 1998. An evaluation of close-cut mechanical harvesting of Eurasian watermilfoil. *J. Aquatic Plant Manage.* 36: 93–100.

USOTA. 1993. Harmful Non-Indigenous Species in the United States. U.S. Congress, Office Tech. Assess., Washington, DC.

van Nes, E.H., M. Scheffer, M.S. van den Berg and H. Coops. 2002. Aquatic macrophytes: restore, eradicate or is there compromise? *Aquatic Bot.* 72: 387–403.

Wade, P.M. 1990. Physical control of aquatic weeds. In: A. Pieterse and K. Murphy (Eds.), *Aquatic Weeds, The Ecology and Management of Nuisance Aquatic Vegetation*. Oxford University Press, Oxford, UK. pp. 93–135.

Welch, E.B., E.B. Kvam and R.F. Chase. 1994. The independence of macrophyte harvesting and lake phosphorus. *Verh. Int. Verein. Limnol.* 25: 2301–2304.

Wile, I. 1974. Lake restoration through mechanical harvesting of aquatic vegetation. *Verh. Int. Verein. Limnol.* 19: 660–671.

Wile, I. 1978. Environmental effects of mechanical harvesting. *J. Aquatic Plant Manage.* 16: 14–20.

Wile, I., G. Hitchin and G. Beggs. 1979. Impact of mechanical harvesting on Chemung Lake. In: J. Breck, R. Prentki and O. Loucks (Eds.), *Aquatic Plants, Lake Management, and Ecosystem Consequences of Lake Harvesting.* Inst. Environ. Stud., University Wisconsin, Madison. pp. 145–159.

Wunderlich, W.E. 1938. Mechanical hyacinth destruction. *Military Eng.* 30: 5–10.

Zimba, P.V., M.S. Hopson and D.E. Colle. 1993. Elemental composition of five submersed aquatic plants collected from Lake Okeechobee, Florida. *J. Aquatic Plant Manage.* 31: 137–140.

15 Sediment Covers and Surface Shading for Macrophyte Control

15.1 INTRODUCTION

Rooted aquatic vegetation control with sand, gravel, or clay has been mostly unsuccessful because root systems remain to produce shoots that penetrate earthen covers, and because many aquatic plants reestablish through fragments carried to the treated site from other lake areas, or by seeds, turions, and rhizomes that survive the treatment. Another option is synthetic sheeting and screening materials. Their effectiveness is higher because shoots may be unable to penetrate them, but they are expensive, application is labor intensive, effectiveness is correlated with application techniques and types of material, and most materials manufactured for this purpose are not widely available. Some are no longer manufactured.

Engel (1982) listed these advantages of sediment covers:

1. Their use is confined to specific lake areas.
2. Screens are usually out of sight and thus create no disturbance on shore.
3. They can be installed in places where harvesters or sprayer boats cannot gain access.
4. No toxic substances are released.
5. They usually require no permit or license.
6. They are easy to install over small areas.
7. They can be removed.

These are their disadvantages (Engel, 1982):

1. They fail to correct the cause of the problem.
2. They are expensive.
3. They are difficult to apply over large areas or on sites with obstructions.
4. They may slip on steep grades or float to the surface after trapping gases beneath them.
5. They can be difficult to remove or relocate.
6. They may rip during application.
7. Some materials are degraded by sunlight.

Reviews include Armour et al. (1979), Cooke (1980), Nichols and Shaw (1983), Perkins (1984), and Newroth and Truelson (1984).

15.2 COMPARISON OF SYNTHETIC SEDIMENT COVERS

15.2.1 POLYETHYLENE

One of the earliest uses of sediment covers was the treatment of 10 ha of Marion Millpond, Wisconsin with 0.1 mm (4 ml) impermeable black polyethylene sheeting (Born et al., 1973; Peterson et al., 1974; Nichols, 1974; Engel, 1982; and Engel and Nichols, 1984). In 1969 to 1970 the pond was drained and the basin cleared of stumps and debris. Sheeting was laid over the sediments and covered with 7 to 15 cm of sand and gravel. Application in soft mud was tedious, so enough water was added to just cover the treatment area and *allow*ed to freeze. Sheeting and the sand and gravel cover were placed on the ice and water withdrawn, cracking the ice. The materials fell to the pond's bottom. The pond was refilled in 1971, and control of macrophytes was achieved in 1971–1972. *Chara* and filamentous algae covered the treated area in 1973 (Nichols, 1974). The screened areas were covered with macrophytes in 1978, although biomass was about half that of untreated areas. The polyethylene sediment cover exerted little control of macrophyte biomass after 1973 because plants like *Najas*, *Myriophyllum*, and some species of *Potamogeton* colonized the sand and gravel layer over the sheeting (Engel, 1982; Engel and Nichols, 1984). The cost was $371 per ha (2002 prices) for materials (Born et al., 1973). Labor was donated by lake users.

An area of 0.43 ha in Skaha Lake, British Columbia was treated, but most of the polyethylene sheeting washed up during storms. Treatments at some sites were effective but plants grew through holes punched for gas escape, or grew on accumulated sediment (Armour et al., 1979).

The Skaha Lake application cost $38,960 per ha (2002 prices), including labor and materials. Negative features of polyethylene sheeting include:

1. It is difficult to apply over an irregular bottom or over a high density of weeds.
2. Gas forms under the sheets, even with 1.2 cm holes.
3. It is not feasible to move and relocate the sheets.
4. It slides down steep inclines.
5. It is deteriorated by sunlight (about 1 year in direct sun).
6. Its buoyancy makes it difficult to handle.

Perhaps one of the greatest negative features of polyethylene sheeting, as well as some of the other covering materials, is that all plants can be eliminated, but regrowth after sheeting removal includes the target exotic plant. For example, polyethylene sheeting, in place for 4–6 weeks in a Wisconsin reservoir, eliminated all species. In the next year, native plants regrew on 40% of the treated area while milfoil (*Myriophyllum spicatum*) appeared on 60%. In contrast, areas treated with 2,4-D (Chapter 16) had native plant recovery in 10–12 weeks over 95% of the area, while milfoil covered only 5% (Helsel et al., 1996).

15.2.2 POLYPROPYLENE

Polypropylene is a black, woven, semi-permeable sheeting often used as a soil stabilizer or "geo-technical" material. It has a specific gravity less than one, thus requiring anchoring to prevent bulges or floating. It is permeable to gases and does not need slits or holes. It was effective in controlling *M. spicatum*, but plant fragments grew on the sediments accumulating over the screen, and also penetrated the screen (Armour et al., 1979). Polypropylene anchored with concrete blocks, did not allow root penetration by *M. spicatum* over three summers, and a plant-free water column was evident, even though plant fragments did appear on the sediments that accumulated on the screen (Lewis et al., 1983). Anchored polypropylene (cement blocks) was completely effective in preventing the growth of *Najas flexilis*, *Potamogeton gramineus*, *P. crispus*, *P. foliosus*, and *P. pusillus* for 1 year in an area that previously had been drawn down and exposed to freezing. Small growths of *N. flexilis* and filamentous algae were evident on the screens (Cooke and Gorman, 1980). Polypropylene may be difficult to remove and clean, so plants can grow on sediment deposits. There can be gas accumulation and "ballooning" beneath polypropylene screens (Engel, 1984). Costs of materials (2002 prices) range from US$14,715 per ha to $65,900 per ha. Commercial availability for use in lakes and ponds is uncertain.

15.2.3 AQUASCREEN

Aquascreen (Menardi-Criswell, Augusta, GA 30913) is a fiberglass screen coated with polyvinyl chloride. It is flexible and dense (specific gravity = 2.54), with a mesh size of 62 apertures/cm^2 (400/in.^{2}). The standard roll size is 7 × 100 ft.

Aquascreen can be effective in eliminating aquatic plants, at least for the application period. Sediment accumulation on the screen eventually allows new plants to root. This is a problem typical of all synthetic screening materials, and users are advised to remove and clean the screens annually. Another problem with Aquascreen, and with similar screens (including window screen), is the development of a community of attached organisms, thus eliminating or severely reducing gas permeability (Pullman, 1990). This promotes gas accumulation and the screen may lift off the sediments.

The relationship between Aquascreen coverage time and effective control of *M. spicatum* was tested in 9 × 24 m plots in shallow (0.5 to 2.0 m) and deep (2 to 3 m) areas of Union Bay, Lake Washington (Perkins et al., 1980; Boston and Perkins, 1982). Panels were removed at 1-month, 2-month, and 3-month intervals. One month of coverage produced decreases of 25% and 35% for shallow and deep plots, respectively. One month after panel removal, plant regrowth was small. Panels in place for 2 months produced decreases of 78% in shallow plots and 56% in deep, with minimal regrowth. Plant biomass and regrowth after 3 months of coverage was small. The screen was most effective where there was good contact with the lake bottom. Plant death in test aquaria was slow enough to prevent DO loss and P accumulation under the typical field conditions (Boston and Perkins, 1982).

Aquascreen's effectiveness, as well as its impact on benthic macroinvertebrates, was investigated in Cox Hollow Lake, Wisconsin. The screen was easily applied, removed, cleaned, and stored. Macrophyte growth was prevented regardless of when Aquascreen was installed during the summer, and few plants could be found under firmly anchored screen, although there was considerable growth under loosely applied screen. Eichler et al. (1995) also found that plants such as *M. spicatum* grew through the screen, or grew under it. There was little control in the second season after application unless the screens were removed, cleaned, and repositioned. This is easily accomplished with Aquascreen (Engel, 1982, 1984).

Macroinvertebrates were eliminated by Aquascreen panels in Cox Hollow Lake, apparently because of poor circulation and low DO during the 1-year application period (Engel, 1982, 1984). Other screening materials also reduce or eliminate macroinvertebrates beneath them (e.g., Bartodziej, 1992; Ussery et al., 1997).

15.2.4 BURLAP

Burlap (340 g/m^2) was applied to two sites in Lake Rockwell Reservoir, Ohio. Burlap at one site was lightly pretreated with Netset (Nichols Net and Twine Co., East St. Louis, Illinois), a sealant and preservative. Despite burlap's porosity, "ballooning" occurred at the site with unconsolidated organic muck sediments due to high benthic metabolism and to difficulty in securing the material in the highly fluid mud. At both locations, plant growth was controlled over the growing season, but treated and untreated burlap rotted during the 3 months of placement (Jones and Cooke, 1984). Untreated burlap applications in British Columbia were cost effective, and plant growth was controlled for 2 to 3 years, after which rotting and sediment accumulation over the burlap curtailed effectiveness (Newroth and Truelson, 1984). The difference with regard to rotting in these studies may be the highly organic, biologically active, warm sediments of the Ohio Lake, where decomposition is faster. Rotting of a bottom barrier could be an advantage. Burlap's cost is lower than most other materials ($7,900 per ha, 2002 prices), plus installation.

Benthic barriers are effective for macrophyte control where they are firmly positioned directly on the sediments at depths where they cannot be dislodged by waves or boat propellers. "Ballooning" of barrier material is a major problem (Gunnison and Barko, 1992), but this can be minimized by placement prior to biomass development. Assuming adequate gas escape, plant control for several years could be possible before new plants root on the accumulated material on top of the barrier. Because of cost, most uses will be for small areas.

15.3 APPLICATION PROCEDURES FOR SEDIMENT COVERS

Application technique is important. Barriers should be applied close to the sediments, without "ballooning" or pockets after installation. In muck-type unconsolidated sediments, this ideal cannot be met because stakes may never be held by the soil, and bricks or cement blocks will anchor only areas where they have been placed. Plant infestations also may prevent application close to the sediment. There should be no gaps between barrier strips, since plants will grow there, and the option of barrier removal and repositioning should be kept in mind.

The first step is to survey the lake bottom for obstructions and to test sediment ability to hold stakes. SCUBA equipment is essential. In fluid, unconsolidated sediment, long stakes will be required, and these should be tested for their holding ability prior to application. If the sediments are too flocculent, bricks or cement blocks can be used, or link chain can be sewn into the edges of the fabric. Steel stakes usually cannot be used in gravel, hard clay, or rock bottoms, because they cannot be pushed deeply enough by the diver to hold the screen.

Stakes can be made from 6- to 7-mm diameter steel reinforcing bar. Bend the bar at one end into an L-shaped "handle." Stake length varies with sediment softness. Sharpen the long end and drive it through a doubled layer of screen until the "handle" end is flush with the sediments.

Screens are efficiently applied from a reel in the stern of a rowboat. The material is unrolled by two applicators, one on each side, and staked on opposite sides every 1 to 2 m. If the screening is applied directly over vegetation, stakes should be placed at 1 m intervals to prevent lifting. In deeper water, where SCUBA is used, another helper in addition to the rower will be needed to hand stakes to the divers and to assist in diving emergencies. Divers will disturb sediments and visibility will be low. Apply barriers perpendicular to the shoreline.

Application can be improved by returning to the site to flatten bulges in 15 to 20 days, when plants beneath the screen have started to decompose. This will increase application costs.

The ideal application time is prior to plant growth. One technique is to place them on top of the ice. With proper weighting, the covering material will sink at ice-out, although it is possible that "rafts" of ice will displace the screens. Placement on ice is much more effective if done with water level drawdown (Chapter 13). Screens in place for about 2 months can be removed and used elsewhere in the lake (Perkins et al., 1980; Engel, 1982), meaning that sites covered in May and June can be uncovered and the screen moved for July and August coverage in another location. Application can be facilitated by lake-level drawdown followed by application of screen to frozen lake sediments, or by harvesting and screen placement. Complete coverage of a dense infestation of a small pond could produce a sharp decline in DO if the screening process is done over a short period.

Sand "blankets" in shallow beach areas create a better bottom surface for wading, and can inhibit some macrophyte growth.

15.4 SHADING OF MACROPHYTES WITH SURFACE COVERS

Reduction of macrophyte biomass through shading has received little attention because a surface cover denies use of the treated area, and because covers can be easily dislodged.

Black polyethylene sheeting was used as a surface cover to control plants in a pond (Mayhew and Runkel, 1962). Polyethylene sheets were floated (specific gravity = 0.92) over 186-m^2 plots, and the corners were anchored to prevent shifting. Eight similar plots, populated by different dominant species, were studied. All species of *Potamogeton* were controlled for the entire summer if a cover period of 15 to 21 days occurred before the plants matured (May in north temperate latitudes of the U.S.). *Ceratophyllum demersum* was controlled by continuous cover of 18 to 28 days. Where plants were controlled, filamentous algae invaded and revegetated the plots. The covers did not control *Chara vulgaris*, *Sagittaria latifolia*, and emergent species.

This procedure needs further evaluation. For example, swimming areas could be covered in early May and the covers allowed to remain in place for 25 to 35 days. This would probably not deny the area to swimming, since water temperatures may not reach the comfortable range until June in northern latitudes. If the sheeting was removed carefully, it could be reused in subsequent seasons.

Shade created by riparian vegetation, especially trees, can reduce littoral submersed plant growth. Property owners should be discouraged from cutting trees along the shoreline.

Dyes to suppress plant growth have been suggested (Eicher, 1947). Aquashade (Aquashade, Inc., Eldrod, New York) is designed specifically to shade plants in hydrologically closed systems such as ponds. The active ingredients are Acid Blue 9 and Acid Yellow, and these dyes control submersed plants by filtering wave lengths of light critical to photosynthesis (Madsen et al., 1999). Aquashade is added as a concentrate, and winds disperse it throughout the pond, turning the pond blue. The manufacturer claims that the material is effective against *Elodea*, *Potamogeton*, *Najas*, *Myriophyllum*, *Hydrilla*, *Chara*, and various filamentous algae, without toxicity to aquatic life, but effectiveness would likely be poor in water depths less than 1 m. Swimming is permitted immediately after application, but the material cannot be used in a potable water source. Aquashade does not reduce transparency to the point that safe swimming standards are violated (Madsen et al., 1999). There is insufficient published information at this time to evaluate commercial dyes. The mode of action is light limitation and not direct toxicity to the plants (Spencer, 1984; Manker and Martin, 1984).

REFERENCES

Armour, G.D., D.W. Brown and K.T. Marsden. 1979. *Studies on Aquatic Macrophytes. Part XV. An Evaluation of Bottom Barriers for Control of Eurasian Watermilfoil in British Columbia.* Water Investigations Branch, Vancouver.

Bartodziej, W. 1992. Effects of a weed barrier on benthic macroinvertebrates. *Aquatics* 14(1): 14–16.

Born, S.M., T.L. Wirth, E.M. Brick and J.P. Peterson. 1973. *Restoring the Recreational Potential of Small Impoundments. The Marion Millpond Experience.* Tech. Bull. No. 71. Wisconsin Department of Natural Resources, Madison.

Boston, H.L. and M.A. Perkins. 1982. Water column impacts of macrophyte decomposition beneath fiberglass screens. *Aquatic Bot.* 14: 15–27.

Cooke, G.D. 1980. Covering bottom sediments as a lake restoration technique. *Water Res. Bull.* 16: 921–926.

Cooke, G.D. and M.E. Gorman. 1980. Effectiveness of DuPont Typar sheeting in controlling macrophyte regrowth after overwinter drawdown. *Water Res. Bull.* 16: 353–355.

Eicher, G. 1947. Aniline dye in aquatic weed control. *J. Wildlife Manage.* 11: 193–197.

Eichler, L.W., R.T. Bombard, J.W. Sutherland and C.W. Boylen. 1995. Recolonization of the littoral zone by macrophytes following the removal of benthic barrier material. *J. Aquatic Plant Manage.* 33: 51–54.

Engel, S. 1982. Evaluating Sediment Blankets and a Screen for Macrophyte Control in Lakes. Office of Inland Lake Renewal, Wisconsin Dept. Nat. Res., Madison, WI.

Engel, S. 1984. Evaluating stationary blankets and removable screens for macrophyte control in lakes. *J. Aquatic Plant Manage.* 22: 43–48.

Engel, S. and S.A. Nichols. 1984. Lake sediment alteration for macrophyte control. *J. Aquatic Plant Manage.* 22: 38–41.

Gunnison, D. and J.W. Barko. 1992. Factors influencing gas evolution beneath a benthic barrier. *J. Aquatic Plant Manage.* 30: 23–28.

Helsel, D.R., D.T. Gerber and S. Engel. 1996. Comparing spring treatments of 2,4-D with bottom fabrics to control a new infestation of Eurasian Watermilfoil. *J. Aquatic Plant Manage.* 34: 68–71.

Jones, G.B. and G.D. Cooke. 1984. Control of nuisance aquatic plants with burlap screen. *Ohio J. Sci.* 84: 248–251.

Lewis, D.H., I. Wile and D.S. Painter. 1983. Evaluation of Terratrack and Aquascreen for control of macro-phytes. *J. Aquatic Plant Manage.* 21: 103–104.

Madsen, J.D., K.D. Getsinger, R.M. Stewart, J.G. Skogerboe, D.R. Honnell and C.S. Owens. 1998. Evaluation of transparency and light attenuation by Aquashade. *Lake and Reservoir Manage.* 15: 142–147.

Manker, D.C. and D.F. Martin. 1984. Investigation of two possible modes of action on the inert dye Aquashade on hydrilla. *J. Environ. Sci. Health A* 19(b): 725–753.

Mayhew, J.K. and S.T. Runkel. 1962. The control of nuisance aquatic vegetation with black polyethylene plastic. *Proc. Iowa Acad. Sci.* 69: 302–307.

Newroth, P.R. and R.L. Truelson. 1984. Bottom barriers to control rooted macrophytes. *LakeLine* 4(5): 8–10.

Nichols, S.A. 1974. *Mechanical and Habitat Manipulation for Aquatic Plant Management. A Review of Techniques.* Wisconsin Dept. Nat. Res., Madison.

Nichols, S.A. and B.H. Shaw. 1983. Review of management tactics for integrated aquatic weed management of Eurasian watermilfoil (*Myriophyllum spicatum*) curly-leaf pondweed (*Potamogeton crispus*) and elodea (*Elodea canadensis*). In: *Lake Restoration, Protection and Management.* USEPA-440/5-83-001. pp. 181–192.

Perkins, M.A. 1984. An evaluation of pigmented nylon film for use in aquatic plant management. In: *Lake and Reservoir Management.* USEPA 440/5-84-001. pp. 467–471.

Perkins, M.A., H.L. Boston and E.F. Curren. 1980. The use of fiberglass screens for control of Eurasian watermilfoil. *J. Aquatic Plant Manage.* 18: 13–19.

Petersen, J.O., S. Born and R.C. Dunst. 1974. Lake rehabilitation techniques and experiences. *Water Res. Bull.* 10: 1228–1245.

Pullman, G.D. 1990. Benthic barriers tested. *LakeLine* 10(4): 4,8.

Spencer, D.F. 1984. Influence of Aquashade on growth, photosynthesis, and P uptake of microalgae, *J. Aquatic Plant Manage.* 22: 80–84.

Ussery, T.A., H.L. Eakin, B.S. Payne, A.C. Miller and J.W. Barko. 1997. Effects of benthic barriers on aquatic habitat conditions and macroinvertebrate communities. *J. Aquatic Plant Manage.* 35: 69–73.

16 Chemical Controls

16.1 INTRODUCTION

Herbicides are chemical pesticides used for plant management. Herbicides kill plants or severely interrupt their normal growth processes. An herbicide formulation consists of an active ingredient, an inert carrier, and possibly other chemicals such as adjuvants that make the herbicide more effective. "Today's modern (herbicide) applicator strives to selectively treat exotic species encouraging native species re-establishment, and to treat other excessive vegetation in more 'direct use' areas leaving less-utilized areas of native species as nutrient and habitat buffers in the ecosystem." This quote by Kannenberg (1997) suggests that the role of herbicides in lake and reservoir management is threefold: (1) eradicate exotic species; (2) change plant community composition; and (3) treat excessive vegetation growth in direct or high-use areas.

The decision to use herbicides should be based on the same criteria — efficacy; cost; health, safety, and environmental impacts; regulatory appropriateness; and public acceptability- that are used for other management techniques (Chapter 11). This was not always the case. Because herbicide (and other pesticide) treatments were fast, relatively cheap, and many times very effective, they were used in inappropriate ways regarding health, safety and environmental impacts. This influenced public perception about the acceptability of using pesticides.

One of the more striking historical cases of overuse of a toxic but very effective aquatic herbicide was the use of sodium arsenite. Between 1950, when the Wisconsin Department of Natural Resources began keeping records and 1970 when it was no longer used, approximately 798,799 kg of sodium arsenite were added to 167 lakes (Lueschow, 1972). The environmental impacts of these treatments were not monitored. However, the use of sodium arsenite causes long-term problems for further management in some lakes where it was heavily used. The sediments in these lakes are a hazardous waste so other lake management options such as dredging become extremely difficult if not impractical (Dunst, 1982).

Herbicides are a useful technique in a lake manager's "tool box." The largest obstacle to using them may be public perception. Poor public perception can be overcome with good demonstration projects, reliable monitoring (Chapter 11), education, full disclosure of known environmental impacts, and responsible use by applicators.

16.2 EFFECTIVE CONCENTRATION — DOSE, TIME CONSIDERATIONS, ACTIVE INGREDIENTS, SITE-SPECIFIC FACTORS, AND HERBICIDE FORMULATION

Aquatic herbicides were originally developed for terrestrial use, mainly for agriculture. In terrestrial systems an effective concentration of active ingredients (a.i.) is applied directly to the plant or the soil. Exposure time is usually not a consideration unless there is a meteorological event like a rainstorm that washes the herbicide off the plant. Similarly, an effective concentration of herbicide can be applied directly to emergent and floating-leaf aquatic species. For submergent species an effective dose is delivered through water so dilution and dispersion are considerations. The water volume treated, currents, drift and micro-stratification (Chapter 11) effect dilution and dispersion.

The success or failure of treating any species is dependent on an effective dose of active ingredient contacting or being taken up by the plant. This is dependent on the concentration/exposure

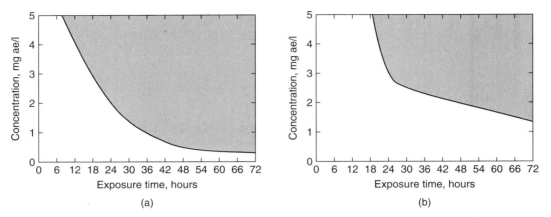

FIGURE 16.1 Examples of concentration/exposure time (CET) relationships using endothall for *Myriophyllum spicatum* (A) and *Hydrilla verticillata* (B) control. The shaded area represents CETs that give 85–100% *M. spicatum* control with very limited regrowth up to 4 weeks post-treatment and 85–100% *Hydrilla* control with very limited or no regrowth up to six weeks post-treatment. The CET relationship is different for each species-herbicide combination. (After Netherland, M.D. et al. 1991. In: *J. Aquatic Plant Manage.* 29: 61–67. With permission.)

time (CET) relationship for controlling the target plant (Getsinger, 1997). An effective concentration can be achieved using a high dose of herbicide and a short contact time or a low dose of herbicide and a long contact time (Figure 16.1). While a low dose of material is more desirable for cost, safety, health, and environmental reasons, an effective CET relationship and thus efficacy is more difficult to achieve for submersed species because any bulk water movements away from the plant affects the CET relationship.

This does not imply that an effective dose is always easily achieved for emergent and floating-leaved species. Accurate application requires that the equipment be well calibrated and that the boat or other application vehicle is moving at a constant speed. This is difficult in heavy vegetation and a boat tends to submerge the plants that it passes over, washing off the herbicide. What it does imply is an effective dose of herbicide is easier to calculate for emergent and floating-leaf species. Application rates are calculated based on the area treated. For submergent species, water depth and velocity also need to be considered.

Understanding active ingredient is critical to proper CET calculations. Active ingredient is the concentration of herbicidally active chemical in a formulation. It can vary tremendously between different formulations or different manufactures of the same product. It is expressed as weight to volume (g/L) for liquid formulations and weight to weight (g/kg) for granular formulation, or it may be represented as a percentage. For example in a liquid formulation the active ingredients could be expressed as 300 g/L or 30%. Active ingredient concentration is given on the herbicide label.

Site-specific treatment factors affect the choice of herbicide formulation, which affects application equipment, techniques, and timing. For example, a surface application of a liquid formulation is appropriate in quiescent, isothermal water. These conditions allow an even distribution and mixing of a surface application. In a dense plant stand that creates a temperature-stratified environment, or in areas of great water movement, a granular or pellet formulation, or subsurface injection of a liquid formulation, will more evenly distribute the herbicide.

16.3 TYPES OF CHEMICALS

There are only six herbicides: copper (Chapter 10), 2,4-D, diquat, endothall, fluridone, and glyphosate, that are registered and commonly used for lake and reservoir management in the

TABLE 16.1
Aquatic Herbicide Characteristics[a]

Compound	Formulation[b]	Contact vs. Systemic[b]	Mode of Action[b]	Half-life in Water (days)[b,c]	Method of Disappearance[c,d]
Complexed copper	Various complexing agents with copper-liquid and granular	Systemic	Plant cell toxicant	3	Precipitation Adsorption
2,4-D	Butoxyethel ester — salt Dimethylamine — liquid Isooctyl ester — liquid	Systemic	Selective plant growth regulator	7–48	Microbial degradation Photolysis Plant metabolism
Diquat	Liquid	Contact	Disrupts plant cell membrane integrity	1–7	Adsorption Photolysis Microbial degradation
Endothall	Liquid and granular	Contact	Inactivates plant protein synthesis	4–7	Plant metabolism Microbial degradation
Fluridone	Liquid and granular	Systemic	Disrupts carotenoid synthesis, causing bleaching of chlorophyll	20–90	Photolysis Microbial degradation Adsorption
Glyphosate	Liquid	Systemic	Disrupts synthesis of phenylalanine	14; Used over but not in water	Adsorption Microbial degradation
Triclopyr[e]	Liquid	Systemic	Selective plant growth regulator	—	—

[a] Herbicides registered by U.S. Environmental Protection Agency.

[e] Experimental use permit only.

Sources: [b]After Madsen, J.D. 2000. Advantages and Disadvantages of Aquatic Plant Management. Tech. Rept. ERDC/EL MP-00-01. U.S. Army Corps of Engineers, Vicksburg, MS. [c]After Langeland, K.A. 1997. In: M.V. Hoyer and D.E. Canfield (Eds.), Aquatic Plant Management in Lakes and Reservoirs, NALMS, Madison, WI and Lehigh, FL. pp. 46–72. [d]After Wisconsin Dept. Nat. Res., 1988. Environmental Assessment Aquatic Nuisance Control (NR 107) Program. Wisconsin Dept. Nat. Res., Madison, WI.

United States. A seventh herbicide, triclopyr, is used under an experimental use permit. Other herbicides may be approved for use in other countries or approved for aquatic uses that are not appropriate to lake and reservoir management because they have a long use restriction time or they are toxic to fish or other aquatic organisms.

These herbicides and other chemicals can be categorized in a number of ways depending on their use, mode of contact, selectivity, and persistence in the environment (Table 16.1).

16.3.1 CONTACT VS. SYSTEMIC

Contact herbicides act quickly and are generally lethal to the plant cells they contact. Because of their rapid action and other physiological reasons, they do not move extensively within the plant and they kill tissue only where they contact the plant. For this reason they are generally more effective on annual plants (see Table 12.10 in Chapter 12 for information regarding annual vs. perennial plants). Perennial plants can be defoliated by contact herbicides but they regrow from unaffected parts, especially parts that are protected beneath the sediment. Contact herbicides are more effective than systemic herbicides on old, slow growing, or senescent plants, so they are preferred later in the growing season for controlling aquatic nuisances where, for lack of time or for physiological reason, systemic herbicides are not effective.

Systemic herbicides are translocated from absorption sites to critical growth points in the plant. They act slowly when compared with contact herbicides, but they are generally more effective for controlling perennial and woody plants. They are also more selective than contact herbicides. Correct application rates are critical. If application rates are too high, systemic herbicides can act like contact herbicides. They stress the plants so much that the herbicides are not translocated to critical plant growth areas (Nichols, 1991).

16.3.2 BROAD-SPECTRUM VS. SELECTIVE HERBICIDES

Broad-spectrum herbicides control all or most of the vegetation they contact. Selective herbicides control certain plants but not others. Selectivity is based on the different response of different species to the herbicide. It is a function of both the plant and the herbicide.

Selectivity can be affected by the CET relationship of the herbicide. For example, water hyacinth (*Eichhornia crassipes*) is selectively controlled amongst spatterdock (*Nuphar* sp.) using the recommended rate of 2,4-D, but spatterdock can be controlled by using higher rates and granular formulations (Langeland, 1997).

Systemic herbicides are the most physiologically selective herbicides. However, as stated above, they must be translocated to the site where they are active. Herbicides may be bound on the outside of the plant or bound immediately after they enter the plant so they cannot move to the activity site. For other reasons, not all understood, herbicides are more readily translocated in some plants than in others, which results in selectivity (Langeland, 1997). Some plants have the ability to alter or metabolize a herbicide so it is no longer active, and some herbicides affect very specific biochemical pathways so they only work on plants or groups of plants with those pathways (Langeland, 1997).

Selectivity is also affected by the physiology of perennial species during their growth cycle. During early stages of growth, energy reserves are translocated upward in the plant so an herbicide taken up by the roots is most effective. Late in the growth cycle, material is translocated downward to the roots so a foliar herbicide is most effective (Langeland, 1997).

16.3.3 PERSISTENT VS. NON-PERSISTENT

Persistent herbicides retain their activity in water for a long time, usually measured in weeks or months. Non-persistent herbicides act only when sprayed directly onto foliage or they lose their phytotoxicity rapidly on contact with soil, particulate matter in the water, or plant cells. Non-persistent herbicides may decay rapidly in water. There is no set time that separates persistent from non-persistent. The half-life of the herbicide in water is a useful measure of persistence (Table 16.1).

16.3.4 TANK MIXES

In addition to single uses, herbicides are mixed to increase efficacy. Diquat and copper chelates are a popular tank mix that provides a broad spectrum of control for aquatic plants plus the convenience of working with a liquid formulation.

16.3.5 PLANT GROWTH REGULATORS (PGRs)

Growth regulators prevent plants from obtaining normal stature. They keep plants short but functional by preventing cell division and elongation. PGR research on aquatic plants has occurred for over 15 years. Unfortunately it has yet to be commercialized so PGRs cannot and have not been used for management purposes.

Laboratory and field tests show that Thiadiazuron and Bensulfuron Methyl maintained milfoil (*Myriophyllum spicatum*), hydrilla (*Hydrilla verticillata*) and *Potamogeton* spp. in short stature (Anderson, 1986, 1987; Anderson and Dechoretz, 1988; Lembi and Netherland, 1990; Nelson and

Van, 1991). Thiadiazuron inhibited tuber and turion production in hydrilla (Klaine, 1986). Bensulfuron methyl inhibited propagule formation in *P. nodosus, P. pectinatus,* and hydrilla (Anderson, 1987).

Growth regulators are a very interesting technology because they have the potential for utilizing the beneficial aspects of aquatic plants without letting them grow to nuisance proportions. There are still many questions to answer regarding product delivery, mode of uptake, mode of action, differential plant responses, efficacy, health, safety, and environmental impacts that probably will not be answered without commercial interest in the technique.

16.3.6 ADJUVANTS

Adjuvants are chemicals added to herbicides to increase their effectiveness. There are activator adjuvants, spray-modifier adjuvants, and utility-modifier adjuvants (Thayer, 1998). They include wetting agents and emulsifiers that allow the herbicide to mix more easily. Spreaders allow herbicides to spread evenly over treated surfaces. Stickers, thickeners, invert emulsifiers, and foaming agents increase the adherence of the herbicide to the treated surface and help control herbicide drift. Penetrants enhance absorption of herbicides by decreasing surface tension or by penetrating through waxy coatings. Many herbicide formulations contain a small percentage of adjuvants and all the categories of adjuvants mentioned may not be used in the aquatic situation. Wetting agents and spreader-stickers are probably the most frequently used adjuvants (Binning et al., 1985).

16.4 INCREASING HERBICIDE SELECTIVITY

Ideally, herbicides should be used to selectively control undesirable species and to change plant community structure to a more desirable type. Past control efforts usually did not take selectivity into consideration and research continues to make herbicides more selective. Some tools for using herbicides selectively are already present and include efficacy information as well as location-selective, time-selective, and dose-selective applications.

Using the differential susceptibility of plants to herbicides is one method of selective control. In a mixed plant community, if the undesirable species are controlled by an herbicide and desirable species are not, there is a basis for selective herbicide control based on herbicide efficacy. An example is using 2,4-D to control Eurasian watermilfoil or coontail (*Ceratophyllum demersum*) in a mixed pondweed (*Potamogeton* spp.) community. 2,4-D effectively controls milfoil and coontail but not pondweeds. As a basis for planning selective management, herbicide efficacy is summarized in Table 16.2. Label instructions for specific efficacy information should be consulted before using any herbicide.

Applications can be selective by carefully placing the herbicide on target plants and avoiding non-target plants. Experienced personnel for example, using a handgun applicator, can control small areas of water hyacinth among bulrushes (*Scirpus* sp.) using 2,4-D and careful placement of the herbicide on the target plant (Langeland, 1997). Likewise, if diquat were used in the above scenario, although it is a broad-spectrum, contact herbicide, it would only kill bulrush stems above the waterline. The extensive underground bulrush roots and rhizomes are not affected and the plant regrows after the initial effect of the herbicide (Langeland, 1997).

Adjuvants that restrict herbicide movement are a way of selectively treating an area. This method is especially appropriate for treating areas that are monotypes of nuisance species while keeping the herbicide from drifting into a valuable plant community. Another method of restricting herbicide movement is to treat in conjunction with a drawdown. The sediments of Lake Ocklawaha, Florida were treated experimentally with fluridone and other chemicals under drawdown conditions to test the efficacy of controlling hydrilla plants and tubers (Westerdahl et al., 1988). Herbicides can be precisely placed in terrestrial areas.

Water temperature and light influence macrophyte growth, physiological status, and phenology. Most herbicides work best when plants are actively growing. Some species, *Elodea canadensis, P.*

TABLE 16.2
Aquatic Plant Response to Herbicides Commonly Used for Lake and Reservoir Management[a]

	Glyphosate	2,4-D	Endothall	Diquat	Fluridone
Emergent and Floating-Leaf Species					
Acorus calamus	—	C	—	—	—
Alternanthera philoxeroides	CC	CC	—	—	CC
Brasenia schreberi	—	C	CC	CC	CC
Eleocharis spp.	—	—	—	—	CC
Glyceria borealis	—	—	—	C	—
Hydrocotyle umbellate	—	CC	—	C	—
Justicia americana	—	C	—	CC	CC
Ludwigia uruguayensis	—	C	CC	CC	CC
Lythrum salicaria	C	—	—	—	—
Nasturtium sp.	—	C	—	—	—
Nelumbo lutea	CC	C	CC	—	—
Nuphar spp.	C	C	CC	—	CC
Nymphaea odorata	C	C	CC	—	CC
Phragmites spp.	CC	—	—	—	—
Polygonum spp.	CC	CC	CC	CC	CC
Pontederia sp.	CC	CC	—	—	—
Salix spp.	C	C	—	—	—
Sagittaria spp.	C	C	—	—	C
Scirpus spp.	C	C	—	CC	C
Sparganium spp.	—	—	C	—	—
Trapa natans	—	CC	—	—	—
Typha spp.	C	CC	—	CC	CC
Floating Species					
Azolla caroliniana	—	CC	—	CC	CC
Eichhornia crassipes	CC	C	CC	C	—
Lemna spp.	—	CC	CC	C	C
Pistia stratiotes	CC	CC	CC	C	—
Salvinia rotundifolia	—	—	CC	C	CC
Spirodela polyrhiza	—	CC	—	C	CC
Wolffia columbiana	—	—	—	CC	CC
Wolffiella floridana	—	—	—	CC	CC
Submergent Species					
Cabomba caroliniana	—	CC	C	CC	CC
Ceratophyllum demersum	—	CC	C	C	CC
Chara spp.[b]	—	—	—	—	—
Egeria densa	—	—	C	CC	CC
Elodea canadensis	—	—	C	C	CC
Hydrilla verticillata[b]	—	—	CC	CC	CC
Myriophyllum aquaticum	—	C	C	C	—
Myriophyllum spicatum	—	C	C	C	CC
Najas spp.	—	CC	C	C	CC
Potamogeton spp.	—	—	C	CC	CC
P. richardsonii	—	—	C	—	C
Ranunculus aquatilis	—	—	CC	C	—
Ruppia maritima	—	—	CC	C	—
Utricularia spp.	—	CC	—	CC	CC

TABLE 16.2 (Continued)
Aquatic Plant Response to Herbicides Commonly Used for Lake and Reservoir Management[a]

	Glyphosate	2,4-D	Endothall	Diquat	Fluridone
Vallisneria americana	—	—	CC	CC	—
Zannichellia palustris	—	—	C	—	—
Zosterella dubia	—	C	C	—	—

Note: C, controlled by the herbicide; CC, conditionally controlled by the herbicide; this could mean that efficacy depends on specific formulation or application techniques, that it was rated as only fair or good control by Westerdahl and Getsinger (1988), or that it is labeled only for partial control. —, not controlled by the herbicide, not registered for use with this species, or information is unknown.

[a] For use as a general guide; read label instructions for details.

[b] Can be controlled by copper or copper complexes.

Source: After Lembi, C.A. and M. Netherland. 1988. Category 5, Aquatic Pest Control. Dept Botany, Purdue University, W. Lafayette, IN; Westerdahl, H.E. and K.D. Getsinger. 1988. Aquatic Plant Identification and Herbicide Use Guide, Volume II: Aquatic Plants and Susceptibility to Herbicides. Aquatic Plant Cont. Res. Prog. Tech. Rept. A-88-9. U.S. Army Corps of Engineers, Vicksburg, MS; Binning, L., B. Ehart, V. Hacker, R.C. Dunst, W. Gojmerac, R. Flashinski and K. Schmidt. 1985. Pest Management Principles for Commercial Applicator: Aquatic Pest Control. University Wisconsin-Ext., Madison; Cooke, G.D. 1988. In: The Lake and Reservoir Guidance Manual. USEPA 1440/5-88-02. pp. 6-20–6-34.

crispus, and *M. spicatum* for example, grow better at low water temperatures and appear earlier in the growing season than many other species. This provides an opportunity to treat these species with a contact or short-lived systemic herbicide before other species are actively growing. Refer to Chapter 11 for a discussion of the importance of phenology and resource allocation patterns when determining management strategies.

A thorough knowledge of CET relationships allows selective management based on varying dose or contact time of the same herbicide. The water hyacinth and spatterdock example was given above. An endothall label suggests that *P. crispus* can be effectively treated at about one-half the concentration needed to control *P. americanus* and many emergent and free-floating species. Adams and Schulz (1987) found that *M. spicatum* and *E. canadensis* were highly sensitive to low concentrations of diquat. "Fine tuning" treatments based on CET relationships constitute a very active area of research. It is not easy because of previously mentioned problems of dispersion and dilution but it is an area that holds great promise for selectively managing plant communities with herbicides and for reducing environmental impacts from herbicide treatments.

16.5 ENVIRONMENTAL IMPACTS, SAFETY AND HEALTH CONSIDERATIONS

16.5.1 HERBICIDE FATE IN THE ENVIRONMENT

Knowing the fate of aquatic herbicides in the environment is important for determining environmental impacts, safety and health. How long do herbicides persist in the environment, what are the breakdown products, where do the herbicides or breakdown products go when they "disappear" are all important questions. Disappearance refers to the removal of the herbicide from a certain part of the environment (Langeland, 1997). Aquatic herbicides disappear by dilution, adsorption to bottom sediments, volatilization, absorption by plants and animals, and by dissipation. Herbicides dissipate by photolysis, microbial degradation, or metabolism by plants and animals. The rate of disappearance (Table 16.1, half-life) depends upon: (1) initial herbicide concentration, (2) water movement, (3) temperature, (4) amount of plant matter, (5) water chemistry, (6) water volume, (7) the presence of decomposing organisms, and (8) the mode of disappearance.

Table 16.1 summarizes the methods of herbicide disappearance. Of the contact herbicides, endothall biodegrades into carbon dioxide and water. Diquat is rapidly taken up by plants or binds tightly to particles in the water or bottom sediments. When bound to clay mineral particles, diquat is not biologically available. When bound to organic matter, microorganisms slowly degrade diquat. It is photo-degraded to some extent when applied to leaf surfaces. Information about the persistence or biological effects of degradation products of diquat was not found (WDNR, 1988).

Microbial action is the primary mode of degradation of 2,4-D and photolysis may be important under alkaline conditions (WDNR, 1988). 2,4-D degrades into naturally occurring compounds. 2,4-D amine for example degrades to carbon dioxide, water, ammonia, and chlorine (Langeland, 1997).

Dissipation of fluridone from water occurs mainly by photo-degradation. Microbial breakdown is probably the most important method of breakdown in bottom sediments. Degradation rate is variable and may be related to the time of year of application. Applications when days are shorter and sun's rays less direct result in longer half-lives. Fluridone usually disappears from water after 3 to 9 months. It usually remains in bottom sediments between 4 months and 1 year (Langeland, 1997).

Although glyphosate is not applied directly to water, when it does enter water, binding to particulate matter and to bottom sediments inactivates it. It is degraded to carbon dioxide, water, nitrogen, and phosphorus over a period of several months (Langeland, 1997).

Complexing is the major means of removing soluble copper ions from water. The copper ion is chemically bound by carbonate and hydroxide ions in natural waters as well as by organic humic acids. This binding is rapid in high alkalinity, hardness, and pH waters. Some lakes received massive doses of copper over an extended period of time. Lakes Kegonsa and Waubesa in Dane County, Wisconsin were treated with 586,750 kg and 692,182 kg, respectively, of copper sulfate between 1950 and 1970 (Lueschow, 1972). Copper sulfate was applied to the five Fairmont Lakes in southern Minnesota at cumulative rates of 1647 kg/ha over a 58-year period (Hanson and Stefan, 1984). Copper concentrations in lake sediments of the Dane County lakes were as high as nearly 1% of total sediment weight (WDNR, 1988). In the Dane County lakes the highest concentration of copper is found in sediments at the greatest water depth and copper concentration decreases toward the top of the sediment, which indicates the sediments with the highest copper concentration are being buried. There appears to be an annual copper cycle in the lakes with greater copper concentrations found in the water during the autumn lake turnover. Increased copper levels are largely in the suspended organic fraction of the water; relatively small increases have been observed in soluble copper (WDNR, 1988). See Chapter 10 for additional details about copper.

The active ingredients are not the only chemicals added to the waters. Inert ingredients, manufacturing contaminants, and adjuvants are also added. The fates of some of these products have been studied but generally their fate is less well known than the fate of the active ingredients. Modeling is becoming an increasingly important tool for characterizing ecological risks of using pesticides in aquatic environments at the individual, population and community levels (Bartell et al., 2000).

16.5.2 TOXIC EFFECTS

In the United States, the United States Environmental Protection Agency (USEPA) registers aquatic herbicides for use. An herbicide can be registered if it does not cause "unreasonable adverse effects" to human health or the environment. Registration does not mean that an herbicide has no health or environmental risks. Herbicide registration decisions balance the risks involved with the benefits.

The USEPA decides whether or not to register an herbicide after considering the ingredients; the manufacturing process; the physical and chemical properties; the mobility, volatility, breakdown rates, and accumulation potential in plants and animals; the toxicity to animals; and the carcinogenic or mutagenic properties. The USEPA can approve or disapprove registration of a new herbicide and may further restrict or cancel the registration of those in use.

An herbicide's capacity to harm fish, plants, and other aquatic life depends on the toxicity of the herbicide, the dose rate used, the exposure time of the affected organism, and the persistence of the herbicide in the environment. Toxic effects may be direct or indirect. Direct effects impact the organism of concern. Direct effects may be lethal if they kill the organism or they can be sub-lethal. Sub-lethal or chronic effects include biomass loss, low resistance to disease, compromised reproduction rates or sterility, loss of attention, low predator avoidance, and deformed body parts. The short-term indirect effects are the ecological effects caused by the death and decay of the target plants. Long-term effects are changes caused by a restructuring of the plant community or broader ecological changes like the change in stable-state from a macrophyte-dominated lake to an algae-dominated lake, or changes in food webs (Chapter 9). The direct and indirect impacts of herbicide use are summarized in Figure 16.2.

16.5.2.1   Direct Effects

The most obvious direct toxic effect is damage to non-target aquatic plants. This can occur to plants present in the targeted treatment area or it can affect plants not in the target area by spray drift or residue movement in water currents. The potential for this impact can be calculated knowing the CET relationship between the non-target species, the herbicide, and the herbicide concentration after considering dissipating factors.

The lethal and sub-lethal effects to invertebrates, fish, and higher animals or humans are not as easily assessed. A variety of tests and extrapolations are performed on aquatic organisms to ascertain herbicide toxicity. Acute toxicity is usually reported as lethal concentration, effective concentration, or tolerance limit (WDNR, 1988). A lethal concentration (LC) is the concentration that kills 50% of the test organisms in a given time period such as 24, 48, or 96 hours. It is one of the most commonly tested and reported parameters for fish and other aquatic organisms. It is reported as LC_{50}, 24, 48, or 96 hours. The effective concentration is the dosage that immobilizes the test organism. It is often used for insects and crustaceans where determining death is difficult. The tolerance limit is an extrapolated or mathematically determined concentration used to estimate the point of toxicity. The "no observable effect" level is another means of reporting toxicity. It is the highest test concentration that shows no observable impact on the test organisms.

Most assays are conducted under laboratory conditions that allow careful control over a wide variety of factors affecting test results. Such simplified tests present obvious difficulties interpreting the impacts of an herbicide on a complicated, dynamic system like a lake. There is also a concern over the species and life stages selected for testing (Paul et al., 1994). It is impossible to test all potentially affected organisms, at all life stages, in all habitat conditions. Many of the test species may not occur in the area where the herbicide is used.

The bulk of the published data on herbicide toxicity to aquatic biota relates to effects on invertebrates and fish but there are effects on phytoplankton, micro-organisms, and higher animals. Many higher animals are not obligate aquatic organisms so less attention has been paid to them. However, some higher animals like frogs and toads are obligate aquatic organisms in early life stages.

Sub-lethal or chronic effects are probably even more difficult to assess than lethal effects. How do you tell if a bluegill is not feeling well today? The main ways are through population, growth, and life-cycle studies that can be extremely complex in a lake or reservoir ecosystem.

The objective of this section is not to review all the toxicological data and do a risk assessment for aquatic herbicides, but to give some idea of the complexity of the task. The information is too voluminous and should be done by a professional toxicologist. To learn more, the best resources

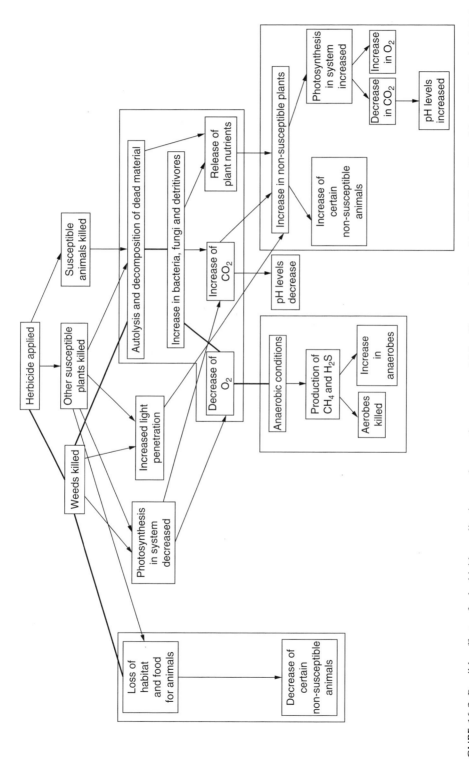

FIGURE 16.2 Possible effects of a herbicide application on the aquatic ecosystem. Main effects are indicated by thick lines. (From Murphy, K.J. and. P.R.F. Barrett. 1990. In: A. Pieterse and K. Murphy (Eds.), *Aquatic Weeds, The Ecology and Management of Nuisance Aquatic Vegetation.* Oxford University Press, Oxford, UK. pp. 136–173. With permission of the original author, David Mitchell.)

are environmental assessments done by governmental agencies that reviewed the toxicology and assessed risk of aquatic herbicides (Shearer and Halter, 1980; WDNR, 1988). Another excellent resource is the Extension Toxicology Network on the WEB. To find it, type Extoxnet in the WEB search function. Extoxnet provides a pesticide information profile (PIP) that summarizes trade names, regulatory status, formulations, toxicological effects, ecological effects, environmental fate, physical properties, and manufactures. It also provides references to further information.

The two herbicides with the greatest potential for direct toxic effects are the monoamine salt of endothall (trade name Hydrothol 191) and copper sulfate. Due to its toxicity, liquid Hydrothol 191 is not recommended for use in water bodies where fish are an important resource (WDNR, 1988). Copper at low levels can produce mortality and sub-lethal toxicity affecting the growth and reproduction of aquatic life on several trophic levels. The concentrations of copper used to control algae are higher than those that have been shown to produce chronic toxicity in a range of aquatic organisms and are above those that produce acute toxicity in particularly sensitive organisms (WDNR, 1988). Trout living in soft water are particularly sensitive to copper.

16.5.2.2   Indirect Impacts

Indirect impacts of herbicide use (Figure 16.2) include changes to water chemistry; detritus accumulation; ecosystem alteration including changes in community structure, food webs, and stable state; and the possibility of accumulating trace contaminants. For managers, Engel (1990) provides a concise literature review of the likely ecosystem impacts of herbicide use. The water chemistry changes are similar to those described in Chapter 11 through natural aquatic plant death and decay. Most water chemistry changes caused by herbicide treatment occur quickly because plant death occurs in days or weeks. If nuisances are great enough to consider herbicidal control, aquatic plant biomass is usually high. Under natural conditions 33–50% of macrophyte biomass may decompose in the first three weeks after death (Adams and Prentki, 1982). Decomposition may occur more quickly after an herbicide treatment, especially if the herbicide disrupts plant tissue. Therefore, there is a large amount of plant material consuming oxygen for decay, releasing nutrients, and adding to bottom detritus over a short time period. Often this occurs during warm months and warm water temperatures do not hold as much oxygen as cold waters. Growing conditions for algae are optimal so released nutrients stimulate algae "blooms."

The oxygen demand caused by decomposition is exacerbated by oxygen loss from photosynthesis as plants die. The main factors involved in oxygen depletion after herbicide treatments are water temperature, turnover rate of the water column, water depth, macrophyte biomass and shoot nitrogen content, and the rate of external oxygen input. Short-term recovery from deoxygenation following an herbicide treatment usually results from a phytoplankton bloom or replacement plant growth (Murphy and Barrett, 1990).

Respiratory CO_2 increases with decay can shift the inorganic carbon equilibrium. In poorly buffered waters this may result in a daytime change of more than one pH unit (Murphy and Barrett, 1990). Plant nutrients released from decaying macrophytes to the water column favors the growth of phytoplankton or free-floating species like *Lemna* sp. (Murphy and Barrett, 1990). If free-floating species dominate, daytime dissolved oxygen levels may not recover to pre-treatment levels for a prolonged period (Murphy and Barrett, 1990). The plant biomass that ends up on the lake bottom as detritus continually consumes oxygen in the decay process. Low oxygen creates reducing conditions in sediments causing further nutrient releases. Loss of canopy foliage can increase sunlight penetration and water temperature. Particulate organic matter from macrophyte decay can temporarily increase turbidity.

Long-term studies show the magnitude of some nutrient and detrital inputs. In Lake Okeechobee, Florida, an estimated 14,281 metric tons (m.t.) of detritus were produced, and 285 m.t. of N and 74 m.t. of P were returned to the water column over a 24-year period from herbicide-treated, freely floating aquatic vegetation (Grimshaw, 2002). In addition 4,472 m.t. of detritus were produced and 88 m.t. of N and 23 m.t. of P were returned to the water column in the Kissimmee River, the main inflow to Lake Okeechobee, over a 15 year period. The nutrient loading from herbicidal control was estimated at 4–49% for P and 1–17% for N of the external nutrient loading to the lake. In addition, some detritus, N, and P from Kissimmee River treatments likely reached Lake Okeechobee. In Lake Istokpoga, Florida, reduction of hydrilla (*H. verticillata*) through herbicide treatments during 1988–1992 resulted in significant increases in total P and chlorophyll *a* concentrations, and a decrease in Secchi depths (O'Dell et al., 1995). These results were expected since nutrients bound in the extensive hydrilla mats were released with herbicidal treatment. Decomposition of the hydrilla mats may also have increased sediment resuspension, and changed the primary producers from macrophytes to algae.

Food chains and food webs are changed with loss of macrophyte habitat. Plant-dwelling invertebrates and epiphyton decline from habitat loss but benthic invertebrates can increase with increased detritus (Hilsenhoff, 1966). Loss of shelter exposes young fish, zooplankton and plant dwelling invertebrates to increased predation. Loss of macrophyte cover can increase bank erosion and suspension of bottom sediment. Water birds can disperse to quiet waters with protective cover and food. When food webs are altered due to loss of macrophytes and associated epiphyton there are "winner" species and "loser" species. For instance, within the carrying capacity for fish in a lake, high aquatic plant abundance favors fish species that are adapted to aquatic plants and low aquatic plant abundance favors fish species that are adapted to open water. A major factor determining the value of aquatic plants to fish species is whether the fish is a predator or prey species. The presence of aquatic plants increases the structural complexity of lake ecosystems that provides refuge for prey species and interferes with the feeding of predator species. Even for single species there are "trade-offs." Herbicides may kill some zooplankton and expose them to increased predation but phytoplankton blooms after an herbicide treatment increases their food supply.

There is some concern that continued use of herbicides will develop herbicide resistant organisms. In the past there was scant evidence for this occurring (WDNR, 1988) the way herbicides were normally used. However, recent evidence indicates that there is a differential susceptibility of hydrilla to fluridone in several aquatic systems in Florida (Netherland et al., 2001). This was unexpected and a significant new development in aquatic plant management. Part of the problem may be related to the low dose rate of fluridone usage. Low doses could exert great selective pressure where there are small differences in susceptibility.

Another concern is the development of herbicide resistant plant communities. Herbicides are selective so the susceptible species are killed and the tolerant species remain. To kill the remaining species a different herbicide may be used. If this scenario is repeated enough times, only species resistant to most herbicides remain. This may be beneficial if the species are desirable, but if not, herbicides will no longer be effective and an aquatic plant management tool is lost. Over the short term, herbicide treatment causes regression to an earlier stage of fresh water plant succession. Opportunistic disturbance-tolerant plants fill the newly vacated niches followed by the seral replacement of opportunists by slower-growing, but more competitive, plant species (Murphy and Barrett, 1990; Newbold, 1976). *Chara* spp., *Najas flexilis*, and *Potamogeton foliosus* are often initial pioneering species and *Chara* spp. and *Vallisneria americana* are persistent species after herbicide treatments (Brooker and Edwards, 1973; Crawford, 1981; Getsinger et al., 1982; Hestand and Carter, 1977). In the longer term a single herbicide treatment may have little effect on macrophyte community structure (Wade, 1981; Wade, 1982 as cited in Murphy and Barrett, 1990). Over the years following treatment, hydroseral processes lead to the re-establishment of the original plant community but repeated treatments may keep the plant community in a hydroserally early stage (Murphy and Barrett, 1990). Windfall Lake, a 23-ha lake with a maximum depth of 9.2 m in northeastern Wisconsin, was an example of the above scenario (Dunst et al., 1974). Three years of extensive treatments with a variety of herbicides reduced a mixed aquatic plant community to dense, monotypic stands of *Chara* over much of the lakes littoral zone. *Chara* growth reached the water surface in 2 m of water in some areas of the lake — a perceived macrophyte problem turned into a real problem for lakeshore residents. Within 3 years of a "doing nothing" (see Chapter 12), *Potamogeton amplifolius*, a much more desirable species in this case, replaced *Chara* over large areas of the lake.

In shallow, eutrophic lakes herbicide treatments may shift the "stable state" (Scheffer et al., 1993) from a macrophyte dominated lake to an algae dominated lake (Moss et al., 1996). Herbicides are not unique in this regard. Other management techniques can also cause this shift. It is very difficult to calculate how much management will cause a shift (van Nes et al., 2002) and once the shift occurs it can be difficult to return to a macrophyte dominated state (Scheffer, 1998).

The case studies later in this chapter provide some information about both direct and indirect environmental effects related to specific treatments. More detailed information is often found in the references related to these treatments.

16.5.2.3 What Should a Lake Manager or Concerned Citizen Do?

Ultimately a lake manager, riparian owner, or governmental agency has to make a decision on whether to use or allow the use of herbicides. Are there risks? — some unanswered (such as the possibility of trace contaminants) — yes. As with any management practice, based on the evidence available, the risks need to be balanced with the benefits. From a practical point of view, currently registered aquatic herbicides have been used for a long time with no known dire consequences to aquatic ecosystems. The majority of the data suggest that the impacts are transient. So far, there is little evidence of any build-up of herbicide residues or chronic toxicity in natural aquatic systems and fish populations appear not to be adversely affected (Murphy and Barrett, 1990). Most problems can be traced to inappropriate use. Currently, no product can be registered for aquatic use if it poses more than a one in a million chance of causing significant damage to human health, the environment, or wildlife resources and, in addition, it may not show evidence of biomagnification, bioavailability, or persistence in the environment (Madsen, 2000). Because of dilution, adsorption by soil particles and organisms, volatilization, and other means of dissipation, organisms are exposed to the applied concentration of herbicide for only a short period of time. Given an escape route, mobile organisms (mainly fish) show an avoidance reaction to some herbicides (Murphy and Barrett, 1990). Can herbicides change aquatic ecosystem functions? The answer again is yes. Sometimes this is the desired result, in other cases the results are known. For purposes of this book it should be noted that there is a big difference between the limited use of herbicides to change aquatic plant community composition or to eradicate an exotic species, and the prolonged use of herbicides to manage an aquatic nuisance without addressing the cause of the nuisance. The Dane County, Wisconsin and Fairmount, Minnesota references given earlier are examples of the latter situation. The next section discusses ways to minimize environmental risks when using herbicides. The more effective the treatment, the longer lasting the impacts are likely to be or the more environmental change that is likely to occur.

16.6 WAYS OF MINIMIZING ENVIRONMENTAL RISKS

The most important means of minimizing environmental risk is to follow the label instructions for the herbicide. Herbicides were tested for safety based on labeled conditions. Not following label procedures is illegal. There are restrictions on the use of herbicide treated water for human drinking, swimming, and fish consumption; for animal drinking; and for irrigation of turf, forage, and food crops. These restrictions are subject to change but are provided on the label so make sure you understand and can abide by them before using the herbicide, and follow them after application. Notifying lake users of herbicide applications prevents inadvertent use of restricted waters and many times is legally required (Figure 16.3). The label also provides information on the efficacy of the product. Applying an herbicide that does not control target species adds unneeded chemicals to the environment and wastes money and effort.

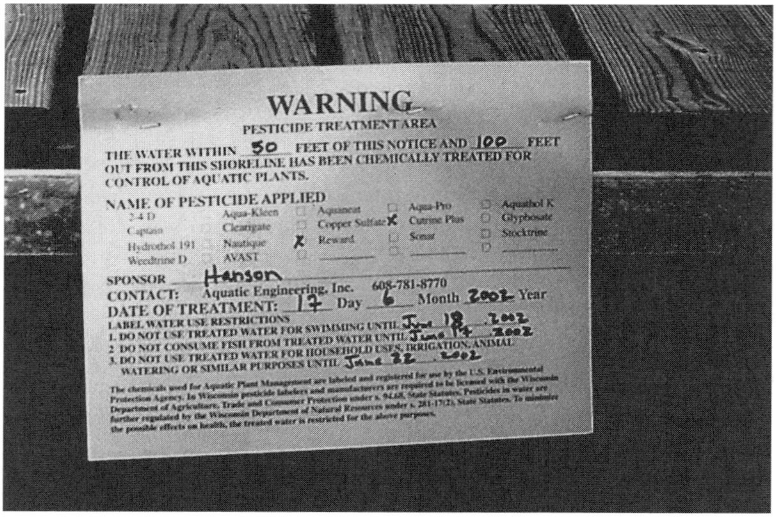

FIGURE 16.3 Posted notice of an herbicide application.

Applying herbicides beginning at the shoreline and working outward provides mobile organisms an avenue of escape. In heavy weed infestations, treat only a portion of the area at one time. Allow 2–3 weeks between treatments. This minimizes dissolved oxygen depletions and nutrient pulses caused by decomposing vegetation. It also allows recruitment of a variety of organisms from untreated refuges.

Treat only the area that needs to be managed. This may seem obvious but, with fluridone a whole lake treatment is recommended. Areas can be isolated for treatment by deploying temporary, non-permeable barrier curtains to reduce water exchange with other part of the lake (McNabb, 2001). This also reduces herbicide cost.

Applicators need to keep current with technology. On-board computers, fathometers, global positioning (GPS) units, and digital flow meters allow applicators to be much more precise with the area treated and treatment doses (Figure 16.4) (Kannenberg, 1997). Low-dose applications of fluridone and endothall and new formulations of 2,4-D and copper chelates are products or techniques that reduce environmental risk (Kannenberg, 1997).

FIGURE 16.4 Typical herbicide application equipment. Notice the GPS antennae and the on-board computer.

Maintenance management is another tool to reduce environmental risk. A maintenance management program controls plants at low levels before they become a problem. It is used effectively in Florida to control water hyacinth. By maintaining water hyacinth to less than 5% coverage, herbicide usage was reduced by a factor as great as 2.6, detritus deposition was reduced by a factor of 4, and reduced depression of dissolved oxygen occurred beneath vegetation mats (Langeland, 1998). By using maintenance management on the St. John River, Florida, the U.S. Army Corps of Engineers reduced the area of *Pistia stratiotes* that needed treatment from 881 ha to 33 ha and the area of water hyacinth that needed treatment from 649 to 28 ha between 1995 and 2000 (Allen, 2001). Maintenance management works well on water hyacinth because it grows rapidly and nearly continually, and it is aerially exposed so it is easily targeted. Maintenance management would probably work well on other floating or emergent species with similar characteristics. Maintenance control of submersed species in lakes is more difficult (Langeland, 1998). Part of the problem is probably the herbicide dilution factor and part is probably that the plants need to be growing to be effectively treated. Plants cannot be treated if they are not there.

Additional governmental regulations may impact the safety of herbicide use. Federal court actions necessitated the issuance of National Pollution Discharge Elimination System (NPDES) permits for applications of aquatic herbicides used for water hyacinth and egeria (*Egeria densa*) control programs in California (Anderson and Thalken, 2001). Permits were issued in 2001 and required extensive environmental monitoring and toxicity testing as well as compliance with conditions imposed by the Endangered Species Act.

16.7 CASE STUDIES

The literature describing herbicide use to control aquatic plants is voluminous. The case studies selected emphasize using species selective herbicides to change plant community structure and/or eradicate exotic species with minimal damage to native aquatic plants. In addition, the herbicide treatment was done only one to a few times in any water body and there were follow-up plant monitoring data for at least 1 year after treatment.

16.7.1 PLANT MANAGEMENT WITH FLURIDONE IN THE NORTHERN UNITED STATES

16.7.1.1 Minnesota Experiences

In 1992 the Minnesota Department of Natural Resources (MNDNR) initiated an evaluation to determine whether application of fluridone to whole bays or lakes can control Eurasian watermilfoil and have minimal effects on native vegetation. Whole lake applications of herbicides to public waters of Minnesota is generally not allowed because it destroys more vegetation than is necessary to provide lake access. Whole lake application of fluridone might be acceptable if it selectively controlled Eurasian watermilfoil. This might be possible using low fluridone concentrations and long contact times. Selective milfoil control was defined as removal of milfoil while causing little reduction in other plants (Welling et al., 1997). Elimination and subsequent re-establishment of native plants was not considered selective control. Parkers, Zumbra, and Crooked Lakes were selected for this evaluation (Table 16.3). All were spring treatments, and targeted whole lake fluridone concentrations were 10 µg/L for Parkers and Zumbra Lakes and 15 µg/L for Crooked Lake.

Fluridone treatment reduced the percentage of sampling stations with vegetation in both Parkers and Zumbra Lakes (Table 16.4). In Lake Zumbra the average number of vascular plants per sampling station declined during the year of treatment to one-quarter of the number observed before treatment and remained at this reduced level through the second year after application (Welling et al., 1997). Eurasian watermilfoil had not reappeared by the second year after application and two native species, coontail and *P. zosteriformis* disappeared (Table 16.5). *Nymphaea* sp., *P. pectinatus*, and

TABLE 16.3

Characteristics of Fluridone-Treated Lakes in the Northern United States[a]

Lake[b]	Treatment Time	Area (ha)	Depth (m)	Target conc. (μg/L)
Parkers, MN	Mid-May, 1994	39	11.3 (max.)	10
Zumbra, MN	Late May, 1994	66	17.7 (max.)	10
Crooked, MN	Early May, 1992	47	8 (max.)	15
Potters, WI	Fall, 1997	66	7.9 (max.)	14
Random, WI	Fall, 1999	85	6.4 (max.)	12
Big Crooked, MI	Mid-May, 1997	65	18.5 (max.)	5 in top 3.05 m
Camp, MI	Mid-May, 1997	65	16.7 (max.)	5 in top 3.05 m
Lobdell, MI	Mid-May, 1997	221	24.4 (max.)	5 in top 3.05 m
Wolverine, MI	Mid-May, 1997	98	17.9 (max.)	5 in top 3.05 m
Burr Pond, VT	Early June, 2000	34.5	4.4 (ave.)	6
Hortonia, VT	Early June, 2000	195	5.8 (ave.)	6

[a] Target species for treatment were *Myriophyllum spicatum* and *Potamogeton crispus* in all lakes except Potters, Random, and Burr Pond where only *M. spicatum* was targeted.

[b] MN, Minnesota; WI, Wisconsin; MI, Michigan; VT, Vermont.

TABLE 16.4

Frequency (%) of Vegetated Sampling Stations in Three Fluridone-Treated Minnesota Lakes

Lake	Pre-treatment[a]	Year of Treatment	First Year after Treatment	Second Year after Treatment	Third Year after Treatment
Zumbra	96	63	43	68	—
Parkers	97	33	77	90	—
Crooked	—	—	—	87	97

[a] Pre-treatment surveys were done in May, the year of treatment. Post treatment surveys were done in August.

Source: After Welling, C. et al. 1997. Evaluation of Fluridone for Selective Control of Eurasian Watermilfoil: Final Report. Minnesota Dept. Nat. Res., Minneapolis.

curly-leaf pondweed (*P. crispus)* became a more dominant part of the vegetation (Table 16.5), although based on absolute frequency *Nymphaea* and *P. pectinatus* both declined.

In Parkers Lake, Eurasian watermilfoil was found at the end of the first year after treatment (Table 16.6) and the frequency of milfoil nearly returned to pre-treatment levels by the end of the second year (Welling et al., 1997). Coontail and *M. sibiricum* were not found in post treatment surveys (Table 16.6) but they were found at other locations in the lake. Sago pondweed, *Zosterella dubia*, *P. foliosus*, and *Chara* sp. were found at greater frequencies after the fluridone treatment (Welling et al., 1997) and became more dominant members of the plant community (Table 16.6). Unfortunately, curly-leaf pondweed also became more dominant.

Secchi disk transparency decreased after fluridone application in Lake Zumbra and reached a minimal value that was 43% of pre-treatment levels during the first year after treatment. Transparency returned to pre-treatment levels the second year after treatment (Welling et al., 1997). Chlorophyll

TABLE 16.5
Relative Frequency (%) of Common[a] Aquatic Plants before and after a Fluridone Treatment in Lake Zumbra, Minnesota

Species	Year Before Treatment[b] (1993)	Year of Treatment (1994)	First Year After Treatment (1995)	Second Year After Treatment (1996)
Ceratophyllum demersum	25.8	5.7	0	0
Myriophyllum spicatum	28.9	5.7	0	0
Nymphaea sp.	17.5	17.1	31.9	33.7
Potamogeton crispus	8.9	30.7	40.7	45.9
P. pectinatus	6.7	0	21.5	18.4
P. zosteriformis	12	5.7	3	0

[a] Only species with a frequency more than 24% are included.

[b] Comparisons are made based on August sampling except for *P. crispus* where a May or June sampling are compared for 1994, 1995, and 1996. This could partially explain the large increase in the relative frequency of *P. crispus* between 1993 and the later years.

Source: After Welling, C. et al. 1997. Evaluation of Fluridone for Selective Control of Eurasian Watermilfoil: Final Report. Minnesota Dept. Nat. Res., Minneapolis.

TABLE 16.6
Relative Frequency (%) of Common[a] Aquatic Plants before and after a Fluridone Treatment in Parkers Lake, Minnesota

Species	Year Before Treatment[b] (1993)	Year of Treatment (1994)	First Year After Treatment (1995)	Second Year After Treatment (1996)
Ceratophyllum demersum	22.8	0	0	0
Myriophyllum spicatum	13.4	0	1.3	11.4
M. sibiricum	13.4	0	0	0
Potamogeton crispus	0	74.5	43.3	35.4
P. foliosus/pusillus	1	0	5.8	11.4
P. pectinatus	0	0	22.3	15.2
P. zosteriformis	33.7	7.4	3.1	2.7
Ranunculus longirostris	10.9	0	3.1	3.8
Zosterella dubia	4.7	18.1	21	20.2

[a] Only species with a frequency more than 24% are included.

[b] Comparisons are made based on August sampling except for *P. crispus* where a May or June sampling are compared for 1994, 1995, and 1996. This could partially explain the large increase in the relative frequency of *P. crispus* between 1993 and later years.

Source: After Welling, C. et al. 1997. *Evaluation of Fluridone for Selective Control of Eurasian Watermilfoil: Final Report.* Minnesota Dept. Nat. Res., Minneapolis.

a levels were also higher the first year after treatment than they were pre-treatment or the year of treatment. In Parkers Lake, Secchi disk transparency did not decrease after the fluridone treatment.

Crooked Lake surveys indicated that in the third and fourth years after treatment vegetation coverage was nearly 100%, values similar to pre-treatment levels (Table 16.4). Eurasian watermilfoil was not discovered in Crooked Lake until the fourth year after treatment. *P. richardsonii* and *M. sibiricum* were not found after the treatment and coontail declined dramatically. *Najas* sp., *Z. dubia,* *P. foliosus,* and sago pondweed all became more dominant members of the plant community by

TABLE 16.7
Relative Frequency (%) of Common[a] Aquatic Plants before and after a Fluridone Treatment in Crooked Lake, Minnesota

Species	Pre-Treatment (May 1992)	First Year After Treatment (July 1993)	Second Year After Treatment (August 1994)	Third Year After Treatment (August 1995)	Fourth Year After Treatment (August 1996)
Ceratophyllum demersum	21.4	0	1.7	2.4	4.6
Myriophyllum sibiricum	17.9	0	0	0	0
M. spicatum	22.6	0	0	0	3.2
Najas sp.	0	0	0	11	17.8
Potamogeton amplifolius	17.9	0	0	7.5	10
P. crispus	9.5	41.8	21.6	19.7	7.8
P. foliosus	4.8	0	14.2	26	18.9
P. pectinatus	0	47.3	39.2	18.5	17.8
P. richardsonii	6	0	0	0	0
Zosterella dubia	0	11	23.3	15	19.9

[a] Only species with a frequency more than 24% are included.

Source: After Welling, C. et al. 1997. Evaluation of Fluridone for Selective Control of Eurasian Watermilfoil: Final Report. Minnesota Dept. Nat. Res., Minneapolis.

the fourth year after treatment (Table 16.7). Initially curly-leaf pondweed became more dominant but by the fourth year after treatment, its importance declined.

Due to degradation by photolysis, adsorption to hydrosoils, plant uptake, and dilution fluridone concentrations are usually less than target values and decrease over time. Fluridone concentrations were equal to or greater than target concentrations for 30 days after application for both Zumbra and Parkers Lakes (Welling et al., 1997). Plant exposure in these lakes was probably more than needed to control milfoil (Welling et al., 1997).

Based on these results the MNDNR concluded that the unavoidable damage to non-target plants and the potential effects on other aspects of the lake ecosystem were great enough so as not to generally permit whole lake fluridone applications (Welling et al., 1997). Criteria considered to permit an application variance are: (1) high potential to eliminate milfoil from a lake, (2) low potential to damage native plants, (3) high potential for the lake to become a source for the spread of milfoil, and (4) low potential for the reintroduction of milfoil into the lake. A hypothetical situation where the MNDNR might issue a variance to allow a whole-lake fluridone treatment is a lake that: (1) has no inlet or outlet, (2) is small (less than 40 ha), and (3) is located in an area with no other milfoil lakes (Welling et al., 1997).

16.7.1.2 Wisconsin Experiences — Potters and Random Lakes

Potters and Random Lakes (Table 16.3) in southeastern Wisconsin were selected for fall fluridone treatments. Eurasian watermilfoil was confirmed present in Potters Lake in 1975, and by 1997 it had a 99% frequency. Native plants were not diverse or abundant. *Chara* sp., coontail, and *Elodea canadensis* were the most common native species (Table 16.8). Potters Lake was treated in October, 1997 with an initial target fluridone concentration of 14 µg/L. Pre- and post treatment aquatic plant, herbicide residue, and water quality data were collected as part of the permit requirements (Toshner et al., 2001).

The FasTest™ for fluridone indicated the chemical was applied evenly and averaged within 0.5 µg/L of the target concentration. Fluridone degraded more slowly than expected with a half-

TABLE 16.8
Relative Frequencies (%) of Aquatic Plants in Potters Lake,
Wisconsin before and after a Fluridone Treatment

	Pre-treatment	Post-treatment			
Species	1997	1998	1999	2000	2001
Ceratophyllum demersum	11.8	35.3	3.0	2.0	0
Chara sp.	19.3	39.7	67.3	45.7	52.5
Elodea canadensis	19.3	0	0	0	0
Myriophyllum spicatum	30.8	0	0	0	0
Najas flexilis	6.9	0	0	0	0
Nymphaea odorata	0	7.8	0.9	1.0	1.1
Potamogeton crispus	5.0	0	7.9	13.1	13.7
P. pectinatus	6.5	16.4	20.8	38.2	32.8
Zanichellia palustris	0.3	0	0	0	0
Zosterella dubia	0.9	0	0	0	0

Source: Data from Scott Toshner and Shelley Garbisch, Wisconsin Dept. Nat. Res.,
Personal communications, 2002.

life of approximately 195 days. The results were concentrations of 4–6 μg/L greater than expected 30 days after treatment and concentrations were still above 2 μg/L in July 1998 (Scott Toshner, WDNR, personal communication, 2002). Based on Secchi depth, total P, and chlorophyll *a* concentrations, post-treatment water quality increased slightly compared to the year before treatment but was similar to long-term average conditions (Scott Toshner, WDNR, personal communication, 2002).

Effectiveness criteria were set before the treatment. The treatment was considered successful if it reduced Eurasian watermilfoil to 20–30% of pre-treatment levels (essentially a frequency of 20–30%) until July 2000, and native plant frequency increased to 50% or greater. The frequency of *M. spicatum* dropped to nothing and was not recorded in the year 2000 sampling. The frequency of native plants went from 62.4% at pre-treatment, to 45.9% in 1998, 68.2% in 1999, and 90.6% in 2000. The frequency of "no plant" sampling points went from 1.2% pre-treatment, to 54.1% in 1998, 31.8% in 1999, and 9.41% in 2000. Both criteria for a successful treatment were met (Scott Toshner, WDNR, personal communication, 2002). In addition to Eurasian watermilfoil, elodea and *Najas flexilis* were eliminated. *Chara* sp. and *Potamogeton pectinatus* were the two dominant members of year 2000 plant community (Table 16.8). The exotic curly-leaf pondweed also increased in dominance by 2000.

The pre-treatment plant community in Random Lake was more diverse than Potters Lake (Table 16.9) but was dominated by Eurasian watermilfoil. The lake was treated in October 1999 at an initial target concentration of 12 μg/L. The same effectiveness criteria and sampling requirements as for Potters Lake were used (Toshner et al., 2001).

Water quality data were not given, but fluridone sampling showed that the initial treatment was right on the target concentration and the decay was slow with a 6 μg/L concentration in February 2000 and a 2 μg/L concentration still available in June 2000 (Scott Toshner, WDNR, personal communication, 2002).

Clearly the plant criteria were met on Random Lake with the frequency of *M. spicatum* dropping from 60% in 1999 to 1% in 2000 then rebounding to 9% in 2001. The native species, *P. pectinatus,* had a frequency of 48% in 2001 (John Masterson, WDNR, personal communication, 2002). The exotic *Najas marina* was not found after treatment and the native *Potamogeton amplifolius* was found only after the treatment. *Chara* sp., *P. pectinatus,* and *P. crispus* became more important

TABLE 16.9
Relative Frequencies (%) of Aquatic Plants in
Random Lake, Wisconsin before and after a
Fluridone Treatment

| | Pre-treatment | Post treatment | |
| | 1999 | 2000 | 2001 |
Species
Chara sp.	20.6	37.5	28.9
Myriophyllum spicatum	36.4	0.6	6.0
Potamogeton pectinatus	20	37.5	32.3
Najas flexilis	0.6	0	0
Potamogeton crispus	0.6	2.6	12.8
P. illinoensis	8.5	11.8	11.4
N. marina	6.1	0	0
Nymphaea odorata	3.0	3.3	0
Nuphar variegata	3.0	3.3	4
Utricularia vulgaris	0.6	0	1.3
P. natans	0.6	3.3	3.4

Source: Data supplied by John Masterson, WDNR. Personal
communications, 2002; from data in report to WDNR from Aron
and Associates.

community members after treatment (Table 16.9). Because of the rebound of Eurasian watermilfoil, spot treatment with 2,4-D was recommended to the Village of Random Lake to protect the longer-term success of the treatment (John Masterson, WDNR, personal communication, 2002).

16.7.1.3 Michigan Experiences

Four Michigan lakes (Table 16.3) were treated with low doses of fluridone as part of a U.S. Army Corps of Engineers, Aquatic Research Program (APCRP) and the Aquatic Ecosystem Restoration Foundation (AERF) research study. The primary study objective was to determine whether submersed plant diversity and frequency were impacted by whole-lake, low-dose fluridone applications in the year of treatment when targeting the control of Eurasian watermilfoil (Getsinger et al., 2001). Secondary objectives included: (1) determining herbicide effects on curly-leaf pondweed; (2) evaluating changes in species diversity at 1 year after treatment; (3) measuring the effect of thermal stratification on the water column distribution of fluridone; (4) verifying laboratory results of fluridone CET relationships with efficacy; and (5) correlating an immunoassay fluridone water residue technique with the conventional high-performance liquid chromatography method (Getsinger et al., 2001).

Observations from previous whole-lake treatments in Michigan indicated that, in many cases, plants growing at depths greater than 3.05 m were not affected by the fluridone application, even though the volume of the entire lake was used to calculate the treatment rate (Getsinger et al., 2002a). Outdoor mesocosm studies on mixed submersed plant communities suggested that fluridone application rates between 5 and 10 µg/L, with an exposure time of greater than 60 days, and with residues remaining above 2 µg/L effectively controlled milfoil with minimal effects on native, non-target species; and an early season fluridone application provided better control of Eurasian watermilfoil and enhanced selectivity than did later season applications (Getsinger et al., 2002a).

Based on the above observations, a treatment strategy was developed utilizing an initial mid-May fluridone application with a targeted concentration of 5 µg/L in the top 3.05 m of the water column. A booster application of fluridone, designed to re-establish the 5 µg/L concentration, followed 2–3 weeks after the initial application. The booster application was used to compensate

TABLE 16.10
Relative Frequency (%) of Common[a] Submergent Species before and after a Fluridone Treatment in Big Crooked Lake, Michigan

	Pre-treatment	Post treatment		
Species	May 1997	August 1997	May 1998	August 1998
Ceratophyllum demersum	7.5	19.5	5.8	8.9
Chara sp.	9.5	18.1	12.1	8.9
Myriophyllum spicatim	19.5	0	0	2.6
Najas guadalupensis	0	0	0	12.3
Potamogeton amplifolius	17.5	15.4	20.4	15.2
P. crispus	12	0.5	22.1	8.9
P. illinoensis	0	6.8	0	0
P. praelongus	20.5	6.8	0	0
P. robbinsii	1	9.5	5.8	4.1
P. zosteriformis	12.5	19.9	21.3	15.2
Zosterella dubia	0	9	0	15.2

[a] Only species with a frequency of 5% or more were considered.

Source: After Getsinger, K.D. et al. 2001. Whole-Lake Applications of Sonar for Selective Control of Eurasian Watermilfoil. Rept. ERD/EL TR-01-07. U.S. Army Corps of Engineers, Vicksburg, MS.

for any low initial fluridone residue and to extend the overall herbicide exposure period in the lakes for at least 60 days. The plant communities in four additional lakes were studied to determine if the results in the treated lakes could be attributed to fluridone treatment or to natural causes (Getsinger et al., 2002a).

Eurasian water-millfoil control was excellent in three of the lakes, with a reduction of milfoil frequency of 100% in Big Crooked, 95% in Camp, and 93% in Lobdell Lakes (Madsen et al., 2002). Eurasian watermilfoil was removed from the water column in these lakes in 8–12 weeks. The slow collapse of the milfoil canopy was likely caused by the low fluridone rates used and the advanced growth stage of the plants at the time of treatment. Fluridone treatment did not reduce total plant species diversity in these lakes and total plant cover and native plant cover remained the same or significantly increased (Madsen et al., 2002). These results may have been related to natural events as similar trends were seen in the non-treated lakes. In all cases, post treatment plant cover was maintained at levels above 60%.

Eurasian watermilfoil was not eliminated in any of the lakes (Tables 16.10–16.13). Only time will tell whether it returns to its former dominance. Curly-leaf pondweed also became more dominant, at least over the short-term, in Big Crooked and Lobdell Lakes (Tables 16.10 and 16.11). However, *Najas guadalupensis* and *Zosterella dubia* were found in Big Crooked Lake; *Potamogeton amplifolius, P. pectinatus, Ranunculus sp., Vallisneria americana,* and *Z. dubia* were found in Camp Lake; and *N. flexilis, N. gracillima, P. pectinatus, and V. americana* were found in Lobdell Lake after, but not before, the fluridone treatments (Tables 16.10–16.12).

In contrast to the above three lakes, the treatment of Wolverine Lake failed to control Eurasian watermilfoil (Madsen et al., 2002). Milfoil frequency was reduced by only 27% in the year of treatment and by August 1988 the frequency was 54%, 8% greater than in the pre-treatment evaluation. However, because of the addition of coontail, *N. gracillima, P. foliosus, P. illinoensis, P. zosteriformis, U. minor, U. vulgaris, and Z. dubia* to the post-treatment community, Eurasian watermilfoil was a much less dominant community member (Table 16.13).

TABLE 16.11

Relative Frequency (%) of Common[a] Submergent Species before and after a Fluridone Treatment in Lobdell Lake, Michigan

| | Pre-treatment | Post-treatment | | |
	May 1987	August 1987	May 1988	August 1988
Species				
Ceratophyllum demersum	3	2.3	0.4	7.1
Chara sp.	34	25.7	24.8	26.1
Myriophyllum spicatum	38	1.4	5.6	5.4
Najas flexilis	0	0	0	8.7
N. gracillima	0	0	0	4.3
Potamogeton amplifolius	8	9	10.9	3.8
P. crispus	14	0	21.3	2.2
P. illinoensis	1	12.4	9.1	2.7
P. pectinatus	0	7.6	6.1	3.3
P. zosteriformis	1	10.5	14.8	2.7
Utricularia vulgaris	1	3.3	4.3	13
Vallisneria americana	0	27.6	2.6	21.2

[a] Only species with a frequency of 5% or more were considered.

Source: After Getsinger, K.D. et al. 2001. Whole-Lake Applications of Sonar for Selective Control of Eurasian Watermilfoil. Rep. ERD/EL TR-01-07. U.S. Army Corps of Engineers, Vicksburg, MS.

TABLE 16.12

Relative Frequency (%) of Common[a] Submergent Species Before and After a Fluridone Treatment in Camp Lake, Michigan

| | Pre-treatment | Post treatment | | |
	May 1987	August 1987	May 1988	August 1988
Species				
Ceratophyllum demersum	2	1.5	0	4.4
Chara sp.	7	24.4	32.8	33.7
Elodea canadensis	16	0.1	5.8	3.3
Myriophyllum spicatum	37	1.5	5.0	5.1
Potamogeton amplifolius	0	0	0.8	2.9
P. crispus	33	12.2	35.7	12.5
P. pectinatus	0	5.6	1.7	0.4
P. praelongus	5.5	10.4	5.4	7.3
Ranunculus sp.	0	0	5.4	0.4
Vallisneria americana	0	17.4	0	16.1
Zosterella dubia	0	26.3	7.5	13.9

[a] Only species with a frequency of 5% or more were considered.

Source: After Getsinger, K.D.. et al. 2001. Whole-Lake Applications of Sonar for Selective Control of Eurasian Watermilfoil. Rept. ERD/EL TR-01-07. U.S. Army Corps of Engineers, Vicksburg, MS.

TABLE 16.13
Relative Frequency (%) of Common[a] Submergent Species before and after a Fluridone Treatment in Wolverine Lake, Michigan

Species	Pre-treatment May 1987	Post-treatment August 1987	Post-treatment May 1988	Post-treatment August 1988
Ceratophyllum demersum	0	4.1	1.2	0.4
Chara sp.	37.9	39.3	24.8	31.9
Myriophyllum spicatum	32.9	17.9	28	21.3
Najas gracillima	0	0	0	9.8
Potamogeton amplifolius	14.3	6.1	11	7.9
P. crispus	12.1	0	14.2	0.4
P. foliosus	0	0	0	7.9
P. illinoensis	0	0	1.2	4.7
P. pectinatus	2.9	18.9	14.2	0
P. zosteriformis	0	3.6	4.7	2.8
Utricularia minor	0	0	0.4	7.1
U. vulgaris	0	7.7	0	5.5
Zosterella dubia	0	2.6	0.4	0.4

[a] Only species with a frequency of 5% or more were considered.

Source: After Getsinger, K.D. et al. 2001. Whole-Lake Applications of Sonar for Selective Control of Eurasian Watermilfoil. Rep. ERD/EL TR-01-07. U.S. Army Corps of Engineers, Vicksburg, MS.

This study found that fluridone was well mixed in the area above the thermocline and it was not found below the thermocline. This has management implications: (1) in the whole-lake treatment of stratified lakes fluridone concentration should be based on the volume of water above the thermocline; and (2) thermocline depths vary as the season progresses so calculating water volumes can be difficult, especially for an herbicide that needs to be active for more than 60 days to achieve desired management. Basing fluridone concentrations on volumes greater than the thermocline depth causes higher than intended fluridone concentrations that could lead to non-target species damage. Basing fluridone concentrations on depths shallower than the thermocline lowers the intended fluridone concentrations that could lead to the lack of target species control. The later situation likely caused the failure to control Eurasian watermilfoil in Wolverine Lake. The thermocline was much deeper than the 3.05 m target depth. Remember, some lakes do not stratify or they mix often (polymictic). This must be known before herbicide dosage can be accurately calculated. Previous studies showed a significant difference in the species-selective properties of fluridone between 5 and 10 µg/L (Getsinger et al., 2002a), so maintaining the proper concentration of fluridone is critical to selective management.

The failure to control curly-leaf pondweed was also disappointing. Madsen et al. (2002) speculated that curly-leaf pondweed growth after treatment may be stimulated by the reduced competition from Eurasian watermilfoil. They suggested a fall or early spring (late March through mid-April) fluridone application at the same rates used on Eurasian milfoil would be more successful. An early season treatment has the added benefit of controlling curly-leaf prior to turion formation.

16.7.1.4 Vermont Experiences — Lake Hortonia and Burr Pond

Lake Hortonia and Burr Pond (Table 16.3) were treated with low doses of fluridone to determine whether submersed plant diversity and frequency were impacted in the year of treatment and beyond when targeting Eurasian water-millfoil control. Both lakes had widespread and diverse aquatic plant

TABLE 16.14
Species Relative Frequencies (%) before and after a Fluridone Treatment in Burr Pond, Vermont

Species	Pre-treatment			Post-treatment		
	June 1999	August 1999	June 2000[A]	August 2000	June 2001	August 2001
Ceratophyllum demersum	0.9	1.6	2.0	1.8	3.3	0.4
Chara sp.	20.7	14.7	18.9	24.6	46.4	32.2
Elodea canadensis	9.0	5.0	7.7	6.4	0	0.4
Myriophyllum sibiricum	1.7	2.1	0.6	0	0	0
M. spicatum	36.1	28.7	37.2	27.8	6.6	8.4
Najas flexilis	2.5	5.3	0	0	2.0	11.7
Nuphar variegata	1.1	2.4	3.0	3.2	6.6	5.6
Nymphaea odorata	4.2	4.5	6.4	7.1	11.9	3.8
Potamogeton amplifolius	5.9	0.5	9.1	1.8	2.7	1.4
P. gramineus	4.5	8.1	0	3.5	0.6	4.2
P. illinoensis	0.5	10.0	2.3	0	0	4.2
P. robbinsii	0.5	1.0	1.4	4.4	5.3	7.1
P. zosteriformis	3.0	3.4	6.4	2.9	3.9	4.2
Utricularia gibba	1.1	1.6	1.0	3.5	0.6	3.3
Vallisneria americana	4.5	8.9	1.7	7.1	2.0	7.5
Zosterella dubia	0.3	0.5	0.3	4.3	2.0	3.3
Other species[b]	3.6	1.8	2.0	1.7	6.0	2.2
Native species	63.9	71.3	62.8	72.2	93.4	91.6

[a] Treatment occurred in June 2000. Because fluridone is slow acting, this date is considered pre-treatment.

[b] Species with a frequency of less than 5% at all sampling times: *Potamogeton natans, P. nodosus, P. pectinatus, Ranunculus longirostris, Equisetum* sp., *Scirpus validus, Sparganium americanum, Utricularia vulgaris,* and *Megalondonta beckii.*

Source: After Getsinger, K.D. et al. 2002. Use of Whole-Lake Fluridone Treatments to Selectively Control Eurasian Watermifoil in Burr Pond and Lake Hortonia. Vermont. Rept. ERDC/EL TR-02-39. U.S. Army Corps of Engineers, Vicksburg, MS.

communities (Tables 16.14 and 16.15). Pre-treatment (June 1999) frequencies for *M. spicatum* were 67.5% for Burr Pond and 58.2% for Lake Hortonia. In addition to Eurasian watermilfoil, the exotic curly-leaf pondweed was found in Lake Hortonia. Predominant submersed native species in both lakes were *Chara* sp., *E. canadensis, P. amplifolius,* and *V. americana* (Getsinger et al., 2002b).

Both lakes were treated on June 4, 2000 at a nominal rate of 6 µg/L fluridone (both lakes were isothermal at the time of treatment except for one basin of Hortonia Lake. For details see Getsinger et al., 2002b). Both lakes were subsequently treated with a booster fluridone application on July 9, 2000 to re-set the whole-lake aqueous fluridone concentration to 6 µg/L. Herbicide residue sampling at 1 day after treatment (DAT) indicated the whole-lake concentration was 9.9 µg/L fluridone in Burr Pond. This concentration declined to 4.3 µg/L by 29 DAT and recovered to 5.6 µg/L after the booster treatment. This level slowly declined to 2.5 µg/L by 102 DAT. The aqueous concentration of fluridone in Lake Hortonia was 6.3 µg/L, 1 DAT. This level declined to 3.8 µg/L by 29 DAT, was raised to 6.1 µg/L by the booster treatment, and slowly declined to 2.8 µg/L by day 116 (Getsinger et al., 2002b).

In Burr Pond Eurasian watermilfoil was significantly reduced to a 40.8% frequency 2 months after treatment and to 9.4% frequency by 14 months after treatment. The milfoil biomass was reduced by 92% within 2 months after treatment and remained extremely low (90% reduction from pre-treatment levels) by August 2001. Eighteen native submergent plant species were found in August 2001 and the relative frequency of native species increased to 91.6% (Table 16.14). The

TABLE 16.15
Species Relative Frequencies (%) before and after a Fluridone Treatment in Lake Hortonia, Vermont[a]

Species	Pre-treatment			Post-treatment		
	June 1999	August 1999	June 2000[a]	August 2000	June 2001	August 2001
Ceratophyllum demersum	1.1	3.7	0.9	2.1	0	0.2
Chara sp.	10.3	7.4	10.9	16.6	33.0	24.1
Elodea canadensis	7.2	6.3	6.7	0.6	1.0	1.2
Myriophyllum spicatum	28.4	23.1	30.5	28.8	5.9	5.9
Nuphar variegata	1.6	0.9	3.1	0.8	2.3	1.9
Nymphaea odorata	7.5	4.0	6.9	6.7	9.6	9.2
Potamogeton amplifolius	10.9	0.7	13.8	1.9	6.9	0.7
P. crispus	0	0	3.1	0.2	5.3	3.0
P. gramineus	1.3	4.7	1.3	3.2	1.7	0.7
P. illinoensis	5.9	16.5	0.2	9.7	0	8.8
P. natans	2.1	0.9	0.5	0.2	0	0
P. praelongus	2.6	1.6	5.1	0.2	0	0.2
P. robbinsii	7.2	5.0	5.3	7.1	15.2	8.1
P. zosteriformis	3.3	1.3	3.5	1.5	5.9	5.0
P. pectinatus	0.1	2.4	1.5	0.4	2.3	4.7
Utricularia gibba	1.5	4.9	0.2	8.2	0.7	3.0
U. vulgaris	1.3	1.0	2.0	2.1	4.9	3.3
Vallisneria americana	4.9	10.1	2.3	6.7	2.3	8.1
Zosterella dubia	0.3	2.7	0.4	2.6	0.3	6.7
Other species[b]	2.4	2.7	1.8	0.4	2.6	5.2
Native species	71.6	69.5	66.4	71.0	88.8	91.1

[a] Treatment occurred in June 2000. Because fluridone is slow acting, this date is considered pre-treatment.

[b] Species with a frequency of less than 5% at all sampling times: *Najas flexilis, Potamogeton nodosus, Ranunculus longirostris, Myriophyllum sibiricum, Polygonum amphibium, Pontederia cordata,* and *Megalodonta beckii.*

Source: After Getsinger, K.D. et al. 2002. Use of Whole-Lake Fluridone Treatments to Selectively Control Eurasian Watermifoil in Burr Pond and Lake Hortonia. Vermont. Rep. ERDC/EL TR-02-39. U.S. Army Corps of Engineers, Vicksburg, MS.

frequency and relative frequency of some species, like *E. canadensis* declined, the frequency of other species like *C. demersum* declined but the relative frequency remained similar. The relative frequency of *N. flexilis* and *P. illinoensis* initially declined and then increased by August 2001. The relative frequency of *Chara* sp. increased dramatically after the fluridone treatment (Getsinger et al., 2002b).

In Lake Hortonia the *M. spicatum* frequency was reduced to 44.8% 2 months after treatment and to 8.4% 14 months after treatment. Pre-treatment milfoil biomass was reduced by 80% within 2 months after treatment and 96% by August 2001. The relative frequency of native plants increased to 91.1% by August 2001. Significant reductions in occurrence (Table 16.15) were found in the same native species in Lake Hortonia that were found in Burr Pond. In addition, curly-leaf pondweed occurrence increased but it had not become a widespread nuisance (Getsinger et al., 2002b).

Getsinger et al. (2002b) considered the treatments a success. An acceptable control of Eurasian watermilfoil was achieved while an overall richness and biomass of native plants was maintained. They attributed the native plant reductions to the direct effects of fluridone combined with a reduction of underwater light levels. They reminded managers that consistent, precise, and selective control of *M. spicatum* using low-dose fluridone requires accurate lake bathymetry, knowledge of the lakes mixing characteristics, pre-treatment thermocline information, rapid water herbicide residue analysis, and plant injury assessment. This information needs to be coupled with established fluridone CET relationships.

16.7.1.5 Increasers and Decreasers

An objective of a selective treatment, such as using fluridone, is to change community structure. Some species should increase relative to other species and others should decrease (thus using relative frequencies to illustrate community structure in Tables 16.5–16.15). A species can become a more dominant community member, thus an increaser, if it decreases less than other members, and a species can become less dominant, or a decreaser, if it increases less than other species. Hopefully, desirable species are increasers and undesirable or nuisance species are decreasers or are eliminated.

From the case studies of fluridone treatments (Tables 16.5–16.15), Eurasian watermilfoil is clearly a decreaser (Table 16.16). This was the desired outcome of the treatments. In some lakes it was not eliminated and was becoming more dominant. This was attributed to missed treatment areas, poor control due to dilution (wrong calculation of volume or inflow areas), or re-introduction of the species. As might be expected from a closely related species, the native *M. sibiricum* also decreased. *Elodea canadensis*, *Najas marina*, and *P. richardsonii* were decreasers in the limited number of lakes where they occurred. There was no effect of herbicide treatment on *C. demersum*, *P. amplifolius*, *P. praelongus*, *P. zosteriformis*, *Ranunculus* sp. and *V. americana*; or they decreased depending on herbicide concentrations. *Potamogeton crispus*, *P. pectinatus*, *P. robbinsii*, *Najas gracillima*, *Nymphaea* sp., *Utricularia vulgaris*, and *Zosterella* showed no effect or they increased with herbicide treatment. *N. guadalupensis*, *P. foliosus*, and *Utricularia minor* also increased. There was little or no effect on *Zannichellia palustris*, *P. illinoensis*, *P. natans*, *P. gramineus*, and *Nuphar variegata*. There was little pattern to the responses of *Chara* sp. and *N. flexilis* (Table 16.16).

Table 16.16 is very preliminary because some species were only found in a small number of the lakes. It can be refined as more case studies become available. However, many results are similar to those found by Smith and Pullman (1997). They found that elodea, *Najas* spp., coontail, and native watermilfoil were very susceptible to fluridone. Their response was very similar to that of Eurasian watermilfoil. *P. zosteriformis*, *V. americana*, and medium and large-leaf *Potamogeton* spp. (e.g., *P. amplifolius*, *P. gramineus*, *P. illinoensis*) exhibited intermediate sensitivity to fluridone. These species were frequently eliminated by high fluridone dose rates but usually survived treatments below 10 µg/L. They reported that the narrow-leaved pondweeds also exhibited intermediate sensitivity to fluridone, whereas Table 16.16 indicates that *P. pectinatus* and *P. foliosus* were not affected or became more dominant with a fluridone treatment. Smith and Pullman (1997) also reported that *Utricularia* spp., *Zosterella dubia*, and *P. robbinsii* were very tolerant of fluridone treatments. *Chara*, they reported, is extremely tolerant and typically develops dense carpets throughout much of the littoral zone following a fluridone treatment. *Nymphaea* sp. and *Nuphar* sp. became somewhat chlorotic after fluridone applications but their abundance did not appear to be affected by dose rates below 20 µg/L. In summary, the plant community changes and damage to non-target species will likely be dictated by the susceptibility of the species in the lake to the dose and contact time used.

16.7.2 2,4-D in Cayuga Lake, New York and Loon Lake, Washington State

16.7.2.1 Cayuga Lake

Eurasian watermilfoil began invading Cayuga Lake in the early 1960s and by the 1970s the northern end of Cayuga Lake was a continuous milfoil bed (Miller and Trout, 1985). Cayuga Lake has a surface area of 172 km². The northern 1600 ha are shallow with a depth less than 4 m. Approximately

TABLE 16.16
Species Response to Fluridone Treatment[a]

	Big Crooked	Camp	Lobdell	Wolverine	Hortonia	Burr	Parkers	Zumbra	Random	Potters	Crooked
Target conc.[b] (µg/L)	5	5	5	5	6	6	10	10	12	14	15
Assess. length[c]	1	1	1	1	1	1	2	2	3	3	4
Ceratophyllum demersum	0	0	0	0	0	0	-	-	-	-	-
Chara sp.		+			+	+			+	+	
Elodea canadensis		-	-	-	-	+	-			-	-
Myriophyllum sibiricum					0	0					
M. spicatum	-	-	-		-	-	0	-	-	-	
Najas flexilis or sp.			+			+	0		0	-	
N. gracillima			0	+							
N. guadalupensis	+										+
N. marina									-		
Nuphar variegata	0	0	-	-	0	0			0		
Nymphaea sp.	+	0			0	0		+	0	0	
Potamogeton amplifolius					-	-			-		
P. crispus	+	0	+	0	+		+	+	+	+	0
P. foliosus					0	0	+				+
P. gramineus	0										
P. illinoensis			0		0	0			0		
P. natans									0		
P. pectinatus		0	0	0			+	+	+	+	+
P. praelongus	-	0			0						
P. richardsonii											-
P. robbinsii	0		0	0	0	+	-				
P. zosteriformis	0				0	0	-	-			
Ranunculus sp.		0			0	0					
Utricularia gibba				+							
U. minor			+	+							
U. vulgaris			-		0	0			0		
Vallisneria americana		0				0				0	
Zannichellia palustris					0		+			0	
Zosterella dubia	+	+			+	0				0	+

[a] Combined results from Tables 16.5 to 16.15. + indicates an increase in relative frequency of more than 5%, — indicates a decrease in relative frequency of more than 5%, 0 indicates a change in relative frequency less than 5%, a blank indicates the species was not recorded in that lake; generally between first and last sampling unless strong seasonality of growth was suspected.
[b] Target concentration may be different than measured concentration, refer to text for differences.
[c] Growing seasons after treatment.

90% of this area was a dense milfoil stand. A 36-ha area within this stand was treated in May of 1975 with a second treatment in May 1977, with 2,4-D (butoxyethel ester granular, 20% a.i.) at the rate of 100 kg/ha. Following the 1975 treatment and for the next 5 years, vegetation was monitored at two locations within the treatment area (designated the north and south treatment location) and two control locations (Miller and Trout, 1985).

No major shifts in the aquatic macrophyte community occurred in the control areas during the study (Miller and Trout, 1985). There was a striking shift in macrophyte dominance in the north treatment location. No significant regrowth of Eurasian milfoil or any other species occurred for the remainder of the growing season following the 1975 treatment. In 1976 *Chara vulgaris* became dominant and remained so in subsequent years of the study. It constituted approximately 83% of the dry-weight biomass of the area (Miller and Trout, 1985). It decreased in June, 1977, perhaps because of the second 2,4-D treatment, but it recovered by mid-August. Species diversity remained low in this area and no growth trends were seen in other species.

In contrast, *Najas flexilis*, coontail, and curly-leaf pondweed increased at the south treatment location. *M. spicatum* also showed a greater recovery than at the north treatment location. Species diversity increased (Miller and Trout, 1985).

Interestingly, while non-treated areas maintained similar macrophyte populations throughout the study and while there was a lack of obvious differences in environmental conditions, the resulting plant communities at the sampling sites were quite different. From this experience and from a literature review, Miller and Trout (1985) classified the plant communities that result from herbicide treatment into three general categories: (1) the regrowth of the target species into dense, monospecific stands, (2) the growth of a community dominated by a species resistant to the herbicide (e.g., the dominance of *Chara* in the North Treatment area), and (3) a mixed macrophyte community containing target species, resistant species, and others (e.g., the higher diversity, mixed plant community of the South Treatment area). It is difficult to predict which plant community will develop after an herbicide treatment. However, Miller and Trout (1985) suggested that the probability of the first type increases where there is a survival of the rootstock of the target plant. The probability of the second type increases where there is a species present that has a high reproductive and growth potential.

They also suggested that the availability of sexual or asexual propagules from peripheral areas influences the composition and succession of the resulting community. Chemical management should consider leaving untreated areas with preferred species adjacent to treated areas as a source of recolonizing material and, where practical, herbicide application should correspond to the maturity of reproductive structures of desirable species in this "reproductive stock" area.

16.7.2.2 Loon Lake

Loon Lake has a 445 ha area, a 30.5 m maximum depth, and a 14 m mean depth. Eurasian watermilfoil was first found in the lake in September 1996. During the summer of 1997 diver hand-pulling and benthic barriers were used in an attempt to control the population. By the end of the summer it was evident that milfoil was continuing to spread beyond the levels that divers could contain, but it was still limited to small patches within 24 ha in the northern half of the lake (Parsons et al., 2001). Water depths in this area were less than 3 m deep. The area was treated on July 8, 1998 with a target 2,4-D (butoxyethel ester granular, 19% a.i.) concentration of 1 to 2 mg/L for 24 to 48 h. The biomass and frequency of plant occurrence were assessed before treatment and at six weeks and one year after treatment in both treated and untreated areas.

Biomasses in the untreated plots showed no significant changes except for *V. americana* that increased significantly in August, 1998 sampling as compared to June 1998 and 1999 samples. A similar change occurred in the treated area. This probably resulted from the seasonal growth pattern of this species (Parsons et al., 2001). Eurasian watermilfoil biomass decreased by 98% by six weeks after treatment. One year after treatment milfoil biomass was still reduced by 87% compared to

TABLE 16.17
Relative Frequencies (%) of Submergent Species Before and After a
2,4-D Treatment in Loon Lake, Washington

Species	Pre-treatment	Post-treatment	
	June 1998	August 1998	June 1999
Chara sp.	15.2	11.8	13.7
Elodea canadensis	3.6	3.6	4.8
Megalodonta beckii/Myriophyllum sibiricum[a]	13.8	11.8	8.2
Myriophyllum spicatum	12.3	3.0	2.7
Najas flexilis	2.2	6.5	2.7
Potamogeton amplifolius	19.6	18.3	18.5
P. gramineus	7.2	7.1	8.2
P. robbinsii	12.3	17.8	13.7
P. zosteriformis	2.2	1.2	4.8
P. pectinatus	5.8	4.1	9.6
Utricularia vulgaris	3.6	4.7	9.6
Vallisneria americana	2.2	10.1	3.4

[a] These two species were not separated in the field.

Source: After Parsons, J.K. et al. 2001. *J. Aquatic Plant Manage.* 39: 117–125.

pre-treatment levels. *Megalodonta beckii* was the only other species where biomass in the treated area changed significantly. *M. beckii* biomass decreased between August 1998 and June 1999. Again, this was probably the result of the seasonal growth pattern of this species (Parsons et al., 2001).

Other than the decline of *M. spicatum*, there was little difference in plant coverage before and after the treatment. The frequency of quadrats with no plant was 12–13%. The relative frequencies of most species (Table 16.17) varied little before and after treatment.

By June 1999, Eurasian watermilfoil was found in areas where it was not found in June 1998, indicating it was still spreading in Loon Lake. The 2,4-D treatment may have slowed milfoil spread, but it had not halted it.

In summary the 2,4-D treatment selectively and significantly reduced the biomass and frequency of Eurasian watermilfoil in Loon Lake both during the year of treatment and one year after treatment without significantly impacting native species. However, one year after treatment Eurasian milfoil was still spreading so continued management will be required to control it.

16.7.3 TRICLOPYR IN PEND OREILLE RIVER, WASHINGTON STATE AND LAKE MINNETONKA, MINNESOTA

Triclopyr has an activity spectrum similar to 2,4-D and other auxin-type growth regulating, phenoxy herbicides. It is toxic to most dicots; monocot species are not as adversely affected by triclopyr applications. Presently it is not a registered herbicide and can only be used under an experimental use permit in the United States.

16.7.3.1 Pend Oreille River

In August 1991 two areas of the Pend Oreille River with very different flow characteristics were treated with triclopyr. The riverine plot varied in depth from 0.3 m to 2.5 m. Flows were non-detectable within the plant bed but the half-life for water exchange was 20 hours. A 6-ha plot within the riverine area was treated with a 2.5 mg/L dose of triclopyr. A cove plot ranged in depth from 0.75 m to 2.8 m. Again, water flow was non-detectable in the plant beds and the half-life for water

TABLE 16.18

Relative Frequency (%) of Submergent Species before and after Triclopyr Treatments in the Pend Oreille River, Washington

Species	Pre-treatment (1991)			One Year after Treatment (1992)			Two Years after Treatment (1993)		
	Control Plot	River Plot	Cove Plot	Control Plot	River Plot	Cove Plot	Control Plot	River Plot	Cove Plot
Ceratophyllum demersum	1.1	4.8	9.4	3	8.2	22.9	3.6	8.2	20.9
Elodea canadensis	1.6	3.7	13.1	5.5	14.7	36	7.3	10.6	27.1
Myriophyllum sibiricum	0	3.7	0	0	0.3	0	0	0	0
M. spicatum	54.1	50	41.8	59.8	16.5	9.7	34.7	25.1	20.2
M. verticillatum	0	0.5	0	0	0.3	0	0.4	1.6	0
Potamogeton crispus	9.2	2.1	3.3	16.6	3.9	5.8	31.8	3.9	10.3
P. nodosus	4.3	0.5	0	3	0.3	0	1.8	0	0
P. obtusifolius	0	0	2.8	0	11.5	2.7	0	2.8	0.3
P. pectinatus	6.5	2.7	5.2	0	2.6	0.4	2.7	2.3	0.7
P. perfoliatus	1.1	1.1	0.5	0	1.8	0.4	1.1	1	0.3
P. praelongus	0	0	0	0	0	0	0	0.3	0.3
P. pusillus	0	0	0	0	0	0	0.4	10.3	0.3
P. vaseyii	0	5.3	3.8	0	0	0.4	0	0.3	0
P. zosteriformis	8.2	4.9	18.8	6.7	18.8	14.0	5.8	24.8	18.2
Ranunculus longirostris	2.7	.4	1.4	4.9	14.7	7.4	7.7	5.1	0.3
Zosterella dubia	1.6	4.3	0	0.6	2.4	0.4	2.9	5.8	1.0

Source: After Getsinger, K.D. et al. 1997. *Reg. Rivers Res. Manage.* 13: 357–375.

exchange was greater than 50 hours. A 4-ha cove plot was treated with a 1.75 mg/L dose of triclopyr. An untreated riverine section was used for comparison (Getsinger et al., 1997).

The biomass of Eurasian watermilfoil at 4 weeks post-treatment was 1% of pre-treatment levels in both the cove and riverine treatments. One year after treatment milfoil biomass in the river plot was 28% of pre-treatment levels and 1% of pre-treatment levels in the cove area. Two years after treatment milfoil biomass was still significantly lower (47–66%) in both plots. This compares to milfoil biomass in the control plot where it remained the same or increased during the same time period (Getsinger et al., 1997).

Total biomass was significantly reduced 4 weeks after the triclopyr treatment. However, there was no effect on total community biomass 1 and 2 years after treatment. The reduction in milfoil biomass was compensated for by an increase in native plant biomass (Getsinger et al., 1997). Native plant biomass remained significantly higher in both treated plots 2 years after treatment. The plant community biomass was not significantly affected over the long term.

Milfoil root crowns were severely damaged or completely destroyed in the treated plots by four weeks after treatment. Therefore, Getsinger et al. (1997) concluded that most of the regrowth was from stem fragments carried into the treated areas from adjacent, non-treated areas.

The relative frequency of milfoil dropped from pretreatment levels of 50% (river plot) and 42% (cove plot) to 16.5% and 9.7% 1 year after treatment and 25.1% and 20.2% 2 years after treatment (Table 16.18) indicating the drop in milfoil importance in the plant community. In contrast, coontail and elodea importance nearly or more than doubled in both treated plots by 2 years after treatment. *Potamogeton obtusifolius* importance increased dramatically the first year after treatment, and *P. pusillus* and *P. zosteriformis* the second year after treatment in the river plot (Table 16.18). In addition *P. obtusifolius, P. praelongus,* and *P. pusillus* were not found before treatment but were found after treatment in one or both of the treated plots (Table 16.18).

Although triclopyr targets dicots, the increase in dicot diversity along with an increase in monocots (primarily *Potamogeton* species and elodea) substantially increased the diversity of the treated plots. The increased diversity probably resulted from the removal of the dense milfoil canopy and a seed/propagule bank sufficient to reestablish native plants. It also appeared that the native plant community that resulted from treatment delayed the reestablishment of problematic milfoil levels for up to three growing seasons (Getsinger et al., 1997) even though the treated areas were relatively small and were adjacent to dense milfoil beds. As might be expected the triclopyr treatment had little effect on the non-native, monocot, curly-leaf pondweed. Its relative frequency tripled in the cove plot by two years after treatment (Table 16.18).

This restoration effort confirmed the efficacy of laboratory CET values. In fact, enhanced field efficacy was observed. Getsinger et al. (1997) attributed this enhanced field efficacy to levels of environmental stress (e.g., wave action, currents, turbidity, microbes, and pathogens) that are lacking or minimized under laboratory conditions. Milfoil was partially controlled for up to 250 m down-stream from the riverine plot. No plant injury was observed for more than 20 m from any other boundaries of either plot. Getsinger et al. (1997) suggested that even lower triclopyr concentrations (0.25 mg/L) could be used with the same results if the exposure time could be increased from 12 to 24 hours. This could occur in some regulated rivers by modifying dam operations for a short time period to reduce flow through the system.

16.7.3.2 Lake Minnetonka

Lake Minnetonka is a 5,801 ha lake with a mean depth of 6.9 m and a maximum depth of 30.8 m. It is composed of 15 morphologically distinct basins. Two test plots, one in Carsons Bay and one in Phelps Bay, of approximately 6.5 ha each were treated at the full label rate of 2.5 mg/L of triclopyr on July 23, 1994. An additional area, Carman Bay, was used as an untreated control plot.

After initial Eurasian watermilfoil biomasses of 57 g/m^2 in Phelps Bay and 42 g/m^2 in Carsons Bay, no milfoil biomass was in either area 6 weeks after treatment (Petty et al., 1998) The milfoil biomass in Carman Bay (270 g/m^2) did not significantly change during this period. At 1 year after treatment, milfoil biomass recovered to approximately 25% of pretreatment levels in Phelps Bay but only low levels were found in Carsons Bay. The milfoil found in these areas was small rooted stem fragments that drifted into the plots from other areas on the lake. Few fragments were found in Carsons Bay because of a restricted water entrance, compared with Phelps Bay, which is open to the lake.

Native plant biomass significantly decreased in Carsons Bay 6 weeks after treatment. The mean biomass of native plants in Phelps Bay was unchanged from pretreatment levels (Petty et al., 1998). The longer exposure time to triclopyr in Carsons Bay probably caused this decreased native plant biomass (Petty et al., 1998). Native plant biomass was not eliminated in Carsons Bay and increased to pretreatment levels by 1 year after treatment. Native plant biomass also increased in Carman Bay 1 year after treatment primarily due to an increase in coontail.

Milfoil frequency also decreased in the treated areas from approximately 70% before treatment to 0% 6 weeks after treatment. One year after treatment, milfoil frequency increased to 50% of pretreatment levels in Phelps Bay but only 15% of pretreatment levels in Carsons Bay (Petty et al., 1998). Again, recovery of Eurasian watermilfoil was due to fragments floating into the treated area from other sites on the lake.

Native plant coverage also decreased at both treated sites by about 5–10%. This apparent contradiction, i.e., an increase in native plant biomass but a decrease in the area of coverage, is explained by a significant increase in the biomass of coontail without a concomitant increase in coverage (Petty et al., 1998). By 1 year after treatment native plant coverage was significantly higher in all three plots, although the untreated reference plot was still dominated by Eurasian watermilfoil. Treatment with triclopyr at the full label rate likely caused the mortality of some native species, particularly in Carsons Bay where slow water exchange reduced triclopyr dissipation.

By 1 year after treatment, native plant diversity recovered to near pretreatment levels in the treated areas and was increasing in Carman Bay (Petty et al., 1998).

16.8 COSTS

Definitive costs are difficult to provide. The variability is high depending on the equipment used, the herbicide, the dose, the size of the treated area, the difficulty treating the area; and depreciation, maintenance, fuel, labor, overhead, administrative and contingency costs. Although costs are often cited, it is difficult to determine what those costs included so it is difficult to determine comparability. Also, the year-to-year cost variability makes comparisons difficult. For the types of treatments discussed in this chapter, that is a single or a few low dose treatments, the best option is probably to work with a commercial applicator. They have the equipment and expertise to do the job. Good applicators have trained personnel, well-calibrated application equipment, experience, and good insurance coverage. For small jobs, like treatments in front of individual properties, they often charge a flat fee, depending on the amount of frontage and the distance from shore treated. The fee covers season-long control, so retreatment is covered in the fee. For large areas, seek competitive bids and have a contract specifying exactly the type of treatment. Monitoring and sampling costs also need to be included (see Chapter 11). The plant community should be monitored to determine the efficacy of the treatment. Herbicide concentrations and residue should be monitored to determine if treatment was adequate based on CET criteria. Factors like dissolved oxygen, nutrient status, water transparency, and some sampling of non-target organisms should be monitored to determine environmental effects of the treatment and for protection against any complaints about the treatment. Although monitoring is an added cost, the educational value for keeping people informed is worth the expense.

16.9 CONCLUDING REMARKS

The herbicides registered for aquatic plant management are limited. Table 16.19 summarizes the advantages and disadvantages of each. Because there is a limited market, new herbicides are not likely in the near future. Advances in chemical control technology will probably be made using currently available herbicides. Promising technologies like plant growth regulators have not been commercialized.

Advances in chemical plant management are likely to be: (1) in the development of new uses for existing compounds (e.g., formulation improvements); (2) application methods that improve the precision of treatments to maximize the impact to target species and minimize the impact to non-target organisms; and (3) determining the compatibility of herbicides with other aquatic plant management methods for use in integrated management (Murphy and Barrett, 1990). Rigorous assessment procedures to determine the environmental safety of aquatic herbicides would raise the "comfort level" for using herbicides.

Selective control based on CET relationships is a very active research area and is one method of improving treatment precision. However, from the case studies above, CET recommendations can be hard to attain under field conditions. The dose needed for some applications is very precise; too large a dose causes collateral damage to non-target species; too low a dose lacks efficacy. Water volumes needed to calculate target herbicide concentrations can be difficult to determine, especially where there is a thermocline or in flowing water.

Systemic herbicides are the most physiologically selective herbicides so they have been used for treating large areas to change plant communities. Contact herbicides are used in a locationally selective manner for spot treatments, especially for emergent, free-floating, and floating-leaf species. Contact herbicides can also be used in a time-selective manner by using phenological information to treat nuisance species when desirable species are not present. There is also evidence that some

TABLE 16.19
Suggested Uses for Registered Herbicides[a]

Compound	Max. Water Conc. (µg/L)	Exposure Time	Advantages	Disadvantages	Where Effective	Plant Response
Complexed copper	1	Intermediate (18–72 h)	Inexpensive, rapid action, approved for drinking water	Not biodegradable, but biologically inactive in sediments	Lakes as algicide, herbicide in high water exchange areas	Broad-spectrum, acts in 7–10 days or up to 4–6 weeks
2,4-D	2	Intermediate (18–72 h)	Inexpensive, systemic	Public perception	Water hyacinth, Eurasian watermilfoil, purple loosestrife (*Lythrum salicaria*) in lakes and slow-flow areas	Selective to dicots, acts in 5–7 days, up to 2 weeks
Diquat	2	Short (12–36 h)	Rapid action, limited drift	No effect on underground portions	Shorelines, spot treatments, high water exchange rate areas	Broad spectrum acts in 7 days
Endothall	5	Short (12–36 h)	Rapid action, limited drift	No effect on underground portions	Shorelines, spot treatments, high water exchange rate areas	Broad spectrum acts in 7–14 days
Fluridone	0.15	Very long (30–60 days)	Very low dosage required, few label restrictions, systemic	Very long contact period	Small lakes, slow flowing systems	Broad spectrum, acts in 30–90 days
Glyphosate	0.2	Not applicable	Widely used, few label restrictions	Very slow action, no submersed control	Nature preserves and refuges; emergent and floating leaved-plants only	Broad spectrum, acts in 7–10 days, up to 4 weeks
Triclopyr	2.5	Intermediate (12–60 h)	Selective, systemic	Not labeled for general aquatic use	Lakes and slow-flow areas, purple loosestrife control	Selective to dicots, acts in 5–7 days, up to 2 weeks

Source: After Madsen, J.D. 2000. Advantages and Disadvantages of Aquatic Plant Management. Tech. Rept. ERDC/EL MP-00-01. U.S. Army Corps of Engineers, Vicksburg, MS.

species are more susceptible to contact herbicides than are other species so there is potential for using them in a dose-selective manner.

The case studies mostly involved Eurasian watermilfoil. Although this may appear to be a limited perspective, it makes an ideal species to test the ability to change community structure with herbicides. Eurasian watermilfoil is: (1) an Eurasian exotic in North America; (2) a serious aquatic nuisance; (3) susceptible to low doses of fluridone; (4) a dicot, so it is susceptible to 2,4-D and triclopyr; and (5) invasive in many lakes that previously had diverse aquatic plant communities.

From the case studies it appears that nuisance plants are usually not eliminated using herbicides. Additional plant management is likely needed although it may not be herbicidal treatment. In some cases regulators felt there was an unacceptable level of collateral damage to non-target species. In other cases the collateral damage was acceptable and non-target species recovered after treatment.

The resulting plant community after herbicide treatment is difficult to predict. If a diverse native plant community returns it appears to slow down the reinvasion or reduce the dominance of the target species. Reinvasion of treated areas with the nuisance species is a problem. Reinvasion can occur from untreated areas of the lake, from areas where the treatment was not successful, or by reintroduction into the lake. Invasion by desirable species is also possible and should be considered in a management plan. A treatment that maintains a desirable plant refugia and is timed when desirable species are most likely to produce reproductive parts enhances the probability of desirable plant invasions.

Timing of the treatment is very important. As mentioned above, timing can be used for selective control and it can be used to increase the probability of invasion by desirable species. Treatments done when the target plant is most physiologically susceptible enhances efficacy (see Chapter 11 on resource allocation and phenology).

REFERENCES

Adams, M.S. and R.T. Prentki. 1982. Biology, metabolism and functions of littoral submersed weedbeds of Lake Wingra, Wisconsin, U.S.: A summary and review. *Arch. Hydrobiol./Suppl.* 62(3/4): 333–409.

Adams, M.S. and K. Schulz. 1987. Concentration Effects of Diquat Herbicide on Selected Aquatic Macrophytes of Wisconsin. Res. Rept., Dept. Botany, University of Wisconsin, Madison.

Allen, N.P. 2001. Aquatic plant management of the St. John River (abstract). In: 41st Annu. Mtg. Aquatic Plant Manage. Soc., Minneapolis, MN.

Anderson, L.W.J. 1986. Annual Report — 1986, Aquatic Weed Control Investigation. USDA, Agric. Res. Serv., Dept. Botany, California, Davis.

Anderson, L.W.J. 1987. Annual Report — 1987, Aquatic Weed Control Investigation. USDA, Agric. Res. Serv., Dept. Botany, University of California, Davis.

Anderson, L.W.J. and N. Dechoretz. 1988. Bensulfuron methyl: a new aquatic herbicide. In: Proc. 22nd Annu. Meet., Aquatic Plant Cont. Res. Prog., Misc. Paper A-88-5. U.S. Army Corps of Engineers, Vicksburg, MS. pp. 224–235.

Anderson, L.W.J. and P. Thalken. 2001. California's water hyacinth and *Egeria densa* control program: compliance with the National Pollution Discharge Elimination System (NPDES) permit requirements and the U.S. Fish and Wildlife "Section Seven" (abstract). In: 41st Annu. Mtg. Aquatic Plant Manage. Soc., Minneapolis, MN.

Bartell, S.M., K. Campbell, C.M. Lovelock, S.K. Nair and J.L. Shaw. 2000. Characterizing aquatic ecological risks from pesticides using a diquat dibromide case study III. Ecological process models. *Environ. Toxicol. Chem.* 19: 1441–1453.

Binning, L., B. Ehart, V. Hacker, R.C. Dunst, W. Gojmerac, R. Flashinski and K. Schmidt. 1985. *Pest Management Principles for Commercial Applicator: Aquatic Pest Control.* University of Wisconsin-Ext., Madison.

Brooker, M.P. and R.W. Edwards. 1973. Effects of the herbicide paraquat on the ecology of a reservoir. *Freshwater Biol.* 3: 157–176.

Cooke, G.D. 1988. Lake and reservoir restoration and management techniques, In L. Moore and K. Thorton (Eds.), The Lake and Reservoir Restoration Guidance Manual. USEPA 440/5-88-02. pp. 6-20–6-34.

Crawford, S.A. 1981. Successional events following simazine applications. *Hydrobiologia* 77: 217–223.

Dunst, R.C. 1982. Sediment problems and lake restoration in Wisconsin. *Environ. Int.* 7: 87–92.

Dunst, R.C., S.M. Born, P.D. Uttormark, S.A. Smith, S.A. Nichols, J.O. Peterson, D.R. Knauer, S.L. Serns, D.R. Winter and T.L. Wirth. 1974. Survey of Lake Rehabilitation Techniques and Experiences. Tech. Bull. 75. Wisconsin Dept. Nat. Res., Madison, WI.

Engel, S. 1990. Ecosystem Responses to Growth and Control of Submerged Macrophytes: A Literature Review. Tech. Bull. 170. Wisconsin Dept. Nat. Res., Madison, WI.

Getsinger, K.D. 1997. Appropriate use of aquatic herbicides. *LakeLine* 17(1): 20.

Getsinger, K.D., G.J. Davis and M.M. Brinson. 1982. Changes in a *Myriophyllum spicatum* L. community following a 2,4-D treatment. *J. Aquatic Plant Manage.* 20: 4–8.

Getsinger, K.D., E.G. Turner, J.D. Madsen and M.D. Netherland. 1997. Restoring native vegetation in an Eurasian watermilfoil dominated plant community using the herbicide triclopyr. *Reg. Rivers Res. Manage.* 13: 357–375.

Getsinger, K.D., J.D. Madsen, T.J. Koschnick, M.D. Netherland, R.M. Stewart, D.R. Honnell, A.G. Staddon and C.S. Owens. 2001. Whole-Lake Applications of Sonar for Selective Control of Eurasian Water-milfoil. Rep. ERD/EL TR-01-07. U.S. Army Corps of Engineers, Vicksburg, MS.

Getsinger, K.D., J.D. Madsen, T.J. Koschnick and M.D. Netherland. 2002a. Whole lake fluridone treatments for selective control of Eurasian watermilfoil: I. application strategy and herbicide residues. *Lake and Reservoir Manage.* 18: 181–190.

Getsinger, K.D., R.M. Stewart, J.D. Madsen, A.S. Way, C.S. Owens, H.A. Crosson and A.J. Burns. 2002b. Use of Whole-Lake Fluridone Treatments to Selectively Control Eurasian Watermifoil in Burr Pond and Lake Hortonia. Vermont. Rep. ERDC/EL TR-02-39. U.S. Army Corps of Engineers, Vicksburg, MS.

Grimshaw, H.J. 2002. Nutrient release and detritus production by herbicide-treated freely floating aquatic vegetation in a large, shallow subtropical lake and river. *Arch. Hydrobiol.* 154: 469–490.

Hanson, M.J. and H.G. Stefan. 1984. Side effects of 58 years of copper sulfate treatment of the Fairmont Lakes, Minnesota. *Water Res. Bull.* 30: 889–900.

Hestand, R.S. and C.C. Carter. 1977. Succession of various aquatic plants after treatment with four herbicides. *J. Aquatic Plant Manage.* 15: 60–64.

Hilsenhoff, W.L. 1966. Effect of diquat on aquatic insects. *J. Econ. Ent.* 59: 1520–1521.

Kannenberg, J.R. 1997. Aquatic pesticide application - past, present, future (An applicator's view). *LakeLine* 17(1): 22.

Klaine, S.J. 1986. Influence on thidiazuron on propagule formation in *Hydrilla verticillata*. *J. Aquatic Plant Manage.* 18: 27–29.

Langeland, K.A. 1997. Aquatic plant management techniques. In: M.V. Hoyer and D.E. Canfield (Eds.), *Aquatic Plant Management in Lakes and Reservoirs*, NALMS and Aquatic Plant Manage. Soc., Madison, WI and Lehigh, FL. pp. 46–72.

Langeland, K.A. 1998. Environmental and public health considerations. In: K.A. Langeland (Ed.), Training Manual for Aquatic Herbicide Applicators in the Southeastern United States, Florida, Cent. Aquatic Invasive Plants (internet edition), Gainesville, FL.

Lembi, C.A. 1998. *Category 5, Aquatic Pest Control*. Dept. Botany, Purdue, W. LaFayette, IN.

Lembi, C.A. and M. Netherland. 1990. Bioassay of Plant Growth Regulator Activity on Aquatic Plants. Tech. Rept. A-90-7. U.S. Army Eng., Waterways Exp. Sta., Vicksburg, MS.

Lueschow, L.A. 1972. Biology and Control of Selected Aquatic Nuisances in Recreational Waters. Tech. Bull. 57. Wisconsin Dept. Nat. Res., Madison, WI.

Madsen, J.D. 2000. Advantages and Disadvantages of Aquatic Plant Management. Tech. Rept. ERDC/EL MP-00-01. U.S. Army Corps of Engineers, Vicksburg, MS.

Madsen, J.D.. K.D. Getsinger, R.M. Stewart and C.S. Owens. 2002. Whole lake fluridone treatments for selective control of Eurasian watermilfoil: II. Impacts on submersed plant communities. *Lake and Reservoir Manage.* 18: 191–200.

McNabb, T. 2001. Using barrier curtains to isolate Eurasian milfoil treatment areas during a sonar herbicide application (abstract). In: *41st Annu. Mtg. Aquatic Plant Manage. Soc.*, Minneapolis, MN.

Miller, G.L. and M. Trout. 1985. Changes in the aquatic plant community following treatment with the herbicide 2,4-D in Cayuga Lake, New York. In L. Anderson (Ed.), *Proc. First Int. Symp. on Watermilfoil (Myriophyllum spicatum) and Related Halogagaceae Species.* Aquatic Plant Manage. Soc., Vancouver, BC. pp. 126–138.

Moss, B., J. Madgwick and G.L. Phillips. 1996. *A Guide to the Restoration of Nutrient-enriched Shallow Lakes.* Broads Authority, Norwich, Norfolk, UK.

Murphy, K.J. and P.R.F. Barrett. 1990. Chemical control of aquatic weeds, In A. Pieterse and K. Murphy (Eds.), *Aquatic Weeds, The Ecology and Management of Nuisance Aquatic Vegetation.* Oxford University Press, Oxford, UK. pp. 136–173.

Nelson, L.S. and T.K. Van. 1991. Growth Regulation of Eurasian Watermilfoil and Hydrilla using Bensulfuron Methyl. Aquatic Plant Cont. Res. Prog. Rep. A-91-1. U.S. Army Corps of Engineers, Vicksburg, MS.

Netherland, M.D., W.R. Green and K.D. Getsinger. 1991. Endothall concentration and exposure time relationships for the control of Eurasian watermilfoil and hydrilla. *J. Aquatic Plant Manage.* 29:61–67.

Netherland, M.D., B. Kiefer and C.A. Lembi. 2001. Use of plant assay techniques to screen for tolerance and to improve selection of fluridone use rates (abstract). In: 41st Annu. Mtg. Aquatic Plant Manage. Soc., Minneapolis, MN.

Newbold, C. 1976. Environmental effects of aquatic herbicides. *Proc. Symp. Aquatic Herbicides, British Crop Prot. Council Monograph* 16: 78–90.

Nichols, S.A. 1991. The interaction between biology and the management of aquatic macrophytes. *Aquatic Bot.* 41: 225–252.

O'Dell, K.M., J. VanArman, B.H. Welch and S.D. Hill. 1995. Changes in water chemistry in a macrophyte dominated lake before and after herbicide treatment. *Lake and Reservoir Manage.* 11: 311–316.

Parsons, J.K., K.S. Hamel, J.D. Madsen and K.D. Getsinger. 2001. The use of 2,4-D for selective control of an early infestation of Eurasian watermilfoil in Loon Lake, Washington. *J. Aquatic Plant Manage.* 39: 117–125.

Paul, E.A., H.A. Simonin, J. Symula and R.W. Bauer. 1994. The toxicity of diquat, endothall, and fluridone to the early life stages of fish. *J. Fresh Water Ecol.* 9(3): 229–239.

Petty, D.G., K.D. Getsinger, J.D. Madsen, J.G. Skogerboe, W.T. Haller, A.M. Fox and B.A. Houtman. 1998. Aquatic Dissipation of the Herbicide Triclopyr in Lake Minnetonka, Minnesota. Aquatic Plant Cont. Res. Prog. Tech. Rept. A-98-1. U.S. Army Corps of Engineers, Vicksburg, MS.

Scheffer, M. 1998. *Ecology of Shallow Lakes.* Chapman Hall, London.

Scheffer, M., S.H. Hosper, M.-L. Meijer, B. Moss and E. Jeppesen. 1993. Alternative equilibria in shallow lakes. *Trends Ecol. Evol.* 8: 275–279.

Shearer, R.W. and M.T. Halter. 1980. Literature Reviews of Four Selected Herbicides: 2,4-D, Dichlobenil, Diquat, and Endothall. Municipality of Metropolitan Seattle, WA.

Smith, C.S. and G.D. Pullman. 1997. Experiences using Sonar[R] A.S. aquatic herbicide in Michigan. *Lake and Reservoir Manage.* 13: 338–346.

Thayer, D.D. 1998. Adjuvants in aquatic plant management. In K.A. Langeland (Ed.), Training Manual for Aquatic Herbicide Applicators in the Southeastern United States. University of Florida, Ctr. Aquatic Invasive Plants (Internet ed.), Gainesville.

Toshner, S., D.R. Helsel and K. Aron. 2001. Selective control of Eurasian watermilfoil using a fall Sonar[TM] treatment in Potters Lake, Walworth County, and Random Lake, Sheboygan County, Wisconsin (abstract). In: *Proc. 21st Int. Symp. North. Amer. Lake Manage. Soc.*, Madison, WI.

van Nes, E.H., M. Scheffer, M.S. van den Berg and H. Coops. 2002. Aquatic macrophytes: restore, eradicate or is there compromise? *Aquatic Bot.* 72: 387–403.

Wade, P.M. 1981. The long-term effects of aquatic herbicides on the macrophyte flora of fresh water habitats-a review. In: *Proc. Assoc. Applied Biol. Conf.: Aquatic Weeds and Their Control.* pp.223–240.

Wade, P.M. 1982. The long-term effects of herbicide treatment on aquatic weed communities. In: *Proc. Eur. Weed Res. Soc., Sixth Symp. on Aquatic Weeds.* pp. 278–285.

WDNR. 1988. Environmental Assessment Aquatic Nuisance Control (NR 107) Program. Wisconsin Dept. Nat. Res., Madison, WI.

Welling, C., W. Crowell and D.J. Perleberg. 1997. Evaluation of Fluridone for Selective Control of Eurasian Watermilfoil: Final Report. Minnesota Dept. Nat. Res., Minneapolis.

Westerdahl, H.E. and K.D. Getsinger. 1988. Aquatic Plant Identification and Herbicide Use Guide, Volume II: Aquatic Plants and Susceptibility to Herbicides. Aquatic Plant Cont. Res. Prog. Tech. Rept. A-88-9. U.S. Army Corps of Engineers, MS.

Westerdahl, H.E., K.D. Getsinger and W.R. Green. 1988. Efficacy of Sediment-Applied Herbicides Following Drawdown in Lake Acklawaha, Florida. Infor. Exchange Bull., Vol. A-88-1. U.S. Army Corps of Engineers, Vicksburg, MS.

17 Phytophagous Insects, Fish, and Other Biological Controls

17.1 INTRODUCTION

Mechanical and chemical methods (Chapters 12, 13, 14, 16, and 20) are the primary management procedures for nuisance aquatic plants. They are often successful, usually expensive, and frequently provide only relatively short-term control. There has been a widespread, sometimes justified, fear of herbicides. Mechanical/physical techniques can be slow, ineffective, subject to breakdowns, and may spread the infestation. Neither type of method is selective, but instead provides temporary elimination of most plants, including the target plant, usually producing habitat removal instead of restoration of the community to a prior and more desirable condition.

Eight exotic aquatic plants have proliferated in lakes of North America and elsewhere. They are: Hydrilla (*Hydrilla verticillata* (L.f.) Royle), Water hyacinth (*Eichhornia crassipes* (Mart.) Solms-Laubach), Alligatorweed (*Alternanthera philoxeroides* (Mart.) Griseb.), Eurasian watermilfoil (*Myriophyllum spicatum* L.), Floating Fern (*Salvinia molesta* D.L. Mitchell), and Waterlettuce (*Pistia stratiotes* L.), curly leafed pondweed (*Potamogeton crispus* L), and Brazilian elodea (*Egeria densa* Planch. (= *Anacharis densa* (Planch.) Vict.). Their success is due to invasions of highly favorable, often disturbed, habitats where biological controls are limited or absent, rather than a response to eutrophication. The problem is acute in southern U.S. states where there is an abundance of shallow, warm, naturally fertile aquatic habitats, and a long growing season.

The widespread economic damage and inconvenience caused by these plants, coupled with dissatisfaction with mechanical and chemical methods, has led to the development of biological controls, including phytophagous insects and fish, plant pathogens such as fungi and viruses, and allelopathy. Biological controls, including food web manipulations (Chapter 9) and use of barley straw for management of algal biomass, are not without problems, including slow response, inability to eradicate the nuisance plant or treat a problem area such as a beach, low predictability, and the potential to create additional problems if the biological control organism has unintended and undesirable impacts.

This chapter describes some of these biological control methods, focusing primarily on aquatic plant management. Their deployment is recent, and there is much to be learned. Our reliance on mechanical and chemical methods has been necessary during the early years of aquatic plant control, and they continue to be important tools. The future may lie with integrating traditional techniques with biological ones, an approach requiring sustained efforts to better understand aquatic ecosystems, and to monitor closely those treated with any of these methods.

Biological control differs substantially from mechanical, and especially chemical, techniques. The objective of biological control is to significantly reduce target plant biomass without eradication (which would also eradicate the biocontrol organism). The goals are to identify a biological agent specific to the target plant, to establish a dynamic equilibrium between this organism and the plant at an acceptable level of plant biomass, and to return the system to an earlier and more desirable community structure. Biocontrol is a suppression technique. There is no goal of plant elimination (Grodowitz, 1998). Plant biomass control will be achieved slowly, and ideally it will be very long lasting, economical, and the biocontrol organism itself will not become a nuisance. The principles

TABLE 17.1
Insect Species Released for Biological
Control of Aquatic Plants

Target Plant	Insect
Alligatorweed	*Amynothrips andersoni* O'Neill
Alligatorweed	*Vogtia malloi* Pastrana
Alligatorweed	*Agasicles hygrosphila* Selman and Vogt
Water lettuce	*Neohydronomus affinis* Hustache
Water lettuce	*Spodoptera pectinicornis* (Hampson)
Hydrilla	*Hydrellia pakistanae* Deonier
Hydrilla	*Bagous affinis* Hustache
Hydrilla	*Bagous hydrillae* O'Brien
Hydrilla	*Hydrellia balciunasi* Bock
Water hyacinth	*Arzama densa* Walker
Water hyacinth	*Sameodes albiguttalis* (Warren)
Water hyacinth	*Neochetina eichhorniae* Warner

of biological control of exotic pests, and the problems and concerns associated with them, continue to be debated (e.g., Hoddle, 2004; Louda and Stiling, 2004).

There are two types of biological control. One is *augmentive,* where a naturally occurring (native or endemic) organism is identified and cultured, and individuals are added to the natural population at a particular site. An example is the milfoil weevil *Euhrychiopsis lecontei* Dietz (Coleoptera: Curculionidae), a herbivore that appears to have switched host preference from the native *Myriophyllum sibiricum* Komar (= *M. exalbescens* Fernald) to the exotic *M. spicatum.* The second approach, *classical* biocontrol, involves the addition of a herbivore or pathogen from the exotic plant's native range. A series of research stages must occur that may end in the release of an exotic organism to control an exotic plant. The target plant is studied in its native range to identify promising species, and to determine whether they feed on or affect closely related and/or economically or ecologically important plants. Host-specific insects are imported under quarantine to a U.S. Department of Agriculture (USDA) facility in Gainesville, Florida. Here, host specificity and potential effectiveness are examined. Insects that prove to be safe for application may then be released from quarantine through authorization from the Animal and Plant Inspection Service (APHIS) of the USDA. Also, the U.S. Department of Interior can restrict the introduction of exotic species for biological control (Hoddle, 2004). Examples of this lengthy procedure are found in Buckingham and Balciunas (1994) and Buckingham (1998). Twelve insects have been released from quarantine in the U.S. for treatment of nuisance aquatic plants (Table 17.1). Plant pathogens from nuisance plant home ranges are still unavailable for application, but may be brought into the U.S. for study at the quarantine facility at Fort Detrick, Maryland (see later section).

The following paragraphs describe the use of insects for control of four of the eight exotic nuisance aquatic plants in U.S. lakes.

17.2 HYDRILLA (*HYDRILLA VERTICILLATA*)

Hydrilla verticillata (L. f.) Royle (= "hydrilla") has caused great ecological and economic damage in the U.S. The dioecious biotype (plants have male or female flowers) was introduced to Florida by an aquarium dealer in about 1950; the monoecious biotype (each plant has male and female flowers) appeared in the late 1970s, possibly from Korea. Eradication is essentially impossible because plants reproduce from tiny fragments that are easily transported to other aquatic habitats, and from seeds, turions and tubers that are resistant to drought, cold, and herbicides. Thick mats form in shallow water, or in clear deep water, whether eutrophic or oligotrophic (Buckingham and Bennett, 1994; Balciunas et al., 2002).

Hydrilla is one of the most troublesome aquatic plants in the southeastern U.S., causing millions of dollars in damage to irrigation operations, hydroelectric power generation, and recreational activities. Infested lakes can become closed to most uses. There is now concern about the northward spread of the monoecious biotype. It is found at 55° N latitude in Europe and could survive in any U.S. state (Balciunas et al., 2002). Newly established infestations of the monoecious biotype in Pennsylvania, Connecticut and Washington states are not new foreign introductions, as demonstrated by randomly amplified polymorphic DNA analysis. The plant is found in at least 16 U.S. states and 185 drainage basins (Madeira et al., 2000). The monoecious biotype has higher production of shoots (source of fragments) at lower temperatures, than the dioecious biotype (Steward and Van, 1987; McFarland and Barko, 1999). Global climate change could be a factor in enhancing its northward spread.

If hydrilla spreads northward, it will be important for lake managers to recognize and attempt to eradicate it immediately. It is difficult to distinguish from other species of Hydrocharitaceae. There are two native members of this family, *Elodea canadensis* and *E. nuttalii* and one exotic, *Egeria densa,* which look like hydrilla. Hydrilla has marginal teeth on the leaves that are visible without a lens, whereas the other species require a hand lens to see the fine marginal teeth (Dressler et al., 1991; Borman et al., 1997).

Hydrilla management typically involves either grass carp (= white amur, see later paragraphs) introduction or herbicide application. However, classical biocontrol agents are also used. Two weevils (Coleoptera: Curculionidae), *Bagous affinis* Hustache and *B. hydrillae* O'Brien, were released in Florida in 1987 and 1991, respectively, but neither was successful (Buckingham and Bennett, 1994; Balciunas et al., 2002). Two ephydrid flies (Diptera: Ephydridae), *Hydrellia pakistanae* Deonier and *H. balciunasi* Bock, were released in 1987 and 1989, respectively. *H. balciunasi* has established at only a few sites, apparently due to high wasp parasitism, poor host plant food quality, and possible genetic differences between hydrilla in the U.S. and hydrilla in Australia, where the flies are native (Grodowitz et al., 1997). *H. pakistani* produced significant decreases in hydrilla, along with recovery of native plants. Successful biocontrol of hydrilla with this insect may be slow. For example, insects were released in 1992 into Lake Seminole, Georgia. Hydrilla declines were noted in 1997 and large-scale decreases were evident in 1999 (Balciunas et al., 2002). The impact on hydrilla may be enhanced by combining insect application with a pathogenic fungus, *Fusarium culmorum* (Shabana et al., 2003). The success of this insect may be influenced by the nutritional status of the hydrilla host. Plants with low tissue N or with tough leaves lead to higher insect mortality and impaired development (Wheeler and Center, 1996), suggesting that host plant adaptation to the insect may be another important factor in unsuccessful biocontrol.

Presently, classical biocontrol of hydrilla is in a developmental stage, and use of grass carp, harvesters, and herbicides remain reliable and effective choices. More research is needed, including overseas surveys, to locate biocontrol agents and to assess factors influencing establishment and growth of biocontrol organisms.

17.3 WATER HYACINTH (*EICHHORNIA CRASSIPES*)

Water hyacinth, introduced to the U.S. in the 1880s, has created much economic and environmental damage and some consider it to be "the world's most troublesome aquatic weed" (Center et al., 1999). This plant is a nuisance throughout tropical and subtropical areas of the Earth, and has posed human life-threatening situations (e.g., trapped boats, collapsed bridges, enhanced mosquito habitat). It is a floating plant with large leaves, an attractive flower, and very high growth rates, leading to a dense, interconnected mat. Under favorable conditions, complete surface coverage of a pond or small lake is possible, and wind-drifted mats trap boats and close dock areas. Water hyacinth can reproduce via seeds that remain viable in aquatic sediments for 15–20 years, but fastest population growth is through vegetative processes (Center et al., 2002). Mechanical and chemical controls have met with varying degrees of success, in part because rapid re-growth follows treatment.

Biocontrol agents were investigated in Argentina in the 1960s and 1970s, leading to importation under quarantine of three insects that were later released after extensive testing. Argentina was chosen because water hyacinth is native to South America and because its climate is similar to the infested areas of North America (Center, 1982). The imported insects are: the moth *Niphograptera* (= *Sameodes*) *albiguttalis* (Warren) (Lepidoptera: Pyralidae), and the beetles *Neochetina eichhorniae* Warner and *N. bruchi* Hustache (Coleoptera: Curculionidae). The mite *Orthogalumna terebrantis* Wallwork (Acarina: Galuminidae), a native North American species, was also suggested. *N. eichhorniae* and *N. bruchi* were released in Florida in 1972 and 1974, respectively, and the moth was released in 1977 (Center et al., 2002).

The beetles are host specific and both adults and larvae affect the plants. Eggs are embedded in plant tissues. Tiny (2 mm) larvae appear in the spring and burrow into leaf petioles, causing wilting and leaf loss from the stems. Mature larvae (8 to 9 mm) enter the stem and attack the apical meristem. Pupae are found attached to roots below the water surface. The adults attack the youngest leaves, eating epidermal cells, which provide sites for microorganisms to augment plant damage. Leaf death occurs slightly faster than leaf renewal, leading to a net loss of leaves. Water hyacinth requires a minimum number of leaves in order to float, and when leaf loss exceeds this limit, plants sink and die (Center et al., 1988).

Classical biocontrol of water hyacinth is highly successful, as illustrated by results from Louisiana, where the infestation averaged 500,000 ha during the fall months of 1974 to 1978. *N. eichhorniae* was released in southeastern states in 1974 to 1976, becoming established by 1978. *N. bruchi* was released in 1975 and *N. albiguttalis* in 1979. By 1980, insect impact was evident, reducing coverage to 122,000 ha. Coverage in 1999 was well below 100,000 ha. Other factors, including herbicide use, saltwater intrusions, and weather do not account for the extent of this decline (Figure 17.1) (Center et al., 2002).

A sustained threshold density of 1.0 insect/plant for 6 months, followed by a peak of 3 or more/plant, is needed to reduce plant coverage. This density is affected by season, plant vigor, and plant pathogens. A natural cycling of plant and insect abundance should develop in which plant

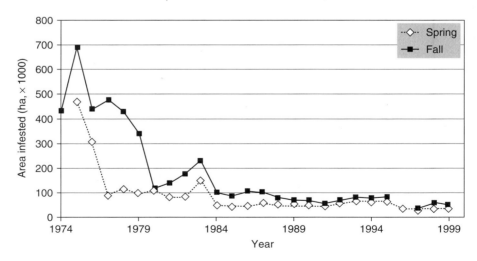

FIGURE 17.1 Data from Louisiana, showing reduced waterhyacinth cover and limited annual growth after introduction of *Neochetina eichhorniae* in 1974, *N. bruchi* in 1975, and *Niphograpta albiguttalis* in 1979. (From Center, T.D. et al. 2002. In: R. Van Driesch et al. (Tech. Coord.), Biological Control of Invasive Plants in The Eastern United States. U.S. Department of Agriculture Forest Service Pub. FHTET-2002-04. Bull. Distribution Center, Amherst, MA. Chapter 4.)

density increases for 2 to 3 years and then declines as the slower growing insect biomass reaches threshold density. Plant biomass then remains low for some period, leading to reduced insect density, plant recovery, and so forth. Plant or insect eradication, except on a small scale, is unlikely. Little is known about other mortality sources (e.g., fish, birds) of insect biocontrol agents and is a major research area (Sanders and Theriot, 1986).

Successful insect use to control water hyacinth illustrates important facts about biocontrol. First, the process is slow, does not produce eradication (e.g., Figure 17.1), and provides long-term, low cost reduction in biomass. Successful biocontrol returns the water resource to all uses. These points are important because 2,4-D, an effective herbicide on water hyacinth, is not available to many tropical and subtropical people. Second, insect control of aquatic plants is not compatible with plant removal via harvesting or herbicides. Chemical and mechanical treatments remove immobile eggs, larvae and pupae so that when plant re-growth occurs from seeds and fragments, few insects remain to suppress the new growth. Long-term control with insects is more likely without intense management (Center, 1987). An integrated approach, where several large lake areas are not sprayed or cut, may allow survival of enough insects to re-infest new growth (Haag, 1986; Haag and Habeck, 1991).

Because there may be public pressure for immediate relief from an infestation, significant research areas are to identify herbicides and adjuvants that are non-toxic to biocontrol insects, and to develop management protocols that allow for treatment of critical lake use areas, but protect the insects for long-term plant suppression (Center et al., 1999). Water hyacinth appears to be spreading northward from southeastern U.S. states, and an important research area is to identify cold tolerant biocontrol agents (Center et al., 2002).

17.4 ALLIGATORWEED (*ALTERNANTHERA PHILOXEROIDES*)

Classical insect control of alligatorweed is very successful. The plant was introduced to the U.S. in the 1880s. It spread rapidly through southeastern states, forming interwoven mats, some as thick as 1 m, sometimes over an entire pond, lake, or canal. Alligatorweed is a rooted, perennial plant that reproduces vegetatively in the U.S. and is capable of becoming terrestrial if a habitat dries (Buckingham, 2002).

Investigations in Argentina, followed by studies under quarantine in the U.S., led to releases of three insects (Maddox et al., 1971): a flea beetle *Agasicles hygrophila* Selman and Vogt (Coleoptera: Curculionidae), a thrip *Amynothrips andersoni* O'Neill (Thysanoptera: Phlaeothripidae), and a moth *Vogtia malloi* (Pastrana) (Lepidoptera: Pyralidae), released in 1964, 1967 and 1971, respectively.

Agasicles has been so successful in controlling alligatorweed that the plant is no longer a nuisance, except in local areas. Five factors led to its success: (1) high reproductive potential, (2) a life history spent on or in alligatorweed, making it less vulnerable to insectivores, (3) complete dependence or specificity on alligatorweed, (4) high mobility and dispersion power, and (5) high tolerance to some chemicals, including certain insecticides (Spencer and Coulson, 1976). Larvae and adults feed on leaves, and larvae bore into the stem to pupate.

Vogtia and *Agasicles* were successfully introduced into Tennessee, southern Alabama, Louisiana, Georgia, North and South Carolina, Texas, and Arkansas. The terrestrial form of alligatorweed is not controlled by these species, though the flightless thrip *Amynothrips* can be locally effective but not widely distributed.

Temperature and water level fluctuations affect the success of *Agasicles*. Greatest effectiveness in controlling alligatorweed occurs where weather permits peak populations to develop by June. The northern limit of effectiveness corresponds roughly with a mean January temperature of 12°C. There is no winter diapause in *Agasicles* so it is eliminated in northern latitudes, or in sites where alligatorweed is frozen back to the shoreline so that beetles cannot feed. The southern limit occurs where summer dormancy to escape intense heat is so extended that no fall population peak occurs

(Spencer and Coulson, 1976). Flooding eliminates insects and droughts stimulate the terrestrial form of the plant, eliminating alligatorweed as a food source for flea beetles and stem borers (Cofrancesco, 1984).

The flea beetle's effectiveness is enhanced by *Vogtia* and *Amynothrips*. There are also possibilities for combining insect use with herbicide pre-treatment (Gangstad et al., 1975) or with plant pathogens or mechanical methods. Unquestionably, insects have been successful in alligatorweed control, eliminating or greatly reducing the need for machines and chemicals, and allowing native plant species to return. Unfortunately, another exotic, such as water hyacinth or hydrilla might replace the controlled species, but insect control of these species, especially water hyacinth, is also possible.

17.5 EURASIAN WATERMILFOIL (*MYRIOPHYLLUM SPICATUM*)

Eurasian watermilfoil ("milfoil," EWM), a native to Asia, Africa and Europe, was introduced to North America between the 1880s and 1940, and spread to nearly every state and three southern Canada provinces. It has displaced native milfoils and other submersed species, in part because it forms a distinct canopy on the lake surface, shading understory species. EWM spreads via fragments, infesting an entire lake or pond, or dispersing to new habitats through lake outflows or human activities. Seeds are formed in spike-like flowers extending above the water surface, but the primary reproduction method is vegetative (Creed, 1998; Johnson and Blossey, 2002). This exotic, perhaps more than any other aquatic plant in North America, has produced extensive biodiversity declines, high treatment costs, and loss of aesthetic and recreational attributes of lakes and reservoirs.

Traditional milfoil management methods (harvesting and herbicides) have not always been satisfactory, in part because plants re-grow rapidly or harvesters spread fragments to uninfested lake areas. Grass carp (see later sections) do not prefer them. Sudden, unexplained declines in heavily infested lakes suggested that biological agents, including insects, could be responsible. While searches for biocontrol organisms in milfoil's native range (for classical biocontrol) have not been successful, native and naturalized insects in North America that consume milfoil were investigated for their potential to provide augmentive control. However, there can be problems with augmentive control, including: (1) native insect populations may not remain at the high densities needed (perhaps due to long-established predator-prey and other density regulation processes), (2) native insect life histories may be "out of phase" with the exotic plant's, and (3) augmentation is expensive (Creed and Sheldon, 1995).

To be an effective augmentive biocontrol agent, the insect must be nearly monophagous on the exotic plant. Otherwise, the insect may prefer and disperse to non-target plants it evolved with. If the exotic plant was not controlled by native insects when it invaded, then use of these insects for augmentive control could be unsuccessful.

Despite these concerns, several native and naturalized insect species have been investigated. *Triaenodes tarda* Milner (Trichoptera: Leptoceridae) and *Cricotopus myriophylii* Oliver n. sp. (Diptera: Chironomidae) damage milfoil in British Columbia lakes, but have not been cultured and used in augmentation (Kangasniemi, 1983; Oliver, 1984; MacRae et al., 1990).The moth *Acentria ephemerella* Denis and Schiffermuller (= *A. nivea* Olivier) (Lepidoptera: Pyralidae), an invader from Europe, is established and ubiquitous in eastern and central North America (Johnson et al., 1998), and is a major source of EWM mortality when larvae reach a density of 6–8 per 10 apical tips. The native weevil *Litodactylus leucogaster* (Marsham), also associated with milfoil, appears to have little potential for biocontrol (Painter and McCabe, 1988; Johnson and Blossey, 2002). The impacts of *Litodactylus,* and especially *Acentria,* on milfoil in a group of Ontario lakes, are illustrated in Figure 17.2. The native milfoil weevil *Euhrychiopsis lecontei* Dietz (Coleoptera: Curculionidae) has been associated with EWM declines (e.g., Kangasniemi, 1983), and recent laboratory and field experiments demonstrated that the association was causal. This insect is available commercially for field augmentations (e.g., Hilovsky, 2002). *A. ephemerella* and *E.*

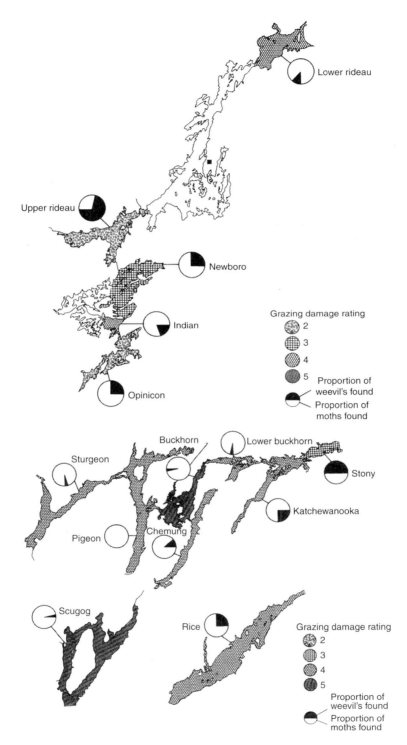

FIGURE 17.2 Insect grazing damage estimates for Ontario lakes and the proportion of weevil larvae (*Litodactylus leucogaster*) and moth larvae (*Acentria nivea*) and cases observed. (From Painter, D.S. and K.J. McCabe. 1988. *J. Aquatic Plant Manage.* 26: 3–12. With permission.)

lecontei have potential as augmentive biocontrol agents for EWM in North America and are discussed further in subsequent sections.

Acentria is the dominant herbivore on EWM in Cayuga Lake, New York. The larvae mine leaflets and feed on the apical meristem, eventually removing the meristem tip as the cocoon is formed, preventing canopy formation and eliminating a competitive advantage over native plants with lower growth forms. The larvae overwinter in *Ceratophyllum demersum* stems (Johnson et al., 1998; Johnson and Blossey, 2002).

One effect of EWM apical tip removal by insects is that this is the site of most intense production of the algicidal substance tellimagrandin II (Gross, 2000). Reduced production of this compound leads to increased epiphyte growth on leaves and possibly to shading and reduced photosynthesis, an effect similar to fish predation on epiphyte-grazing snails (Chapter 9).

The effectiveness of augmenting *Acentria* populations is unknown, although there have been experimental releases in New York state. The larvae are generalist feeders in the laboratory but select for and do serious damage to EWM in the field (Johnson et al., 1998). Earlier field observations (Creed and Sheldon, 1995) indicated that *Acentria* was associated with milfoil declines in Brownington Pond, Vermont. *Acentria* exhibits reduced growth on milfoil, compared to *Potamogeton*, possibly due to the high phenolic content of milfoil leaves (Choi et al., 2002). Additional research is needed, mainly with methods to grow large quantities of *Acentria* for field augmentation, and with observations of effectiveness.

E. lecontei apparently evolved with the North American native milfoil *Myriophyllum sibiricum* Kom. (= *M. exalbescens* Fern.), but the weevil prefers EWM in host specificity tests (Newman et al., 1997; Solarz and Newman, 2001). Females lay eggs on apical meristems. While adults feed on leaves, the larvae have the greatest negative effects, eating about 15 cm of the meristem, and eventually mining the stem and destroying vascular tissue. Larvae move about 0.5 to 1.0 m from the apical meristem, burrow into the stem, and pupate. The plant's leaf-stem-root connection may be eliminated leading to nutrient deficiencies and less carbohydrate storage in roots. The larvae may also create optimum conditions for fungal and bacterial infections of the plant. Normally there can be 4 to 5 generations per summer. Adults crawl or fly to the shore in autumn, overwintering in drier leaf litter, up to 6 m from shore. Adults return to the lake, beginning at ice-out (Creed, 2000; Mazzei et al., 1999; Newman et al., 2001; Johnson and Blossey, 2002; Newman, 2004;). Attempts to eliminate plants with harvesting, herbicides, or grass carp usually reduce insect density to ineffective low levels (i.e., Sheldon and O'Bryan, 1996).

R.P. Creed Jr., S.P. Sheldon, and co-workers (e.g., Creed et al., 1992; Creed and Sheldon, 1993, 1995) were among the first to examine weevil impacts on EWM. Laboratory and field enclosure experiments demonstrated that *Acentria* and especially *E. lecontei* reduced EWM growth. Field observations showed an association of the insects with milfoil declines, and suggested that the weevil was most damaging.

The decline of EWM in Cenaiko Lake, Minnesota appears to be the first demonstration that it was caused by the presence of *E. lecontei*, because there was no evidence of fungal infection and *A. ephemerella* and the midge *Cricotopus myriophylli* were associated with other plants. *Acentria* may have prevented milfoil resurgence at this lake (Newman and Biesboer, 2000).

A key feature of successful insect biocontrol is host specificity. *E. lecontei* evolved with North American milfoils, but has very high preference for the exotic EWM. Weevils distinguish between exotic and native milfoil, possibly because adult weevils can detect a substance in EWM at distances up to 10 cm in still water, inducing preference for EWM. *E. lecontei* has higher egg-laying and development rates on EWM, and greater adult mass than on other species (Solarz and Newman, 2001; Newman, 2004). No-choice experiments with nine non-milfoil submersed species demonstrated that the weevil did not damage these plants, laid no eggs, and survived poorly (Sheldon and Creed, 1995). Thus *E. lecontei* is host-specific, having abandoned native milfoils where choice is possible. An effective density of *E. lecontei* is in the range of 50–100/m^2, about two adults, larvae, eggs or pupae/stem (Creed and Sheldon, 1995; Newman and Biesboer, 2000).

Factors regulating weevil density are poorly known. In a Minnesota lake, black crappie (*Pomoxis nigromaculatus*) and perch (*Perca flavescens*) consumed no life stage, while bluegills (*Lepomis macrochirus*) consumed adults and larvae, but not pupae. Bluegills could be a major mortality source with low insect and high fish densities. Odonate larvae are apparently unsuccessful larval predators (Sutter and Newman, 1999). More research is needed on weevil predators. Adults could be especially vulnerable in the fall as they move to shore to overwinter (Newman et al., 2001). Undisturbed shoreline areas, with no insecticide residuals, are apparently essential for successful overwintering. Lawns manicured to the lake's edge are unlikely to provide suitable overwintering sites, though this has not been investigated.

Acentria and *E. lecontei* clearly have negative impacts on milfoil. They rarely occur as co-dominants, suggesting competition (Johnson et al., 1998) and their use for biocontrol depends on which species can be easily cultured. At this time, only the weevil is being cultured for control purposes. Another question concerns the efficacy of the weevil in southern U.S. lakes and reservoirs, well away from their established range (Creed, 2000). High summer temperatures (> 35°C) in southern lakes and low temperatures (< 18°C) in more northern lakes may limit effectiveness to mid-latitude North America (Mazzei et al., 1999).

Currently, *E. lecontei* is used to augment natural populations, but there are few long-term evaluations. There were no milfoil declines in Vermont that could be attributed to widespread augmentations with the weevil (Crosson, 2000 in Madsen et al., 2000), but preliminary data from 12 Wisconsin lakes suggest some control in the first year of augmentation (Jester et al., 2000).

In summary, insects are effective, but they are slow and do not lead to eradication of target plants. Severe infestations can be reduced with insects, and when used with herbicides in a way that preserves an insect "reservoir," there can be longer-term control. What other native insects could be used for aquatic plant control? Basic lake ecological research must continue.

17.6 GRASS CARP

17.6.1 HISTORY AND RESTRICTIONS

The grass carp, or white amur (*Ctenopharyngodon idella* (Val.) (Cyprinidae) is native to the large rivers of China and Siberia. The controversy in the U.S. over this exotic fish for aquatic plant control stems from the history of its introduction, its subsequent escape to North American rivers, and its expected impacts on lakes and reservoirs. It was shipped to the Fish Farming Experimental Station in Arkansas, and to Auburn University, from Malaysia in 1963. Between 1970 and 1976, 115 lakes and ponds in Arkansas were stocked, including Lake Conway, a hydrologically open system. Free-ranging fish were discovered outside of Arkansas in 1971, all from the 1966 age class (Guillory and Gasaway, 1978).

Unlike the introduction of exotic insects to U.S. waters for plant control, grass carp were introduced without rigorous preliminary studies under quarantine. It should have been predicted that this "generalist" herbivore would have many negative features. It is likely that grass carp importation to the U.S. would not receive authorization by the U.S. Department of Agriculture if permission had been requested in more recent times. A scientific effort to understand the beneficial and harmful effects was launched after their broadcast to the waters of North America, a classic example of the "stock and see" mentality (Bain, 1993) so common with importation of exotic plants and animals. There have been many concerns about impacts on aquatic habitats where plants are desirable, and about their potential to enrich lake waters or to interfere with game fish or other biota.

Some states prohibit their use, or have restricted use to the sterile triploid fish (Table 17.2). There has been a general restriction on importation and release in Canada, although triploids are under investigation in some provinces.

Grass carp are popular, largely because they can provide low cost, long-term plant control, with acceptable negative impacts for some lake users. For example, a lake can become completely

TABLE 17.2
State Regulations on Possession and Use of Grass Carp

A. Diploid (Able to Reproduce) and Triploid (Sterile) Permitted

Alabama	Hawaii	Kansas	Oklahoma
Alaska	Iowa	Mississippi	New Hampshire
Arkansas	Idaho	Missouri	Tennessee

B. Only 100% Triploids permitted

California	Illinois	New Jersey	South Dakota
Colorado	Kentucky	New Mexico	Texas
Florida	Lousiana	North Carolina	Virginia
Georgia	Montana	Ohio	Washington
	Nebraska	South Carolina	West Virginia

C. 100% Triploids Permitted for Research Only

New York	Oregon	Wyoming

D. Grass Carp Prohibited

Arizona	Maryland	North Dakota	Vermont
Connecticut	Massachusetts	Pennsylvania	
Indiana	Minnesota	Wisconsin	
Maine	Nevada	Utah	

accessible for boating and swimming, though this may be at the expense of many lake and lake shore species, and an increase in trophic state.

The purpose of this section is to provide lake managers with the information to make informed decisions about grass carp use.

17.6.2 BIOLOGY OF GRASS CARP

Grass carp exhibit an unusual metabolic strategy. Their aerobic metabolic rate is about half that of many fish, but their average consumption rate (at 21°C or higher) as adults is about 50–60% of body weight/day, and may equal body weight/day in small (< 300 g) fish (Osborne and Riddle, 1999). This rate is two to three times that of carnivorous fish. Their low metabolism and high consumption rates offset their low assimilation efficiency, which is about one third that of carnivorous fish (Wiley and Wike, 1986). Young grass carp are omnivorous, perhaps as a means of obtaining adequate protein (Chilton and Muoneke, 1992). Food assimilation decreases with increasing fish size and increases with increasing temperature. Up to 74% of ingestion is defecated, providing a significant load of partially digested organic matter and nutrients to the sediments. An energy budget for adult triploid carp is (Wiley and Wike, 1986):

$$100\,I = 21\,M + 67\,E + 12\,G$$

where I = ingestion, M = metabolism, E = egestion, and G = growth

The feeding rate is temperature dependent. They apparently do not feed at temperatures below 3°C, while active feeding begins at 7–8°C, and peak feeding is at 20–26°C (Chilton and Muoneke, 1992; Opuszynski, 1992). There may be regional acclimation so that fish in temperate climates, for example, begin feeding at lower temperatures, an important factor in stocking models (Leslie and Hestand, 1992). Triploid fish have a consumption rate that is about 90% of diploid fish. Average growth rates are 9–10 cm/year as juveniles, decreasing to 2–5 cm/year as adults (Chilton and

Muoneke, 1992). Common adult weights exceed 9–10 kg, and 30–40 kg fish occur in Florida (Leslie and Hestand, 1992).

Grass carp exhibit feeding preferences, varying somewhat among U.S regions. This fact has important implications for stocking rates (see later section). Table 17.3, modified from Cooke and Kennedy (1989), is a feeding preference list for triploid grass carp in Florida, Illinois, and Oregon-Washington. Other state and regional preference lists are available (Florida, Colorado, California, Pacific Northwest United States, and New Zealand) (Chapman and Coffey, 1971; Swanson and Bergerson, 1988; Pine and Anderson, 1991; Leslie and Hestand, 1992).

Eurasian watermilfoil (*Myriophyllum spicatum*) is not a preferred food plant. It has a high protein and gross energy content, but the lower stem is tough and fibrous, leading to rejection by the fish. Only when the more tender upper, new growth can be reached will grass carp eat this plant (Pine et al., 1989), suggesting that accessibility and ease of mastication may be more important than nutritional quality in determining grass carp preferences. Control of milfoil may be deferred until stocked fish are larger and preferred (often native) plants have been eliminated.

Regional differences in food preferences have management implications. *Ceratophyllum demersum* is a preferred plant in Florida, variably eaten in Oregon-Washington, but not eaten by Illinois grass carp (Table 17.3). Triploid grass carp also rejected *C. demersum* during experiments in northern California (Pine and Anderson, 1991). The question remains whether palatability varies from region to region, whether there is a genetic basis to grass carp feeding behavior, or whether further studies will demonstrate that these geographical differences are due to experimental design. One approach is to test palatability of nuisance plants for each water body prior to stocking (Chapman and Coffey, 1971; Bonar et al., 1987). Major nuisance exotic species, including water hyacinth and alligatorweed, are not eaten or are non-preferred. Additional research is needed about grass carp feeding preferences.

Feeding preferences mean that grass carp may allow non-preferred plants to become abundant, particularly when fish are under-stocked or when fish escape or die. At low fish density, only palatable species are consumed (e.g., Fowler and Robson, 1978; Fowler, 1985). For example, in Deer Point Lake, Florida (Van Dyke et al., 1984; Leslie et al., 1987; J.M. Van Dyke, Florida Department of Natural Resources, personal communication), a large reservoir stocked in 1975–1978 (see case history), *M. spicatum* became a problem after a native plant (*Potamogeton illinoiensis*) was eliminated and grass carp density declined from escape and death. In some lakes, grass carp feed on detritus and animals after plant eradication (Edwards, 1973).

Plant preference rankings (e.g., Table 17.3) may be an oversimplification of the palatability problem. Consumption rates of *Egeria densa* and *Elodea canadensis*, taken from Pacific Northwest lakes with varying chemical content, were significantly correlated with lake-to-lake variations in plant tissue composition. Feeding rates were positively correlated with calcium content and negatively correlated with cellulose (Bonar et al., 1990).

17.6.3 REPRODUCTION OF GRASS CARP

An issue with grass carp is whether they will escape from a stocked lake, reproduce, and invade non-target habitats where vegetation is desirable. The criteria for successful reproduction are stringent (Stanley et al., 1978; Chilton and Muoneke, 1992), and it was assumed by importers that reproduction would be unlikely outside the native range. Spawning occurs in rivers, and is elicited by a sharp rise in water level and by temperatures above 17°C. The eggs must remain in suspension, and it was assumed that currents of about 0.6 m/s were needed. However, Leslie et al. (1982) found that a velocity of only 0.23 m/s was sufficient to transport eggs in a Florida river. Thus, in a warm Florida river at this or greater current velocity, only 28 km would be required for incubation and hatching of eggs, a much shorter distance than previously reported. Stream length required for hatching of eggs increases with decreasing temperature. Larvae develop in quiescent areas (oxbows, sloughs) where they feed on zooplankton.

TABLE 17.3
Feeding Preference List, in Approximate Order of Preference, for Triploid Grass Carp in Florida, Illinois, and Oregon–Washington Studies

Florida	Illinois[a]	Oregon-Washington
	Preferred Plants	
Hydrilla verticillata (hydrilla)	*Najas flexilis* (brittle naiad)	*Potamogeton crispus* (curly-leafed pondweed)
Potamogeton illinoiensis (Illinois pondweed)	*Najas minor* (naiad)	*Potamogeton pectinatus* (sago pondweed)
Potamogeton spp. (pondweeds)	*Chara* (muskgrass)	*Potamogeton zosteriformis* (flat-stemmed pondweed)
Najas guadalupensis (southern naiad)	*Potamogeton foliosus* (pondweed)	*Elodea canadensis* (elodea)
Egeria densa (Brazilian elodea)	*Elodea canadensis* (elodea)	*Vallisneria* sp. (tapegrass)
Elodea canadensis (elodea)	*Potamogeton pectinatus* (sago pondweed)	*Egeria densa* (Brazilian elodea)
Chara spp. (muskgrass)		
Lemna spp. (duckweed)		
Nitella spp. (stonewort)		
Ceratophyllum demersum (coontail)		
Eleocharis acicularis (needle rush)		
Pontederia lanceolata (pickerelweed)		
Wolffiella spp. (bog mat)		
Wolffia spp. (watermeal)		
Typha spp. (cattail)		
Azolla spp. (azolla)		
Spirodela (duckweed)		
	Variable Preference — May Eat	
Myriophyllum spicatum (EWM)	*Potamogeton crispus* (curly-leafed pondweed)	*Myriophyllum spicatum* (Eurasian watermilfoil)
Bacopa spp. (bacopa)		*Ceratophyllum demersum* (coontail)
Polygonum spp.(smartweed)		*Utricularia vulgaris* (bladderwort)
Utricularia spp. (bladderwort)		*Polygonum amphibium* (amphibious smartweed)
Cabomba spp. (fanwort)		*Myriophyllum exalbescens* (native milfoil)
Fuirena spp. (umbrellagrass)		
Nymphaea spp.(water lilies)		
	Variable Preference — May Eat	
Brasenia schreberi (watershield)		
Hydrocotyl spp. (pennywort)		
Panicum repens (torpedograss)		
Stratiotes aloides (water aloe)		
	Non-preferred — Does Not Eat	
Nuphar luteum (spatterdock)	*Ceratophyllum demersum* (coontail)	*Potamogeton natans* (floating leaf pondweed)
Vallisneria americana (tapegrass)	*Myriophyllum* spp.	*Brasenia schreberi* (watershield)
Myriophyllum brasiliense (parrotfeather)	*Ranunculus longirostris*	
Eichhornia crassipes (water hyacinth)	*Ranunculus flabellaris* (buttercup)	

TABLE 17.3 (Continued)
Feeding Preference List, in Approximate Order of Preference, for Triploid Grass Carp in Florida, Illinois, and Oregon–Washington Studies

Florida	Illinois[a]	Oregon-Washington
Alternanthera philoxeroides (alligatorweed)		
Nymphoides spp. (floating heart)		
Pistia stratiotes (waterlettuce)		
Phragmites spp. (reed)		
Carex spp. (sedge)		
Scripus spp. (bulrush)		
Ludwigia octovalis (water primrose)		
Colocasia esculentum (elephant-ear)		

[a] Diploid carp.

Sources: Data based on Hestand, R.S. and C.C. Carter. 1978. *J. Aquatic Plant Manage.* 16; Osborne, J.A. 1978. Final Report to Florida Department of Natural Resources. University of Central Florida, Orlando; Nall, L.E. and J.D. Schardt. 1980; Van Dyke, J.M. et al. 1984. *J. Aquatic Plant Manage.* 22; Miller, A.C. and J.L. Decell. 1984; Sutton, D.L. and V.V. Van Diver. 1986. Grass Carp: A Fish for Biological Management of Hydrilla and Other Aquatic Weeds in Florida. Bull. 867. Florida Agric. Exper. Sta., University of Florida, Gainesville; Bowers, K.L. et al. 1987. In: G.B. Pauley and G.L. Thomas (Eds.), *An Evaluation of the Impact of Triploid Grass Carp (Ctenopharyngodon idella) on Lakes in the Pacific Northwest.* Washington Cooperative Fisheries Unit, University of Washington, Seattle; Leslie, A.J., Jr. et al. 1987. Unpublished Report; Pauley, G.B. et al. 1994; Van Dyke, J.M. 1994; Murphy, J.E. et al. 2002. *Ecotoxicolgy* 11.

Despite the assumption that reproduction would not occur outside the native range, there have been many instances — in areas of diverse topography and latitude, ranging from the former USSR to Japan, Taiwan, the Philippines, and Mexico — where introduced grass carp have spawned successfully (Stanley et al., 1978). There is direct evidence that grass carp have reproduced in the Missouri, Mississippi, Lower Trinity (Texas), and Atachafalaya (Florida) Rivers, and in their tributaries and adjoining bays (Connor et al., 1980, Brown and Coon, 1991; M.A. Webb et al., 1994; Raibley et al., 1995). It is unknown if grass carp populations will disperse, but they do spawn in smaller river systems and farther north than previously documented (Brown and Coon, 1991). Because many escaped fish are diploids, wild grass carp populations may expand in distribution, with unknown impacts. The continued sale and use of diploid fish in North America should cease.

Sterile grass carp were developed to solve the reproduction problem. Early attempts to use sterile fish involved hybrids, but these had lower feeding efficiencies and fertile diploids could occur. A solution involved the production of pure (unhybridized) triploid (three members of each chromosome in cells) fish, using hydrostatic pressure or high temperature techniques that produce nearly 100% triploids (Cassani and Caton, 1986).

No known procedure produces 100% triploidy consistently, and diploids and triploids cannot be accurately separated by sight. Fish producers must verify that fish sold are triploid. One technique is to examine a blood sample with a Coulter Counter with a channelizer. Triploid red blood cells are larger than those in diploids, and are verified with the Counter. Three workers can examine 2000 to 3000 fish/day, with 100% accuracy. Triploids are functionally sterile, with a very low probability of being a source of reproducing diploids (Allen et al., 1986; Allen and Wattendorf, 1987). The production and verification of 100% sterile fish prompted several states to permit stocking (Table 17.2).

17.6.4 STOCKING RATES

Stocking density is important in successful use of grass carp. Feeding activity, and its impact on vegetation, is affected by water temperature, length of the warm-water season, type of plants, size of fish stocked, mortality or escapement, and pre-stocking plant control activities. Overstocking may occur when the dominant plant species is highly palatable (e.g., hydrilla), leading to plant eradication. Stocking rates must be higher if unpalatable or non-preferred plants dominate (e.g., milfoil), and palatable (often native) plants will be eliminated first. Preferential feeding means that the target plant could remain a nuisance for some time and lake user dissatisfaction may be high, possibly leading to further over-stocking. Problems with over- or understocking are more likely when lake managers are advised to use a fixed stocking rate (same rate state-wide) often recommended by state agencies. Stocking models were developed to provide stocking rates appropriate for each of several regions of the U.S. Models in use include (1) the White Amur Stocking Rate Model (Miller and Decell, 1984; Stewart and Boyd, 1994), (2) the Illinois Herbivorous Fish Stocking Simulation System (Wiley and Gorden, 1985), and (3) the Colorado model (Swanson and Bergersen, 1988). Reservoir and lake managers should consult the appropriate model, or see Leslie et al. (1987) and Wiley et al. (1987).

For example, the Illinois stocking model (Wiley et al., 1987) requires the following data: lake area, percent of area less than 2.4 m (8 ft) in depth, percent of area heavily vegetated at peak biomass, specific identity of dominant plants (adjusts for feeding preferences), and the climatic region (adjusts for water temperature and length of growing season). The model assumes fish 25 cm (10 in.) in length will be stocked in the spring season, and considers whether all fish will be stocked at once (batch stocking) or whether serial stocking will be used (e.g., fish added every 5 years) as long as control is desired. The latter strategy uses fewer fish.

The Illinois model emphasizes an attempt to maintain 40% plant coverage in littoral areas after stocking, an amount optimal for largemouth bass in that state (Wiley et al., 1984), although optimal coverage apparently varies from region to region. For example, in 56 Florida lakes, ranging greatly in area, depth, trophic state, and macrophyte abundance, adult largemouth bass density was not related to macrophyte abundance, but was positively correlated with trophic state. Younger bass density was weakly correlated with macrophyte abundance (Hoyer and Canfield, 1996a, b). But, when submersed vegetation fell below 20% of total lake coverage in 30 Texas reservoirs, bass standing crop and recruitment decreased (Durocher et al., 1984).

Figure 17.3 illustrates the application of the Illinois model to three plant communities, dominated respectively by unpalatable (milfoil), palatable (pond-weed) and very palatable (*Chara*) plant species. The figure compares stocking recommendations with the fixed stocking rate, showing that with the fixed rate, the number of fish will be too high when littoral zone coverage is low and palatable plants dominate, and too low when coverage is high and unpalatable plants are the nuisance.

The significance of palatability and latitude in stocking rates is illustrated with the Illinois model. Consider a pond or lake near Chicago, Illinois (approx. latitude 42°N). If the lake is dominated by palatable species like *Chara* and naiads, the stocking rate would be 40 25-cm fish/ha followed in 6 years with a second stocking of 30/ha. However, if this had been a milfoil-dominated lake, the stocking rate would be 170/ha followed by 69/ha 7 years later. An identical pair of lakes in southern Illinois (approximately latitude 36°N) would have an initial stocking of 20/ha, followed by another 20 fish/ha in 5 years for the lake with palatable plants, and 151 fish/ha followed by 79/ha 7 years later for the lake with unpalatable plants (Wiley et al., 1987).

Fish size is important. Stocking of fingerlings may result in high mortality, possibly from bass predation. Fish at least 25 cm (10 in.) in total length are recommended in northern latitudes and at least 30 cm (12 in.) in Florida (Shireman et al., 1978; Canfield et al., 1983).

Stocking to achieve an intermediate density of plants, while ideal, is difficult in practice. While there are cases of partial plant control (e.g., Lake Conway, Florida; Miller and King, 1984), they may be the exception (Bauer and Willis, 1990; Hanlon et al., 2000). Stocking rates for an optimal

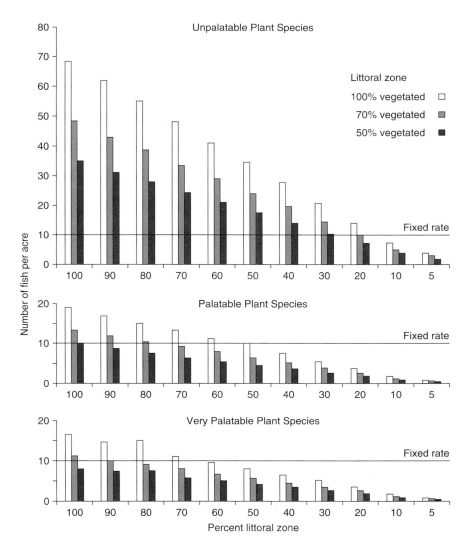

FIGURE 17.3 A comparison of fixed rate (10 fish per acre) recommendations with recommendations from the Illinois Stocking Model for three categories of plant palatability. Each comparison shows rate for northern Illinois when littoral zones are 50, 70, and100% vegetated. Graphs give stocking rate, in number of 10-in, fish, as a function of percentage of lake in littoral zone. (From Wiley, M.J. et al. 1987. Controlling Aquatic Vegetation with Triploid Grass Carp. Circular 57. Illinois Natural History Survey, Champaign.)

plant density are difficult to calculate due to variable rates of plant re-growth, water temperature, fish growth, and fish mortality (or escape) (Mitchell, 1980). Eradication of plants or failure to control them are the usual outcomes of attempts to obtain intermediate plant biomass (Bonar et al., 2001).

An integrated control approach, utilizing low stocking densities, combined with initial chemical or mechanical control, may circumvent the ecologically disruptive use of high densities followed by plant eradication (Shireman and Maceina, 1981; Shireman et al., 1983). This strategy is difficult for two reasons. First, some lake users are dissatisfied if plant control is not rapid and complete. In Washington state lakes, for example, grass carp took 2 years or more to produce effects (Bonar et al., 2001), leading lake users to add more carp which produced an overstocking. The integrated approach requires patience and still may lead to plant eradication (Shireman et al., 1983). Secondly, herbicide-treated plants (e.g., diquat and fluridone) may have lower palatability to grass carp because residues can persist (Kracko and Noble, 1993). This can lead to slower plant control, especially when lower grass carp densities are used.

Containment is an important part of stocking. Most states require an escapement barrier at the lake's outlet and grass carp should not be added to a lake or impoundment unless an adequate barrier is in place. As demonstrated at Deer Point Lake, Florida (Leslie et al., 1987; J.M. Van Dyke, Florida Department of Natural Resources, personal communication), containment is essential to maintaining enough grass carp to bring about plant control (see case history). In reality, barriers are costly, and may impede water outflow if blocked with debris. Grass carp also jump over barriers. Therefore, escape is common and the fish become pollutants.

Once stocked with grass carp, lake users are committed. There is no effective method of selectively removing them, and plant control may persist for 15 or more years. Fish Management Bait, a rotenone-laced pellet (Prentiss Inc, Floral Park, New York, 11001), has some potential for grass carp removal (Mallison et al., 1995). Bonar et al (1993) investigated several methods. Earlier work showed that fyke, gill, and trammel nets, and electroshocking, were ineffective. Grass carp could be lured to traps with lettuce (*Latuca sativa*) when submersed plants had been eradicated or the lake had non-preferred plants. Other baits (e.g., bread, cabbage, spinach, alfalfa, soybeans) were far less effective. Angling, using lettuce tied to a #8 hook with > 9 kg test line, was somewhat successful (0.0–0.14 fish/man hour) in calm weather where attractant lettuce bundles were not blown away and the lake was devoid of submersed plants. Other angling baits (e.g., doughballs, bread, catfish power bait, crappie jigs) were unsuccessful. An effective technique (0.17–0.56 fish/man hour) was herding fish into nets. Angling and herding would be ineffective in large, deeper lakes. The most effective options involve lake draining (with a high escape barrier) or application of rotenone. All fish should be eliminated, and this may have other beneficial effects for the lake (Chapter 9).

Grass carp cannot be stocked into one area of the lake with the expectation that they will remain there. Unlike harvesting and herbicide treatments, grass carp choose where and when to feed unless barriers to movement are used, as demonstrated in Lake Seminole, Georgia where grass carp were prevented from leaving a 365 ha embayment. Non-electrified barriers were ineffective, but an electrified one prevented escape from the bay, demonstrating that it is possible to treat a selected area. Cost of the barrier was $72,000 (Maceina et al., 1999).

17.6.5 CASE HISTORIES

17.6.5.1 Deer Point Lake, Florida

Deer Point Lake, a 1900-ha reservoir built in 1961, is the water supply for Panama City, and a recreational area. By 1975, *Potamogeton illinoiensis* and milfoil covered large areas, interfering with lake use and drinking water intakes. The previous edition of this text (Cooke et al., 1993) stated that pesticides were used on Deer Point Lake from 1972 to 1975. That statement was incorrect. Instead, grass carp were stocked in 1975 into fenced-off, predator-free grow-out areas at 43 fish/ha of lake area. The fish were released to the open lake in 1976. Additional grass carp were added between 1976 and 1978, bringing stocking density to 61/ha by 1978 (Van Dyke et al., 1984; Van Dyke, 1994).

P. illinoiensis, a preferred plant by grass carp, was selectively grazed and eliminated by 1977–1978. Milfoil, a non-preferred plant, remained abundant until 1979, then declined. In 1981, milfoil increased again, although native and preferred plants remained scarce (Figures 17.4 and 17.5). In 1985, there was a new stocking (21/ha) that maintained plant control until 1993, when non-preferred native plants (*Bacopa caroliniana, Vallisneria americana*) increased, providing new waterfowl and fish habitat. When preferred native species (e.g., *Najas guadalupensis, Nitella* spp.) began to increase, lake managers concluded that the lake could be susceptible to the expanding

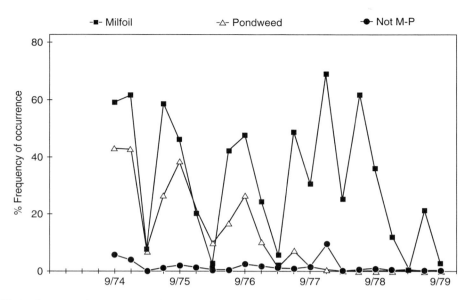

FIGURE 17.4 Deer Point Lake, Florida vegetation transect data from 1974 to 1979. (From Van Dyke, J.M. 1994. In: *Proceedings, Grass Carp Symposium.* U.S. Army Corps Engineers, Vicksburg, MS. pp. 146–150.).

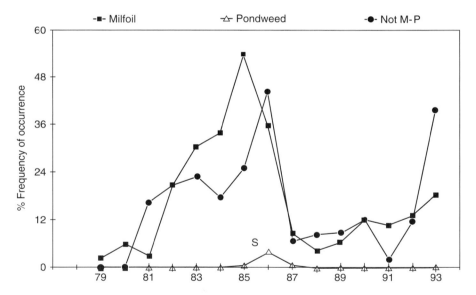

FIGURE 17.5 Deer Point Lake, Florida vegetation transect data from 1979 to 1993. (From Van Dyke, J.M. 1994. In: *Proceedings, Grass Carp Symposium.* U.S. Army Corps Engineers, Vicksburg, MS. pp. 146–150.)

hydrilla problem in northern Florida, and therefore added more grass carp to control the native species and to prevent hydrilla establishment (Van Dyke, 1994).

The Deer Point Lake project illustrates that preferred plants will be chosen, allowing the target, non-preferred plant to expand. An identical response occurred in Guntersville Reservoir, Tennessee (D.H. Webb et al., 1994). Only after the preferred species are eliminated will the target plant be consumed. High stocking density may reduce the delay time before the target plant is controlled, but will likely assure plant eradication.

Eradication may be consistent with management goals at lakes with certain recreational activities, and where an exotic plant with no natural controls other than light, space, and nutrients, has curtailed most lake uses. Plant eradication must be carefully considered before choosing it as a management goal because it will mean long-term elimination of habitat for many lake species and changes in lake water quality (see Section 17.6.5.3).

17.6.5.2 Lake Conway, Florida

Lake Conway, a 730-ha, 5-pool, urban reservoir near Orlando was stocked with diploid monosex (female) grass carp in 1977, at low but different rates (7.5 to 12.5/ha) in the different pools, to control hydrilla. This lower stocking rate was sufficient to nearly eliminate hydrilla, but *Nitella megacarpa* and *Potamogeton illinoiensis* were not greatly affected, and *Vallisneria americana* increased. Triploids were stocked in 1986 and 1988 at low doses (2.4 and 1.5 fish/ha, respectively) in response to increasing hydrilla. These low stocking rates controlled hydrilla but did not affect other species (Leslie et al., 1994).

17.6.5.3 Lake Conroe, Texas

The goal for stocking grass carp in Lake Conroe, an 8,100 ha reservoir used for recreation, shoreline housing, and water supply for Houston, was plant eradication. Hydrilla was first recorded there in 1975. By 1980, 34% of the reservoir area was infested, primarily with hydrilla (80% of infested area) but also with milfoil and coontail (*Ceratophyllum demersum*), decreasing recreational activities and shoreline property values. Despite angler protests, 75 diploid fish/vegetated ha (270,000 fish > 250 mm) were stocked, a stocking rate double that required. In two years, submersed vegetation was eliminated, although coontail growth continued in 1982 because it is non-preferred (Figure 17.6) (Noble et al., 1986; Martyn et al., 1986).

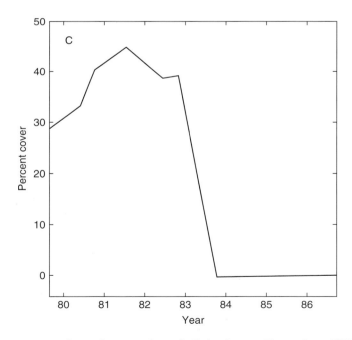

FIGURE 17.6 Percent cover of aquatic macrophytes in Lake Conroe, Texas, from 1979 to 1987. Diploid grass carp were stocked at 33 fish/ha (74/vegetated ha). (From Maceina, M.J. et al., 1992. *J. Fresh Water Ecol.* 7: 81–95. With permission.)

FIGURE 17.7 Mean monthly chl *a* in Lake Conroe, Texas. Grass carp were stocked in 1982. (From Maceina, M.J. et al., 1992. *J. Fresh Water Ecol.* 7: 81–95. With permission.)

Significant changes occurred following plant eradication. Chlorophyll (chl) *a* increased from 12 mg/m³ to 19–22 mg/m³ (Figure 17.7), and transparency declined. The Carlson (1977) trophic state index increased from 55 to 60 for chlorophyll (chl) *a* and transparency, blue-green algae became dominant, and Cladocera relative abundance fell from 22% to 3%. By 1992, largemouth bass (*Micropterus salmoides*) and crappie (*Pomoxis nigromaculatus, P. annularis*) became uncommon, whereas threadfin shad (*Dorosoma pretense*), white and yellow bass (*Morone chrysops, M. mississippiensis*) and channel catfish (*Ictalurus punctatus*) increased. Submersed vegetation remained absent through 1994, with grass carp feeding on filamentous algae, terrestrial leaves, detritus, and presumably on benthic invertebrates (Noble et al., 1986; Maceina et al., 1991, 1992; M.A. Webb et al., 1994). Plant eradication is not an environmentally sound management objective, especially for natural lakes.

17.6.5.4 Smaller Lakes and Ponds

Grass carp treatments of golf course ponds, farm ponds, real estate lakes and other smaller waterbodies are more environmentally sound than application of tens of thousands of fish to large, hydrologically open systems. Grass carp can be more easily removed from ponds, their escape can be prevented, and plant eradication will have little impact on waterfowl and other lake species. The "all-or-none" response to stocking occurs in small lakes, but plant elimination may not produce the extensive negative impacts of eradication in large multi-use lakes.

17.6.6 WATER QUALITY CHANGES

Impacts on non-target species and habitats, as well as on lake water quality and trophic state, are major concerns (as they are for other aquatic plant control techniques). But unlike herbicide or harvesting applications, grass carp treatments remain effective until fish die or escape, a variable period with persistent effects ranging from 5–9 years in Santee Cooper Reservoirs, South Carolina (largest grass carp release in North America; Kirk and Socha, 2003), to at least 15 years in Florida lakes (Colle and Shireman, 1994).

The Lake Conway, Florida study (Miller and Potts, 1982; Miller and Boyd, 1983; Miller and King, 1984) is a detailed examination of grass carp impacts. Mean BOD, and filterable and TP concentrations decreased, and ammonia and chl *a* increased, compared to pre-stocking baseline data. Algal populations were double those of comparable months before stocking.

In Lake Conroe, Texas, major water quality and fish community changes (see case history) were observed (Maceina et al., 1992; M.A. Webb et al., 1994). Submersed macrophytes were quickly eliminated, transparency declined, nutrient levels increased, and average annual chl *a* doubled and the filamentous algae *Oscillatoria* dominated.

Lakes Baldwin (80 ha) and Pearl (24 ha), Florida were dominated (80–95% coverage) by hydrilla. Each was herbicide-treated and stocked with grass carp. All submersed vegetation was eliminated, and remained so for at least 15 years. Chl *a* and nutrients increased, and transparency decreased, indicating a switch to an alternative trophic state (Chapter 9) maintained by grass carp. Fish standing crop declined in Lake Pearl and six fish species were apparently eliminated (Shireman et al., 1985; Colle and Shireman, 1994).

Major changes in sports fishing have been reported, though effects on fish are not well understood. In Lake Conroe, vegetation-dependent fish like bluegills declined as plants were eradicated and were replaced by cyprinids. Largemouth bass shifted to fewer, but larger individuals, though total biomass declined and fewer bass were caught per hour of fishing (Noble et al., 1986; Maceina et al., 1992). Lakes Baldwin and Pearl, Florida have been without macrophytes since about 1980 (Colle and Shireman, 1987). Food base in these lakes shifted from phytophilous insects and zooplankton to insects that do not require vegetation. There were immediate decreases in non-game species such as golden shiners and chubsuckers, and these species did not recover. Bluegill and redear sunfish were unaffected or increased, and largemouth bass did not change because their food base (bluegills, redear) remained intact. Negative impacts to bluegill and largemouth bass in Lake Conway, Florida were not apparent, probably because vegetation was not eliminated, and angler success with bass increased dramatically (Miller and King, 1984). In Lake Marion, South Carolina (Santee Cooper Reservoir) there was no change in fish species abundance over the 8 years after stocking, again because submersed plants remained after hydrilla was reduced by 90% (Killgore et al., 1998).

Waterfowl density and diversity declined in Lake Conway, though this change may also be related to the urbanization of the area. Herbivorous turtles, and turtles that feed on snails, were negatively affected by grass carp (Miller and King, 1984).

Another adverse effect of grass carp stocking occurs when plants are eradicated. Even casual observations reveal shoreline and littoral zone erosion. Prior to stocking, vegetation damped wave action. In some Florida lakes, wind and powerboat-generated waves produce enough shoreline erosion to cause trees to fall after submersed plants were eliminated (J. Van Dyke, Florida Department of Natural Resources, personal communication).

Grass carp are difficult to remove and plant eradication, with its undesirable side effects, is common, meaning that desirable native plant species have no chance to replace the target plants, often because native species are highly preferred. While many lake users desire plant eradication, this should not be an option for multi-use lakes. Anglers, an important lake user group that make significant contributions to local economies, understand the need for vegetation (Henderson et al., 2003).

Table 17.4 presents a cost comparison of herbicides, harvesting, and grass carp. Several factors are important in this comparison. First, grass carp are ineffective with plants such as alligatorweed and water hyacinth, and insects and/or herbicides are the least costly, most effective approach for these plants. Harvesters are too slow. Secondly, grass carp stocking rates must be higher in northern climates because of lower water temperatures, shorter growing season, and the need for greater numbers of fish in lakes dominated by the non-preferred plant, Eurasian watermilfoil. Chemical and mechanical methods may be needed more than once per season, depending upon the herbicide or the harvesting technique employed (mowing vs. root crown removal), resulting in higher overall annual costs. Finally, initial grass carp costs are amortized over the effective life of the fish, whereas other methods must be used at least every year. For example, the cost of chemically treating 15,000 ha of hydrilla in Florida in 1977 was about $9.1 million, whereas grass carp stocked at 35/ha would have cost $1.71 million. The most important point, however, is that the state would have had the $9.1 million cost every year, assuming no inflation, and the $1.71 million would provide control for several years (Shireman, 1982).

TABLE 17.4
Ranges of Costs for Grass Carp, Harvesting, and Herbicide Treatments for Aquatic Plant Management

	Midwest ($)	Florida ($)
Harvesting	508–1,423 (206–577)	1,137–55,102[a] (461–22,315)[a]; 1,137–4,500[b] (461–1,823)[b]
Herbicides	771–1,406 (289–570)	574–1,377 (232–592,157)
Grass carp	264 (107)	70–119[c] (37–63)

Note: Calculations assume use of 25 cm fish costing $8 each and with an 8-year longevity. Stocking rates for Florida lakes ranged from 59 to 101 fish/ha (24 to 41/acre) of hydrilla, and the rate for Illinois was 170 fish/ha (69/acre) of EWM (*M. spicatum*). Costs are for a single treatment and are in dollars/hectare for harvesting and herbicides, and in dollars/hectare per year for grass carp. Corrected to 2002 dollars. Costs in parentheses are per acre.

[a] Dense infestation of water hyacinth.

[b] Dense infestation of hydrilla.

[c] Costs amortized over 8 years. The estimated minimum annual cost (Florida) for grass carp is therefore $92/ha; for one harvest it is $1,137, and for one herbicide application it is $574.

Sources: Data from Cooke, G.D. and R.H. Kennedy. 1989. U.S. Army Corps Engineers, Vicksburg, MS; Leslie, A.J., Jr. et al. 1987. *Lake and Reservoir Manage*. 3; Wiley, M.J. et al. 1987. Controlling Aquatic Vegetation with Triploid Grass Carp. Circular 57. Illinois Natural History Survey. Champaign.

In summary, grass carp are powerful, long-term, cost-effective agents for macrophyte control. Appropriate stocking rates are critical to achieving control without eradication of desirable vegetation. These stocking rates are often difficult to achieve. Many case histories report plant eradication, and this has been associated with major adverse water quality changes, including a switch to higher algal biomass. Elimination of plants lasts for many years, and constitutes habitat elimination for littoral species. While long-term observations are still needed, it appears that sport fishing has improved in some cases, and declined in others. Plant eradication may produce major negative impacts to amphibians, reptiles, especially to waterfowl. It is difficult to assess this factor because macrophyte-free lakes tend to attract shoreline development leading to lake enrichment and to an ecosystem less attractive to native fauna and flora. Grass carp treatments have these characteristics (Van Dyke, 1994): (1) they are like an inexpensive, powerful, persistent, moderately selective "herbicide" that produces a slow rather than rapid nutrient release, (2) they are effective but somewhat unpredictable, and (3) they should be used as supplements to other plant control methods, with effective barriers to prevent escape.

17.7 OTHER PHYTOPHAGOUS FISH

Fish of the genus *Tilapia* (Cichlidae), native to India, Africa, South America, and other warm water climates, were suggested for algae and macrophyte control where water temperature does not fall below 10°C (Florida, or in lakes or reservoirs receiving a heated discharge) (Schuytema, 1977). For example, Hyco Reservoir, North Carolina (1,760 ha), received a heated water and fly ash discharge from a coal-fired power plant. *T. zilli* were accidently introduced in 1984. Winter temperature near the discharge did not fall below 14°C and selenium pollution eliminated largemouth bass and severely reduced bluegills, both tilapia predators. Tilapia increased rapidly. *Egeria densa*, the dominant plant, and other macrophytes, were eliminated by the end of 1985. Alkalinity and NO_3–NO_2 increased, but other nutrients and transparency apparently remained unchanged through

1988. *T. zilli* switched to detritus, benthic invertebrates, and zooplankton after eliminating macrophytes, thus maintaining its population (Crutchfield et al., 1992).

Like grass carp, high densities of *T. zilli* may eradicate macrophytes, but tilapia have significant predators and in most lakes their potential to eradicate plants is limited. Their ability to switch to other food ensures continued plant control as long as water temperature and predation are controlled. The apparent absence of algal blooms in Hyco Reservoir, and other responses to macrophyte elimination, may be related to the selenium contamination. Other water bodies may not respond in this manner.

Filter feeding species of Tilapia (e.g., *T. aurea* or *T. galilaea*) are size-selective and suppress populations of large celled algae such as dinoflagellates or intermediate size nanoplankton, as well as planktonic crustaceans and rotifers. Reduction of zooplankton density may allow increases in non-grazed phytoplankton (McDonald, 1985; Drenner et al., 1987; Vinyard et al., 1988). *T. aurea* has little potential for algal control and spread rapidly in Florida waters, becoming a nuisance. *T. melanopleura* was studied for macrophyte control potential in Florida and was effective, but because there were negative aspects (high reproductive potential and interference with game fish), it should not be used (Ware et al., 1975). Other *Tilapia* with some potential for algae and macrophyte control include *T. mossambica* and *T. nilotica* (Schuytema, 1977), and *T. rendalli* (Chifamba, 1990).

T. mossambica, like *T. zilli*, could be ideal in ponds or small lakes where no vegetation is desired and aesthetics rather than fishing is the primary purpose. An initial stocking of 140/ha multiplied to over 26,000/ha in one growing season when no bass were present. About 2,500/ha were sufficient to eliminate macrophytes and keep the water clear. The fish could be removed in the fall and a small supply kept over winter (at 20 to 27°C) for restocking in the next summer (Childers and Bennett, 1967).

The use of tilapia will remain insignificant in the U.S. due to the requirement for warm waters (unless fish are restocked annually) and because their impact on the pond or lake may cause more problems than are solved. They may be valuable to use in ponds where fishing is not desired.

17.8 DEVELOPING AREAS OF MACROPHYTE AND ALGAE MANAGEMENT

There are several aquatic plant and algae biocontrol methods in early stages of development. The following brief discussion may interest lake management researchers.

17.8.1 FUNGAL PATHOGENS

Fungi have possibilities for aquatic plant control. These characteristics make them desirable biocontrol agents: (1) numerous and diverse, (2) often host specific, (3) easily disseminated and self-maintaining, (4) capable of limiting populations without eliminating the species, and (5) non-pathogenic to animals (Zettler and Freeman, 1972; Freeman, 1977). Research continues, but currently there are no operational biocontrols of aquatic plants involving fungi. Reviews by Theriot (1989), Theriot et al. (1996), Joye (1990), and Shearer (1994) are used to summarize the status of using pathogens to control water hyacinth, hydrilla, and Eurasian watermilfoil. Imports of plant pathogens from native habitats of nuisance plants are now possible through the U.S. Department of Agriculture's Foreign Disease-Weed Science Laboratory at Ft. Detrick, Maryland, and classical biocontrol with plant pathogens may be developed (Shearer, 1997).

17.8.2 WATER HYACINTH

A new species of *Cercospora*, *C. rodmanii* Conway, was isolated from a declining population of water hyacinth in Rodman Reservoir, Florida, and described by Conway (1976a,b). Studies were carried out in quarantine and it was determined to be a strong pathogen of water hyacinth without major detrimental effects to other plants. A closely related species, *C. piaropi*, caused the decline of water hyacinth in a Texas reservoir (Martyn, 1985).

C. rodmanii may be useful for water hyacinth management (Theriot, 1989), but it may be restricted to certain types of lakes. Under situations of high nutrient concentrations, water hyacinth can outgrow the progress of the disease, limiting the use of *C. rodmanii* to conditions favoring slow host growth. Better results are obtained when the pathogen is used in combination with the insect *Neochetina* (Sanders and Theriot, 1986; Charudattan, 1986). *Acremonium zonatum* is another endemic fungal pathogen with potential for use on water hyacinth (Martinez-Jimenez and Charudattan, 1998).

17.8.3 HYDRILLA

Joye and Cofrancesco (1991) isolated *Mycoleptodiscus terrestris* (Gerdemann) Ostazeski, an endemic fungus that was non-pathogenic to 44 of 46 other plant species within 22 families. The fungus reduced hydrilla biomass during field tests, but did not create a disease epidemic in the pond, allowing hydrilla to re-grow. When the fungus was combined with low doses of fluridone (Chapter 16), plant control occurred and plant susceptibility to the herbicide increased (Shearer, 1996; Netherland and Shearer, 1996; Nelson et al., 1998). There is significant progress in the use of plant pathogens for hydrilla control, providing new possibilities to reduce hydrilla's spread, especially if classical biocontrol agents become available. A new approach, involving the U.S. Army Corps of Engineers (Vicksburg, Mississippi), the U.S. Department of Agriculture (Peoria, Illinois) and SePro (Carmel, Indiana, U.S.) involves fermentation methods to concentrate *M. terrestris* propagules into a low cost "bioherbicide" (Balciunas et al., 2002).

17.8.4 EURASIAN WATERMILFOIL

Several fungi have been isolated from *Myriophyllum spicatum* (Andrews and Hecht, 1981; Andrews et al., 1982, 1990; Sorsa et al., 1988). Of these, *Colletotrichum gloeosporioides* (Penz.) Sacc. was considered promising, but later was found to have little potential (Smith et al., 1989). Surveys in the U.S. for pathogens of milfoil are underway.

The effective use of fungal pathogens presents problems, among them the ability of aquatic plants to multiply and overwhelm the infection. One action of fungi is to fragment plants, which may increase the plant's distribution in the lake. Conditions in a macrophyte bed range from high DO, temperature, and pH in the lighted canopy, to dark, cooler, and possibly anaerobic conditions near the sediments. A successful pathogen may have to thrive over this entire range. Another problem is the level of inoculum, which usually has to be large. Dilution rates in the littoral zone can be high, limiting contact time of the fungal inoculum with the plants. The high doses could affect other organisms by increasing turbidity or oxygen demand. There is also a paucity of destructive diseases of aquatic plants, making the isolation and development of a successful pathogen even more difficult (Charudattan et al., 1989; Joye, 1990).

17.8.5 ALLELOPATHIC SUBSTANCES

The production and release of a substance by one plant or algal species that interferes with the growth and reproduction of another species (allelopathy) may have promise for aquatic plant management (e.g., Szczepanski, 1977). Some angiosperm and algal species release allelopathic materials (Gross, 2003), but questions remain about identity of these compounds, and how to maintain their concentrations in the littoral zone at levels sufficient to control target plants. Is it possible to enhance the growth of desirable native species that have allelopathic properties? As examples, *Ceratophyllum demersum* reduces phytoplankton growth, even with abundant nutrients in the water (Mjelde and Faafeng, 1997), and *Chara* appears to have some negative effects on certain phytoplankton (van Donk and van de Bund, 2002). Eelgrass (*Vallisneria americana*) is a candidate for hydrilla and milfoil control (Elakovich and Wooten, 1989), but high concentrations of the allelopathic material may be required. Allelopathy has promise and should be of interest to lake management researchers.

17.8.6 Plant Growth Regulators

Another approach is the application of gibberellin synthesis inhibitors to nuisance plants to limit stem development so that plants do not fill the water column. Some biomass production is preserved, and oxygen production, littoral soil stabilization, and other functions of rooted plants can continue (Lembi et al., 1990; Nelson, 1990; Lembi and Chand-Goyal, 1994). Studies with plant growth regulators need additional attention.

17.8.7 Barley Straw

Barley straw (but apparently not oat or wheat straw) appears to have algistatic properties when allowed to decompose in oxygen-rich waters. The first reports or tests of this came from England (Welch et al., 1990; Gibson et al., 1990). The effect appears to be inhibitory rather than toxic (e.g., Newman and Barrett, 1993), meaning it does not work against a current algal problem but may inhibit future (weeks or months later) problems.

The active substance(s) from rotting barley straw is not known, although it appears to be from the straw rather than from the flora of decomposition (i.e., not an antibiotic from fungi), and is associated with lignin oxidation and solubilization (Ridge and Pillinger, 1996; Barrett et al., 1996).

Field trials in the United Kingdom produced good results. In early spring (April), 3.5 tons were applied to Linacre Reservoir, using six anchored booms across the surface at intervals from the inlet to mid-reservoir. Water was released at a steady rate from the upper (control) reservoir. Phytoplankton reduction occurred within 12 days (Everall and Lees, 1997). The first potable water supply was treated in 1993 (Aberdeen, Scotland; Barrett et al., 1996) using tubular, high density, polyethylene netting, 0.5 m in diameter with 10–12 mm mesh, to contain the straw. Each tube contained 20 kg of loosely packed straw, and was floated on the surface to assure an aerobic environment. Complete decomposition occurred in 4–6 months. Diatom and cyanobacteria density declined to less than half of pre-treatment levels, taste and odor complaints were fewer, and filter backwash frequency declined to the minimum (Barrett et al., 1999). The barley straw technique is used throughout England, Scotland, and Ireland.

Reports of applications in the U.S. are mainly anecdotal. Experiments in confined areas such as tubs or limnocorrals (e.g., Boylan and Morris, 2003) have not been successful, possibly because these systems did not have the oxygenation and mixing required for barley straw decomposition and release of algistatic materials.

The dose recommended for water clarification by McComas (2003) is 22–24 g/m^2 (200–250 lb/acre) for phytoplankton and 2 to 3 times this amount for filamentous algae. Straw should be added by late spring (earlier is better), packed loosely in mesh bags, and floated on the surface to assure oxygenated conditions. Aeration may be required with stagnant habitats.

Nuisance algae are "pests," and the U.S. Environmental Agency (USEPA) regards any substance added to control a "pest" to be a "pesticide." As a result, barley straw is not registered as an algicide, unlike copper sulfate, a broad-spectrum, highly toxic material (Chapter 10). Barley straw cannot be sold for algae control and commercial applicators and lake managers cannot legally recommend it or apply it for this purpose. An owner of a private lake may apply it, but it cannot be used legally on public waters to control algae (Lembi, 2001; C. Mayne, Ecosystem Consulting, Inc., Coventry, Connecticut, U.S., personal communication).

17.8.8 Reducing Algae Growth with Bacteria

There are several commercial formulations of "microbial products" advertised as effective, non-toxic preparations said to "out-compete" algae for nutrients, leading to reduced algae growth. They are not advertised as algicides, avoiding requirements of state agencies and the USEPA for disclosure of data on efficacy and impacts to non-target organisms. Five commercial bacterial formulations were tested in laboratory and greenhouse settings and did not control algae (Duvall and Anderson, 2001). Experimental pond studies with three commercial bacterial products also failed to control planktonic or filamentous algae, and at least one product did not increase bacterial density. In every case, bacterial density returned to control levels within days (Duvall et al., 2001). There appears to be no evidence from peer-reviewed journals that these products are effective, and caution is suggested.

17.8.9 VIRUSES FOR BLUE-GREEN ALGAE MANAGEMENT

Safferman and Morris (1963) discovered the first blue-green algal virus or cyanophage, a virus they named LPP-1 after its ability to infect *Lyngbya, Phormidium, and Plectonema*. These are the properties of cyanophages: (1) selective and specific, (2) non-toxic to other microorganisms, (3) harmless to animals, (4) without direct effect on water quality, and (5) increase during use rather than decrease. Their effect in natural systems appears to be one of preventing an algal bloom from developing rather than eliminating an already formed bloom (Desjardins, 1983).

There have been few field studies of cyanophages. While it appears unlikely that they will become practical blue-green algae controls, they might be effective when used in conjunction with other lake management activities, like artificial circulation, which enhances phage activity.

Successful biological management of plants and algae requires far more research, including greatly increased research in basic limnology.

REFERENCES

Allen, S.K., Jr. and R.J. Wattendorf. 1987. Triploid grass carp: status and management implications. *Fisheries* 12: 20–24.

Allen, S.K., Jr., R.G. Thiery and N.T. Hagstrom. 1986. Cytological evaluation of the likelihood that triploid grass carp will reproduce. *Trans. Am. Fish. Soc.* 115: 841–848.

Andrews, J.H. and E.P. Hecht. 1981. Evidence for pathogenicity of *Fusarium sporotrichoides* to EWM, *Myriophyllum spicatum. Can. J. Bot.* 59: 1069–1077.

Andrews, J.H., E.P. Hecht and S. Bashirian. 1982. Association between the fungus *Acremonium curvulum* and Eurasian watermilfoil, *Myriophyllum spicatum. Can. J. Bot.* 60: 1216–1221.

Andrews, J.H., R.F. Harris, C.S. Smith and T. Chand. 1990. Host Specificity of Microbial Flora from Eurasian Watermilfoil. Tech. Rept. A-90-3. U.S. Army Corps Engineers, Vicksburg, MS.

Bain, M.B. 1993. Assessing impacts of introduced aquatic species — grass carp in large systems. *Environ. Manage.* 17: 211–224.

Balciunas, J.K., M.J. Grodowitz, A.F. Cofrancesco and J.F. Shearer. 2002. Hydrilla. In: R. Van Driesche et al. (Tech. Coord.), Biological Control of Invasive Plants in the Eastern United States. U.S. Department of Agriculture Forest Service Pub. FHTET-2002-04. Bull. Distrib. Center, Amherst, MA. Chapter 7.

Barrett, P.R.F., J.C. Curnow and J.W. Littlejohn. 1996. The control of diatom and cyanobacterial blooms in reservoirs using barley straw. *Hydrobiologia* 340: 307–312.

Barrett, P.R.F., J.W. Littlejohn and J. Curnow. 1999. Long-term algal control in a reservoir using barley straw. *Hydrobiologia* 415: 309–314.

Bauer, D.L. and D.W. Willis. 1990. Effects of triploid grass carp on aquatic vegetation in two South Dakota lakes. *Lake and Reservoir Manage.* 6: 175–180.

Bonar, S.A., G.L. Thomas and G.B. Pauley, 1987. The efficacy of triploid grass carp (*Ctenopharyngodon idella*) for plant control. In: G.B. Pauley and G.L. Thomas (Eds.), An Evaluation of the Impact of Triploid Grass Carp (*Ctenopharyngodon idella*) on Lakes in the Pacific Northwest. Cooperative Fisheries Unit, University of Washington, Seattle. pp. 98–178.

Bonar, S., H.S. Sehgal, G.B. Pauley and G.L. Thomas. 1990. Relationship between the chemical composition of aquatic macrophytes and their consumption by grass carp, *Ctenopharyngodon idella. J. Fish. Biol.* 36: 149–157.

Bonar, S.A., G.L. Thomas, S.L. Thiesfeld, G.B. Pauley and T.B. Stables. 1993. Effect of triploid grass carp on the aquatic macrophyte community of Devils's Lake, Oregon. *North Am. J. Fish. Manage.* 13: 757–765.

Bonar, S.A., B. Bolding and M. Divens. 2001. Effects of triploid grass carp on aquatic plants, water quality and public satisfaction in Washington state. *North Am. J. Fish. Manage.* 21: 96–105.

Borman, S., R. Korth and J. Temte. 1997. *Through the Looking Glass. A Field Guide to Aquatic Plants.* University of Wisconsin, Stevens Point, Wisconsin.

Bowers, K.L., G.B. Pauley and G.L. Thomas 1987. Feeding preference of the triploid grass carp (*Ctenopharyngodon idella*) on Pacific Northwest aquatic macrophytes. In: G.B. Pauley and G.L. Thomas (Eds.), An Evaluation of the Impact of Triploid Grass Carp (*Ctenopharyngodon idella*) on Lakes in the Pacific Northwest. Washington Cooperative Fisheries Unit, University of Washington, Seattle. pp. 70–97.

Boylan, J.D. and J.E. Morris. 2003. Limited effects of barley straw on algae and zooplankton in a midwestern pond. *Lake and Reservoir Manage.* 19: 265–271.

Brown, D.J. and T.G. Coon. 1991. Grass carp larvae in the lower Missouri River and its tributaries. *North Am. J. Fish. Manage.* 11: 62–66.

Buckingham, G.R. 1998. Surveys for Insects that Feed on Eurasian Watermilfoil, Myriophyllum spicatum, and hydrilla, *Hydrilla verticillata*, in The People's Republic of China, Japan, and Korea. Tech. Rept. A-98-5. U.S. Army Corps of Engineers, Vicksburg, MS.

Buckingham, G.R. 2002. Alligatorweed. In: R. Van Driesch et al. (Tech. Coord.), Biological Control of Invasive Plants in the Eastern United States. U.S. Department of Agriculture Forest Service Pub. FHTET-2002-04. Bull. Distribution Ctr., Amherst, MA. Chapter 1.

Buckingham, G.R. and J.K. Balciunas. 1994. Biological Studies of *Bagous hydrillae*. Tech. Rept. A-94-6. U.S. Army Corps of Engineers, Vicksburg, MS.

Buckingham, G.R. and C.A. Bennett. 1994. Biological and Host Range Studies with Bagous affinis, an Indian Weevil that Destroys Hydrilla Tubers. Tech. Rept. A-94-8. U.S. Army Corps of Engineers, Vicksburg, MS.

Canfield, D.E., Jr., M.J. Maceina and J.V. Shireman. 1983. Effects of hydrilla and grass carp on water quality in a Florida lake. *Water Res. Bull.* 19: 773–778.

Carlson, R.E. 1977. A trophic state index for lakes. *Limnol. Oceanogr.* 22: 361–369.

Cassani, J.R. and W.E. Caton. 1986. Efficient production of triploid grass carp (*Ctenopharyngodon idella*) utilizing hydrostatic pressure. *Aquaculture* 55:43–50.

Center, T.D. 1982. The waterhyacinth weevils *Neochetina eichhorniae* and *N. bruchi. Aquatics* 4: 8, 16, 18–19.

Center, T.D. 1987. Insects, mites, and plant pathogens as agents of waterhyacinth (*Eichhornia crassipes* (Mart.) Solms) leaf and ramet mortality. *Lake and Reservoir Manage.* 3: 285–293.

Center, T.D., A.F. Cofrancesco and J.K. Balciunas. 1988. Biological control of aquatic and wetland weeds in the Southeastern U.S. *Proc. VII. International. Symposium. Control of Weeds.* Rome. pp. 239–262.

Center, T.D., F.A. Dray, G.P. Jubinsky and M.J. Grodowitz. 1999. Biological control of water hyacinth under conditions of maintenance management: Can herbicides and insects be integrated? *Environ. Manage.* 23: 241–256.

Center, T.D., M.P. Hill, H. Cordo and M.H. Julien. 2002. Waterhyacinth. In: R. Van Driesch et al. (Tech. Coord.), Biological Control of Invasive Plants in The Eastern United States. U.S. Department of Agriculture Forest Service Pub. FHTET-2002-04. Bull. Distribution Center, Amherst, MA. Chapter 4.

Chapman, Y.J. and D.J. Coffey. 1971. Experiments with grass carp in controlling exotic macrophytes in New Zealand. *Hydrobiologia* 12: 313–323.

Charudattan, R. 1986. Integrated control of waterhyacinth (*Eichhornia crassipes*) with a pathogen, insects, and herbicides. *Weed Sci.* 34(Suppl): 26–30.

Charudattan, R.S.B., J.T. DeValerio and V.J. Prange. 1989. Special problems associated with aquatic weed control. In: R. Baker and E. Dunn (Eds.), *New Directions in Biological Control.* Alan Liss, New York.

Chifamba, P.C. 1990. Preference of *Tilapia rendalli* (Boulenger) for some species of aquatic plants. *J. Fish. Biol.* 36: 701–705.

Childers, W.F. and G.W. Bennett. 1967. Experimental vegetation control by largemouth bass — Tilapia combinations. *J. Wildlife Manage.* 31: 401–407.

Chilton, E.W. and M.I. Muoneke. 1992. Biology and management of grass carp (*Ctenopharyngodon idella*, Cyrpinidae) — a North American perspective. *Rev. Fish Biol. Fisheries* 2: 283–320.

Choi, C.C. Bareiss, O. Walenchiak and E.M. Gross. 2002. Impact of polyphenols on growth of the aquatic herbivore *Acentria ephemerella. J. Chem. Ecol.* 28: 2245–2256.

Cofrancesco, A.F., Jr. 1984. Alligatorweed and its Biocontrol Agents. Information Exchange Bulletin A-84-3. U.S. Army Corps Engineers, Vicksburg, MS.

Colle, D. and J.V. Shireman. 1987. Bass, grass carp, and hydrilla. *Aquaphyte* 7: 12.

Colle, D.E. and J.V. Shireman. 1994. Use of grass carp in two Florida lakes, 1975 to 1994. *Proceedings, Grass Carp Symposium*. U.S. Army Corps of Engineers, Vicksburg, MS. pp. 111–120.

Connor, J.V., R.P. Gallagher and M.F. Chatry. 1980. Larval evidence for natural reproductions of the grass carp (*Ctenopharyngoden idella*) in the Lower Mississippi River, In: *Proceedings 14th Annual Larval Fish Conference*. U.S. Fish Wildlife Service, Biol. Science Program, Natl. Power Plant Team, Ann Arbor, MI. FWS/UBS-80/43.

Conway, K.E. 1976a. *Cercospora rodmanii*, a new pathogen of water hyacinth with biological control potential. *Can. J. Bot.* 54: 1079–1083.

Conway, K.E. 1976b. Evaluation of *Cercospora rodmanii* as a biological control of water-hyacinths. *Phytopathology* 66: 914–917.

Cooke, G.D. and R.H. Kennedy. 1989. Water Quality Management for Reservoirs and Tailwaters. Report I. In-reservoir Water Quality Management Techniques. Tech. Rept. A-89-1. U.S. Army Corps Engineers, Vicksburg, MS.

Cooke, G.D., E.B. Welch, S.A.Peterson and P.R. Newroth. 1993. *Restoration and Management of Lakes and Reservoirs*. 2nd ed. Lewis Publishers, Boca Raton, FL.

Creed, R.P., Jr. 1998. A biogeographic perspective on Eurasian watermilfoil declines: Additional evidence for the role of herbivorous weevils in promoting declines? *J. Aquatic Plant Manage.* 36: 16–22.

Creed, R.P., Jr. 2000. The weevil–watermilfoil interaction at different spatial scales: What we know and what we need to know. *J. Aquatic Plant Manage.* 38: 78–81.

Creed, R.P., Jr. and S.P. Sheldon. 1992. The effect of herbivore feeding on the buoyancy of Eurasian watermilfoil. *J. Aquatic Plant Manage.* 30: 75–76.

Creed, R.P., Jr. and S.P. Sheldon. 1993. The effect of feeding by a North American weevil, *Euhrychiopsis lecontei*, on Eurasian watermilfoil (*Myriophyllum spicatum*). *Aquatic Bot.* 45: 245–256.

Creed, R.P., Jr. and S.P. Sheldon. 1994. The effect of two herbivorous insect larvae on Eurasian watermilfoil. *J. Aquatic Plant Manage.* 32: 21–26.

Creed, R.P., Jr. and S.P. Sheldon. 1995. Weevils and watermilfoil: Did a North American herbivore cause the decline of an exotic plant? *Ecol. Appl.* 5: 1113–1121.

Creed, R.P., Jr., S.P. Sheldon and D.M. Cheek. 1992. The effect of herbivore feeding on the buoyancy of Eurasian watermilfoil. *J. Aquatic Plant Manage.* 30: 75–76.

Crutchfield, J.U., Jr., D.H. Schiller, D.D. Herlong and M.A. Mallen. 1992. Establishment and impact of redbelly tilapia in a vegetated cooling reservoir. *J. Aquatic Plant Manage.* 30: 28–35.

Desjardins, P.R. 1983. Cyanophage: History and likelihood as a control. In: *Lake Restoration, Protection and Management*. USEPA 440/5-83-001. USEPA. pp. 242–248.

Drenner, R.W., K.D. Hambright, G.L. Vinyard, M. Gophen and U. Pollingher. 1987. Experimental study of size-selective phytoplankton grazing by a filter-feeding cichlid and the cichlid's effects on plankton community structure. *Limnol. Oceanogr.* 32: 1138–1144.

Dressler, R.L., D.W. Hall, K.D. Perkins and N.H.Williams 1991. *Identification Manual for Wetland Plant Species of Florida*. University of Florida. Gainesville.

Durocher, P.P., W.C. Provine and J.E. Kraai. 1984. Relationship between abundance of largemouth bass and submerged vegetation in Texas reservoirs. *North Am. J. Fish. Manage.* 4: 84–88.

Duvall, R.J. and W.J. Anderson. 2001. Laboratory and greenhouse studies of microbial products used to biologically control algae. *J. Aquatic Plant Manage.* 39: 95–98.

Duvall, R.J., W.J. Anderson and C.R. Goldman. 2001. Pond enclosure evaluations of microbial products and chemical algicides in lake management. *J. Aquatic Plant Manage.* 39: 99–106.

Edwards, D.J. 1973. Aquarium studies on the consumption of small animals by 0-group grass carp, *Ctenopharyngodon idella* (Val.). *J. Fish Biol.* 5: 599–605.

Elakovich, S.D. and J.W. Wooten. 1989. Allelopathic potential of sixteen aquatic and wetland plants. *J. Aquatic Plant Manage.* 27: 78–84.

Everall, N.C. and D.R. Lees. 1997. The identification and significance of chemicals released from decomposing barley straw during reservoir algal control. *Water Res.* 31: 614–620.

Fowler, M.C. 1985. The results of introducing grass carp, *Ctenopharyngodon idella*, into small lakes. *Aquacult. Fish. Manage.* 16: 189–201.

Fowler, M.C. and T.O. Robson. 1978. The effects of food preferences and stocking rates of grass carp (*Ctenopharyngodon idella* Val.) on mixed plant communities. *Aquatic Bot.* 5: 261–276.

Freeman, T.E. 1977. Biological control of aquatic weeds with plant pathogens. *Aquatic. Bot.* 3: 175–184.

Gangstad, E.O., N.R. Spencer and J.A. Forest. 1975. Towards integrated control of alligatorweed. *Hyacinth Control J.* 13: 30–33.

Gibson, M.T., I.M.Welch, P.R.F. Barrett and I. Ridge. 1990. Barley straw as an inhibitor of algal growth. II: Laboratory studies. *J. Appl. Phycol.* 2: 241–248.

Grodowitz, M.J. 1998. An active approach to the use of insect biological control for the management of non-native aquatic plants. *J. Aquatic Plant Manage.* 36: 57–61.

Grodowitz, M.J., A.F. Confrancesco, J.E. Freedman and T.D. Center. 1997. Release and establishment of *Hydrellia balciunasi* (Diptera: Ephydridae) for the biological control of the submersed aquatic plant *Hydrilla verticillata* (Hydrocharitaceae) in the United States. *Biol. Control* 9: 15–23.

Gross, E.M. 2000. Seasonal and spatial dynamics of allelochemicals in the submersed macrophyte *Myriophyllum spicatum* L. *Verh. Int. Verein. Limnol.* 27: 2116–2119.

Gross, E.M. 2003. Allelopathy of aquatic autotrophs. *Crit. Rev. Plant Sci.* 22: 313–339.

Guillory, V. and R.D. Gasaway. 1978. Zoogeography of the grass carp in the U.S. *Trans. Am. Fish. Soc.* 107: 105–112.

Haag, K.H. 1986. Effective control of waterhyacinth using *Neochetina* and limited herbicide application. *J. Aquatic Plant Manage.* 24: 70–75.

Haag, K.H. and D.H. Habeck. 1991. Enhanced biological control of waterhyacinth following limited herbicide application. *J. Aquatic Plant Manage.* 29: 24–28.

Hanlon, S.G., M.V. Hoyer, C.E. Cichra and D.E. Canfield Jr. 2000. Evaluation of macrophyte control in 38 Florida lakes using triploid grass carp. *J. Aquatic Plant Manage.* 38: 48–54.

Henderson, J.E., J.P. Kirk, S.D. Lambrecht and W.E. Hayes. 2003. Economic impacts of aquatic vegetation to angling in two South Carolina reservoirs. *J. Aquatic Plant Manage.* 41: 53–56.

Hestand, R.S. and C.C. Carter. 1978. Comparative effects of grass carp and selected herbicides on macrophyte and phytoplankton communities. *J. Aquatic Plant Manage.* 16: 43–50.

Hilovsky, M. 2002. Invasive Eurasian watermilfoil can be controlled naturally. *Land Water* 46(2): 46–50.

Hoddle, M.S. 2004. Restoring balance: Using exotic species to control invasive exotic species. *Cons. Biol.* 18: 38–49.

Hoyer, M.V. and D.E. Canfield, Jr. 1996a. Lake size, macrophytes, and largemouth bass abundance in Florida lakes: A reply. *J. Aquatic Plant Manage.* 34: 48–50.

Hoyer, M.V. and D.E. Canfield, Jr. 1996b. Largemouth bass abundance and aquatic vegetation in Florida lakes: An empirical analysis. *J. Aquatic Plant Manage.* 34: 23–32.

Jester, L.L., M.A. Bozek, D.R. Helsel and S.P. Sheldon. 2000. *Eurhychiopsis lecontei* distribution, abundance, and experimental augmentation for Eurasian watermilfoil in Wisconsin lakes. *J. Aquatic Plant Manage.* 38: 88–97.

Johnson, R.L. and B. Blossey. 2002. Eurasian Watermilfoil. In:. R. Van Driesch et al. (Tech. Coord.), Biological Control of Invasive Plants in the Eastern United States U.S. Department of Agriculture Forest Service Pub. FHTET-2002-04. Bull. Distribution Center, Amherst, MA. Chapter 6.

Johnson, R.L., E.M. Gross and N.G. Hairston. 1998. Decline of the invasive submersed macrophyte *Myriophyllum spicatum* (Halgoraceae) associated with herbivory by larvae of *Acentria ephemerella* (Lepidoptera). *Aquatic Ecol.* 31: 273–282.

Joye, G.F. 1990. Biocontrol of the aquatic plant *Hydrilla verticillata* (L.f.) Royce with an endemic fungal disease. Unpublished Report. U.S. Army Corps Engineers, Vicksburg, MS.

Joye, G.F. and A.F. Cofrancesco, Jr. 1991. Studies on the Use of Fungal Pathogens for Control of *Hydrilla verticillata* (L.f.) Royle. Tech. Rept. A-91-4. U.S. Army Corps Engineers, Vicksburg, MS.

Kangasniemi, B.J. 1983. Observations on herbivorous insects that feed on *Myriophyllum spicatum* in British Columbia. In: *Lake Restoration, Protection and Management*. USEPA-440/5-83-001. USEPA. pp. 214–219.

Killgore, K.J., J.P. Kirk and J.W. Folz. 1998. Response of littoral fishes in Upper Lake Marion, South Carolina following hydrilla control by triploid grass carp. *J. Aquatic Plant Manage.* 36: 82–87.

Kirk, J.P. and R.C. Socha. 2003. Longevity and persistence of triploid grass carp stocked into the Santee Cooper Reservoirs of South Carolina. *J. Aquatic Plant Manage.* 41: 90–92.

Kracko, K.M. and R.L. Noble. 1993. Herbicide inhibition of grass carp feeding on hydrilla. *J. Aquatic Plant Manage.* 31: 273–275.

Lembi, C.A. 2001. Barley straw for algae control. *Aquatics* 23: 13–18.

Lembi, C.A. and T. Chand-Goyal. 1994. Plant Growth Regulators as Potential Tools in Aquatic Plant Management: Efficacy and Persistence in Small-Scale Tests. Contract Rep. A-94-1. U.S. Army Corps Engineers, Vicksburg, MS.

Lembi, C.A., T. Chand and W.C. Reed. 1990. Plant growth regulator effects on submersed aquatic plants. In: *Proceedings 24th Annual Meeting Aquatic Plant Control Research Program.* U.S. Army Corps Engineers, Vicksburg, MS.

Leslie, A.J., Jr., R.S. Hestand, III. 1992. Managing aquatic plants with grass carp: A practical guide for natural resources managers. Large impoundments. Unpublished Report. Florida Department of Natural Resources, Tallahassee, FL and Florida Game and Fresh Water Fish Commission, Eustes.

Leslie, A.J., Jr., J.M. Van Dyke and L.E. Nall. 1982. Current velocity for transport of grass carp eggs. *Trans. Am. Fish. Soc.* 111: 99–101.

Leslie, A.J., Jr., J.M. Van Dyke, R.S. Hestand, III and B.Z. Thompson. 1987. Management of aquatic plants in multi-use lakes with grass carp (*Ctenopharyngodon idella*). *Lake and Reservoir Manage.* 3: 266–276.

Leslie, A.J., L.E. Nall, G.P. Jubinsky and J.D. Schardt. 1994. Effects of grass carp on the aquatic vegetation in Lake Conway, Florida. In: *Grass Carp Symposium.* U.S. Army Corps Engineers, Vicksburg, MS. pp. 121–128.

Louda, S.M. and P. Stiling. 2004. The double-edged sword of biological control in conservation and restoration. *Conserv. Biol.* 18: 50–53.

Maceina, M.J., P.W. Bettoli, W.G. Klussmann, R.K. Betsill and R.L. Noble. 1991. Effect of aquatic macrophyte removal on recruitment and growth of black crappies and white crappies in Lake Conroe, Texas. *North Am. J. Fish. Manage.* 11: 556–563.

Maceina, M.J., M. F, Cichra, R.K. Betsill and P.W. Bettoli. 1992. Limnological changes in a large reservoir following vegetation removal by grass carp. *J. Fresh Water Ecol.* 7: 81–95.

Maceina, M.J., J. Slipke and J.M. Grizzle. 1999. Effectiveness of three barrier types for confining grass carp in embayments of Lake Seminole, Georgia. *North Am. J. Fish. Manage.* 19: 968–976.

MacRae, I.V., N.N. Winchester and R.A. Ring. 1990. Feeding activity and host preference of the milfoil midge, *Cricotopus myriophylli* Oliver (Diptera: Chironomidae). *J. Aquatic Plant Manage.* 28: 89–92.

Maddox, D.M., L.A. Andres, R.D. Hennessey, R.D. Blackburn and N.R. Spencer. 1971. Insects to control alligatorweed, an invader of aquatic ecosystems in the U.S. *BioScience* 21: 985–991.

Madeira, P.T., C.C. Jacono and T.K. Van. 2000. Monitoring hydrilla using two RAPD procedures and the nonindigenous aquatic species database. *J. Aquatic Plant Manage.* 38: 33–40.

Madsen, J.D., H. Crosson, K.S. Hamel, M.A. Hilovsky and C.H. Welling. 2000. Panel Discussion. Management of Eurasian watermilfoil in the United States using native insects: State regulatory and management issues. *J. Aquatic Plant Manage.* 38: 121–124.

Mallison, C.T., R.S. Hestand III and B.Z. Thompson. 1995. Removal of triploid grsss carp with an oral rotenone bait in two central Florida lakes. *Lake and Reservoir Manage.* 11: 337–342.

Martinez-Jimenez, M. and R. Charudattan. 1998. Survey and evaluation of Mexican native fungi for potential biocontrol of waterhyacinth. *J. Aquatic Plant Manage.* 36: 145–148.

Martyn, R.D. 1985. Waterhyacinth decline in Texas caused by *Cercospora piaropi*. *J. Aquatic Plant Manage.* 23: 20–32.

Martyn, R.D., R.L. Noble, P.W. Bettoli and R.C. Maggio. 1986. Mapping aquatic weeds with aerial color infrared photography and evaluating their control by grass carp. *J. Aquatic Plant Manage.* 24: 46–56.

Mazzei, K.C., R.M. Newman, A. Loos and D.W. Ragsdale. 1999. Developmental rates of the native milfoil weevil, *Euhrychiopsis lecontei,* and damage to Eurasian watermilfoil at constant temperatures. *Biol. Control* 16: 139–143.

McComas, S. 2003. *Lake and Pond Management Guidebook.* Lewis Publishers, Boca Raton, FL.

McDonald, M.E. 1985. Growth of a grazing phytoplanktivorous fish and growth enhancement of the grazed alga. *Oecologia* 67: 132–136.

McFarland, D.G. and J.W. Barko. 1999. High temperarture effects on growth and propagule formation in hydrilla biotypes. *J. Aquatic Plant Manage.* 37: 17–25.

Miller, H.D. and J. Boyd. 1983. Large-Scale Management Test of the Use of the White Amur for Control of Problem Aquatic Plants; Report 4. Third Year Poststocking Results. Vol VI: The Water and Sediment Quality of Lake Conway, Florida. Tech. Rept. A-78-3. U.S. Army Corps Engineers. Jacksonville, Florida.

Miller, A.C. and J.L. Decell. 1984. Use of White Amur for Aquatic Plant Management. Instruct. Rep. A-84-1. U.S. Army Corps Engineers, Vicksburg, MS.

Miller, A.C. and H.R. King. 1984. Large-scale Operations Management Test for Use of the White Amur for Control of Problem Plants. Report 5. Synthesis Report. Tech. Rept. A-78-2. U.S. Army Corps Engineers, Vicksburg, MS.

Miller, H.D. and R. Potts. 1982. Large-Scale Operations Management Test of the Use of the White Amur for Control of Problem Aquatic Plants; Report 3. Second Year Poststocking Results. Vol VI: The Water and Sediment Quality of Lake Conway, Florida. Tech. Rept. A-78-2. U.S. Army Corps Engineers, Vicksburg, MS.

Mitchell, C.P. 1980. Control of water weeds by grass carp in two small lakes. *J. Mar. Fresh Water Res.* 14: 381–390.

Mjelde, J. and B.A. Faafeng. 1997. *Ceratophyllum demersum* hampers phytoplankton development in some small Norwegian lakes over a wide range of phosphorus concentrations and geographical latitude. *Freshwater Biol.* 37: 355–366.

Murphy, J.E., K.B. Beckmen, J.K. Johnson, R.B. Cope, T. Lawmaster and V.R. Beasley. 2002. Toxic and feeding deterrent effects of native aquatic macrophytes on exotic grass carp (*Ctenopharyngodon idella*). *Ecotoxicology* 11: 243–254.

Nall, L.E. and J.D. Schardt. 1980. Large-scale operations management test using the white amur at Lake Conway, Florida. Aquatic macrophytes. In: *Proceedings 14th Annual Meeting, Aquatic Plant Control*. Res. Plan. Oper. Rev. Misc. Paper A-80-3. U.S. Army Corps Engineers, Vicksburg, MS. pp. 249–272.

Nelson, L.S. 1990. Plant growth regulators for aquatic plant management. In: *24th Annual Meeting, Aquatic Plant Control*. Res. Program. Misc. Paper A-90-3. U.S. Army Corps Engineers, Vicksburg, MS. pp. 115–118.

Nelson, L.S., J.F. Shearer and M.D. Netherland. 1998. Mesocosm evaluation of integrated fluridone-fungal pathogen treatment on four submersed plants. *J. Aquatic Plant Manage.* 36: 73–77.

Netherland, M.D. and J.F. Shearer. 1996. Integrated use of fluridone and a fungal pathogen for control of hydrilla. *J. Aquatic Plant Manage.* 34: 4–8.

Newman, J.R. and P.R.F. Barrett. 1993. Control of *Microcystis aeruginosa* by decomposing barley straw. *J. Aquatic Plant Manage.* 31: 203–206.

Newman, R.M. 2004. Invited Review. Biological control of Eurasian watermilfoil by aquatic insects: Basic insights from an applied problem. *Arch. Hydrobiol.* 159: 145–184.

Newman, R.M. and D.D. Biesboer. 2000. A decline of Eurasian watermilfoil in Minnesota associated with the milfoil weevil *Euhrychiopsis lecontei*. *J. Aquatic Plant Manage.* 38: 105–111.

Newman, R.M., M.E. Borman and S.W. Castro. 1997. Developmental performance of the weevil *Euhrychiopsis lecontei* on native and exotic watermilfoil host plants. *J. North Am. Benthol. Soc.* 16: 627–634.

Newman, R.M., D.W. Ragsdale, A. Milles and C. Oien. 2001. Overwinter habitat and the relationship of overwinter to in-lake densities of the milfoil weevil *Euhrychiopsis lecontei*, a Eurasian watermilfoil control agent. *J. Aquatic Plant Manage.* 39: 63–67.

Noble, R., P.W. Bettoli and R.K. Betsill. 1986. Considerations for the use of grass carp in large, open systems. *Lake and Reservoir Manage.* 2: 46–48.

Oliver, D.R. 1984. Description of a new species of *Cricotopus* van der Wulp (Diptera: Chironomidae) associated with *Myriophyllum spicatum*. *Can. Entomol.* 116: 1287–1292.

Opuszynski, K. 1992. Are herbivorous fish herbivorous? *Aquaphyte* 12: 1,12–13.

Osborne, J.A. 1978. Management of Emergent and Submergent Vegetation in Stormwater Retention Ponds using Grass Carp. Final Report to Florida Department of Natural Resources. University of Central Florida, Orlando.

Osborne, J.A. and R.D. Riddle. 1999. Feeding and growth rates for triploid grass carp as influenced by size and water temperature. *J. Fresh Water Ecol.* 14: 41–46.

Painter, D.S. and K.J. McCabe. 1988. Investigation into the disappearance of Eurasian watermilfoil from the Kawartha lakes. *J. Aquatic Plant Manage.* 26: 3–12.

Pauley, G.B. et al. 1994. An overview of the use and efficacy of triploid grass carp *Ctenopharyngodon idella* as a biological control of aquatic macrophytes in Oregon and Washington state lakes. In: *Proceedings, Grass Carp Conference*. U.S. Army Corps Engineers, Vicksburg, MS.

Pine, R.T. and W.J. Anderson. 1991. Plant preferences of triploid grass carp. *J. Aquatic Plant Manage*. 29: 80–82.

Pine, R.T., L.W.J. Anderson and S.S.O. Hung. 1989. Effects of static versus flowing water on aquatic plant preferences of grass carp. *Trans. Am. Fish. Soc*. 118: 336–344.

Raibley, P.T., D. Blodgett and R.E. Sparks. 1995. Evidence of grass carp (*Ctenopharyngodon idella*) reproduction in the Illinois and Upper Mississippi Rivers. *J. Freshwater Ecol*. 10: 65–74.

Ridge, I. and J.M. Pillinger. 1996. Towards understanding the nature of algal inhibitors from barley straw. *Hydrobiologia* 340: 301–306.

Safferman, R.S. and M.E. Morris. 1963. Algal virus: isolation. *Science* 140: 679–680.

Sanders, D.R. and E.A. Theriot. 1986. Large-Scale Operations Management Test (LSOMT) of Insects and Pathogens for Control of Waterhyacinth in Louisiana. Vol. II. Results for 1982–1983. Tech. Rept. A-85-1. U.S. Army Corps Engineers, Vicksburg, MS.

Schuytema, G.S. 1977. *Biological Control of Aquatic Nuisances — A Review*. USEPA-600/3-77-084.

Shabana, Y.M., J.P. Cuda and R. Charudattan. 2003. Combining plant pathogenic fungi and the leaf-mining fly, *Hydrellia pakistanae,* increases damage to hydrilla. *J. Aquatic Plant Manage*. 41: 76–81.

Shearer, J.F. 1994. Potential role of plant pathogens in declines of submerged macrophytes. *Lake and Reservoir Manage*. 10: 9–12.

Shearer, J.F. 1996. Field and Laboratory Studies of the Fungus *Mycoleptodiscus terrestris* as a Potential Agent for Management of the Submersed Aquatic Macrophyte *Hydrilla verticillata*. Tech. Rept. A-96-3. U.S. Army Corps Engineers, Vicksburg, MS.

Shearer, J.F. 1997. Endemic Pathogen Biocontrol Research on Submersed Macrophytes: Status Report 1996. Tech. Rept. A-97-3. U.S. Army Corps Engineers, Vicksburg, MS.

Sheldon, S.P. and R.P. Creed. 1995. Use of a native insect as a biological control for an introduced weed. *Ecol. Appl*. 5: 1122–1132.

Sheldon, S.P. and L.M. O'Bryan. 1996. The effects of harvesting Eurasian watermilfoil on the aquatic weevil *Euhrychiopsis lecontei. J. Aquatic Plant Manage*. 34: 76–77.

Shireman, J.V. 1982. Cost analysis of aquatic weed control: fish versus chemicals in a Florida lake. *Prog. Fish. Cult*. 44: 199–200.

Shireman, J.V. and M.J. Maceina. 1981. The utilization of grass carp, *Ctenopharyngodon idella* Val., for hydrilla control in Lake Baldwin, Florida. *J. Fish. Biol*. 19: 629–636.

Shireman, J.V., D.E. Colle and R.W. Rottman. 1978. Size limits to predation on grass carp by largemouth bass. *Trans. Am. Fish. Soc*. 107: 213–215.

Shireman, J.V., W.T. Haller, D.E. Colle, C.E. Watkins, D.F. Durant and D.E. Canfield. 1983. Ecological Impact of Integrated Chemical and Biological Aquatic Weed Control. USEPA-660/3-83–098. USEPA.

Shireman, J.V., M.V. Hoyer, M.J. Maceina and D.E. Canfield. 1985. The water quality and fishing of Lake Baldwin, Florida: 4 years after macrophyte removal by grass carp. *Lake and Reservoir Manage*. 1: 201–206.

Smith, D.W., S.J. Slade, J.H. Andrews and R.F. Harris. 1989. Pathogenicity of the fungus *Colletotrichum gloeosporioide*s (Penz.) Sacc. to EWM (*Myriophyllum spicatum* L.). *Aquatic Bot*. 33: 1–12.

Solarz, S.L. and R.M. Newman. 2001. Variation in host plant preference and performance by the milfoil weevil, *Euhrychiopsis lecontei* Dietz, exposed to native and exotic watermilfoils. *Oecologia* 126: 66–75.

Sorsa, K.K., E.V. Nordheim and J.H. Andrews. 1988. Integrated control of Eurasian watermilfoil, *Myriophyllum spicatum,* by a fungal pathogen and a herbicide. *J. Aquatic Plant Manage*. 26: 12–17.

Spencer, N.R. and J.R. Coulson. 1976. The biological control of alligator weed *Alternanthera philoxeroides*, in the United States of America. *Aquatic Bot*. 2: 177–190.

Stanley, J.G., W.W. Miley, II and D.L. Sutton. 1978. Reproductive requirements and likelihood for naturalization of escaped grass carp in the U.S. *Trans. Am. Fish. Soc*. 107: 119–128.

Steward, K.K. and T.K. Van. 1987. Comparative studies of monoecious and dioecious hydrilla (*Hydrilla verticillata*). *Weed Sci*. 35: 204–210.

Stewart, R.M. and W.A. Boyd. 1994. Simulation model evaluation of sources of variability in grass carp stocking requirements. In: *Proceedings, Grass Carp Symposium*. U.S. Army Corps Engineers, Vicksburg, MS. pp. 85–92.

Sutter, T.J. and R.M. Newman. 1997. Is predation by sunfish (*Lepomis* spp.) an important source of mortality for the Eurasian watermilfoil biocontrol agent *Euhrychiopsis lecontei*? *J. Fresh Water Ecol.* 12: 225–234.

Sutton, D.L. and V.V. VanDiver. 1986. Grass Carp: a Fish for Biological Management of Hydrilla and Other Aquatic Weeds in Florida. Bull. 867. Florida Agric. Exper. Sta., University of Florida, Gainesville.

Swanson, E.D. and E.P. Bergersen. 1988. Grass carp stocking model for coldwater lakes. *North Am. J. Fish. Manage.* 8: 284–291.

Szczepanski, A.J. 1977. Allelopathy as a means of biological control of water weeds. *Aquatic Bot.* 3: 193–197.

Theriot, E.A. 1989. Biological control of aquatic plants with plant pathogens. In: *Proceedings, Workshop on Management of Aquatic Weeds and Mosquitoes in Impoundments*. Water Resources Research Institute, University of North Carolina, Charlotte.

Theriot, E.A., S.L. Kees and H.B. Gunner. 1996. *Specific Association of Plant Pathogens with Submersed Aquatic Plants*. Tech. Rept. A-96-9. U.S. Army Corps Engineers, Vicksburg, MS.

van Donk, E. and W.J. van de Bund. 2002. Impact of submerged macrophytes including charophytes on phyto- and zooplankton communities: Allelopathy versus other mechanisms. *Aquatic Bot.* 72: 261–274.

Van Dyke, J.M. 1994. Long-term use of grass carp for aquatic plant control in Deer Point Lake, Bay County, Florida. In: *Proceedings, Grass Carp Symposium*. U.S. Army Corps Engineers, Vicksburg, MS. pp. 146–150.

Van Dyke, J.M., A.J. Leslie, Jr. and L.E. Nall. 1984. The effects of the grass carp on the aquatic macrophytes of four Florida lakes. *J. Aquatic Plant Manage.* 22: 87–95.

Vinyard, G.L., R.W. Drenner, M. Gophen, U. Pollingher, D.L. Winkelman and K.D. Hambright. 1988. An experimental study of the plankton community impacts of two omnivorous filter-feeding cichlids, *Tilapia golidaea* and *Tilapia aurea. Can. J. Fish. Aquatic Sci.* 45: 689–690.

Ware, F.D., R.D. Gasaway, R.A. Martz and T.F. Drda. 1975. Investigations of herbivorous fishes in Florida. In: P.L. Brezonik and J.L. Fox (Eds.), *Water Quality Management through Biological Control*. Department of Environmental Engineering Sciences, University of Florida, Gainesville. pp. 79–84.

Webb, D.H., L.N. Mangum, A.L. Bates and H.D. Murphy. 1994. Aquatic vegetation in Guntersville Reservoir following grass carp stocking. In: *Proceedings, Grass Carp Symposium*. U.S. Army Corps Engineers, Vicksburg, MS. pp. 199–209.

Webb, M.A., H.S. Elder and R.G. Howells. 1994. Grass carp reproduction in the Lower Trinity River, Texas. In: *Proceedings, Grass Carp Symposium*. U.S. Army Corps Engineers, Vicksburg, MS. pp. 29–32.

Welch, I.M., P.R.F. Barrett, M.T. Gibson and I. Ridge. 1990. Barley straw as an inhibitor of algal growth. I. Studies in the Chesterfield Canal. *J. Appl. Phycol.* 2: 231–239.

Wheeler, G.S. and T.D. Center. 1996. The influence of hydrilla leaf quality on larval growth and development of the biological control agent *Hydrellia pakistanae* (Diptera: Ephydridae). *Biol. Control* 7: 1–9.

Wiley, M.J. and R.W. Gorden. 1985. Biological Control of Aquatic Macrophytes by Herbivorous Carp. Part 3. Stocking Recommendations for Herbivorous Carp and Description of the Illinois Herbivorous Fish Stocking Simulation System. Aquatic Biology Tech. Report. 1984 (12). Illinois Natural History Survey, Champaign.

Wiley, M.J. and L.D. Wike. 1986. Energy balances of diploid, triploid, and hybrid grass carp. *Trans. Am. Fish. Soc.* 115: 853–863.

Wiley, M.J., S.W. Waite and T. Powless. 1984. The relationship between aquatic macrophytes and sport fish production in Illinois ponds: a simple model. *North Am. J. Fish Manage.*, 4: 111–119.

Wiley, M.J., P.P. Tazik and S.T. Sobaski. 1987. Controlling Aquatic Vegetation with Triploid Grass Carp. Circular 57. Illinois Natural History Survey, Champaign.

Zettler, F.W. and T.E. Freeman. 1972. Plant pathogens as biocontrols of aquatic weeds. *Annu. Rev. Phytopathol.* 10: 455–470.

Section IV

Multiple Benefit Treatments

18 Hypolimnetic Aeration and Oxygenation

18.1 INTRODUCTION

Depletion of dissolved oxygen (DO) in the hypolimnia of stratified eutrophic lakes is one of the first signs of eutrophication. Anoxia occurs if respiration of organic matter in hypolimnetic water and sediments is sufficient to exhaust most or all of the hypolimnetic DO before autumn destratification. Anoxia produces undesirable changes in lake quality, including accelerated internal recycling of nutrients, solubilization of metals that are undesirable in water supplies and limitation of fish distribution, especially cold-water species. Chapter 2 describes the expected seasonal changes in temperature and DO in stratified lakes and reservoirs.

Hypolimnetic aeration, used first in Lake Bret, Switzerland (Mercier and Perret, 1949), is a lake management technique designed to counteract hypolimnetic anoxia and its associated problems. The specific objectives of hypolimnetic aeration are threefold. The first, and usually most attainable, is to raise the oxygen content of the hypolimnion without destratifying the water column or warming the hypolimnion. The second, which is largely dependent on the first, is to provide an increased habitat and food supply for coldwater fish species. The third, if sediment-to-water exchange of P is controlled by iron redox, is to reduce sediment P release by establishing oxic conditions at the sediment-water interface. Other constituents that reach high and possibly undesirable concentrations under anoxic conditions, such as NH_4^+, Mn, and Fe, should also be diminished by hypolimnetic aeration. This technique and its effects were described by Pastorak et al. (1981, 1982). However, many device modifications and improvements have occurred since that review.

18.2 DESCRIPTION AND OPERATION OF UNITS

There are several designs for hypolimnetic aerators. Fast and Lorenzen (1976) reviewed 21 designs and grouped them into three categories: (1) mechanical agitation, involving removal, treatment, and return of the hypolimnetic water; (2) injection of pure oxygen; and (3) injection of air, either through a full or partial air-lift design or through a down-flow injection design.

Mechanical agitation involves drawing water from the hypolimnion, aerating it on shore or on the lake surface by means of a splash basin, and returning the water to depth with minimal increase in temperature. It is not a popular system because of poor gas exchange efficiency (Pastorak et al., 1982). To improve efficiency of gas exchange, pure oxygen has been used to aerate hypolimnetic water drawn from depth, treated at the shore or on the surface, with or without pressure, and returned. Pure oxygen also can be introduced into the hypolimnion and forced downward with a pump or released at depth and allowed to rise through the hypolimnion. The latter is referred to as a deep oxygen injection system (DOIS; Figure 18.1). Although gas exchange is more favorable with pure oxygen than with aeration (bubbles are 85–100% instead of 20% O_2), there are potential problems, especially with N_2-filled bubbles escaping to the lake surface and the associated mixing of hypolimnetic water with epilimnetic water (Fast and Lorenzen, 1976). If bubbles are small enough (≤ 1 mm radius) and rising plumes weak enough, bubbles should completely dissolve in the hypolimnion (Wüest et al., 1992). In sufficiently deep lakes (≥ 30 m), the pure oxygen bubbles

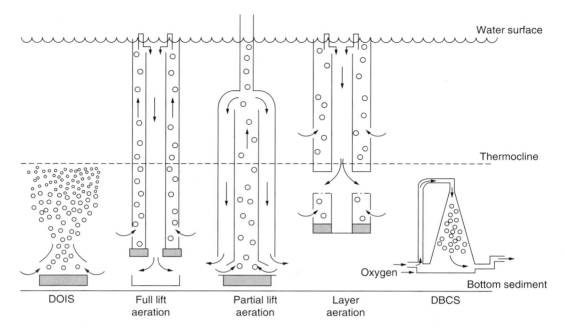

FIGURE 18.1 Comparison of units to aerate and oxygenate hypolimnia. DOIS = deep oxygen injection system, DBCS = double bubble contact system. (Modified from NALMS, 2001. *Managing Lakes and Reservoirs*. 3rd ed. USEPA 841-B-01–006. NALMS, Madison, WI.)

should completely dissolve before reaching the epilimnion (Gächter, 1987; Prepas et al., 1997). Complete dissolution of small bubbles may occur within 8 m (Babin et al., 1999). A DOIS system used in Europe is called "Tanytarsus," after a midge that prefers aerobic conditions (Jungo, 1993). The potential destratifying effect of pure oxygen bubbles in relatively shallow lakes is overcome by using a down flow double bubble contact system (DBCS), in which water is circulated by pumping (Speece, 1971; Doke et al., 1995; Beutel and Horne, 1999).

Injection of air via air-lift systems has been the most popular for hypolimnetic aeration. Full air-lift brings bottom water to the surface by forcing compressed air into the bottom of an outer cylinder (Figure 18.1). The rising bubbles drive the air–water mixture to the surface, exposing water to the atmosphere, and then returns it to the hypolimnion via an inner cylinder after first venting the air bubbles (Figure 18.1). The system can be composed of separate pipes as well (Figure 18.2). Partial air-lift units aerate hypolimnetic water in place, with water and air bubbles being separated at depth with the excess air being discharged at the surface (Figure 18.1). A hypolimnetic aeration unit that is commercially available is shown in Figure 18.3. In both air-lift systems, the contained air forces the oxygenated water to distribute horizontally into the hypolimnion (Figure 18.1). Pure O_2 is used instead of air to increase gas transfer efficiency, but this provides less distribution force than air.

Fast et al. (1976) and Lorenzen and Fast (1977) found that the full air-lift design is least costly and more efficient at delivering oxygen than the other systems (Pastorak et al., 1982). The partial air-lift design, however, is probably the most frequently used system, possibly because of its commercial availability (Verner, 1984). The flexible "Limno," constructed mainly of noncorrosive PVC-coated polyester fabric, was upgraded by a system composed of more substantial hard PVC (Figure 18.3). Also, partial airlift systems avoid the forces of wind, wave, and pendulum torque that confront full airlift systems, therefore requiring less sturdy and costly structure.

A different type of full air-lift hypolimnetic aerator ("Tibean") uses a submerged centrifugal pump and electromotor rather than compressed air (Jaeger, 1990). The pumped water entrains air

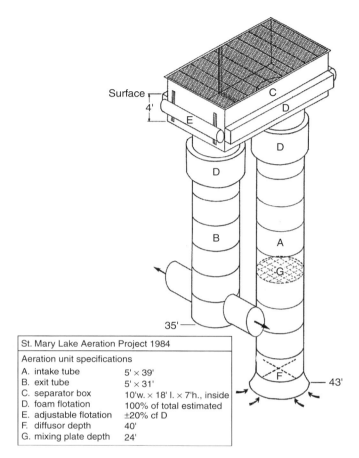

Surface

4'

C

D

E

D

D

B

A

G

35'

F

43'

St. Mary Lake Aeration Project 1984

Aeration unit specifications

A. intake tube	5' × 39'
B. exit tube	5' × 31'
C. separator box	10'w. × 18' l. × 7'h., inside
D. foam flotation	100% of total estimated
E. adjustable flotation	±20% cf D
F. diffusor depth	40'
G. mixing plate depth	24'

FIGURE 18.2 A full air-lift hypolimnetic aerator. (Designed by K.I. Ashley, Fisheries Research and Technical Services Section, Fish and Wildlife Branch, Ministry of Environment, Vancouver, BC.)

bubbles brought from the surface through a pipe, and the air-water mixture is injected at the bottom of a vertical tube much like other units. A similar technique, called MIXOX, has been used in 70 Finnish lakes since 1981 (Lappalainen, 1994; Keto et al., 2004). However, rather than entraining air bubbles (DOIS), high DO epilimnetic water is simply pumped to depth creating an internal circulation. While the lighter epilimnetic water initially rises, it soon mixes with hypolimnetic water and does not destratify the lake.

All units produce a circular field with a DO gradient diminishing distal to the unit. The number of units necessary to reach an acceptable hypolimnetic DO concentration depends on the amount of DO that must be delivered to meet the DO demand in the hypolimnion. Airflow, and thus DO output, to meet the determined demand must be carefully estimated prior to installation of any units. Insufficient aerator sizing is common, due to both insufficient oxygen supply and induced current mixing.

In some instances, hypolimnetic aeration resulted in a metalimnion with near zero DO, accompanied by high P and hydrogen sulfide, despite of an aerobic hypolimnion (Steinberg and Arzet, 1984; McQueen and Lean, 1986). A metalimnetic DO minimum does not promote complete nutrient mineralization of settling organic matter (Gächter, 1987). To minimize metalimnetic DO minima, expand the oxygenated environment, and minimize diffusion of nutrients to the photic zone, a layer aeration system was developed and tested in two Connecticut Utility Reservoirs (Kortmann et al, 1988, 1994; Kortmann, 1989). The system redistributes heat and oxygen, while

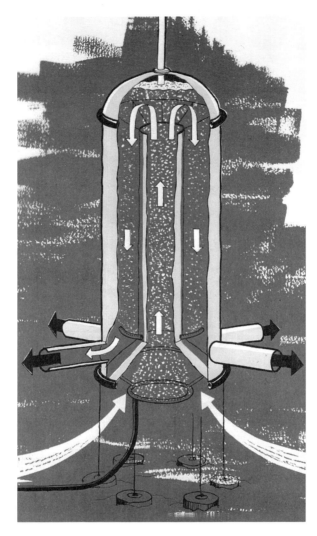

FIGURE 18.3 Limno partial air-lift hypolimnetic aerator. Arrows indicate directions of air and water flows. (Courtesy of R. Geney, General Environmental Systems, Summerfield, NC.) Standard length is 10 m; placement is 1.2 m above the bottom, making the unit similar to a full air-lift in reservoirs and lakes 12 m deep.

maintaining a deep barrier to a reduced volume of anoxic, high-P water (Figure 18.1). Results of increased DO in one of those reservoirs are shown in Figure 18.4. Suggested power and air-flow requirements are lower for layer aeration than for hypolimnetic aeration. With a similar purpose, Stefan et al. (1987) designed a metalimnetic aerator, as well as system to oxygenate the thermocline region only while maintaining ice cover in winterkill threatened lakes (Ellis and Stefan, 1990).

Hypolimnetic aeration may not operate satisfactorily if the water body is too shallow. Although stratification may exist, the density gradient may not be sufficient to resist thermocline erosion and complete mixing even if the hypolimnion is mildly circulated. Such slow circulation, if the water body destratifies, may result in low DO throughout and formation of scums of blue-green algae in a warmed surface layer (McQueen and Lean, 1986; Cooke and Carlson, 1989). Therefore, hypolimnetic aeration is not recommended if maximum depth is less than 12 to 15 m and/or the hypolimnetic volume is relatively small (Cooke and Carlson, 1989). However, oxygenation with a Speece cone (Figure 18.1) can be used in

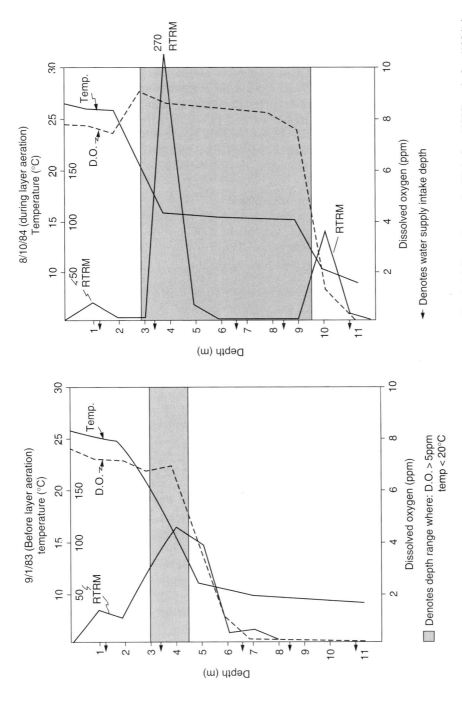

FIGURE 18.4 Temperature, oxygen, and RTRM (relative thermal resistance to mixing) profiles in Mulberry Reservoir, CT, before (1983) and after (1984) layer aeration. (From Kortmann, R.W. et al. 1988. *Lake and Reservoir Manage.* 4: 35–50. With permission.)

lakes with less depth (Doke et al., 1995). Also, air separation delivering ~80% O_2 with a partial airlift system did not destratify a shallow lake (max. depth 7.9 m; Gibbons, pers. comm.).

18.3 UNIT SIZING

The effectiveness of aerators to oxygenate a hypolimnion depends on the DO transport efficiency between air bubbles and pumped water (Ashley, 1985; Ashley et al., 1992). An adequate sizing of the aeration system, even oversizing it, is important to compensate for unpredicted variation in oxygen consumption and hypolimnetic volume, and for mechanical breakdown (Pastorak et al., 1982). If hypolimnetic DO decreases in late summer, with an aeration system operating, the cause is usually an undersized aerator (McQueen and Lean, 1986). Procedures for sizing hypolimnetic aerators to oxygenate hypolimnia without destratifying the water column were outlined in detail by Lorenzen and Fast (1977), Ashley (1985) and Ashley et al. (1987). Ashley (1985) developed and tested an empirical sizing method outlined in the following steps:

1. Determine the maximum hypolimnetic volume.
2. Estimate the hypolimnetic oxygen consumption rate in kg/day by calculating the slope for the relationship between average hypolimnetic DO and time (Chapter 3).
3. Calculate water flow necessary to meet the DO demand rate. This depends on the aerator input DO concentration, which is variable and usually less than saturation.
4. Calculate aerator flow and outflow tube size (radius) from water flow and estimated velocity. Tube length should be as long as possible to optimize oxygen transfer efficiency. Outflow tube radius should exceed that of the inflow tube so water flow is not restricted.
5. Determine friction losses at the aerator entrance and exit, as well as in the line, to estimate total head loss.
6. Determine density of the air-water mixture assuming the theoretical head generated is due to the density difference between ambient water and the air-water mixture.
7. Calculate the air-flow rate to transport the required water flow to satisfy the oxygen demand, considering length and diameter of the line.
8. Estimate the pressure requirements at the compressor (from the pressure drops) and the power demand.

Ashley (1985) presented equations and example calculations for each of these steps as an appendix, which the reader is urged to review for a more complete understanding of the sizing procedure. Little (1995) developed a model to estimate induced water flow rate and O_2 transfer necessary to meet O_2 demand, as well as the extent of the oxygenated zone.

The principal factor that determines the air flow rate required is the hypolimnetic oxygen demand, usually expressed as the areal hypolimnetic oxygen deficit rate (AHOD; Chapter 3). The AHOD is determined from a plot of mean hypolimnetic DO concentrations (volume-weighted) over time and estimating the rate from the slope of a least squares regression line. Beginning the computation with a well-oxygenated hypolimnion is important to ensure that the maximum depletion rates are obtained (Ashley and Hall, 1990). Therefore, sampling should begin prior to thermal stratification in the spring. However, data for analysis should not include DO concentrations < 1–2 mg/L, because DO demand rate greatly declines below that level (McQueen et al., 1984). Although less accurate, a rate may also be estimated from two observations of widely differing mean hypolimnetic DOs, prior to depletion:

$$\text{AHOD (mg/m}^2 \text{ per day)} = \frac{\bar{X}\text{DO}_{t_1} - \bar{X}\text{DO}_{t_2}}{t_2 - t_1} \cdot \bar{Z}_h \qquad (18.1)$$

where DO_{t_1} = dissolved oxygen at the beginning of stratification, DO_{t_2} = DO prior to any concentrations less than 1–2 mg/L, $t_2 - t_1$ = elapsed time in days, and Z_h = mean depth of hypolimnion.

Hypolimnetic BOD determinations in water samples may not realistically estimate the demand because (1) sediment demand is not included, and (2) dissolved organic matter may accumulate in an anoxic hypolimnion, producing bottle BODs many times greater than the actual AHOD (Sehgal and Welch, 1991).

From the AHOD, the required air-flow rate can be determined:

$$\text{Air Flow, m}^3/\text{day} = \frac{AHOD \cdot A_h \cdot I \cdot 10^{-6}}{1.205 \cdot 0.2} \tag{18.2}$$

where A_h = area of hypolimnion, in m²; I = factor to allow for unmeasured and induced O_2 demand that occurs with increased DO concentration and mixing across the sediment surface when aerated (McQueen et al., 1984; Ashley et al. 1987; Moore et al., 1996; Beutel, 2003). An additional factor of 2–4 per 10°C to I may be necessary if temperature increases significantly (Steinberg and Arzet, 1984); 0.2 = fraction of air as O_2; 10^{-6} = kg/mg, 1.205 = kg/m³ air at 1 atm and 20°C.

The induced O_2 demand factor (I) was observed to range from about 1.5 to 4.0 (Soltero et al., 1994; Moore et al., 1996; Prepas et al., 1997; Beutel, 2003). Increased O_2 demand with aeration results from increased DO and water velocity at the sediment surface. The increased velocity reduces the diffusional boundary overlying the sediment, allowing more O_2 to contact sediment organic matter (Moore et al., 1996; Beutel, 2003). The effect increases with the sediment demand relative to that in the water column. A low I (1.5) would apply to deep and a high I (3.3) to shallow hypolimnia, according to the observed relative importance of sediment demand to total AHOD (15% and 75%, respectively; Beutel, 2003). Livingstone and Imboden (1995) concluded that water column volumetric demand is rather constant, while the sediment demand accounts for most of the difference among lakes and the effect of eutrophication. Induced sediment demand can be estimated for unit sizing purposes from the expected above-sediment DO concentration and water velocity, which also can be estimated with hydrographic data (Moore et al., 1996). Designing systems to enhance induced sediment demand increases the depth of O_2 penetration and organic matter consumption (Beutel, 2003).

Because oxygen transfer from the gas to solute phase is rather inefficient, Kortmann (1989) inserted an additional factor in the numerator of Equation 18.2 (100/depth × 2.5) to allow for 2.5% gas transfer per meter (after Bernhardt, 1967). From this estimate of air flow rate, one can proceed to determining friction losses and pressure requirements.

Other important design factors that affect oxygen transfer are the depth of air injection, the orifice (pore) size in the diffuser tube, and the surface area in the surface separator box of full air-lift systems. Field and laboratory experiments showed that depth of injection had a small, but significant, effect on oxygen transfer, while a small orifice size (140 μm diameter) substantially increased transfer over that of larger sizes (≥ 794 μm; Ashley and Hall, 1990; Ashley et al., 1992). Retrofitting a full air-lift aerator in St. Mary Lake (BC, Canada) with 140 μm orifice size diffuser, in place of one with 3,175 μm, increased late summer hypolimnetic DO from 0.4 to 3.0–4.0 mg/L (Ashley, 2000).

18.4 BENEFICIAL EFFECTS AND LIMITATIONS

Although not inclusive, 28 cases of aeration/oxygenation are listed in Table 18.1. Three other unlisted lakes are cited in the text; 70 Finnish lakes with the MIXOX technique are not discussed here, and many lakes are aerated with little or no published record.

Oxygen is the target of hypolimnetic aeration/oxygenation and results are nearly always positive. All but one of the lakes (Mirror) listed in Table 18.1 showed an increase in hypolimnetic DO, and often to at least 7 mg/L, following aeration.

TABLE 18.1
Lakes Receiving Hypolimnetic Aeration[1] or Oxygenation[2] and Associated Characteristics

Lake	Depth Max.	Depth Mean	Depth Device	Volume (m³ × 10⁶)	Area (ha)	Q air/gas (m³/min)	Ref.
Brunsviken, Sweden	—	—	13[1]	—	100	15.5	Atlas Copco, 1976
Caldonazzo, Italy	50	—	11[1]	—	700	44	Atlas Copco, 1980
Hemlock, Michigan	18.6	—	18.6[1]	—	2.4	2.8	Fast, 1971a
Jarlasjön, Sweden	24	9.3	24[1]	7.8	84	22.8	Bengtsson and Gelin, 1975
Kolbotnnvatn, Norway	18.5	10.3	18.5[1]	3.1	30.3	5.5	Atlas Copco, 1980; Holton and Holton, 1978
Larson, Wisconsin	11.9	4.0	11.9[1]	0.188	4.8	0.45	Smith et al., 1975
Mirror, Wisconsin	13.1	7.6	12.8[1]	0.4	5.3	0.45	Smith et al., 1975
Ottoville Quarry, Ohio	18	8.6	18[2]	0.063	0.73	0.11	Fast, 1973; Overholtz et al., 1977
Spruce Run, New Jersey	13.1	—	12.2[2]	—	—	0.15	Whipple et al., 1975
Tegel, Germany	16	6	12/16[1]	24.6	400	63	Atlas Copco, 1980
Waccabuc, New York	13	7.5	13[1]	4.053	53.6	7.93	Fast et al., 1975 Garrell et al., 1977
Medical, Washington	18.3	9.8	18[1]	6.2	63	4.5	Soltero et al. 1994
Camanche, California	31	17	~30[2]	511	3000	4.6	Beutel and Horne, 1999
Amisk, Alberta	34	10.8	33[2]	25.1	233	0.3–0.6	Prepas et al., 1997
Fenwick, Washington	7.9	4.0	6.7[1]	0.42	10.4	—	Gibbons, pers. Commun.
Baldegg, Switzerland	65	32.7	~60[2]	170	520	5.6	Jungo, 1993; Gächter and Wehrli, 1998
Sempach, Switzerland	85	21.5	~80[2]	670	1440	3.3	Jungo, 1993; Gächter and Wehrli, 1998
Hallwil, Switzerland	45	28.4	~40[2]	290	1020	—	Jungo, 1993
Stevens, Washington	46	20.5	43[1]	194	421	44	Gibbons, pers. Commun.
Black, British Columbia	9	—	7[1]	0.18	—	1.13	Ashley and Hall, 1990; Hall et al., 1994
Newman, Washington	9.1	5.8	~9[2]	28.4	490	1.15	Beutel and Horne, 1999
Gross-Glienicker, Germany	11	6.5	~11	4.2	68	—	Wolter, 1994
Vadnais, Minnesota	16.5	8.1	~16[1]	12.5	155	—	Walker et al. 1989
Wahnbach, W. Germany	43	19.2	40[1]	41.63	214.5	9	Bernhardt, 1967, 1974
Spruce Knob, W. Virginia	5.7	2.1	5.2[1]	0.224	10.5	1.3	Hess, 1977; LaBaugh, 1980
Ghirla, Italy	14.0	8.0	14.0[2]	2.0	24.5	—	Bianucci and Bianucci, 1979
Tory, Ontario	10.0	4.5	9.0[1]	0.055	1.23	3.54	Taggart and McQueen, 1981
Irondequoit Bay, New York	23.7	3.5	14–23[2]	23.4	679	0.6	Babin et al. 1999

Source: Modified from Pastorak, R.A. et al. 1981. *Evaluation of Aeration/Circulation as a Lake Restoration Technique.* USEPA 600/3-81-014. .

The principal benefit of hypolimnetic aeration/oxygenation is that anoxic hypolimnia can be readily switched to an oxic state through effective aeration while maintaining a normal coldwater environment. With few exceptions, there was either no increase in hypolimnetic temperature or the increase was < 2–3°C. In only two cases did the hypolimnion increase in temperature by 4°C or more, but in one of those cases, the design was inappropriate (Fast, 1971a; Fast et al., 1973). Considerable disturbance of the thermocline and entrainment in the epilimnion occurred in Lake Tegel, Germany, due largely to the lake's shallowness (mean depth 6 m, maximum 16 m; Lindenschmidt and Chorus, 1997). Beneficial biological effects are associated with enlargement of the oxic environment.

A secondary benefit of hypolimnetic aeration/oxygenation is a potential decrease in internal loading of P, Fe and Mn, ammonium, and hydrogen sulfide. A decrease in Fe, Mn, and hydrogen sulfide could result in improved water supply quality. By changing the sediment-water interface from reducing to oxidizing, the release of dissolved forms of P, Fe and Mn should decrease, as well as NH_4 due to increased nitrification (McQueen and Lean, 1984). Fe decreased in all lakes where it was monitored (Jarlasjön, Spruce Run, Stevens and Wahnbach; Table 18.1) and Mn decreased in three of those lakes in addition to Baldegg (Beutel and Horne, 1999). Ammonia decreased in 10 of the 11 lakes cited in Table 18.1 (Jarlasjön, Larson, Spruce Run, Waccabuc, Wahnbach, Tory). Beutel and Horne (1999) cited 60–95% reduction in ammonia in Amisk, Baldegg, Sempach and Camanche and 100% loss of H_2S in Amisk and Camanche.

Improvement in P, however, has not always reached expectations following aeration. Although P was reduced in the hypolimnion during aeration, the effect was not as great or as permanent as with other techniques such as P inactivation with alum or hypolimnetic withdrawal. The response was usually a 30–50% reduction in hypolimnetic P with a rather high residual concentration. For example, while TP in the hypolimnion of Jarlasjön during the stratified period decreased by about one half, 400 µg/L TP still remained. As soon as aeration stopped, P rapidly returned to pre-aeration levels (Bengtsson and Gelin, 1975). The P decrease was only 30% in the hypolimnion of Lake Waccabuc following aeration (Garrell et al., 1977).

Results from Brunsviken were similar. While the spring minimum P content decreased after aeration began in 1972, there was little change in the autumn maximum, which remained at approximately 200 to 300 µg/L. Phosphorus content near the sediment-water interface in Kolbotnvatn decreased from 1,600 to 500 µg/L during winter stagnation and summer hypolimnetic P was still high (600 µg/L), as with the other examples, even after a 33% decrease. Epilimnetic P content reached 300 µg/L and chlorophyll (chl) at times exceeded 100 µg/L. In this case, however, high epilimnetic P and chl *a* were maintained by high external loading to the lake (Holton and Holton, 1978; Holton et al., 1981). Soluble reactive P in a 15-m deep aerated enclosure decreased about 30% (before iron addition; McQueen and Lean, 1984). On the other hand, Ravera (1990) found no reduction in TP or soluble reactive P (SRP) in a 6-m deep, 40-m diameter enclosure with an aerated hypolimnion. In other cases the decrease was greater; hypolimnetic P decreased by 55% in Lake Wesslinger, Germany (Steinberg and Arzet, 1984), 56% in Tory Lake, 75% in Lake Stevens and 61% in Medical Lake (Table 18.1).

The marginal effect on hypolimnetic P as a result of aeration/oxygenation may be partly due to a lack of iron to bind all the P released (Lean et al., 1986). In contrast to hypolimnetic P reduction on the order of ~ 50% following aeration, reductions were ≥ 90% in three lakes where iron addition was coupled with aeration (Black, Vadnais and Gross-Glienicker; Table 18.1) and in Lake Södra Hörken, Sweden, a lake that for many years received effluent from an iron mining operation as well as municipal wastewater, although iron content was less than P (Björk, 1985; Verner, 1984). When iron was added to a 15-m deep, aerated enclosure, SRP was sharply decreased by about two thirds (McQueen and Lean, 1984; Lean et al., 1986). Reduction in hypolimnetic P in Vadnais Reservoir increased from 30% with aeration alone to 93% with iron addition plus aeration (Walker et al., 1989).

Adding other binding agents along with hypolimnetic aeration/oxygenation has enhanced control of internal P loading. Newman Lake in eastern Washington received injections of alum through a DBCS (Moore, personal communication). Alum was added because P concentrations increased during oxygenation only (Thomas et al., 1994). Injection of $Ca(OH)_2$ was employed, along with a Tibean aerator, in Lake Schmaler Luzin, Germany, to increase calcite precipitation and promote P precipitation and sediment retention (Koschel et al., 2001). One basin (max. depth 33 m) was treated while the other (max. depth 34 m) was untreated. Hypolimnetic pH increased from 7.3–7.7 to 8.2, calcite precipitated in the treated basin and TP decreased by 33%.

Gächter (1987) suggested that the effectiveness of hypolimnetic oxygenation in reducing P depends on the P retention capacity of the sediment. If gross sedimentation (due to external loading) exceeds the retention capacity, rendering the hypolimnion oxic will likely not decrease lake P concentration. Phosphorus decreased in three deep Swiss lakes, Sempach, Baldegg and Halwill, which received DOIS ("Tanytarsus") for several years (Gächter and Wehrli, 1998; Jungo, 1993; Table 18.1). Over 15 years of observation in these lakes showed that TP decreased, but much of the decrease was due to reduced external inputs prior to and during oxygenation. Despite the DOIS's effectiveness in maintaining an oxic hypolimnion, the sediment surface in Sempach and Baldegg remained anoxic due to high sedimentation rates of organic matter, and P was not effectively retained in the sediment (Gächter and Wehrli, 1998). Moreover, low P retention was due partly to low Fe:S ratios in sediment, which results in the iron being sequestered in anoxic sediment, thus allowing high sediment P release even if the surficial sediment is oxic (Gächter and Müller, 2003).

Designing aeration/oxygenation systems to enhance induced sediment DO demand should increase O_2 penetration into the sediment while maintaining increased hypolimnetic DO. That should in turn increase the thickness of the oxidized layer below the sediment-water interface and oxidize reduced forms of iron to the ferric state thereby increasing P retention. However, evidence suggests that aeration, in this case complete circulation, did not increase oxidation of and loss of organic sediment (Engstrom and Wright, 2002). Long-term sediment response in five aerated and five non-aerated lakes in Minnesota was evaluated by lead-210 dating. Neither historic patterns of sediment accumulation rate or sediment organic matter concentration declined over two decades, with or without aeration. However, circulation may have actually increased production in some of those lakes due to insufficient air flow, as was apparently the case in Crystal Lake (Chapter 19). While induced AHOD is a possible indicator of greater O_2 penetration and increased depth of organic matter and iron oxidation, direct evidence of this phenomenon in aerated/oxygenated lakes is generally lacking, although a MIXOX unit reportedly reduced sediment organic content in Lake Särkinen, Finland (Sandman et al., 1990).

Hypolimnetic aeration/oxygenation has both improved and worsened trophic state. In cases where hypolimnetic P was not controlled, little change in epilimnetic quality was expected (Smith et al., 1975). Although hypolimnetic P declined with aeration, there was no effect on epilimnetic chl *a* in 18-m deep Medical Lake, Washington (Soltero et al., 1994). However, Beutel and Horne (1999) observed a 75% reduction in peak hypolimnetic SRP and a proportional decrease in peak summer chl *a* following operation of a DBCS in Comanche Reservoir, California. As with expectations from alum treatments in stratified lakes, the effect on trophic state depends on the availability of hypolimnetic P to the epilimnion, which can be predicted prior to treatment (Chapters 3 and 8), and on external loading.

Trophic state can worsen if stratification is disturbed. Fifteen aerators operating in Lake Tegel from 1980 to 1992 resulted in entrainment of hypolimnetic water into the epilimnion at times, due partly to the air release located at 2 m and the force of circulation from the aerators in the shallow hypolimnion, especially during windy conditions (Lindenschmidt and Chorus, 1997; Lindenschmidt and Hamblin, 1997). During a year without aeration, the water column was stable with no entrainment and epilimnetic TP was 40–100 µg/L. In contrast, with aeration, bottom temperature in June was 2–3.5°C greater, sediment P release began in June (rather than July without aeration) and epilimnetic TP was 100–180 µg/L. The increased turbulence from aeration favored *Aphanizomenon* and *Microcystis*. Oxygenation (DOIS) of Lake Baldegg apparently increased algal biomass and favored *Planktothrix* over *Microcystis* (Buergi and Stadelmann, 2000). TP decreased, but was still 100 µg/L.

There could be a significant indirect effect on phytoplankton control, however, by furnishing a greatly increased aerobic habitat as a refuge for herbivorous zooplankton. The hypolimnion is dark to dimly lit, which gives further protection against sight-feeding predators during daylight, in addition to having a greater dilution of the populations (Fast, 1979; Shapiro, 1979). However, if the lake is highly eutrophic, metalimnetic minima may occur despite an aerated hypolimnion, and could serve as a barrier to zooplankton diurnal migration (Taggart and McQueen, 1981).

Investigation of increased zooplankton grazing following aeration is limited to a few cases. Large *Daphnia pulex* became abundant (90 times greater) in Hemlock Lake following hypolimnetic aeration (Fast, 1971b). The smaller cladocerans (*Bosmina* and *Diaphanasoma*) also increased, although to a lesser extent than *Daphnia*, and there was little change in *Diaptomus*. Before aeration, all zooplankton were excluded from depths between 11 m and the bottom (18.5 m) by anaerobic conditions. Large *Daphnia* increased following layer aeration in Lake Shenispit, Connecticut (Kortmann et al., 1994). McQueen and Post (1988) observed an increased abundance of large-bodied *Daphnia* in an enclosure with an aerobic hypolimnion compared to one with an anoxic hypolimnion. Both contained planktivorous fish. The aerobic enclosure also had higher transparency. Encouragement of the large zooplankters could lead to greater phytoplankton loss rates and lower biomass (Chapter 9). If that effect could be consistently realized, hypolimnetic aeration could offer much broader benefits in restoring lakes than are presently expected.

A DOIS was evaluated for 2 years in one basin of Amisk Lake (Table 18.1) and two species of *Daphnia* increased in the oxygenated hypolimnion (Field and Prepas, 1997; Prepas et al., 1997). However, there was little change noted in the epilimnion even though vertical migration increased. Epilimnetic TP and chl *a* decreased by 87% and 45% in response to a 57% decrease in hypolimnetic TP to 42 µg/L in the treated basin, showing that hypolimnetic P had been available to the epilimnion (Prepas and Burke, 1997). Cyanobacterial blooms were reduced and delayed in favor of extended spring diatom blooms (Webb et al., 1997).

Growth and survival of coldwater fishes, as well as reproduction of some lake-spawning species (e.g., *coregonus*), can be severely limited in eutrophic lakes with anaerobic hypolimnia. Preferred temperatures for coldwater species are usually less than 18°C (Welch and Jacoby, 2004). When epilimnetic temperatures exceed that level, as is common even in cool climates, growth is reduced and fish seek cooler waters. If the cool hypolimnetic waters are devoid of DO, that habitat is not available and coldwater fish production may be impaired. In addition, the production of benthic fish-food organisms, especially chironomids, *Chaoborus*, and tubificids, may increase following a change from anaerobic to aerobic conditions (Fast, 1971a).

In three of the lakes listed in Table 18.1 (Hemlock, Ottoville Quarry, and Waccabuc), planted rainbow trout distributed themselves throughout the hypolimnion (Fast, 1973; Garrell et al., 1977; Overholtz et al., 1977). Trout stomachs analyzed before aeration showed that they utilized food organisms that subsequently increased in abundance in the hypolimnion as a result of the aerobic conditions (e.g., *Chaoborus* and chironomids) or, like *Daphnia*, migrated into the hypolimnion during daylight hours (Fast, 1973; Garrell et al., 1977).

Oxygenation of the Newman Lake (Figure 18.1) hypolimnion expanded the aerobic habitat, and subsequent increased trout stocking decreased chaborids while the higher DO increased chironomid and oligochaete populations (Doke et al., 1994). Chironomid density and biomass increased many fold in the oxygenated hypolimnion of one basin of Amisk Lake, compared with the untreated basin (Dinsmore and Prepas, 1997). Also, cisco expanded their distribution by 2–8 m in depth in the treated basin (Prepas et al., 1997).

18.5 UNDESIRABLE EFFECTS

Supersaturation of hypolimnetic water with N_2 was suggested as a possible problem that might lead to gas bubble disease in fish. Although this apparently has not occurred as a result of hypolimnetic aeration, N_2 content can reach potentially damaging levels; 150% saturation relative to surface temperature and pressure occurred after 80 days of aeration in Lake Waccabuc (Fast et al., 1975) and similar levels have been shown by others (Bernhardt, 1974; McQueen and Lean, 1983). Kortmann et al. (1994) suggested that increased N_2 levels resulting from hypolimnetic aeration might be a concern, but only in the deepest of treated lakes.

Hypolimnetic aeration may increase eddy diffusion of nutrients into the epilimnion even though stratification is maintained. Metalimnetic P content increased by a factor of two in lake Wesslinger during hypolimnetic aeration, which doubled the mostly blue-green phytoplankton biomass (Steinberg and Arzet, 1984). Similar entrainment of hypolimnetic water increased blooms in Lake Tegel, as discussed above. One of the principal reasons for maintaining stratification during aeration is to prevent the recirculation of sediment-released P. If this is not accomplished, hypolimnetic aeration loses some of its appeal over complete circulation as an enhancement to fisheries-usable habitat. Metalimnetic DO minima have also been observed (Bengtsson and Gelin, 1975; Garrell et al., 1977; Walker et al., 1989). Such minima could restrict fish movement. Some of these real or potential problems with hypolimnetic aeration may be eliminated or minimized with layer aeration (Kortmann et al., 1988, 1994) or with a DBCS or DOIS (Beutel and Horne, 1999).

18.6 COSTS

The cost of hypolimnetic aeration depends primarily on the O_2 demand to be satisfied. However, the amount of air required (i.e., compressor size) to deliver the O_2 to the hypolimnion depends on the distance from the compressor to the discharge site and depth of the unit. The extent to which frictional losses due to pipe length and size and head loss due to depth are optimized with compressor size can account for large variations in energy efficiency and project costs.

Costs for aeration/oxygenation are seldom reported. Aqua Technique (now General Environmental Systems; R. Geney, personal communication) provided project costs for seven partial air-lift systems installed in the 1970s–1980s. These showed average energy efficiency in operation of 1.4 kg O_2/kwhr. Using $0.12/kwhr (2002 U.S. dollars), that represents an average cost efficiency of $0.86 ($\pm$ 36%)/kg O_2. The operating cost plus the average installed cost, spread over an assumed 10-year longevity at 160 days per year, was $0.39/kg O_2 per day. Using cost efficiency and the average air flow reported for the 15 aeration projects listed in Table 18.1 (15.3 m³/min or 3.7 kg O_2/min), gives about $340,000 for 160 days per year for installation and operation. That is about $3,000/ha per year, based on the average area of the 15 lakes. Many of these were small projects so there is little economy of scale in this cost.

Two custom-built units have delivered 15.5 metric tons O_2/day to the 17×10^6 m³ hypolimnion of Lake Stevens for the past 10 years (Table 18.1). The operating and capital costs of that project for 160 days/year were $0.21/kg O_2 per day or $1,240/ha per year. The 15 Limno units in Lake Tegel (Table 18.1) delivered 4.5 tons O_2/day starting in 1980 (Verner, 1984). The initial cost of that system was $3,770,160 (2002 dollars). Including 10 years of operation (they actually operated for 12 years), an operation cost of $0.09/kg O_2 per day and 160 days/year ($64,800) gives about $442,000 per year or $1,052/ha per year.

18.7 SUMMARY

If aerators are adequately sized and the water body is sufficiently deep (> 12 to 15 m), hypolimnetic aeration will provide greatly increased DO, decreased iron, manganese, and ammonium concentrations and often moderately decreased P concentrations in the hypolimnion. Oxygenation devices (DOIS, DBCS) also have been successful. Additionally, the expanded aerobic environment should enhance growth and distribution of coldwater fish and the abundance of large-bodied zooplankton. Although not specifically demonstrated, the latter could significantly increase the grazing loss rate of algae. If aerators are inadequately sized, metalimnetic low DO and increased diffusion of P into the epilimnion can result, with enhanced blooms of algae. The addition of iron along with aeration will further reduce P in lakes with low Fe/P.

Long-term costs for installation and operation (6 months) of hypolimnetic aeration spread over ten years can be expected to range from $0.20 to $0.40/kg O_2 per day (2002 dollars). Sizing the aeration/oxygenation system depends mostly on the oxygen demand and energy losses in delivering O_2 to the hypolimnetic treatment depth. The cost for DOIS is at or less than the lower end of that range; $0.15/kg O_2 per day for the same projected time period (Babin et al., 1999).

REFERENCES

Ashley, K.I. 1985. Hypolimnetic aeration: Practical design and application. *Water Res.* 19: 735–740.

Ashley, K.I. 2000. Recent advances in hypolimnetic aeration design. *Verh. Int. Verein. Limnol.* 27: 2256–2260.

Ashley, K.I. and K.J. Hall. 1990. Factors influencing oxygen transfer in hypolimnetic aeration Systems. *Verh. Int. Verein. Limnol.* 24: 179–183.

Ashley, K.I., S. Hay and G.H. Schoeten. 1987. Hypolimnetic aeration: Field test of the empirical sizing method. *Water Res.* 21: 223–227.

Ashley, K.I., D.S. Mavinic and K.J. Hall. 1992. Bench-scale study of oxygen transfer in coarse bubble diffused aeration. *Water Res.* 26: 1289–1295.

Atlas Copco. 1976. *Aeration of Lake Brunsviken.* Communications Dept., Wilrijk, Belgium.

Atlas Copco. 1980. Communications Dept., Wilrijk, Belgium.

Babin, J.M., J.M. Burke, T.P. Murphy, E.E. Prepas and W. Johnson. 1999. Liquid oxygen injection to increase dissolved oxygen concentration in temperate zone lakes. In: T. Murphy and M. Munawar (Eds.), *Aquatic Restoration in Canada.* Backhuys Publ., Leiden, The Netherlands. pp. 109–125.

Bengtsson, L. and C. Gelin. 1975. Artificial aeration and suction dredging methods for controlling water quality. In: *Proc. Symposium on Effects of Storage on Water Quality.* Water Res. Center, Medmenham, England.

Bernhardt, H. 1967. Aeration of Wahnbach Reservoir without changing the temperature profile. *J. Am. Water Works Assoc.* 9: 943–964.

Bernhardt, H. 1974. Ten years experience of reservoir aeration. In: *Seventh Conf. on Water Pollution Research,* Paris.

Beutel, M.W. 2003. Hypolimnetic anoxia and sediment oxygen demand in California drinking water reservoirs. *Lake and Reservoir Manage.* 19: 208–221.

Beutel, M.W. and A.J. Horne. 1999. A review of the effects of hypolimnetic oxygenation on lake and reservoir water quality. *Lake and Reservoir Manage.* 15: 285–297.

Bianucci, G. and E.R. Bianucci. 1979. Oxygenation of a polluted lake in northern Italy. *Effluent Water Treat. J.* 19: 117–128.

Björk, S. 1985. Scandinavian lake restoration activities. In: *Lake Pollution and Recovery.* Int. Comp. European Water Pollut. Cont. Assoc. pp. 293–301.

Buergi, H.R. and P. Stadalmann. 2000. Change in phytoplankton diversity during long-term restoration of Lake Baldegg (Switzerland). *Verh. Int. Verein. Limnol.* 27: 574–581.

Cooke, G.D. and R.E. Carlson. 1989. *Reservoir Management for Water Quality and THM Precursor Control.* Am. Water Works Assoc. Res. Found., Denver, CO.

Dinsmore, W.P. and E.E. Prepas. 1997. Impact of hypolimnetic oxygenation on profundal macroinvertebrates in a eutrophic lake in central Alberta. I. Changes in macroinvertebrate abundance and diversity. *Can. J. Fish. Aquatic Sci.* 54: 2157–2169.

Doke, J.L., W.H. Funk, S.T.J. Juul and B.C. Moore. 1995. Habitat availability and benthic invertebrate population changes following alum treatment and hypolimnetic oxygenation in Newman Lake, Washington. *J. Fresh Water Ecol.* 10: 87–102.

Ellis, C.R. and H.G. Stephan. 1990. Hydraulic design of winter lake aeration system. *J. Environ. Eng. Div. ASCE* 116: 376–393.

Engstrom, D.R. and D.I. Wright. 2002. Sedimentological effects of aeration-induced lake circulation. *Lake and Reservoir Manage.* 18: 201–214.

Fast, A.W. 1971a. *The Effects of Artificial Aeration on Lake Ecology.* Water Pollut. Cont. Res. Ser. 16010 Exe 12/71. USEPA.

Fast, A.W. 1971b. Effects of artificial destratification on zooplankton depth distribution. *Trans. Am. Fish. Soc.* 100: 355–358.

Fast, A.W. 1973. Effects of artificial hypolimnion aeration on rainbow trout (*Salmo gairdneri* Richardson) depth distribution. *Trans. Am. Fish. Soc.* 102: 715–722.

Fast, A.W. 1977. Artificial aeration and oxidation of lakes as a restoration technique. In: J. Cairns, Jr., K.L. Dickson and F.E. Herricks (Eds.), *Recovery and Restoration of Damaged Ecosystems.* University Press of Virginia, Charlottesville.

Fast, A.W. 1979. Artificial aeration as a lake restoration technique. In: *Proc. Natl. Conf. Lake Rest.* USEPA 440/5-79-001. pp. 121–132.

Fast, A.W. and M.W. Lorenzen. 1976. Synoptic survey of hypolimnetic aeration. *J. Environ. Eng. Div. ASCE* 102: 1161–1173.

Fast, A.W., B. Moss and R.G. Wetzel. 1973. Effects of artificial aeration on the chemistry and algae of two Michigan lakes. *Water Resour. Res.* 9: 624–647.

Fast, A.W., V.A. Dorr and R.J. Rosen. 1975. A submerged hypolimnion aerator. *Water Resour. Res.* 11: 287–293.

Fast, A.W., M.W. Lorenzen and J.H. Glenn. 1976. Comparative study with costs of hypolimnetic aeration. *J. Environ. Eng. Div. ASCE* 1026: 1175–1187.

Field, K.M. and E.E. Prepas. 1997. Increased abundance and depth distribution of pelagic crustacean zooplankton during hypolimnetic oxygenation in a deep, eutrophic Alberta lake. *Can. J. Fish. Aquatic Sci.* 54: 2146–2156.

Gächter, R. 1987. Lake restoration. Why oxygenation and artificial mixing cannot substitute for a decrease in the external P loading. *Schweiz. Z. Hydrol.* 49: 170–185.

Gächter, R. and B. Müller. 2003. Why the phosphorus retention of lakes does not necessarily depend on the oxygen supply to their sediment surface. *Limnol. Oceanogr.* 48: 929–933.

Gächter, R. and B. Wehrli. 1998. Ten years of artificial mixing and oxygenation: No effect on the internal phosphorus loading of two eutrophic lakes. *Environ. Sci. Technol.* 32: 3659–3665.

Garrell, M.H., J.C. Confer, D. Kirchner and A.W. Fast. 1977. Effects of hypolimnetic aeration on nitrogen and P in a eutrophic lake. *Water Resour. Res.* 13: 343–347.

Geney, R.S. Personal communication. General Environmental Systems, Oak Ridge, TN.

Gibbons, H. Personal communication. Tetra Tech, Inc., Seattle, WA.

Hall, K.J., T.P.D. Murphy, M. Mawhinney and K.I. Ashley. 1994. Iron treatment for eutrophication control in Black Lake, British Columbia. *Lake and Reservoir Manage.* 9; 114–117.

Hess, L. 1977. Lake Destratification Investigations. Job 1–3: Lake Aeration June 1, 1972 to June 30, 1977, Final Rpt. West Virginia Dept. Nat. Res., D-J Proj. F-19-R.

Holton, H. and G. Holton. 1978. Sammerstilling av undersdøkelsesresulates 1972–1977. Report No. 0-5/70. Norwegian Institute for Water Research, Oslo.

Holton, H., P. Brettum, G. Holton and G. Kjellberg. 1981. Kolbotnvatn med tillop: Sammerstilling av undersdøkelsersesultates 1978–1979. Report No. 0-78007. Norwegian Institute for Water Research, Oslo.

Jaeger, D. 1990. TIBEAN — a new hypolimnetic water aeration plant. *Verh. Int. Verein. Limnol.* 24: 184–187.

Jungo, E. 1993. Ten years internal measures in Swiss lakes — experiences and results. In: G. Giussani, and C. Callieri (Eds.), *Proc. 5th Int. Conf. Conserv. and Manage. of Lakes.* Pallanza, Italy.

Keto, A., A. Lehtinen, A. Mäkelä and I Sammalkorpi. 2004. Lake Restoration. In: P. Eloranta (Ed.). *Inland and Coastal Waters of Finland.* University of Helsinki and Palminia Centre for Cont. Education.

Kortmann, R.W. 1989. Aeration technologies and sizing methods. *LakeLine* 9: 6–7, 18–19.

Kortmann, R.W., M.E. Conners, G.W. Knoecklein and C.H. Bonnell. 1988. Utility of layer aeration for reservoir and lake management. *Lake and Reservoir Manage.* 4: 35–50.

Kortmann, R.W., G.W. Knoecklein and C.H. Bonnell. 1994. Aeration of stratified lakes: Theory and practice. *Lake and Reservoir Manage.* 8: 99–120.

Koschel, R.H., M. Dittrich, P. Casper, A. Hoiser and R. Rossberg. 2001. Induced hypolimnetic calcite precipitation — ecotechnology for restoration of stratified eutrophic hardwater lakes. *Verh. Int. Verein. Limnol.* 27: 3644–1649.

LaBaugh, J.W. 1980. Water chemistry changes during artificial aeration of Spruce Knob Lake, West Virginia, *Hydrobiologia* 20: 201–216.

Lappalainen, K.M. 1994. Positive changes in oxygen and nutrient contents in two Finnish lakes induced by Mixox hypolimnetic oxygenation method. *Verh. Int. Verein. Limnol.* 25: 2510–2513.

Lean, D.R.S., D.J. McQueen and V.R. Story. 1986. Phosphate transport during hypolimnetic aeration. *Arch. Hydrobiol.* 108: 269–280.

Lindenschmidt, K.E. and I. Chorus. 1997. The effect of aeration on stratification and phytoplankton populations in Lake Tegel, Berlin. *Arch. Hydrobiol.* 139: 317–346.

Lindenschmidt, K-E. and P.F. Hamblin. 1997. Hypolimnetic aeration in Lake Teget, Berlin. *Water Res.* 31: 1619–1628.

Little, J.C. 1995. Hypolimnetic aerators: Predicting oxygen transfer and hydrodynamics. *Water Res.* 29: 2475–2482.

Livingstone, D.M. and D.M. Imboden. 1996. The prediction of hypolimnetic oxygen profiles: A plea for a deductive approach. *Can. J. Fish. Aquatic Sci.* 53: 924–932.

Lorenzen, M.W. and A.W. Fast. 1977. A Guide to Aeration/Circulation Techniques for Lake Management. Ecol. Res. Ser. USEPA 600/3-77-004.

McQueen, D.L. and D.R.S. Lean. 1983. Hypolimnetic aeration and dissolved gas concentrations. *Water Res.* 17: 1781–1790.

McQueen, D.J. and D.R.S. Lean. 1984. Aeration of anoxic hypolimnetic water: Effects on nitrogen and P concentrations. *Verh. Int. Verein. Limnol.* 22: 267–276.

McQueen, D.J. and D.R.S. Lean. 1986. Hypolimnetic aeration: An overview. *Water Pollut. Res. J. Can.* 21: 205–217.

McQueen, D.L. and J.R. Post. 1988. Limnocorral studies of cascading trophic interactions. *Verh. Int. Verein. Limnol.* 23: 739–747.

McQueen, D.L., S.S. Rao and D.R.S. Lean. 1984. Hypolimnetic aeration: Change in bacterial population and oxygen demand. *Arch. Hydrobiol.* 99: 498–514.

Mercier, P. and J. Perret. 1949. Aeration station of Lake Bret. *Schweiz. Ver. Gas. Wasserfach. Monatsbull.* 29: 25–30.

Moore, B.C., P.-H. Chen, W.H. Funk and D. Yonge. 1996. A model for predicting lake sediment oxygen demand following hypolimnetic aeration. *J. Am. Water Res. Assoc.* 32: 723–731.

NALMS. 2001. Managing Lakes and Reservoirs, 3rd ed. USEPA 841-B-01–006. NALMS, Madison, WI.

Overholtz, W.J., A.W. Fast, R.A. Tubb and R. Miller. 1977. Hypolimnion oxygenation and its effects on the depth distribution of rainbow trout (*Salmo gairdneri*) and gizzard shad (*Dorosoma cepedianum. Trans. Am. Fish. Soc.* 106: 371–375.

Pastorak, R.A., T.C. Ginn and M.W. Lorenzen. 1981. Evaluation of Aeration/Circulation as a Lake Restoration Technique. USEPA 600/3-81-014.

Pastorak, R.A., M.W. Lorenzen and T.C. Ginn. 1982. Environmental Aspects of Artificial Aeration and Oxygenation of Reservoirs: A Review of Theory, Techniques, and Experiences. Tech. Report No. E-82-3, U.S. Army Corps Engineers, Vicksburg, MS.

Prepas, E.E. and J.M. Burke. 1997. Effects of hypolimnetic oxygenation on water quality in Amisk Lake, Alberta, a deep, eutrophic lake with high internal phosphorus loading rates. *Can. J. Fish. Aquatic Sci.* 54: 2111–2120.

Prepas, E.E., K.M. Field, T.P. Murphy, W.L. Johnson, J.M. Burke and W.M. Tonn. 1997. Introduction to the Amisk Lake Project: Oxygenation of a deep, eutrophic lake. *Can. J. Fish. Aquatic Sci.* 54: 2105–2110.

Ravera, O. 1990. The effects of hypolimnetic oxygenation in a shallow and eutrophic Lake Comabbio (Northern Italy) studied by enclosure. *Verh. Int. Verein. Limnol.* 24: 188–194.

Sandman, O., K. Eskonen and A. Liehu. 1990. The eutrophication history of Lake Särkinen, Finland and effects of lake aeration. *Hydrobiologia* 214: 191–199.

Sehgal, H.S. and E.B. Welch. 1991. A case of unusually high oxygen demand in a eutrophic lake. *Hydrobiologia* 209: 235–243.

Shapiro, J. 1979. The need for more biology in lake restoration. In: *Proc. Natl. Conf. on Lake Restoration.* USEPA 440/5-79-001. pp. 161–168.

Smith, S.A., D.R. Knauer and T.L. Wirth. 1975. Aeration as a Lake Management Technique. Tech. Bull. No. 87. Wisconsin Dept. Nat. Res., Madison.

Soltero, R.A., L.M. Sexton, K.I. Ashlen and K.O. McKee. 1994. Partial and full lift hypolimnetic aeration of Medical Lake, WA to improve water quality. *Water Res.* 28: 2297–2308.

Speece, R.E. 1971. Hypolimnion aeration. *J. Am. Water Works Assoc.* 63: 6–9.

Stefan, H.G., M.D. Bender, J. Shapiro and D.I. Wright. 1987. Hydrodynamic design of a metalimnetic lake aerator. *ASCE J. Environ. Eng.* 113: 1239–1264.

Steinberg, C. and K. Arzet. 1984. Impact of hypolimnetic aeration on abiotic and biotic conditions in a small kettle lake. *Environ. Tech. Lett.* 5: 151–162.

Taggart, C.T. and D.J. McQueen. 1981. Hypolimnetic aeration of a small eutrophic kettle lake: Physical and chemical changes. *Arch. Hydrobiol.* 91: 151–180.

Thomas, J.A., W.H. Funk, B.C. Moore and W.W. Budd. 1994. Short term changes in Newman Lake following hypolimnetic aeration with the Speece Cone. *Lake and Reservoir Manage.* 9: 111–113.

Verner, B. 1984. Longterm effect of hypolimnetic aeration of lakes and reservoirs with special consideration of drinking water quality and preparation cost. In: *Lake and Reservoir Management.* USEPA-440/5-84-001. pp. 134–138.

Walker, W.E., C.E. Westerberg, D.J. Schuler and J.A. Bode. 1989. Design and evaluation of eutrophication control measures for the St. Paul water supply. *Lake and Reservoir Management.* 5: 71–83.

Webb, D.J., R.D. Roberts and E.E. Prepas. 1997. Influence of extended water column mixing during the first 2 years of hypolimnetic oxygenation on the phytoplankton community of Amisk Lake, Alberta. *Can. J. Fish. Aquatic Sci.* 54: 2133–2145.

Welch, E.B. and J.M. Jacoby. 2004. *Pollutant Effects in Fresh Water: Applied Limnology.* SPON Press, London and New York.

Whipple, W., Jr., J.V. Hunter, F.B. Trama and T.J. Tuffey. 1975. Oxidation of Lake and Impoundment Hypolimnia. Water Resources Res. Inst., Proj. No. B-050-NJ, final report.

Wolter, K-D. 1994. Restoration of eutrophic lakes by phosphorus precipitation – Lake Gross, Glienicker, Germany. In: *Restoration of Lake Ecosystems — A Holistic Approach.* IWRB Publ. 32. pp. 109–118.

Wüest, A., N.H. Brooks and D.M. Inboden. 1992. Bubble plume modeling for lake restoration. *Water Res.* 28: 3235–3250.

19 Artificial Circulation

19.1 INTRODUCTION

Artificial circulation, also referred to as destratification, and hypolimnetic aeration/oxygenation (Chapter 18) are two general techniques for aerating lakes. Circulation has been achieved by pumps, jets, and diffused air. Complete lake circulation is usually the objective, and in the majority of cases examined either stratification was prevented or destratification occurred. Unlike hypolimnetic aeration/oxygenation, the temperature of the whole lake is raised with complete circulation; the greatest increase in temperature occurs at depths that were previously part of the cooler hypolimnion.

The principal improvements in water quality caused by complete circulation are oxygenation and chemical oxidation of substances in the entire water column (Pastorak et al., 1981, 1982). Similar to hypolimnetic aeration, its main benefit is enlarging the suitable habitat for aerobic animals. Complete circulation may reduce internal loading of P, if the principal P-release mechanism was due to iron reduction in anoxic profundal sediments (Chapter 18). Complete circulation may also reduce algal biomass by increasing the mixed depth, thereby reducing available light, and by subjecting mixed algal cells to rapid changes in hydrostatic pressure (Lorenzen and Mitchell, 1975; Fast, 1979; Forsberg and Shapiro, 1980). Although reduced internal P loading and decreased phytoplankton biomass may be reasonable expectations, other factors such as nutrient availability in the photic zone, may be more important to P availability, and actually be enhanced with circulation. In some instances, phytoplankton biomass and P content either did not change or were increased following circulation.

Artificial circulation has been employed as a management technique since at least the early 1950s (Hooper et al., 1953). Initially it was used to prevent winter fish kills in shallow, ice-covered lakes (Halsey, 1968). Although not discussed here, refinements to winterkill prevention were proposed recently (McCord et al., 2000; Miller et al., 2001; Miller and Mackey, 2003). Nearly all of the reported applications of the technique to control eutrophication effects and to improve water quality occurred later than the mid 1960s. Complete circulation has been the most frequently used technique to improve water quality (except for algicides and herbicides).

19.2 DEVICES AND AIR QUANTITIES

Introduction of compressed air through a diffuser or perforated pipe located at depth employs the air-lift method of circulating lakes and reservoirs, in which water is welled up by the rising plume of air bubbles (Pastorak et al., 1981, 1982). Although techniques using pumps and water jets have been used successfully to circulate lakes, the air-lift method, through diffusion of compressed air, is apparently the least expensive and is easiest to operate (Lorenzen and Fast, 1977). However, high efficiencies of oxygenation have been reported from pumped jets in some cases (Stefan and Gu, 1991; Michele and Michele, 2002).

If the lake is already stratified, mixing is usually achieved only above the depth of air injection. If the lake is not stratified however, injection near the surface can prevent stratification (Pastorak et al., 1981, 1982). The effect of an unconfined rising plume of air bubbles on water circulation in an already stratified lake is illustrated in Figure 19.1. As the plume rises, the mixture becomes heavy, upward water flow ceases and the water plume spreads laterally or sinks to a neutral

TABLE 19.1
Lakes Receiving Treatment by Artificial Circulation with Associated Characteristics

Lake	Depth Max.	Depth Mean	Depth Device	Volume (10⁶/m³)	Area (ha)	Q Air/m³/min	Q Air/m³ × 10⁶	Q Air/km²	Reference
Clines Pond, OR	4.9	2.5	4.9	0.003	0.13	0.028[a]	10.2	21.6	Malueg et al., 1973
Parvin, CO	10.0	4.4	10.0	0.849	19.0	2.1[a]	2.5	11.18	Lackey, 1972
Section 4, MI	19.1	9.8	18.3	0.110	1.1	2.21[a]	20.0	200.0	Fast, 1971a
Boltz, KY	18.9	9.4	18.9	3.614	39.0	3.17[a]	0.88	8.17	Symons et al., 1967, 1970; Robinson et al., 1969
University, NC	9.1	3.2	9.1	2.591	80.9	0.40[a]	0.15	0.49	Weiss and Breedlove, 1973
Kezar, NH	8.2	2.8	8.2	2.008	73.0	2.83[a]	1.41	3.88	Anon., 1971; Haynes, 1973
Indian Brook, NY	8.4	4.1	2.2	0.302	7.3	4.53[a]	15.0	62.06	Riddick, 1957
Prompton, PA	10.7	3.7	10.7	0.193	112.0	4.53[a]	1.08	4.04	McCullough, 1974
Cox Hollow, WI	8.8	3.8	8.8	1.480	38.8	2.04[a]–	1.38–	5.26–	Wirth and Dunst, 1967
						4.08	2.76	10.53	Wirth et al., 1970
Stewart, OH	7.5	3.4	7.0	0.090	2.6	0.25[b]	2.83	9.80	Barnes and Griswold, 1975
Wahnbach, 1961–1962	43.0	19.2	43.0	41.618	214.0	2.01[b]	0.048	0.94	Bernhardt, 1967
West Germany 1964						5.95[b]	0.143	2.78	
Starodworskie, Poland	23.0		23.0		7.0	0.27[a]		3.81	Lossow et al., 1975
Roberts, NM	9.1	4.4	9.1	1.233	28.3	3.54[a]	2.87	12.5	USEPA, 1970
						2.26[a]	1.84	8.00	McNally, 1971
Falmouth, KY	12.8	6.1	12.8	5.674	91.0	3.26	0.58	3.58	Symons et al., 1967, 1970; Robinson et al., 1969
Test II, U.K.	10.7	9.4	10.7	2.405	25.4	2.01[a]	0.84	7.92	Knoppert et al., 1970
Test I, U.K.	10.7	9.4	10.7	2.097	22.7	2.01[a]	0.96	8.86	Knoppert et al., 1970
Mirror, WI	13.1	7.6	12.8	0.40	5.3	0.45[a]	1.13	8.55	Smith et al., 1975; Brynildson and Serns, 1977
Växjosjön, Sweden	6.5	3.5	6.0	3.1	87.0	7.2[a]	2.32	8.28	Bengtsson and Gelin, 1975
Buchanan, ON	13	4.9	13	0.42	8.9	0.28[a]	0.67	3.17	
Corbett, BC	19.5	7.0	19.5	1689	24.2	4.5[a]	2.66	18.52	Halsey, 1968; Halsey and Galbraith, 1971
Maarsseveen, U.K.	29.9	14.0	19.0, 29.9	8.018	60.7	2.49[a]	0.31	4.10	Knoppert et al., 1970
Casitas, CA	82.0	26.8	39.0, 55.0	308.0	1100.0	17.84[b]	0.06	1.62	Barnett, 1975
Hyrum, UT	23.0	11.9	15.2	23.1	190.0	2.83[b]	0.17	1.49	Drury et al., 1975
Waco, TX	23.0	10.7	23.0	128.0	2942.0	3.11[b]	0.02	0.10	Biederman and Fulton, 1971

Location									Reference
Catharine, IL	11.8	5.0	8.5	3.034	59.5	0.76[c]	0.25	1.27	Kothandaraman et al., 1979
El Capitan, CA 1965-	62.0	9.8	21.3	17.99	183.9	6.09[b]	0.34	3.31	Fast, 1968
1966		9.4	28.3	21.05	222.0	6.09[b]	0.29	2.74	
Calhoun, MN	27.4	10.6	23.0	18.01	170.4	2.83[b]–3.54	0.16–0.20	1.66–2.08	Shapiro and Pfannkuch, 1973
Eufaula, OK	27.0	16.2	27.0	703.1	414.8×10^2	33.98[b]	0.05	0.06	Leach et al., 1980
Pfaffikersee, Switzerland	35.0	18.0	28.8	56.5	325.0	6.0[b]	0.11	1.85	Thomas, 1966; Ambuhl, 1967
Wahiawa, HI	26.0	8.0	2.7	1.7	20.0	2.4[b]	1.4	12.0	Devick, 1972
Trasksjön, Sweden	4.0	3.0	4	0.365	12.1	[a]			Karlgren and Lingren, 1963
Altoona, GA 1968–1969	46.0	9.4	42.7	453	4800	21.6[b]–27.7	0.05–0.86	0.45–0.58	USAE, 1973; Raynes, 1975
						27.7[b]	0.06	0.58	
Lafayette, CA	24.0	9.1	18.0	5.243	53	1.68[c]	0.32	3.17	Laverty and Nielsen, 1970
Hot Hole, NH	13.3	5.7	13.3	0.733	12.9	0.59[a]	0.80	4.57	NHWSPCC, 1979
Heart, Ontario	10.4	2.7	10.0	0.392	14.5	0.23[a]–0.92	0.58–2.34	1.56–6.33	Nicholls et al., 1980; Nicholls[d]
Clear, CA	15.0	10.2	14.0	115.9	1217	17[a]	6.82	114	Rusk[d]
Kremenchug, Poland	3.0	2.0	2.6	0.002	0.12	4.38[a]	1750	3500	Ryabov et al., 1972; Sirenko et al., 1972
Tarago, Australia	23.0	10.5	14.0	27.6	360	3.0–9.0	0.08–0.24	0.83–2.50	Bowles et al., 1979
						3.0–7.50	0.08–0.20	0.83–2.08	
Silver, OH	12.0	4.22	10.0	1.68	38.44	3.37[b]	2.01	8.77	Brosnan, 1983
East Sydney, NY	15.7	4.9	15	4.17	0.85	1.8[b]	0.43	2.1	Barbiero et al., 1996a
Crystal, MN	10.4	3.0	10	0.93	0.31	1.44[b]	1.55	4.6	Osgood and Stiegler, 1990
King George VI, U.K.	16.0	14.0	10.0	20.0		142.0 Water jet[c]			Ridley et al., 1966
Queen Elizabeth II, U.K.	17.5	15.3	17.5			128.0 Water jet[c]			Ridley et al., 1966
Ham's, OK	10.0	2.9	1.2	115.0		40.0 Axial-flow pump[a]			Stichen et al., 1979; Toetz, 1977a,b
Stewart Hollow, OH	7.6	4.6	7.6	0.148		3.2 Axial-flow pump[a]			Garton et al., 1978
Cladwell, OH	6.1	3.0	6.1	0.123		4.0 Axial-flow pump[b]			Irwin et al., 1966
Pine, OH	5.2	2.1	5.2	0.121		5.7 Axial-flow pump[b]			Irwin et al., 1966
Vesuvius, OH	9.1	3.6	9.1	1.554		42.5 Axial-flow pump[c]			Irwin et al., 1966
Arbuckle, OK 1975; 1977	24.7	9.5	6.0; 2.0	89.3×10^2		951.0 Axial-flow pump[c]			Toetz, 1977a, b, 1979
West Lost, MI	12.8	6.2	11.9	0.089		1.4 Pump[c]			Hooper et al., 1953

[a] Flow rate produced destratification.
[b] Partly mixed.
[c] Flow rate inadequate to destratify.
[d] R.A. Pastorak, personal communication.

Source: From Pastorak, R.A. et al. 1981; Pastorak, R.A. et al. 1982. Tech. Rept. No. E-82-3. U.S. Army Corps of Engineers; with additions.

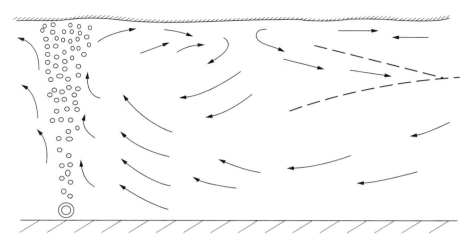

FIGURE 19.1 The process of destratification as a result of entrainment of water by a rising plume of air bubbles. Cooler, hypolimnetic waters from elsewhere replace the volume entrained near the plume, ultimately eroding away the thermocline. (From Davis, J.M. 1980. *Water Serv.* 84: 497–504. With permission.)

buoyancy level. However, the bubbles continue to rise with increased buoyancy having expanded due to reduced hydrostatic pressure at shallow depth, repeating the water-entrainment process, until they reach the surface. Assuming air flow is adequate, the process continues until the density difference above the diffuser is zero (Zic and Stefan, 1994; Sahoo and Luketina, 2002). The overall effect is that water is pulled from the hypolimnion into the epilimnion, breaking up the thermocline, producing generally homothermous, completely mixed conditions near the plume. As mixing and entrainment continue, erosion of the thermocline proceeds away from the plume so long as the energy applied through the airlift system exceeds the energy of resistance due to thermal (density) stability.

Injection of compressed air at maximum depth usually affords the greatest rate of mixing, because flow of the entrained water is a function of depth of release and air-flow rate. Lorenzen and Fast concluded that an air-flow rate per lake surface area of 9.2 m^3/km^2 per min (1.33 ft^3/acre per min) should provide adequate surface reaeration and other benefits of circulation. However, the areal air-flow rates approached or exceeded that critical value in only 42% of the cases cited in Table 19.1. Effectiveness of that flow rate is substantiated by the cases in Table 19.1 where before and after temperature data were provided (Pastorak et al., 1982). Figure 19.2 is a plot of the degree of destratification (percent reduction in Δt in the water column) related to air-flow rate per unit area. Except for three observations, areal air-flow rates approaching or exceeding 9.2 m^3/km^2 per min produced complete mixing, or 100% decrease in the surface to bottom Δt. In two of the three exception lakes to the right of the line in Figure 19.2, the final Δt was < 3°C, which was used as the criterion for satisfactory destratification (Pastorak et al., 1982). In 30 of the 45 cases cited for the airlift technique, where temperature data were available, the presented air-flow rates were adequate to destratify or prevent stratification (Table 19.2).

The Lorenzen and Fast areal air-flow rate criterion has been more reliably followed in more recent commercially installed systems. The average areal air-flow rate for 21 systems installed by General Environmental Systems in reservoirs and lakes > 23 ha during 1991–2002 was 7.8 m^3/km^2 per min (Geney, personal communication). Delivering the air to as much of the deep area of the water body as possible is also important to attain and maintain destratification (Geney, 1994).

The basis for the areal air-flow rate criterion of 9.2 m^3/km^2 per min is a relationship among air-flow rate, depth, and flow rate of up-welled water above an orifice (Lorenzen and Fast, 1977; Pastorak et al., 1982):

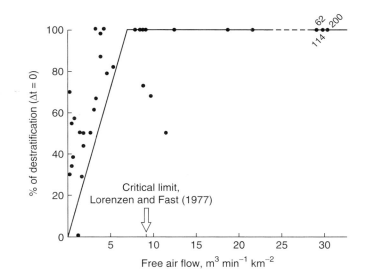

FIGURE 19.2 Percent destratification, based on surface to bottom temperature differences (Δt) before and after circulation, related to free air flow. (Data from Pastorak, R.A. et al. 1982. Environmental Aspects of Artificial Aeration and Oxygenation of Reservoirs: A Review of Theory, Techniques, and Experiences. Tech. RepT. No. E-82-3, U.S. Army Corps of Engineers, Vicksburg, MS; from Cooke et al. 1993. With permission.)

$$Q_w(X) = 35.6C(X + 0.8) \left(\dfrac{-V_o \ln \left(1 - \dfrac{X}{h + 10.3} \right)}{\mu_b} \right) \tag{19.1}$$

where $Q_w(X)$ = water flow rate in m³/s, $C = 2V_o + 0.05$ m³/s, X = height above orifice in m, V_o = air flow in m³/s at 1 atm, h = depth of orifice in m, and $\mu_b = 25V_o + 0.7$ m³/s.

Using this estimated water flow rate, the effect of various air-flow rates on hypothetical lake and reservoir morphometry was studied (Chen and Orlob, 1975). Results from 38 airlift cases over a range of lake reservoir areas, volumes, and depths, indicated an air-flow rate approaching or greater than the 9.2 m³/km² per min level (midpoint of a range 6.1–12.3 m³/Km² per min) consistently achieved destratification (Table 19.1).

The diffuser should be a pipe with multiple orifices, usually located at the deepest point in the lake, but suspended sufficiently well off the bottom (1 to 2 m) to minimize sediment entrainment. Orifice spacing should be about 0.1 times the depth of air release, because the rising water plume will spread horizontally at 0.05 m/m of rise (Lorenzen and Fast, 1977).

Another approach to designing an air-lift system to destratify lakes and reservoirs was described in detail by Davis (1980). This approach requires the following information/steps:

1. Obtain surface area and volume as function of depth.
2. Determine or assume temperature or density profile.
3. Existing stability and added heat input and theoretical energy required to overcome it are calculated.
4. Calculate free air-flow rate at the compressor.
5. Calculate perforated (diffuser) pipe length (50 m suggested as minimum).

6. Select diffuser pipe and hole diameters (0.8 mm suggested) and hole spacing (0.3 m suggested).
7. Determine internal pipe diameter and air pressure at compressor considering losses due to hydrostatic pressure, excess pressure at pipe end, friction in the pipe, the pipe bends, valves, etc..
8. Recheck diffuser length, considering pressure losses and free air flow through a single hole.
9. Calculate anchor weight.

Stability is calculated first, as the difference between the unmixed, existing density gradient, and the mixed condition:

$$S = g \sum_{i=1}^{n} \rho_{im} V_i h_i - \sum_{i=1}^{n} \rho_{is} V_i H_i \tag{19.2}$$

where S = stability, joules (kg m²/s²), g = acceleration due to gravity, m/s², ρ_i = density of layer i, kg/m³, V_i = volume of layer i, m³, h_i = height of centroid of layer i, m, m = mixed, and s = stratified. The energy required for destratification is calculated by

$$E = S + R - W \tag{19.3}$$

where S = stability, R = heat input, and W = wind energy, all in joules. Wind is neglected, as a conservative approach, so that mixing is possible without wind. R can be approximated as 5 J/m² per day.
    The required air-flow rate (Q) in L/s is

$$Q = \frac{0.196E}{T \ln\left(1 + \dfrac{D}{10.4}\right)} \tag{19.4}$$

where E = energy input required; 20 times the theoretical level (that assumes isothermal conditions and bubble pressure slightly in excess of the hydrostatic head) is factored into the equation, T = time to achieve destratification, D = depth of diffuser in m, and 10.4 = depth of water equivalent to atmospheric pressure.
    The volume of water entrained by the air bubbles from a perforated pipe is recommended to be 2.5 times the volume of the lake or reservoir to be destratified and can be calculated according to

$$V_e = 0.486LT \left(\frac{gQ}{L}\right)^{1/3} \left(1 + \frac{D}{10.4}\right)^{-1/3} \ln\left(1 + \frac{D}{10.4}\right) \tag{19.5}$$

From Equation 19.4, and knowing the volume to be destratified (m³) and the required air flow (L/s), the length of perforated pipe (diffuser) in m can be calculated:

$$L = 3.73 \left(\frac{V^3 \left(1 + \dfrac{D}{10.4}\right)}{T^3 Q \left(\ln\left(1 + \dfrac{D}{10.4}\right)\right)^3} \right)^{1/2} \tag{19.6}$$

Pastorak et al. (1982) compared the calculated flow rates required by the two procedures, using an example from Davis (1980) for a body of water with a volume of 20×10^6 m^3, a maximum depth of 20 m, and an area of 1.2×10^6 m^2. The flow rate recommended by the Davis procedure would be 70 L/s (3.5 m^3/km^2 per min). By the Lorenzen and Fast (1977) procedure the rate would be 6 m^3/km^2 per min, or 120 L/s, nearly twice the Davis rate. The rate used here is the the lower end of the range (6.1 to 12.3 m^3/km^2 per min) because deeper lakes generally require less air to mix than do shallow lakes (Pastorak, personal communication).

According to Equation 19.6, the diffuser pipe length needed to destratify is inversely related to air-flow rate. Thus, pipe length would be 216 m, based on a 70 L/s air-flow rate and 182 m based on 120 L/s for destratification to occur in 5 days. For the example lake, Davis (1980) selected a high-density polyethylene pipe of diameter 50.8 mm, perforated with 1-mm diameter holes spaced at 0.3 m. An air pressure of 5.3 bar (5.5 kg/cm^2) at the compressor was calculated by summing the hydrostatic pressure represented by the water depth over the pipe, mean excess pressure above the hydrostatic pressure at the end of the pipe (related to pipe length), friction loss in the pipe (related to pipe diameter) and pressure drop from bends in the pipe. An air-flow rate of 108 L/s was recalculated for pipe length and pore size and number of holes with that compressor pressure (5.3 bar). That exceeded the calculated 70 L/s so the nominal pipe length of 250 m was considered adequate. A longer pipe length than the minimum calculated facilitates destratification with greater air distribution. These estimates can be obtained from nomographs in Davis (1980).

While calculation of required free air-flow rate at the diffuser end and the initial estimate of minimum diffuser length to accommodate that rate are relatively straightforward, determining the required pressure at the compressor, and a more precise estimate of diffuser length incorporating all the pressure losses, is not straightforward and involves an iterative process (Meyer, 1991). Consistent with the above procedure, first obtain an initial estimate of diffuser length (Equation 19.6). Then, determine hydrostatic and internal pipe pressures to obtain a new estimate of free air flow from a single diffuser hole. From that air flow and knowing the diffuser hole-spacing and total air flow required, a new pipe length can be determined. With that pipe length, pressures can be recalculated and the process repeated until the optimum diffuser length is obtained. To simplify the process, Meyer (1991) incorporated the equations and charts from Davis (1980) into a spreadsheet, which allowed an iterative process of changing variables and formulas to arrive at an optimum diffuser length. Results using the spreadsheet procedure for a hypothetical reservoir are summarized as follows:

- Surface area: 1,011,750 m^2
- Diffuser depth: 10 m
- Volume above diffuser: 10,117,500 m^3
- Time to destratify 5 days: 432,000 s
- Temp. range from 30°C @ surface to 21.8°C @ 25 m
- Theoretical energy required (E) = stability (S) + solar input (R) (Equation 19.3):

$$1.9 \times 10^8 \text{ J} + 0.25 \times 10^8 \text{ J} = 2.15 \times 10^8 \text{ J}$$

Air flow required (from Equation 19.4):

$$Q = \frac{(0.196)(2.15 \times 10^8 \text{ J})}{432 \times 10^3 \text{ s ln}\left(1 + \dfrac{10 \text{ m}}{10.4}\right)} = 144.5 \text{ L/s}$$

Diffuser length, initial calculation (from Equation 19.6):

$$L = 3.73 \left(\frac{(10.1175 \times 10^6 \text{ m}^3) \left(1 + \frac{10 \text{ m}}{10.4} \right)}{(432 \times 10^3 \text{ s})^3 (144.51/\text{s}) \left(\ln \left(1 + \frac{D}{10.4} \right) \right)} \right) = 89 \text{ m}$$

Selected:

- Supply line: 500 m
- Internal diameter supply line: 45 mm
- Internal diameter diffuser: 35 mm

Through iteration, an optimum diffuser length of 339 m and compressor pressure of 9.7 kg/cm^2 (135 psi) were determined.

The iterative approach was used to estimate air-flow pressure and diffuser length for East Sidney Lake, New York, 85 ha, 15.7 m maximum depth and 4.9 m mean depth (Meyer et al., 1992). The respective values by using the Davis nomographs were 1.53 m^3/min, 3.4 kg/cm^2, and 107 m. Those using the iterative process were 2.19 m^3/min, 3.9 kg/cm^2, and 135 m. A destratifying time of 5 days was used with both procedures.

To gain flexibility and control over the long and narrow reservoir, 244 m of total diffuser length was installed, with 8 separate 30-m lines spread through the reservoir. A 15 hp compressor was used to deliver 1.8 m^3/min air flow at 3.6 kg/cm^2 pressure.

The system operated satisfactorily during 1989–1990 to maintain destratified conditions (< 2°C difference surface to bottom) in the near field, but temperature difference was greater in the far field or whole lake, despite the extended lines. Also, bottom DO levels dropped below 3 mg/L. Use of a diffuser longer than calculated, i.e., "underloading" the diffuser, may have accounted for and restricted destratification capacity. However, the total air delivery per area to the reservoir, which was 2.1 m^3/km^2 per min, relative to the Lorenzen and Fast criterion, was not discussed. That rate for East Sidney Lake was well below their median criterion and probably accounted for some of the less-than-expected water quality response, discussed later in this chapter. While a successful outcome for complete circulation depends on the size and length of diffuser pipes, results indicate that for best results in improvement of water quality, as well as achieving destratification, adherence to the Lorenzen and Fast criterion is also advisable.

Mechanical mixing devices have been used less frequently than compressed air (Table 19.1). Two types of pumps have been developed for destratifying reservoirs: (1) axial-flow pumps with a large propeller (6 to 15 ft diameter) that generates a low velocity jet (Punnet, 1991), and (2) direct drive mixer with a small propeller (1 to 2-ft diameter) that generates a high velocity jet (Stefan and Gu, 1991; Price, 1988, 1989). Design of a pumping system to destratify a lake or reservoir depends on the desired time to destratify (or rates of circulation) and depth of hydraulic jet penetration. Time to destratify in turn depends on the degree of stratification or resistance to mixing. The number of pumps needed to achieve a given depth of penetration and time to mixing can be calculated (Holland, 1984; Gu and Stephan, 1988; Stefan and Gu, 1991). Destratification was complete ($\Delta t < 3$°C) for 4 of the 10 cases for pumps and jets cited in Table 19.1.

Mixing devices powered by solar and wind energy are available commercially, but published results of effectiveness were unavailable for inclusion here.

19.3 THEORETICAL EFFECTS OF CIRCULATION

19.3.1 DISSOLVED OXYGEN (DO)

The principal, and probably the most reliable, effect of circulation is to raise the dissolved oxygen (DO) content throughout the lake over time. If the lake is destratified, the DO content in what was the hypolimnion will increase, and that in the epilimnion will decrease, at least at first. This can occur from simple dilution. Additional reasons why the surface water DO may decrease are the transfer of oxygen-demanding substances toward the surface and a decrease in photosynthesis in the photic zone due to increased mixing depths (Haynes, 1973; Ridley et al., 1966; Thomas, 1966). DO will continue to increase as circulation is maintained, largely because water undersaturated with oxygen is brought into contact with the air. While the vertical transport of water is achieved by entraining water through releasing compressed air at some depth, little oxygen increase is achieved through direct diffusion from bubbles (King, 1970; Smith et al. 1975).

19.3.2 NUTRIENTS

Internal loading of P theoretically can be decreased through increased circulation. This would occur in situations where the dominant mechanism of P release was from iron-bound P in anoxic hypolimnetic sediments. By aerating the sediment-water interface of lakes where iron is controlling P solubility, P should be adsorbed from solution by ferric-hydroxy complexes (Mortimer, 1941, 1971; Stumm and Leckie, 1971; Chapters 8, 18, 20). Thus P would be prevented from migrating from high concentrations in sediment interstitial water to the overlying water. Calcium may control P solubility in hardwater lakes, rather than iron, or the iron/phosphorus ratio may be too low to control P release (Jensen et al., 1992), in which case the release rate could be due largely to a function of aerobic decomposition of organic matter (Kamp-Nielsen, 1975). In that event, internal P loading may actually increase as temperature at the sediment-water interface is raised in the circulation process. Also, some sediments with a low Fe:P ratio have a high organic and water content and are very flocculent, and may have a high loosely bound P fraction (Boström, 1984). In that latter situation as well, internal loading could actually increase from such sediments following circulation. P exchange rates are dependent upon circulation at the sediment-water interface and that process could be enhanced by mixing (Lee, 1970). Degree of wind mixing had a dominant effect on summer internal loading of P in shallow Moses Lake, Washington (Jones and Welch, 1990).

Internal loading of P may be high in unstratified, shallow, eutrophic lakes in which the sediment-water interface is usually oxic (Jacoby et al., 1982; Kamp-Nielsen, 1975; Søndergaard et al., 1999). Therefore, reduced internal P loading probably cannot be expected to result from artificial circulation. Internal loading and whole-lake TP may decrease in shallow stratified lakes following circulation (Ashby et al., 1991), but the concentration available for growth in the photic zone may increase, as has been observed (Brosnan and Cooke, 1987; Osgood and Stiegler 1990). Thus, depth is an important criterion in determining the candidacy of shallow lakes for complete circulation from not only phytoplankton production related to available light, but also internal P loading. Unless oxic conditions will substantially reduce P internal loading, maintaining stratified conditions may be preferable for limiting P availability in the photic zone.

Other potential changes in chemical content resulting from complete circulation are the conversion of ammonium to nitrate and the complexation and sedimentation of trace metals such as manganese and iron. Ammonium decrease can largely be attributed to increased nitrification, which requires aerobic conditions (Brezonik et al., 1969; Toetz, 1979). This effect will be greater the longer that duration and completeness of hypolimnetic deoxygenation proceeded prior to circulation. The decrease in trace metals like manganese and iron should also be greater in lakes with larger oxygen deficits prior to aeration increases. Because these metals diffuse from the sediment in their reduced, soluble forms, aeration will promote their oxidation and subsequent complexation and precipitation. This can be an important benefit in lakes used for drinking water supplies.

19.3.3 PHYSICAL CONTROL OF PHYTOPLANKTON BIOMASS

Circulation can reduce phytoplankton biomass through light limitation, brought about by providing a greater depth of mixing of plankton cells in the water column so that the total light received during their brief period in the photic zone is insufficient for net photosynthesis (photosynthesis in excess of respiration) and thus any growth or increase in cell mass. This is known as the "critical depth" concept, first formulated to predict the timing of the spring diatom bloom in the ocean (Sverdrup, 1953). By knowing light at the surface, compensation depth, and the extinction coefficient, the critical depth can be calculated as the point above which net production is possible; when that calculated depth exceeds the mixed-layer depth, a bloom can occur. This model is dependent upon some relationship between light intensity and gross photosynthesis, assuming a constant rate of respiration.

The same concept applies in lakes (Talling, 1971). The combination of low surface light intensity and deep mixing prevented net photosynthesis during winter in relatively deeper lakes (> 30 m) of the English Lake District, but not in the shallower lakes (10 m). Growth rate during the spring phytoplankton maximum was directly related to light intensity in a long-term data series (Neale et al., 1991). Normally, lakes are shallow enough to allow some net photosynthesis even in winter, but decreasing mixing depth, as stratification develops and surface light intensity increases in the spring, usually accounts for the large increase in net photosynthesis and the spring diatom bloom in deeper lakes.

Light can limit maximum phytoplankton biomass even in shallow eutrophic lakes (Sheffer, 1998). A 35-year data base from Lake Võrtsjär (270 km², mean depth 2.8 m), Estonia, showed that the water level change produced a 2.5 times difference in mean depth resulting in biomass levels significantly lower in high water level years (Nõges and Nõges, 1999; Nõges et al., 2003). Thus, artificial circulation may produce light-limiting benefits in shallow, eutrophic lakes with normally high particulate matter concentrations and light extinction.

The concept of physical control of phytoplankton growth was extended to the effects of artificial circulation in eutrophic lakes (Lorenzen and Mitchell, 1975; Murphy, 1962; Oskam, 1978). Forsberg and Shapiro (1980) and Shapiro et al. (1982) integrated the effects of nutrients with those of physical factors. By increasing the depth of mixing, a lake potentially can be returned to a winter condition where light is limiting, assuming maximum depth and light attenuation are sufficient. Increasing mixing depth would not be great enough in most cases to prevent net biomass production completely, which is not expected. This effect of mixing depth is clearly shown in results from Kezar Lake (Figure 19.3; Lorenzen and Mitchell, 1975). Increased mixing depth though complete circulation is expected to substantially reduce algal biomass due to light limitation alone. However, nutrients may initially be limiting in the epilimnion, so that a slight increase in mixed depth may entrain water with higher nutrient content from below and biomass may increase (point A to point B in Figure 19.3). At some point light will limit and productivity and biomass will decrease (point C to point D). Note that biomass is plotted as mass per area (g/m²), which was expected to decrease by only 38% for a mixing-depth increase of 2 to 6 m. Biomass concentration (g/m³), however, was expected to decrease by 80%, which would also include the effect of water column dilution. This model predicted only the potential productivity without nutrient limitation and included no losses from sinking, grazing, parasitism, or washout. Actual values may therefore fall below the line in Figure 19.3, as was the case for Kezar Lake.

Little change in biomass may occur in oligotrophic lakes following circulation, because the slope of the ascending line in Figure 19.3 (nutrient limitation) would be less for such lakes (Pastorak et al., 1981, 1982). Because that line represents the maximum nutrient-limited biomass, any displacement of the biomass vertically by circulation would bring about a smaller change in biomass concentration in oligotrophic than in eutrophic lakes. That was not the case in experiments in deep

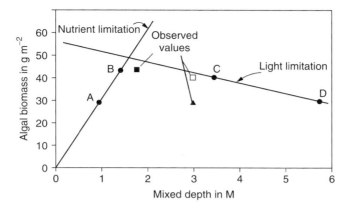

FIGURE 19.3 Theoretical and observed peak biomass of algae in Kezar Lake (see text for explanation of points A–D). Solid circles: theoretical values; solid square: 1968, stratified; triangle: 1969, destratified; open square: 1970, destratified. (From Lorenzen, M.W. and R. Mitchell. 1975. *J. Am. Water Works Assoc.* 67: 373–376. With permission; Pastorak, R.A. et al. 1981. *Evaluation of Aeration/Circulation as a Lake Restoration Technique.* 600/3-81-014. USEPA; Pastorak, R.A. et al. 1982. Environmental Aspects of Artificial Aeration and Oxygenation of Reservoirs: A Review of Theory, Techniques, and Experiences. Tech. Rept. No. E-82-3. U.S. Army Corps of Engineers, Vicksburg, MS.)

plastic bags in an oligotrophic lake (15 µg/L TP) in which biomass increased with mixing depth up to 15 m so long as background turbidity was low (Diehl et al., 2002). Results verified the hypothesis that increased mixing depth reduces growth rate, but at the same time reduces cell and nutrient loss. However, light attenuation should be greater and nutrient conservation less important under eutrophic conditions, as was demonstrated with increased background turbidity; i.e., biomass decreased beyond a mixing depth of 6 m.

Oskam (1973, 1978) developed a model to express the effect of mixing-depth change on productivity and maximum biomass. Because net productivity (P_{net}, mg C/m² per day) is the difference between gross productivity and respiration in the mixed layer, the following equation should hold:

$$P_{net} = CP_{max}\left(\frac{F(i)\lambda}{\varepsilon_w + C\varepsilon_c} - 24\ rZ_m\right) \qquad (19.7)$$

where C = chlorophyll (chl) a in mg/m³, P_{max} = maximum photosynthetic rate in mgC/mg chl per hour, $F(i)$ = dimensionless function of light intensity (expands P_{max} to total areal rate), λ = daylight hours, ε_w = extinction for water in 1/m, ε_c = specific extinction coefficient per unit algae in m²/mg chl, Z_m = depth of mixing in m, 24 = 24 h/d, r = respiration/P_{max}.

According to this equation, as the depth of mixing increases, assuming uniform distribution of algae, net productivity decreases. The mixing depth can be increased by artificial circulation. Critical depth can be calculated without knowing P_{max} by setting $P_{net} = 0$ and solving for Z_m:

$$Z_m = \frac{F(i)\lambda}{24r\left(\varepsilon_w + C\varepsilon_c\right)} \qquad (19.8)$$

Maximum biomass (mg chl/m³) can also be estimated from Equation 19.7) as a function of mixing depth by setting $P_{net} = 0$, and solving for C_{max}:

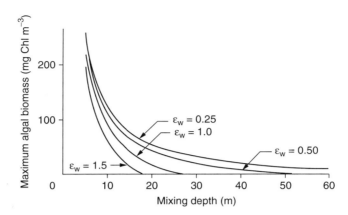

FIGURE 19.4 Relation of maximum chlorophyll concentration to mixing depth for different levels of nonalgal attenuation of light. (From Oskam, G. 1978. *Verh. Int. Verein. Limnol.* 20: 1612–1618. With permission.)

$$C_{\max} = \frac{1}{\varepsilon_c}\left(\frac{F(i)\lambda}{24rZ_m} - \varepsilon_w\right) \tag{19.9}$$

The maximum biomass possible is plotted for four different water extinction coefficients (Figure 19.4), assuming that $\varepsilon_c = 0.02$, $F(i) = 2.7$, $\lambda = 12$, and $r = 0.05$. Further important assumptions are that nutrients are not limiting and there are no significant losses other than respiration. Accordingly, the maximum biomass concentration attainable in a lake with mixing depth = 5 m would be 220 mg/m³ chl *a*. If either nutrient limitation, grazing, sinking, or washout were significant, then the maximum would be correspondingly less. These relationships show the sensitivity of potential maximum biomass to mixing depth in shallow lakes and may offer a first approximation of the feasibility for circulation to reduce algae in a particular, non-nutrient-limited lake.

Forsberg and Shapiro (1980) and Shapiro et al. (1982) developed an expanded model to include nutrient limitation and losses. Their equation for maximum biomass in the mixed layer is

$$C = \frac{\ln(Io/Iz')P_{\max}^{sat} - D\theta Z_m\varepsilon_w}{\varepsilon_c D\theta Z_m + [\ln(Io/Iz')P_{\max}^{sat}Kq]/TP} \tag{19.10}$$

where C = chl *a* concentration in mg/m³, Io = incident radiation, Iz' = radiation at a depth one-half the photosynthesis saturated light intensity (I_k in Talling, 1971), $P_{\max}^{sat}$ = maximum specific rate of photosynthesis under saturated nutrient concentration in mg C/mg chl per day, D = loss rate through sinking, grazing, parasitism, washout, etc. in 1/d, θ = c/chl *a* ratio, Z_m = depth of mixing in m, ε_w = extinction for water in 1/m, ε_c = extinction coefficient for chl *a* in m²/mg chl *a*, Kq = subsistence quota of TP in mg TP/mg chl and TP = TP concentration in mg/m³.

The basis for the nutrient effect in Equation 19.10 is an expression of cell nutrient quota, which is approximated by the ratio of TP to chl *a:*

$$P_{\max} = P_{\max}^{sat}\left(1 - \frac{Kq'}{TP/chl\ a}\right) \tag{19.11}$$

where $P_{\max}$ = maximum specific daily rate of photosynthesis at saturating nutrient level and Kq' = minimum ratio of TP/chl a required for photosynthesis to occur (1.8 in Forsberg and Shapiro, 1980).

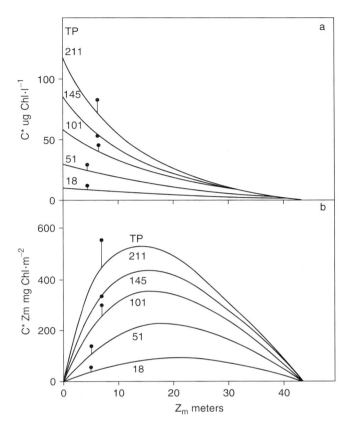

FIGURE 19.5 The effect of changes in the mixed depth and TP on: (a) the maximum concentration of chl *a* and (b) the maximum aerial standing crop of chl *a* in the mixed layer of Twin Lake, Minnesota, as predicted by the model (closed circles indicate observations and connecting lines indicate the deviation between predicted and observed results). (From Shapiro, J. et al. 1982. *Experiments and Experiences in Biomanipulation — Studies of Biological Ways to Reduce Algal Abundance and Eliminate Blue Greens.* USEPA-600/3-82–096.)

The relationships between maximum biomass (chl) per unit volume and per unit area in the mixed layer, and the depth of mixing, based on this model, are shown in Figure 19.5. Clearly, the concentration of limiting nutrient determined the maximum biomass at any depth of mixing. This is an important point, and should be considered with predictions of improvements following circulation because of the great potential for increasing the nutrient available to algae following destratification. Of course, if nutrient content is relatively high already and increases do not occur with mixing, then biomass concentration should decrease, with the greatest decrease occurring at mixing depths less than 10 m (Figure 19.5a).

As Pastorak et al. (1982) indicate, there are several problems with application of this model. The most serious would appear to be difficulties in estimating loss rates, which would decrease with mixing depth increase (Diehl et al., 2002), as well as the effects of shifts in species composition (see below) and the nutrient history on the growth-rate response of algae. Nevertheless, rather good agreement between the model predictions and experimental results were observed (Figure 19.5; Forsberg and Shapiro, 1980). The low level of complexity in this model makes it appealing as a tool to guide the application of the circulation technique. However, a separate prediction for TP is necessary.

19.3.4 EFFECTS ON PHYTOPLANKTON COMPOSITION

There are several hypotheses to explain the dominance of blue green algae (cyanobacteria) in eutrophic lakes (Welch and Jacoby, 2004). There are three that may explain a shift from dominance by bloom-forming blue-greens to dominance by more desirable diatoms or green algae as a result of complete circulation. These involve changes in (1) CO_2 and pH, (2) distribution of buoyant cells, and (3) grazing by zooplankton, all of which could be results from increased circulation.

Blue-green algae-dominated cultures shifted to dominance by green algae in response to decreased pH and associated increases in free CO_2 concentration (King, 1970, 1972; Shapiro, 1973, 1984, 1990; Shapiro and Pfannkuch, 1973; Shapiro et al., 1975). Blue-greens apparently absorb CO_2 at lower concentrations, compared to green algae, giving them an advantage at higher pH. Green algae may have a competitive advantage over blue-greens with respect to nutrients at lower pH. The observed rapid die-off of blue-greens following pH decrease, however, may have been caused by lysing of the blue-greens by viruses that were favored by low pH (Shapiro et al., 1982). King introduced the CO_2 hypothesis based on comparisons of algal populations and chemical conditions existing in sewage lagoons, and suggested that the potential for lakes to promote blue-green dominance increases as alkalinity (buffering capacity) decreases at any given P loading. Shapiro (1984, 1990) was able to shift dominance from blue-greens to greens in bag experiments *in situ* with either HCl or CO_2 addition, but the shift was more complete if nutrient additions were also included.

Increased circulation can cause CO_2 to increase and pH to decrease in the euphotic zone by vertical transport of bottom water, in which CO_2 content is high due to respiration in the absence of photosynthesis, as well as by increased contact with the atmosphere. For circulation to promote the shift from blue-greens, the surface waters should not be nutrient-limited, because high content of N and P also exists in bottom water, that, with vertical entrainment, could increase blue-green biomass already present.

A large-scale experiment in Squaw Lake, Wisconsin during summer 1993 produced some doubt about the role of the CO_2/pH hypothesis (Shapiro, 1997). The lake is naturally divided into two basins; south (9.1 ha, 2.55 m mean depth) and north (16.8 ha, 2.92 m mean depth). The south basin was artificially circulated and enriched with CO_2. The pH exceeded 10 in the north basin, but remained steady at around 7 in the enriched south basin during circulation. Despite the contrasting pH/CO_2 condition, populations of *Aphanizomenon* and *Anabaena* reached levels exceeding 300 µg/L chl *a* in both basins.

Because the blue-green algae have gas vacuoles that permit buoyancy, the increased stability brought about by thermal stratification and calm weather will allow blue-greens to produce surface "scums" and a decreased light environment for non-buoyant algae. Therefore, increased circulation can favor non-buoyant algae, which otherwise tend to sink rapidly (especially diatoms) under stable conditions. *Anabaena* succeeded *Tabellaria* as thermal stratification developed in summer, although specific growth rate of the blue-green was not different from that of a spring dominating diatom (Knoechel and Kalff, 1975). This change was explained as a physical effect based on sinking-rate difference, rather than one based on growth-rate differences related to nutrient changes. Had stratification been prevented by artificial circulation the diatom may have persisted. Artificial destratification in a Thames River reservoir in late July promoted a second bloom of the diatom *Asterionella*, which had previously bloomed in the spring and had subsequently declined (Taylor, 1966). Such a decline in *Asterionella* has been attributed to the combination of nutrient limitation and sinking losses (Lehman and Sandgren, 1978). One likely explanation for the second bloom in the Thames reservoir was a inoculation of high-Si bottom water to the lighted zone, but conditions must have likewise been more favorable for the large diatom because of a probable decreased sinking rate. Subsequent destratification in the fall, when blue-green algae were very abundant, did not result in a third *Asterionella* bloom. Destratification did not seem to affect blue-greens in this case. Increases in *Asterionella* following induced circulation have been observed elsewhere (Bern-hardt, 1967; Fast et al., 1973).

The advantage that blue-greens have in stable water, through their buoyancy regulation, is negated by increased circulation. Gas vacuole adjustment of buoyancy in blue-greens is probably controlled by a combination of light, pH, CO_2, N, and P, allowing their movement between the more lighted surface waters and the more nutrient-rich intermediate depths (Reynolds, 1975; Walsby and Reynolds, 1975; Klemer et al., 1982; Reynolds, et al., 1987) if the water column is stable. Increased circulation, however, can prevent that pattern. An *Oscillatoria* population, located in the metalimnion, was dispersed following circulation (Bernhardt, 1967). Likewise, circulation of 313-ha T. Howard Duckett Reservoir, Virginia, eliminated summer blue-green blooms (Robertson et al., 1988). Cyanobacteria (*Ctkubdrisoerniosus* and *Anabaena*) were replaced by diatoms during summer following circulation (8 h/day) of an Australian reservoir (13.4 m max. depth) at an air flow exceeding the Lorenzen and Fast criterion (Hawkins and Griffiths, 1993).

Artificial circulation experiments that tested a combination of the CO_2/pH and buoyancy regulation hypotheses were carried out in 1-m diameter by 7-m deep plastic bags in two lakes (Forsberg and Shapiro, 1980; Shapiro et al., 1982). To a great extent, results from these *in situ* mixing experiments verified the above hypotheses regarding change in pH, CO_2, nutrients, and species composition. At the slowest mixing rates, TP and chl *a* increased, with blue-green algae dominating the plankton. At intermediate and high rates of mixing, diatoms and greens tended to dominate the plankton, with biomass and TP increasing at the intermediate rate of mixing, but decreasing at the high rate. Green algae did not dominate unless pH was low and nutrients were high. Overall, abundance of algae was more related to nutrient content than to light availability, which was controlled by mixing up to 7 m (Figure 19.5).

Mixing rate and neutralization of buoyancy regulation may have been more important in the Forsberg and Shapiro experiments than CO_2/pH, in view of the failure of CO_2/pH alone to alter blue-green dominance in the Squaw Lake experiment, described above. Results from whole lake mixing investigations in Lake Nieuwe Meer (1.32 km², 18 m mean depth, 30 m maximum depth), The Netherlands, support that view (Van der Veer et al., 1995; Visser et al., 1996b). *Microcystis* dominance shifted to a mixed community of flagellates and diatoms during two summers of complete circulation (1993–1994). *Microcystis* decreased from 90% to < 5% of the biomass. Buoyancy loss, due to entrainment through the water column, increased with greater distance from the diffuser plumes. That is, a higher percentage of sinking cells (determined microscopically) meant more carbohydrate stored due to more light received, because photosynthesis stores carbo-hydrate. A lower percentage of sinking cells meant less carbohydrate stored and a greater neutral-ization of buoyancy. If mixing was not continuous, *Microcystis* was able to reach a higher biomass by spending more time in the illuminated zone.

Designing the system in Leke Nieuwe Meer for mixing rate velocity (~ 1 m/h) was the key to controlling blue greens, because that rate exceeded the mean flotation velocity of *Microcystis* (0.11 m/h) and approached its maximum of 2.6 m/h. That velocity (~ 1 m/h) was achieved with an overall air-flow rate of 9.9 m³/km² per min — similar to the Lorenzen and Fast criterion. Nutrient content was high before and during mixing (TP, 420–450 µg/L and SRP, 350–380 µg/L), showing that circulation can restrict the abundance of buoyant cyanobacteria despite nutrient level, and shift communities to less objectionable taxa that are more successful in mixed conditions (i.e., low compared to high $Z_{eu}:Z_{mix}$).

While Lake Nieuwe Meer was relatively deep, mixing may effectively reduce cyanobacteria in much shallower lakes if mixing rate is adequate. The light gradient even in a shallow pond (1.8 m) was sufficient to promote strong buoyancy/sinking behavior in *Anabaena*, as indicated by the proportion of vacuolated cells (Spencer and King, 1987).

These potential benefits of artificial circulation are illustrated together in Figure 19.6. Aeration may reduce trace elements and P internal loading and, consequently, algal biomass. Algal biomass may also decrease because mixed depth increases and because silt stirred up by circulation could cause light to limit photosynthesis. This would be most likely in nutrient-rich lakes, because light is more likely to limit than nutrients.. Circulation may increase the mixing rate enough to neutralize the buoyancy/sinking advantage of blue greens. Epilimnetic CO_2 may increase and pH decline as a result of mixing bottom waters enriched with CO_2. The lowered pH may stimulate cyanophage activity, lysing blue-greens, while increased free CO_2 concentration could provide green algae with a growth advantage. Thus more edible size phytoplankton (small green algae and diatoms), plus the enlarged aerobic, dimly lit habitat serving as a refuge for large zooplankton from zooplanktivorous fishes, could result in greater loss rates of phytoplankton through grazing (Figure 19.6).

19.4 EFFECTS OF CIRCULATION ON TROPHIC INDICATORS

The four indicators that have improved most consistently following artificial circulation are DO, ammonium, epilimnetic pH, and the trace metals iron and manganese (Table 19.2). DO increased and trace metal decreased in a very high percentage of cases studied, while favorable changes in ammonium and pH were less frequent. Changes in all four variables were statistically significant, and are a result of increased contact of a mixed water column with the atmosphere.

Increased circulation usually results in the complexation and precipitation of Fe and Mn. However, upon close examination, Chiswell and Zaw (1991) found poorer control of Mn than Fe and increases in both metals over time despite continued circulation in two Australian reservoirs. The lack of particulate oxides of Mn (IV) was considered undesirable for water supplies if insoluble Mn (II and III) has adsorbed Mn^{2+}, which could be desorbed during treatment with alum.

Oxic conditions should also be effective in decreasing P content, because P should sorb to oxidized iron complexes. The results for P, however, are much less impressive than for Fe and Mn (Table 19.2). The cases where P increased or did not change following circulation were more frequent (65%) than those where decreases occurred. For many of the cases examined, there may have been other sources of internal loading that could be more significant than release from pelagic sediments into anoxic overlying water. These could include aerobic release from littoral sediment, plant decomposition, or littoral release from photosynthetically caused high pH, although midlake pH usually decreased with circulation. Also, external loading may have represented most of the input to the water column. However, if P was not controlled by iron, then aerobic release through microbial decomposition or exchange of loosely sorbed P could become the principal mechanism for P internal loading, as often occurs in unstratified lakes (Kamp-Nielsen, 1974, 1975; Boström, 1984). In that event, destratification, along with increased water exchange at the mudwater interface and increased temperature, could have resulted in greater release of P than occurred before circulation.

Surface waters in a stratified lake usually cool slightly during the destratification process, and the bottom waters heat up as much as 15 to 20°C, to approach a temperature similar to the surface water (Pastorak et al., 1982). If the air-flow rate is much less than the Lorenzen and Fast criterion, which was true for about 58% of the cases listed in Table 19.1, then microstratification may develop at the surface (Fast, 1973). That would provide light conditions that are highly desirable for phytoplankton production, because the ratio of effective mixing depth to "critical depth" would be rather small. On the other hand, if the air-flow rate is too high, sediment can be suspended in the water column.

Worst-case results from inadequate air-flow rates have been described in detail. Destratification of shallow Crystal Lake, Minnesota ($\bar{Z}$ = 3 m; Z_{max} = 10.4) was resumed after aerators had been shut off for 2 years as a control period (Osgood and Stiegler, 1990). Internal P loading substantially increased following resumed circulation, as evidenced by a 2–3 fold increase in summer epilimnetic TP, TN and chl a, with a proportional decrease in transparency. Circulation apparently increased the rate and availability (to the lighted zone) of P from internal loading. While the lake may represent an example where iron is not controlling P, circulation was nonetheless inadequate to provide continuous oxic conditions at the sediment-water interface. The air-flow rate employed was only 4.7 m^3/km^2 per min — only about one half the Lorenzen and Fast criterion. The conclusion that

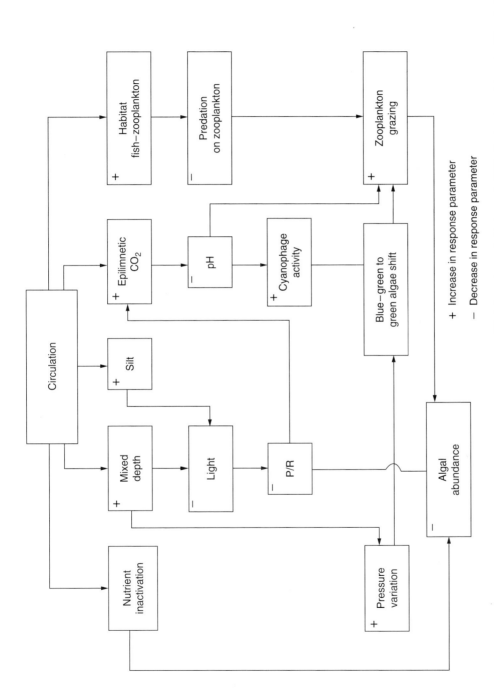

FIGURE 19.6 Potential beneficial effects of artificial circulation on phytoplankton. (Modified from Pastorak, R.A. et al. 1981. *Evaluation of Aeration/Circulation as a Lake Restoration Technique.* USEPA-600/3-81-014; Shapiro, J. 1979. In: *Lake Restoration.* USEPA-440/5-79-001.)

TABLE 19.2
Summary of Lake Responses to Artificial Circulation, Diffused-Air Systems Only

Parameter	N		Lake Responses +	−	0	?	χ^2
Δt After[a]	45	No.	15	30			5.0^b
		%	33	67			
SD	19	No.	4	10	2	3	6.50^b
		%	21	53	11	16	
DO	41	No.	33	1	2	5	55.2^d
		%	80	1	5	12	
Phosphate	17	No.	3	5	7	2	1.60
		%	18	29	41	12	
TP	20	No.	5	6	8	1	0.74
		%	25	30	40	5	
Nitrate	20	No.	7	8	3	2	2.33
		%	35	40	15	10	
Ammonium	20	No.	3	13	3	1	10.5^c
		%	15	65	15	5	
Iron/manganese	22	No.		20	2		33.1^d
		%		91	9		
Epilimnetic pH	21	No.	1	9	8	3	6.33^b
		%	5	43	38	14	
Algal density	33	No.	6	14	8	5	3.71
		%	18	42	24	15	
Biomass/chlorophyll	23	No.	5	6	6	6	0.12
		%	22	26	26	26	
Green algae	18	No.	7	4	7		1
		%	39	22	39		
Blue-green algae	25	No.	5	13	5	2	5.57
		%	20	52	20	8	
Ratio of green to blue-green algae	21	No.	11	3	6	1	4.90
		%	52	14	29	5	

[a] Temperature differential between surface and bottom water during artificial mixing: + means $\Delta t > 3°C$; − means $\Delta t < 3°C$.
[b] $p < 0.05$. Goodness-of-fit test to uniform frequency distribution for +, −, 0 responses only.
[c] $p < 0.01$.
[d] $p < 0.001$.

Source: From Pastorak, R.A. et al. 1982. Environmental Aspects of Artificial Aeration and Oxygenation of Reservoirs: A Review of Theory, Techniques, and Experiences. Tech. Rept. No. E-82-3. U.S. Army Corps of Engineers, Vicksburg, MS.

circulation caused the worsened quality of the lake was challenged by Laing (1992) and subsequently defended by Osgood and Stiegler (1992).

A similar experience occurred in East Sydney Lake, New York, which was essentially destratified according to the $\Delta t < 3°C$ criterion and bottom DO was increased (see earlier discussion of this lake). However, weak stratification still occurred intermittently (Barbiero et al., 1996a, b). This recurring condition allowed low bottom DO (but still oxic), continued internal P loading and entrainment of P into the photic zone to levels as high as 60 µg/L with or without mixing. As a result, circulation did not reduce algal biomass or increase transparency and cyanobacteria blooms continued to occur during summer. Although the aeration system was carefully designed (Meyer et al., 1992), overall air-flow rate was only 2.1 m^3/km^2 per min, less than one fourth the Lorenzen and Fast criterion. Similar adverse effects of inadequate circulation occurred in Silver Lake, Ohio (Brosnan and Cooke, 1987).

Another interesting case of worsened lake quality due to complete circulation occurred in Lake Wilcox, southern Ontario, during the normally unstratified period (Nürnberg et al., 2003). That practice promoted blooms of the cyanophyte *Planktothrix rubescens*, which previously did not produce large blooms. Circulation produced increased blooms from continued entrainment of P and algae throughout the water column and exposure to higher light conditions.

Transparency (SD) worsened more often than it improved (53% vs. 21%) in cases examined following circulation (Table 19.2). Transparency may decrease following treatment, if: (1) the photic zone is initially nutrient-limited such that phytoplankton content increases following circulation of entrained nutrients, (2) circulation is too weak, resulting in microstratification that favors buoyant blue-green algae with a more favorable light climate for productivity, or (3) circulation is so intense that particulate matter becomes resuspended. These complicating effects were probably responsible for SD decrease being common in most cases. Also, decreased transparency was frequently reported from artificially circulated water supply reservoirs (AWWA, 1971).

The CO_2/pH mechanism may be important in controlling the blue-green to green algae shift, which may occur if mixing is fast enough to allow sufficient CO_2 transport from the atmosphere (or hypolimnion) to drop the pH substantially (Shapiro, 1984, 1990). However, the failure of the whole-lake experiment in Squaw Lake, Wisconsin, casts some doubt on the role of CO_2/pH in cyanobacteria dominance (Shapiro, 1997). Affecting the buoyancy/sinking characteristic of bloom-causing cyanobacteria probably offers more promise in restricting their dominance with circulation. Maintaining an air-flow rate near the Lorenzen and Fast criterion produced a mixing velocity that neutralized the buoyancy/sinking advantage of *Microcystis* (Visser et al., 1996a). Heretofore, designers of circulation systems had not considered the mixing rate criterion with respect to buoyancy, although the effect of mixing was recognized.

This buoyancy effect was shown in enclosure experiments, where intermittent mixing reduced the total biomass as well as that of blue-greens because of interference in the growth of fast-growing species and the population build-up of slower growing species (Reynolds et al., 1984). Others have demonstrated that intermittent destratification was effective at reducing summer blue-green algal blooms (Steinberg and Zimmerman, 1988). However, while blue-greens, especially *Limnothrix redakei* (a thin, long-filament type like *Oscillatoria*), were controlled in a Bavarian kettle lake for 3 years following destratification, that species reappeared in bloom proportions in autumn of the fourth year at six times the pretreatment level (Steinberg and Tille-Backhous, 1990). The algal biomass in general was higher the fourth year and the authors suggested that the resulting low CO_2 may have favored the blue-green. P content was apparently not an explanation for the increased algal production. In the example of Crystal Lake mentioned earlier, phytoplankton (principally *Microcystis*) doubled following destratification, but blue-green control may have been hampered by the lower than recommended air-flow rate.

Circulation can increase the aerobic environment, as noted earlier, which in turn can greatly influence the depth distribution of zooplankton. If allowed to distribute themselves to greater but more poorly lit depths, zooplankton should be able to avoid predation by fish (Zaret and Suffern, 1976). *Daphnia* increased from 5–8 times and was distributed to greater depths in an incomplete destratification of Lake Calhoun (Shapiro et al., 1975). McQueen and Post (1988) have shown similar positive effects of circulation this regard. Fast (1971b) also found that circulation resulted in most of the zooplankton occupying water below 10 m following circulation, while most occupied depths less than 10 m before circulation. In 8 of the 13 cases examined, circulation increased the depth distribution of some or all zooplankters (Pastorak et al., 1981, 1982). In others, zooplankton was distributed to depth before circulation. Abundance increased in 10 of 15 cases examined. Such abundance increases could be due to reduced predation from planktivorous fishes (Shapiro et al., 1975; Andersson et al., 1978; Kitchell and Kitchell, 1980; McQueen and Post, 1988). High nutrient content (i.e., less removal by algae) in Thames Valley reservoirs (e.g., Queen Elizabeth II, Table 19.1) is due to light limitation promoted by deep mixing and grazing by high abundances of large *Daphnia* (Duncan, 1990). High rates of herbivory by filter-feeding zooplankton resulted in a lower biomass of phytoplankton with mixing in 6-m plastic bags in an eutrophic lake, compared to higher biomass with low herbivory (Weithoff et al., 2000).

There have been few studies of macroinvertebrates and fish following circulation (Pastorak et al., 1981, 1982). In most instances (six of eight), the expanded aerobic habitat resulted in increased abundance of macroinvertebrates, while increased diversity occurred in seven of eight cases examined. In all cases, fish expanded their depth distribution following circulation, but growth rate seldom increased possibly because study duration was insufficient to detect changes. Higher trophic levels are usually slow to respond to manipulation. Fish kills were averted in several cases, and where temperature remained satisfactory (temperatures $> 20°C$ should be avoided), salmonids survived in eutrophic lakes.

19.5 UNDESIRABLE EFFECTS

There are several potential adverse effects of artificial circulation, some more likely to occur than others; 13 are illustrated in Figure 19.7. If nutrients are limiting productivity in the epilimnion, then circulation may increase particulate P, which could be mineralized to the usable form, or highly dissolved P itself may be transferred to the lighted zone. Water transparency (SD) could then become worse than before circulation, due to increased silt as well as algal biomass. Increased algal abundance and photosynthesis would lower epilimnetic CO_2, raise pH, and prevent the succession from cyanobacteria (blue-greens) to greens. Because less edible cyanobacteria would tend to dominate as a result of the increased productivity and concomitant chemical changes, zooplankton would have a lesser effect on algal loss rates through grazing. The existing cyanobacteria crop could then increase.

Figure 19.7 suggests that reduced algal sinking may cause increased algal abundance, but as indicated under the description of benefits, such an increase could be represented by diatoms. In that case, it would be recognized as a benefit because diatoms would be less objectionable than cyanobacteria. The diagram also assumes that the air-flow rate is adequate to attain complete mixing. If air flow is not adequate, then partial stratification may occur during periods of summer heating, and algal abundance (especially cyanobacteria) would increase, because of increased nutrients, decreased grazing, decreased sinking, and increased available light or any combination of these factors (Brosnan and Cooke, 1987).

Temperature increase is also omitted from Figure 19.7. The increase from 15 to 20°C in the hypolimnetic waters as a result of complete circulation may be the most adverse effect. This is especially true where coldwater fish species are involved.

A negative effect on fish from supersaturated N_2 has been suggested (Chapter 18).

19.6 COSTS

Cost information for artificial circulation on a project basis is usually not included in published articles. Lorenzen and Fast (1977) cite an annual cost of $202,000 (2002 U.S. dollars) for two air compressors producing an air flow rate of 34.3 m^3/min (1200 ft^3/min) at standard conditions. The cost included pipes and air diffusers. At the recommended rate of 9.2 m^3/km^2 per min, this represents $540/ha for first year of operation, which is modest relative to other restoration techniques. That cost is at the lower end of the range for 13 projects in Florida; $400 to $4,700/ha and $120 to $2,265/ha for initial and annual costs, respectively (Dierberg and Williams, 1989). Median values for initial and annual costs, respectively, were $991 and $442/ha (2002 dollars).

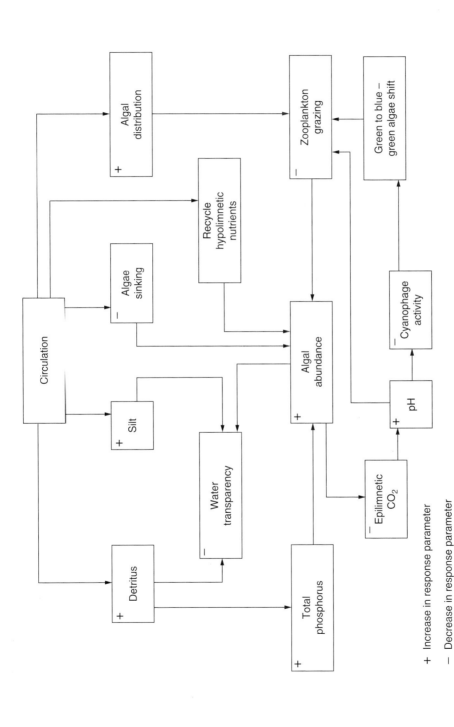

FIGURE 19.7 Potential adverse effects of artificial circulation, including the promotion of blue-green algal blooms (From Pastorak, R.A. et al. 1981. *Evaluation of Aeration/Circulation as a Lake Restoration Technique.* 600/3-81-014. USEPA; Shapiro, J. 1979. In: *Lake Restoration.* USEPA-440/5-79-001. pp. 161–167; with modification.)

For the example presented in the design section, costs for a compressor, pipe, and 1-year operation for that system at 6 m³/min were about $56,600 (2002 dollars), or about $470/ha, including installation (Davis, 1980). Another similar example installed was about $77,300 (2002 dollars) or about $640/ha more.

There is usually an economy of scale for circulation projects. Costs for 33 projects installed by General Environmental Systems during 1991–2002 averaged $588/ha ($n = 17$), $1,295/ha ($n = 4$), and $5,960/ha ($n = 12$) for water bodies of > 53 ha, 23–35 ha and < 10 ha, respectively (Geney, personal communication).

19.7 SUMMARY AND RECOMMENDATIONS

Artificial circulation has been recommended as an inexpensive management technique (Pastorak et al., 1981). The technique should be most applicable in lakes that are not nutrient-limited and where oxygen depletion is a threat to warmwater fish and the quality (metal content) of water supplies. The best record of improvement has been observed with DO, Fe and Mn content, ammonium, and pH. The principle of increasing the depth of mixing, thereby decreasing available light to plankton algae, however, may also have a good chance of working in sufficiently deep lakes where nutrients usually are not limiting. Furthermore, increased mixing may discourage cyanobacteria while encouraging diatoms and green algae. If the lake is marginally nutrient-limited and mixing is sufficient, then more algae, and even more cyanobacteria (especially if pH is raised) may result. That is probably why algal abundance and cyanobacteria have decreased following circulation in only about half of the cases cited, and increased in others.

The depth at which compressed air is released is critical to the problem of preventing the persistence of anaerobic bottom water. Likewise, the sizing of the system for flow rate, whether using compressed air or a pump, is critical to achieving complete mixing and preventing micros-tratification at the lake surface and achieving a mixing rate that discourages cyanobacteria. To avoid water quality problems, placing air diffusers at the lake bottom and using a combination of compressed air and a surface pump for very deep lakes is recommended (Pastorak et al., 1981, 1982). For best results, systems should be designed to produce an air-flow rate of about 9.2 m³/km² per min and within the range of 6.1 to 12.3 m³/km² per min. Nutrient transport to the surface lighted zone should be minimized and sedimentation maximized if circulation is begun prior to stratification or done gradually when started after stratification.

Circulation would probably be best used alone as a management technique and not combined with other methods designed to reduce P content. This is because the benefits for algal control would be best achieved in non-nutrient-limited situations, and because increased circulation may encourage internal loading of P and counteract other efforts to lower P. The exception may be diversion in which a large internal loading persists (see Gächter, 1987).

REFERENCES

Ambuhl, H. 1967. Discussion of impoundment destratification by mechanical pumping. By W.H. Irwin, J.M. Symons and G.G. Robeck. *J. San. Eng. Div. ASCE* 93: 141–143.

American Water Works Association (AWWA). 1971. Artificial Destratification in Reservoirs. *Committee Rept.* 63: 597–604.

Andersson, G., H. Berggren, G. Cronbert and C. Gelin. 1978. Effects of planktivorous and benthivorous fish on organisms and water chemistry in eutrophic lakes. *Hydrobiologia* 59: 9–15.

Anonymous. 1971. *Algae Control by Mixing.* New Hampshire Water Supply and Pollut. Cont. Commun., Concord.

Ashby, S.L., R.H. Kennedy, R.E. Price and F.B. Juhle. 1991. Water Quality Management Initiatives in East Sidney Lake. New York. Tech. Rept. E-91-3, U.S. Army Corps Engineers, Vicksburg, MS.

Barbiero, R.P., B.J. Speziale and S.L. Ashby. 1996a. Phytoplankton community succession in a lake subjected to artificial circulation. *Hydrobiologia* 331: 109–120.

Barbiero, R.P., S.L. Ashby and R.H. Kennedy. 1996b. The effects of artificial circulation on a small northeastern impoundment. *Water Res. Bull.* 32: 575–584.

Barnes, M.D. and B.L. Griswold. 1975. Effect of artificial nutrient circulation on lake productivity and fish growth. In: *Proc. Conf. on Lake Reaeration Research.* Amer. Soc. Civil Eng., Gatlinburg, TN.

Barnett, R.H. 1975. Case study of reaeration of Casitas Reservoir. In: *Proc. Conf. on Lake Reaeration Research.* Amer. Soc. Civil Eng., Gatlinburg, TN.

Bengtsson, L. and C. Gelin. 1975. Artificial aeration and suction dredging methods for controlling water quality. In: *Proc. Symposium on Effects of Storage on Water Quality.* Water Resource Center, Medmenham, England.

Bernhardt, H. 1967. Aeration of Wahnbach Reservoir without changing the temperature profile. *J. Am. Water Works Assoc.* 59: 943–964.

Biederman, W.J. and E.E. Fulton. 1971. Destratification using air. *J. Am. Water Works Assoc.* 63: 462–466.

Boström, B. 1984. Potential mobility of phosphorus in different types of lake sediments. *Int. Rev. ges. Hydrobiol.* 69: 454–474.

Bowles, B.A., I.V. Powling and F.L. Burns. 1979. Effects on Water Quality of Artificial Aeration and Destratification of Tarago Reservoir. Tech. Paper No. 46. Australian Water Resources Council, Australian Govt. Publ. Serv., Canberra.

Brezonik, P.L., J. Delfino and G.F. Lee. 1969. Chemistry of N and Mn in Cox Hollow Lake, Wisconsin, following destratification. *J. San Eng. Div. ASCE* 95: 929–940.

Brosnan, T.M. 1983. Physical, chemical and biological effects of artificial circulation on Silver Lake, Summit Co., Ohio. M.S. Thesis, Kent State University, Kent, OH.

Brosnan, T.M. and G.D. Cooke. 1987. Response of Silver Lake trophic state to artificial circulation. *Lake and Reservoir Manage.* 3: 66–75.

Brown, D.J., T.G. Brydges, W. Ellerington, J.J. Evans, M.F.P. Michalski, G.G. Hitchin, M.D. Palmer and D.D. Veal. 1971. Progress Report on the Destratification of Buchanan Lake. Ontario Water Research Commission, Aid for Lakes Program.

Brynildson, O.M. and S.I. Sterns. 1977. Effects of Destratification and Aeration of a Lake on the Distribution of Planktonic Crustacea, Yellow Perch and Trout. Tech. Bull. No. 99. Wisconsin Dept. Nat. Res., Madison.

Chen, C.W. and G.T. Orlob. 1975. Ecological simulation for aquatic environments. In: *Systems Analysis and Simulation in Ecology,* Vol. III. Academic Press, New York. pp. 475–588.

Chriswell, B. and M. Zaw. 1991. Lake destratification and speciation of iron and manganese. *Environ. Monit. Assess.* 19: 433–447.

Cooke, G.D., E.B. Welch, S.A. Peterson and P.R. Newroth. 1993. *Restoration and Management of Lakes and Reservoirs,* 2nd ed. Lewis Publishers and CRC Press, Boca Raton, FL.

Davis, J.M. 1980. Destratification of reservoirs — a design approach for perforated-pipe compressed-air systems. *Water Serv.* 84: 497–504.

Devick, W.S. 1972. Limnological Effects of Artificial Aeration in the Wahiawa Reservoir. Job Completion Rept. Proj. F-9-2. Job 2. Study IV. Honolulu, HI.

Diehl, S., S. Berger, R. Ptacnik and A. Wild. 2002. Phytoplankton, light, and nutrients in a gradient of mixing depths: field experiments. *Ecology* 2002: 399–411.

Dierberg, F.E. and V.P. Williams. 1989. Lake management techniques in Florida, USA: Costs and water quality effects. *Environ. Manage.* 13: 729–742.

Drury, D.D., D.B. Porcella and R.A. Gearheart. 1975. The Effects of Artificial Destratification on the Water Quality and Microbial Populations of Hyrum Reservoir. Utah Water Res. Lab. Proj. EW 011-1.

Duncan, A. 1990. A review: limnological management and biomanipulation in the London reservoirs. *Hydrobiologia* 200/201: 541–548.

Fast, A.W. 1968. Artificial Destratification of El Capitan Reservoir by Aeration. Part I: Effects on Chemical and Physical Parameters. Fish. Bull. No. 141. Calif. State Dept. Fish and Game.

Fast, A.W. 1971a. The Effects of Artificial Aeration on Lake Ecology. Water Pollut. Contr. Res. Ser. 16010 EXE. 12/71/USEPA.

Fast, A.W. 1971b. Effects of artificial destratification on zooplankton depth distribution. *Trans. Am. Fish. Soc.* 100: 355–358.

Fast, A.W. 1973. Effects of artificial destratification on primary production and zoobenthos of El Capitan Reservoir, California. *Water Resour. Res.* 9: 607–623.

Fast, A.W. 1979. Artificial aeration as a lake restoration technique. In: *Lake Restoration*. USEPA 440/5-79-001. pp. 121–132.

Fast, A.W., B. Moss, R.G. Wetzel. 1973. Effects of artificial aeration on the chemistry and algae of two Michigan lakes. *Water Resour. Res.* 9: 624–647.

Forsberg, B.R. and J. Shapiro. 1980. Predicting the algal response to destratification. In: *Restoration of Lakes and Inland Waters*. USEPA 440/5-81-010. pp. 134–139.

Gächter, R. 1987. Lake restoration. Why oxygenation and artificial mixing can not substitute for a decrease in the external phosphorus loading. *Schweiz. Z. Hydrol.* 49: 170–185.

Garton, J.E., R.G. Strecker and R.C. Summerfelt. 1978. Performance of an axial-flow pump for lake destratification. In: W.A. Rogers (Ed.), *Proc. 13th Annu. Conf. S.E. Assoc. Fish. and Wildlife Agencies.* pp. 336–346.

Geney, R.S. 1994. Successful diffused aeration of lakes/reservoirs using the Lorenzen-Fast 1977 sizing criteria. Presented at the North Amer. Lake Manage. Soc. Conf. Orlando, FL.

Geney, R.S. Personal communication. General Environmental Systems Inc., Oak Ridge, TN.

Gu, R. and H.G. Stephan. 1988. Mixing of temperature — stratified lakes and reservoirs by buoyant jets. *J. Environ. Eng.* 114: 898–914.

Halsey, T.G. 1968. Autumnal and overwinter limnology of three small eutrophic lakes with particular reference to experimental circulation and trout mortality. *J. Fish. Res. Board Can.* 25: 81–99.

Halsey, T.G. and D.M. Galbraith. 1971. Evaluation of Two Artificial Circulation Systems Used to Prevent Trout Winter-Kill in Small Lakes. B.C. Fish and Wildl. Br., Fish Manage. Publ. No. 16.

Hawkins, P.R. and D.J. Griffiths. 1993. Artificial destratification of a small tropical reservoir: effects upon the phytoplankton. *Hydrobiologia* 254: 169–181.

Haynes, R.C. 1973. Some ecological effects of artificial circulation on a small eutrophic lake with particular emphasis on phytoplankton. I. Kezar Lake experiment. *Hydrobiologia* 43: 463–504.

Holland, J.P. 1984. Parametric Investigation of Localized Mixing in Reservoirs. Tech. Rept. E-84-7. U.S. Army Corps Engineers, Vicksburg, MS.

Hooper, F.F., R.C. Ball and H.A. Tanner. 1953. An experiment in the artificial circulation of a small Michigan lake. *Trans. Am. Fish. Soc.* 82: 222–241.

Irwin, W.H., J.M. Symons and G.G. Robeck. 1966. Impoundment destratification by mechanical pumping. *J. San. Eng. Div. ASCE* 92: 21–40.

Jacoby, J.M., D.D. Lynch, E.B. Welch and M.A. Perkins. 1982. Internal phosphorus loading in a shallow eutrophic lake. *Water Res.* 16: 911–919.

Jensen, H.S., P. Kristensen, E. Jeppsen and A. Skytte. 1992. Iron: phosphorus ratio in surface sediment as an indicator of phosphate release from aerobic sediments in shallow lakes. *Hydrobiologia* 235/236: 731–743.

Jones, C.A. and E.B. Welch. 1990. Internal phosphorus loading related to mixing and dilution in a dendritic, shallow prairie lake. *J. Water Pollut. Control Fed.* 62: 847–852.

Kamp-Nielsen, L. 1974. Mud-water exchange of phosphate and other ions in undisturbed sediment cores and factors affecting the exchange rates. *Arch. Hydrobiol.* 73: 218–237.

Kamp-Nielsen, L. 1975. Seasonal variation in sediment-water exchange of nutrient ions in Lake Esrom. *Verh. Int. Verein. Limnol.* 19: 1057–1065.

Karlgren, L. and O. Lindren. 1963. Luftningastudier; Traksjön. *Sartryck ur Vattenygien* 3: 67–69.

Klemer, A.R., J. Feuillade and M. Feuillade. 1982. Cyano-bacterial blooms: Carbon and nitrogen limitation have opposite effects on the buoyancy of *Oscillatoria*. *Science* 215: 1629–1631.

King, D.L. 1970. The role of carbon in eutrophication. *J. Water Pollut. Control Fed.* 42: 2035–2051.

King, D.L. 1972. Carbon limitation in sewage lagoons. In: *Nutrients and Eutrophication. Special Symposium*, Vol. 1. Am. Soc. Limnol. and Oceanogr., Michigan State University, W.K. Kellogg Biol. Sta., East Lansing. pp. 98–110.

Kitchell, J.A. and J.F. Kitchell. 1980. Size-selective predation, light transmission, and oxygen stratification. Evidence from the recent sediments of manipulated lakes. *Limnol. Oceanogr.* 25: 389–402.

Knoechel, R. and J. Kalff. 1975. Algal sedimentation: The cause of a diatom-blue-green succession. *Verh. Int. Verein. Limnol.* 19: 745–754.

Knoppert, P.L., J.J. Rook, T. Hofker and G. Oskan. 1970. Destratification experiments at Rotterdam. *J. Am. Water Works Assoc.* 62: 448–454.

Kothandaraman, V., D. Roseboom and R.L. Evans. 1979. Pilot Lake Restoration Investigations: Aeration and Destratification in Lake Catherine, IL. Illinois State Water Survey, Springfield.

Lackey, R.T. 1972. Response of physical and chemical parameters to eliminating thermal stratification in a reservoir. *Water Res. Bull.* 8: 589–599.

Laing, L.L. 1990. The effects of artificial circulation on a hypereutrophic lake: Discussion. *Water Res. Bull.* 28: 409–412.

Laverty, G.L. and H.L. Nielsen. 1970. Quality improvements by reservoir aeration. *J. Am. Water Works Assoc.* 62: 711–714.

Leach, L.E., W.R. Duffer and C.C. Harlin, Jr. 1980. Induced Hypolimnion Aerations for Water Quality Improvement of Power Releases. Water Pollut. Cont. Res. Ser. 16080. USEPA.

Lee, G.F. 1970. *Factors Affecting the Transfer of Materials between Water and Sediments.* Water Res. Center, University of Wisconsin, Madison.

Lehman, J.T. and C.D. Sandgren. 1978. Documenting a seasonal change from phosphorus to nitrogen limitation in a small temperate lake, and its impact on the population dynamics of *Asterionella. Verh. Int. Verein. Limnol.* 20: 375–380.

Lorenzen, M.W. and A.W. Fast. 1977. *A Guide to Aeration/Circulation Techniques for Lake Management.* Ecol. Res. Ser. USEPA-600/3-77-004.

Lorenzen, M.W. and R. Mitchell. 1975. An evaluation of artificial destratification for control of algal blooms. *J. Am. Water Works Assoc.* 67: 373–376.

Lossow, K., A. Sikorowa, H. Drozd, A. Wuckowa, H. Nejranowska, M. Sobierajska, J. Widuto and I. Zmyslowska. 1975. Results of research on the influence of aeration on the physico-chemical systems and biological complexes in the Starodworski Lake obtained hitherto. *Pol. Arch. Hydrobiol.* 22: 195–216.

Malueg, K.W., J.R. Tilstra, D.W. Schultz and C.F. Powers. 1973. Effect of induced aeration upon stratification and eutrophication processes in an Oregon farm pond. *Geophys. Monogr. Ser.* 17: 578–587.

McCord, S.A., S.G. Schladow and T.G. Miller. 2000. Modeling artificial aeration kinetics in ice-covered lakes. *J. Environ. Eng.* 126: 1–11.

McCullough, J.R. 1974. Aeration revitalized reservoir. *Water Sewage Works* 121: 84–85.

McNally, W.J. 1971. *Destratification of Lakes.* Federal Aid to Fisheries, Proj. Comp. Rept., State of New Mexico, F-22-R-11, J of C-8.

McQueen, D.J. and J.R. Post. 1988. Limnocorral studies of cascading trophic interactions. *Verh. Int. Verein. Limnol.* 23: 739–747.

Meyer, E.B. 1991. Pneumatic Destratification System Design using a Spreadsheet Program. Water Operations Tech. Support E-91-1. U.S. Army Corps Engineers, Vicksburg, MS.

Meyer, E.B., R.E. Price, S.C. Wilhelms. 1992. Destratification System Design for East Sidney Lake, New York. Misc. Paper. W-92-2. US Army Corps of Engineers, Vicksburg, MS.

Michele, J. and V. Michele. 2002. The free jet as a means to improve water quality: Destratification and oxygen enrichment. *Limnologica* 32: 329–337.

Miller, T.G. and W.C. Mackay. 2003. Optimizing artificial aeration for lake winterkill prevention. *Lake and Reservoir Manage.* 19: 355–363.

Miller, T.G., W.C. Mackay and D.T. Walty. 2001. Under ice water movements induced by mechanical surface aeration and air injection. *Lake and Reservoir Manage.* 17: 263–287.

Mortimer, C.H. 1941. The exchange of dissolved substances between mud and water in lakes. Parts 1 and 2. *J. Ecol.* 29: 280–329.

Mortimer, C.H. 1971. Chemical exchanges between sediments and water in the Great Lakes - speculations on probable regulatory mechanisms. *Limnol. Oceanogr.* 16: 387–404.

Murphy, G.I. 1962. Effects of mixing depth and turbidity on the productivity of fresh water impoundments. *Trans. Am. Fish. Soc.* 91: 69–76.

NHWSPCC. 1979. Effects of Destratification upon Temperature and Other Habitat Requirements of Salmonid Fishes 1970–1976. Staff Rept. No. 100, New Hampshire Water Supply and Pollution Control Comm., Concord.

Neale, P.J., S.I. Heaney and G.H.M. Jaworski. 1991. Long time series from the English Lake District: Irradiance-dependent phytoplankton dynamics during the spring maximum. *Limnol. Oceanogr.* 36: 751–760.

Nicholls, K.H., W. Kennedy and C. Hammett. 1980. A fish-kill in Heart Lake, Ontario, associated with the collapse of a massive population of *Ceratium hirundinella* (Dinophyceae). *Freshwater Biol.* 10: 553–561.

Nõges, T. and P. Nõges. 1999. The effect of extreme water level decrease on hydrochemistry and phytoplankton in a shallow eutrophic lake. *Hydrobiologia* 408/409: 277–283.

Nõges, T., P. Nõges and R. Laugaste. 2003. Water level as the mediator between climate change and phytoplankton composition in a large shallow temperate lake. *Hydrobiologia* 506–509: 257–263.

Nürnberg, G.K. and B.D. LaZerte. 2003. An artificially induced *Planktothrix rubescens* surface bloom in a small kettle lake in southern Ontario compared to blooms worldwide. *Lake and Reservoir Manage.* 19: 307–322.

Osgood, R.A. and J.E. Stiegler. 1990. The effects of artificial circulation on a hypereutrophic lake. *Water Res. Bull.* 26: 209–217.

Osgood, R.A. and J.E. Stiegler. 1992. The effects of artificial circulation on a hypereutrophic lake: Reply to discussion by R.L. Laing. *Water Res. Bull.* 28: 413–415.

Oskam, G. 1973. A kinetic model of phytoplankton growth and its use in algal control by reservoir mixing. *Geophys. Monogr. Ser.* 17: 629–631.

Oskam, G. 1978. Light and zooplankton as algae regulating factors in eutrophic Biesbosch reservoirs. *Verh. Int. Verein. Limnol.* 20: 1612–1618.

Pastorak, R.A., T.C. Ginn and M.W. Lorenzen. 1981. Evaluation of Aeration/Circulation as a Lake Restoration Technique. USEPA-600/3-81-014.

Pastorak, R.A., M.W. Lorenzen and T.C. Ginn. 1982. Environmental Aspects of Artificial Aeration and Oxygenation of Reservoirs: A Review of Theory, Techniques, and Experiences. Technical Rept. No. E-82-3. U.S. Army Corps of Engineers.

Pastorak, R.A. Personal communication. Tetra Tech. Inc.

Price, R.E. 1988. Applications of Mechanical Pumps and Mixers to Improve Water Quality. Water Operations Tech. Support E-88-2, U.S. Army Corps Engineers, Vicksburg, MS.

Price, R.E. 1989. Evaluating Commercially Available Destratification Devices. Water Operations Tech. Support E-89-2, U.S. Army Engineers Waterways Exp. Sta., Vicksburg, MS.

Punnett, R.E. 1991. Design and Operation of Axial Flow Pumps for Reservoir Destratification. Instruct. Rept. W-91-1. U.S. Army Corps of Engineers. Vicksburg, MS.

Raynes, J.J. 1975. Case study — Altoona Reservoir. In: *Symp. on Reaeration Research*, Am. Soc. Civil Eng., Gatlinburg, TN.

Reynolds, C.S. 1975. Interrelations of photosynthetic behavior and buoyancy regulation in a natural population of a blue-green alga. *Freshwater Biol.* 5: 323–338.

Reynolds, C.S., S.W. Wiseman and M.J.O. Clarke. 1984. Growth and loss rate responses of phytoplankton to intermittent artificial mixing and their potential application to the control of planktonic algal biomass. *J. Appl. Ecol.* 21: 11–39.

Reynolds, C.S., R.L. Oliver and A.E. Walsby. 1987. Cyanobacterial dominance: The role of buoyancy regulation in dynamic lake environments. *N.Z.J. Mar. Fresh Water Res.* 21: 379–390.

Riddick, T.M. 1957. Forced circulation of reservoir waters yields multiple benefits at Ossining, New York. *Water Sewage Works* 104: 231–237.

Ridley, J.E. 1970. The biology and management of eutrophic reservoirs. *Water Treat. Exam.* 19: 374–399.

Ridley, J.E., P. Cooley and J.A.P. Steel. 1966. Control of thermal stratification in Thames Valley reservoirs. *Proc. Soc. Water Treat. Exam.* 15: 225–244.

Robertson, P.G. et al. 1988. Effect of Artificial Destratification on the Water Quality of an Impoundment. Maryland Dept. Environment, Baltimore, MD and Washington Suburban Sanitary Comm., Laurel, MD.

Robinson, E.L., W.H. Irwin and J.M. Symonds. 1969. Influence of artificial destratification on plankton populations in impoundments. *Trans. Ky. Acad. Sci.* 30: 1–18.

Ryabov, A.K., B.I. Nabivanets, Zh.M. Argamova, Ye.M. Palamarchuk and I.S. Kozlova. 1972. Effect of artificial aeration on water quality. *J. Hydrobiol.* 8: 49–52.

Sahoo, G.B. and D. Luketina. 2003. Bubbler design for reservoir destratification. *Mar. Fresh Water Res.* 54: 271–285.

Shapiro, J. 1973. Blue-green algae: why they become abundant. *Science* 197: 382–384.

Shapiro, J. 1979. The need for more biology in lake restoration. In: Lake Restoration. USEPA-440/5-79-001. pp. 161–167.

Shapiro, J. 1984. Blue green dominance in lakes: The role and management significance of pH and CO_2. *Int. Rev. ges. Hydrobiol.* 69: 765–780.

Shapiro, J. 1990. Current beliefs regarding dominance by blue greens: The cases for the importance of CO_2 and pH. *Verh. Int. Verein. Limnol.* 24: 38–54.

Shapiro, J. 1997. The role of carbon dioxide in the initiation and maintenance of blue-green dominance in lakes. *Freshwater Biol.* 37: 307–323.

Shapiro, J. and H.O. Pfannkuch. 1973. The Minneapolis Chain of Lakes: A Study of Urban Drainage and Its Effects. Int. Rept. No. 9 Limnol. Res. Center, University of Minnesota, Minneapolis.

Shapiro, J., V. Lamarra and M. Lynch. 1975. Biomanipulation: An ecosystem approach to lake restoration. In: P.L. Brezonik and J.L. Fox (Eds). *Proc. Symp. Water Quality Management through Biological Control.* University of Florida, Gainesville and USEPA. pp. 85–95.

Shapiro, J., B. Forsberg, V. Lamarra, G. Lindmark, M. Lynch, E. Smeltzer and G. Zoto. 1982. *Experiments and Experiences in Biomanipulation — Studies of Biological Ways to Reduce Algal Abundance and Eliminate Blue Greens.* USEPA-600/3-82–096.

Sheffer, M. 1998. *Ecology of Shallow Lakes.* Chapman and Hall, London.

Sirenko, L.A., N.V. Avil'tseva and V.M. Chernousova. 1972. Effect of artificial aeration on pond water on the algal flora. *J. Hydrobiol.* 8: 52–58.

Smith, S.A., D.R. Knauer and T.L. Wirth. 1975. Aeration as a Lake Management Technique. Tech. Bull. No. 87. Wisconsin Dept. Nat. Res., Madison.

Søndergaard, M. J.P. Jensen and E. Jeppesen. 1999. Internal phosphorus loading in shallow Danish lakes. *Hydrobiologia* 408/409: 145–152

Spencer, C.N. and D.L. King. 1987. Regulation of blue-green algal buoyancy and bloom formation by light, inorganic nitrogen, CO_2, and trophic interactions. *Hydrobiologia* 144: 183–192.

Stefan, H.G. and R. Gu. 1991. Conceptual design procedure for hydraulic destratification systems in small ponds, lakes, or reservoirs for water quality improvement. *Water Res. Bull.* 27: 967–978.

Stichen, J.M., J.E. Garton and C.E. Rice. 1974. The effect of lake destratification on water quality. *J. Am. Water Works Assoc.* 71: 219–225.

Steinberg, C. and R. Tille-Backhaus. 1990. Re-occurrence of filamentous planktonic cyanobacteria during permanent artificial destratification. *J. Plankton Res.* 12: 661–664.

Steinberg, C. and G.M. Zimmerman. 1988. Intermittent destratification: a therapy measure against cyanobacteria in lakes. *Environ. Technol. Lett.* 9: 337–350.

Stumm, W. and J.O. Leckie. 1971. Phosphate exchange with sediments: its role in the productivity of surface waters. In: *Proc. 5th Int. Conf. Water Pollut. Res.*, London. III-26/1–16.

Sverdrup, H.U. 1953. On conditions for the vernal blooming of phytoplankton. *J. Cons. Int. Explor. Mer.* 18: 287–295.

Symons, J.M., W.H. Irwin, E.L. Robinson and G.G. Robeck. 1967. Impoundment destratification for raw water quality control using either mechanical- or diffused-air pumping. *J. Am. Water Works Assoc.* 59: 1268–1291.

Symons, J.M., J.K. Carswell and G.G. Robeck. 1970. Mixing of water supply reservoirs for quality control. *J. Am. Water Works Assoc.* 62: 322–334.

Talling, J.F. 1971. The underwater light climate as a controlling factor in the production ecology of fresh water phytoplankton. *Mitt. Int. Verein. Limnol.* 19: 214–243.

Taylor, E.W. 1966. Forty-Second Report on the Results of the Bacteriological Examinations of the London Waters for the Years 1965–66. Metropolitan Water Board, New River Head, London.

Thomas, E.A. 1966. Der Pfaffikersee vor, wahrand, und nach kunstlicher durchmischung. *Verh. Int. Verein. Limnol.* 16: 144–152.

Toetz, D.W. 1977a. Biological and Water Quality Effects of Whole Lake Mixing, Tech. Rept. A-068-OKLA. Water Resour. Res. Inst.

Toetz, D.W. 1977b. Effects of lake mixing with an axial flow pump on water chemistry and phytoplankton. *Hydrobiologia* 55: 129–138.

Toetz, D.W. 1979. Biological and water quality effects of artificial mixing of Arbuckle Lake, Oklahoma, during 1977. *Hydrobiologia* 55: 129–138.

U.S. Army Corps of Engineers. 1973. Alatoona Lake, Destratification Equipment Test Rept. U.S. Army Corps Engineers, Savannah, Georgia.

U.S. Environmental Protection Agency (USEPA). 1970. *Induced Aeration of Small Mountain Lakes*. Water Pollut. Cont. Res. Ser., 16080-11/70.

Vandermeulen, H. 1992. Design and testing of a propeller aerator for reservoirs. *Water Res.* 26: 857–861.

Van der Veer, B., J. Koedood and P.M. Visser. 1995. Artificial mixing: a therapy measure combating cyano-bacteria in Lake Nieuwe Meer. *Water Sci. Technol.* 31: 245–248.

Visser, P.M., H.A.M. Ketelaare, L.W.C.A. van Breemen and L.R. Mur. 1996a. Diurnal buoyancy changes of *Microcystis* in an artificially mixed storage reservoir. *Hydrobiologia* 331: 131–141.

Visser, P.M., B.W. Ibelings, B. van der Veer, J. Koedood and L.R. Mur. 1996b. Artificial mixing prevents nuisance blooms of the cyanobacterium *Microcystis* in Lake Nieuwe Meer, the Netherlands. *Freshwater Biol.* 36: 435–450.

Walsby, A.E. and C.S. Reynolds. 1975. Water blooms. *Biol. Rev.* 50: 437–481.

Weiss, C.M. and B.W. Breedlove. 1973. Water Quality Changes in an Impoundment as a Consequence of Artificial Destratification. Rept. No. 80. North Carolina Water Resour. Res. Inst., Chapel Hill.

Weithoff, G., A. Lorke and N. Walz. 2000. Effects of water-column mixing on bacteria, phytoplankton, and rotifers under different levels of herbivory in a shallow eutrophic lake. *Oecologia* 125: 91–100.

Welch, E.B. and J.M. Jacoby. 2004. *Pollutants Effects in Fresh Water: Applied Limnology*, 3rd ed. Spon Press, London/New York.

Wirth, T.L. and R.C. Dunst. 1967. Limnological Changes Resulting from Artificial Destratification and Aeration of an Impoundment. Fish. Res. Rept. No. 22. Wisconsin Conserv. Dept., Madison, WI.

Wirth, T.L., R.C. Dunst, P.D. Uttormark and W. Hilsenhoff. 1970. Manipulation of Reservoir Waters for Improved Quality and Fish Population Response. Fish. Res. Rep. No. 22. Wisconsin Conserv. Dept., Madison.

Zaret, T.M. and J.S. Suffern. 1976. Vertical migration in zooplankton as a predator avoidance mechanism. *Limnol. Oceanogr.* 21: 804–813.

Zic, K. and H.G. Stefan. 1994. Destratification Induced by Bubble Plumes. Tech. Rept. W-94-3. U.S. Army Corps of Engineers, Vicksburg, MS.

20 Sediment Removal

20.1 INTRODUCTION

Dredging, due to some poor past practices, has received a bad reputation. However, properly conducted, sediment removal is an effective, but expensive, lake management technique. New to this chapter is an extensive case history concerning contaminated sediment removal and the realization that formerly named "special purpose" dredges are becoming more common to lake restoration, at least in Europe. This chapter describes objectives, environmental concerns, dredging depths, removal techniques, lake conditions, dredge selection, disposal area designs, some case histories, and costs associated with sediment removal (adjusted for inflation to June 2002). Sediment removal, while common, is very limited in documentation concerning the success or failure of most projects. Thus, material in this chapter is not exhaustive, but rather representative of various lake sediment removal procedures.

20.2 OBJECTIVES OF SEDIMENT REMOVAL

20.2.1 DEEPENING

When recreational activities are impaired due to shoaling, the only practical means of restoration is lake deepening through sediment removal. According to the United States Department of Agriculture (USDA, 1971), lakes must have a water volume sufficient to exceed water loss by seepage and evaporation, and sufficient depth to prevent complete freezing. In the latter case that means a depth anywhere from 1.5 to 4.5 m, depending on the region of the country. A depth of at least 4.5 m is usually required to avoid winterkill of fish in colder parts of the U.S. (Toubier and Westmacott, 1976). These and other factors, such as intended lake use, availability of a suitable dredged material disposal area, and available funds, must be considered when designing and implementing any lake-deepening project. The reasons for deepening and the means of measuring the success of such a project are the most direct aspects of the sediment removal objectives. Modern dredging equipment efficiently moves large volumes of sediment. Therefore, nearly all dredging projects are considered successful at the time of their completion (Pierce, 1970). However, more recent information from Wisconsin shows that lake deepening can be reversed by sedimentation in 10 years or less (Wisconsin Department of Natural Resources, 1990). Specific examples include the millponds of Bugle Lake and Lake Henry. Therefore, sedimentation rates must be determined before dredging is recommended.

Success in terms of deepening is not the only criterion for determining success of a dredging project. Deepening might be accomplished while the overall condition of the lake is actually worsened due to poor dredging techniques (Gibbons and Funk, 1983). Therefore, dredging procedure is a critical aspect of the dredging project.

20.2.2 NUTRIENT CONTROL

Many shallow, eutrophic lakes do not stratify thermally (polymictic or amictic) making them susceptible to continual or periodic nutrient inputs from the sediment. Deeper stratified lakes might become destratified when a passing summer, cold weather front depresses the thermocline pushing nutrient rich water into the photic zone of the epilimnion (Stauffer and Lee, 1973). Power boat wakes and bottom fish also are problematic in shallow lakes. Thus, obnoxious algal blooms occur most frequently during peak summer recreation periods.

Sediment-regenerated P amounted to approximately 45% of the P loading to Linsley Pond, CT (Livingston and Boykin, 1962). Welch et al. (1979) estimated P inputs to Long Lake, Washington were 200 to 400 kg/yr, or about 25% to 50% of the external loading. Shagawa Lake, MN, experienced summer sediment P pulses of approximately 2000 to 3000 kg during June, July, and August. This compares to an annual P loading from the City of Ely, MN, of 5000 to 5500 kg before advanced waste treatment (AWT) and about 1000 to 1500 kg after AWT (Larsen et al., 1981). Before AWT, sediment P loading to Shagawa Lake was about 28% to 35% of the total loading. The sediment portion of the TP loading to the lake increased to 66% following AWT, even though the total loading decreased considerably after AWT (Peterson, 1981). Sediment-recycled P in Shagawa Lake has been sufficient to produce large summer algal blooms, thus slowing the lake's predicted rate of recovery (Larsen et al., 1981; Chapter 4).

In cases where a significant nutrient loading from sediment can be documented, sediment removal might be expected to reduce the rate of internal nutrient recycling, thus improving overall lake and water quality conditions. However, while dredging rich surface sediments will reduce internal nutrient recycling, this effect might be temporary if external sources are shut off. Kleeberg and Kohl (1999) demonstrated that trophic state in Lake Muggelsee, Germany is controlled more by photic zone production and its associated sedimentation than by nutrient release from the sediment if surface inputs of P are not cut off. Additionally, Sondergaard et al. (1996) found that surface sediment TP in Danish lakes was highly correlated to the external P loading, but only weakly related to other sediment parameters. This strongly reinforces the idea that P input reduction is the first line of defense in lake management and restoration.

Consideration of nutrient inactivation is another option for shallow lakes that might not need deepening per se. It is easier, less expensive, and likely to be more successful in terms of nutrient control per se (Welch and Cooke, 1995).

20.2.3 Toxic Substances Removal

Toxic substances are a common concern among industrialized nations. Large-scale surveys and improved analytical techniques demonstrate that toxicants are more common to fresh water sediments than previously suspected (Bremer, 1979; Horn and Hetling, 1978; Matsubara, 1979). Many toxicants are recycled from the sediment to the overlying water, where they bioaccumulate in aquatic organisms. Perhaps the most infamous incident of this type (marine water) was mercury pollution of Minimata Bay, Japan, first discovered in 1956 (Fujiki and Tajima, 1973). Other incidents, in the U.S., have involved kepone contamination of the James River, VA (Mackenthun et al., 1979), and PCB contamination of Waukegan Harbor in Lake Michigan (Bremer, 1979). Few occurrences of toxic problems like the one for mercury at Gibraltar Lake, CA, were reported in the past (Spencer Engineering, 1981). However, that has changed in recent times as PCBs and heavy metals, particularly mercury, have been recognized as a more prevalent fish tissue bioaccumulation problem (Gullbring et al., 1998; Peterson et al., 2002).

The most obvious solution to contaminated sediment is removal, but contaminated sediment removal frequently is complicated by pollution of the overlying water column, through sediment agitation. Most conventional dredges can cause massive resuspension of fine sediment (Suda, 1979; Barnard, 1978). Sediment resuspension while dredging toxic substances must be minimized to prevent secondary environmental damage. Proper selection and design of dredging equipment becomes more important when removing toxic sediment (see the Lake Järnsjön case history in this chapter).

20.2.4 Rooted Macrophyte Control

Some rooted aquatic plants in a lake are desirable since they provide habitat for young fish and reduce beach erosion. However, an overabundance of plants may interfere with fishing, boating, and swimming and may be aesthetically displeasing. Respiration by large plant masses in the littoral zone during hours of darkness might significantly reduce dissolved oxygen concentrations. In addition, there is increasing literature concerning the effects of macrophytes on internal nutrient cycling. Their role in this process may be an important reason for attempting to control macrophytes by selectively removing them from a lake. Wetzel (1983) indicated that most of the organic matter found in small lakes may be derived from their littoral zones.

Fresh Water aquatic plants extract nutrients chiefly from the sediment (Schults and Malueg, 1971; Twilley et al., 1977; Carignan and Kalff, 1980), but they do not excrete large quantities of nutrients to the surrounding water while in the active growth phase (Barko and Smart, 1980). They do tend, however, to concentrate sediment-supplied nutrients in their tissues. These nutrients are recycled to the lake when plants fruit and during the senescence, death, and decay stages (Barko and Smart, 1979; Lie, 1979; Welch et al., 1979) (see also Chapter 11). Barko and Smart (1979) estimated that in-lake mobilization of P by *Myriophyllum* in Lake Wingra, WI, might amount to 62% of the annual external P loading. Welch et al. (1979) indicated that much of the "sediment" P loading in Long Lake, Washington probably was due to rapid plant die-off and decay. Current information indicates that any long range lake restoration project concerned with in-lake nutrient controls needs to focus on both macrophytes and sediment (Barko and Smart, 1980; Carignan and Kalff, 1980).

20.3 ENVIRONMENTAL CONCERNS

20.3.1 In-Lake Concerns

Sediment resuspension during dredging is the primary in-lake concern (Herbich and Brahme, 1983). One of the most common problems is nutrient liberation. Phosphorus is of particular concern because of its high concentration in sediment interstitial waters of eutrophic lakes. Dredge agitation and wind action move nutrient-laden sediment into the euphotic zone of the lake, creating the potential for algal blooms. Churchill et al. (1975) reported increased P concentration in Lake Herman, SD, coincident with cutterhead hydraulic dredging, but no increased algal production was noted. This lack of algal increase presumably was due to the high turbidity level. Dunst (1980), on the other hand, found increased algal production in Lilly Lake, WI, when hydraulic dredging began, but it was short lived and never posed a nuisance. While nutrient enrichment due to dredging can become a problem, in most cases the effects are short term and negligible relative to the long-term benefits.

Another, and potentially greater, concern associated with resuspended sediments is the liberation of toxic substances. Small-lake toxic sediment removal projects are relatively uncommon, but a few have been undertaken (Bremer, 1979; Matsubara, 1979; Sakakibara and Hayashi, 1979; Spencer Engineering, 1981). Fine particles pose the major concern. Murakami and Takeishi (1977) showed that up to 99.7% of the polychlorinated biphenyls (PCBs) associated with marine sediments are attached to particles less than 74 µm in diameter. This could pose a particular problem for fresh water dredging projects, where particle-settling times are significantly greater than for marine waters. Therefore, added precautions need to be taken when dredging contaminated sediments. Such precautions might include special dredges (see Sediment Removal Techniques section of this chapter and case histories) and special disposal and treatment techniques (Barnard and Hand, 1978; Matsubara, 1979).

A common dredging concern among fisheries managers is the destruction of benthic fish-food organisms. If the lake basin is dredged completely, 2 to 3 years may be required to reestablish the benthic fauna (Carline and Brynildson, 1977). However, if portions of the bottom are left undredged, reestablishment can vary from almost immediate (Andersson et al., 1975; Collett et al., 1981) to 1 to 2 years (Crumpton and Wilbur, 1974). Lewis et al. (2001) concluded that small scale dredging impacts on benthos in shallow water bayous were "counteracted" by beneficial effects to other biota due to the removal of sediments and the increase in depth and circulation. In any case, the effect on benthic communities appears to be short lived and generally acceptable relative to the longer term benefits derived. However, partial dredging fisheries benefits must be weighed against the increased potential for nutrient liberation from poorly executed partial dredging projects (Gibbons and Funk, 1993).

These concerns are associated primarily with dredging as a sediment removal technique. Another technique for sediment removal involves lake drawdown (lowering the water level) to expose the littoral sediments, or in some cases (Born et al., 1973) the entire lake basin, followed by removal of sediment with earth moving equipment after it has dried sufficiently. Drawdown accompanied by bulldozer operation is more destructive of the benthic community than dredging. It may also pose additional nuisance problems such as noise, dust, and truck traffic. The section on sediment removal techniques addresses dredging techniques that minimize many of these concerns.

20.3.2 Disposal Area Concerns

The major non-lake impact of sediment removal concerns the area chosen for dredged materials disposal. The problem of finding disposal sites in urban areas has become more acute in the U.S. with the promulgation of Section 404 of Public Law 92–500 (The Clean Water Act); this law prohibits the dredging or filling of any wetland area exceeding 4.0 ha (10 acres) without a federal permit. However, Section 404 of the Law was challenged and reversed by a Supreme Court ruling in 2000 that said in effect only those wetlands contiguous with navigable waters are protected by fill permitting. This makes many small wetlands vulnerable to draining, filling and wanton destruction.

Flooding of wooded areas with dredged material should be avoided. Flooding kills trees, providing unsightly evidence of improper disposal. Disposal areas may become attractive nuisances in the legal sense and can be extremely dangerous. They tend to form thin dry crusts that, like thin ice, break easily when subjected to the weight of a person or vehicle. Even dewatered and apparently dried disposal areas can be deceiving. Those with strong surface crusting, deep cracking, and vegetation can swallow earth-moving equipment if excavation is attempted too early. Disposal areas covered to depths greater than 1 m should be tested thoroughly to determine their ability to support heavy equipment before any rework on the disposal areas is attempted. It is advisable to fence and post disposal areas for safety.

A disposal method used frequently in recent years employs diking in upland areas. A common problem with these sites is dike failure accompanied by flooding of adjacent areas (Calhoun, 1978). Groundwater contamination near upland disposal sites has been identified as a potential problem, however, there are no documented contamination cases involving lake sediment disposal even where monitoring was extensive (Dunst et al., 1984). Upland disposal areas are commonly used for a variety of purposes once they are closed and dewatered.

Another lake dredging problem is under-design of the disposal area capacity. Unfortunately, these failings usually become apparent only after the project is fully operational. The problem may be caused by the slow settling rate of suspended sediment in fresh water (Wechler and Cogley, 1977) and reduced ponding depth as the project proceeds. This may result in failure to meet the requirements of suspended solids discharge permits. If that happens there are two choices: shut down until seepage and evaporation allows additional filling, or treat the discharge water. Either alternative adds additional cost to the project. However, increasingly stringent requirements for dredged material return flow waters require innovative settling techniques. A dredging project at Lake Tahoe, CA required that dredge water return flows to the lake be no more than 5 Nephelometric Turbidity Units (NTU), a standard that could not be met by any known technology (Macpherson et. al., 2003). A compromise was reached that allowed discharge at no more than 20 NTU into an adjacent dry marsh. However, even this standard could not be met and the use of polyacrylamides, polymines, aluminum, and iron-based coagulants were discouraged because of potential environmental problems. Therefore, a low toxicity, non-contaminant, biodegradable coagulant (chitosan) was tested and used. This product is derived from shellfish shells and marketed under the name of Gel-Floc®. Gel-Floc placed in the 2,000 gpm recirculation flow consistently reduced dredge water turbidity from 1,000 NTU to an average of 17 NTU. Conductivity, pH, and temperature of the treated water remained unaffected.

Disposal areas must be designed for end-of-project efficiency, not average discharge requirements over the entire use period. Palermo et al. (1978) along with a later section of this chapter summarize important technical information that assists with the proper design, construction, and maintenance of disposal areas for dredged material. Barnard and Hand (1978) describe when and how to treat disposal area discharges if standards cannot be met. Brannon (1978), Chen et al. (1978), Gambrell et al. (1978), and Lunz et al. (1978) provide valuable information that help minimize environmental problems at disposal sites.

20.4 SEDIMENT REMOVAL DEPTH

When restoring a lake for sailing, power boating, and associated activities, the deepening requirements are relatively straightforward. When deepening to control internal nutrient cycling and macrophyte growth, the criteria are less clearly defined.

Lake Trummen, Sweden, is perhaps the most thoroughly documented case of sediment removal to control internal nutrient cycling and macrophyte encroachment. Sediment removal depth in Lake Trummen was determined by mapping both the horizontal and the vertical distribution of nutrients in the sediment. Digerfeldt (1972), as cited by Björk (1972), determined that approximately 40 cm of fine surface sediment accumulated from 1940 to 1965. Aerobic and anaerobic release rates of $PO_4 - P$ and $NH_4^+ - N$ from sediment surface layers were markedly greater than for the underlying sediment (see the Lake Trummen case study in this chapter). Based on these differences, a plan was developed to remove the upper 40 cm of sediment.

Another approach to determine sediment removal depth was proposed by Stefan and Hanson (1979) and by Stefan and Ford (1975). This approach is similar to that developed by Stauffer and Lee (1973), which described thermocline erosion by wind in northern temperate lakes. Stefan and Hanson (1979) used their model to predict the depth to which Hall Lake, MN, must be dredged to control adverse nutrient exchange from the sediment during the summer. In other words, to determine what depth was necessary to establish permanent summer thermal stratification (dimictic condition).

The Stefan and Hanson (1979) model assumes stable summer stratification is necessary to prevent enriched hypolimnetic waters from mixing into the epilimnion. Based on that assumption, they calculated that Hall Lake (one of the Fairmont, MN, lakes) would require dredging to a maximum depth of 8.0 m to change it from a polymictic to a dimictic lake. Dredging volume to obtain the 8.0 m depth would be enormous, given Hall Lake's 2.25 km² surface area and 2.1 m mean depth.

There was little apparent chemical or physical distinction between shallow and deep sediments in Hall Lake. Phosphorus concentration was relatively uniform from the sediment surface to a depth of 8.5 m (737 to 1412 mg/kg for 37 samples, with a mean of 1,097 mg/kg). It is possible, however, that the P release rates from deeper sediment could be less than those of surface sediments (they were not measured). Nutrient release from the deeper sediment could be slow enough to significantly reduce the adverse impact of nutrients on the overlying water, even though stratification might not be permanent (Bengtsson et al., 1975). If that is the case, surface sediment skimming might produce nearly the same result as deep dredging, and at a considerable saving. Therefore, it would be advisable to conduct incremental nutrient release rate experiments prior to adopting a lake temperature modeling approach to determine dredging depth for nutrient control.

Dredging will remove rooted macrophytes from the littoral zone of lakes, but there have been few detailed studies to determine the depths necessary to prevent regrowth of nuisance plants. Factors influencing the areas in which rooted macrophytes grow include temperature, sediment texture, nutrient content, slope, and light level (see Chapter 11).

Using field data developed by Belonger (1969) and Modlin (1970), the Wisconsin Department of Natural Resources developed a guide to prescribe dredging depths necessary to control the regrowth of macrophytes. The guide was developed by regression of the maximum depth of plant growth in several Wisconsin lakes against the average summer Secchi disc transparency of the lakes. The relationship is described by the equation

$$Y = 0.83 + 1.22 \, X \tag{20.1}$$

where Y = maximum plant growth depth (m) and X = average summer water transparency (m).

Wisconsin lakes with a mean Secchi disc transparency of 1.5 m have few macrophytes growing beyond a depth of 2.7 m. According to Dunst (1980), this relationship was used in Wisconsin as a rough guide to develop dredging plans for macrophyte control. Dunst indicated, however, that dredging depths do not always need to exceed the predicted Y value to achieve control since slight deepening frequently changes plant speciation to less objectionable forms (see Lilly Lake, WI, case study in this chapter and Chapter 11, Table 11.3, for other regression equations for different geographic areas).

Work by Collett et al. (1981) attempted to establish the depth of dredging necessary to prevent plant regrowth in the usually turbid Tuggarah Lakes of New South Wales. They bracketed the light compensation depth by dredging three 30 m² test plots 1.0 m, 1.4 m, and 1.8 m deep in a 30 × 180 m rectangular area parallel to and about 300 m from the lake shore. Three control plots of the same size (30 m²) were left undredged. Results indicated rapid recolonization (within 4 months) in the plot dredged to 1.0 m. One year after dredging, macrophyte biomass in the 1.0-m plot was about 60% of the pre-dredging level. Macrophytes had not reestablished in the 1.4 m and 1.8 m test plots during the same year. Sediment nutrient levels were found to be similarly high in all test plots, so nutrient deficiency was ruled out as a probable cause of reduced growth. The authors speculated that reduced light penetration at the 1.4 m and 1.8 m depths limited regrowth, but they also noted that deeper plots tended to fill with plant debris and lake detritus, altering the texture of the substrate. Unfortunately, no quantitative measurement of light level or sediment particle size was reported to corroborate their speculations.

That macrophytes ordinarily grow to depths up to 2 m (Higginson, 1970) in the Tuggarah Lakes seemed to imply that light alone should not have prevented regrowth at 1.4 m and 1.8 m. The more flocculent sediments in deeper plots may have had a greater influence than indicated by Collett et al. (1981). Their study did not answer conclusively the question of the influence of light on regrowth of plants. It may even raise some question about the rationale for using light level to determine dredging depth. This seems, however, to be a reasonable approach given what we know about macrophyte growth characteristics and light requirements. The maximum depth of autotrophic plant growth depends upon water transparency (Hutchinson, 1975; Maristo, 1941).

Canfield et al. (1985) reevaluated the relationship between macrophyte maximum depth of colonization (MDC) and Secchi disc transparency. Duarte and Kalff (1987) confirmed the work of Canfield et al. using several variables from Canadian and U.S. lake data sets. The subject of macrophyte growth characteristics in lakes was addressed briefly in Chapter 2 and covered in much greater detail in Chapter 11. In addition, Duarte and Kalff (1990) is an excellent reference for in-depth coverage on the subject.

20.5 SEDIMENT REMOVAL TECHNIQUES

There are two major techniques for sediment removal from freshwater lakes and reservoirs. The first one, lake drawdown followed by bulldozer and scraper excavation, has limited application. It has been used most successfully in small reservoirs (Born et al., 1973). The obvious limitation of this technique is that water must be drained or pumped from the basin. A second drawback is that the basin must be allowed to dewater sufficiently before earth-moving equipment can operate.

Despite these problems, plus the added concern of truck traffic to transport the removed sediment, this approach has been used successfully at Steinmetz Lake, NY (Snow et al., 1980).

The second, and most common, sediment removal technique is dredging. Huston (1970) reviewed the many types of dredges in use. This chapter addresses only dredges commonly used in lakes and those with special features that minimize adverse dredging effects. Dredges are divided into mechanical and hydraulic types. A third category, "special purpose dredges," is included to highlight low-turbidity systems for dredging fine-grained and toxic sediments, both of which are relatively common in fresh water lakes and reservoirs.

20.5.1 MECHANICAL DREDGES

Grab-type mechanical dredges are used commonly in lake restoration (Figure 20.1). Figure 20.1A shows a clamshell bucket dredge in operation. Figure 20.1B shows a typical Sauerman grab bucket set-up. A limitation of all grab bucket dredges is that they must discharge in the immediate vicinity of the sediment removal area or into barges or trucks for transportation to the disposal area. Their normal reach is no more than 30 to 40 m. Another disadvantage is the rough, uneven bottom contours they create. Production rates are relatively slow due to the time-consuming bucket swing, drop, close, retrieve, lift, and dump operating cycle. Grab dredges commonly create very turbid water conditions due to bucket drag on the bottom as it pulls free from the sediment, dragging an open bucket through the water column, bucket leakage once it clears the water surface, and the occasional intentional overflow of receiving barges to increase their solids content. Another disadvantage is that many lake sediments are highly flocculent, reducing the pickup efficiency of a grab bucket.

Grab-bucket dredges have at least two advantages over the other dredge types: they can be transported with ease from one location to another and they can work in relatively confined areas. Thus, their chief use in lake restoration and management is shoreline modification, particularly around docks and marinas. They are readily operated around stumps and trash frequently found in these areas. A grab bucket operates most efficiently in near-shore areas that contain soft to stiff mud. Depth is no impedance, but efficiency drops rapidly with depth, because of the time consuming operating cycle.

Silt curtains reduce some of the turbidity-associated problems mentioned above. A silt curtain is a continuous polyethylene sheet (skirt) buoyed at the surface and weighted at the bottom so it hangs perpendicular to the water surface. It may be used to encircle an open water dredging operation or to isolate a length of shoreline (Figure 20.1). The purpose of the silt curtain is to isolate turbidity within the immediate dredging area, protecting clean surface water areas downstream. Silt curtains, while effective in controlling surface turbidity, are open at the bottom and permit the escape of turbid water near the sediment–water interface.

Another means of minimizing turbidity from grab bucket dredging is to use a covered, watertight unit (Figure 20.2). Watertight buckets range in sizes from 2 to 20 m^3. Manufacturers claim turbidity reductions from 30% to 70% compared to open buckets of comparable size. The dredging process with watertight buckets is cleaner than with conventional buckets, but production is still relatively inefficient compared to hydraulic dredges.

20.5.2 HYDRAULIC DREDGES

There are many variations of hydraulic dredges, including the suction dredge, the hopper, the dustpan, and the cutterhead suction dredge. Hopper dredges are impractical for dredging small inland lakes. Cutterless suction dredges have not been used extensively. Attempts to use one at Lilly Lake, WI, in 1978 were abandoned when it was discovered that the partially decomposed plant material in the sediment prevented it from "flowing" to the suction head (Dunst, 1982). A cutterhead suction dredge subsequently was employed.

Dustpan dredges are not commonly used in lake restoration, although a "dustpan-like" dredge was used to remove flocculent sediment from Green Lake, Washington in 1961 and 1962 (Pierce,

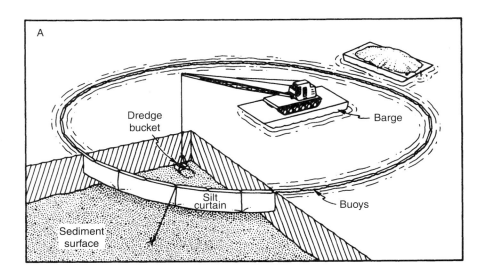

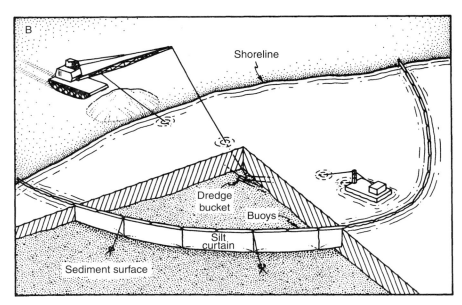

FIGURE 20.1 (A) Silt-curtain encirclement of an open-water grab dredge operation. (B) Shoreline isolation of a bucket dredge operation, using a silt curtain. (Cooke et al., 1993. With permission.)

1970). The device consisted of a 15.25 m suction manifold with slot openings. The total size of the inlet ports was designed to produce inlet velocities of at least 300 cm/s. As sediment consistency increased with depth, some of the inlet ports were sealed to increase flow velocity in the open ones. The dustpan-like suction head was barge mounted and designed to swing in a full 180° arc and discharge into a 50.8 cm diameter pipeline. The discharge distance was about 792 m. This dredge successfully removed 917,500 m³ of sediment. Björk (1974) indicated that the dredge head used at Lake Trummen, Sweden had a specially designed "nozzle." The positive experience at Green Lake and at Lake Trummen indicates that dustpan types and other variations of conventional hydraulic suction heads should receive additional consideration for dredging highly flocculent fresh water lake sediments.

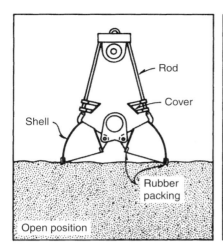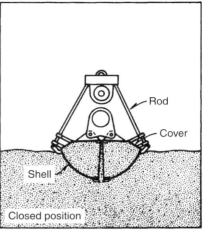

FIGURE 20.2 Open and closed positions of the watertight bucket. (Redrawn from Barnard, W.D. 1978. Prediction and Control of Dredged Material Dispersion Around Dredging and Open-water Pipeline Disposal Operations. Tech. Rept. DS-78-13. U.S. Army Corps Engineers, Vicksburg, MS.)

Inland lake sediment removal is most commonly accomplished with a cutterhead hydraulic pipeline dredge. Small, portable, cutterhead hydraulic dredges are the dominant equipment used for inland lake dredging. The primary components of any cutterhead dredge system include the hull, cutter head, ladder, pump, power unit, and a pipeline to distribute dredged material (Figure 20.3).

The hull is made of steel and constructed to withstand the constant vibration created by the cutterhead. The hull is the working platform that houses the main power plant, pump, lever room, and the assemblage of winches, wires, "A" frames, etc., that comprise the dredge.

At the bow is a steel boom or ladder with a cutter mounted at its distal end. Ladder length determines the practical dredge depth limitations. The ladder also supports the suction pipe and the cutter drive motor and shaft. In some cases, there may be a submersible auxiliary suction pump mounted on the ladder. The ladder is raised and lowered by suspension cables attached at the outer end and to a hull-mounted winch.

The cutter or cutterhead typically consists of three to six smooth or toothed conical blades that rotate at 10 to 30 rpm to loosen compacted sediment (Bray, 1979). Cutterheads may be open nose, closed nose, straight vane, ribbon screw shape, or auger-like. Most cutters have been designed specifically to loosen sand, silt, clay, or even rock material. Few, if any conventional hydraulic cutterheads have been designed to remove soft, flocculent lake sediment, so most of them are less efficient than they could be for lake dredging.

Spuds, vertically mounted pipes ranging from 25.4 cm to 127 cm in diameter, depending on the dredge size, are located at the stern of the hull on both sides (Figure 20.4). They are used to "walk" the dredge forward by alternately raising and lowering them into the sediment.

Operationally, sediment loosened by the cutter moves to the pickup head by suction from the dredge pump, usually a centrifugal type. The sediment slurry is then discharged by pipeline to a remote disposal area. Cutterhead dredges are described by the diameters of their discharge pipes. Hydraulic dredges used for inland lake work usually range in size from 15 to 35 cm, although the one used at Vancouver Lake, Washington was 66 cm (Raymond and Cooper, 1984). Figure 20.4 shows how the cutterhead is moved from side to side, and how pulling alternately on port and starboard swing wires creates the cut path. A major advantage of hydraulic cutter suction dredges over bucket types is that they are not confined in operation by the limitation of cable reaches. Another advantage is their continuous operating cycle. This cycle permits hydraulic dredges to

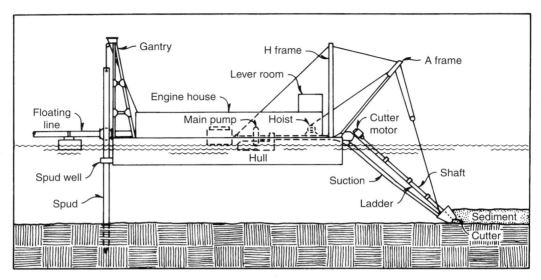

FIGURE 20.3 Configuration of a typical cutterhead dredge. (From Barnard, W.D. 1978. Prediction and Control of Dredged Material Dispersion Around Dredging and Open-water Pipeline Disposal Operations. Tech. Rept. DS-78-13. U.S. Army Corps Engineers, Vicksburg, MS.)

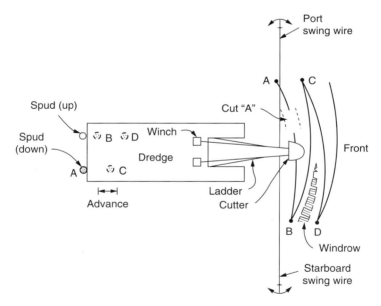

FIGURE 20.4 Spud-stabbing method for forward movement, and resultant pattern of the cut. (From Barnard, W.D. 1978. Prediction and Control of Dredged Material Dispersion Around Dredging and Open-water Pipeline Disposal Operations. Tech. Rept. DS-78-13. U.S. Army Corps Engineers, Vicksburg, MS.)

produce large volumes of dredged material. This advantage, however, is not without its downside. Most hydraulic dredge slurries contain only 10% to 20% solids and 80% to 90% water. This means that relatively large disposal areas, with adequate residence times, are needed to precipitate solids from the dredge slurry. Also, it means that the large pumping capacity of hydraulic dredges might produce unplanned lake drawdowns, unless disposal-area overflow water is returned to the lake.

The amount of sediment supplied to the suction head is controlled by cutter rotation rate, thickness of the cut, and the swing rate (Barnard, 1978). Improper combination of any of these

FIGURE 20.5 The Mud Cat® dredge features a unique auger-type cutterhead. The size of the dredge makes it extremely portable. (Photo courtesy of Ellicott, Division of Baltimore Dredges, LLC, Baltimore, MD.)

might result in excessive turbidity. Therefore, not only the configuration of the dredge equipment, but the skill of the operator is important to minimizing turbidity. New computer technology on special purpose dredges has reduced this problem considerably.

20.5.3 SPECIAL-PURPOSE DREDGES

Portable cutterhead dredges are essentially miniatures of large coastal waterway dredges. The cutterheads of coastal dredges were designed for cutting sand, clay, and silt; they were not intended for use in fine, flocculent, organic lake sediments (frequently 40% to 60% organics). Consequently, soft lake sediments have challenged the dredging industry that responded with several dredging innovations. Among them is the cutter head used on Mud Cat® dredges. These dredges utilize a horizontal auger to dislodge and move sediment to the center of a 2.4 m wide, shielded, dredge head where it is sucked up by the pump and transported through a 20.3 cm discharge pipeline. Mention of the Mud Cat dredge is to illustrate their auger type cutter head and the mobility of small dredges (Figure 20.5). There are several others that are just as portable (see Clark, 1983).

Note in Figure 20.5 the mud shield, which can be raised or lowered over the auger head to minimize sediment resuspension. Nawrocki (1974) reported that turbidity plumes due to dredging with a Mud Cat machine were confined to an area no more than 6 m from the dredge, though operating conditions were not clearly defined. Suspended solids in the area of increased turbidity ranged from 39 to 1,260 mg/L. Those near the bottom averaged approximately 100 mg/L. More turbidity is created by forward motion of the dredge than by backward motion. This appears to be caused by raising the mud shield while moving forward, but lowering it when moving backward. Mallory and Nawrocki (1974) indicated that the Mud Cat dredge should be capable of producing slurry containing 30% to 40% solids. This represents nearly a doubling of the solids content commonly produced by conventional cutterhead dredges.

The Mud Cat guidance system is well suited to work on small water bodies. The dredge operates on a cable anchored at both shorelines. The guidance system permits uniform dredging of the bottom, with few missed strips. Mud Cat dredges have been used successfully at Collins Park and several other small lakes in New York State. The portability, guidance system, reduced turbidity, and increased solids content resulting from use of these dredges makes them ideally suited to small lake restoration projects. New and improved guidance and operating systems on Mud Cat® dredges have been instrumental in successful dredging of lakes in Europe (see case histories in this chapter).

Clark (1983) reported on a survey of portable hydraulic dredges available for use in the U.S. The survey identified 46 models of portable equipment available from several different manufacturers. No attempt was made to critically analyze the features of one dredge relative to another, but tables are presented that describe the general dredge specifications, the pump characteristics, suction and discharge diameters, cutter type, and working capacity. The information should be useful to engineers for selecting dredges, since it includes dredging depth ranges from 3 to 18 m, production rate ranges from 15 to 1375 m^3/h, and a wide variety of cutterhead types.

Equipment that removes water from hydraulically dredged material by centrifugal force exists, but we are not aware of any published evaluations. While this technique would reduce pond holding times for sediment settling, the high volume of water (typically 80% to 85%) in dredged material would still need to be managed.

20.5.4 PNEUMATIC DREDGES

Pneumatic (air-driven) dredge systems might have several advantages over conventional dredge systems relative to removal of fine grain lake sediment (Cooke et al., 1993). All of the pneuma systems (Oozer®, Cleanup®, Pneuma®) are Japanese. To our knowledge, the only use of one of these systems was the Ooozer-like (Figure 20.6) pneuma pump used at Gibraltar Lake, CA in 1981 (Spencer Engineering, 1981) to remove mercury-contaminated sediments.

After major modification of the valving material in the pump body, the pneumatic system performed satisfactorily (Spencer Engineering, 1981). Goldman et al. (1981) confirmed these findings and reported there were no elevated mercury levels in the water column at any station or at any depth during dredging. The dredging was so clean that no bathing beach areas in the 110.8 ha lake were forced to close during any phase of the dredging. Despite these positive findings, pneumatic dredging systems have not been used widely in the United States and, therefore, will not be discussed further in this text.

20.6 SUITABLE LAKE CONDITIONS

Peterson (1981, 1982a) described some sediment problems to consider when assessing dredging feasibility. Lake size, except for total cost, is not a dredging constraint. Peterson's (1979) examination of 64 lake-dredging projects showed that size ranged from less than 2 to over 1,050 ha, and that sediment volume removed ranged from a few hundred to over 7 million cubic meters.

One factor that might limit dredging of a large inland lake is the requirement for a commensurately large disposal area. Restoration most frequently is sought for lakes in high use areas, where sediment disposal space is scarce, but also where the greatest user benefits will be derived (JACA, 1980). Therefore, it is important that disposal alternatives be explored for these situations.

Various productive uses of dredged material have been examined (Lunz et al., 1978; Spaine et al., 1978; Walsh and Malkasian, 1978). At Nutting Lake, MA, 153 × 103 m^3 of sediment was sold as soil conditioner at $1.40/$m^3$. This reduced the total dredging cost by $215,000 and per unit dredging cost to about $1/$m^3$ (Worth, 1981). However, the final Nutting Lake report refutes this information saying that no substantial income was realized from the sale of dredged material

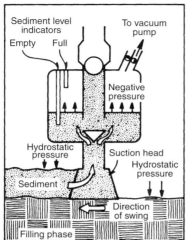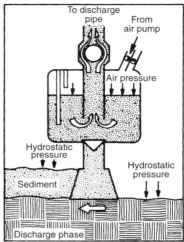

FIGURE 20.6 Schematic diagram of Oozer® dredge system. (Cooke et al., 1993. With permission.)

(Baystate Environmental Consultants, 1987). This was attributed to excavation difficulties fostered by the slow drying of material in the disposal basins. But, the containment area subsequently was sold for $450,000, nearly recovering the invested project costs. In Japan, sediment disposal areas are commonly sold for industrial development or converted to parks (Matsubara, 1979).

To be cost effective, a sediment removal project should have reasonable assurance of longevity. An estimate of sedimentation rates helps determine the infilling rate and, thus the duration of sediment removal effectiveness. Although dredging is expensive per unit of dredged material, where costs are amortized over the life expectancy of the project they may look much more reasonable. All other conditions being similar, lakes with relatively small watershed-to-surface ratios (nominally 10:1) will have lower sedimentation rates than those with large watersheds. Thus, a large lake with a small watershed should benefit more from dredging than will the reverse situation.

Depth, size, disposal area, watershed area, and sedimentation rate described above are all physical features. What about the influence of sediment chemistry on lake biota? Current information demonstrates that lakes with highly enriched surface sediments relative to underlying sediment (see Lake Trummen case history) might benefit from shallow dredging (Andersson et al., 1975; Bengtsson et al., 1975). Lake Trummen, Sweden, showed marked changes in water chemistry and biota when 40 cm of rich surface sediments were removed (Björk, 1978). Similar changes were observed in Steinmetz Lake, NY, when 25 cm of organic sediments were removed and replaced by the same amount of clean sand (Snow et al., 1980). In both cases, extensive sediment surveys before dredging revealed that surface sediment was disproportionately rich in P and N relative to the deeper sediment. In lakes, open water sediment is usually more important in sediment surveys than littoral zones, since sediment is transported toward the deeper zones of lakes. Surface inflow areas also need to be considered. Littoral zones tend to be cleaned by wave action and, in the temperate zone, by spring ice scouring. Reservoirs accumulate sediment quickly at their inflows due to their extensive watersheds. Sediment surveys should, at the minimum, determine the area of sediment to be removed and the depth (see the next section). Horizontal sediment characteristics normally are more uniform than vertical sediment profiles. Sediment depth may vary considerably, depending on the basin configuration at the time of the lake formation or the transport of sediment to the lake via stream inlets. Vertical variation in the survey is important to note. Sediment profiles can be developed with the assistance of a Livingston piston corer. It is important to note sediment color and texture differences with depth and to chemically characterize (P and N) surface (0 to approximately 10 cm) sediment relative to deeper sediment if nutrient control is the intent (Peterson,

1981). Beyond this it is quite useful to know sediment particle size, settling rate, sediment volume, etc., to properly select a dredge for the job and design an adequate disposal area.

Several variables determine the suitability of a lake for dredging, but generally the most suitable lakes have shallow depths, low sedimentation rates, organically rich sediments, relatively small (10:1) watershed-to-surface ratios, long hydraulic residence times, and the potential for extensive use following dredging.

20.7 DREDGE SELECTION AND DISPOSAL AREA DESIGN

This section draws heavily from the work of Pierce (1970). Implementation of lake dredging requires several decisions. The most important ones are what dredging equipment to use and what factors to consider in the disposal area design. Equipment selection depends on several variables, including availability, project time constraints, slurry transport distances, discharge head, and the physical and chemical characteristics of the dredged material.

The primary factor controlling the disposal area design is the amount of dredged material that must be contained. A second factor is the need to meet the discharge permit suspended solids requirements. Therefore, sediment grain size, specific gravity, plasticity, and settling characteristics of the dredged material must be considered when designing the disposal area.

To illustrate these considerations an example is offered. A feasibility study conducted on hypothetical Dead Lake, located in a rural area of the glaciated upper mid-western U.S. reveals these characteristics:

- Lake area = 120 ha
- Maximum depth = 5.5 m
- Average depth = 2.0 m
- Normal water level = 245 m above sea level
- Sediment water content = 30% to 60%

Since total project cost is usually based on a measure of actual material removed, it is necessary to estimate the amount and type of sediment contained in the basin. The usual procedure is to collect hydrographic data suitable to developing a lake-bottom map that describes the configuration of the original basin. The accuracy of this map depends on the sampling interval and the original basin relief. Even relatively shallow glacial lakes may have deep holes, reinforcing the need for sediment depth mapping.

Sediment sampling frequency to determine volume varies depending on basin configuration and desired survey accuracy. Preliminary sampling stations should be broadly spaced to provide a rough estimate of the solid bottom relief of the lake. This helps define and limit the required number of stations for final mapping. Pierce (1970) suggested that small to medium sized (< 40.5 ha) sediment removal projects should be mapped routinely by laying out sampling locations in a 15.25 m grid pattern. Pattern layout can be done by survey or using GPS units. He also suggested that, for lakes with surface areas > 40.5 ha, the sample station grid size could be increased to 30.5 m without significant loss of accuracy. He noted further that there will be far less variance horizontally than there will be vertically in lake sediment quality. Individual lake characteristics ultimately dictate the required station frequency.

A common procedure for obtaining the necessary data is to make sediment depth/lake hard bottom measurements at stations prescribed by the chosen grid size and relating the measurements to known elevation datum points on shore (topographic map, U.S. Geological Survey bench mark, etc.). The measurements can then be converted to elevations, thereby permitting the development of hydrographic maps and calculation of sediment volume.

A simple means of obtaining the required data is to measure, at each station, the water depth to the sediment–water interface and the distance (depth) to which a probe can be pushed into the lake sediment before contacting hard bottom. Both measurements can be made at the same time by using a graduated probing ("sounding") rod. Lake sediment probes usually are steel rods measuring 0.95 cm to 1.6 cm in diameter. If the rods are forced they can be bent and accuracy is reduced. The investigator needs to develop "a feel" for the degree of resistance that determines hard lake bottom. Sediment depth is determined by calculating the difference between the rod interval reading at "hard bottom" and the reading at the sediment–water interface. Distinction of the sediment–water interface may be difficult in lakes with flocculent, highly organic sediments. In these cases, it is advisable to use a lightweight disc or foot at the tip of the probing rod to establish water depth to the sediment surface. Alternatives to this are the use of a graduated line and Secchi disk, or an electronic depth sounder, some of which are extremely accurate.

Depth determination is easiest during calm periods on open waters and pontoon boats are great platforms for doing this work. In cold climates the work can be accomplished even more easily by making the measurements through holes drilled in the ice. Winter lake mapping makes it much easier to locate your position accurately, particularly when using GPS. Pierce (1970) indicated that a properly equipped crew working efficiently should be able to collect water and sediment data in this manner over 4 to 8 ha of lake surface per day. Efficiency is enhanced if data are collected during early winter; before ice has thickened to more than 15 or 20 cm. Sediment depth measurement is critical. Miscalculations in the sediment volume leads to errors in projecting cost estimates and to selecting proper dredging equipment, so accuracy should be stressed.

Sediment mapping of Dead Lake indicated deposits of highly organic silt material (muck). Water content of surface sediment averaged about 60%, while that at mid-depth and beyond ranged from 30% to 40%. Mapping data showed that sediment thickness was nearly 3.6 m at the south end, near the inlet, and that it decreased to about 1.8 m on the north end. These sediment conditions are well suited to the use of a hydraulic cutterhead dredge. Three sediment disposal areas were located around the lake. The desire to minimize pumping distances made it convenient to divide the lake surface area into three pieces; each one identified with the nearest disposal area. Figure 20.7 shows how the lake might be divided to best utilize the available upland disposal areas.

The feasibility study shows that sedimentation rates in the lake have been reduced significantly over the past 15 years as a result of shifts from row crop to small grain and hay crop farming in the watershed. The accumulated sediment is not contaminated, and recent accumulations result mostly from autochthonous organic material decomposition. Therefore, it appears that deepening at least 15% of the lake to about 6.0 m, while leaving a fish spawning and wildlife area intact, will have a positive effect toward restoring the fishery and other beneficial uses. The study indicated further that water depth 60 m from shore should be a minimum of 2.5 m, and that the bottom should then slope at a 5% grade to a depth of 3.5 m. Reconfiguring the lake in this manner will provide adequate water volume and depth to maintain adequate dissolved oxygen (DO) levels to avoid fish winterkills (Toubier and Westmacott, 1976).

The maximum depth calculations based on these recommendations indicate that approximately 1,530,000 m^3 of sediment needs to be removed. It is desirable to complete the project as rapidly as possible, to minimize lake use disruption, so project duration is targeted for 2 years (mid-April through mid-November: ice-free months, over two consecutive seasons).

20.7.1 Dredge Selection

Proper selection and use of hydraulic dredging equipment will implement feasibility recommendations. The remainder of this section presents a series of considerations for selecting a cutterhead dredge (Pierce, 1970).

20.7.1.1 Plan to Optimize the Available Disposal Area

Long pumping distances to disposal areas should be minimized, since energy requirements increase with pumping distances. Disposal area No. 1 is the closest, at 750 m, when pumping from lake

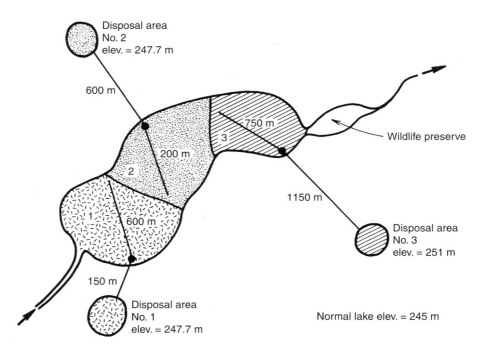

FIGURE 20.7 Dead Lake (hypothetical), showing the planned dredging areas, pipeline distances to disposal areas, and the wildlife area that will remain undredged (not to scale). (Cooke et al., 1993.With permission.)

area No. 1 (Figure 20.7). Disposal area No. 2 is 800 m and disposal area No. 3 is 1,900 m, when pumping from the respective lake areas. It was calculated that areas 1, 2, and 3 will hold 574,000, 413,000, and 918,000 m³ of dredged material, respectively. Therefore, areas 1 and 2 would receive 574,000 and 413,000 m³, respectively, with area 3 receiving the remainder of the dredged material (543,000 m³), to optimize disposal efficiency by minimizing pipeline length.

20.7.1.2 Analyze the Production Capacity of Available Dredging Equipment

It is necessary to analyze the production of various sized dredges to determine which equipment might complete the job within the planned 2-year period. A survey of equipment reveals that 20-cm, 25-cm, and 30-cm dredges are available, so production analysis is limited to these sizes.

Dredge pump production rates usually are listed as ranges since dredging conditions, and thus production rates, vary considerably. Production ranges for the available dredges (20, 25, and 30 cm) are taken from Figure 20.8 to illustrate the method. Similar dredge capacity charts are available from various dredge pump manufacturers. Charts for the specific equipment in question should be used whenever they are available. Figure 20.7 and the feasibility study for Dead Lake showed that the greatest sediment volume is located near the center of the lake and that transport from this area to the disposal cells will require pipeline transport distances in excess of 600 m. Based on that information, the following dredge pump production range analysis was developed, using the minimum, a medium, and the maximum pipeline lengths:

300-m length of discharge pipeline:

> 20-cm pump = 50 to 110 m³/h, average 80 m³/h
> 25-cm pump = 80 to 190 m³/h, average 135 m³/h
> 30-cm pump = 310 to 420 m³/h, average 365 m³/h

600-m length of discharge pipeline:

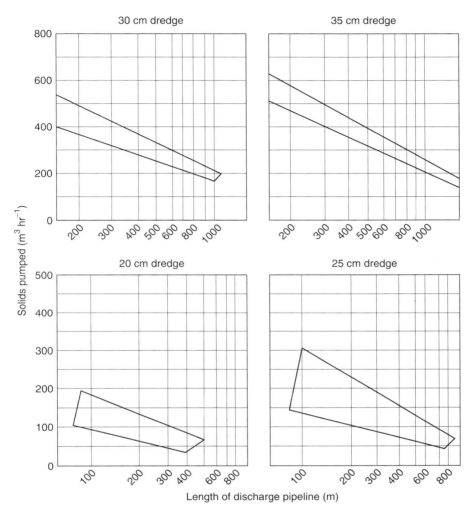

FIGURE 20.8 Representative production characteristics for various sizes of dredge systems. (Modified from Pierce, N.D. 1970. Inland Lake Dredging Evaluation. Tech. Bull. 46. Wisconsin Dept. Nat. Res., Madison.)

20-cm pump = beyond effective discharge length; booster pump required
25-cm pump = 60 to 120 m³/h, average 90 m³/h
30-cm pump = 220 to 290 m³/h, average 255 m³/h

800-m length of discharge pipeline:

20-cm pump = beyond effective discharge length; booster pump required
25-cm pump = 50 to 80 m³/h, average 65 m³/h
30-cm pump = 190 to 250 m³/h, average 220 m³/h

The analysis reveals that use of the 25 cm system for distances of 600 to 800 m is marginally efficient, based primarily on the dredge pump characteristics and its power (kilowatts). As pipeline length increases pipeline friction increases and solids transport efficiency decreases. A pipeline discharge velocity of 3 to 4 m/s must be maintained to transport solids. Thus, discharge pipeline length must be limited to that which permits the velocity to be maintained at 3 to 4 m/s. Longer pipes can be used with booster pumps. The analysis indicated that disposal in cell No. 3 from lake area No. 3, even with the largest system available (30 cm), would require a booster pump.

20.7.1.3 Compute Dredging Days Required to Complete the Job

Approximately 1,530,000 m³ of sediment must be removed from Dead Lake. For efficiency, a hydraulic dredge normally operates 24 h/d unless noise is a problem. Noise could be a concern on urban or small lakes.

There is always some down time for maintenance and pipeline relocation, so for this example we will assume a 24 h/d operation schedule with a normal productive dredge time of approximately 20 h/d, 6 d/week.

Previously, it was observed from the above production analysis that the 20 cm dredge would not operate efficiently without a booster pump at discharge distances exceeding 600 m. Since most of the pumping will require discharge at 600 m and beyond, the 20 cm pump should not be considered. Using the average (rounded down for illustrative convenience) discharge rates of the 25 cm and 30 cm systems for 600 m and 800 m pipeline lengths (Figure 20.8 or the discharge pipeline summary above) the number of days required to complete the Dead Lake dredging project can be calculated.

25 cm pump system:

$$\frac{90 \text{ m}^3/\text{h} + 65 \text{ m}^3/\text{h}}{2} = 77.5 \text{ m}^3/\text{h} \cong 75 \text{ m}^3/\text{h}$$

$$\frac{1,530,000 \text{ m}^3}{(75 \text{ m}^3/\text{h})(20 \text{ h/d})} = 1020 \text{ d}$$

30 cm pump system:

$$\frac{255 \text{ m}^3/\text{h} + 220 \text{ m}^3/\text{h}}{2} = 237.5 \text{ m}^3/\text{h} \cong 230 \text{ m}^3/\text{h}$$

$$\frac{1,530,000 \text{ m}^3}{(230 \text{ m}^3/\text{h})(20 \text{ h/d})} = 333 \text{ d}$$

Dead Lake is located in the northern U.S. so it is frozen over from approximately mid-November until mid-April. This reduces annual open water workdays to about 185. If the 25 cm pump system is used, completion time would exceed 5 years (1,020 ÷ 185 = 5.5 year). If the project is to be completed within the 2 year target time (two open water seasons), the 30 cm dredge is required (333 ÷ 185 = 1.8 year). Disposal in area No. 3 requires use of an appropriately placed booster pump, since the pipeline length to that area exceeds the efficient pumping distance (approximately 1000 m) of the 30 cm pump system.

20.7.1.4 Determine the Required Head Discharge Characteristics of the Main Pump When Pumping Material with the Specific Gravity of Lake Sediment (Approximately 1.20)

The required head-discharge characteristics of a pump depend on the discharge pipe length, i.e., the longer the pipeline, the higher the total head required. Pump head discharge characteristics must be analyzed for both minimum and maximum discharge distances. In the case of Dead Lake, the minimum is about 300 m (150 m from shore to disposal area plus 150 m off-shore in the lake) since dredging to the shoreline is seldom done when pumping from lake area 1 to disposal area 1, and the maximum is about 1,900 m when pumping from lake area 3 to disposal area 3.

The sum of the total suction lift and total discharge head is the total dynamic head against which a pump works (Pierce, 1970). Heads commonly are computed from basic hydraulic formulae, corrected for specific gravity of the pumped material. Suction lift incorporates suction elevation head, suction velocity head, and friction head in the suction pipe. The total discharge head is calculated by summing the pump velocity head, the discharge elevation head, and the friction head in the pipeline. Minor head losses usually are not considered.

20.7.1.4.1   Suction Head

Since the weight of dredged material (specific gravity of lake sediment is approximately 1.20) is greater than water, the surface of a column of water equal to the depth of Dead Lake would always have a greater elevation than the surface of an equal sized (diameter) column of dredged material of the same weight. The resultant difference in column heights is the suction elevation head. The suction elevation head always refers to the horizontal center line of the main pump and is computed as

$$h_{ss} = S_1 A - S_2 B \qquad (20.2)$$

where: h_{ss} = the suction elevation head (meters of fresh water), S_1 = specific gravity of lake water (1.0), S_2 = specific gravity of material being pumped (1.2), A = distance from the bottom of the cut to the water surface (m), and B = distance from the pump center to the bottom of the cut (m).

Assuming that the dredge pump is mounted on the hull at lake level and that maximum dredged depth is 8.5 m, the static suction head is

$$h_{ss} = 1.00(8.5) - 1.20(8.5)$$

$$h_{ss} = -1.7 \text{ m}$$

The minus sign indicates that a suction head exists. This number must be added positively to other heads computed for the suction system.

The suction velocity head is the energy required to start the movement of dredge material into the suction pipe. It can be computed as

$$h_{sv} = S_2 \frac{Vs^2}{2_g} \qquad (20.3)$$

where: h_{sv} = velocity head (meters of fresh water), S_2 = specific gravity of the material being pumped, Vs = velocity of the mixture in the suction pipe (m/s), and g = acceleration rate of gravity (m/s²).

The acceleration rate of gravity is 9.82 m/s². Normal suction pipe velocity should be maintained at 3.0 to 4.0 m/s to assure that solids are carried into the pump. If we assume an upper midrange suction pipe velocity of 3.6 m/s, the velocity head in the suction pipe will be

$$h_{sv} = (1.20) \frac{(3.6)^2}{2(9.82)}$$

$$h_{sv} = 0.8 \text{ m}$$

Friction head losses caused by aqueous flow characteristics in pipes create the major portion of the head that a dredge pump must overcome. Pipeline friction loss is influenced by several variables. Among them are the type and diameter of pipe, flow velocity in the pipe, pipeline length and configuration, and the percentage and type of solids in the pumped mixture. Since friction losses are magnified as the diameter of the suction pipe decreases, many small dredges utilize suction pipes one size (usually 5 cm increments) larger than the discharge pipe. For example, a 30-cm dredge (size of discharge) might have a 35 cm suction line. Velocity in the discharge pipe (30 cm) will be greater than that in the suction pipe (35 cm), since the volume entering the larger suction pipe must be squeezed through the smaller diameter discharge pipe. The velocity in the discharge pipe varies as the ratio of the square of the diameter of the larger pipe divided by the square of the diameter of the smaller one $[(35)^2 \div (30)^2 = 1.36]$. Therefore, the 3.6 m/s velocity in the suction will be increased to approximately 4.9 m/s in the discharge. All influences affecting pipeline friction losses (*suction friction head*) must be considered and applied to an acceptable equation for formulating friction losses. Suction friction head is the energy required to overcome friction losses in the pump suction line (Pierce, 1970). The suction friction head can be computed from the Darcy–Weisbach formula:

$$h_{sf} = f \left[\frac{1 + (P - 10)}{100} \right] \frac{LV_s^2}{2gD} \tag{20.4}$$

where: h_{sf} = friction head (meters of fresh water), f = the friction factor, P = solids in dredge slurry (% by volume), L = equivalent length of suction pipe (m), Vs = velocity of the mixture in the suction pipe (m/s), g = acceleration rate of gravity (m/s²), and D = inside diameter of the suction pipe (m).

The friction factor (f) is a dimensionless number that is a function of the Reynolds number and the relative roughness (absolute roughness ÷ diameter of pipe in m) of different types of pipe. The functions have been obtained experimentally for clear water and expressed graphically (Figure 20.9) by Moody (1944). Use of f as described by the Moody diagram for computing dredged material pipeline transport necessarily becomes an approximation at best, since solids in the slurry will affect the number. Despite this apparent problem, Moody (f) values are commonly used to estimate various hydraulic, pipeline dredging figures. The Reynolds number can be calculated from the formula

$$R = \frac{VD}{v} \tag{20.5}$$

where: V = velocity in the pipeline (m/s), D = inside diameter of the pipeline (m), and v = temperature-corrected kinematic viscosity of water (m²/s × 10⁻⁶) (see Table 20.1).

As stated above, the velocity of suction pipeline slurries commonly ranges from 3.0 to 4.0 m/s or greater to maintain the suspension of solids (turbulent flow). If we use the previous assumed slurry velocity of 3.6 m/s and a kinematic viscosity of water at 20°C (1.0×10^6 m²/s) and apply these figures to a 35 cm (0.35 m) suction pipe, the following Reynolds number can be calculated from Equation 20.5

$$R = \frac{3.6 \, (0.35)}{1.0 \times 10^{-6}}$$

$$R = 1.3 \times 10^6$$

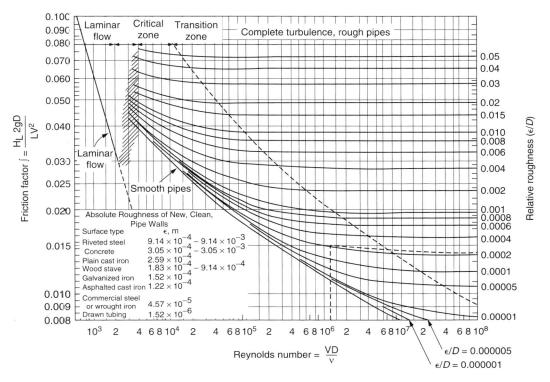

FIGURE 20.9 Moody diagram showing friction factors for pipe flows. (Redrawn from Moody, L.F. 1944. *Trans. ASME* 66: 51–61. With permission.)

If we assume a pipe roughness of 8.7×10^{-5} m, the relative roughness will be

$$rr = \frac{e}{D} \qquad\qquad (20.6)$$

where: rr = relative roughness, e = absolute roughness (m), D = inside diameter of pipe (m). Then

$$rr = \frac{8.7 \times 10^{-5}}{0.35}$$

$$rr = 2.5 \times 10^{-4}$$

Applying relative roughness and the Reynolds number to the Moody diagram (Figure 20.9), the resultant friction factor is 0.015 for these conditions. Note that the intersection of these two variables falls within the transition zone between laminar flow and complete turbulence in rough pipes.

Friction-head losses in suction pipes vary with configuration, valving, suspended solids concentration, and cutterhead types. Cutterhead losses are highly variable and losses associated with fine grain dredge material are not well defined. (Note the comment above concerning the use of f values.)

Correction factors for these variables are not readily available in tabular form. Therefore, engineering best judgment based on a combination of practical experience and laboratory tests frequently is applied to actual suction pipe lengths to calculate "the equivalent length of suction

TABLE 20.1
Selected Physical Properties of Water at Various Temperatures

Temperature T (°C)	Density p (g/cm³)	Viscosity μ (g/cm/s ×10²)	Kinematic Viscosity v (cm²/s × 10²)[a]
0	0.9999	1.787	1.787
5	1.0000	1.514	1.514
10	0.9997	1.304	1.304
15	0.9991	1.137	1.138
20	0.9982	1.002	1.004
25	0.9971	0.891	0.894
30	0.9957	0.798	0.802
35	0.9941	0.720	0.725
40	0.9923	0.654	0.659
50	0.9881	0.548	0.554
60	0.9832	0.467	0.475
70	0.9778	0.405	0.414
80	0.9718	0.355	0.366
90	0.9653	0.316	0.327
100	0.9584	0.283	0.295

[a] cm²/s × 10⁴ = m²/s.

Source: Modified from Montgomery, R.L. 1978. *Methodology for Design of Fine-Grained Dredged Material Containment Areas for Solids Retention.* Tech. Rept. D-78-56. U.S. Army Corps Engineers, Vicksburg, MS.

pipe." In effect, the equivalent length is a correction for suction pipe head loss. The suction pipe "correction factor" commonly is within the range of 1.3 to 1.7 (Hayes, 1980). To dredge to a depth of 8.5 m (maximum lake depth after dredging), the dredge ladder (suction pipe length) will need to be approximately 15 m long. Applying a suction-pipe-equivalency correction factor of 1.7, the equivalent suction pipe length is 25.5 m (15 × 1.7 = 25.5). Substituting the required figures (assume 20% solids) into Equation 20.4 determines the suction friction head

$$h_{sf} = 0.015 \left[1 + \frac{(20-10)}{100} \right] \frac{25.5\,(3.6)^2}{2\,(9.82)\,(0.35)}$$

$$h_{sf} = 0.8 \text{ m}$$

The total suction head (H_s) on the dredge pump is the sum of the suction elevation head (-1.7, added positively), the velocity head (0.8) and the friction head (0.8)

$$H_s = h_{ss} + h_{sv} + h_{sf}$$

$$= 1.7 + 0.8 + 0.8$$

$$H_s = 3.3 \text{ m}$$

20.7.1.4.2 Discharge Head

Discharge elevation head is represented by the difference in elevation (vertical distance) between the pump centerline and the end of the discharge pipe, corrected for the specific gravity of the dredge slurry. As mentioned previously, the specific gravity of dredge slurry for this example is 1.20. The pump centerline of dredges being considered for this job is at the water line of the dredge hull (from Figure 20.7, normal water level is 245 m). The top of the dike elevation at disposal sites 1 and 2 is 247.7 m. This information yields the discharge elevation head, using the equation

$$h_{de} = S_2 \, (E_D - E_p) \tag{20.8}$$

where: h_{de} = discharge elevation head (meters of fresh water), S_2 = specific gravity of the mixture being pumped, E_D = elevation of the center line of the discharge pipe at the point of discharge (m), and E_p = elevation of the center line of the dredge pump (m).

Therefore, when pumping to disposal areas 1 and 2, the discharge elevation head will be

$$h_{de} = 1.20 \, (247.7 - 245.0)$$

$$h_{de} = 3.2 \text{ m}$$

The *discharge friction head* is the energy needed to overcome friction losses in the discharge pipe; it can be computed using Equation 20.4. The dredge pump will have to overcome maximum friction head when pumping from lake area 2 to disposal area 2 (greatest discharge distance without a booster pump). The pipeline length in this case is about 200 m of floating pipe and 600 m of shore pipe. The two pipes differ considerably in joint configuration, since the floating pipe must be flexible enough to accommodate wave action and relocation of the dredge. Therefore, the factor applied to the two types of pipe to calculate the equivalent length is different. Pierce (1970) indicates that the floating pipe factor typically ranges from 1.35 to 1.5 (more bends than shore pipe), while that for shore pipe is usually between 1.1 and 1.25. If we use the maximum factor of 1.5 for floating pipe (200 m) and the minimum of 1.1 for shore pipe (600 m) the factors will tend to normalize the pipeline equivalent lengths. Therefore,

- Floating pipe length = 200 (1.5) = 300 m
- Shore pipe length = 600 (1.1) = 660 m
- Total equivalent length = 960 m

This total equivalent length, substituted into Equation 20.4 with the calculated discharge pipeline velocity (4.9 m/s), results in a discharge pipeline friction-head loss of

$$h_{df} = 0.015 \left[1 + \frac{(20 - 10)}{100} \right] \frac{960(4.9)^2}{2(9.82)(0.30)}$$

The same value can be obtained from Figure 20.10 by entering the velocity scale at 4.9 m/s and reading vertically to the 0.30 m pipeline intersection and then reading left to 2.05 on the friction-head loss scale. The friction-head loss scale is in meters per 30.5 m of pipe, so the scale reading must be multiplied by the number of times that 30.5 can be divided into the equivalent pipe length (960 ÷ 30.5 = 31.47; 2.05 × 31.47 = 64.5 ≅ 65 m).

The pump velocity head is the energy required to increase the pump suction line velocity to the discharge pipeline velocity. It is computed from Equation 20.9:

$$h_{dv} = S_2 \, \frac{V_d^2 - V_s^2}{2(g)} \tag{20.9}$$

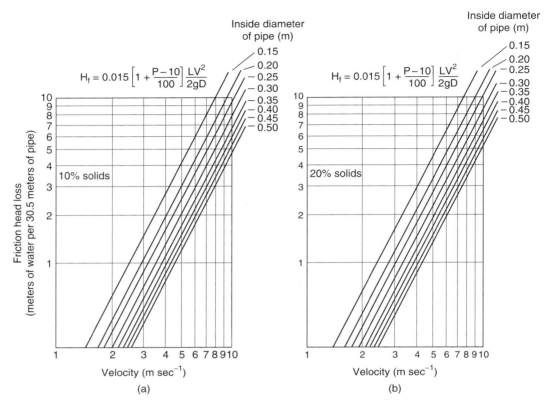

FIGURE 20.10 Friction-head loss for 10% and 20% solids in various diameter pipelines as a function of slurry velocity. (Modified from Pierce, N.D. 1970. Inland Lake Dredging Evaluation. Tech. Bull. 46. Wisconsin Dept. Nat. Res., Madison.)

where: h_{dv} = pump velocity head (meters of fresh water), S_2 = specific gravity of the dredged material, V_d = velocity of dredged material; in the discharge pipeline (m/s), V_s = velocity of dredged material in the suction pipeline (m/s), and g = acceleration rate of gravity (m/s^2).

The suction velocity for this example is 3.6 m/s, and the discharge velocity is 4.9 m/s. Therefore, the pump velocity head is

$$h_{dv} = (1.20)\frac{(4.9)^2 - (3.6)^2}{2(9.82)}$$

$$h_{dv} = 0.7 \text{ m}$$

The total discharge head on the main pump is the sum of the discharge heads, per Equation 20.10:

$$H_d = h_{de} + h_{df} + h_{dv}$$

$$= 3.2 + 65.0 + 0.7 \qquad (20.10)$$

$$H_d = 68.9 \text{ m}$$

The total dynamic head on the main pump is the sum of the total suction head and the total discharge head, per Equation 20.11:

$$H_{TDH} = H_s + H_d$$

$$= 3.3 + 68.9 \tag{20.11}$$

$$H_{TDH} = 72.2 \text{ m}$$

Once the total dynamic head is known, the power necessary to operate the pump against the resistance in the system can be calculated. First, however, it is necessary to know the theoretical pump output, which can be calculated as follows:

$$Q = \frac{\pi}{4} \times 3600 \; D^2 V_d \tag{20.12}$$

where: Q = output of the dredge pump (m³/h), D = inside diameter of the discharge pipe (m), and V_d = velocity of slurry in the discharge pipe (m/s).

The 30 cm dredge pump output, when pipeline velocity is 4.9 m/s, is:

$$Q = \frac{3.14}{4} \times 3600(0.30)^2 (4.9)$$

$$Q = 1246 \text{ m}^3 / \text{h}$$

Therefore, a dredge pump should be selected that most nearly meets the required head discharge characteristics of 1246 m³/h at a total dynamic head of 72.2 m. The performance curve for the dredge pump shown in Figure 20.11 meets these requirements at Point C.

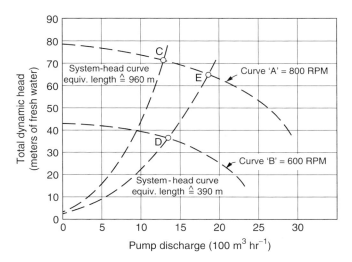

FIGURE 20.11 System head curve for a 30 cm dredge pump. (Modified from Pierce, N.D. 1970. Inland Lake Dredging Evaluation. Tech. Bull. 46. Wisconsin Dept. Nat. Res., Madison.)

In addition, the dredge power plant must be sufficiently powerful to force the pump output through the pipeline. The required power is usually specified by engineers in terms of brake hp (BHP) and can be computed from Equation 20.13:

$$BHP = \frac{QH_{TDH}S_2}{2.737\,E}$$ (20.13)

where: BHP = continuous brake hp at the pump, Q = dredge output (m³/h), H_{TDH} = total dynamic head the pump works against (meters of fresh water), S_2 = specific gravity of the dredged material, and E = dredge pump efficiency (%).

Pump efficiency of smaller dredges ranges from 50% to 65% and decreases with wear (Pierce, 1970). Thus, it is recommended that a conservative figure be adopted for efficiency. Pierce (1970) recommends an efficiency figure of 55%. Therefore,

$$BHP = \frac{(1248)(72.2)(1.20)}{(2.737)(55.0)}$$

$$BHP = 718 \text{ hp}$$

The manufacturer's rated continuous duty capacity at any rpm should be discounted by at least 10% (Pierce, 1970). This assures that the power plant is adequately sized to rotate the pump at the required rpm (800 in this case), and at the same time it will provide a longer, more trouble free engine lifetime. In this case, $718 \times 1.10 = 790$ hp at 800 rpm will be required. Engine selection should be made based on a 1.5- to 1.0-reduction gear between the engine and the pump. Therefore, an engine of at least 790 hp at 1,200 rpm should be used (BHP $\times$ 0.7457 = kilowatts (kw); thus, 790 hp $\times$ 0.7457 = 589 kw).

Conclusions from the above dredge pump system analysis are:

1. The minimum dredge size for this job is 30 cm, with sufficient ladder length to dredge to a depth of 8.5 m.
2. The average production rate of a 30 cm dredge pumping distances of 600 m will be approximately 230 m³/h.
3. The job can be completed in two summer seasons using a 30 cm dredge.
4. The closest disposal sites should be filled first.
5. There will be a maximum head of 72.2 m on the dredge pump when pumping from lake area 2 to disposal area 2.
6. The dredge pump power plant should have a minimum continuous rating of 790 hp or 589 kw at 1,200 rpm.

This represents an analysis of maximum head conditions on the pump.

20.7.1.5 Determine Minimum Head Conditions When Pumping to the Nearest Disposal Area

The total head on the pump decreases as the pumping distance decreases. This means the pump output increases and the velocity of dredged material in the pipeline increases. As noted previously, a minimum pumping distance of about 300 m (150 m of shore pipe and 150 m of floating pipe) will be encountered when dredging and disposing in area 1. By doing the same series of computations as were done for the 800 m pipeline, the following can be concluded for the 300 m pipeline:

1. The average production rate for a 30 cm dredge with a 300 m discharge pipeline length is about 360 m³/h.
2. The system head curve for the 300 m pipeline is shown in Figure 20.11 as having an equivalent length of approximately 390 m. Point E in Figure 20.11 shows that at 800 rpm, the pump discharge exceeds 1,800 m³/h. This increases the discharge pipeline velocity beyond 6.5 m/s, creating excessive pump and pipeline wear, possible pump cavitation, and extreme taxing of the engine, all of which create an inefficient operation.
3. There are two possible solutions to the problem: install a smaller pump impeller or reduce the engine speed. Figure 20.11 shows that if the 30 cm pump is operated at 600 rpm in conjunction with a 390 m equivalent pipeline length (point D), the pump delivers about 1,340 m³/h at a head of 36 m. According to Pierce (1970), at this capacity the discharge velocity is reduced to an acceptable 5.1 m/s.
4. The continuous hp required for the 36 m head and 600 rpm operating conditions is reduced to 385 hp. Using the 1.5 to 1 reduction would require the engine to operate at 900 rpm, to turn the pump at 600 rpm.

It is apparent that the head discharge curves for a pump over its recommended speed range is very helpful in selecting a dredge system and in determining the optimum pump speed at various discharge distances, to maximize production. It should be noted that pump system performance changes as the system components wear through use, and it is good policy to periodically check the performance against the manufacturer's rating curves. This is most easily accomplished by running tests on clear water when the system is first mobilized and rerunning the tests periodically after the pump has been in service. These tests permit the dredge operator to modify operating procedures as necessary to maintain optimum production.

20.7.1.6 Analyze Booster Pump Requirements for Pumping to Distances Beyond the Capacity of the Main Pump

The pipeline transport distance (1,900 m) from lake area 3 to disposal site 3 (elevation = 251 m) exceeds the efficient capacity of the 30 cm dredge (see Figure 20.8) due to the increased friction head and reduced output from the pump. The booster pump selected must be capable of increasing the total discharge rate to maintain a minimum discharge pipeline velocity of 4.9 m/s.

Figure 20.12 shows the head discharge for the main pump operating alone and the curve for the main pump and an identical booster pump operating in series. The "operating point" on the two-pump curve can be determined by calculating the system head curve for the 1,900 m discharge pipeline. For this system, the equivalent length of the discharge line is

- Floating pipe length = 760 (1.5) = 1,140 m
- Shore pipe length = 1160 (1.1) = 1,276 m
- Total equivalent length= 2,416 m

The system head curve, as shown in Figure 20.12 for the 1,900 m discharge pipeline, can be computed as

$$h_{ss} = 1.0 \ (8.5) - 1.20(8.5)$$

$$h_{ss} = -1.7 \ m$$

$$h_{sv} = 1.2 \ \frac{(3.6)^2}{2(9.82)}$$

$$h_{sv} = 0.8 \ m$$

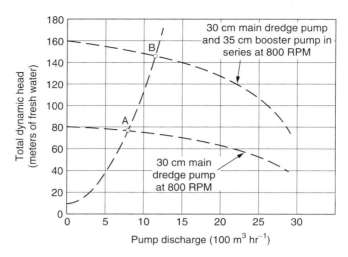

FIGURE 20.12 Head discharge relationships for a 30 cm dredge pump and a 30 cm booster pump. (Modified from Pierce, N.D. 1970. Inland Lake Dredging Evaluation. Tech. Bull. 46. Wisconsin Dept. Nat. Res., Madison.)

$$h_{sf} = 0.015 \left[1 + \frac{(20-10)}{100} \right] \frac{25.5\,(3.6)^2}{2\,(9.82)\,(0.35)}$$

$$h_{sf} = 0.8 \text{ m}$$

$$H_s = 1.7 + 0.8 + 0.8$$

$$H_s = 3.3 \text{ m}$$

$$h_{de} = 1.2\,(251 - 245)$$

$$h_{de} = 7.2$$

$$h_{dv} = 1.20 \frac{(4.9)^2 - (3.6)^2}{2\,(9.82)}$$

$$h_{dv} = 0.7 \text{ m}$$

$$h_{df} = 0.015 \left[1 + \frac{(20-10)}{100} \right] \frac{2,416\,(4.9)^2}{2\,(9.82)\,(0.30)}$$

$$h_{df} = 162 \text{ m}$$

$$H_d = 7.2 + 0.7 + 162$$

$$H_d = 169.9 \ m$$

$$H_{TDH} = H_s + H_d$$

$$= 3.3 + 169.9$$

$$H_{TDH} = 173.2 \text{ m}$$

$$Q = 2,826 \ (0.30)^2 \ (4.9)$$

$$Q = 1,246 \text{ m}^3/\text{h}$$

The calculated head discharge relationship determines one point on the pump system head curve (Figure 20.12). Other points on the curve can be calculated to develop the system curve for plotting. Point A in Figure 20.12 shows that the dredge pump alone produces about 815 m³/h at a head of 75 m. Rearranging Equation 20.12 to

$$V_d = \frac{Q}{\dfrac{\pi}{4} \times 3,600 \ D^2}$$

determines that the discharge pipeline velocity under these conditions is reduced to slightly more than 3 m/s, which is at the lower end of the efficient operating range (3.0 to 4.0 m/s). Point B in Figure 20.12 shows that the dredge pump operating in series with a second identical booster pump would increase the discharge to about 1180 m³/h at a head of 145 m. Under these conditions, velocity in the discharge pipeline increases to a more acceptable 4.6 m/s.

Use of a 30 cm booster pump is quite acceptable. What if the 30 cm pump is unavailable? Assume that the only pump available is a 35 cm, high head model. The head discharge curve for this booster pump, together with the 30 cm dredge pump curve, is shown in Figure 20.13. Figure

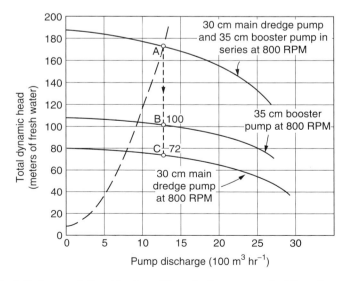

FIGURE 20.13 Head discharge relationships for a 30 cm dredge pump and a 35-cm booster pump. (Modified from Pierce, N.D. 1970. Inland Lake Dredging Evaluation. Tech. Bull. 46. Wisconsin Dept. Nat. Res., Madison.)

20.13 also shows the 30 and 35 cm series pump curve and the system head curve for the 1900-m discharge pipeline. Point A on Figure 20.13 represents the capacity of the two pumps in series when delivering through the 30 cm diameter, 1,900 m long discharge pipeline. The discharge at point A is about 1281 m³/h, at a head of 172 m. The discharge pipeline velocity at this discharge rate is about 4.9 m/s.

What portion of the total head comes from each pump can be determined from Figure 20.13. Simply construct a vertical line from Point A downward. The total head for the main pump is 72 m (Point C), and that for the booster pump is 100 m (Point B). Added together, the two heads equal the total head of 172 m at operating point A. The continuous horsepower (hp or BHP – brake horsepower) needed to operate the two pumps is for the dredge pump

$$BHP = \frac{(1,281)\ (72.0)\ (1.2)}{(2.737)\ (55.0)}$$

$$BHP = 735 \times 1.10 = 809\ \text{hp}$$

and for the booster pump

$$BHP = \frac{(1,281)\ (100.0)\ (1.2)}{(2.737)\ (55.0)}$$

$$BHP = 1,021 \times 1.10 = 1,123\ \text{hp}$$

The hp requirement for the dredge pump is slightly more than the 790 hp that was computed previously, but the difference is not enough to pose any problem for operating, if one considers the 10% factor that was used in the initial selection of the power plant. The booster pump power plant should be selected in conjunction with a speed reduction gear, so that the booster pump runs at 800 rpm.

An energy diagram (Figure 20.14), depicting the heads developed throughout the length of the pipeline is useful in determining the maximum and minimum allowable distances between the dredge pump and the booster pump. Since there are several variables in the dredging process, a positive suction head (H_s) of 10.6 m at the booster pump will be assumed. Also, since there is a positive suction head (H_s) of about 3.3 m on the main dredge pump, that amount must be subtracted from the total dynamic head (H_{TDH}) on the dredge pump to obtain the dredge pump head (H_d): 72.0 − 3.3 = 68.7. From this and the discharge friction-head loss per 30.5 m, for 30-cm discharge pipe at a velocity of 4.9 m/s (from Figure 20.10), the maximum pump spacing in pipe equivalent length is

$$\frac{(68.7 - 10.6)\ 30.5}{2.05} = 864\ \text{m}\ (\textit{equivalent length})$$

The pipeline equivalent length divided by the floating pipeline equivalent length correction factor (1.5) results in the actual pipeline length (576 m). Since the floating pipeline length is about 760 m, this means the booster pump will have to be barge mounted on the lake at a maximum distance not to exceed 576 m from the dredge. Plus and minus signs used with H_s in Figure 20.14 indicate the presence of a suction head or suction lift, respectively. Therefore, it can be seen that a booster pump located at a distance greater than L_{max} from the dredge pump will operate under a suction lift. This condition should be avoided.

The minimum spacing between the dredge pump and the booster pump can be computed if the discharge pipeline working pressure is known. Assume a working pressure of 1.40×10^5

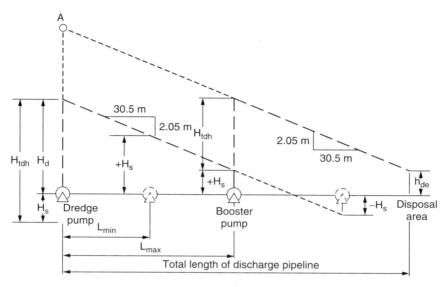

FIGURE 20.14 Energy diagram for a 30 cm dredge pump operating in series with a 35-cm booster pump. (Modified from Pierce, N.D. 1970. Inland Lake Dredging Evaluation. Tech. Bull. 46. Wisconsin Dept. Nat. Res., Madison.)

kg/m², which is equivalent to 140 m of water. The two pumps placed immediately adjacent to one another produce a discharge head of 168.7 m [(dredge pump head H_d − dredge pump suction head H_s) + booster pump total dynamic head H_{TDH}], indicated by point A in Figure 20.14. By proportion, the 168.7 m head results in a discharge pipeline pressure of 1.68×10^5 kg/m², which exceeds the required working pressure of 1.40×10^5 kg/m². Therefore, it will be necessary to locate the booster pump some minimum distance from the dredge, such that pipeline friction will reduce the pipeline pressure to a value below the working pressure. This distance should be calculated so that the discharge pressure of the dredge pump plus the booster pump, minus the pipeline friction-head loss between the two pumps, is less than the pipeline working pressure. The slope of the energy gradient in Figure 20.14 is nearly constant throughout the length of the discharge pipe. The friction-head loss created by the minimum distance between pumps can be computed from Equation 20.14:

$$H_1 = H_{TDH} + H_d - W_p \tag{20.14}$$

where: H_1 = head loss in the discharge pipe between the dredge pump and the booster pump (meters of water), H_{TDH} = booster pump total dynamic head (m), H_d = dredge pump total head minus the positive suction head (m), and W_p = discharge pipe working pressure (meters of water). Therefore,

$$H_1 = 100 + 68.7 - 140$$

$$H_1 = 28.7 \text{ m of water}$$

The pipeline length necessary to create this friction-head loss can be computed from information above. Recall from Figure 20.10 that 30 cm discharge pipeline head loss is 2.05 m per 30.5 m of pipeline length at a discharge velocity of 4.9 m/s. Therefore,

$$\frac{(30.5)\ 28.7}{2.05} = 427 \text{ m equivalent pipeline length}$$

The actual pipeline length equals the equivalent pipeline length divided by the equivalent pipeline length correction factor (1.5):

$$\frac{427}{1.5} \cong 284 \text{ m}$$

To operate within the prescribed discharge pipeline pressure limits, Pierce (1970) developed two formulas for calculating the maximum and minimum distances required between the dredge pump and the booster pump. The formulas are quite useful in determining these distances when working with the smaller dredges commonly employed in lake restoration. Pierce pointed out that the approach described above can be used to determine spacing between pumps if more than one booster pump is required. The two formulas are:

$$L_{\max} = \frac{30.5 \,(H_d - H_s)}{1.5 h_{df}} \tag{20.15}$$

$$L_{\min} = \frac{30.5 \,(H_d + H_{TDH} - W_p)}{1.5 h_{df}} \tag{20.16}$$

where: $L_{\max}$ = maximum distance between the main dredge pump and the booster pump (meters of discharge pipeline), $L_{\min}$ = minimum distance between the main dredge pump and the booster pump (meters of discharge pipeline), H_d = dredge pump discharge head (meters of fresh water, which equals the discharge pressure gauge reading at the dredge pump), H_s = booster pump suction head (m), h_{df} = friction loss in the discharge pipeline (meters per 30.5 m of discharge pipeline; see Figure 20.10), H_{TDH} = booster pump total dynamic head (meters of water; equals the discharge pressure gauge reading at the dredge pump), and W_p = discharge pipe working pressure (meters of water).

From this analysis, conclusions are drawn for pumping from lake area 3 to disposal site 3:

1. The disposal area is beyond the efficient pumping capacity of the dredge pump, so a booster pump is required.
2. Discharge pipeline velocities can be increased to an acceptable level by employing either an identical 30 cm booster pump or the available 35 cm model.
3. Head discharge characteristic graphs are highly desirable, for booster pump selection, since the required head discharge characteristics of the booster are highly dependent on those of the dredge pump.
4. Locating the booster pump too close to or too far away from the dredge pump must be avoided. If the booster is too close excessive pipeline pressures will result. If the booster is too distant, it may operate under a suction lift that can cause pump cavitation, reduced output, and excessive equipment wear.

This information on hydraulic dredge selection should be helpful to the dredge-plan designer for selecting the proper equipment, and to the lake manager, to assure that the designer has selected the proper equipment. How dredging is actually conducted depends on the type of equipment selected and on site-specific conditions, all of which must be considered when developing the dredge operating plan. The example above can be used as a general guide, but when choosing a dredge for use there is no substitute for actual dredge pump discharge relationship curves for the dredges being considered (Pierce 1970). Pump suction heads can be obtained from these relationships or from pump manufacturers' specifications.

20.7.2 DISPOSAL AREA DESIGN

Once dredge equipment is selected, the other critical concern is the disposal area(s). Upland disposal is common. The challenge is to design and construct containment and disposal areas of adequate size and retention time to hold the dredged material volume and to reduce suspended solids concentrations to meet effluent requirements. Comprehensive guidance for design, operation, and management of upland confined disposal areas is available (USACOE, 1987). This guidance contains procedures for designing disposal areas for retention of suspended solids based on the settling characteristics of fresh water sediments (Montgomery, 1978, 1979, 1982). These procedures apply directly to lake dredging and are summarized in the following paragraphs, but the above references should be consulted for design details.

Field investigation of the dredge site must be conducted, to obtain disposal area design information. A field estimate of the in-place sediment volume is basic. Two laboratory tests also are needed. The first characterizes sediment, including natural water content, Atterberg limits, organic content, and specific gravity for fine grain sediments. Grain size analyses are adequate for coarse-grained sediment. The second determines sedimentation rate. Montgomery (1978) demonstrated that most fresh water dredge slurries could be characterized by flocculent settling tests, where particles agglomerate during settling, with different physical properties and settling rates. Hydraulically dredged lake sediments are characterized by this test procedure.

Montgomery (1978) prefaced the flocculent settling test procedures with a caution. He noted that an interface forming near the top of the settling column during the first day of the test indicates that sedimentation is governed by zone settling, and that a zone settling test should be conducted. Also, he indicated that zone settling, where the flocculent suspension forms a lattice structure and settles as a mass, might prevail at high solids concentrations or if sediments are contaminated with high levels of organics (lake sediment commonly contains 30% to 40% organics). The zone settling process transitions to a compression settling process in which settling occurs by compression of the lattice structure. Compression settling behavior governs the initial storage volume occupied by dredged sediment in a confined disposal site.

20.7.2.1 Flocculent Settling Procedure

1. A settling column is used (Figure 20.15). The test column depth should approximate the effective settling depth of the proposed containment area. A practical test depth is 2 m. The column should be at least 20 cm in diameter, with sample ports at 0.3 m intervals. The column should have provisions to bubble air from the bottom, to keep the slurry mixed during the column filling period.
2. Mix the sediment slurry to the desired suspended solids concentration in a container with sufficient volume to fill the test column.
3. Pump or pour the slurry into the test column, using air to maintain a uniform concentration during the filling period.
4. While the column is completely mixed, draw off samples at each sample port, determine the suspended solids concentration, average these values, and use the results as the initial concentration. After the initial samples are taken, stop the air bubbling and begin the test.
5. While the slurry is settling, withdraw samples from each sampling port at regular time intervals, and determine the suspended solids concentrations. Sampling intervals depend on the settling rate of the solids — usually at 30 min intervals for the first 3 h and then at 4 h intervals until the end of the test. Continue the test until the interface of solids can be seen near the bottom of the column and the suspended solids level in the fluid above the interface is <1 g/L. Test data are tabulated as in Table 20.2.
6. If an interface has not formed within the first day on any previous tests, run one additional test with suspended solids concentration high enough to induce zone-settling behavior.

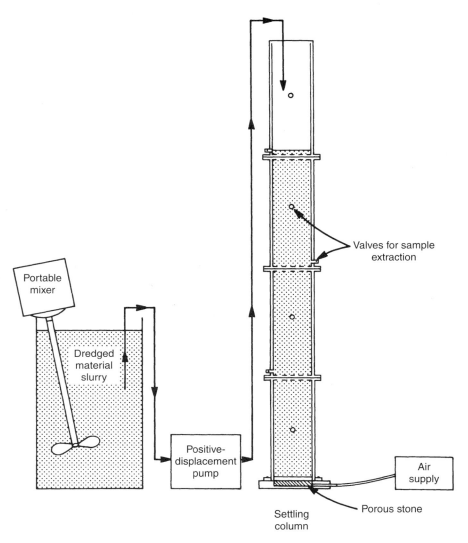

FIGURE 20.15 Schematic of fine grained sediment settling test equipment. (Modified from Montgomery, R.L. 1978. Methodology for Design of Fine-Grained Dredged Material Containment Areas for Solids Retention. Tech. Rept. D-78-56. U.S. Army Corps Engineers, Vicksburg, MS.)

This test should be carried out according to the procedures outlined below. The exact concentration at which zone settling behavior occurs depends on the sediments being used to estimate the volume required for dredged material storage.

20.7.2.2 Zone/Compression Settling Test Procedure

This test consists of placing slurry in a column similar to that in Figure 20.15 and recording the fall of the liquid–solid interface over time. The depth to the interface is then plotted as a function of time. From this plot, the slope of the constant settling zone of the curve represents the zone settling velocity, which is a function of the initial test slurry concentration. Information needed to design a containment area when zone settling characteristics prevail can be obtained by using the following procedure from Montgomery (1978).

TABLE 20.2
Observed Flocculent Settling Concentrations (g/L) with Depth

Time (min)	Depth From Top of Settling Column (m)					
	0.3	**0.6**	**0.9**	**1.2**	**1.5**	**1.8**
0	132.00	132.00	132.00	132.00	132.00	132.00
30	46.00	99.00	115.00	125.00	128.00	135.00
60	25.00	49.00	72.00	96.00	115.00	128.00
120	14.00	20.00	22.00	55.00	78.00	122.00
180	11.00	14.00	16.00	29.00	75.00	119.00
240	6.80	10.20	12.00	18.00	64.00	117.00
360	3.60	5.80	7.50	10.00	37.00	115.00
600	2.80	2.90	3.90	4.40	14.00	114.00
720	1.01	1.60	1.90	3.10	4.50	110.00
1020	0.90	1.40	1.70	2.40	3.20	106.00
1260	0.83	1.14	1.20	1.40	1.70	105.00
1500	0.74	0.96	0.99	1.10	1.20	92.00
1740	0.63	0.73	0.81	0.85	0.94	90.00

Source: Data from actual test on fresh water sediments (initial concentration = 132 g/L). Modified from Montgomery, R.L. et al. 1983. *J. Environ. Eng. Div. ASCE* 109: 466–484. With permission.

1. Use a settling column such as that shown in Figure 20.15. It is important that the column diameter be sufficient to reduce wall effects, and that the slurry column depth for the test is the same (or nearly the same) as disposal area slurry depths expected in the field.
2. Mix the slurry to the desired concentration and pump or pour it into the test column while mixing with air to maintain suspension concentrations between 60 and 200 g/L.
3. Record the depth to the solid–liquid (sediment–water) interface with respect to time. Observations must be made at regular intervals to gain data for plotting the curve of depth to interface vs. time. It is important to make enough observations to clearly define this curve for each test (see Palermo et al., 1978: pp. A3).
4. Continue the readings until sufficient data are available (tests should be repeated at least 8 times) to define the maximum point of curvature of the depth to interface vs. time for each test. These data are used to develop a zone settling velocity versus concentration curve.
5. The tests should be performed on sediment slurries at a concentration of about 145 g/L and continued for a period of at least 15 d to provide data for estimating volume requirements.

20.7.2.3 Design Procedures

Montgomery (1978) described a complete procedure for designing dredged material disposal areas, based on field and laboratory observations. He described the methods for both saltwater sediments and fresh water sediments, based on flocculent settling properties and zone/compression settling properties. Since this text is concerned with fresh water lake sediments that usually are flocculent, only that portion of Montgomery's design procedure is described below.

20.7.2.3.1 Estimate In Situ Sediment Volume

The initial step in any proposed dredging project is to estimate the amount of sediment to be removed. Methods for doing this were described earlier in this chapter. Any procedure that provides an accurate estimate for computing disposal volume and cost is satisfactory.

20.7.2.3.2 Determine Sediment Physical Characteristics

The information required is the same as that mentioned above in the section on disposal area design. Montgomery (1978) recommended use of the methods described by Palermo et al. (1978) for these determinations. However, Palermo does not include methods for all of the analyses. He refers to OCEDA (1970): a laboratory soils testing manual. Other standard soils testing manuals will provide most of the same information required for the following calculations. Fine-grained sediment samples obtained with a Petersen dredge are adequate for defining the *in situ* water content of sediment. This determination is critical to the disposal area design, because sediment water content of representative samples is used to determine the *in situ* void ratio (e_i) using the following equation

$$e_i = \frac{(w/100)\,G_s}{S_d/100} \tag{20.17}$$

where: w = water content of sediment (%), G_s = specific gravity of sediment solids, and S_d = the degree of saturation (equal to 100% for sediment) (see OCEDA, 1970).

20.7.2.3.3 Analyze the Proposed Dredging and Disposal Data and Laboratory Sedimentation Tests

Once tentative dredge selection is made, this information is used for disposal area design. For example, the designer must estimate the containment area influent rate, influent suspended solids concentration, effluent rate (for weir sizing), effluent suspended solids concentration allowed, and time required to complete the disposal activity. If previously described information is unavailable, assume the largest size dredge that "could be used" on the project to assure the disposal area is not undersized. Montgomery (1980a) recommended using a suspended solids concentration of 145 g/L (13% by weight) for design purposes if the actual concentrations are unknown. That figure, however, is derived from maintenance dredging experience on fresh water waterways, not from dredging lake-type sediments, so it should be used with caution when dealing with flocculent lake sediments. It is strongly recommended that the flocculent and/or the zone/compression sedimentation settling tests for fresh water be conducted according to the methods described above.

20.7.2.3.4 Design Method for Fresh Water Sediments

According to Montgomery (1980b), the following design method should be sufficient to remove suspended solids from fresh-water disposal area discharges, down to a level of < 1 g/L. If that level is insufficient to meet discharge requirements it will be necessary to treat the discharge with flocculents.

Flocculents commonly are used to settle fine particles in dredge material disposal areas. Flocs form complex matrices that settle slowly in water. The slow settling increases contact time with suspended particles to increase sediment removal efficiency. Therefore, greater ponding depth in the disposal area provides greater probability for contact among sediment and floc particles, thus improving settling efficiency.

Laboratory settling rate tests produced the data in Table 20.2 (Montgomery, 1978). Then

TABLE 20.3
Percent Initial Concentration[a] with Time

Time (min)	Depth From Top of Settling Column (m)		
	0.3	0.6	0.9
0	100.0	100.0	100.0
30	35.0	75.0	87.0
60	19.0	37.0	55.0
120	11.0	15.0	17.0
180	8.0	11.0	12.0
240	5.0	8.0	9.0
360	3.0	4.0	6.0
600	2.0	2.2	3.0
720	2.0	1.2	1.4

[a] Initial suspended solids concentration = 132 g/L.

Source: Modified from Montgomery, R.L. et al. 1983. *J. Environ. Eng. Div. ASCE* 109: 466–484. With permission.

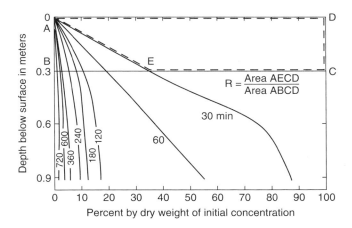

FIGURE 20.16 Profile of percent initial concentration vs. depth. (Modified from Montgomery, R.L. et al. 1983. *J. Environ. Eng. Div. ASCE* 109: 466–484. With permission.)

1. Arrange the data in Table 20.2 into the form of Table 20.3 (as a percentage of the initial conc. (132 g/L), determined as: Column concentration at test start = 132 g/L; concentration at 0.3 m, at time = 30 min, is 46 g/L (Table 20.2); percent of initial concentration is 46 ÷ 132 = 0.35 = 35%; calculate for each time and depth to create Table 20.3.
2. Plot Table 20.3 data as in Figure 20.16. The solid curved lines represent the concentration depth profile at various times during settling. Letters appearing on the horizontal depth lines (0 m and 0.3 m) indicate area boundaries. Labeling the other concentration/depth intersects (0.6 m and 0.9 m) in Figure 20.16 will assist with computing other area boundaries.

The design concentration, CD, is the average concentration of the dredged material in the containment area at the end of the disposal activity. Use the following steps to compute CD:

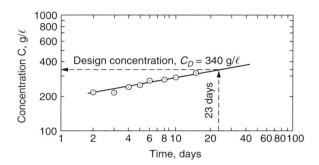

FIGURE 20.17 Concentration determined from column test vs. time. (From Montgomery, R.L. et al. 1983. *J. Environ. Eng. Div. ASCE* 109: 466–484. With permission.)

1. Compute concentration vs. time for the 15 d settling test and organize the data in tabular form. Assume zero solids in the water above the solids interface, to simplify calculation.
2. Plot the tabular form concentrations vs. time on log-log paper (Figure 20.17).
3. Draw a straight line through the data points representing the consolidation zone.
4. Estimate the time of dredging by dividing the dredge production rate into volume of sediment to be dredged.
5. Estimate the concentration at time $t_{1/2}$ (half the time required for the disposal activity, determined in the previous step), using the figure developed in Steps 2 and 3. This time is an approximation of the average time of residence for the dredged material in the containment area. Since concentration is a function of time, one half the dredging time would represent a period during which one half of the dredged material would have been in the area longer than, and the other half less than, a time equal to one half the dredging time. This value is the design solids concentration, CD.

Laboratory tests indicate 15% of the sediment to be dredged is coarse grained (>200 sieve); thus, from Dead Lake the volume of coarse-grained material (V_{sd}) is

$$V_{sd} = 1,530,000(0.15) = 229,500 \text{ m}^3$$

and the volume of fine-grained material (V_i) is

$$V_i = 1,530,000 - 229,500 = 1,300,500 \text{ m}^3$$

20.7.2.3.5   Compute Detention Time Required for Sedimentation

1. Calculate removal percentage at depths of 0.3, 0.6, and 0.9 m for various times, using the plot illustrated in Figure 20.16. The removal percentage for depth d_1 (0.3 m) and t = 30 min is computed from Formula 20.18:

$$R = \frac{\text{Area A, E, C, D}}{\text{Area A, B, C, D}} \times 100 \tag{20.18}$$

where: R = the removal percentage and the area letters correspond to the letters used in Figure 20.16 to indicate the area boundaries for the total area down to depth 0.3 m (A, B, C, D), and the area to the right of the t = 30 min line (A, E, C, D). Calculations are repeated for each depth as a function of time.

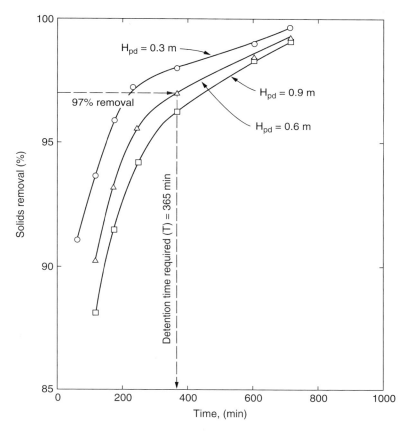

FIGURE 20.18 Solids removal vs. time, as a function of depth. (Modified from Montgomery, R.L. et al. 1983. *J. Environ. Eng. Div. ASCE* 109: 466–484. With permission.)

2. Plot the solids removal percentages vs. time calculations as in Figure 20.18.
3. Theoretical detention time (T) can be selected from Figure 20.18 for various solids removal percentages. Average ponding depth (H_{pd}) is 0.6 m. Initial sediment concentration is 132 g/L. The effluent suspended solids (SS) requirement is 4 g/L. Therefore, the effluent SS requirement is $(C_i - C_e) \div C_i$, or $132 - 4 \div 132 = 0.97$ or 97%. Then: enter Figure 20.18 on the x axis at 97%, read over 0.6 m, and read time (T) on the y axis: $T = 365$ min.
4. Apply a correction factor of 2.25 to the theoretical detention time T, to correct for short-circuiting and dispersion of flows through the disposal basin according to Equation 20.19 (Montgomery, 1978):

$$T_d = 2.25T \qquad (20.19)$$

where: T_d = design detention time, and T = theoretical detention time.

$$T_d = 2.25 \times 365 = 822 \text{ min}$$

The pond detention time required for dredged material to settle to an acceptable level of 4 g/L is 822 min. The above procedures are aimed at providing sedimentation basins with sufficient areas and detention times to accommodate continuous hydraulic dredge disposal activities while also meeting requirements for effluent suspended solids. Therefore, the basins must also be designed to meet the volume requirements of the job. The sedimentation basin total volume requirement includes volume for storage of dredged material, volume for sedimentation (ponding depths), and freeboard volume (volume above water surface). Coarse-grained material storage volume (> 200 sieve) must be determined separately, since this material behaves differently from the fine grained (< 200 sieve) material.

20.7.2.3.6 Estimate the Sedimentation Basin Dredged Material Volume

1. Compute the average void ratio of fine grained dredged material in the sedimentation basin at the completion of the dredging operation, using the design concentration determined in earlier steps as dry density of solids. Use the following equation to determine void ratio

$$e_o = \frac{G_s \gamma_w}{\gamma_d} - 1 \qquad (20.20)$$

where: e_o = average void ratio of dredged material in the sedimentation basin at the completion of the dredging operation, G_s = specific gravity of sediment solids, γ_w = density of water (g/L), and γ_d = dry density of solids (g/L) at design concentration (CD = γ_d) (from Figure 20.18).

2. Compute the changes in volume of fine-grained sediments after disposal in the sedimentation basin

$$\Delta V = V_i \frac{e_o - e_i}{1 + e_i} \qquad (20.21)$$

where: ΔV = change in volume of fine grained sediments after disposal in the sedimentation basin (m³), e_i = in situ void ratio computed as; specific gravity of sediment solids ÷ degree of saturation (equal to 100% of sediments), and V_i = volume of fine grained sediments (m³).

3. Compute the volume required in the sedimentation basin for dredged material

$$V = V_i + \Delta V + V_{sd} \qquad (20.22)$$

where: V = volume of dredged material in the sedimentation basin at the end of the dredging operation (m³) and V_{sd} = volume of sand (compute using 1:1 ratio) (m³).

20.7.2.3.7 Estimate Basin Depth

The above procedures provide a design detention time, T_d, required for sedimentation of fine-grained dredged material. Equations 20.20 to 20.22 estimate the volume and corresponding depth requirements for solids storage in the containment area. Topography and surficial geology are significant relative to the average containment area depth. The following procedures estimate the thickness of dredged material at the end of disposal (fresh water sediments) (Montgomery, 1978).

1. Compute the volume required for sedimentation

$$V_B = Q_i T_d \qquad (20.23)$$

where: V_B = sedimentation basin volume needed to meet requirements for suspended solids in effluent (m³), Q_i = influent rate (m³/h)[$Q_i = A_p V_d$, assuming V_d = 4.6 m/s in the absence of data, convert Q_i from m³/s to m³/h, A is the cross-sectional area of the dredge discharge pipe (m²), V_d is the velocity of dredge discharge (m/s)], and T_d = required design detention time, from Equation 20.19.

2. Consult with soils design engineers to determine maximum confining dike height (D) allowances.

3. Compute the required design area as a minimum required surface area for solids storage.

$$A_d = \frac{V}{H_{dm(max)}} \tag{20.24}$$

where: A_d = the design basin surface area (m²) and V = volume of dredged material in the sedimentation basin at the end of the dredging operation (m³) from Equation 20.22, $H_{dm(max)} = D - H_{pd} - H_{fb}$ (m), where D = dike height (m), H_{pd} = average ponding depth over the area (m), and H_{fb} = free board above basin water surface to prevent wave overtopping and subsequent damage to the confining earth dikes (m). $H_{dm(max)}$ is the thickness of dredged material in the basin corresponding to a known volume of material. Thickness decreases as surface area increases. Minimums of 0.6 m are recommended for H_{pd} and H_{fb}, to account for fetch and wind.

4. Evaluate volume available for sedimentation near the end of the disposal operation:

$$V^* = H_{pd} A_d \tag{20.25}$$

where: V^* = volume available for sedimentation near the end of the disposal operation (m³), H_{pd} = average ponding depth over the area (m), and A_d = design basin surface area (m²).

5. Compute V^* and V_B. If the volume required for sedimentation is larger than V^*, the sedimentation basin will not meet the suspended solids effluent requirements throughout the disposal operation. If that is the case, one of the following measures can be taken to ensure that effluent suspended solids concentrations are met:
   a. Increase the design area A_d.
   b. Operate the dredge intermittently whenever the calculated V^* becomes less than V_B, or use a smaller dredge.
   c. Provide treatment of disposal pond effluent to remove solids (see Barnard and Hand, 1978).

Sediment that exhibits zone-settling characteristics requires a different disposal area design method. If zone settling is encountered, refer to USACOE (1987). According to Palermo (2003) the 1987 USACOE Engineering Manual is being revised, but it will not be available until late 2004. It remains to be seen if previous criticism of the manual to adequately address flocculent organic sediment found in lakes will be considered in the revision. Dunst et al. (1984) reported that the specific gravity of sediment from Lilly Lake, WI, was only 1.02. Therefore, one should be extremely cautious about relying completely on the above disposal area design procedures. When possible, it would be wise to oversize disposal areas by 5% to 10% to alleviate most concerns associated with their operation.

20.8 CASE STUDIES

In 1974, Dunst et al. (1974) identified 33 lake-dredging projects completed or in progress. Seven years later, Peterson (1981) listed twice that number. Both the 1974 and 1981 listings probably were incomplete, since many small lake dredging projects are completed, but never reported in the literature. A quick survey for the rewriting of this book found similar circumstances in 2002. Even where unpublished records are available little data are collected beyond the dredging completion date, so few well documented, long-term effect assessments are available. The following case studies represent relatively well-documented projects that illustrate various approaches to (or purposes for) sediment removal. Peterson (1981) described additional case studies.

Not all dredging case studies to control internal nutrient recycling are successful. There have been failures or marginal successes that can usually be traced to inaccurate pre-dredging assessment, to inadequate amounts of sediment removed (Brashier et al., 1973; Churchill et al., 1975; George et al., 1981; Ryding, 1982), to poor dredging technique (Gibbons and Funk, 1983) and/or to lack of proper watershed erosion control measures (Garrison and Ihm, 1991). The need for accurate assessment of the problem cannot be overemphasized. For example, Garrison and Ihm (1991) found that the infilling rate for Lake Henry, Wisconsin, was 11,000 m³/yr rather than the predicted 4,600 m³/yr causing a gross underestimate of the post dredging infiltration rate.

20.8.1 Lake Trummen, Sweden

Lake Trummen, Sweden has the most thoroughly documented, long-term evaluation of lake dredging. Located near the city of Växjö, in south-central Sweden, Lake Trummen began receiving domestic waste discharge from the kitchen of St. Sigfrid's Hospital in 1895 (Sjön Trummen i Växjö, 1977). Modern toilets, served by septic tanks near the lakeshore, were installed in 1936. In 1943, a flax mill began discharging to the 100 ha lake. From that time, winter fish kills became common and lake quality declined rapidly. Sewage discharge to the lake stopped in 1959, after the flax dressing plant closed and wastewater from the surrounding area was connected to a municipal treatment system. Despite the wastewater diversion, the lake showed no sign of recovery. In fact, conditions became so bad in the early 1960s that citizens of Växjö considered filling the basin (Björk, 1974). Internal nutrient recycling from the extremely rich upper 1 m of sediment, a process later recognized as rather common in shallow lakes, was discovered (Björk, 1974).

One half meter of sediment was dredged from the main lake basin in 1970, and another one half meter was removed in 1971 (Björk, 1974). This removal increased the mean depth of the lake from 1.1 to 1.75 m and the maximum depth from 2.1 to 2.5 m. The total volume of sediment removed was about 30×10^5 m³. By any standard, the lake was still extremely shallow and could have remained eutrophic, except that the sediment skimming treatment reduced the total P content in the surface layer of sediment from approximately 0.78 to 0.03 mg/kg (Figure 20.19; Sjön Trummen i Växjö, 1977).

Part of the dredged material was disposed in shallow, diked-off bays. The remainder went to upland diked ponds, where return flow was treated with aluminum sulfate to reduce the TP concentration from about 1 mg/L to 30 µg/L. Dried dredge material was sold as topsoil dressing for about $5.73 (2002 U.S.)/m³.

The role of the sediment in recycling nutrients was reduced substantially following dredging. TP in the surface water of Lake Trummen after dredging was reduced approximately 90% (from 600 µg/L to a range of 70 to 100 µg/L). Phosphorus reductions during summer were especially noticeable. Total nitrogen concentrations also were reduced approximately 80%, from 6.3 to 1.3 mg/L (Andersson, 1984). There was a temporary increase of P in 1975 that was associated with a large influx of planktivorous cyprinid fish (Andersson et al., 1978). This type of fish population is generally undesirable and associated with reduced water quality. During 1976, about 2 metric tons (30 kg/ha) of fish were taken from the lake. The practice continued through 1979 and P concentrations remained low. When fishing stopped in 1979, the P levels began to rise again and Andersson (1988) suggested that prolonged or repeated fishing is probably necessary to maintain the low nutrient levels in Lake Trummen. Shapiro et al. (1975) were correct when they suggested that eutrophication symptoms in shallow lakes might be controlled by fish population manipulations (see Chapter 9).

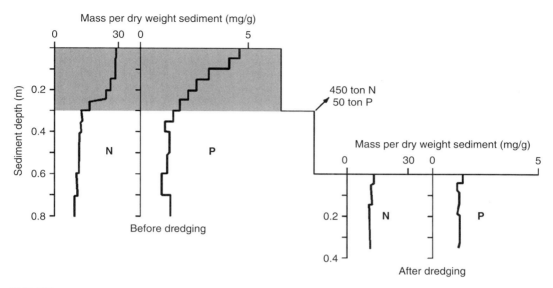

FIGURE 20.19 Lake Trummen sediment nutrient concentrations before and after dredging. (Redrawn from Sjön Trummen i Växjö, 1977. *Förstörd, Restaurerad, Pånyttfödd. Länsstyrelsen I Kronobergs Län.* Växjö Kommun.)

The significant nutrient reduction in the lake resulted in equally significant biological changes. The Shannon diversity index for phytoplankton rose from 1.6 in 1968 to 3.0 in 1973 (Cronberg et al., 1975). SD transparency went from 23 to 75 cm during the same period. The blue-green algal biomass was reduced dramatically, and the nuisance species *Oscillatoria agardhii* disappeared completely. Phytoplankton productivity decreased from 370 g C/m^3 in 1968–1969 to about 225 g C/m^3 in 1972–1973. The latter is demonstrated by the phytoplankton biomass figure (Figure 20.20) of Cronberg (2004). Figure 20.20 also shows reduced phytoplankton biomass from 1976 through 1979 while the cyprinid fish harvest was conducted. Phytoplankton biomass increased again from 1980 through 1987, when fishing was not done, but it never reached the levels that had persisted before dredging in 1970.

The effect of dredging on the benthic community of Lake Trummen was negligible. A year after dredging, tubificid oligochaetes and chironomids were more numerous than before dredging, but the total number of benthic organisms changed little (Andersson et al., 1975). Rapid recolonization was attributed to the mobility and constant swarming of chironomids.

A photo of Lake Trummen during dredging in 1970 shows how it was choked with aquatic weeds, had severe algal blooms, and suffered from winter fish kills (Figure 20.21). A 1972 photo (Figure 20.22), immediately following restoration, shows Lake Trummen as an environmental asset to the urbanized setting by supporting swimming, wind surfing, and sport fishing (Björk, 1985). A subsequent description of the Lake Trummen experience by Andersson (1988) indicated there had been no significant deterioration in lake quality from the restored level.

A Lake Trummen photo by Andersson in 2001 attests to the continued high quality of the lake (Figure 20.23). The location of the building spire on the distant shore in Figures 20.21 through 20.23 serves as an orientation landmark. At a NALMS symposium in Madison Wisconsin in 2001 one of us (Peterson) was asked, "Why do you keep talking about Lake Trummen? It is a very old example!" In this case, the question was answered by the comment, "Indeed, it is a very old example of a very successful project. That is precisely why we continue to talk about it." The photos in Figures 20.21 through 20.23, personal communication from Gunnar Andersson (2001), and Cronberg's biomass figure (Figure 20.20) are strong indicators that the effects of dredging at Lake Trummen have lasted 35 years. Figure 20.20 is probably the longest continuous record of lake

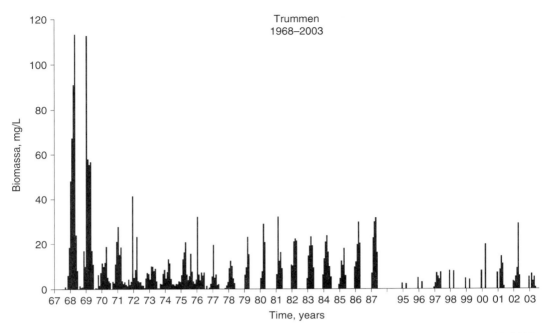

FIGURE 20.20 Biomassa (biomass) in Lake Trummen, Sweden, before and after dredging: 1968–2003. (Courtesy of Gertrud Cronberg, Department of Limnology, University of Lund, Sweden.)

FIGURE 20.21 Lake Trummen, Sweden at the time of dredging in 1970. Note location of far-shore steeple. (Photo courtesy of Gunnar Andersson, Department of Limnology, University of Lund, Sweden.)

water quality in relationship to a lake restoration project. While data from 1988 through 1994 are missing from this figure, it is evident from the 1995 through 2003 records that water quality in Lake Trummen remained at levels similar to that immediately following the 1970 dredging. Cronberg (2004) provided no explanation for the noticeably reduced biomasses from 1995 through 1999, but did note that she feared water quality was beginning to decline in 2002 and 2003. As a result,

FIGURE 20.22 Lake Trummen, Sweden immediately after the completion of dredging in 1971. Note location of far-shore steeple. (Photo courtesy of Gunnar Andersson, Department of Limnology, University of Lund, Sweden.)

FIGURE 20.23 Lake Trummen, Sweden 30 years after (2001) the completion of dredging. Note location of far-shore steeple: (Photo courtesy of Gunnar Andersson, Department of Limnology, University of Lund, Sweden.)

wetland filter systems were being installed to treat increased urban runoff. If dredging (and associated fish removal) worked this well on a shallow lake like Trummen, there is little reason to think that it will not work on other lakes where nutrient and sediment inputs are controlled and where low P level sediments can be exposed.

There are, however, two key factors that must not be overlooked in the success at Lake Trummen. One is continued fish management following dredging (removal of rough fish like roach and bream) from 1976 through 1978. The second is associated with the "special nozzle" suction dredge that permitted precision removal of soft sediment in two successive one-half meter increments. Precision dredging technology might play a much larger role in the long-term success story at Lake Trummen than was thought previously (see the Lake Järnsjön case history). It is becoming evident with more in-depth case histories that "how" a lake is dredged (equipment used) is every bit as important, if not more important, to success of the project as is the dredging plan (what sediment to remove) itself. Precision dredging equipment has great advantages over more conventional equipment that can actually make a bad situation worse (Gibbons and Funk, 1983).

Van der Does et al. (1992) compared lake eutrophication recovery from external P reduction without and with accompanying dredging. They concluded that in-lake P reductions can be achieved with or without dredging of rich surface sediments, i.e., by diversion alone (Chapter 4). However, they presented no clear differences and concluded that the choice between dredging and not dredging depended on depth, residence time, and sediment characteristics.

20.8.2 Lilly Lake, Wisconsin

20.8.2.1 Initial Diagnosis and Results

Lilly Lake is a 37 ha closed basin lake in southeastern Wisconsin. Its agricultural watershed is only 155 ha. The lake suffered several years of infilling by aquatic plants. By 1977, shoaling had reduced the lake to a maximum depth of 1.8 m and a mean depth of only 1.4 m. The basin contained more than 10 m of partially decomposed plant materials. The Wisconsin Department of Natural Resources (WDNR, 1969) reported that chemical eradication and restocking of centrarchid and northern pike failed due to severe winterkill problems. Infilling by plant material had reached 0.5 cm/year (Dunst, 1981).

Restoration of Lilly Lake for fish management purposes was recommended by the WDNR (1969). The plan called for deepening at least 10% of the basin to a depth of 6 m. This required removing 665×10^3 m^3 of sediment. Hydraulic suction dredging was proposed, with the idea that sediment would flow to the dredge head as a cone of depression was created. The lightweight sediment (specific gravity equaled 1.02), however, was much more cohesive than the pre-dredging assessment indicated, and a cutterhead had to be employed.

Dredging began in July 1978 and continued until near freeze-up at the end of October. It began again in May 1979 and was completed at the end of August. Most of the dredged material was pumped approximately 3 km to an abandoned gravel pit. Although the groundwater table rose temporarily due to this disposal, it resumed normal levels shortly after disposal ceased in both 1978 and 1979, and monitoring wells revealed no adverse impact due to chemicals (Dunst et al., 1984). Prior to dredging in 1977, the total inorganic nitrogen (TIN) concentrations were near detection limits (Figure 20.24). Although TP increased to more than 40 µg/L during July, August, and September before dredging (Figure 20.25), the increase did not produce excessive phytoplankton concentrations (Figure 20.26). Increased TP was likely caused by decomposition of sloughed macrophyte parts and filamentous algae, while shading from sediment resuspension probably controlled phytoplankton populations (chlorophyll *a*). Much of the TP increase during dredging was due to resuspended sediment. Despite previously reported winterkills of fish, the DO concentrations remained above 6 mg/L throughout the study, except for February and March of 1979 (Figure 20.27). These DO declines were preceded by large increases in TIN and TP late in the fall of 1978. The DO depletion may have resulted from oxygen demand from organics suspended as a result of late fall dredging. There was no similar oxygen decline the following year, when dredging stopped in August. The water quality *per se* of Lilly Lake was not exceptionally poor prior to dredging, but there was very little water in the lake (mean depth = 1.4 m), due to infilling by macrophytes and partially decomposed macrophytes.

Short-term changes that were adverse, but not severe, occurred during active dredging. TIN increased rapidly (Figure 20.24) when dredging began in July 1978, due mostly to liberation of ammonium nitrogen from the sediment (Dunst, 1981). It remained in excess of 3.5 mg/L during

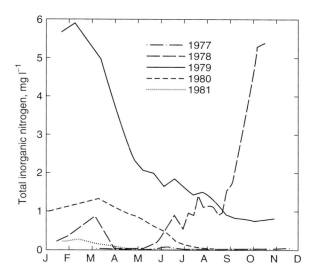

FIGURE 20.24 Total inorganic nitrogen in Lilly Lake, WI. (Data courtesy of Russell Dunst, Wisconsin Department of Natural Resources, Madison.)

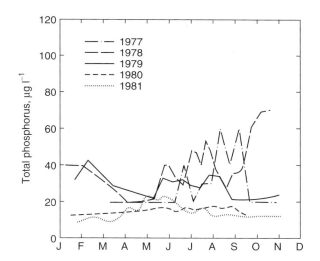

FIGURE 20.25 Total P in Lilly Lake, WI. (Data courtesy of Russell Dunst, WDNR, Madison.)

most of the 1978–1979 ice-cover period, then declined steadily throughout the spring and summer, even though dredging ended in November 1978 and began again in May 1979.

The TP concentrations behaved similar to TIN, except there were minor peaks (lower than pre-dredging, 1977) in TP during the summer of 1978. TP remained relatively high during the fall and winter of 1978–1979, probably due to increased levels of suspended solids in the water column due to dredging. This is confirmed in part by elevated turbidity and BOD levels during the same period (Dunst et al., 1984).

Phytoplankton responded in a predictable manner to the increased nutrient concentrations when dredging began. The July to September average gross primary productivity increased from 185 mg C/m^3 per day in 1976 and 140 mg C/m^3 per day in 1977 to over 1,000 mg C/m^3 per day in 1978 (Dunst, 1981).

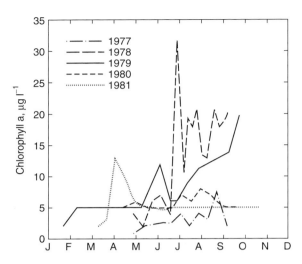

FIGURE 20.26 Chlorophyll *a* in Lilly Lake, WI. (Data courtesy of Russell Dunst, WDNR, Madison.)

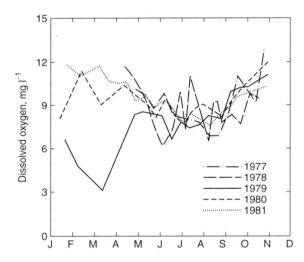

FIGURE 20.27 DO in Lilly Lake, WI. (Data courtesy of Russell Dunst, WDNR, Madison.)

With the completion of dredging in September 1979, all of the above variables returned to near or below pre-dredging levels. The 1980 and 1981 TP concentrations were significantly lower than concentrations prior to or during dredging. Chl *a* concentration was slightly higher in 1980 and 1981 than in the pre-dredging year of 1977, but the increase was relatively insignificant (approximately 3 to 4 μg/L). More importantly, the water storage capacity (basin volume) of the lake increased 128% (Dunst, 1981), and the overall objective of increasing the recreational potential of Lilly Lake appeared to be realized. The two-summer inconvenience of dredging resulted in what lake users hoped would be a long-term benefit.

20.8.2.2 Long-Term Effects

From April 1989 through September 1989, Chl *a* in Lilly Lake ranged from 0.9 to 5.3 μg/L (mean = 2.6 μg/L) (Garrison, 1989). While the lake never demonstrated algal problems (it was dominated by macrophytes), these continued low chlorophyll values in combination with a diverse macrophyte community attest to the continued long-term success of this project. Secchi disk (SD)

TABLE 20.4
Summary of Selected Limnological Parameters Immediately following Dredging and a Decade Later (Means for June–Sep. 1981; Annual Mean 1989) in Lilly Lake, WI

	Year	
Parameter	1981	1989
TP (µg/L)	14	9
Total N (mg/L)	1.1	0.8
Total inorganic N (mg/L)	0.02	0.02
Chl *a* (µg/L)[a]	<5	3.3
SD (m)	2.2	4.0

[a] Summer mean.

Source: Garrison, P.J. and D.M. Ihm. 1991. Final Annual Report of Long Term Evaluation of Wisconsin Clean Lake Projects: Part B: Lake Assessment. WDNR,, Madison.

depths ranged from 5.9 m in the spring to 3.2 m in July, further confirming the overall high quality of the lake. TP concentrations in May and August 1989 were 9 and 12 µg/L, respectively, versus the 30 to 60 µg/L ranges found in the lake prior to dredging. TIN reported for the same dates was < 20 µg/L (Table 20.4).

Dredging did not create a completely weed-free lake. High water transparency (SD = 2.2 m) in 1981 permitted plant growth to a depth of approximately 3.7 m, close to the growth depth (3.5 m) predicted by Equation 20.1 (Dunst et al., 1984). Plant growth over 75% of the lake basin, in 1981, was dominated by *Chara* and *Myriophyllum* spp, and was particularly offensive because it grew to the surface in many areas. In 1982, plant growth covered the same area, but remained 1.2 and 1.8 m below the water surface except in the near-shore areas. Similar outcomes could be expected in other shallow lakes, which might make the one time expenditure associated with dredging more acceptable relative to yearly expenditures for harvesting, herbicides, etc., that usually provide few if any long-term solutions to lake deterioration.

While the 1980 and 1981 macrophyte biomass was considerably less (about 100 g/m²) than during the pre-dredging period (685 and 335 g/m² in 1976 and 1977, respectively), Cooke et. al. (1986) posed concern about the longer term effectiveness of dredging to control nuisance macrophytes due to their rapid reinvasion once dredging stopped. Dunst et al. (1984), indicated that two years after dredging a mixture of rooted macrophytes replaced the original *Chara* sp. that invaded the lake the first year after dredging.

A more recent report (WDNR, 1990) indicated the Lilly Lake macrophyte control project has been a long-term success. Eleven years after dredging there was no measurable sediment infilling (Figure 20.28). The macrophyte community is highly diverse. More importantly, macrophytes did not clog the lake surface as they did prior to dredging. In fact, results indicate that plant surfacing up to 1991 was rare. Garrison and Ihm (1991) summarized the macrophyte biomass changes in Lilly Lake from 1976 through 1990 (Table 20.5).

Ten years after dredging, macrophyte biomass was nearly as great in deeper water as it was over the entire shallow lake prior to dredging (Table 20.5). However, as noted previously, these plants pose little nuisance problem since they seldom reach the water surface. Plant succession in this system is interesting. Prior to dredging the population was dominated by *Potomogeton robbinsii* on both muck and sand substrates. During the first full year following dredging (1980) the dominant plant was *Chara* sp. In 1981, dominance shifted to *Myriophyllum* spp. on muck substrates, but

Pre-dredge

Post-dredge

1990

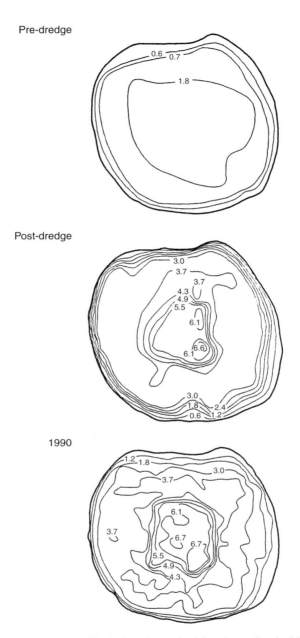

FIGURE 20.28 Morphometric maps of Lilly Lake prior to dredging, immediately following dredging, and a decade later. (From Garrison, P.J. and D.M. Ihm. 1991. Final Annual Report of Long Term Evaluation of Wisconsin Clean Lake Projects: Part B: Lake Assessment. WDNR, Madison.)

Chara sp. remained on sandy substrates. By 1990, the dominance reverted to *P. robbinsii*, the pre-dredging condition, on both muck and sand substrates.

Prior to dredging, Lilly Lake suffered from severe DO depletion and fish kills in the winter months. Since dredging this problem disappeared completely. Winter oxygen levels just below the ice are typically 8 mg/L while those in deep waters are not below 4 mg/L. A final testimonial to project success is that the lake supports as much recreational use today as it did immediately after dredging was completed (Garrison and Ihm 1991; Garrison, 2002).

TABLE 20.5
Macrophyte Biomass in Lilly Lake before and after Dredging (g/m² Dry Weight)

		Depth (m)				
		0–0.5	0.5–1.5	1.5–2.5	2.5–3.5	3.5–4.5
1976	Muck	790	898	659	—	—
	Sand	38	71	—	—	—
1977	Muck	345	432	335	—	—
	Sand	76	125	—	—	—
1980	Muck	—	283	—	9	—
	Sand	194	485	—	—	—
1981	Muck	—	309	292	94	—
	Sand	49	237	224	—	—
1990	Muck	—	—	770	834	730
	Sand	118	315	—	—	—

Source: Garrison, P.J. and D.M. Ihm. 1991. Final Annual Report of Long Term Evaluation of Wisconsin Clean Lake Projects: Part B: Lake Assessment. Wisconsin Dept. Nat. Res., Madison.

Before Lilly Lake, control of algal productivity by dredging appeared to be more successful than control of macrophytes. However, Lilly Lake changed that perception. The Lilly Lake experience demonstrates the utility of using the relationship between Secchi depth and the maximum depth of plant growth (Equation 20.1) or Chapter 11; Table 11.3) when designing a dredging project for macrophyte management. The Lilly Lake experience demonstrates macrophyte control, not eradication. Significant amounts of plants remain in the lake, but their nuisance factor is noticeably reduced due to increased water depth, some species shifts, subsequent increased open water, and thus, more general lake use.

20.8.2.3 Other WDNR Dredging Experiences

While Lilly Lake is a notable success story, not all WDNR dredging projects have been successful. Bugle Lake, Lake Henry, and the Upper Willow Flowage (all small, shallow impoundments, Table 20.6) in Wisconsin all received similar treatments. The treatments included stream bank stabilization via riprapping, resloping and seeding of stream banks, selective fencing, and construction of cattle crossings. All of the impoundments were then dredged (Garrison and Ihm, 1991).

In 1971, stream-bank stabilization was implemented along 19 km of the 29 km of Elk Creek, upstream from Bugle Lake. Riprapping was done on 3 km of stream bank and 1.1 km was resloped and seeded. In 1980, 4.4 km more riprapping was done, along with 136 m of resloping and seeding. Three cattle crossings, one cattle ramp and 2,488 m of fencing were constructed along the creek. In 1980 and 1981, 109,800 m³ of sediment was dredged from the lake, increasing its volume from 74,760 to 209,600 m³. By 1989, much of the 3 to 4 m increased dredging depth decreased to 2 m due to infilling. The entire project proved to be for naught when spring ice flows in 1990 burst the dam and drained Bugle Lake. Perhaps this amounts to stream restoration.

Lake Henry had 2.6 km of stream bank (Trempealeau River) riprapped, resloped, and seeded in 1977 and 1978. Also, 12 km of fencing was installed and 11 stream crossings constructed. In 1979, 176,000 m³ of sediment was removed from the lake, increasing the volume from 88,000 to 264,000 m³. By 1991, the lake volume was reduced by 105,000 m³ (Garrison and Ihm, 1991). This represented infilling at nearly twice the predicted rate. The mean annual infilling rate from 1984 to 1991 was 76,000 m³ giving the reservoir a life expectancy of approximately 21 years.

TABLE 20.6
Selected Features of Three Wisconsin Impoundments

Parameter	Bugle Lake	Lake Henry	Upper Willow Flowage
Lake area	14 ha	17 ha	81 ha
Watershed area	29,000 ha	47,000 ha	45,000 ha
Land use			
Agriculture	70%	65%	80%
Forest	—	35%	—
Other	30%	—	20%
Estimated annual soil loss to lake	90,000 m^3	18,000 m^3	?
Dredging purpose	Deepen entire lake	Deepen entire lake	Create sediment trap at inflow

Source: Garrison, P.J. and D.M. Ihm. 1991. Final Annual Report of Long Term Evaluation of Wisconsin Clean Lake Projects: Part B: Lake Assessment. WDNR,, Madison.

At the Upper Willow Flowage, WI, 25 of the 165 sections of stream bank susceptible to erosion were resloped, seeded, and riprapped. A total of 4,200 m^2 of riprap was placed and 3,500 m^2 of eroded banks were stabilized by sloping and reseeding. In 1982, dredging removed 153,000 m^3 of sediment from the lake, which included a 23,000 m^3 sediment trap near the inflow. By 1990, only a small portion of the dredged area refilled, however, the sediment trap was full. Garrison and Ihm's (1991) description of the Upper Willow Flowage indicated that sediment trapping protected the lake in general, but that the sediment traps needed cleaning at about 8 year intervals. This project appears to be more successful than the other two, but demonstrates that, in most cases, considerable maintenance is required. These small flowage projects also demonstrate that infilling probably occurs much more rapidly than predicted. Many predictions are based on miscalculations of infilling rates, producing a hidden cost for what are likely already expensive dredging projects.

20.8.3 LAKE SPRINGFIELD, ILLINOIS

This case history provides detailed information on feasibility studies and permit requirements. Lake Springfield is a 1,635 ha water supply reservoir in central Illinois (Figure 20.29). It provides cooling water for two coal fired power plants and drinking water for the city of Springfield. Since impoundment in 1935 it accumulated nearly 5.9 metric tons of sediment from its 689 km^2 watershed (Hinsman and Skelly, 1987). This reduced the reservoir storage capacity by 9.5 million m^3 (13.26% of its original capacity), promoted nuisance aquatic plant growth in the upper reaches of the lake, and reduced the recreational utility. Agricultural runoff (88% of the watershed) reduced fish habitat and contaminated several species of fish with toxic substances to the extent that the Illinois Environmental Protection Agency (IEPA) recommended against human consumption.

Buckler et al. (1988) reported on the history and three-phase (dredging, shoreline stabilization, and soil conservation in the watershed) remediation of Springfield Lake. About 50% of the sedimentation in the lake accumulated in a relatively small area of the lake. Highway and railroad bridges across upper areas of the lake constricted inflow and inadvertently trapped sediment (Figure 20.30). Sediment loading to the lake threatened the long-term water supply for the city, that caused the following expenses: (1) $12 million to acquire an alternate water supply, (2) $46,000/year to remove increased turbidity, and (3) approximately $1,117,500 in annual deferred maintenance cost due to the accumulation of approximately 289,909 m^3 of sediment per year. These factors prompted the city of Springfield to address the sedimentation problem.

The Springfield Lake feasibility study began in April, 1985 ($60,000 budgeted) and was designed to: (1) evaluate and prioritize proposed dredging sites, (2) survey and estimate quantities

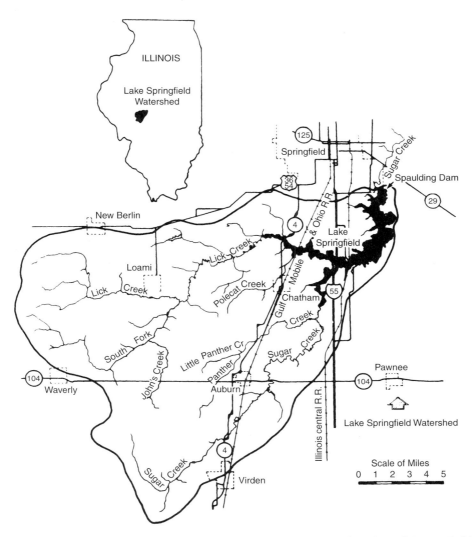

FIGURE 20.29 Lake Springfield watershed. (From Buckler, J.H. et al. 1988. *Lake and Reservoir Manage.* 4(1). With permission.)

of sediment to be removed, (3) determine cost estimates to remove the sediment, (4) identify potential sediment disposal sites, and (5) provide general engineering assistance for permit applications and environmental assessments.

20.8.3.1   Sediment Removal Guidelines

The study developed general guidelines for sediment removal. The primary objective was to remove sediment in a manner that conformed closely with the original lakebed configuration since cutting into lakebed material would be difficult, costly and thus not justifiable. Dredging in any of the original inflow streams was not recommended since it would not provide any added recreational benefit. Also, the dredging guidelines recommended that logical lake basin drainage patterns be maintained along with the integrity of the littoral zone of the lake.

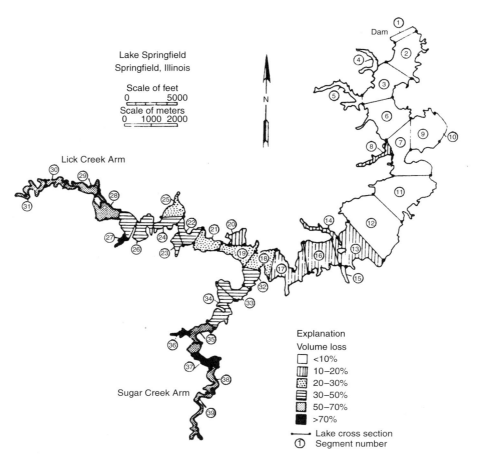

FIGURE 20.30 Percent loss of original storage volume for various areas of Lake Springfield. (From Buckler, J.H. et al. 1988. *Lake and Reservoir Manage.* 4(1): 143–152. With permission.)

20.8.3.2 Sediment Removal Techniques and Disposal Site Selection

The study examined several sediment removal options including draglines and dry mechanical removal (draw-down and bulldoze). The recommended option was hydraulic dredging because of its efficiency and proven record.

Nineteen agricultural land tracts were evaluated as potential disposal areas. The primary criterion for disposal site selections was proximate locations relative to the proposed dredging locations so that pipeline transportation costs were minimized. Secondary site selection criteria included (1) availability of suitable soil for levee material, (2) flat or gently rolling terrain, (3) potential for more than one weir to permit use of flow-through rather than a plug system, (4) little elevation difference between the disposal area and the high-water level of the lake to minimize total dynamic head (pumping to heights is expensive), and (5) reasonable isolation to minimize concerns for safety, odors, aesthetics, and need for right of ways for piping.

The final consideration concerning the disposal sites was that they be kept low profile, both figuratively and factually. Therefore, the average levee height of 2.4 m allowed a final pooled sediment depth of 0.9 m (0.6 m of clarifying water and 0.3 m of freeboard). Based on these criteria, and a recommended sediment removal volume of 2.06 million m^3, the city bought 199 ha of farmland in 1985 for sediment disposal at a cost of $1,890,201 (2002 prices).

The large recommended removal volume prompted authorities to investigate the potential for reclamation of the disposed sediment. The addition of lake sediment to eroded central Illinois farmland produced greater corn yields (Lembke et al., 1983) and addition to Sudan grass under laboratory conditions increased their yields (Olson and Jones, 1987). Thus, disposal sites could be reclaimed for farmland following project completion. Dunst et al. (1984) found similar results for corn, grain, and Sudan grass, due to increased nitrogen availability, when Lilly Lake, WI sediment was added to test plots. They concluded the applications were valid up to, and probably in excess of, 89.6 metric tons/ha.

20.8.3.3 Permits

Be aware of permit requirements so they can be obtained in time for the project to proceed as planned. When preparing for a dredging project, it might seem that dredging is easy compared to the feasibility analyses, permit requirements, and the public meetings one must attend. Obtaining permits is no small task: A variety of Federal, State, and Local permits must be obtained. The Springfield Lake project is an excellent example of the permit workload.

At the Federal and State levels, the U.S. Army Corps of Engineers (USACOE) reviews proposed projects under Section 404 of the Clean Water Act (which covers the navigable waters of the U.S.). However, since Lake Springfield was not considered navigable under Section 10 of the USA Rivers and Harbors Act of 1899, and dredged material was to be disposed upland rather than in wetlands, a nationwide permit was required for the return flow from the disposal site pursuant to 33 CRF 330.5 (a) (16) in lieu of a Section 404 permit. This permit was issued by the USACOE only after receipt of a Section 401 water quality certification and a construction and operating permit from the IDOT. Construction of the disposal site berms required construction permits from the Illinois Department of Transportation's Division of Water Resources, Dam Safety Section, since the berms were classified as Class III, low-hazard dams. In addition to these formal permits, the City was required to conduct an archeological survey of the disposal area to determine that there were no significant cultural features that might be impacted due to the construction (Wells, 1986). The archaeological survey was negative and construction began.

Local zoning proved to be the most controversial aspect of the entire project. The City, as well as Sangamon County, required the disposal site property to be rezoned from "A-1 agricultural use" to "conditional permitted use." Rezoning by the County Board was fine with citizens as long as it was not close to them. Residents living up to 2 miles away objected to the disposal sites because of chlordane and dieldrin in the sediment. Fears included depressed real estate values, health risks, and the possibility of later conversion of the disposal site to a general landfill. Odors and nuisance insects also concerned the surrounding residents. In addition, it was alleged by the citizens that the sediments were subject to disposal in accordance with the Resource Conservation and Recovery Act (RCRA) since they contained dieldrin that was listed as an acute hazardous substance under that statute. If so, disposal of 2.06 million m^3 of sediment according to RCRA criteria would result in exorbitant costs. Therefore, each of these public concerns had to be addressed before the public would support permitting.

In response to the concerns, the City issued statements indicating that offensive odors were not expected as a result of the settling tests they conducted. However, to demonstrate support of their statements, the City announced that a malodor control program was developed for the site as an operating contingency. Water level control of the disposal ponds was proposed as a means for minimizing nuisance insects.

Dieldrin and chlordane in the sediment were more serious concerns. The city tested sediment and sediment elutriates and found that both chemicals were routinely below detection limits. These results were confirmed by the IEPA based on sediment surveys from 1977. Additional analyses of sediment samples by the IEPA in 1984 showed average concentrations of chlordane and dieldrin were 18.5 and 12 µg/kg, respectively.

Health concerns relative to the sediment contaminants were considered by examining the sediment concentrations, the potential exposure pathway, and the probability of excessive exposure. The IEPA estimated that a person would need to ingest more than 400 kg of lake sediment for dieldrin to be lethal. In addition, it was determined the farmland itself, in the disposal area, contained an average of 82 and 313 μg/kg of dieldrin and chlordane, respectively.

The IEPA and U.S. Environmental Protection Agency (USEPA) advised the City that the dredged material would not qualify to be regulated under RCRA since it did not meet the hazardous waste criteria for ignitability, corrosivity, reactivity, or toxicity under the law. In addition, the definition of solid waste specifically excludes silt, dissolved or suspended solids, or other significant pollutants in water resources (40 CFR 24.101(v)).

20.8.3.4 Disposal Site

The IEPA's operating permit for the disposal site required detention times sufficient to meet the 15 mg/L total suspended solids requirement. Monitoring requirements included total suspended solids, ammonium-nitrogen, oil and grease, chlordane, dieldrin, pH, temperature, nitrate nitrogen, TP and DO.

Large-scale sediment settling tests were conducted to determine the settling pond retention time to meet the required 15 mg/L suspended solids discharge allowed from the disposal ponds. The estimated retention time was calculated based on the slurry pumping rate per hour, a 10% to 20% solids concentration in the slurry, and a 16 to 20 h/d operating schedule. This together with a series of large-scale settling tests, conducted according to USACOE (1978; now upgraded to USACOE 1987), determined that a 7 d retention time was required. The retention ponds were constructed accordingly with an approximate 20% oversize safety factor. This safety margin is very desirable since many disposal areas end up being undersized. An undersized disposal area can be very costly to a project since dredging equipment, once mobilized cannot be shut down to wait for seepage and evaporation in the disposal ponds to reestablish adequate retention times. The actual, as-constructed, retention time for the Lake Springfield disposal cells was 8.7 days. Adjustable weirs at each of the three outfall structures controlled discharge allowing the IDOT dewatering and flooding standards to be met (50% of the total storage volume in 30 days) and passing a 25 year flood event, respectively.

The per unit price of the disposal site was $185/m³ based on the estimates for engineering expenses ($110,000), construction costs ($732,900), and land acquisition costs ($1,426,700). Completion time for the disposal sites was less than 75 days. Operation of the site began in June, 1987.

20.8.3.5 Sediment Removal

Actual dredging in this project was anticlimactic following the enormous effort to obtain the necessary permits. Seven pre-qualified dredging contractors submitted bids ranging in price from $2.50/m³ to $6.65/m³. The average bid was $4.07/m³. These costs included all ancillary costs such as mobilization, bonds, insurance, and cash authorization allowance. Direct per unit costs for dredged material ranged from $2.10/m³ to $5.72/m³ and averaged $3.20/m³ (equivalent to $2.69/m³ to $7.32/m³ and $4.10/m³ in 2002 dollars). A contract was let that resulted in a total unit project cost (development of disposal site and dredging work) of $4.53/m³ of dredged sediment. The dredger worked without incident for 12 weeks to remove approximately 363,000 m³ (half of phase I) of sediment by October 1987. There was no apparent adverse impact on lake water quality and the retention ponds worked to maintain compliance with the relevant effluent limits. Project plans called for completion of the Phase II dredging in 1989.

This case study points out a major aspect of any large-scale environmental project. That is the need to inform the public of intent, to listen to their concerns, and to subsequently do everything possible to alleviate those concerns.

20.8.4 Lake Järnsjön, Sweden

No case study involving contaminated sediments was included in Cooke et al. (1993) because there were no well-documented cases at the time. The most thoroughly documented success story to date for contaminated sediment removal is that of Lake Järnsjön, in the river Eman (southeastern Sweden) (Bremie and Larsson, 1998; Blom et al., 1998; Elander and Hammar, 1998; Gullbring et al., 1998). The primary contaminants in Lake Järnsjön were Cd, Cu, and Pb from a Ni/Cd battery factory that discharged directly to the river and from leaching slag piles along side the river. The Ni/Cd factory closed in the mid-1970s. Measures were taken several years later to reduce leaching from the slag piles. Both measures reduced the flow of heavy metals to the lake, but it still contained high concentrations of heavy metals following the abatements.

Another, and perhaps more perplexing, problem was that of Hg and PCBs that accumulated in the lake from paper pulp processing discharges over a 100 year period. Mercury was prohibited from use in the Swedish paper pulp industry in 1972. Also, prior to 1972, copy paper containing PCBs was recycled into the pulp production process. Thus, large quantities of PCBs and Hg-containing paper fibers settled to the bottom of Lake Järnsjön. Elevated levels of PCBs were first detected in surface foam on the lake in the early 1980s. About the same time, high PCB levels were found in the fatty tissue of fish (140 mg/kg) (Gullbring et al., 1998). A detailed sediment analysis revealed that approximately 400 kg of PCBs were contained in the upper 0.4–1.6 m of lake sediment (Bremie and Larsson, 1998). It was also revealed that PCB concentrations were greater downstream from Lake Järnsjön than they were upstream, indicating that the lake was acting not only as a sediment trap for PCB, but also as a source for downstream contamination.

Because this is a reservoir, increasing downstream levels of PCB due to remediation posed a potential problem. As indicated previously, conventional cutterhead dredging can liberate considerable amounts of turbidity and associated contaminants to overlying water. Therefore, creation of excessive turbidity had to be avoided in the remediation process. Contaminated sediment removal from Lake Järnsjön required the use of a special purpose dredge. The Ellicott-designed, "Mud Cat®-like," MILMAN® dredge was capable of high-precision depth and cutting control, minimum turbidity, and low water discharge (see the section on Mud Cat dredges). Mud Cat dredges have a unique shielded auger cutter head mounted horizontal to the dredge hull so that cuts are made as the dredge moves forward or backward via its cable guidance system. The MILMAN dredge employed a Mud Cat type auger (Figure 20.31), but it was mounted perpendicular to the dredge hull so that cutting was accomplished in the more conventional fashion of a cutter head dredge by swinging from side to side as the dredge was advanced via cables anchored on land. This dredge differed considerably from a portable Mud Cat dredge in size, dredging capacity, and operational mode. Specifications of the dredge are shown in Table 20.7.

The MILMAN Mud Cat dredge employs an auger and shield like the conventional Mud Cat dredge and like the Japanese Clean-Up® dredge. While not stated in Ellicott literature, it appears the MILMAN II Mud Cat auger head shields are used in both swing directions based on the swing pattern of operation rather than the forward and backward movement of small Mudcat units and from Figure 20.31. This greatly improves the overall efficiency by reducing excessive turbidity regardless of the direction of cut. In addition to the auger and shield, the MILMAN II Mud Cat dredge employs other sophisticated technologies. These include both positioning and depth sounding equipment that shows exactly where ($\pm < 2.5$ cm) the cutter head is relative to the bottom. Computers on board store data concerning the appearance of the bottom, flow, dry material content and other information useful to the dredge operator. Systems equipped like this do an excellent job of removing unconsolidated sediment with minimum ancillary sediment disturbance and turbidity. Part of the reason is that the auger/shield configuration loosens and removes sediment at near *in-situ* density, with water introduction held to a minimum. According to Ellicott (McKegg, 2001), this results in production of nearly 30% suspended solids (SS) in the dredged material — approximately twice the average SS concentration produced by conventional cutter head dredges. However,

FIGURE 20.31 MILMAN II Mud Cat® dredge used to remove contaminated sediment from Lake Järnsjön, Sweden. (Photo courtesy of Ellicott, Division of Baltimore Dredges, LLC, Baltimore, MD.)

TABLE 20.7
Specifications of the MILMAN II Mud Cat® Suction Dredge Used at Lake Järnsjön, Sweden

Characteristic	Specification
Length overall	33.0 m
Beam	4.3 m
Draft	0.8 m
Displacement	40 T
Engine	Scania 282 kw @ 1800 rpm
Pump	Gould 25.4/20.24 cm
	600 m^3 @ 60 m head
Auger width	3.5 m
Auger diameter	0.5 m
Working depth	1–14 m
Swing width	0–200 m
Swing speed	0–12 m/min
Depth of cut	0.1–0.5 m
Removal rate	100–200 m^3/h (dry material)
Pump range	200 m @ 10 m head; with booster pump >2000 m

Source: Elander, P. and T. Hammar. 1998. *Ambio* 27(5).

it is only about half of that claimed by owners of the Japanese Clean-Up dredging system (pneumatic pumping system) (Sato, 1978).

Positioning of the MILMAN II Mud Cat dredge was accomplished with the help of a 140T Geodimeter that utilizes infrared light and radio signals. Dredging progress was monitored continuously with scanning echo depth sounders ahead and behind the auger. On board monitoring included sediment removal and environmental effects. The position of the dredge auger relative to the bottom was monitored continuously by the computer system with the help of precision positioning, with vertical accuracy of about 10 cm and horizontal accuracy of about 5 cm. Continuous echo depth soundings were made to assure that sediment removal was proceeding properly and that no ridged areas or mounds were left behind. This is an important consideration in that ridges and valleys in sediment created by inefficient dredging increases the bottom surface area exposed to overlying water and has the potential for making a bad situation even worse (Gibbons and Funk, 1983).

The dredging system used at Lake Järnsjön was designed for soft sediment and did not work well in sand and gravel. A bucket dredger surrounded by a silt curtain was used in these areas. The eastern part of the lake was most heavily contaminated with from 0.4 to 1.6 m accumulated sediment. Dredging in this area occurred from May through November 1993. The highly contaminated sediment was removed in successive 0.4 m layers, in a manner similar to that used in Lake Trummen, Sweden for nutrient removal. Dredging ceased during periods of high sensitivity to aquatic life (December to April), thus, the western part of the lake was dredged during the summer of 1994. Dredging in mixed sand/soft sediment conditions caused problems that required more water introduction (lower TSS) into the dredge slurry (Elander and Hammar, 1998). This caused greater dewatering problems at the disposal end. However, this was not a major problem and dredging removed about 150,000 m³ of material at completion. A follow-up study revealed that 394 kg of PCB, or 97% of that estimated to be contained in the lake sediment, was removed.

The Swedish Environmental Protection Agency monitored turbidity, TSS, and PCBs upstream, within the enclosed dredging area (Geotextile screen), and downstream during dredging. Suspended Solids within the enclosed area varied between 2 and 6 mg/L while those upstream varied between 2 and 5 mg/L and those downstream varied between 1 and 4 mg/L. This meant dredge efficiency was good at minimizing resuspension of sediment (Elander and Hammar, 1998). However, even though TSS concentrations were low within the enclosed area, PCB concentrations during dredging were relatively high (average of 60 ng/L). This compares with 1 to 4 µg PCB/L upstream and 10 to 15 ng PCB/L downstream of the dredging operation. Figure 20.32 summarizes PCB concentration, water temperature, and water flow in the lake during 1993–1994.

The dredged material disposal area was carefully selected (250 m from the lake) and designed to avoid adverse impacts on other lakes in the area. Sand and gravel was first removed from the

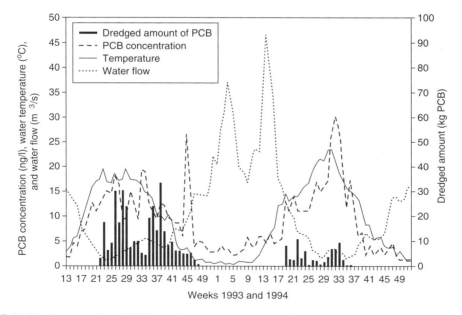

FIGURE 20.32 Concentration of PCB (ng/L), water temperature (°C), water flow (m³/s) in water leaving the lake, and the amount of dredged PCB (kg) at Lake Järnsjön, Sweden. (From Elander, P. and T. Hammar, 1998. *Ambio* 27(5). With permission.)

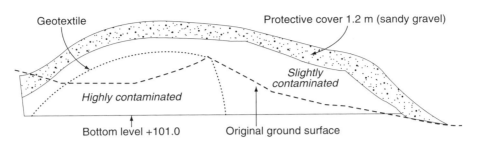

FIGURE 20.33 General design of the landfill for disposal of PCB contaminated sediment at Lake Järnsjön, Sweden. (From Elander, P. and T. Hammar, 1998. *Ambio* 27(5). 1998. With permission.)

area to minimize groundwater mounding as dredge fill was added. The resultant disposal area was about 32,000 m² with a maximum height of 6.5 m and a volume of about 90,000 m³. In the landfill, heavily contaminated sediments were separated from less contaminated sediments by a geotextile screen so that the most contaminated materials could be easily located some time in the future when decontamination technology might become available to break down the PCBs. The general design of the landfill is shown in Figure 20.33.

Blom et al. (1998) provide full details of the sublethal effects of Lake Järnsjön dredging on rainbow trout. Forlin and Norrgren (1998) describe the physiological and morphological effects of dredged material on perch. All results indicate that the project was a complete success. This is one more example of how high-tech, precision dredging can be done in an environmentally friendly manner such that desired results are achieved with minimal disturbance to, and around the target water body.

Peterson (1982a) addressed additional case studies. Most are not as complete as those described above. Because dredging is expensive and mistakes are costly, the objectives of any dredging project need to be clearly defined. The plans should be explicit and need to be executed in the best manner possible.

20.9 COSTS

Costs throughout the chapter are 1991 costs unless indicated otherwise or converted as indicated to 2002 costs by applying a 1.28 inflation factor to the 1991 costs. Project-to-project cost comparisons for sediment removal are difficult because of the large number of variables that affect dredging costs. Variables include equipment types used, project size (volume of material to be removed), disposal site availability, density of the material being removed, distance to the disposal area, and ultimate use of the removed material. Data for 64 sediment removal projects in the U.S. indicated a cost range from $0.36/m³ to $21.00/m³ ($0.46 to $26.88 in 2002 dollars) (Peterson, 1981). Given such wide variability, an average cost figure is not helpful in estimating single project costs. However, hydraulic dredging costs commonly range from $2.25/m³ to $5.65/m³ ($2.88 to $7.23 in 2002 dollars) and probably can be considered reasonable. A 1996 bid for hydraulic removal of muck from Lake Macy in Lake Helen, FL quoted a price of $4.00/yd³ ($5.23 m³) for removal of up to 20,000 yd³ (15,292 m³) of muck. Muck removal beyond 20,000 yd³ was quoted at a rate of $3.50 yd³ ($4.58 m³). Additional costs included a bid of $57,000 to remove 19 acres (7.69 ha) of aquatic vegetation and $60,000 for mobilization and installation of temporary pipeline (McDougal Construction, 1996). Mobilization is not a trivial part of dredging costs and can be a significant part of small jobs. If sediment contaminated with toxic substances is encountered and special dredges or treatment methods are required, dredging costs may exceed $52.00/m³ (Barnard and Hand, 1978; Koba et al., 1975; Matsubara, 1979). In general, the per unit volume cost of sediment removal is inversely related to the total volume of material removed. Any time dredged material can be used as potting soil or topsoil dressing, as was the case at Lake Trummen, Sweden (Sjön Trummen i Väjö, 1977), Lilly Lake, Wisconsin (Dunst et al., 1984), and Paradise Lake, Illinois (Lembke et al., 1983), the overall project costs might be reduced significantly.

With the difficulty of comparing costs among sediment removal projects (Peterson, 1981), it may seem impossible to make realistic cost comparisons between sediment removal and other lake treatment techniques. Peterson (1982b) attempted to compare dredging and P inactivation costs by assuming that both treatments are aimed at controlling in-lake P cycling and that both treatments are based on criteria thought necessary to achieve reduced P cycling. Thus, treatment costs should reflect the amount of P removed (dredging) or bound into the lake system (P inactivation) to prevent excessive internal cycling.

Direct cost comparisons of the two treatments cannot be made, since P inactivation costs usually are based on the materials and labor required to treat a hectare of lake surface (area), while dredging costs are based on the cost of removing a cubic meter of sediment (volume). Peterson (1982b) used per-cubic-meter dredging cost, lake area, percent of basin area dredged, and dredge volume to calculate the per hectare dredging cost. If the assumption is correct that only sufficient sediment was removed to control internal P cycling (lake deepening projects *per se* were not included in the calculation), then estimated dredging cost and P inactivation costs can be compared. Calculations were made using P inactivation costs from Cooke and Kennedy (1981), Cooke et al. (1982), and dredging costs from Peterson (1981) (see Table 20.8).

The P inactivation costs in Table 20.8 do not include equipment costs that generally are small relative to other costs (about 9% of total costs at Medical Lake, WA). Adding equipment costs to the Medical Lake project increases the total by $384/ha, bringing the most expensive (per hectare) P inactivation project cost close to the least expensive dredging project cost. Labor costs for P inactivation in Table 20.8 assumed an 8 h workday and a conservative $5/h (1982 cost) labor charge. Even if doubled, however, labor cost would not make dredging competitive with P inactivation, if considering cost alone. However, P inactivation longevity has not exceeded 15 years (Chapter 8).

To put the Table 20.8 costs in perspective, it should be noted that Medical Lake project costs are abnormally high. They represent real costs, but are high because the lake has steep banks, it is relatively deep (mean depth = 10 m; maximum depth = 18 m), has a large volume (6.2×10^6 m^3), and it has an uncommonly high total alkalinity (approximately 750 mg/L as $CaCO_3$). The large volume required and the high alkalinity permitted the application of large amounts of aluminum sulfate to reduce the soluble reactive P concentration by 87%. These factors make Medical Lake unusual among the alum-treated lakes. Also, the 59th Street Pond dredging costs were unusually high, for the reasons noted in Table 20.8. If these two outliers are eliminated and the median cost per hectare of the remaining projects is computed, P inactivation cost ($564/ha) is quite attractive relative to dredging ($17,894/ha).

It might be argued that dredging and P inactivation costs should not be compared, since dredging addresses one set of lake problems while P inactivation addresses another. However, Peterson (1982b) found considerable overlap between the lake conditions suitable to dredging and those suited to P inactivation. A possible exception to the similarities was depth and rooted plant growth. There was some question about the effectiveness of P inactivation in shallow lakes, which are generally ideal candidates for dredging. However, Welch et al. (1982) found, at Long Lake, Washington that P inactivation was very effective in a shallow lake. Today there is little question that P inactivation in shallow lakes can be even *more* effective than in deeper lakes (Welch and Cooke, 1995; Chapter 8).

The costs in Table 20.8 were calculated as "end-of-treatment" costs, i.e., what it actually cost to complete the treatment; which is misleading. More meaningful, are costs that are amortized over the effective life expectancy of a project. Chapter 8 and Welch and Cooke (1999) show that a P inactivation effectiveness of 10 to 15 years is common. This suggests that, based on cost and long-term effectiveness, P inactivation has advantages over dredging. Dredging does have the advantage that it adds no "foreign" substance to the lake. However, the "foreign" substance added

TABLE 20.8
Per-Hectare Cost Comparison of Dredging and Alum Treatment to Control In-Lake P Dynamics in 20 Treated Lakes

Lake	Treatment Type	Physical and Chemical Data	Chemical and Dose	Sediment Removed (m^3)	Treatment Cost ($/ha)
Horseshoe Lake, Wisconsin	Liquid alum	$A0 = 8.9$ ha $V = 3.6 \times 10^5$ m^3 $Z_{max} = 16.7$ m $Z = 4.0$ m Alk = 218-278 mg/L pH = 6.8–8.9 Dimictic	2.6 g Al/m^3	—	150
Lake San Marcos, California	Liquid alum	$A0 = 18.2$ ha $V = 4.3 \times 10^5$ m^3 $Z_{max} = 2.3$ m $Z = 2.3$ m Alk = 190–268 mg/L pH = 7.3–9.1	6.0 g Al/m^3	—	189
Welland Canal, Ontario,	Liquid alum, surface application	$A0 = 74$ ha $V = 6.2 \times 10^6$ m^3 $Z_{max} = 9.0$ m $Z = 9.0$ m Alk = 109 mg/L Dimictic	2.5 g Al/m^3	—	306
Mirror Lake, Wisconsin	Liquid alum and aeration	$A0 = 5.1$ ha $V = 4 \times 10^5$ m^3 $Z_{max} = 13.1$ m $Z = 7.8$ m Alk = 222 mg/L pH = 7.6 Monomictic	6.6 g Al/m^3	—	600[a]
Shadow Lake, Wisconsin	Liquid alum	$A0 = 17.1$ ha $V = 9.1 \times 10^5$ m^3 $Z_{max} = 12.4$ m $Z = 5.3$ m Alk = 188 mg/L pH = 7.4 Dimictic	5.7 g Al/m^3	—	600[a]
Cline's Pond, Oregon	Liquid sodium aluminate and HCl	$A0 = 0.4$ ha $V = 9600$ m^3 $Z_{max} = 4.9$ m $Z = 2.4$ m Alk = 30–50 mg/L pH = 7.0–7.7 Monomictic	10.0 g Al/m^3	—	630
West Twin Lake, Ohio	Liquid alum	$A0 = 34$ ha $V = 14.2 \times 10^4$ m^3 $Z_{max} = 11.5$ m $Z = 4.4$ m Alk = 102–149 mg/L Dimictic	26.0 g Al/m^3	—	638

TABLE 20.8 (Continued)
Per-Hectare Cost Comparison of Dredging and Alum Treatment to Control In-Lake P Dynamics in 20 Treated Lakes

Lake	Treatment Type	Physical and Chemical Data	Chemical and Dose	Sediment Removed (m^3)	Treatment Cost ($/ha)
Dollar Lake, Ohio	Liquid alum	$A0 = 2.2$ ha $V = 0.86 \times 10^5$ m^3 $Z_{max} = 7.5$ m $Z = 3.9$ m Alk = 101–127 mg/L pH = 6.7–8.6 Dimictic	20.9 g Al/m^3		756
Medical Lake, Washington	Liquid alum	$A0 = 64$ ha $V = 6.4 \times 10^6$ m^3 $Z_{max} = 18$ m $Z = 10$ m Alk = 750 mg/L pH = 8.5–9.5 Dimictic	12.2 g Al/m^3		2,610[b]
Half Moon Lake, Wisconsin	Dredging (~30% of basin)	$A0 = 53.4$ ha $V = 8.9 \times 10^5$ m^3 $Z_{max} = 2.7$ m $Z = 1.7$ m	—	25×10^3	3,205
Lilly Lake, Wisconsin	Dredging (100% of basin)	$A0 = 35.6$ ha $V = 5.3 \times 10^8$ m^3 $Z_{max} = 1.8$ m $Z = 1.4$ m	—	680×10^3	6,876
Commonwealth Lake, Oregon	Dredging (100% of basin)	$A0 = 2.6$ ha $Z = 0.9$ m	—	19×10^3	8,653
Steinmetz, New York	Draining and bulldozing (75% of basin)	$A0 = 1.2$ ha $V = 8.1 \times 10^5$ m^3 $Z_{max} = 2.1$ m $Z = 1.5$ m	—	2×10^3	10,849
Carnegie Lake, New Jersey	Dredging(75% of basin)	$A0 = 110$ ha 765×10^3 $Z_{max} = 3$ m	18,081		
Lenox Lake, Iowa	Dredging (100% of basin)	$A0 = 13.4$ ha $Z_{max} = 3.4$ m $Z = 0.9$ m	—	76×10^3	19,992
Nutting Lake, Massachusetts	Dredging (56% of basin)	$A0 = 31.6$ ha $Z_{max} = 2.1$ m $Z = 1.3$ m	—	275×10^3	28,288
Sunshine Springs, Wisconsin	Dredging (100% of basin)	$A0 = 0.4$ ha $V = 1.7 \times 10^3$ m^3 $Z_{max} = 1.2$ m $Z = 0.5$ m	—	5.1×10^3	40,248
Krause Springs, Wisconsin	Dredging (100% of basin)	$A0 = 0.3$ ha $V = 0.98 \times 10^3$ m^3 $Z_{max} = 1.0$ m $Z = 0.34$ m	—	4.9×10^3	47,631

TABLE 20.8 (Continued)
Per-Hectare Cost Comparison of Dredging and Alum Treatment to Control In-Lake P Dynamics in 20 Treated Lakes

Lake	Treatment Type	Physical and Chemical Data	Chemical and Dose	Sediment Removed (m^3)	Treatment Cost ($/ha)
Collins Park Lake, New York	Dredging (15% of basin)	$A0 = 24.3$ ha $V = 6.5 \times 10^5$ m^3 $Z_{max} = 9.75$ m $Z = 2.7$ m	—	52×10^3	58,767
59th Street Pond, New York	Draining and bulldozing (100% of basin)	$A0 = 1.8$ ha $Z = 0.2$ m	—	13×10^3	150,907[c]

Note: Costs were calculated from selected data of Peterson (1981) and Cooke and Kennedy (1981), and adjusted to June 1991 costs using inflation factors (1.28 × table costs = 2002 U.S. dollar costs).

[a] Cost per hectare, spread over Mirror Lake and Shadow Lake as a unit (5.1 ha ++ 17.1 ha = 22.2 ha).

[b] From Gasperino et al. (1981).

[c] All sediment from 59th Street Pond had to be trucked out of New York City, thus escalating the costs.

Sources: Peterson, S.A. 1981. Sediment Removal as a Lake Restoration Technique. USEPA-600/3-81-013. Cooke, G.D. and R.H. Kennedy, 1981. Precipitation and Inactivation of Phosphorus as a Lake Restoration Technique. USEPA-600/3-81-012. Gasperino, A.F. et al. 1981. USEPA-440/5-81-010.

in most P inactivation projects (alum) is commonly used to treat municipal water supplies destined for human consumption.

If amortization is applied to Lake Trummen dredging costs (not shown in Table 20.8), they are much more reasonable than the end-of-project costs. The 1971 cost of the entire project was $572,222 (times 3.36 = 1991 cost) (using a 1971 conversion factor of 4.5 Swedish Kronor per dollar). The cost was approximately $5,722/ha. However, when the cost is amortized over the post-treatment years the lake has shown benefits (now 35 years), the costs are reduced to approximately $163/ha per year. This figure continues to decline as long as the lake maintains its current quality. Cost/ha per year is a reasonable approach to compare the "cost effectiveness" of different treatment techniques.

Comparing the amortized dredging cost of treatment effectiveness for Lake Trummen (about $163/ha) with that for P inactivation in West Twin Lake, Ohio (Cooke et al., 1981), P inactivation remains very attractive. Phosphorus levels in West Twin Lake in 1991 were still close to those just after treatment in 1975 (Welch and Cooke, 1999) when treatment cost was $425/ha or $26.56/ha per year (16 years of effectiveness). Few records of Twin Lake quality are available after 1991. However, assuming that another 16 years of improved lake quality could be bought for another treatment at $425/ha, resulting in the same $26.56/ha cost, that would be very attractive; even to the lowest amoritized dredging cost, ($163/ha) at Lake Trummen.

From this brief analysis, P inactivation is much more cost effective than dredging, but dredging introduces no foreign materials into the lake and actually removes them. Dredging is still the only practical method (short of drawdown/scraping or damming) for deepening a shoaled lake, or if littoral zone removal of macrophytes is targeted, as in Lake Trummen, Sweden and Lilly Lake, Wisconsin, or if toxic contamination of sediment is a problem.

20.10 SUMMARY

There are four major reasons for removing sediment from a lake: (1) deepening, (2) limiting nutrient recycling, (3) reducing of macrophyte nuisances (by deepening and subsequent light limitation), and (4) removing toxic sediments. Environmental concerns associated with lake sediment removal appear to be less negative than might be expected. Case studies reveal that in-lake adverse impacts are not severe and are short lived. Dredged material disposal can be problematic. However, if disposal area planning and management is properly done, the area can enhance the overall project.

Dredging depth depends on the purpose of the project and the sediment characteristics of the lake. Two different approaches can be used for determining dredging depth for the control of in-lake sediment nutrient cycling. A third method is useful for controlling macrophytes. An important aspect of any sediment removal project is a thorough pretreatment evaluation of a nutrient mass balance, including sediment behavior and a lake history from sediment core analyses (see Chapter 2), that helps determine dredging feasibility and the expected longevity of effectiveness .

Almost all projects designed to deepen a lake are successful. Those designed to control internal nutrient cycling show mixed results, but most failures can be traced to incomplete pre-evaluation, removing too little sediment, or to poor dredge operation. Macrophyte control projects have been limited, but the experience from Lilly Lake, WI makes deepening to limit light penetration appear quite promising. Toxic sediment removal projects have been completed successfully, but generally require use of special dredges and disposal practices that increase costs dramatically.

Cost comparisons among dredging projects are difficult due to the large number of variables involved. Cost comparisons of dredging with other treatment techniques are not commonly made because of the many variables and the non-uniform units for measuring costs — dollars per hectare of treated area for P inactivation; dollars per cubic meter of sediment dredged. A few basic assumptions permit conversion of the different costing units into common terms. It appears that dredging is several times more expensive than P inactivation to accomplish the same short-term results (minimizing internal nutrient recycling) if cost alone is considered. "Real cost" determination, based on amortized cost over the duration of treatment effectiveness, can be made based on 10 to 30 year case histories. When done, it appears that P inactivation clearly is less expensive. However, where addition of chemicals to the lake is prohibited, dredging can be a reasonable (but costly) alternative.

There is little doubt that dredging can be a successful lake restoration technique if there is a thorough pre-implementation evaluation of the lake setting, if proper equipment is selected, if the disposal areas are designed for end-of-treatment effectiveness, and if the dredging is conducted by conscientious and competent operators. Successful examples of both surface-sediment skimming and deep (6 to 8 m) dredging are cited. Thorough pre-implementation evaluation followed by end-of-project assessment and periodic long-term monitoring assures better decisions on future projects.

REFERENCES

Andersson, G. 1984. Personal communication. Limnological Research Institute, University of Lund, Lund, Sweden.

Andersson, G. 1988. Restoration of Lake Trummen, Sweden: effects of sediment removal and fish manipulation. In: G. Balvay (Ed.), *Eutrophication and Lake Restoration. Water Quality and Biological Impacts*. Thonon-les-Bains. pp. 205–214.

Andersson, G. 2001. Personal communication. Limnological Research Institute, University of Lund, Lund, Sweden.

Andersson, G., H. Berggren and S. Hambrin. 1975. Lake Trummen restoration project III. Zooplankton macrobenthos and fish. *Verh. Int. Verein. Limnol.* 19: 1097.

Andersson, G., H. Berggren, G. Cronberg and C. Gelin. 1978. Effects of planktivorous and benthivorus fish on organisms and water chemistry in eutrophic lakes. *Hydrobiologia* 59: 8–15.

Barko, J.W. and R.M. Smart. 1979. The role of *Myriophyllum spicatum* in the mobilization of sediment phosphorus. In: J.E. Breck. R.J. Prentki and O.L. Loucks (Eds.), *Aquatic Plants, Lake Management, and Ecosystem Consequences of Lake Harvesting*. Institute for Environmental Studies, Center for Biotic Systems, University of Wisconsin, Madison. pp. 177–190.

Barko, J.W. and R.M. Smart. 1980. Mobilization of sediment phosphorus by submerged fresh water macrophytes. *Freshwater Biol.* 10: 229–238.

Barnard, W.D. 1978. Prediction and Control of Dredged Material Dispersion Around Dredging and Openwater Pipeline Disposal Operations. Tech. Rept. DS-78-13. U.S. Army Corps Engineers, Vicksburg, MS.

Barnard, W.D. and T.D. Hand. 1978. Treatment of Contaminated Dredged Material. Tech. Rept. DS-78-14. U.S. Army Corps of Engineers, Vicksburg, MS.

Baystate Environmental Consultants, Inc. 1987. Evaluation of the Nutting Lake Dredging Program. Baystate Environmental Consultants, Inc., East Longmeadow, MA.

Belonger, B. 1969. Aquatic Plant Survey of Major Lakes in the Fox-Illinois Watershed. Res. Rep. No. 39. Wisconsin Dept. Nat., Madison.

Bengtsson, L., S. Fleischer, G. Lindmark and W. Ripl. 1975. Lake Trummen restoration project. I. Water and sediment chemistry. *Verh. Int. Verein. Limnol.* 19: 1080.

Björk, S. 1972. Ecosystem studies in connection with the restoration of lakes. *Verh. Int. Verein. Limnol.* 18: 379–387.

Björk, S. 1974. *European lake rehabilitation activities*. Plenary Lecture of the Conference on Lake Protection and Management, Madison, WI.

Björk, S. 1978. *Restoration of degraded lake ecosystems*. Lecture at MAB Project 5 Regional Workshop, Land Use Impacts on Lake and Reservoir Ecosystems, Warsaw, Poland, May 26–June 2, 1978. CODEN LUNBDS/(NBLI-3008)/1-24 (1978)/ISSEN 0348-0798. Lund, Sweden.

Björk, S. 1985. Scandinavian lake restoration activities. In: *Lakes Pollution and Recovery*. Proc. European Water Pollut. Control Assoc. Int. Congr., Rome, April 15–18. pp. 293–301.

Blom, S., L. Norrgren and L. Forlin. 1998. Sublethal effects in caged rainbow trout during remedial activities in Lake Jarnsjon. *Ambio* 27(5): 411–418.

Born, S.M., T.L. Wirth, E.M. Brick and J.O. Peterson. 1973. Restoring the Recreational Potential of Small Impoundments. Tech. Bull. No. 70. Wisconsin Dept. Nat. Res., Madison.

Brannon, J.M. 1978. Evaluation of Dredged Material Pollution Potential. Tech. Rept. DS-78-6. U.S. Army Corps Engineers, Vicksburg, MS.

Brashier, C.K., L. Churchill and G. Leidahl. 1973. *Effect of Silt Removal in a Prairie Lake*. USEPA Ecol. Res. Series R3-73-037. Corvallis, OR.

Bray, R.N. 1979. *Dredging: A Handbook for Engineers*. Arnold Press, London.

Bremer, K.E. 1979. PCB contamination of the Sheboygan River, Indiana Harbor, and Saginaw River and Bay. In: S.A. Peterson and K.K. Randolph (Eds.), *Management of Bottom Sediments Containing Toxic Substances*. Proc. 4th U.S./Japan Experts Meeting. USEPA-600/3-79-102. Corvallis, OR.

Bremie, G. and P. Larsson. 1998. PCB in Eman River ecosystem. *Ambio* 27(5): 384–392.

Buckler, J.H., T.M. Skelly, M.J. Luepke and G.A. Wilken. 1988. Case study: The Lake Springfield sediment removal project. *Lake and Reservoir Manage.* 4(1): 143–152.

Calhoun, C.C., 1978. Personal communication. U.S. Army Corps of Engineers, Waterways Exp. Sta., Vicksburg, MS.

Canfield, D.E., Jr., K.A. Langeland, S.B. Linda and W.T. Haller. 1985. Relations between water transparency and maximum depth of macrophyte colonization in lakes. *J. Aquatic Plant Manage.* 23: 25–28.

Carline, R.F. and O.M. Brynildson. 1977. Effects of Hydraulic Dredging on the Ecology of Native Trout Populations in Wisconsin Spring Ponds. Tech. Bull. No. 98. Wisconsin Dept. Nat. Res., Madison.

Carignan, R. and J. Kalff. 1980. Phosphorus sources for aquatic weeds: water or sediment? *Science* 207: 987–989.

Chen, K.Y., B. Eichenberger, J.L. Mang and R.E. Hoeppel. 1978. Confined Disposal Area Effluent and Leachate Control (Laboratory and Field Investigations). Tech. Rept. DS-78-7. U.S. Army Corps Engineers, Vicksburg, MS.

Churchill, C.L., C.K. Brashier and D. Limmer. 1975. Evaluation of a Recreational Lake Rehabilitation Project. OWRR Comp. Rep. No. B-028-SDAK. Water Resources Inst., South Dakota State University, Brookings.

Clark, G.R. 1983. Survey of Portable Hydraulic Dredges. Tech. Rept. HL-83-4. U.S. Army Corps Engineers, Vicksburg, MS.

Collett, L.C., A.J. Collins, P.J. Gibbs and R.J. West. 1981. Shallow dredging as a strategy for the control of sublittoral macrophytes: a case study in Tuggerah Lakes, New South Wales. *Aust. J. Mar. Fresh Water Res.* 32: 563–571.

Cooke, G.D. and R.H. Kennedy, 1981. *Precipitation and Inactivation of Phosphorus as a Lake Restoration Technique.* USEPA-600/3-81-012. Washington, DC.

Cooke, G.D., R.T. Heath, R.H. Kennedy and M.R. McComas. 1982. Change in lake trophic state and internal phosphorus release after aluminum sulfate application. *Water Res. Bull.* 18: 699–705.

Cooke, G.D., E.B. Welch, S.A. Peterson and P.R. Newroth. 1986. *Lake and Reservoir Restoration,* 1st ed. Butterworth, Stoneham, MA.

Cooke, G.D., E.B. Welch, S.A. Peterson and P.R. Newroth. 1993. *Restoration and Management of Lakes and Reservoirs,* 2nd ed. Lewis Publishers, Boca Raton, FL.

Cronberg, G. 2004. Figure 20.20. Unpublished biomass figure authorized for book publication. Personal communication. University of Lund, Lund, Sweden.

Cronberg, G., Gelin, C. and Larsson, K. 1975. Lake Trummen restoration project II. Bacteria, phytoplankton, and phytoplankton productivity. *Verh. Int. Verein. Limnol.* 19: 1088.

Crumpton, J.E. and R.L. Wilbur. 1974. Habitat Manipulation. Dingell-Johnson Job Completion Report, Proj. No. F-26-5. Florida Game and Fresh Water Fish Comm., Tallahassee.

Digerfeldt, G. 1972. The post-glacial development of Lake Trummen, regional vegetation history, water level changes, and paleolimnology. *Folia Limnol. Scand.* 16: 1.

Duarte, C.M. and J. Kalff. 1987. Latitudinal influences on the depths of maximum colonization and maximum biomass of submerged angiosperms in lakes. *Can. J. Fish. Aquatic Sci.* 44: 1759–1764.

Duarte, C.M. and J. Kalff. 1990. Patterns in the submerged macrophyte biomass of lakes and the importance of the scale of analysis in interpretation. *Can. J. Fish. Aquatic Sci.* 47: 357–363.

Dunst, R. 1980. Sediment problems and lake restoration in Wisconsin. In: S.A. Peterson and K.K. Randolph (Eds.), *Management of Bottom Sediments Containing Toxic Substances,* Proc. 5th U.S./Japan Experts Meeting. Ecol. Res. Ser. Rep. USEPA-600/9-8-044.

Dunst, R.C. 1981. Dredging activities in Wisconsin's lake renewal program. In: *Restoration of Lake and Inland Waters: International Symposium on Inland Waters and Lake Restoration.* USEPA-440/5-81-010.

Dunst, R. 1982. Sediment problems and lake restoration in Wisconsin. *Environ. Int.* 7: 87–92.

Dunst, R.C., S.M. Born, P.O. Uttormark, S.A. Smith, S.A. Nichols, J.O. Peterson, D.R. Knauer, S.R. Serns, D.R. Winters and T.L. Wirth. 1974. Survey of Lake Rehabilitation Techniques and Experiences. Tech. Bull. 75. Wisconsin Dept. Nat. Res., Madison.

Dunst, R.C., J.G. Vennie, R.B. Corey and A.E. Peterson. 1984. *Effect of Dredging Lilly Lake, Wisconsin.* USEPA-600/3-84-097.

Elander, P. and T. Hammar. 1998. The remediation of Lake Jarnsjon: Project implementation. *Ambio* 27(5): 393–398.

Forlin, L. and L. Norrgren. 1998. Physiological and morphological studies of feral perch before and after remediation of a PCB contaminated Lake: Jarnsjon. *Ambio* 27(5): 418–424.

Fujiki, M. and S. Tajima. 1973. The pollution of Minamata Bay and the neighbouring sea by factory wastewater containing mercury. In: F. Coulston, F. Korte and M. Goto (Eds.), *New Methods in Environmental Chemistry and Toxicology,* Papers presented at the Int. Symp. On Ecol. Chem., International Academic Printing Co., Susono, Totsuka, Tokyo.

Gambrell, R.P., R.A. Kincaid and W.H. Patrick, Jr. 1978. Disposal alternatives for Contaminated Dredged Material as a Management Tool to Minimize Adverse Environmental Effects. Tech. Rept. DS-78-8. U.S. Army Corps Engineers, Vicksburg, MS.

Garrison, P. 1989. Personal communication. Wisconsin Dept. Nat. Res., Madison.

Garrison, P. 2002. Personal communication. Wisconsin Dept. Nat. Res., Madison.

Garrison, P.J. and D.M. Ihm. 1991. Final Annual Report of Long Term Evaluation of Wisconsin Clean Lake Projects: Part B: Lake Assessment. Wisconsin Dept. Nat. Res., Madison.

Gasperino, A.F., M.A. Beckwith, G.R. Keizur, R.A. Saltero, D.G. Nichols and J.M. Mires. 1981. Medical Lake improvement project: success story. In: *Restoration of Lakes and Inland Waters.* USEPA-440/5-81-010.

George, C., P. Tobiessen, P. Snow and T. Jewell. 1981. The Monitoring of the Restorational Dredging of Collins Lake, Scotia, New York. Final Project Report. Grant No. R804572. USEPA/CERL, Corvallis, OR.

Gibbons, H.L., Jr. and W.H. Funk. 1983. A Few Pacific Northwest Examples of Short-Term Lake Restoration Successes and Potential Problems with Some Techniques. In: *Lake Restoration and Management,* Second Annual Conference NALMS. USEPA-440/5-83-001.

Goldman, C.R., R.M. Gersberg and R.P. Axler. 1981. Gibraltar Lake Restoration Project, City of Santa Barbara. Final Report on Limnological Monitoring During Dredging. Ecological Research Associates, Davis, CA.

Gullbring, P., T. Hammar, A. Helgee, B. Troedsson, K. Hansson and F. Hansson. 1998. Remediation of PCB-contaminated sediments in Lake Jarnsjon: investigations, considerations and remedial actions. *Ambio* 27(5): 374–384.

Hayes, D. 1980. Personal communication. U.S. Army Corps of Engineers, Vicksburg, MS.

Herbich, J.B. and S.B. Brahme. 1983. Literature Review and Technical Evaluation of Sediment Resuspension During Dredging. Rep. No. COE-266. Ocean and Hydraulic Engineering Group, Texas A & M University, College Station.

Higginson, F.R. 1970. Ecological effects of pollution in Tuggereh Lakes. *Proc. Ecol. Soc. Aust.* 5: 143–152.

Hinsman, W.J. and T.M. Skelly. 1987. Clean Lakes Program Phase I Diagnostic/Feasibility Study for the Lake Springfield Restoration Plan. Springfield City Water, Light, and Power, Springfield, IL.

Horn, E. and L. Hetling. 1978. Hudson River PCB study description and detailed work plan. In: S.A. Peterson and K.K. Randolph (Eds.), *Management of Bottom Sediments Containing Toxic Substances.* Proc. 3rd U.S./Japan Experts Meeting. USEPA-600/3-78-084.

Huston, J. 1970. *Hydraulic Dredging: Theoretical and Applied.* Cornell Maritime Press, Cambridge, MD.

Hutchinson, G.E. 1975. *A Treatise on Limnology — Vol. III — Limnological Botany.* Wiley, New York.

JACA Corp. 1980. *Economic Benefits Assessment of the Section 314 Clean Lakes Program.* USEPA Office of Water, Washington, DC.

Kleeberg, A. and J.G. Kohl. 1999. Assessment of the long-term effectiveness of sediment dredging to reduce benthic phosphorus release in shallow Lake Muggelsee (Germany) *Hydrobiologia* 394: 153–161.

Koba, H., K. Shinohara and E. Sato. 1975. Management techniques of bottom sediments containing toxic substances. Paper presented at 1st U.S./Japan Experts Meeting on the Management of Bottom Sediments Containing Toxic Substances. Nov. 17–21, 1975. USEPA, Corvallis, OR.

Larsen, D.P., D.W. Schults and K.W. Malueg. 1981. Summer internal phosphorus supplies in Shagawa Lake, Minnesota, *Limnol. Oceanogr.* 26: 740–753.

Lembke, W.D., J.K. Mitchell, J.B. Fehrenbacher, M.J. Barcelona, E.E. Garske and S.R. Heffelfinger. 1983. Dredged Sediment for Agriculture: Lake Paradise. Res. Rept. No. 175. Water Resources Center, University of Illinois, Champaign-Urbana.

Lewis, M.A., D.E. Weber, R.S. Stanley and J.C. Moore. 2001. Dredging impact on an urbanized Florida bayou: effects on benthos and algal-periphyton. *Environ. Pollut.* 115: 161–171.

Lie, G.B. 1979. The influence of aquatic macrophytes on the chemical cycles of the littoral. In: J.E. Breck, R.T. Prentki and O.L. Loucks (Eds.), *Aquatic Plants, Lake Management, and Ecosystem Consequences of Lake Harvesting.* Center for Biotic Systems, University of Wisconsin, Madison.

Livingston, D.A. and J.C. Boykin. 1962. Distribution of phosphorus in Linsley Pond mud. *Limnol. Oceanogr.* 7: 57–62.

Luntz, J.D., R.J. Diaz and R.A. Cole. 1978. Upland and Wetland Habitat Development with Dredged Material; Ecological Considerations. Tech. Rept. DS-78-15. U.S. Army Corps Engineers, Vicksburg, MS.

Mackenthun, K.M., M.W. Brossman, J.A. Kohler and C.R. Terrell. 1979. Approaches for mitigating the kepone contamination in the Hopewell/James River area of Virginia. In: S.A. Peterson and K.K. Randolph (Eds.), *Management of Bottom Sediments Containing Toxic Substances.* Proc. 4th U.S./Japan Experts Meeting. USEPA-600/3-79-102.

Macpherson, J., M. LeMaster, A. (Sandy) Jack, A. Kilander and R. Carreau. 2003. Dredging and Water Quality: Tahoe Keys Marina Project. *Land and Water* May/June: 10–12.

Mallory, C.W. and M.A. Nawrocki. 1974. Containment Area Facility Concepts for Dredged Material Separation, Drying, and Rehandling. DMRP Contract Rep. D-74-6. U.S. Army Corps Engineers, Vicksburg,
MS.

Maristo, L. 1941. Die Seetypen Finnlands auf floristicher und vegetations-physiognomischer Grundlage, *Ann.
Bot. Soc. Zool. Bot. Vanamo* 15: 314–310.

Matsubara, M. 1979. The improvement of water quality at Lake Kasumigaura by the dredging of polluted
sediments. In: S.A. Peterson and K.K. Randolph (Eds.), *Management of Bottom Sediments Containing
Toxic Substances*. Proc. 4th U.S./Japan Experts Meeting. USEPA-600/3-79-102.

McDougal Construction. 1996. Bid letter to City of Lake Helen. McDougal Construction, 8800 NW 112th
Street, Suite 300, Kansas City, MO 64153.

McKegg, D. 2001. Personal communication. Ellicott, Division of Baltimore Dredges, LLC, Baltimore, MD.

Modlin, R. 1970. Aquatic Plant Survey of Major Lakes in the Milwaukee River Watershed. Research Report
No. 52. Wisconsin Dept. Nat. Res., Madison.

Montgomery, R.L. 1978. Methodology for Design of Fine-Grained Dredged Material Containment Areas for
Solids Retention. Tech. Rept. D-78-56. U.S. Army Corps Engineers, Vicksburg, MS.

Montgomery, R.L. 1979. Development of a methodology for designing fine-grained dredged material sedimentation basins. Ph.D. Thesis, Vanderbilt University, Nashville, TN.

Montgomery, R.L. 1980a. Containment area sizing for sedimentation of fine-grained dredged material. In:
S.A. Peterson and K.K. Randolph (Eds.), *Management of Bottom Sediments Containing Toxic Substances*. Proc. 5th U.S./Japan Experts Meeting. USEPA-600/9-80-044.

Montgomery, R.L. 1980b. Personal communication. U.S. Army Corps Engineers, Vicksburg, MS.

Montgomery, R.L. 1982. Containment area sizing for disposal of dredged material. *Environ. Int.* 7: 151–161.

Montgomery, R.L., E. Thackston and F.L. Parker. 1983. Dredged material sedimentation basin design. *J.
Environ. Eng. Div. ASCE* 109: 466–484.

Moody, L.F. 1944. Friction factors for pipe flow. *Trans. ASME* 66: 51–61.

Murakami, K. and K. Takeishi. 1977. Behavior of heavy metals and PCBs in dredging and treating of bottom
deposits. In: S.A. Peterson and K.K. Randolph (Eds.), *Management of Bottom Sediments Containing
Toxic Substances*. Proc. 2nd U.S./Japan Experts Meeting. USEPA-600/3-77-083.

Nawrocki, M.A. 1974. *Demonstration of the Separation and Disposal of Concentrated Sediments*. USEPA-
660/2-74-072.

OCEDA, Office, Chief of Engineers, Department of the Army. 1970. Laboratory Soils Testing. Engineer
Manual EM 1110-2-1906, Nov. 1970. Washington, DC.

Olson, K.R. and R.L. Jones. 1987. Agronomic use of scrubber sludge and soil as amendments to Lake
Springfield sediment dredgings. *J. Soil Water Conserv.* 421: 57–60.

Palermo, M.R. 1980. Personal communication. U.S. Army Corps Engineers, Vicksburg, MS.

Palermo, M.R. 2003. Personal communication. U.S. Army Corps Engineers, Vicksburg, MS.

Palermo, M.R., R.L. Montgomery and E. Poindexter. 1978. Guidelines for Designing, Operating, and Managing
Dredged Material Containment Areas. Tech. Rept. DS-78-10. U.S. Army Corps Engineers, Vicksburg,
MS.

Peterson, S.A. 1979. Dredging and lake restoration. In: *Lake Restoration: Proceedings of a National Conference*. USEPA-400/5-79-001.

Peterson, S.A. 1981. *Sediment Removal as a Lake Restoration Technique*. USEPA-600/3-81-013.

Peterson, S.A. 1982a. Lake restoration by sediment removal. *Water Res. Bull.* 18: 423–435.

Peterson, S.A. 1982b. Dredging and nutrient inactivation as lake restoration techniques: a comparison. In:
Management of Bottom Sediments Containing Toxic Substances. Proc. 6th U.S./Japan Experts Meeting.
Feb. 1981, Tokyo, Japan. U.S. Army Corps Engineers, Dredging Operations Technical Support Pro-
gram, Vicksburg, MS.

Peterson, S.A., A.T. Herlihy, R.M. Hughes, K.L. Motter and J.M. Robbins. 2002. Level and extent of mercury
contamination in Oregon, USA lotic fish. *Environ. Toxicol. Chem.* 21: 2157–2164.

Pierce, N.D. 1970. Inland Lake Dredging Evaluation. Tech. Bull. 46. Wisconsin Dept. Nat. Res., Madison.

Randall, R.E. 1977. *Notes from the Fifth Dredging Engineering Short Course*. Texas A & M University,
College Station.

Raymond, R.B. and F.C. Cooper. 1984. Vancouver lake: dredged material disposal and return flow management
in a large lake dredging project. In: *Lake Restoration and Management*. USEPA 440/5-83-001. pp.
284–292.

Reimold, R.J. 1972. The movement of phosphorus through the salt marsh cord grass, *Spartina alterniflora* Loisel. *Limnol. Oceanogr.* 17: 606–611.

Ryding, S.O. 1982. Lake Trehörningen restoration project. Changes in water quality after sediment dredging. *Hydrobiologia* 92: 549–558.

Sakakibara, A. and O. Hayashi. 1979. Lake Suwa water pollution control projects. In: S.A. Peterson and K.K. Randolph (Eds.), *Management of Bottom Sediments Containing Toxic Substances*. Proc. 4th U.S./Japan Experts Meeting. USEPA-600/3-79-102.

Sato, H. 1978. Personal communication. Japan Bottom Sediments Association, Tokyo.

Schults, D.W. and K.W. Malueg. 1971. Uptake of radiophosphorus by rooted aquatic plants. In: *Proceedings of the 3rd National Symposium on Radioecology*, Oak Ridge, TN, May 10–12, 1971. pp. 417–424.

Shapiro, J., V. LaMarra and M. Lynch. 1975. Biomanipulation: an ecosystem approach to lake restoration. In: P.L. Brezonik and J.L. Fox (Eds.), *Proceedings of a Symposium on Water Quality Management Through Biological Control*. University of Florida, Gainesville.

Sjön Trummen i Växjö. 1977. Förstörd, Restaurerad, Pånyttfödd. Länsstyrelsen i Kronobergs Län. Växjö Kommun. (Lake Trummen in Växjö. 1977. Destroyed, Restored, Regenerated. County Commission in Kronoberg County. Växjö Municipality).

Snow, P.D., W. Cook and T. McCauley. 1980. The Restoration of Steinmetz Pond, Schenectady, New York. USEPA Final Project Report, Grant No. NY-57700108. USEPA, Washington, DC.

Sondergaard, M., J. Windolf and E. Jeppesen. 1996. Phosphorus fractions and profiles in the sediment of shallow Danish lakes as related to phosphorus load, sediment composition and lake chemistry. *Water Res.* 30: 992–1002.

Spaine, P., L. Llopis and E.R. Perrier. 1978. Guidance for Land Improvement Using Dredged Material. Tech. Rept. DS-78-21. U.S. Army Corps Engineers, Vicksburg, MS.

Spencer Engineering. 1981. Gibraltar Lake Restoration Project Final Report. Spencer Engineering, Santa Barbara, CA.

Stauffer, R.E. and G.F. Lee. 1973. The role of thermocline migration in regulation algal blooms. In: E.J. Middlebrooks, D.H. Falkenborg and T.E. Maloney (Eds.), *Modeling the Eutrophication Process*. Utah State University, Logan.

Stefan, H. and D.E. Ford. 1975. Temperature dynamics in dimictic lakes. *J. Hydraul. Div. ASCE* 101 (HY1), Proc. Paper 11058: 97–114.

Stefan, H. and M.J. Hanson. 1979. Fairmont Lakes Study: Relationships between Stratification, Phosphorus Recycling, and Dredging. Proj. Rep. No. 183. St. Anthony Fall's Hydraulic Laboratory, University of Minnesota, St. Paul.

Suda, H. 1979. Results of the investigation of turbidity generated by dredges at Yokkaichi Port. In: S.A. Peterson and K.K. Randolph (Eds.), *Management of Bottom Sediments Containing Toxic Substances*. Proc. 4th U.S./Japan Experts Meeting. USEPA-600-3-79-102.

Toubier, J. and Westmacott. 1976. Lakes and Ponds. Tech. Bull. No. 72. Urban Land Institute, Washington, DC

Twilley, R.R., M. Brinson and G.J. Davis. 1977. Phosphorus absorption, translocation, and secretion in *Nuphar luteum*. *Limnol. Oceanogr.* 22: 1022–1032.

U.S. Army Corps of Engineers, Headquarters (USACOE). 1987. Confined Disposal of Dredged Material. Engineers Manual 1110-2-5027. Washington, DC

U.S. Department of Agriculture. 1971. Ponds for Water Supply and Recreation. Handbook No. 387. U.S. Government Printing Office, Washington, DC.

Van der Does, J.,P. Verstraelen, P. Boers, J. Van Roestel, R. Roijackers and G. Moser. 1992. Lake restoration with and without dredging of phosphorus-enriched upper sediment layers. *Hydrobiologia* 233: 197–210.

Walsh, M.R. and M.D. Malkasian. 1978. Productive Land Use of Dredged Material Containment Areas: Planning and Implementation Consideration. Tech. Rept. DS-78-020. U.S. Army Corps of Engineers, Vicksburg, MS.

Wechler, B.A. and D.R. Cogley. 1977. Laboratory Study Related to Predicting the Turbidity-Generation Potential of Sediments to be Dredged. Tech. Rept. D-77-14. U.S. Army Corps Engineers, Vicksburg, MS.

Welch, E.B. and G.D. Cooke. 1995. Internal phosphorus loading in shallow lakes: Importance and control. *Lake and Reservoir Manage.* 11(3): 273–281.

Welch, E.B. and G.D. Cooke. 1999. Effectiveness and longevity of phosphorus inactivation with alum. *Lake and Reservoir Manage.* 15(1): 5–27.

Welch, E.B., P.D. Lynch and D. Hufschmidt. 1979. Internal phosphorus related to rooted macrophytes in a shallow lake. In: J.E. Breck, R.T. Prentki and O.L. Loucks (Eds.), *Aquatic Plants, Lake Management, and Ecosystem Consequences of Lake Harvesting.* Center for Biotic Systems, University of Wisconsin-Madison.

Welch, E.B., J.T. Michaud and M.A. Perkins. 1982. Alum control of internal phosphorus loading in a shallow lake. *Water Res. Bull.* 18: 929–936.

Wells, C.L. 1986. An Archaeological Reconnaissance of a Sediment Retention Area, Lake Springfield, Sangamon County, Illinois. Prepared for the City of Springfield. Southern Illinois University at Edwardsville.

Wetzel, R.G. 1983. *Limnology,* 2nd ed. Saunders, Philadelphia.

Wisconsin Department of Natural Resources (WDNR). 1969. Lilly Lake, Kenosha County, Wisconsin. Lake Use Rep. No. FX-34. Wisconsin Dept. Nat. Res., Madison.

Wisconsin Department of Natural Resources (WDNR). 1990. Abstracts and Publications for Current Projects. Bureau of Research, Water Resources Research, Madison, WI (Contact Paul Garrison).

Worth, D.M., Jr. 1981. Nutting Lake Restoration Project: a case study. In: *Restoration of Lakes and Inland Waters and Lake Restoration.* USEPA-440/5-81-010.

Index

A

Acid neutralizing capacity (ANC), 52, 53

Acid-volatile sulfide (AVS), 264

Acorus calamus, 306, 307, 310, 313, 388

Acremonium zonatum, 443

Active planting, 317

Advanced wastewater treatment (AWT), 87, 88, 105, 227, 502

Advanced wastewater treatment, diversion and, 87–109

    costs, 104–105

    general, 87–88

    in-lake treatment following diversion, 105–106

    Lake Norrviken, Sweden, 98

    Lake Sammamish, Washington, 96–98

    Lake Søbygaard, Denmark, 103–104

    Lake Washington, Washington, 93–96

    Lake Zürich, Switzerland, 102–103

    Madison Lakes, Wisconsin, 101–102

    recovery of world lakes, 89–93

    Shagawa Lake, Minnesota, 99–101

    techniques for reducing external nutrient loads, 88

AERF, *see* Aquatic Ecosystem Restoration Foundation

Aerial photography

    boat tracks, 295

    close-cut channel persistence analyzed using, 350

AF, *see* Anoxic factor

Agasicles hygrosphila Selman and Vogt, 422, 425

Agricultural field drainage, 127

AHOD, *see* Areal hypolimnetic oxygen deficit rate

Algae

    growth, fertilized lawns and, 130

    identification of, 261

    management, viruses for, 445

    mat-forming filamentous, 261

    natural mortality of, 265

    odor producing, 262

    toxic form of copper to, 259

Algal abundance, 490

Algal biomass

    control, 85–86

    lake P concentration and, 107

    rebound of, 264

Algal bloom(s), 32, 158

    control of, 259

    nuisance, 175

    occurrence of after reflooding, 327

    prevention of in Moses Lake, 155

    reduced, 113

    released nutrients stimulating, 393

Algal problems, 71, 74

Algal productivity, calcite precipitation and, 211

Algicides, 175, 259, 264, 265

Alimagnet Lake, 349

Alkalinity procedure, 181

Alosa pseudoharengus, 236

Alternanthera philoxeroides, 322, 328, 361, 388

Alum, 75, 214

Aluminum, reaction of in water, 176

Ameiurus melas, 314

Amisk Lake, 462, 465

Amynothrips andersoni O'Neill, 422, 425

Anabaena, 90, 265

Anabaena circinalis, 209, 239

Anabaenopsis, 265

Anacystis cyanea, 209

ANC, *see* Acid neutralizing capacity

Animal and Plant Inspection Service (APHIS), 422

Animal waste facilities, 117

Annabessacook Lake, 184

    characteristics and alum doses of, 194

    P inactivation dose, 221

    reduction in mean summer epilimnetic TP, 197

Anoxia, 60, 99

Anoxic factor (AF), 66

Aphanizomenon, 90, 155

Aphanizomenon flos-aquae, 212

APHIS, *see* Animal and Plant Inspection Service

APIRS, *see* Aquatic, Wetland, and Invasive Plant Information Retrieval System

Aqua-Beach Comber, 363

Aquascaping, 303

Aquascreen, 379

Aquatic Ecosystem Restoration Foundation (AERF), 402

Aquatic habitat destruction, 5

Aquatic plant(s)

    groups, 276

    methods for removal, 345

    restoration, categories, 317

Aquatic plant management plan recommendations, 272–276

    chemical treatment, 274

    education and information, 273–274

    finding of feasibility, 276

    harvesting, 275

    plant reassessment, 275

    riparian controls, 274–275

Aquatic Research Program (ARP), 402

Aquatic, Wetland, and Invasive Plant Information Retrieval System (APIRS), 271

Areal air-flow rate criterion, 474

Areal hypolimnetic oxygen deficit rate (AHOD), 460

ARP, *see* Aquatic Research Program

Artificial circulation, 72, 75, 471–498

    costs, 490–492

    devices and air quantities, 471–478

effects of circulation on trophic indicators, 486–490
lake responses to, 488
theoretical effects of circulation, 479–486
    dissolved oxygen, 479
    effects of phytoplankton composition, 484–486
    nutrients, 479–480
    physical control of phytoplankton biomass, 480–483
undesirable effects, 490
Artificial islands, 303
Arzama densa Walker, 422
Asterionella sp., 259
AVS, *see* Acid-volatile sulfide
AWT, *see* Advanced wastewater treatment
Azolla caroliniana, 277, 388

B

Backus Lake, 327
Bacopa caroliniana, 436
Bacteria, 285
    reducing algae growth with, 444
    sulfate-reducing, 170
Bagous
    affinis, 330
    affinis Hustache, 422
    hydrillae O'Brien, 422
Bank
    fishing, 292
    recession, 133
Barleber See, 197
Bathing beach areas, killing of snails in, 263
BATHTUB, 71
Baumé number, 182
Bautzen Reservoir, 211, 249, 251
Behavioral cascade, 237
Below sediment harvesting, 352
Benthic zone, 28
Best management practices (BMPs), 75, 112, 115
Bidens
    beckii, 305
    sp., 322
Big Crooked Lake, 398, 403
Big Green Lake, 296
Big Muskego Lake, 297, 298, 327
Biochemical oxygen demand (BOD), 52, 53
Biofilters, 75
Biological control(s), 73
    insect species released for, 422
    objective of, 421
Biological limnology, 28
Biomagnification, 7
Biomanipulation, 235–257
    biomanipulation, 240–241
    costs, 251
    deep lakes, 248–251
        Bautzen Reservoir, 249–251
        Lake Mendota, 248–249
    over-, 250
    shallow lakes, 241–248

Cockshoot Broad, 243–244
Lake Christina, 246–248
Lake Vaeng, 246
Lake Zwemlust, 244–246
trophic cascade, 235–238
trophic cascade research, 239–240
Biosphere, water in, 4
Bird River, 348
Black Lake
    addition of iron to, 189
    hypolimnetic aeration of, 462
Blooms, 30
BMPs, see Best management practices
Boat
    launch sites
        removal of nuisance plants at, 344
        warning sign at, 343
    tracks, 295, 297
Boating ordinance, slow-no-wake, 294, 295
BOD, see Biochemical oxygen demand
Boom and bust phenomenon, 151
Booster pump requirements, 527
Borderless world, 12
Bottom-feeding fish, biomanipulation and, 249
Bottom sediments, method of breakdown in, 390
Bottom-up top-down hypothesis, 251
Bow reciprocating cutters, 346
Bow rotary cutting machines, 346
Brasenia
    schreberi, 305, 306, 307, 310, 322, 329, 388
    sp., 276
Breakwaters, 299, 302
Broad-spectrum herbicides, 386
Bronmark–Weisner hypothesis, 246
Browns Lake, 366
BrunsvikenLake, hypolimnetic aeration of, 462
Buckhorn Lake, 353, 357
Buffalo Lake, 343
Buffering compounds, 178
Buffer zone, effectiveness of, 131
Bugle Lake, 501, 551, 552
Bureau of Reclamation, 159
Burr Pond, 398, 405, 406
Bussey Reservoir, 329

C

Cabomba caroliniana, 322, 342, 388
CAFOs, *see* Confined animal operations
Calcite precipitation, 211
Calcium
    application of to hardwater lakes, 211
    compounds, P concentration and, 180
Caldonazzo Lake, hypolimnetic aeration of, 462
Calla palustris, 313
Camanche Lake, hypolimnetic aeration of, 462
Campbell Lake, 207
    characteristics and alum doses of, 195
    longevity of phosphorus inactivation in, 204
    reduced TP in, 205

Camp Lake, 398, 404
Candlewood Lake, 333
Canopy-forming plants, 33
Carbon dioxide diffusion, 279
Carex spp., 306, 307, 310, 322
Carlson equation, 93
Carlson index, misuse of, 65
Carman Bay, 414
Carnegie Lake, dredging and alum treatment costs, 563
Carsons Bay, 413
Cascades, realized biomass, 240
Casitas Reservoir, 261, 263, 266
Cassadaga Lakes, 293
Castle Lake, 239
Castor canadensis, 312
Cattle grazing, negative effects of, 131
Cayuga Lake, 345, 408, 428
Cenaiko Lake, 428
Cephalanthus occidentalis, 313, 322, 328
Ceratium hirundinella, 259–261
Ceratophyllum
    demersum, 245, 275, 294, 298, 300, 305, 307, 316, 330,
        340, 353, 387, 399, 400, 401, 404, 405, 438,
        443
    echinatum, 305
    spp., 310
Ceriodaphnia dubia, 264
CET relationship, *see* Concentration/exposure time
        relationship
Chain Lake, 151, 171
    characteristics of withdrawal, 167
    hypolimnetic withdrawal of, 166
Chara
    sp., 294, 297, 302, 316, 405, 409
    spp., 307, 310, 388
    vulgaris, 322, 329, 381
Chatauqua Lake, 354
Chautauqua Lake, 344
Chemical controls, 383–419
    case studies, 397–414
        2,4-D in Cayuga Lake, New York and Loon Lake,
            Washington state, 408–411
        plant management with fluridone in northern United
            States, 397–408
        triclopyr in Pend Oreille River, Washington state
            and Lake Minnetonka, Minnesota, 411–414
    costs, 414
    effective concentration, 383–384
    environmental impacts, safety and health
            considerations, 389–395
        herbicide fate in environment, 389–390
        toxic effects, 390–395
    herbicide selectivity, 387–389
    types of chemicals, 384–387
        adjuvants, 387
        broad-spectrum vs. selective herbicides, 386
        contact vs. systemic, 385–386
        persistent vs. non-persistent, 386
        plant growth regulators, 386–387
        tank mixes, 386
    ways of minimizing environmental risks, 395–397

Chemung Lake, 353, 359
Chesapeake Bay, 369
Chicot Lake, 327
Cladwell Lake, artificial circulation of, 473
Clarke–Bumpus sampler, 54
Classroom Aquatic Plant Nursery Program, 313
Clean Lakes Program (CLP), 15, 74
Clean Water Act, 15, 122, 504, 555
Clear Lake, 127, 147, 154, 239, 473
Clear water lakes, littoral zones, 277
CLIMEX simulation model, 342
Cline's Pond
    artificial circulation of, 472
    dredging and alum treatment costs, 562
Close-cut harvester, modified, 350
CLP, *see* Clean Lakes Program
Coarse woody debris (CWD), 130
Cochnewagon Lake
    characteristics and alum doses of, 194
    P inactivation dose, 221
    reduction in mean summer epilimnetic TP, 197
    treatment cost, 222
Cockshoot Broad, 243, 244
Colletotrichum gloeosporioides, 443
Collins Park Lake, dredging and alum treatment costs, 564
Comanche Reservoir, 464
Combined sewer systems, 112
Commonwealth Lake, dredging and alum treatment costs,
        563
Computerized navigational device, 190
Concentration/exposure time (CET) relationship, 383–384
Confined animal operations (CAFOs), 111
Constructed wetland(s), 122, 128
    case histories, 125
    plan-profile view of, 128
Contact herbicides, 414
Contingency planning, 339
Continuously stirred tank reactors (CSTR), 55
Cootes Paradise Marsh, 313
Copper, 384
Copper Basin Reservoir, 264
Copper sulfate, 259–268, 393
    application guidelines, 261–262
    applications, 259–261
    costs, 265–266
    effectiveness, 262–263
    negative effects, 263–265
    treatment, 72, 263
Coregonus artedi, 249
Cox Hollow Lake, 379, 472
Cricotopus myriophylii Oliver n. sp., 426
Critical depth concept, 480
Critical P source areas (CSAs), 115
Crooked Lake, 397, 398, 399
Cross Lake, 335
Crystal Lake, 277, 464, 473, 486
CSAs, *see* Critical P source areas
CSTR, *see* Continuously stirred tank reactors
Ctenopharyngodon idella, 12, 312
Cultivating equipment, 364
CWD, *see* Coarse woody debris

Cylindrospermopsis raciborskii, 92, 265
Cyperus spp., 306, 307, 310, 322
Cyprinus carpio, 12, 250, 264, 292, 334

D

2,4-D, 384
  suggested uses, 415
  treatment, Loon Lake, 411
Dairy wastewater, 104
Daphnia
  ambigua, 264
  galeata, 247, 249
  magna, 244, 264
  parvula, 264
  pulex, 244, 247, 264
Dead Lake, 514, 515, 538
Decodon verticillatus, 305, 307, 313
Deep lakes, 87
  biomanipulation, 248
  characteristics of, 33, 34
  control of algal biomass in, 240
Deep oxygen injection system (DOIS), 455, 456
Deep-water channels, long-term persistence of, 351
Deer Point Lake, 431, 436
Delavan Lake, 297
  alum treatment of, 214
  aquatic flora of, 299
  frequency of aquatic plants in, 300
  nutrient budget, 355
Demotechnic growth, 5
Depth-area hypsographs, 29
Derooting treatments, 364
Destratification, *see* Artificial circulation
Detritus, 30
Devil's Lake, 170, 344
  characteristics of withdrawal, 167
  hypolimnetic withdrawal of, 166
  system installation costs, 171
DHM, *see* Diel horizontal migration
Diamond Lake, 349
DIC, *see* Dissolved inorganic carbon
Diel horizontal migration (DHM), 244
Dilution and flushing, 147–162
  case studies, 149–159
    Green Lake, 156–158
    Lake Veluwe, 158–159
    Moses Lake, 150–156
  effects, applications, and precautions, 159–160
  theory and predictions, 148–149
Dip-In volunteers, 38
Diquat, 384, 415
Discharge elevation head, 522
Discharge friction head, 523
Disposal area design, 533
Dissolved inorganic carbon (DIC), 279
Dissolved organic carbon (DOC), 264
Dissolved organic matter (DOM), 215
Dissolved oxygen (DO), 26, 52, 178
  circulation effects and, 479

concentrations, decline of, 179
daily changes of, 285
depletion(s), 13, 53
  fish survival and, 286
  hypolimnia, 455
  Lilly Lake, 550
  severe, 262
fish winterkills and, 515
problems, summer drawdown and, 328
transport efficiency, 460
Diver dredging
  advantages of, 362
  limitations of, 362
  Potomac River, 363
Diversion, *see* Advanced wastewater treatment, diversion and
DO, *see* Dissolved oxygen
DOC, *see* Dissolved organic carbon
DOIS, *see* Deep oxygen injection system
Dollar Lake, 60, 190
  characteristics and alum doses of, 194
  dredging and alum treatment costs, 563
  P inactivation dose, 221
  reduction in mean summer epilimnetic TP, 197
DOM, *see* Dissolved organic matter
Dorosoma
  cepedianum, 250
  pretense, 439
Dredge(s)
  grab-bucket, 507
  hydraulic, 507
  mechanical, 507
  pneumatic, 512
  pump
    distance between booster pump and, 532
    energy diagram for, 531
    output, 525
    system analysis, 526
  selection, 514
  slurry, specific gravity of, 523
  special nozzle suction, 545
  special purpose, 501, 511
Dredging
  days, computation of required, 518
  diver, 362
  equipment, production capacity of, 516
  sediment resuspension during, 503
Dreissena
  bugensis, 12
  polymorpha, 12
Dulichium arundinaceum, 305
Dustpan-like dredge, 507
Duwamish River, 104
Dyes, plant growth suppression using, 381

E

East Sydney Lake, 473, 488
East Twin Lake (ETL), 106, 199
  mean surface total P concentration in, 201

reduction in mean summer epilimnetic TP, 197
species diversity of planktonic microcrustacea, 216
trophic state improvement of, 213
Eau Galle Reservoir, 50, 184, 188, 195, 209
Ecological engineering, 112
Ecosystem(s)
functions, herbicides and, 395
most polluted, 7
processes, impact of harvesting on, 360
ED pond, *see* Extended detention pond
Egeria densa, 220, 278, 308, 327, 342, 388, 397, 441
Eichornia crassipes, 277, 308, 329, 339
Elatine minima, 305
Eleocharis
acicularis, 284, 305, 307, 322
baldwinii, 322
cellulosa, 285
coloradoensis, 284
interstincta, 285
palustris, 305, 307
robbinsii, 305
spp., 276, 310, 388
Elodea
canadensis, 275, 300, 305, 314, 316, 322, 340, 363, 387–389, 401, 409, 423
densa, 322
nuttalli, 244
sp., 322
spp., 310
Emergent macrophytes, 276
Endangered Species Act, 397
Endothall, 384, 415
English Lake District, 480
Environmental energy, 292
Environmental resources, aggregate of, 38
Environmental risk, tool to reduce, 397
Equisetum spp., 293, 307
Erie Lake, 25, 207
characteristics and alum doses of, 195
longevity of phosphorus inactivation in, 204
reduced TP in, 205
Eriocaulon aquaticum, 305
Esox lucius, 246, 248
ETL, *see* East Twin Lake
Euhrychiopsis lecontei, 350
Eutrophication, 243
control measures, 77
process, 31
surface drinking water supplies and, 10
symptoms of, 13
Evaporation (EVP), 48
EVP, *see* Evaporation
EXF, *see* Exfiltration
Exfiltration (EXF), 48
Extended detention (ED) pond, 120

F

Fairmont Lakes, 37, 390
Fall drawdown, 334

Farm equipment, outdated, 344
FAS, *see* Ferric aluminum sulfate
FasTest[a], 400
Fathometers, 396
Feed grain, P in, 118
Fe–P redox cycle, 179
Ferric aluminum sulfate (FAS), 211
Fertilization, impact of on stream, 115
Fertilized lawns, algae growth and, 130
Fertilizer controls, 75
59th Street Pond, dredging and alum treatment costs, 564
Figure Eight Lake, 212
Fish
attractors, method of adding, 334
kills, 13
mortality, 218
removals, 241
survival, DO depletion and, 286
winterkills, 34, 240, 515
Fish Lake, 349, 350, 359
Floating-leaved macrophytes, 276
Flocculent settling
concentrations, 535
procedure, 533
Fluridone, 384, 390
FasTest[a] for, 400
plant management with, 397, 398
suggested uses, 415
treatment
Big Crooked Lake, 403
Burr Pond, 406
Camp Lake, 404
Lake Hortonia, 407
Lobdell Lake, 404
species response to, 409
Wolverine Lake, 405
Flushing, *see* Dilution and flushing
FLUX, 71
Fontinalis spp., 276
Food web
functions, copper stress and, 265
manipulations, 72
Founder colony, 311
Foxcote Reservoir, 113, 210
Fox Lake, 330
Fraxinus lanceolata, 328
Free-floating macrophytes, 277
French Lake, 349
Fresh water, status of in United States, 7–11
Friction head losses, 520
Frisken Lake, 212
Fulica atra, 245
Fungal pathogens, 442
Fungi, 285, 442
Fusarium culmorum, 423

G

GAC, *see* Granular activated carbon
Game fish biomass, decline of, 130

Geographical information systems (GIS), 45
Ghirla Lake, hypolimnetic aeration of, 462
Gibralter Lake, 502, 512
GIS, *see* Geographical information systems
Global fresh water runoff, 3, 4
Global positioning (GPS) units, 396, 515
Gloeotrichia, 208, 209
Gloeotrichia echinulata, 157
Glyceria borealis, 322, 388
Glyphosate, 384
GPS units, *see* Global positioning units
Grab-bucket dredges, 507
Grafenheim Lakes, 250
Granular activated carbon (GAC), 265
Grass carp
    costs, 441
    feeding preference list, 432
    reproduction of, 431
    state regulations on possession and use of, 430
    stocking rates, 434
Gratiola aurea, 305
Great Lakes, 6, 12
Great Salt Lake, 284
Green Lake, 147, 149, 178, 208, 218, 507, 508
    dilution of, 157
    dose determination for, 188
    example of benefits of dilution, 156
Ground reconnaissance, 46
Groundwater (GW), 46, 48
Guntersville Reservoir, 312, 437
GW, *see* Groundwater

H

Habitat
    alteration, 294
    buffers, 383
Half Moon Lake, 212, 563
Hall Lake, 505
Halverson Lake, 352, 353, 356, 358
Ham's Lake, artificial circulation of, 473
Harp Lake, 47, 49
Harvesting, 73
    budget calculation sheet, 365, 366
    chemical control and, 370
    impact of on ecosystem processes, 360
    impact of on spawning fish, 360
    shredding and crushing, 361
    suction, 365, 367
Heart Lake, artificial circulation of, 473
Hechtsee, 169
    characteristics of withdrawal, 167
    hypolimnetic withdrawal of, 166
Hemlock Lake, 462, 465
Herbicide(s)
    application
        equipment, 396
        posted notice of, 396
    characteristics, 385
    comfort level for using, 414

decision to use, 383
disappearance, 385, 390
effectiveness, 281
efficacy, 387
indirect impacts of, 393
movement, adjuvants restricting, 387
plant response to, 388
registered, 384
registration, 390, 395
residue, monitoring of, 414
resistant plant communities, 394
secondary effects of, 17
selectivity, 387
suggested uses, 415
treatment costs, 441
Hoover Reservoir, 259
Horseshoe Lake, 189
    characteristics and alum doses of, 195
    dredging and alum treatment costs, 562
    P inactivation costs, 222
    P inactivation dose, 221
Human drinking water, herbicides and, 395
Hyallela azecta, 264, 247
Hyco Reservoir, 441
Hydraulic dredges, 507
Hydraulic washing, 363, 368
Hydrellia
    balciunasi Bock, 422
    pakistanae Deonier, 422
Hydrilla management, 423
Hydrilla verticillata, 278, 281, 308, 322, 342, 384, 386, 422
Hydrocharis morsus-ranae, 342
Hydrochloa caroliniensis, 322
Hydrocotyle umbellate, 388
Hydrodictyon, 261
Hydrologic cycle and quantity of fresh water, 3–7
Hydrothol 191, 393
Hydrotrida caroliniana, 322
Hypolimnetic aeration, 61, 73
Hypolimnetic aeration and oxygenation, 455–470
    beneficial effects and limitations, 461–465
    costs, 466
    description and operation of units, 455–460
    undesirable effects, 466
    unit sizing, 460–461
Hypolimnetic aerator, 456, 457
Hypolimnetic anoxia, depth of, 165
Hypolimnetic withdrawal, 72, 163–173
    adverse effects, 171
    costs, 171
    test cases, 165–171

I

Ice caps, 3
Ictalurus
    nebulosus, 246
    punctatus, 439
Ictiobus cyprinellus, 246, 299
Indian Brook, artificial circulation of, 472

In-lake nutrient control measures, 45
Interlake plant dispersal, 340
Intermediate trophic state hypothesis, 239, 240
Internal loading
    control of non-point, 12
    enhanced, 92
    principal source of, 75
Intertidal communities, strong interactors in, 236
Iron
    forms of in lake water, 179
    inactivation of P by, 210
Irondoquoit Bay
    characteristics and alum doses of, 194
    hypolimnetic aeration of, 462
    reduction in mean summer epilimnetic TP, 197
Irrigation water, 159
Isoete
    echinospora, 305
    lacustris, 305

J

James River, kepone contamination of, 502
Jesenice Reservoir, 129
Juncus spp., 306, 307, 310
Jussiaea diffusa, 322, 328
Justicia americana, 388

K

Kepone contamination, 502
Keystone species, omnivorous shad as, 250
Kezar Lake, 480
    artificial circulation of, 472
    characteristics and alum doses of, 194
    P inactivation dose, 221
    reduction in mean summer epilimnetic TP, 197
    water softness of, 202
Kissimmee lakes, 330
Kissimmee River, 394
Klamath Indian Tribe's fish production, 6
Klamath Lake, 6
Kleiner Montiggler See
    aeration of, 165
    characteristics of withdrawal, 167
    hypolimnetic withdrawal of, 166

L

Lac de Paladru
    characteristics of withdrawal, 167
    hypolimnetic withdrawal of, 166
Lac Léman, 90
LaDue Reservoir, 352
Lafourche Reservoir, 329
Lagarosiphon major, 340
Lago Maggiore, 90
Lake(s)
    air-lift system to destratify, 475

annual mean TP concentration, 90
    artificial circulation of, 472–473
    biotic communities, 28
    -bottom map, 514
    clear water, littoral zones, 277
    conditions, 8, 9–10, 34
    difference between reservoirs and, 14, 23
    dredging and alum treatment costs, 562–564
    fluridone-treated, 398
    hardwater
        calcium applications to, 211
        lime application to, 212
    hypolimnetic aeration of, 462
    -level drawdown, 73
    management, herbicides used for, 384
    monitoring, voluntary, 342
    nutrient concentration, reduction in, 91
    plants
        habitat preferences of, 305
        wildlife and environmental values of, 307
    pollutant sources, 13
    primary interactions in, 32
    problems, sources of, 11–13
    restoration and management, 13–15, 15–17
    state, forces determining, 241
    stratified, 59, 471
    trophic state, 51, 63
    weed-free, 287
    world, recovery of, 89
Lake Altoona, artificial circulation of, 473
Lake Annabessacook, 178
Lake Apopka, 127, 326
Lake Aratiatia, 352, 354
Lake Arbuckle, artificial circulation of, 473
Lake Arendsee, 212
Lake Balaton, 92
Lake Baldegg, hypolimnetic aeration of, 462
Lake Baldwin, 440
Lake Ballinger, 163, 164, 169
    characteristics of withdrawal, 167
    discharge from, 170
    hypolimnion in, 166, 172
Lake Bistineau, 352
Lake Biwa, 5, 6
Lake Bled, 147, 165
    characteristics of withdrawal, 167
    hypolimnetic withdrawal of, 166
Lake Boltz, artificial circulation of, 472
Lake Bomoseen, 327
Lake Buchanan, artificial circulation of, 472
Lake Burgaschi, 165
    characteristics of withdrawal, 167
    hypolimnetic withdrawal of, 166
Lake Butte des Morts, 300
Lake Calhoun, 473, 489
Lake Casitas, artificial circulation of, 472
Lake Catharine, artificial circulation of, 473
Lake Champlain, 131, 349
Lake Christina, 246, 247
Lake Cidra, 343
Lake Conroe, 312, 438

Lake Conway, 429, 434, 438, 440
Lake Corbett, artificial circulation of, 472
Lake Dagowsee, 212
Lake Delavan
    Al added:Al–P formed ratio, 185
    dosing procedures, 186, 187, 188
    phosphorus mobilized in, 286
Lake Donten, 147
Lake El Capitan, artificial circulation of, 472
Lake Eucha, 128
Lake Eufaula, artificial circulation of, 473
Lake Evaluation Index (LEI), 64
Lake Falmouth, artificial circulation of, 472
Lake George, 240, 277, 363
Lake Glum Sø, 90
Lake Groot Vogelenzang, 189, 211
Lake Gross-Glienicker, 210
Lake Haines, 53
Lake Haugatjern, 249
Lake Hecht, 165
Lake Henry, 501, 542, 551, 552
Lake Herman, 503
Lake Hjälmaren, 92
Lake Hortonia, 398, 405, 407
Lake Hot Hole, artificial circulation of, 473
Lake Hyrum, artificial circulation of, 472
Lake the Isles, 186
Lake Istokpoga, 362, 394
Lake Järnsjön, 557, 558
Lake Keesus, 359
Lake Kegonsa, 101, 390
Lake King George VI, artificial circulation of, 473
Lake Kleiner Montiggler, 165
Lake Klopeiner
    characteristics of withdrawal, 167
    hypolimnetic withdrawal of, 166
Lake Kremenchug, artificial circulation of, 473
Lake Lafayette, artificial circulation of, 473
Lake Lagog, 60
Lake Långsjön, 176, 189, 205
Lake Larson, hypolimnetic aeration of, 462
Lake Leba, 197
Lake Lewisville, 312
Lake Lillesjön, 223
    interstitial P content, 223
    treatment cost, 226
Lake Loenderveen, 113
Lake Lyng, 227
Lake Lyngby, 205
Lake Maarsseveen, artificial circulation of, 472
Lake Macy, 560
Lake Mälern, 92
Lake Marion, 440
Lake Mathews, 264
Lake Mauen, 165, 166
Lake McCarron, 127
Lake McDonald, 60
Lake Mead, 33
Lake Meerfelder Maar, 165, 166
Lake Memphremagog, 278, 281
Lake Mendota, 31, 50, 101, 221, 248, 344, 352

Lake Michigan, 220, 502
Lake Minnetonka, 341, 353, 411, 413
Lake Monona, 101, 347
Lake Morey, 178
    alum treatment of, 216
    characteristics and alum doses of, 194
    hypolimnetic Fe:P ratio, 202
    inorganic complexing agents, 219
    P inactivation dose, 221
    reduction in mean summer epilimnetic TP, 197
    treatment of hypolimnion, 218
Lake Muggelsee, 502
Lake Nieuwe Meer, 485
Lake Norrviken, 87, 90, 93, 107, 148
    availability of internal P in, 100
    characteristics of, 95
    diversion cost, 104, 105
    response of lake TP concentration in, 99
    successful diversion in, 98
Lake Ocklawaha, 387
Lake Ohakuri, 352, 354
Lake Okeechobee, 50, 67, 88, 117, 327, 393, 394
Lake Onalaska, 283
Lake Onondaga, 59, 60
Lake Ontario, Al treatment of, 193
Lake Parvin, artificial circulation of, 472
Lake Pattison
    characteristics and alum doses of, 195
    longevity of phosphorus inactivation in, 204
    reduced TP in, 205
Lake Pearl, 440
Lake Pend Oreille, 335
Lake Pfaffikersee, artificial circulation of, 473
Lake Piburger, 165
Lake Poygan, 299
Lake Prompton, artificial circulation of, 472
Lake Queen Elizabeth II, artificial circulation of, 473, 490
Lake and reservoir diagnosis and evaluation, 45–83
    data evaluation, 55–71
    diagnosis/feasibility studies, 45–55
        in-lake, 51–55
        watershed, 45–51
    guidelines for choosing lake restoration alternatives,
        74–76
    lake improvement restoration plan, 76–78
    selection of lake restoration alternatives, 71–74
        algal problems, 71–72
        macrophyte problems, 72–74
Lake Ripley, 296, 297
Lake Roberts, artificial circulation of, 472
Lake Rockwell Reservoir, 379
Lake Sallie, 353, 355, 356, 357
Lake Sammamish, 46, 50, 58, 59, 60, 62, 67, 93, 97, 107
    availability of internal P in, 100
    characteristics of, 95
    diversion cost, 105
    internal loading in, 98
    sewage and dairy plant effluent diversion, 96
    urban sub-watershed and, 122
    wastewater diverted from, 104
Lake San Marcos, dredging and alum treatment costs, 562

Lake Särkinen, 464
Lakescaping, 135
Lake Schmaler Luzin, 212, 464
Lake Seminole, 423, 436
Lake Sharpe, 133
Lakeshore rehabilitation, 135
Lake Søbygaard
    characteristics of, 95
    recovery, 103
Lake Sonderby, 186
Lake Spavinaw, 129
Lake Springfield, 552
    disposal cells, retention time, 556
    watershed, 553
Lake Starodworskie, artificial circulation of, 472
Lake Stevens, 462, 463, 466
Lake Stewart, artificial circulation of, 472
Lake St. George, 239
Lake Susan, 186
Lake Tahoe, 239, 277, 504
Lake Tarago, artificial circulation of, 473
Lake Tegel, 114, 462, 463, 466
Lake Tohopekaliga, 326, 330, 334
Lake Trasksjön, artificial circulation of, 473
Lake Trekanten
    iron naturally present in, 223
    treatment cost, 226
Lake Trummen, 87, 106, 505, 508, 513, 542
    biomass, 544
    post-dredging, 545
    sediment nutrient concentrations, 543
Lake Vaeng, 246
Lake Vallentunasjon, response of lake TP concentration in, 99
Lake Växjosjön, artificial circulation of, 472
Lake Veluwe, 147, 150, 158, 160
Lake Vesuvius, artificial circulation of, 473
Lake Victoria, 12
Lake Waccabuc, 463, 466
Lake Waco, artificial circulation of, 472
Lake Wahiawa, artificial circulation of, 473
Lake Wahnbach, artificial circulation of, 472
Lake Wapeto, characteristics and alum doses of, 195
Lake Waramaug, 165, 170, 169
    characteristics of withdrawal, 167
    hypolimnetic withdrawal of, 166
Lake Washington, 67, 87, 93, 94, 98, 379
    characteristics of, 95
    dilution of, 158
    diversion cost, 105
    recovery to pre-enrichment conditions, 103
    sewage and dairy plant effluent diversion, 96
    summer lake TP, 157
    trophic state of, 175
    wastewater entering, 104
Lake Waubesa, 101, 390
Lake Wesslinger, 463
Lake Whakamarino, 363
Lake Wilcox, 489
Lake Wiler
    characteristics of withdrawal, 167

hypolimnetic withdrawal of, 166
Lake Wingra, 285, 286, 352, 353, 503
    do nothing approach, 293
    internal P loading in, 356
    phosphorus removal from, 357
    species relative frequencies, 294
    vegetation recovery in, 293
Lake Wononscopomuc, 165, 169
    characteristics of withdrawal, 167
    hypolimnetic withdrawal of, 166
    withdrawal water from, 170
Lake Zumbra, 397, 398, 399, 400
Lake Zürich, 93, 102, 105
Lake Zwemlust, 244, 245, 246
Land management procedures, 112
Latuca sativa, 436
Lawrence Lake, 278–279
LC, see Lethal concentration
Leaf photosynthesis, 277
Leersia oryzoides, 322
LEI, see Lake Evaluation Index
Lemna
    minor, 298, 307, 322
    sp., 322
    spp., 277, 310, 316, 388
    trisulca, 307
Lenox Lake, dredging and alum treatment costs, 563
Lepomis macrochirus, 250, 263, 314, 359
Lethal concentration (LC), 391
Liberty Lake, 216
Lilly Lake, 503, 507, 541, 546, 561
    chlorophyll a in, 548
    DO depletion of, 550
    dredging and alum treatment costs, 563
    macrophyte control project, 549
    total inorganic nitrogen in, 547
Lime, addition of to hardwater lakes, 212
Limno, 456
Limnobium spongia, 277, 322
Limnology, 23–45
    biological limnology, 28–30
    characteristics of shallow and deep lakes, 33–34
    ecoregions and attainable lake conditions, 34–41
    eutrophication process, 31–33
    lakes and reservoirs, 23–25
    limiting factors, 30–31
    physical–chemical limnology, 26–28
    schools, emergence of, 23
Linacre Reservoir, 444
Linsley Pond, 502
Littoral zone(s), 277
    erosion, 358
    slope, 281
Livestock, riparian zone destruction by, 138
Loading
    model, 40
    year-to-year variation in, 62
Lobdell Lake, 398, 404
Lobelia dortmanna, 305
Long Lake, Minnesota
    optimum dose of calcium nitrate for, 223

plant growth depth, 277
Long Lake, Washington, 45, 130, 326, 353, 502, 561
    alum application to, 193
    characteristics and alum doses of, 195
    longevity of phosphorus inactivation in, 204
    reduced TP in, 205
    sodium carbonate treatment of, 184
    soft water treatment, 178
Long Lake, Wisconsin, 295, 296
Long Lake South, 206
Loon Lake, 410, 411
LORAN navigation system, 222
Lough Neagh, 90
Ludwigia uruguayensis, 342, 388
Lyngbya, 261
Lythrum salicaria, 12, 298, 388

M

Macrophyte(s)
    biomass of submerged, 65
    colonization of, 283
    death and decay, 286
    distribution, satellite imagery and, 54
    growth–death–decay cycles, 14
    harvesting, phosphorus removal by, 355
    herbivorous control of, 283
    interspecific competition between, 284
    macroconsumers of, 359
    nuisance, herbicide applications for, 326
    nutrient cycles and, 285
    problems, 72
    shading of with surface covers, 380
    zonation of, 277
Macrophyte ecology and lake management, 271–290
    aquatic plant growth and productivity, 277–281
        dissolved inorganic carbon, pH, and oxygen,
            279–280
        light, 277–278
        nutrients, 278–279
        substrate, 280
        temperature, 280–281
    effects of macrophytes on environment, 285–287
    planning and monitoring for aquatic plant management,
        271–276
    plant distribution within lakes, 281
    relationships with other organisms, 283–285
    reproduction and survival strategies, 282–283
    resource allocation and phenology, 281–282
    species and life-form considerations, 276–277
Maintenance management, 397
Manure
    applications, 12, 111, 116
    management, 117
Marsilea quadrifolia, 342
Maximum depth of colonization (MDC), 506
MDC, *see* Maximum depth of colonization
Mechanical derooting, 364
Mechanical dredges, 507

Mechanical management techniques, characteristics of,
    368–369
Mechanical removal methods, cost estimates for, 367
Mechanical shredding, 349
Medical Lake, 217, 463, 464, 561
    alum treatment of, 193
    dredging and alum treatment costs, 563
    hypolimnetic aeration of, 462
    P inactivation dose, 221
    treatment cost, 222
Megalodonta beckii, 323
Melosira sp., 259
Mercury, deposition of in streams, 7
Metal salts, control of P concentration using, 180
Methane production, 213
Microcystis, 90, 265
Microcystis aeruginosa, 244, 249
Micropterus
    dolomieu, 217
    salmoides, 247, 264, 314, 439
Microzone, sediment surface, 179
Mill Pond Reservoir, 259
MILMAN dredge, 557
Minimata Bay, 502
Minnesota ecoregions, land use and water quality data for,
    37
Mirror Lake, 198, 199
    artificial circulation of, 472
    characteristics and alum doses of, 194
    dredging and alum treatment costs, 562
    hypolimnetic aeration of, 462
    trophic state improvement of, 213
Mixing
    devices, 478
    wind-caused, 61
MIXOX unit, 464
Model
    CLIMEX, 342
    loading, 40
    Nürnberg's, 58
    steady state mass balance, 57
    top-down bottom-up, 239
    TP-blue-green algal probability, 102
Modeling, phosphorus, 45
Mondeaux Flowage, 327
Monoamine salt of endothall, 393
Moody diagram, 520, 521
Morone chrysops, 439
Moses Lake, 61, 66, 147, 149, 154, 479
    addition of dilution water to, 150
    dilution rates, 153
    irrigation water added to, 159
    prevention of algal blooms in, 155
    shape of, 152
Motorboat activity, 296
Mount St. Helens ashfall, 151, 153
Mud Cat guidance system, 512
Murphy Flowage, 330
Mycoleptodiscus terrestris, 443
Myocastor coypus, 312

Myriophyllum
    aquaticum, 388
    brasiliense, 322, 329
    exalbescens, 244, 307, 322
    farwellii, 305
    heterophyllum, 275, 305, 322, 342
    scabratum, 328
    sibiricum, 247, 298, 305, 354, 409
    sp., 302, 323
    spicatum, 12, 33, 135, 272, 278, 293, 300, 322, 340,
        388, 399, 404, 405, 443
    spp., 276, 310
    tenellum, 305
    verticillatum, 305

N

Najas
    flexilis, 275, 285, 294, 305, 307, 316, 323, 330, 363,
        394, 409
    gracillima, 305, 403, 405, 409
    guadalupensis, 307, 323, 403, 409, 36
    marina, 297, 298, 342, 401, 409
    minor, 342
    sp., 302
    spp., 310, 388
Nasturtium sp., 388
National Pollution Discharge Elimination System
    (NPDES), 397
National Urban Runoff Program, 120
Natural wetlands, 122
NAWDB, *see* North American Wetland Data Base
Necturus maculosus, 359
Nelumbo lutea, 306, 307, 309, 310, 323, 388
Neochetina eichhorniae Warner, 422
Neohydronomus af.nis Hustache, 422
Neomysis mercedis, 94
Nephelometric Turbidity Units (NTU), 504
Newman Lake, 216, 462, 464, 465
Nitella megacarpa, 438
Nitrate removal, 124
Non-persistent herbicides, 386
Non-point pollution, protection from, 111–145
    constructed wetlands case histories, 125–128
    in-stream phosphorus removal, 112–114
    lakeshore rehabilitation, 135–137
    manure management, 117–119
    non-point nutrient source controls, 114–117
    ponds and wetlands, 120–125
        constructed wetlands, 122–125
        dry and wet extended detention ponds, 120–122
    pre-dams, 128–129
    reservoir shoreline rehabilitation, 133–135
    riparian zone rehabilitation, 129–130
    riparian zone rehabilitation methods, 130–133
Non-point sources (NPS), 11
North American Wetland Data Base (NAWDB), 125
North Lake, 312
NPDES, *see* National Pollution Discharge Elimination
    System

NPS, *see* Non-point sources
NRCS, *see* U.S. Natural Resources Conservation Service
NTU, *see* Nephelometric Turbidity Units
Nuisance algal blooms, 175
Nuisance phytoplankton blooms, 262
Nuisance plants, 351, 416
Nuisance species, control of, 271
Nuphar
    advena, 323
    luteum, 323
    macrophyllum, 323
    polysepalum, 323
    sp., 323
    spp., 276, 306, 310, 388
    variegata, 297, 298, 305, 307, 314, 316, 409
Nürnberg's model, 58
Nutrient(s)
    algae blooms and, 393
    budget, 357
    buffers, 383
    control, 45, 501
    cycles, macrophytes and, 285
    diversion, 71
    inputs, world lakes, 89
    loading, 32, 111
    removal, 354
    retention, 131
    source, substrates as, 280
    transfers, 286
Nutting Lake, 512, 563
Nymphaea
    odorata, 297, 314, 316, 329, 352, 366, 388, 401
    sp., 409
    spp., 276, 306, 307, 310
    tuberosa, 294, 323
Nymphoides peltata, 342
Nyssa aquatica, 328

O

ODR, *see* Oxygen deficit rate
Ogallala Aquifer, 5
OI, *see* Osgood Index of mixing
Okanagan Lakes, 359
Oligotrophication, 32, 243
Olszewski tube, 163
Once through costs, 365
Ondatra zibethica, 312
One Gram Assimilative Capacity Rule, 125, 126
Oozer® dredge system, 513
Open water pelagic zone, 28
Orange Lake, 361
Organization for Economic Cooperation and Development
    (OECD), 15
Oscillatoria, 90, 261
    agardhii, 199, 543
    limosa, 262
    rubescens, 103
Osgood Index of mixing (OI), 59
Outstanding resource waters, 170

Over-biomanipulation, 250
Oxygenation, *see* Hypolimnetic aeration and oxygenation
Oxygen deficit rate (ODR), 68

P

P, *see* Phosphorus
Panicum
    sp., 323
    virgatum L., 131
Paradise Lake, 561
Parker Horn, 154
Parkers Lake, 397, 398, 399, 400
Paul Lake, 352
PCBs, *see* Polychlorinated biphenyls
Pelican Horn, 153
Pend Oreille River, 411, 412
PEP, *see* Phosphorus elimination plant
Perca
    flavescens, 218, 246, 249
    fluviatilis, 249
Persistent herbicides, 386
PGR, *see* Plant growth regulators
Phosphorus (P), 11
    assimilative capacity, 125
    elimination plant (PEP), 113
    inactivation, 71, 73, 180, 213
    index (PI), 114
    loading, control of internal, 176
    modeling, 45
    removal, 112, 180, 355
    runoff, 119
    sedimentation rate, 69
    wind-caused resuspension, 204
Phosphorus inactivation and sediment oxidation, 175–236
    chemical background, 176–180
        aluminum, 176–178
        iron and calcium, 178–180
    costs, 222
    dose determination and application techniques,
        180–193
        aluminum, 180–189
        application techniques for alum, 189–193
        iron and calcium, 189
    effectiveness and longevity of P inactivation,
        193–213
        calcium applications to hardwater lakes, 211–213
        iron applications, 210–211
        ponds, 209–210
        reservoirs, 209
        shallow, unstratified lake cases, 204–209
        stratified lake cases, 193–204
    negative aspects, 214–221
    problems limiting effectiveness of P inactivation,
        213–214
    sediment oxidation, 222–227
        costs, 226
        equipment and application rates, 223
        lake response, 223–226
        prospectus, 226–227

Photosynthesis, 33
    most readily used carbon form for, 279
    in terrestrial plants, 279
    water temperature and, 280
Photosynthetic rate, 277, 481
Phragmites
    australis, 299, 306, 307, 310
    spp., 276, 388
Physical–chemical limnology, 26
Phytophagous insects, fish, and other biological controls,
        421–452
    alligatorweed, 425–426
    developing areas of macrophyte and algae management,
        442–445
        allelopathic substances, 443–444
        barley straw, 444
        Eurasian watermilfoil, 443
        fungal pathogens, 442
        hydrilla, 443
        plant growth regulators, 444
        reducing algae growth with bacteria, 444–445
        viruses for blue-green algae management, 445
        water hyacinth, 442–443
    Eurasian watermilfoil, 426–429
    grass carp, 429–441
        biology, 430–431
        case histories, 436–439
        history and restrictions, 429–430
        reproduction, 431–433
        stocking rates, 434–436
        water quality changes, 439–441
    hydrilla, 422–423
    other phytophagous fish, 441–442
    water hyacinth, 423–425
Phytoplankton
    analysis, 54
    blooms, 262
    composition, circulation and, 484
    grazer control of, 244
    productivity, 356
PI, *see* P index
Piburger See, 166, 168
Pickerel Lake, 205
    characteristics and alum doses of, 194
    longevity of phosphorus inactivation in, 204
Pimephales promelas, 113, 264
Pine Lake
    artificial circulation of, 473
    characteristics of withdrawal, 167
    hypolimnetic withdrawal of, 166
    restoration of, 170
    system installation costs, 171
Pistia stratiotes, 330, 339, 388, 397
Pithophora, 261
Plankton algae, control of, 85–86
Plant(s)
    community restoration, 291–320
        aquascaping, 303–310
        do nothing approach, 292–294
        founder colony, 311–317
        habitat alteration approach, 294–303

establishment of, 309
growth, dyes to suppress, 381
growth regulators (PGRs), 386, 444
lake
    habitat preferences of, 305
    wildlife and environmental values of, 307
nuisance, 416
pathogens, 285
propagation methods, 310
pruning effect, 354
restoration
    categories, 317
    doing nothing, 292
    remediation techniques, 292
volume, methods of reducing, 369
Pneumatic dredges, 512
Politics of scarcity, 7
Pollution, *see* Non-point pollution, protection from
Polychlorinated biphenyls (PCBs), 503
Polygonum
    amphibium, 275, 305, 306, 307, 314
    coccineum, 323
    natans, 323
    spp., 310, 388
Pomoxis nigromaculatus, 439
Pond design, short-circuiting of, 122
Pontederia
    cordata, 305, 306, 307, 310, 323
    sp., 388
Potamogeton
    americanus, 323
    amplifolius, 275, 298, 305, 307, 310, 323, 363, 395,
        400, 401, 403, 404, 405, 409
    berchtholdii, 245
    capillaceus, 307
    crispus, 272, 293, 294, 298, 300, 302, 308, 323, 354,
        378, 399, 409
    diversifolius, 305, 307, 323
    epihydrus, 305, 307, 323
    filiformis, 305
    foliosus, 294, 300, 307, 310, 323, 378, 394, 400, 405,
        409
    friesii, 307
    gramineus, 305, 307, 310, 323, 341, 363, 378
    illinoensis, 279, 293, 298, 305, 306, 307, 341, 405, 409,
        431, 436, 438
    natans, 294, 305, 307, 310, 323, 409
    nodosus, 294, 305, 307, 310, 323
    obtusifolius, 305, 307
    pectinatus, 247, 275, 284, 294, 298, 299, 300, 305, 307,
        310, 323, 341, 400, 401, 403, 405, 409
    praelongus, 305, 306, 307, 409
    pusillus, 247, 302, 305, 307, 310, 378
    richardsonii, 275, 294, 305, 306, 307, 310, 314, 323,
        388, 400, 409
    robbinsii, 323, 305, 307, 330, 363, 409, 549
    sp., 294, 299
    spirillus, 310
    spp., 276, 310, 324, 388
    strictifolius, 305, 307

zosteriformis, 275, 294, 300, 305, 307, 310, 316, 324,
    405, 409
Potomac River, diver dredging in, 363
Potters Lake, 398, 400, 401
Powerboating, 292
Precipitation, lake surface, 47
Pre-dams, 128, 129
Predator population, 292
Predatory game fish, restocking of, 299
Preventive, manual, and mechanical methods, 339–375
    manual methods and soft technologies, 344–345
    mechanical methods, 345–370
        costs and productivity, 365–370
        cutting, 348–351
        diver-operated suction dredges, 362–363
        harvesting, 351–361
        hydraulic washing, 363
        machinery and equipment, 346–348
        materials handling problem, 345
        mechanical derooting, 364–365
        shredding and crushing, 361–362
        weed rollers, 363–364
    preventive approaches, 339–344
        barriers and sanitation, 342–344
        education, enforcement, and monitoring, 342
        probabilities of invasion, 340–342
Prey species, refuge for, 394
Priority Watershed Project, 101–102
PROFILE, 71
Pruning effect, 354

R

Rain garden, 135
Random Lake, 398, 400, 402
Ranunculus
    aquatilis, 388
    flammula, 305
    longirostris, 275, 298, 305, 306, 316, 399
    sp., 403, 409
    spp., 310
    trichophyllus, 305, 306, 324
RCRA, *see* Resource Conservation and Recovery Act
RCWP, *see* Rural Clean Water Projects
Rehabilitation, 14
Reither See, 169
    characteristics of withdrawal, 167
    hypolimnetic withdrawal of, 166
Relative thermal resistance to mixing (RTRM), 61, 153,
    154, 155
Reservoir
    air-lift system to destratify, 475
    basin design, 25
    biotic communities, 28, 29
    conditions of U.S., 8, 9–10
    deep zone of, 24
    difference between lakes and, 14, 23
    fringe wetlands, 326
    inflow, addition of copper to, 262
    management, herbicides used for, 384

plunge point, 25
primary interactions in, 32
problems, sources of, 11–13
pumps for destratifying, 478
rehabilitation projects, 14
restoration and management of, 13–15
shape, 27
shoreline erosion, 133
shoreline rehabilitation, 133
Tennessee Valley Authority, 321, 328
thermally stratified, 28
U.S. Army Corps of Engineers, 69, 133
watershed area, average, 24
Residential development, retrofitted, 120
Residual plants, 292
Resource Conservation and Recovery Act (RCRA), 555
Restoration, definition of, 291
Rhizoclonium, 261
Rice Lake, 313, 314, 316
Riparian zone
destruction, 129
recovery, 131
rehabilitation, 129, 130
vegetation removal, 130
Riplox, 45, 71, 222
Rivers and Harbors Act, 555
Rocky Coulee Wasteway, 150, 151
Root
biomass reduction, 364
removal, 345
Rooted aquatic vegetation control, 377
Rooted plants, primary productivity, 27
Rotary cutting cookie cutter, 362
Rototilling, 364, 369
RTRM, *see* Relative thermal resistance to mixing
Ruppia maritima, 247, 310, 388
Rural Clean Water Projects (RCWP), 117
Rutilus rutilus, 250, 359

S

Safe Drinking Water Act, 15
Sagittaria
cuneata, 308
graminea, 305, 324
latifolia, 305, 308, 310, 313, 324, 381
rigida, 305, 310
spp., 306, 388
subulata, 284
Saidenbach Reservoir, 129
Salix
interior, 324
nigra, 328
spp., 388
Salmo gairdneri, 217
Salvinia
rotundifolia, 388
sp., 277
Sameodes albiguttalis (Warren), 422
Sand blankets, 380

Santee Cooper Reservoirs, 439
Saratoga Lake, 352, 359
Scardinius erythrophthalmus, 244
Schindler–Patalas trap technique, 54
Scirpus
acutus, 296, 308
americanus, 306, 324
californicus, 324
spp., 276, 298, 306, 388
subterminalis, 279
validus, 293, 306, 308, 324
SCUBA equipment, 380
SDT, *see* Secchi disk transparency
Secchi depths, 292
Secchi disk transparency (SDT), 39, 156, 227, 506
Sediment(s)
characteristics, 292
copper accumulation in, 264
covers, 73
advantages of, 377
disadvantages of, 377
density, increased, 298
nutrient levels, 506
organic matter, 281
oxidation, *see* Phosphorus inactivation and sediment
oxidation
phosphorus
inactivation of, 181
release, 178
physical characteristics, 536
release rates, 53–54, 58, 87
resuspension, 241, 503
seed banks, 292
Sedimentation
basin dredged material volume, 540
detention time required for, 538
Sediment covers and surface shading for macrophyte
control, 377–382
application procedures, 380
comparison of synthetic sediment covers, 377–380
Aquascreen, 379
burlap, 379–380
polyethylene, 377–378
polypropylene, 378
shading of macrophytes with surface covers,
380–381
Sediment removal, 73, 501–571
case studies, 541–560
Lake Järnsjön, 557–560
Lake Springfield, 552–556
Lake Trummen, 542–546
Lilly Lake, 546–552
costs, 560–564
depth, 505
dredge selection and disposal area design, 514–541
disposal area design, 533–541
dredge selection, 515–533
environmental concerns, 503–505
disposal area concerns, 504–505
in-lake concerns, 503–504
guidelines, 553

objectives, 501–503
  deepening, 501
  nutrient control, 501–502
  rooted macrophyte control, 502–503
  toxic substances removal, 502
  sediment removal depth, 505–506
  sediment removal techniques, 506–512
    hydraulic dredges, 507–511
    mechanical dredges, 507
    pneumatic dredges, 512
    special-purpose dredges, 511–512
  suitable lake conditions, 512–514
  technique, 504, 506
Selective herbicides, 386
SEM, *see* Simultaneously extracted metal
Sewage, diversion of secondary treated, 104
Shade plants, 277
Shadow Lake, 198, 199
  characteristics and alum doses of, 194
  dredging and alum treatment costs, 562
  trophic state improvement of, 213
Shagawa Lake, 36, 37, 58, 59, 88, 93, 99, 107, 502
  characteristics of, 95
  diversion cost, 105
  external loading and, 175
  internal nutrient loading, 355
  total P concentration in, 100
Shallow lakes
  biomanipulation, 243
  characteristics of, 33, 34
  control of algal biomass in, 240
  forward switches, 241
  herbicide treatments and, 395
  macrophyte growth in, 27
  P inactivation in, 214
  poor recovery in, 93
  recovery of, 87
  sediment release rates in, 87
  thermal stability in, 179
Shoreline
  erosion, 279
  lawns, 130
Shredding, by-catch of, 362
Silt
  curtains, 507
  loading, 32
Silver Lake, 54, 277, 473, 489
Simultaneously extracted metal (SEM), 264
Size–efficiency hypothesis, zooplankton grazing impacts,
  236
Skaha Lake, 378
Slow-no-wake boating ordinance, 294, 295
Sluice Pond, P inactivation dose, 221
Snake Lake, characteristics and alum doses of, 195
Soil erosion, 111, 116
Soil Test Phosphorus concentration (STP), 113
Soluble reactive P (SRP), 90, 112, 186, 463
South Lake, 150, 152, 153, 154
Sparganium
  chlorocarpum, 306, 308, 324
  eurycarpum, 306, 314

spp., 306, 310, 388
Spawning fish, impact of harvesting on, 360
Special purpose dredges, 501, 511
Spirodela polyrhiza, 308, 310, 324, 388
Spodoptera pectinicornis (Hampson), 422
Spruce Knob Lake, hypolimnetic aeration of, 462
Spruce Run Lake, hypolimnetic aeration of, 462
Squaw Lake, 185, 484, 485
SRP, *see* Soluble reactive P
SS, *see* Suspended solids
Steady state mass balance model, 57
Steinmetz Lake, 507, 513, 563
Stephanodiscus sp., 259
Stewart Hollow Lake, artificial circulation of, 473
Stizostedion
  lucioperca, 249
  vitreum, 247, 248
St. John River, 397
St. Mary Lake, 461
Stormwater diversion, 159
STP, *see* Soil Test Phosphorus concentration
Submergent macrophytes, 276
Suction friction head, 520
Suction harvesting, 365, 367
Sulfate-reducing bacteria, 170
Sun plants, 277
Sunshine Springs, dredging and alum treatment costs, 563
Surface drinking water supplies, eutrophication and, 10
Surface hydraulic loading, 56
Surface shading, *see* Sediment covers and surface shading
    for macrophyte control
Suspended solids (SS), 52, 557
Swimmers' itch, 263
Synthetic sediment covers
  Aquascreen, 379
  burlap, 379
  polyethylene, 377
  polypropylene, 378
Systemic herbicides, 414

T

Tank mixes, herbicide, 386
Tanytarsus, 456
Tanytarsus dissimilis, 216
Taxodium distichum, 328
Tennessee Valley Authority (TVA), 321, 328
Thames River, 484
Thames Valley reservoirs, 490
Thermocline erosion, 505
Third Sister Lake, 60
THM production, *see* Trihalomethane production
Three Mile Pond
  P inactivation dose, 221
  treatment costs, 222
Tibean, 456
TIN, *see* Total inorganic nitrogen
TMDL, *see* Total maximum daily load
TNC, *see* Total nonstructural carbohydrate
TOC, *see* Total organic carbon

Top-down bottom-up model, 239
Tory Lake, 462, 463
Total inorganic nitrogen (TIN), 546, 547
Total maximum daily load (TMDL), 131
Total nonstructural carbohydrate (TNC), 282, 342, 351, 353
Total organic carbon (TOC), 219
Total phosphorus (TP), 37, 341
    concentrations, stream, 40
    maps, 40
    reduction, European lake response to, 92
Total suspended solids (TSS), 52, 122, 127, 559
Toxicity, means of reporting, 391
Toxic substances removal, 502
TP, *see* Total phosphorus
TP-blue-green algal probability model, 102
Trachemys scripta elegans, 312
Trapa natans, 308, 342, 348, 388
Trempealeau River, 551
Triaenodes tarda Milner, 426
Triclopyr, 385, 411
    suggested uses, 415
    treatment, Pend Oreille River, 412
Trihalomethane (THM) production, 85
Trophic cascades, 236, 238, 239
Trophic indicators, 486
Trophic state index (TSI), 63, 341
Trophic state indicator, TP used as, 65
TSI, *see* Trophic state index
TSS, *see* Total suspended solids
Tuggarah Lakes, 506
TVA, *see* Tennessee Valley Authority
Twin Lake, 483
Typha
    angustifolia, 293, 313, 351
    glauca, 282
    latifolia, 264, 293, 298, 306, 324
    spp., 276, 306, 310, 388

U

Underwater dam, 129
United States Department of Agriculture, water volume
        required by, 501
United States Environmental Protection Agency (USEPA),
        8, 390
    Clean Lakes Program, 15, 74
    herbicide registration, 390
    pesticides regarded by, 444
    Science Advisory Board, 41
University Lake, artificial circulation of, 472
Upper Klamath Lake, 62
Upper Willow Flowage, 551, 552
Urban runoff, 72, 111
U.S. Army Corps of Engineers (USCOE), 16, 23
    Aquatic Ecosystem Restoration Foundation, 402
    Aquatic Plant Control Research Program, 271
    Aquatic Research Program, 402
    CLIMEX, 342
    cutting of water chestnut in the Potomac River, 348
    reservoirs, 69

U.S. Clean Water Act, 7
USCOE, *see* U.S. Army Corps of Engineers
USDA, *see* U.S. Department of Agriculture
U.S. Department of Agriculture (USDA), 422
USEPA, *see* United States Environmental Protection
        Agency
U.S. Natural Resources Conservation Service (NRCS), 131
Utricularia
    biffa, 328
    geminiscapa, 277, 306
    gibba, 306, 409
    intermedia, 306
    minor, 405, 409
    purpurea, 308, 324
    sp., 324
    spp., 306, 388
    vulgaris, 275, 306, 324, 404, 405, 409

V

Vadnais Lake, 113, 210, 462
Vallisneria americana, 275, 293, 300, 316, 324, 354, 363,
        389, 394, 403, 404, 409, 436, 438, 443
Vancouver Lake, 509
Vegetarianism, 112
Vegetation
    density, 133
    recovery, 293
Vegetative reproduction, 282
Viruses, 285
Vogtia malloi Pastrana, 422
Vollenweider steady state, mass balance P model, 148

W

Wapato Lake, 160, 204, 206
Wastewater (WW), 47, 104
Water
    agricultural use of, 5
    biosphere, 4
    budget construction, 46
    clarity, 292
    physical properties of, 522
    primary sources of, 7
    rake, 363
    retention time, 68
    scarce countries, 5
    stressed countries, 5
    temperature, photosynthesis and, 280
    transparency, 52, 490
    treatment residual (WTR), 118
    war, 6
Water level drawdown, 321–338
    case histories, 333–335
    case studies, 328–333
        Connecticut, 333
        Florida, 329–330
        Louisiana reservoirs, 329
        Oregon, 333

TVA reservoirs, 328–329
Wisconsin, 330
fish management, 333
methods, 321–325
positive and negative factors, 326–328
responses of aquatic plants to, 322–324, 325
Watershed
areal ratio (WWAR), 127
characteristics, 45
drainage, 30
Lake Springfield, 553
TP yield coefficients, 50, 51
Watervliet Reservoir, 348
Water Witch, 363
Weaver Lake, 349
Weed
cutting, 359
-free lake, 287
problem, reproductive ability and, 282
rollers, 363
Welland Canal
dredging and alum treatment costs, 562
P inactivation dose, 221
West Lost Lake, artificial circulation of, 473
West Twin Lake (WTL), 106, 196, 199
characteristics and alum doses of, 194
dredging and alum treatment costs, 562
mean surface total P concentration in, 201
P inactivation dose, 221
reduction in mean summer epilimnetic TP, 197
species diversity of planktonic microcrustacea, 216
trophic state improvement of, 213
Wetland(s)
artificially constructed floating, 300
constructed, 122, 125, 128
destruction, 8
-littoral zone, 28
losses, 8
natural, 122
nitrate removal and, 124
rehabilitation, 122
reservoir fringe, 326

sizing, 127
Weyauwega Lake, 343
White Lough, 211, 226
White River Lake, 272, 273, 276
aquatic vegetation of, 275
harvesting, 276
plant populations of, 275
sampling transect locations, 274
Wind
-caused mixing, 61
disturbance, 33
Windfall Lake, 394
Wintergreen Lake, 130
Winter lake mapping, 515
Wolffia
columbiana, 308, 388
spp., 310
Wolverine Lake, 398, 403, 405
World lakes, recovery of, 89
WTL, *see* West Twin Lake
WTR, *see* Water treatment residual
WW, *see* Wastewater
WWAR, *see* Watershed areal ratio

Z

Zannichellia palustris, 300, 306, 308, 310, 389, 401, 409
Zizania
aquatica, 306, 308, 354
palustris, 314
spp., 279, 306, 310
Zone/compression settling test procedure, 534
Zooplanktivores, top-down control of, 240
Zooplankton
grazing
assessment of, 244
reduced, 236
sampling, 54
Zosterella dubia, 272, 300, 306, 308, 312, 354, 363, 389, 398, 401, 404, 409